NEURAL-NETWORK MODELS OF COGNITION
Biobehavioral Foundations

ADVANCES IN PSYCHOLOGY

121

Editors:

G. E. STELMACH

P. A. VROON

ELSEVIER
Amsterdam – Lausanne – New York – Oxford – Shannon – Tokyo

NEURAL-NETWORK MODELS OF COGNITION

Biobehavioral Foundations

edited by

John W. DONAHOE

and

Vivian PACKARD DORSEL

Department of Psychology
Neuroscience and Behavior Program
University of Massachusetts, Amherst, U.S.A.

1997

ELSEVIER
Amsterdam – Lausanne – New York – Oxford – Shannon – Tokyo

NORTH-HOLLAND
ELSEVIER SCIENCE B.V.
Sara Burgerhartstraat 25
P.O. Box 211, 1000 AE Amsterdam, The Netherlands

ISBN: 0 444 81931 2

This book is printed on acid-free paper.

Printed and bound by Antony Rowe Ltd, Eastbourne

Transferred to digital printing 2005

ACKNOWLEDGMENTS

The editors wish to express their appreciation to Professors Neil R. Carlson and Jerrold S. Meyer for informal consultations about various aspects of this project and to Alana Perocchi for assistance in the preparation of the bibliography. Finally, the editors thank Professors George E. Stelmach and Peter R. Killeen, both of Arizona State University, for their roles in initiating this project.

This book was set in Times PC type using WordStar version 7.0 and a Hewlett-Packard LaserJet IIIP printer.

LIST OF CONTENTS

LIST OF CONTRIBUTORS

DERMOT BARNES, D.Phil.
Department of Applied Psychology
University College
Cork, Ireland
stay8028@bureau.ucc.ie

ANDREW G. BARTO, Ph.D.
Department of Computer Science
Lederle Graduate Research Center
University of Massachusetts
Amherst, MA 01003-4610
barto@cs.umass.edu

NEIL E. BERTHIER, Ph.D.
Department of Psychology
Tobin Hall
University of Massachusetts
Amherst, MA 01003-7710
berthier@psych.umass.edu

ANNA M.T. BOSMAN, Ph.D.
Katholieke Universiteit Nijmegan
Vakgroep Orthopedagogiek
Postbus 9104
6500 HE Nijmegan
The Netherlands
a.bosman@ped.kun.nl

DEAN V. BUONOMANO, Ph.D.
P.O. Box 0732
Keck Center for Integrative Neuroscience
University of California
San Francisco, CA 94143
dbuono@phy.ucsf.edu

JOSÉ E. BURGOS, Ph.D.
Apartado 47562, Los Chaguaramos
Caracas 1041-A
Venezuela
cmunoz@usb.ve

JUNE-SEEK CHOI, M.A.
Neuroscience and Behavior Program
Middlesex House
University of Massachusetts
Amherst, MA 01003-7710
jschoi@titan.umass.edu

JEFFERY A. CLOUSE, Ph.D.
Department of Computer Science
Graham Hall
North Carolina A&T State University
Greensboro, NC 27411
clouse@ncat.edu

JOHN W. DONAHOE, Ph.D.
Department of Psychology
Tobin Hall
University of Massachusetts
Amherst, MA 01003-7710
jdonahoe@psych.umass.edu

WILLIAM T. FARRAR, IV, Ph.D.
Cognitive Systems Group
Department of Psychology
Arizona State University
Tempe, AZ 85287-1104
atwtf@research1.asu.edu

UWE FREY, Ph.D.
Institute for Neurobiology
Gene Regulation & Plasticity
Brenneckestrasse 6
P.O. Box 1860
39008 Magdeburg
Germany
frey@ifn-magdeburg.de

JACK J. GELFAND, Ph.D.
Department of Psychology
Princeton University
Princeton, NJ 08544
jjg@princeton.edu

APOSTOLOS P. GEORGOPOULOS, M.D.
Brain Sciences Center
Veterans Affairs Medical Center
One Veterans Drive
Minneapolis, MN 55417
omega@maroon.tc.umn.edu

MARK A. GLUCK, Ph.D.
Center for Molecular and Behavioral Neuroscience
197 University Avenue
Rutgers University
Newark, NJ 07102-1814
gluck@pavlov.rutgers.edu

STEPHEN D. GOLDINGER, Ph.D.
Cognitive Systems Group
Department of Psychology
Arizona State University
Tempe, AZ 85287-1104
goldinger@asu.edu

VIJAYKUMAR GULLAPALLI, Ph.D.
10 Village Court
Lawrenceville, NJ 08648
vijay@princeton.edu

PETER J. HAMPSON, Ph.D.
Department of Applied Psychology
University College
Cork, Ireland
stay@8026bureau.ucc.ie

JOHN E. HUMMEL, Ph.D.
Psychology Department
University of California
Los Angeles, CA 90095-1563
jhummel@psych.ucla.edu

MICHAEL I. JORDAN, Ph.D.
Department of Brain and Cognitive Sciences
E25-229
Massachusetts Institute of Technology
Cambridge, MA 02139
jordan@psyche.mit.edu

MARLA B. LUSKIN, Ph.D.
Department of Anatomy and Cell Biology
Emory University School of Medicine
Atlanta, GA 30322
luskin@anatomy.emory.edu

MICHAEL M. MERZENICH, Ph.D.
Keck Center for Integrative Neuroscience
University of California
San Francisco, CA 94143
merz@phy.ucsf.edu

JERROLD S. MEYER, Ph.D.
Department of Psychology
Tobin Hall
University of Massachusetts
Amherst, MA 01003-7710
jmeyer@psych.umass.edu

P. READ MONTAGUE, Ph.D.
Division of Neuroscience
Center for Theoretical Neuroscience
Baylor College of Medicine
1 Baylor Plaza
Houston, TX 77030
read@bcm.tmc.edu

JOHN W. MOORE, Ph.D.
Department of Psychology
Middlesex House
University of Massachusetts
Amherst, MA 01003-7710
jmmoore@cs.umass.edu

CATHERINE E. MYERS, Ph.D.
Center for Molecular and Behavioral Neuroscience
197 University Avenue
Rutgers University
Newark, NJ 07102-1814
myers@pavlov.rutgers.edu

DAVID C. PALMER, Ph.D.
Department of Psychology
Smith College
Northampton, MA 01060
dcpalmer@science.smith.edu

MARCUS E. RAICHLE, M.D.
Department of Radiology and Neurology
Washington University School of Medicine
Campus Box 8225
4525 Scott Avenue, Room 2116
St. Louis, MO 63110
marc@npg.wustl.edu

WOLFRAM SCHULTZ, M.D.
Institute of Physiology
University of Fribourg
CH-1700
Fribourg, Switzerland
wolfram.schultz@unifr.ch

STEPHEN L. SENFT, Ph.D.
Neuroengineering and Neuroscience Center
5 Science Park North
Yale University
New Haven, CT 06511
senft-stephen@yale.edu

WOLF J. SINGER, Ph.D.
Max-Planck-Institut für Hirnforschung
Deutschordenstrasse 46
D-60528 Frankfurt/Main
Federal Republic of Germany
singer@mpih-frankfurt.mpg.de

RICHARD S. SUTTON, Ph.D.
Department of Computer Science
Lederle Graduate Research Center
University of Massachusetts
Amherst, MA 01003-4610
rich@cs.umass.edu

KEIJI TANAKA, Ph.D.
The Institute of Physical and Chemical Research (RIKEN)
2-1 Hirosawa, Wako-shi,
Saitama, 351-01, Japan
keiji@postman.riken.go.jp

ARNOLD TREHUB, Ph.D.
Department of Psychology
Tobin Hall
University of Massachusetts
Amherst, MA 01003-7710
trehub@psych.umass.edu

GUY C. VAN ORDEN, Ph.D.
Cognitive Systems Group
Department of Psychology
Arizona State University
Tempe, AZ 85287-1104
guy.van.orden@asu.edu

Neural-Networks Models of Cognition
J.W. Donahoe and V. Packard Dorsel (Editors)

CHAPTER 1

THE NECESSITY OF NEURAL NETWORKS

John W. Donahoe
Department of Psychology
Neuroscience and Behavior Program
University of Massachusetts, Amherst

Introduction

The goal of this volume is to provide a foundation for a natural science-based understanding of human behavior, including the complex phenomena of "mind"—perception, thought, memory, and language. Clearly, the attainment of this goal is very much a task for the future, and remains so at the conclusion of this volume. Nonetheless, knowledge has advanced to the point that genuine progress appears on the near horizon. This knowledge includes not only specific scientific findings, as essential as they are, but also general conceptual insights into the origins of complex phenomena. The present chapter identifies some of the conceptual insights that motivate and inform the search for a natural-science approach to complex human behavior.

The Darwinian contribution

Arguably, the work of Charles Darwin provides us with the deepest and most general insights into how we may understand complex phenomena. How did Darwin go about his task of uncovering the origins of complex structure, and how does his example instruct us to proceed in our own task—uncovering the origins of complex human behavior?

Two major conceptual contributions may be discerned. First, Darwin's example teaches us that structure and function are inseparably linked. For Darwin, complex structure—morphology—arose as the cumulative effect of natural selection, a functional concept: Whatever morphological characteristics an organism possessed, those characteristics endured into the next generation to the degree that they enhanced or, at a minimum, did not detract from reproductive fitness. In short, the function of structure determined the fate of structure. Natural selection decreed that great minds well concealed could not be favored or, as Thomas Huxley put it, "The great end of life is not knowledge, but action." Darwin's second major contribution was the view that complex structure/function arose as the emergent product of lower-level processes acting over time. No appeal to higher-level principles was required to understand the origins of complex structure/function. Instead, complexity emerged as a byproduct of fundamental biological processes whose effects were largely captured by the principle of natural selection. These emergent byproducts of selection were then subjected to still further selections with complexity as a possible outcome. (See **Figure 1**.) The notion that complexity arose through the accumulation of prior favored selections rather than through the playing out of some a priori plan or set of superordinate principles had a further, and disquieting, implication: Since selection could always be traced ultimately to the environment at the moment of selection, the products of selection reflected conditions that existed only in the past. That is, strictly speaking, selection "prepared" us to live in the past, not the future. Thus, those conventional attributes of mind—foresight, intelligence, reason, and the like—are, at best, comforting illusions except insofar as future conditions are similar to past conditions. The relative constancy of selecting conditions over the lifespan of an individual allows these illusions to endure, but they remain fictions nevertheless.

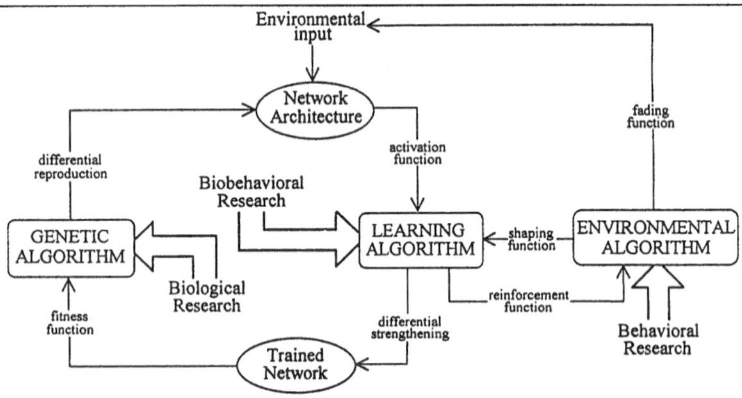

FIGURE 1. A schematic representation of the process of selection. Some initial variations are favored over others by selecting factors in the environment. These favored variations are retained and contribute to the variation available for subsequent selection. Complexity is a possible but not a necessary consequence of the three-step process.

The triumph of Darwinism

As can be easily appreciated, Darwin's selectionist proposals about the origins of complex structure were met with strong resistance from his contemporaries. In 1859, when *On the origin of species* was published, most of his audience rejected the idea that the tremendous diversity and complexity of life could have arisen by natural selection from a few or, perhaps, a single progenitor. Even most scientists, although they agreed that some evolutionary process was responsible for the similarities among living creatures, did not accept natural selection as an adequate account of the origins of that similarity. Most strikingly, Alfred Wallace, who independently proposed the principle of natural selection, explicitly demurred in its application to our species, man. Indeed, approximately 70 years passed before most scientists, let alone most laypersons, endorsed natural selection as the central insight into the evolution of species. (Today, surveys indicate that only about 50% of laypersons in the United States "believe in" natural selection—less than in any other educated society!)

Requirements for acceptance. Acceptance of a natural-science account of human behavior lags far behind that of human morphology. To understand why this might be so, it is useful to examine the history of the acceptance of natural selection to isolate the reasons for its relative success. If the critical events can be identified and their counterparts achieved, then—perhaps—acceptance of a natural-science account of human behavior can be realized. What were the critical events that led most scientists and many laypersons to accept natural selection with, of course, the provisionality that accompanies all principles in science?

Natural selection, as proposed by Darwin and Wallace, was largely based on naturalistic observations of the relation between the possession of certain morphological characteristics (e.g., sturdiness of beak in Darwin's finches) and reproductive fitness (e.g., the likelihood of leaving surviving offspring given the availability of seeds of certain size and hardness of shell). Thus, as initially formulated, the principle of natural selection stated a purely functional relation between morphological antecedents and their consequences for reproductive fitness. The biological mechanisms that underlay these observed relations were not known to Darwin. What was asserted was simply that, insofar as the characteristics were heritable, those that benefited reproductive fitness would become more numerous in subsequent generations because of competition for resources between different organisms located in common environments.

The lack of general acceptance of natural selection by Darwin's scientific contemporaries indicates that a purely functional relation was not persuasive, even when buttressed by a large volume of naturalistic observations and quasi-experimental work with artificial breeding. Realizing this, Darwin proposed a hereditary mechanism based on inferences from his naturalistic observations rather than direct experimental work in genetics. Darwin's theory of *gemmules* was a so-called "blending" theory of heredity in which the characteristics of the offspring took intermediate forms; i.e., "blends" of the characteristics of their parents. However, the Scots engineer Fleeming Jenkin soon showed mathematically that such a theory of heredity was inconsistent with evolution through natural selection; the rare favorable characteristic would be "blended out" and never predominate in the manner prescribed by natural selection.

What transpired between 1859 and the 1930s that led most scientists to accept Darwin's proposal that natural selection was the chief engine of evolution? Two achievements seem paramount. First, in the early 1900s, independent experimental work established a science of genetics. In place of Darwin's blending theory of heredity, a "particulate" theory was devised beginning with the rediscovery of Mendel's experimental work with peas. (Mendel's decision to study certain characteristics of peas was a happy choice. Had he chosen other characteristics or species, he might have observed polygenic characteristics that did not have the simple dominant-recessive relations required to make the particulate nature of heredity apparent.) When Mendel's observations were coupled with other work, notably that revealing mutations, a powerful set of mechanisms became available to implement the functional relations Darwin had observed. The second achievement occurred in the 1920s and 30s when mathematical biologists—Fisher, Haldane, and Wright—devised formal techniques: population genetics. These techniques were capable of tracking the flow of genes over generations. Now the proposed cumulative effects of natural selection could be rigorously traced over time. Together, Darwin's original functional proposal of natural selection, in conjunction with the genetic mechanisms to implement the principle and the mathematical techniques to trace its implications, constitute the *synthetic theory of evolution*. All three components—a functional principle, a biological mechanism, and a formal technique—were necessary preludes to the general acceptance of evolution through natural selection. (See the left column of **Table 1**.)

	Complex Morphology	**Complex Behavior**
Functional Principle	Natural Selection	Behavioral Selection
Biological Mechanism Implementing the Functional Principle	Genetics (Changes in Gene Frequency)	Neuroscience (Changes in Synaptic Efficacy)
Formal Techniques for Tracing the Cumulative Effects of the Functional Principle	Population Genetics	Neural Networks

TABLE 1. Parallels between the evolution of complex morphology via natural selection and complex behavior via behavioral selection.

Toward a Natural-Science Account of Human Behavior

If the Darwinian parallel holds, then a generally accepted, natural-science account of human behavior depends on the confluence of three events: (1) a functional selectionist principle for modifying behavior within the lifetime of the individual, analogous to the principle of natural selection, (2) biological mechanisms for implementing the principle, analogous to genetic mechanisms, and (3) quantitative techniques for tracing the cumulative effects of the functional principle acting over time, analogous to population genetics. Are these requirements currently met, and—if so—what form do they take?

A behaviorally based principle of selection

In the early 1900s, at about the time that the mechanisms of heredity were being rediscovered, work began that was explicitly addressed toward uncovering a principle of behavioral selection. Even casual observations indicated that individual experience changed behavior: We recall the names of persons with whom we are acquainted and we remember incidents that occurred many years before. But precisely what conditions must be present for experience to change behavior? We recall some incidents clearly, some dimly, and others not at all. What characteristics distinguish those situations that change behavior from those that do not?

Ivan Pavlov, the Russian physiologist, and Edward Thorndike, the American psychologist, began the search for a principle of behavioral selection quite independently. However, each explicitly described his work as following the trail that Darwin had blazed. Unlike Darwin, each relied principally on experimental rather than naturalistic observation. The more well controlled conditions of the laboratory permitted fundamental relations to be more readily identified. Also, unlike natural selection, behavioral selection occurred over a time span short enough to be investigated in the laboratory within the lifetime of a single scientist.

Beginning with these laboratory investigations, two fundamental factors have been identified that, together, allow experience to change behavior. First, the critical events affected by behavioral selection must occur very close in time to be effective—only a few seconds apart at most. For example, Pavlov found that the behavioral change of salivating to a tone was acquired only if the dog experienced the tone followed immediately by food, the stimulus that initially elicited salivation. Similarly, the behavioral change of increased leverpressing occurred only if the rat experienced the eliciting stimulus of food immediately after the leverpress. Stimuli that already elicited behavior could change the way in which the environment guided behavior, but only if the eliciting stimulus was separated from the affected events by no more than a few seconds. The elicitation of salivation by food could select salivation to a tone or leverpressing to the sight of a lever only if the food immediately followed these events. Otherwise, behavioral selection did not occur no matter how reliably or frequently these events were followed by the eliciting stimulus of food. Thus, the first factor promoting the selection of behavior was *temporal contiguity*—a close temporal relation between the affected environmental and behavioral events and the eliciting stimulus. (For specialists in the field of learning, a caveat is noted here: Some events, such as gustatory/olfactory stimulation and the gastric effects of ingestion, may be separated by *hours* without preventing behavioral selection. In such cases, the stable relation between these events over evolutionary time has permitted the natural selection of specialized neural circuitry that bridges the long time interval between gustatory/olfactory stimuli and gastric consequences. The toleration of long intervals between events is the great exception not the rule, and in all other respects the findings demonstrate the same phenomena as behavioral selection operating over the normative shorter intervals; cf. Domjan, 1983.)

In addition to temporal contiguity of the environmental and behavioral events with an eliciting stimulus, a second crucial factor was uncovered more recently. The discovery was prompted by the observation that, under some circumstances, an eliciting stimulus that occurred immediately after a stimulus or a behavior would not produce a change in the environmental guidance of behavior. For example, a tone might not acquire the ability to evoke salivation and a lever would not be pressed more frequently even though each had been followed immediately by food. What was responsible for the failure of an eliciting stimulus to support behavioral selection even when the requirement of temporal contiguity was met?

Beginning with the experimental work of Leon Kamin (1968), it became apparent that an eliciting stimulus would foster behavioral selection only if it evoked behavior that was not already occurring at that moment. For example, a tone that was immediately followed by food would not acquire the ability to evoke salivation if the tone was accompanied by a light that had previously been paired with food. Since the light already evoked salivation because of prior light-food pairings, the presentation of food—which also evoked salivation—did not produce a *change* in behavior. Without a change in behavior (in this case an increase in salivation), selection of a relation between the tone and salivation could not occur. The second fundamental factor—that an eliciting stimulus supports behavioral selection only if it evokes a *change* in ongoing behavior—is known as *behavioral discrepancy*. Under controlled experimental conditions, temporal contiguity and behavioral discrepancy have been shown to be necessary and sufficient for behavioral selection to occur; i.e., for experience to change the way the environment guides behavior. (For the major theoretical treatment of these findings, see Rescorla & Wagner, 1972. For related theoretical treatments, see Donahoe, Crowley, Millard, & Stickney, 1982, and for a selective review, see Donahoe & Palmer, 1994.)

A principle of behavioral selection seeks to define the necessary and sufficient conditions whereby experience changes behavior. As such, it is a functional principle specifying the relation between certain conditions—contiguity and discrepancy—and their consequences—behavioral change. Consistent with the Darwinian precedent, even most scientists do not accept the view that the cumulative effects of such a relatively simple functional principle can yield anything approaching the complexity and diversity of human behavior. And, of course, even the staunchest advocates of such a functional principle cannot yet claim such an accomplishment, only its promise.

If the history of ontogeny recapitulates the history of phylogeny, a functional principle of behavioral selection must be supplemented both by the biological mechanisms that implement the relations summarized by the principle—the mechanisms of synaptic plasticity—and by formal techniques that trace the cumulative effects of the principle—neural networks. (See **Table 1**.) Steps toward both of these goals are taken in this volume.

The volume is divided into six major sections, each treating a core component of any natural-science based understanding of human behavior. A brief overview introduces each section and is followed by presentations of both experimental work that describes relevant biobehavioral processes and neural-network research related to those processes. The sections are:

One: **Neural development**. Even a well formulated principle of behavioral selection whose biological mechanisms were completely known would be severely limited in its ability to generate complex behavior if it acted upon an embodied nervous system of inappropriate structure. Behavioral selection operates on a nervous system whose interconnected neural systems are the product of a long history of natural selection with a concurrently selected skeleto-muscular system. Mind and body co-evolved: No history of behavioral selection can lead a pigeon to speak or a person to fly.

Two: **Neural plasticity**. When the requirements for behavioral selection are met—contiguity and discrepancy—cellular processes are initiated that alter synaptic efficacies between neurons. What are those processes, and what are their implications for the functioning of systems of neurons?

Three: **Perceiving**. Complex behavior is the product of a prolonged history of selection—both natural and behavioral—by complex environments. How do organisms extract from the varied combinations of stimuli with which they are incessantly bombarded just those combinations that validly guide behavior? Consider the complex combinations of stimuli that permit a young child to distinguish the faces of its parents from the faces of others. By what biological mechanisms are these stimuli integrated, and how might they be simulated by neural networks?

Four: **Behaving**. The fluidity and coherence of behavior is so conspicuous as to be overlooked. Our fingers move rapidly and deftly as we write a letter; our speech involves the concerted activity of a number of response systems—the tongue, mouth, lips, and muscles of respiration. But what we usually think of as unitary responses, such as writing the letter "a" or uttering the speech sound (phoneme) /b/, are—in truth—the temporally coordinated activity of thousands of muscle fibers. What are the neuromuscular mechanisms that underlie this integrated activity, and can they be simulated in biologically faithful ways?

Five: **Reinforcement learning**. The preceding comments on perceiving and behaving make clear that a central requirement of any adequate learning principle is that it be able to integrate diverse sensory and behavioral activity and, in turn, to coordinate them with each other. By such means, complex combinations of stimuli may come to guide the complex combinations of muscular activity that constitute complex behavior. Both behavioral and neuroscientific research point to one approach to learning—reinforcement learning—as the means whereby this is accomplished. Normative cognitive science typically takes other approaches to learning, such as back-propagation (e.g., Rumelhart, Hinton, & Williams, 1986) or production systems (e.g., Anderson, 1983). Although there are some points of contact between these approaches and reinforcement learning (see the discussion of learning algorithms later in this chapter and at other points in the volume), reinforcement learning is most broadly consistent with experimental findings from biobehavioral research. Other approaches may be useful for purposes of artificial intelligence or engineering, but living organisms appear to exploit the iterative application of reinforcement learning as the means for achieving complex behavior.

Six: **Complex behavior**. The sixth, and final, section of the volume applies biobehavioral findings as interpreted by neural networks to aspects of that most complex of human behaviors—verbal behavior.

Biological mechanisms of behavioral selection

I shall now introduce some of the basic terminology and concepts used in a biobehavioral approach to cognition (i.e., complex behavior) and in its interpretation via neural networks.

Synaptic processes. For the environment to change the way in which behavior is guided, appropriate physical energies must stimulate sensory receptors, the receptors must activate sets of neural pathways, synapses along these pathways must be modified by experience, and these modified pathways must, in turn, initiate effector activity. The modification of synapses changes the ability of neurons to communicate with one another, and neurons communicate chiefly by means of neurotransmitters. Neurotransmitters are liberated by "upstream" (presynaptic) neurons and migrate across the synapse to activate "downstream" (postsynaptic) neurons. Since, by definition, behavioral selection produces changes in the manner in which the environment guides behavior, the biological processes whereby neurons communicate with one another must be affected. **Jerrold Meyer**'s chapter on **neurotransmitters** indicates some of the major characteristics of communication via neurotransmitters and provides constraints that must be honored by formal techniques that attempt to trace the effects of behavioral selection.

A presynaptic neuron does not generally have a constant effect on a postsynaptic neuron. The changing effects of behavioral selection are mediated through changes in the release of neurotransmitter by the presynaptic neuron and/or the response of the postsynaptic neuron to the neurotransmitter. A change in the ability of presynaptic neurons to initiate activity in postsynaptic neurons is known as a change in synaptic efficacy. **Uwe Frey**'s chapter describes cellular processes underlying a type of change in synaptic efficacy known as **long-term potentiation**, or **LTP**. The changes occurring during LTP are thought to provide the most general model of the neural processes mediating behavioral change throughout the nervous system. The dependence of behavioral change on temporal contiguity is, in part, a reflection of the time relations between events occurring at the synapse.

Neural systems. Changes in synaptic efficacy necessarily occur because of events taking place locally at the synapse. However, these local events may be affected by prior events that have occurred at more remote sites in the nervous system and, ultimately, in the environment itself. Thus, in order to understand fully the changes in synaptic efficacies that mediate behavioral change, it is necessary to consider the sets of pathways, or neural systems, of which those synapses are members. **Marla Luskin** describes some of the techniques and processes whereby **neural development** occurs. Of particular importance for learning are those neural systems that project widely throughout the brain. These relatively nonspecific systems can modify synaptic efficacies in large areas of the brain in a coordinated fashion. Dopamine is a neurotransmitter that affects the functioning of many synapses and, for that reason, is often referred to as a neuromodulator rather than simply a neurotransmitter. Studies of the **electrophysiology of dopaminergic neurons** by **Wolfram Schultz** indicate that their activity implements the discrepancy requirement of a functional principle of behavioral selection.

Electrophysiology provides essential information about the activity of single neurons or small ensembles of neurons. However, guidance of even relatively simple behavior requires the coordinated activity of many hundreds of thousands of neurons. To provide more global information about the functioning of many neurons, techniques have been developed for imaging the activity of entire neural systems. **Marcus Raichle** describes research on **neural imaging** during **verbal processing**.

Sensory and motor processes. Synaptic processes within neural systems cannot be modified to implement a principle of behavioral selection unless the rich environmental input to the organism is sufficiently appreciated and the behavioral output of the organism sufficiently integrated to produce an organized response. As examples, recognition of faces requires an appreciation of the relations between visual stimuli that specify form, color, and perhaps motion and depth as well. And, a response such as reaching requires the coordinated contraction of thousands of muscle fibers reflecting the concerted activity of thousands of neurons. **Wolf Singer** describes mechanisms whereby the activity of neurons in the visual system are integrated to permit the **visual processing** of sensory inputs. **Keiji Tanaka** describes **higher-level visual processing** and **Mark Gluck and Catherine Myers** describe the likely role of **hippocampal function** in bringing about the integration of sensory and motor processes to "represent" extremely complex combinations of events. **Apostolos Georgopoulos** describes the means by which the activity of neurons in the motor systems is integrated to specify the **control of movement**. Nonspecific neural systems also play a critical role in coordinating the modification of synaptic efficacies underlying sensory and motor integration.

Behavioral constraints. The biological mechanisms that mediate behavioral selection can be known only through the experimental methods of neuroscience. They cannot be inferred solely from behavioral observations because, in general, an indefinitely large number of underlying mechanisms could mediate any given environment-behavior relation (Donahoe & Palmer, 1994; cf. Smolensky, 1986; Townsend, 1972). Although behavioral observations do not provide a sufficient basis from which to infer biological mechanisms, they do assist the search for such mechanisms: Biological mechanisms must yield outcomes that are consistent with functional relations discovered at the behavioral level. For example, a person with brain damage that produces a certain form of aphasia might not be able to name an object by looking at it, but could name it after picking it up. Neuroscience must accommodate this fact—that a vocal response that cannot be guided by a visual stimulus can nevertheless be guided by a tactile stimulus. There is also that most fundamental contribution of behavioral processes to biological mechanisms: If there were no consequences of the biological mechanisms for the behavior of the organism, then there would be no basis on which the environment could naturally select the biological mechanisms. Behavior and behavioral change are the pacemakers of evolution (Wilson, 1975). The chapter by **David Palmer** indicates some of the important **behavioral constraints** on whatever biological mechanisms mediate behavioral selection.

Formal techniques for tracing behavioral selection

Even if a functional principle of behavioral selection and the biological mechanisms that implement it were completely known, something else is required before a natural-science account of complex behavior can gain general acceptance. As previously noted, a functional principle and its biological implementation must be supplemented by formal techniques that trace the selection process over time.

Formal techniques are required for two primary reasons. First, any truly complex behavior is the end product of a long and incompletely known history of natural and behavioral selection spanning many years. Consider the complex behavior of reading English prose. The visual patterns that distinguish different letters and words from one another must be appreciated, the articulatory movements accompanying those visual stimuli must be acquired and the sounds produced by those movements coordinated with the visual stimuli, the complex motor patterns that make up the "meanings" of the words must be established, and so on. The full history of such a prolonged and complex series of events cannot be studied under experimental conditions, and it is unlikely that it ever will be. Second, even if the selection history were fully known and the various biological mechanisms engaged by that history were fully characterized, the simultaneous occurrence of multiple selecting events and their simultaneous effects on multiple neural pathways in diverse portions of the brain over long periods of time is far too complex to be tracked without formal techniques for doing so. For both of these reasons, some formal means is required to follow the course of selection for the sequence of many simultaneously acting events that yields complex behavior.

The chapters on neural-network simulations present a number of examples of formal techniques for tracing the effects of selection over time. All of these techniques are implemented in neural networks that simulate the actions of one or more of the neural systems found in the brains of living organisms. If, as we claim, formal techniques are necessary to keep track of simple processes acting over time, and if neural networks serve this purpose with respect to the effects of behavioral selection, then neural networks are an inescapable component of a natural science-based account of human behavior. In short, neural networks are *necessary* if the cumulative products of behavioral selection are to be understood.

Interpreting Complex Behavior Using Neural Networks

Neural networks

Network architecture. A neural network is an interconnected set of units, each unit simulating a single neuron or coherent ensemble of neurons and each connection simulating an axon or set of axons communicating activity from one unit to another. The entire set of *units* and *connections* constitutes the *architecture* of the network. (See **Figure 2**.) The units within a network are conventionally divided into three types based on their positions within the architecture. *Input units* sense events in the environment of the network, usually the external environment of the organism. But, if the network is intended to simulate some "module" within a larger network, then the input units sense only those events that affect the particular module. After the input units are stimulated, connections from these units may then activate "hidden" or *interior units* within the network. Connections from interior units may then activate either other interior units or, ultimately, output units of the network. Interior units that are activated by other interior units are said to be in different *layers* of the network. When a network has only input and output units (i.e., no interior units), the network is said to have a perceptron architecture (Rosenblatt, 1962). Most biologically plausible neural networks have a number of hidden layers. *Output units* simulate the behavior of the organism or, in the case of a module, the events that are passed to the input units of another module within a larger network.

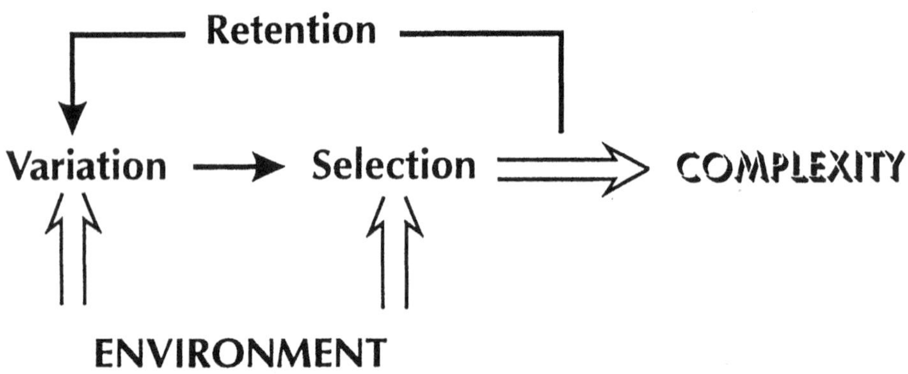

FIGURE 2. A neural-network architecture showing input, hidden (or interior), and output units. Connections between units are also shown with arrows indicating recurrent as opposed to feedforward connections.

Connections are classified into two types. Those connections that go from units in one layer to units in "deeper" layers are called *feedforward connections*. Connections that go from units in "deeper" layers to units in more "superficial" layers are called *recurrent connections*. A recurrent connection allows the activity of the unit from which it arises to affect its own activity at a later time in the operation of the network. This occurs either when the recurrent connection directly reactivates the originating unit or when it activates a more superficial unit in the network that, in turn, reactivates the originating unit. Feedforward connections convey activations from input toward output units. Recurrent connections permit the activity of interior units to affect the functioning of the network at a later point in time. In the nervous system, feedforward and recurrent connections occur with approximately equal frequency. For example, feedforward projections from thalamus to cortex are complemented by recurrent projections from cortex to thalamus. Similarly, feedforward intracortical pathways extend from sensory-association to motor-association cortex and are complemented by recurrent connections in the other direction.

Activation function. Given a network architecture with input, interior, and output units and associated connections, how does the network operate? Events in the environment of the network stimulate its input units and thereby initiate a pattern of activity that ripples through the network until, possibly, output units are activated. The process whereby presynaptic units affect the activity of postsynaptic units via the connection between them is described by an *activation function.* In its simplest form, the level of activation of a postsynaptic unit is determined by three variables—whether the effect of the presynaptic unit is to increase (an *excitatory effect*) or decrease (an *inhibitory effect*) the activation of the postsynaptic units, the level of activation of the presynaptic unit, and the strengths of the connections, or *connection weights* between pre- and postsynaptic units. If a_i symbolizes the activation of the *i*th presynaptic unit and w_{ij} symbolizes its connection weight to the *j*th postsynaptic unit, then the activation level of unit *j* at time *t* is the product of a_i and w_{ij} at time *t*-1. When the connection is from an excitatory unit, a_i has a positive sign; when a_i is from an inhibitory unit, it has a negative sign. If there are a number, *n*, of connections from multiple presynaptic units, each with its own connection weight to the *j*th postsynaptic unit, then the total activation is the sum (with respect to sign) of the products of the presynaptic activations and their associated connection weights (i.e., the inner product in linear algebra).

$$a_j(t) = \sum_{i=1}^{n} a_i(t-1)w_{ij}(t-1) \tag{1}$$

In neural networks, as in the nervous system, the effective activation of the *j*th unit, a'_j, is a nonlinear function of the total activation. In the nervous system, there are a number of sources of nonlinearity; e.g., saturation of ionic flow across concentration gradients as described by the Goldman equation and the firing threshold of neurons. In neural-network research, the nonlinear function is referred to as a "squashing" function and commonly takes the form of the logistic function, Λ, which resembles a cumulative normal distribution. The effective activation (a') after "squashing" by the logistic function is

$$a'_{j,t} = \Lambda(a_{j,t}) \tag{2}$$

If the activation function were linear, the addition of interior units to the network architecture would have no effect on the computational capability of the network: A computationally equivalent network could be constructed that lacked interior units altogether. However, it is known that nonlinearities are necessary to mediate some behaviorally observed input-output relations. For example, in order for a learner to respond when either of two events is present but not when both are present or neither is present, nonlinearly activated hidden units are required. This task is called a patterning problem in the field of learning and an exclusive-or problem in artificial intelligence. It is the simplest nonlinearly separable pattern discrimination. (Nonlinear input-output relations can also be mediated by networks having what are called "higher-order" input units that are designed to respond to only the conjunction of events, and not to either event separately. However, such input units must be present for all possible conjunctions of events that might conceivably be useful to the operation of the network, an unrealistic strategy that has many of the same virtues with respect to nonlinear hidden units as theft with respect to honest toil. Accordingly, most biologically plausible neural networks have multiple layers and nonlinear activation functions.)

Learning algorithm. In a neural network, learning is conceptualized as a change in the connection weights between units. As the values of the w_{ij}s change, the same events acting on the input units of the network produce different activations of the output units in an otherwise constant network architecture. Thus, as with living organisms, the behavior of the network changes as its learning history changes. The process whereby connection weights are modified is described by a *learning algorithm.*

It is useful to distinguish between two general types of learning algorithms—unsupervised and supervised. In *unsupervised learning*, the learning algorithm is unaffected by feedback from the activations of other units or by any environmental consequences of that pattern of activation. That is, there are no means by which the performance of the network affects the learning process. On the neural level, unsupervised learning corresponds to changes in synaptic efficacy that result only from events that are local to the affected synapse with no direct or indirect influence of remote events. In the second type, *supervised learning*, the learning algorithm is affected not only by local events but also by more remote events taking place at output units or in the environment as a consequence of the activity of output units. On the neural level, if the occurrence of food following a response led to the release of a neuromodulator that affected synaptic efficacies in the network, then a supervised-learning algorithm would be required to simulate the behavioral selection process.

Two classes of supervised-learning algorithms are commonly distinguished—reinforcement learning and instructed learning. In *reinforcement learning*, feedback takes the form of a single value that varies with some global measure of remote events. Consider a rat pressing a lever for food. If feedback took the form of a single value that varied directly with the amount and inversely with the delay in the receipt of food, then a reinforcement-learning algorithm would be appropriate. Technically, in reinforcement learning, feedback is a scalar quantity. In *instructed learning*, feedback takes the form of multiple values that reflect different aspects of the remote event. Returning to the lever-pressing example, if the rat received separate feedback concerning the height to which the paw was raised prior to pressing, which paw pressed the lever, and the force with which the lever was pressed, then an instructed-learning algorithm would be required. Technically, in instructed learning, feedback takes the form of a vector quantity.

Note that a supervised-learning algorithm for instructed learning requires the learner to be able to distinguish between different aspects of its performance prior to the beginning of learning. Without prior learning, the different feedback values that correspond to each aspect of performance could not be appreciated. As an illustration, consider the following advice to a novice tennis player after a poor backhand return. "Your return was poor because you failed to draw back your racket soon enough and you hit the ball off your rear foot instead of your front foot." The novice could not benefit from this feedback unless he had already learned to discriminate the words as well as the positions of the ball relative to various parts of his body. For such reasons, instructed-learning algorithms depend on a history of prior learning to be effective and, accordingly, are unlikely to describe the learning process in inexperienced organisms. In a genuine sense, supervised learning by means of instructed learning assumes the prior effects of reinforcement-learning algorithms. (The most common approach in cognitive psychology employs an instructed-learning algorithm, the generalized delta rule; Rumelhart et al, 1986; Werbos, 1974. The delta rule is conceptually related to discrepancy-based reinforcement learning; Sutton & Barto, 1981; but its implementation is problematic in a biologically plausible neural network.)

Neural networks and neuroscience

In normative psychology, underlying structures and processes—including neural networks and learning algorithms—are inferences from observations at the behavioral level constrained only by logical/mathematical arguments. The approach taken in this volume differs in that the structures and processes that underlie behavioral observations are the result of direct observations at the level of the neurosciences. Thus, network architectures and learning algorithms are based on *independent* experimental observations at the same levels as their counterparts in the nervous system (cf. Donahoe & Palmer, 1994). The choice of the neurosciences as the foundation for neural-network research, instead of inferences from behavior and logical/mathematical arguments alone, does not deny contributions from these other sources. Any neuroscience-based treatment of neural networks must yield results that are consistent with behavioral observations, and must satisfy standard criteria of parsimony and logical consistency. In addition, the use of neural-network technology to solve engineering problems may benefit from relaxing constraints based on the characteristics of biological systems.

Levels of analysis. Given that the neural-network research in this volume is informed and constrained by findings from the neurosciences, other issues present themselves. The neurosciences include observations at levels that range from biophysics and molecular biology to neural systems and animal behavior. What is the "appropriate" observational level(s) for a biologically grounded approach to neural networks?

No a priori answer to this question can be given, just as no answer can be given to a comparable question about the appropriate level of analysis for experimental work. What is the level at which questions about neuroanatomy should be answered—electron or light microscopy? What is the level at which questions about synaptic plasticity should be answered—patch-clamp or biochemical measurements? Aside from the nature of the question itself, the choice of a level of analysis in experimental science depends on whether orderly relations are found between independent and dependent variables at that level. If orderly relations are found, then the level of analysis is, for the moment, appropriate. However, if the relations between manipulated and measured variables are not orderly, then either the wrong variables at that level are being studied or the analysis must move to a lower level, at which the effects of previously unconsidered variables are investigated. In general, science typically encounters unruly relations with continued study of an empirical phenomenon at a given level, and then the level of experimental analysis shifts downward.

At the same time that a reductionist course is pursued experimentally, understanding the ordinary world of the unaided senses advances as formal techniques are developed that trace the complex effects of our ever-deepening understanding of fundamental processes. Neural-network research interprets the cumulative effects of the biobehavioral processes uncovered through experimental research and, like experimental research, exploits those levels of analysis that yield orderly relations. In the case of neural networks, the relations are between the activations of input and output units, and their orderliness is evaluated by the degree to which they simulate the relations between independent and dependent variables observed in living organisms.

In short, the appropriate levels of analysis for experimental science and neural-network research are governed by essentially similar considerations. For example, if the operation of a neural network accurately simulates the target phenomenon with activation functions that simply sum the excitatory and inhibitory inputs to a postsynaptic unit, then that is an appropriate level of analysis for the phenomenon. However, it is known experimentally that some inhibitory synapses (those on dendrites) produce graded responses in neurons, whereas other inhibitory synapses (those on the soma or axon hillock) completely negate excitatory dendritic inputs. Therefore, it is likely that a phenomenon will ultimately be encountered that cannot be adequately simulated with a neural network using simple summation for its activation function. At that point, the differential effects of dendritic and somatic inhibitory synapses may need to be incorporated into the simulation. Biologically plausible neural-network simulations must not contain features that are contrary to experimental findings, but they need include only those features that are necessary to accurately simulate the phenomena of interest.

A Conceptual Scheme

The various levels of analysis that contribute to neural-network research are schematically presented in **Figure 3**.

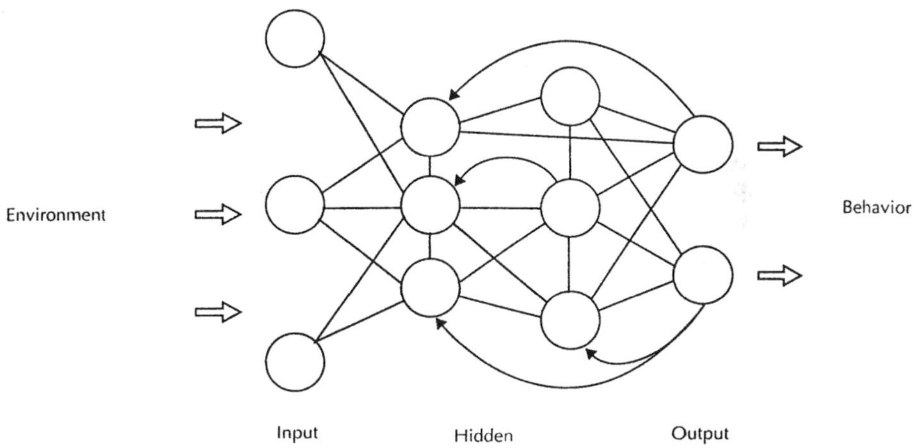

FIGURE 3. A conceptual framework for neural-network research. (See the text for a discussion.)

Learning algorithm

We begin the selection process with the network architecture shown in the oval in the middle of the upper portion of **Figure 3**. The input units for this architecture are stimulated by environmental events that activate interior and output units according to an activation function. A learning algorithm then modifies connection weights within the network. The learning algorithm operates on the connection weights in a manner that is determined by the particular history of inputs and feedback experienced as a consequence of the environmental algorithm.

Environmental algorithm

The environmental algorithm determines the input to the network by means of a fading function. The fading function defines the particular sequence of environmental events that stimulate the input units of the network, and this sequence may be affected by the output of the network. For example, if the network is performing poorly, training could be restricted to more easily discriminated input patterns. The shaping function defines the criteria that the output pattern must satisfy for the network to receive environmental feedback in supervised learning. For example, feedback might be given for partially correct output patterns early in training. Interactions between the learning and environmental algorithms also govern the reinforcement function. The reinforcement function determines the form of the feedback given in supervised learning. Feedback might take the form of either a scalar or a vector signal that is some function of the output pattern occurring at that point in training. For example, the strength of the reinforcement signal might be greater for closer approximations to the target output pattern. In brief, the environmental algorithm implements the contingencies between environmental, behavioral, and reinforcing events that are present in the training environment—technically, a triplet of input, output, and reinforcement vectors. Contingencies of this sort are studied in the experimental analysis of behavior.

Once the connections between units in the network have been differentially strengthened through the learning algorithm as governed by the environmental algorithm, the result is a trained network that mediates the target input-output relations to varying degrees. If a founding population of networks of different architectures is modified by a learning algorithm, the members of that population will, in general, vary in the adequacy with which they perform. These variations reflect to some degree the particular sequence of contingencies to which the networks were exposed by the environmental algorithm. But, they are also importantly influenced by any differences in the network architectures at the beginning of the simulation. Just as not all nervous systems acted upon by the same set of experiences will mediate the same behavior, so not all networks modified by the same learning algorithm will mediate the same input-output relations. Chimpanzees and deaf humans exposed to the same sign-language environment behave differently. And, different network architectures may also behave differently—even after identical histories of behavioral selection.

Genetic algorithm

Since the architecture of a neural network is a crucial determinant of the input-output relations that it can mediate, a comprehensive program of simulation research includes some means for modifying the architecture as well as the connection weights of networks. The structure of the nervous system is a joint product of influences of two major sources—genetics and neurodevelopmental processes. Accordingly, a biologically plausible genetic algorithm simulates mechanisms of these two origins to differentially reproduce network architectures in proportion to their fitness; i.e., their ability to mediate the target input-output relations. The product of the genetic algorithm is a new population of network architectures to be acted upon by the interaction of the learning and environmental algorithms. The culmination of the selection process is a population of neural networks in which connection weights are modified by the learning algorithm and network architectures by the genetic algorithm to simulate complex performance by means of biobehaviorally plausible mechanisms.

The neural-network simulations in this volume illustrate major aspects of the conceptual scheme depicted in **Figure 3**. **Andrew Barto and Richard Sutton** provide a general conceptual framework that exposes fundamental issues in **reinforcement learning**. In separate contributions, **John Donahoe**, **Read Montague**, and **John Moore and June-Seek Choi** propose biologically plausible realizations of reinforcement **learning algorithms**. As already noted, the ability of any learning algorithm to modify connection weights so that the network can mediate input-output relations depends on the architecture of the network. In keeping with that view, the chapter by **Stephen Senft** describes **simulations of neurodevelopmental processes** and the one by **José Burgos** employs a **genetic-developmental algorithm** to produce network architectures. **Neil Berthier**, **Jeffery Clouse**, and **Vijaykumar Gullapalli** separately describe simulation research that employs techniques such as shaping that permit networks to mediate complex, sequential output patterns (i.e., **complex control problems**). **Dean Buonomano and Michael Merzenich**, and **John Hummel** and **Arnold Trehub** in their individual contributions, indicate how neural networks may come to appreciate **complex sequences of input patterns**. Finally, chapters by **Dermot Barnes and Peter Hampson**, **Vijaykumar Gullapalli and Jack Gelfand**, **Michael Jordan**, and **Guy Van Orden, Anna Bosman, Stephen Goldinger, and William Farrar** indicate how neural networks may be used to simulate aspects of that most complex human behavior, **language**. No one simulation exploits all aspects of the conceptual scheme, embodies all potentially relevant biological processes, or accounts for all aspects of the simulated phenomena. Nevertheless, these simulations collectively represent a promising beginning toward the achievement of the ultimate goal of understanding complex human behavior in terms of basic biobehavioral processes through the use of adaptive neural networks.

ACKNOWLEDGMENTS

The author expresses his appreciation to David Palmer and José Burgos for discussions that led to the present formulation. The work was supported by a Faculty Research Grant from the Graduate School, University of Massachusetts, and by grants from the National Institute of Mental Health and the National Science Foundation.

PART ONE: NETWORK ARCHITECTURE AND NEUROANATOMY

This first part of the volume illustrates the use of simulations to explore the implications of neuroscience for the structure of the real nervous system and, reciprocally, the use of neuroscience to constrain the architectures of artificial neural networks. (Note: These introductions were prepared by the editors without consulting the authors of the chapters, and should be regarded as expressing the views of the editors alone.)

As noted in the preceding chapter, the effect of the environment on behavior depends, in part, on the *structure* of the organism on which the environment acts. Consider language. The cellular mechanisms involved in learning (neural plasticity) appear to be identical throughout the class of mammals and, indeed, substantially the same across the entire animal kingdom. And yet, whereas almost all members of our species acquire vocal speech when exposed to a human social environment, chimpanzees do not (Hayes, 1951). The absence of the effectors and neural structures necessary for vocal speech precludes their acquisition of language. Generalizing from this example, even if we knew with perfection the cellular mechanisms of learning and the environmental history required for language acquisition and we were able to faithfully implement that knowledge in a simulation, we could not simulate language acquisition in an artificial neural network lacking the necessary architecture.

Luskin's chapter provides an overview of some of the modern methods available to study the architecture of the real nervous system and some major findings obtained with those methods. Several general conclusions may be drawn from this work. First, modern neuroanatomical methods allow the development of the nervous system to be traced with a precision and completeness that was inconceivable even a few years ago. Thus, the information required for detailed simulations of neural development is becoming available. Second, evolution has devised multiple "strategies" whereby the genome guides neural development. For example, the basic structure of the neocortex appears to result from the migration of incipient neurons (neuroblasts) along glial strands arising from structures deep within the brain, whereas—in other regions of the brain—neurons find their paths through less structural, more purely chemical mechanisms.

Senft's chapter makes the general point that the complexity of neurodevelopmental processes is so great that simulation techniques are *required* if the competence of those processes to produce neural architectures is to be evaluated. Furthermore, graphical presentations produced by the simulations permit the concerted effects of those processes to be appreciated. A functionally comparable problem is confronted when tracing the development of galaxies and solar systems—although the temporal and spatial scales of such processes are much greater, of course. As with the nervous system, the complexity of the resulting structure requires simulation to determine if basic physical processes—e.g., gravity, adhesion, etc.—can mimic galactic evolution as their cumulative product (e.g., Rasio & Ford, 1996). To evaluate the adequacy of simulations of neural development, **Senft** proposes a kind of Turing test; namely, do experts mistake the graphical results of the simulations for microscopic views of the real nervous system? This test is analogous to evaluating the adequacy of cosmological simulations by judging their similarity to telescopic views of the universe. As indicated by the reactions of my colleagues to **Senft**'s graphical representations, some simulations already pass muster for limited regions of the nervous system. Although we are clearly at the beginning of the enterprise that **Senft** has pioneered, the portents are auspicious.

Burgos's chapter describes the selection of architectures for artificial neural networks using a genetic algorithm (see Chapter 1) that is inspired by experimental work on neurodevelopment. In normative cognitive science and artificial intelligence (e.g., McClelland & Rumelhart, 1986), network architectures are designed (and redesigned) such that the input-output relation of interest can be mediated by the network when its connection weights are modified by learning. That is, the network architectures are "handcrafted" rather than produced by a set of principles. The genetic algorithm devised by **Burgos** begins with a founder population of simulated chromosomes and then, by simulating neurodevelopment, produces a population of network architectures whose connections are modified by the learning algorithm. The chromosomes that lead to these networks through neurodevelopment are then reproduced in proportion to the performance of the networks on the task. In this way, the genetic algorithm simulates the process of natural selection to determine network architecture. **Burgos** demonstrates that a single founder population of chromosomes can lead to subsequent populations of chromosomes that generate networks capable of performing a variety of tasks. For example, networks produced in this manner can learn to be sensitive to different temporal relations between events due to interactions between units in the network—i.e., timing occurs without a "timer" (see also Buonomano & Merzenich, this volume). Hybrid genetic-learning algorithms promise a principled approach to selecting neural-network architectures that permit the simulation of environment-behavior relations of indefinite complexity.

Neural-Networks Models of Cognition
J.W. Donahoe and V. Packard Dorsel (Editors)
1997 Elsevier Science B.V.

CHAPTER 2

PROGENITOR CELLS OF THE MAMMALIAN FOREBRAIN: THEIR TYPES AND DISTRIBUTION

Marla B. Luskin
Department of Anatomy and Cell Biology
Emory University School of Medicine

ABSTRACT

Important events underlying the proper functioning of the central nervous system include the production, assembly, and differentiation of appropriate types and numbers of cells. The mechanisms that control these phenomena have been difficult to unravel in the mammalian forebrain because of its diverse cellular composition and because of the displacement of its cells from their site of origin to their permanent location during development. Nonetheless, headway has been made in our understanding of how cells of the developing forebrain "decide" where to go and what to become. The availability in the past few years of replication-defective recombinant retroviruses that encode heritable and easily detectable marker genes, as well as a class of fluorescent lipophilic markers (i.e., molecules that associate with lipids in membranes), has made it possible to tag progenitor cells and subsequently identify their progeny either morphologically or immunohistochemically. This approach has been used to determine the properties and location of the progenitor cells of the forebrain and to analyze the distribution and phenotype of lineally related cells in the developing and mature forebrain. The findings obtained by these studies support the notion that the progeny of individual telencephalic germinal-zone cells generally remain in relatively close proximity because they traverse similar paths. Furthermore, progenitor cells of the prenatal and postnatal brain are a composite of specialized cell types, and the progeny of individual progenitors share a common phenotype. This chapter focuses on the similarities and differences between how the cerebral cortex and the olfactory bulb develop, as a way of deducing the range of mechanisms responsible for generating the diversity of cell types characteristic of each structure.

Generation of Cells Destined for the Cerebral Cortex

The development of the mammalian cerebral cortex proceeds by an orderly sequence of events that has been elucidated by numerous studies utilizing several experimental approaches (for review see McConnell, 1988; Rakic, 1988b). During prenatal development, cells that form the cerebral cortex arise from a layer of neuroepithelial cells surrounding the cerebral ventricles. Postmitotic neurons separate from this ventricular (germinal) zone and form the cortical plate (future cellular layers of the cerebral cortex) by migrating through the intermediate zone toward the outer pial surface (**Figure 1**). This migration of immature neurons takes place predominantly in association with a specialized population of glial cells whose processes extend the full width of the developing cerebral cortex (for review see Rakic, 1990). These radial glial fibers may serve as a road map to help guide migrating cortical neurons to their final destinations. The postmitotic neurons complete their differentiation in the cortical plate and organize into layers; the earliest-born permanent

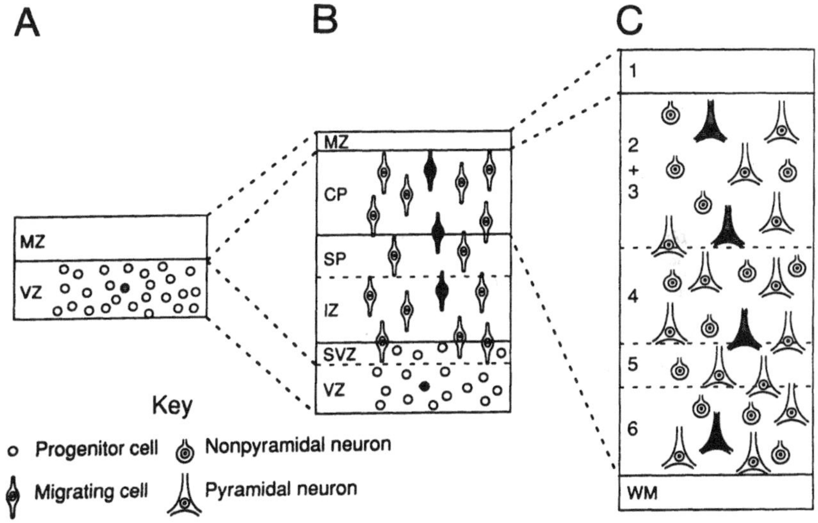

FIGURE 1. Sequential stages in the development of the cerebral cortex, and hypothetical formation of a cortical clone. **A.** Schematic appearance of the cerebral mantle before the generation of postmitotic neurons. The ventricular zone (*VZ*) consists of a densely-packed layer of proliferating cells, and the marginal zone (*MZ*) contains relatively few cell somata. **B.** Schematic drawing of the cerebral mantle midway through cortical neurogenesis. The ventricular zone and subventricular zone (*SVZ*), another layer of dividing cells, generate neurons and glia destined for the cerebral cortex. Most postmitotic neurons migrate radially away from the ventricular zone. The postmitotic neurons migrate through the overlying intermediate zone (*IZ*) and subplate (*SP*), a transient layer of early-born neurons, to the cortical plate (*CP*), where they differentiate. **C.** Schematic drawing showing the layers of the mature cerebral cortex and the position of postmitotic, postmigratory cortical neurons in layers 2-6. The *dark gray cells* in each drawing represent the hypothetical arrangement of clonally related cells in the developing cerebral cortex. The presumed sequence of events is for a progenitor cell in the ventricular zone to divide a number of times (A), casting out postmitotic neurons, which migrate to the cortical plate (B) and differentiate in relative proximity to each other (C). *WM* is white matter containing axons projecting to and from the cortex.

neurons are situated deepest in the cerebral cortex and the latest born are the most superficial (e.g., Rakic, 1974; Luskin & Shatz, 1985a; **Figure 1**). After neurogenesis has occurred and the depletion of the ventricular zone has been completed, glial cells (astrocytes and oligodendrocytes) continue to be generated postnatally in the subventricular zone, which subsequently surrounds the cerebral ventricles (Privat, 1975; LeVine & Goldman, 1988).

Despite our basic understanding of the spatial and temporal patterns of proliferation and migration of cells destined for the cerebral cortex, our understanding of how, when, and where a cortical cell's identity is established is limited. Given that the cells of the cerebral cortex can be phenotypically dis-

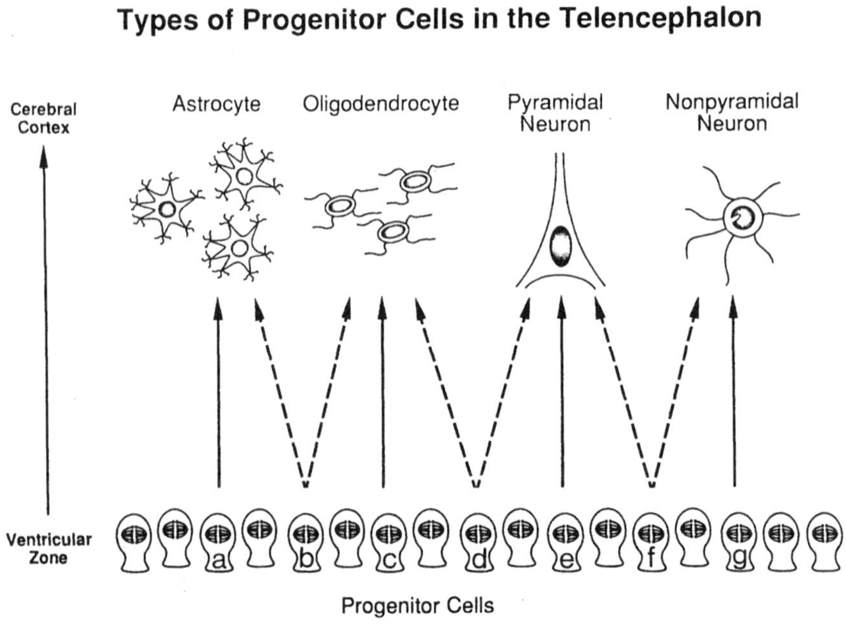

FIGURE 2. Possible types of progenitor cells in the germinal ventricular zone of the developing rat cerebral cortex. Our lineage studies suggest that by the onset of cortical neurogenesis most progenitor cells (*cells a, c, e, and g*) of the prenatal ventricular zone generate a homogeneous population of cells; an individual progenitor cell generates either all astrocytes, all oligodendrocytes, all pyramidal neurons, or all nonpyramidal neurons. Although we cannot rule out the presence of bipotential glial progenitor cells (*cell b*) in the ventricular zone (or subventricular zone) which give rise to astrocytes and oligodendrocytes, they were not detected. Nor did we obtain compelling support for multipotential progenitor cells which give rise to neurons and glia (*cell d*) or to pyramidal and nonpyramidal neurons (*cell f*). Collectively our results suggest that the lineages for the major subtypes of cells in the cerebral cortex have diverged by the onset of cortical neurogenesis.

tinguished by their morphology, connectivity, laminar position, and transmitter candidates, it is likely that a series of "decisions" progressively restricts the fate of a progenitor cell or its progeny. These decisions must be coordinated but need not occur at the same time or place. Moreover, it is important to distinguish which phenotypic features are imposed by inherited factors and are therefore irreversible, and which are sculpted by environmental factors, such as responses to extracellular matrix-bound cues or diffusible factors.

The types of progenitor cells in the prenatal telencephalic germinal zone

Recently, several investigators have investigated which, if any, of the phenotypic properties of a cell in the cerebral cortex are dictated by genetic factors (Luskin, Pearlman, & Sanes, 1988; Price & Thurlow, 1988; Walsh & Cepko, 1988, 1992; Luskin, Parnavelas, & Barfield, 1993; Temple, 1989; Grove, Williams, Li, Hajhosseini, Friedrich, & Price, 1993; Parnavelas, Barfield, Franke, & Luskin, 1991; Williams, Read, & Price, 1991; Mione, Danevic, Boardman, Harris, & Parnavelas, 1994). These studies have addressed the question of whether there are separate progenitor cells for the major subtypes of cells in the cerebral cortex (**Figure 2**). If separate progenitor cells exist for each major subtype, then all the cells derived from any given progenitor should have the same phenotype. This problem is more amenable to examination when the cells of the cerebral cortex are subdivided into the two main types of glia, astrocytes and oligodendrocytes, and into the two main types of neurons, pyramidal and nonpyramidal cells. Pyramidal or projection neurons have axons that project relatively long distances and act to excite the cells upon which they synapse, whereas virtually all nonpyramidal cells or interneurons have locally ramifying axons and are inhibitory neurons.

The most direct approach currently available for revealing whether individual progenitor cells of the ventricular zone generate a heterogeneous or a homogeneous population of cells with respect to the major cell classes is to use retroviral-mediated gene transfer to introduce a marker gene into the DNA of dividing progenitor cells. The integrated marker gene is subsequently inherited by all progeny of the infected cell (Sanes, 1989). The most commonly used retroviral lineage tracers contain the *Escherichia coli lacZ* (ß-galactosidase) gene, the expression of which can be detected histochemically and immunohistochemically in the offspring of infected cells by light and electron microscopy (**Figures 3** and **4**). In our studies, we have relied on well-established ultrastructural features to distinguish between astrocytes, oligodendrocytes, pyramidal neurons, and nonpyramidal neurons (Peters, Palay, & Webster, 1991; Parnavelas, Luder, Pollard, Sullivan, & Lieberman, 1983). We have turned to the ultrastructural level because the identification of ß-galactosidase-positive [lacZ(+)] cells at the light-microscope level can be inconclusive.

Because pyramidal and nonpyramidal cells of the cerebral cortex are generated concurrently throughout the period of cortical neurogenesis (Luskin &

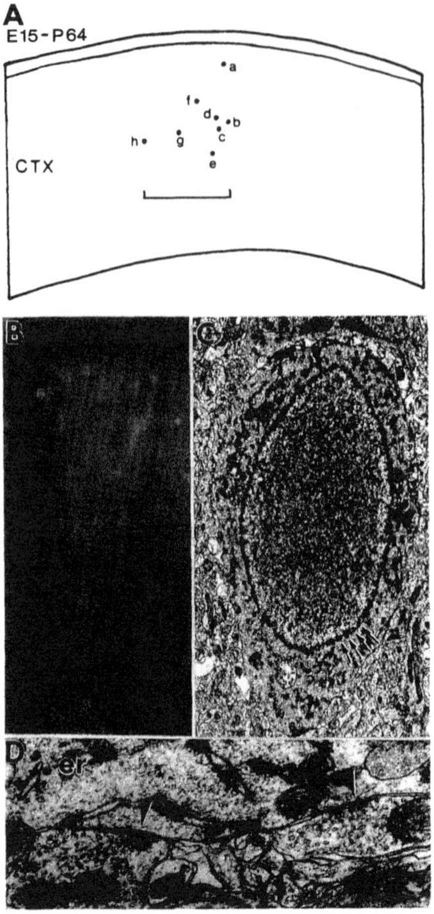

FIGURE 3. Representative example of a homogeneous clone of pyramidal neurons in the mature cerebral cortex from a rat brain injected at the onset of neurogenesis with a replication-defective recombinant retrovirus encoding the E. coli lacZ gene. **A.** The distribution in the cerebral cortex of the eight constituent pyramidal neurons, situated in layers 2-5, constituting the clone. The *bracket*, measuring 550 μm, delimits the clone's maximum dimension. **B.** Bright-field photomicrograph (200x) of two of the lacZ-positive histochemically stained neurons (*b and d*). **C.** and **D.** Electron micrographs (3,700x and 21,100x, respectively) of lacZ-positive cells. At the ultrastructural level, the ß-galactosidase histochemical reaction product is electron dense, and is associated predominantly with the nuclear membrane (*arrows* in C) and granular endoplasmic reticulum (*er* in D). The presence of exclusively symmetrical axosomatic synapses (*arrowheads* in D) was used to conclusively identify pyramidal cells. E15, embryonic day 15; P64, postnatal day 64. (Modified with permission from Luskin, 1993b.)

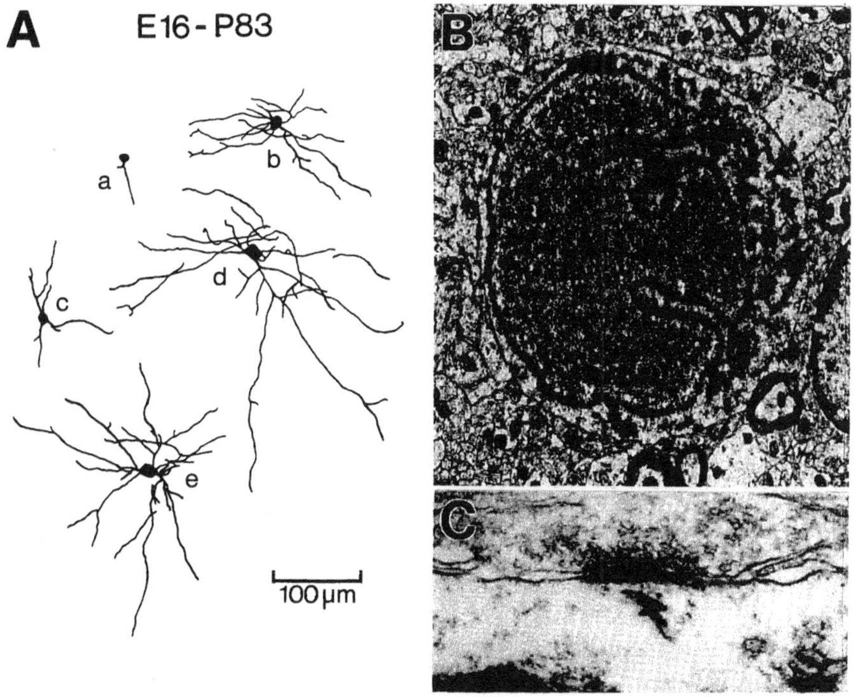

FIGURE 4. Representative example of a homogeneous clone of nonpyramidal neurons in the adult rat cerebral cortex resulting from the injection of a retroviral lineage tracer encoding the lacZ gene at E16. **A.** *Camera lucida* drawings illustrating the similarities at the light-microscopic level of the five clonally related cells constituting the clone. With the exception of *cell a*, the least well-stained member of the clone, the lacZ-positive cells of the clone exhibit characteristics associated with multipolar nonpyramidal neurons. **B.** At the ultrastructural level *cell a* also displayed the features of a nonpyramidal neuron, including an invaginated nucleus of irregular shape, as well as darkly stained nucleoplasm and nuclear membrane. **C.** An example of an asymmetrical axosomatic synapse on a lacZ-positive nonpyramidal cell soma. Note that the presynaptic thickening is greater than the postsynaptic thickening. The presence of both symmetrical and asymmetrical axosomatic synapses, one of the criteria used to classify a cell as a nonpyramidal neuron, were detected on *cell a*, as well as on *cells b-e*. B, x6,370; C, x11,000. (Modified with permission from Luskin et al, 1993.)

Shatz, 1985a), retroviral lineage tracers can be administered at the onset of cortical neurogenesis to determine whether these two lineages are distinct from each other. Experiments employing lineage tracers in rats yielded clear-cut results, provided that the discrete clusters of lacZ(+) cells found in the cerebral cortex were treated as groups of clonally related cells (Luskin et al, 1988, 1993; Parnavelas et al, 1991). (The justification for this definition of clonality is discussed below.) In the developing cerebral cortex, discrete collections of lacZ(+) cells were often observed to be radially arrayed (Luskin et al, 1988; Austin & Cepko, 1990; Price & Thurlow, 1988; Walsh & Cepko, 1992; Grove et al, 1993), and in the mature brain, well-isolated clusters of cells were also apparent although not always as tightly aligned. When the phenotypes of the cells within clusters, or clones, were established ultrastructurally, we found that they were composed wholly of astrocytes, oligodendrocytes, pyramidal neurons, or nonpyramidal neurons (Luskin et al, 1988, 1993; Parnavelas et al, 1991). That is, by the onset of neurogenesis, virtually all the progeny of an individual progenitor cell in the ventricular zone were of the same type. Similar conclusions were reported by Grove et al, 1993 and Mione et al, 1994. Collectively, these results provide strong evidence that the lineages for the major subtypes of cells have diverged by the onset of cortical neurogenesis. However, the possibility that a small proportion of the progenitor cells are bipotential or multipotential cannot be ruled out.

Distribution of clonally related cells in the cerebral cortex. As indicated, the proposition that there are separate progenitor cells in the developing ventricular zone for the major cellular subtypes is based on the notion that a cluster of lacZ(+) cells constitutes a clone. Several observations support this contention. First, there is indirect evidence at the ultrastructural level: Cells of a putative clone have a similar pattern and intensity of ß-galactosidase histochemical staining (Luskin et al, 1993). From clone to clone, however, a wide range of expression exists. If the level of the expression of ß-galactosidase is a function of the insertion site of the viral DNA, consistent expression from cell to cell within a cluster implies that the relationship between the cells is lineal. Second, Austin and Cepko (1990) showed that the number of cerebral cortical clusters after introduction of the retrovirus at the onset of neurogenesis was dependent on the amount of virus injected, as would be expected if the progeny of each progenitor cell were restricted in their distribution. Furthermore, on average, the number of cells per cluster did not vary as a function of the amount of virus administered, although injections at later developmental times produced clusters of fewer cells (Luskin et al, 1988, 1993). Taken together, these results are consistent with the interpretation that cells within discrete lacZ(+) clusters are clonally related.

On the other hand, Walsh and Cepko (1992) showed that the progeny of a contingent of progenitor cells in the ventricular zone become widely dispersed in the cerebral cortex, occasionally over the length of the entire anterior-posterior axis, and that some groups of closely spaced cells are the products of more than one progenitor cell. Walsh and Cepko injected a solution composed of a library of retroviruses—each engineered to contain the lacZ gene and one of 100 genetically distinct viral inserts—into the embryonic brain of rats, with the aim of infecting a small number of ventricular-zone cells. After waiting several days for proliferation and migration to ensue, Walsh and Cepko mapped the distribution of lacZ(+) cells in the brain and determined the specific viral construct present in each lacZ(+) cell by amplifying the sequence of inherited viral DNA with the polymerase chain reaction. They found that clustering of cells bearing the identical insert did occur. Furthermore, some cells possessing identical DNA inserts were separated by many millimeters or centimeters; this was presumed to be the consequence of widespread migration of cells derived from the same infected progenitor cell. For an indisputable argument to be made for such widespread dispersion, however, one must accept a statistical proof that the pool of retroviruses injected contained a maximum of only one copy of each construct and, therefore, that the occurrence of disparately placed cells in the cortex possessing the same insert is not the result of two identical viruses infecting two different cells that produced progeny residing in two different parts of the cortex. The statistical arguments used by Walsh and Cepko have been contested by other investigators (Kirkwood, Price, & Grove, 1992), complicating the interpretation of their results. However, the results of Luskin et al (1993) are not necessarily contradictory to those of Walsh and Cepko (1992), since the majority of clonally related cells remain together in the cerebral cortex. Walsh has recently devised a model to reconcile the seemingly disparate results concerning the extent of dispersion of clonally related cells in the cortex (Reid, Liang, & Walsh, 1995), which can account for the findings of a primate-lineage study of the cells of the cerebral cortex (Kornack & Rakic, 1995).

Another way to evaluate whether a cluster of labeled cells is truly a clone is to determine the probability that homogeneous clusters of neurons and glia arise by chance alone. We found that the probability was extremely low ($p < 0.0001$; Luskin et al, 1993). Similar statistical considerations argue for the existence of separate progenitor cells for pyramidal and nonpyramidal neurons by the time cortical neurogenesis begins. Even though we cannot be certain that the clusters we analyzed constituted complete clones in every case, our general conclusion that clusters represent groups of lineally related cells remains valid.

Radial and nonradial migration of immature cortical neurons. Because of the ongoing uncertainty about the extent to which clonal cohorts disperse, several studies have addressed the issue of whether the predominant form of cell migration in the developing cerebral cortex is radial or nonradial. The excursions of migrating cells into the intermediate zone (**Figure 1**) of the neonatal ferret cortex have been charted by time-lapse microscopy of cells labeled with a nontoxic, lipophilic fluorescent dye, DiI, by O'Rourke, Dailey, Smith, & McConnell (1992). The vast majority of cells traveled in a radial or near-radial direction during the observation period; migration in the nonradial direction, orthogonal to the pial surface, was recorded for a small proportion of the cells.

In addition, some migrating cells changed their orientation mid-course. Whether the amount of nonradial migration observed is sufficient to account for the widespread clones reported by Walsh and Cepko (1992) remains an open question. However, using different techniques, other investigators have reported markedly less net horizontal migration in the intermediate zone (Tan & Breen, 1993; Nakatsuji, Kadokawa, & Suemori, 1991) than in the cortical plate.

As observed by Tan and Breen (1992), X-chromosome inactivation in transgenic mice carrying the lacZ gene on the X chromosome results in a mosaic lacZ expression pattern in the cerebral cortex. This has then been used as a means to evaluate the extent of nonradial migration that occurs during development of the cerebral cortex. Tan and Breen concluded, based on the pattern of ß-galactosidase histochemical staining exhibited by cells in the cortical plate, that clonally related cells seem to initially traverse predominantly the same radial route from the ventricular zone to their destination in the cortical plate. Alternating bands of lacZ(+) cells, referred to as developmental modules, appeared first in the ventricular zone and then extended into the overlying cortical plate, remaining in register with the ventricular-zone labeling. These results were interpreted as providing support for the radial-unit hypothesis of Rakic (1988b). This hypothesis states that the ventricular zone is composed of proliferative units that constitute a "protomap," which presages the cytoarchitectonic layout of the mature cortex. The small amount of mixing of lacZ(+) and lacZ(-) cells in the developing mouse cortex is perhaps the most convincing evidence to date for a limited degree of nonrandom, widespread dispersion of clonally related cortical cells. Tan and Breen (1992) found that bands or columns of lacZ(+) cells were also present in the adult cortex, although labeled cells punctuated the adjacent unlabeled columns, suggesting that some cells had migrated secondarily in a horizontal direction.

The results of Nakatsuji et al (1991) provide further evidence that the majority of clonally related cells do not undergo widespread dispersion during migration to the cortical plate. Radial bands of lacZ(+) cells extend across the developing cerebral cortex in lacZ(+) chimeric mice. Once again, at later stages in cortical development, less distinct columns of lacZ(+) cells were present in the cortex, presumably as a consequence of cell mixing within the cortical plate.

Remaining uncertainties. We are left with a conundrum about the degree of dispersion that takes place among the progeny of individual progenitor cells of the telencephalic ventricular zone. Although it seems judicious to conclude that the overriding direction is radial, perhaps paralleling the organization of radial glial fibers, there appear to be notable exceptions (Fishell, Mason, & Hatten, 1993). It would be intriguing if the horizontally migrating cells in the intermediate zone and cortical plate constituted a unique class of neurons or glial cells, as has been shown for the cells that undergo a horizontal displacement in the optic tectum (Gray & Sanes, 1991). Along these lines, Menezes and Luskin (1994) have recently demonstrated a band of horizontally aligned neurons at the interface between the ventricular and subventricular telencephalic zones during embryogenesis in mice. Although the final destination and fate of these cells has not been determined, they have a number of features that distinguish them from most of the immature neurons bound for the cortical plate (Menezes & Luskin, 1994).

A more direct way to potentially resolve the issue of dispersion by immature cortical neurons is to visualize directly the dynamic aspects of the proliferation and migratory behavior of individual progenitor cells and their progeny in cultured slices of developing cerebral cortex; this has now been achieved for a single division (Chenn & McConnell, 1995). In this way, the extent to which the members of a clone migrate along the same path from the ventricular zone to the cortical plate and the amount of dispersion occurring among clonally related cells can be assessed. However, it still must be determined how closely proliferation and migration *in situ* in cultured slices approximates the same events occurring *in vivo*.

In addition to questions about the undetermined extent of dispersion of clonally related cells, there are several outstanding questions about the range of progenitor cell types in the ventricular zone of the developing telencephalon. For example, how and when do separate progenitor cells for astrocytes, oligodendrocytes, pyramidal cells, and nonpyramidal cells arise? In particular, is there a common progenitor cell for pyramidal and nonpyramidal neurons that then gives rise to distinct progenitor cells, or are there bipotential or multipotential cells of other types? Injections of retrovirus at earlier times would address this question. It is technically quite difficult, however, to make intraventricular injections of retrovirus at earlier embryonic stages than has already been done. On the other hand, the possibility of lineages that are even more restricted is ripe for investigation. For example, do the rather unusual subplate cells, which form a transient layer underlying the cortical plate and appear before the cells of the cortical plate (Luskin & Shatz, 1985b), arise from a lineage distinct from that of cortical neurons destined for the overlying cellular layers? The ultimate answer to each of these questions requires an understanding at the molecular level of how differential gene expression is regulated and of how it governs lineage restrictions.

Generation of Cells Destined for the Olfactory Bulb

Unexpected insight into fundamental aspects of the proliferation and differentiation of neural progenitor cells has emerged from our recent studies concerning a zone of proliferating cells in the postnatal brain (Altman, 1969; Kishi, 1987; Luskin, 1993a). After the cessation of cortical neurogenesis, the subventricular zone of the forebrain continues to supply the postnatal brain with glial cells (Privat, 1975). We and others have shown that injections of a

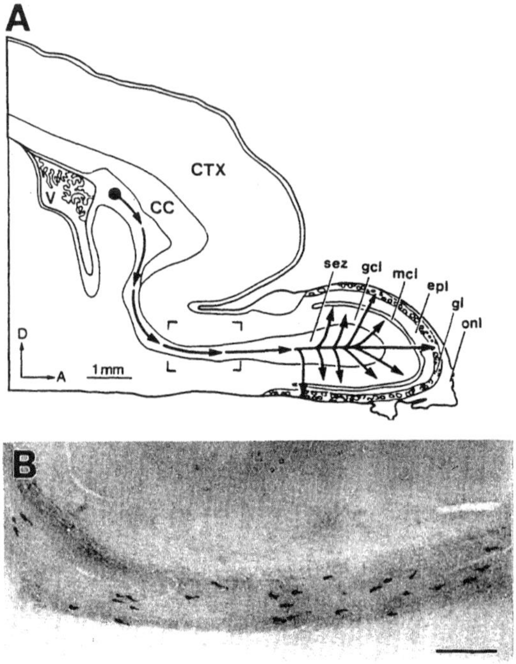

FIGURE 5. Site of origin and path of migration of interneurons destined for the granule-cell and glomerular layers of the olfactory bulb that arise in the postnatal subventricular zone. **A.** A schematic illustration of the site of generation, stereotypical pathway and final destination of cells originating in the anterior part of the subventricular zone (*large black dot*) or SVZa, containing essentially all neuronal progenitor cells. Initially the SVZa-derived cells follow an extended tangential (longitudinal) pathway (*disconnected arrows*) from the anterior edge of the lateral ventricle to the subependymal zone in the middle of the olfactory bulb. Secondarily the SVZa-derived cells ascend in a radial direction to their final positions in either the granule-cell or glomerular layers of the olfactory bulb, where they complete their differentiation. Unlike the migration of most cerebral cortical neurons, the migration of the SVZa-derived cells in the neonatal brain does not appear to involve radial glial fibers. *Abbreviations*: A, anterior; CC, corpus callosum; CTX, cerebral cortex; D, dorsal; epl, external plexiform layer; gcl, granule-cell layer; gl, glomerular layer; mcl, mitral cell layer; onl, olfactory nerve layer; sez, subependymal zone; V, lateral ventricle. **B.** Bright-field photomicrograph corresponding to the bracketed area in A of a 100μm sagittal section of the postnatal day 7 (P7) forebrain embedded in plastic and counterstained with osmium tetroxide, and demonstrating the highly stereotyped pattern of migration followed by SVZa-derived cells. The labeled cells (*black irregularly shaped spots*) resulting from an injection into the SVZa at P2 of a retrovirus encoding the lacZ gene into in the SVZa (not shown) are confined to the pathway (*band of dark staining*). Neither the cell body nor the leading process of lacZ-positive cells extends beyond the limits of the demarcated pathway. Scale bar = 200 μm. Dorsal is up, anterior is to the right. (Modified from Luskin, 1993a with permission.)

retroviral lineage tracer into the subventricular zone normally resulted in homogeneous clones of astrocytes or oligodendrocytes (Levison & Goldman, 1993; Luskin & McDermott, 1994). Thus, like cells arising from the prenatal ventricular and subventricular zones, cells arising from the postnatal subventricular zone maintain distinct lineages. Rather surprisingly, though, we also found that progenitor cells situated in the anterior part of the subventricular zone (SVZa) generate neurons rather than glia (Luskin, 1993a). The SVZa-derived cells are destined exclusively for the olfactory bulb (**Figure 5A**), an anterior extension of the forebrain (Luskin, 1993a; Zigova et al, 1996), and not the cerebral cortex. Furthermore, we showed that cells derived from the SVZa may possess positional information. They migrate a substantial distance toward the olfactory bulb along a highly restricted pathway (Luskin, 1993a; **Figures 5A** and **5B**). Moreover, the migrating neurons travel in a direction perpendicular to the surrounding radial glial fibers (Kishi, Peng, Kakuta, Murakami, Kuruda, Yokota, Hayakawa, Kuge, & Asayama, 1990), which suggests that these cells advance to the olfactory bulb by a different mechanism from that used by cortical neurons in their ascent from the ventricular zone to the cortical plate. We have performed experiments to determine whether the progenitor cells of the prenatal telencephalon can decipher the cues read by the SVZa-derived cells in order to reach the bulb. We found that telencephalic ventricular-zone cells transplanted into the neonatal SVZa fail to undergo migration away from their site of implantation. This suggests that the transplanted ventricular-zone cells lack the ability to migrate in the absence of radial glia (Zigova, Betarbet, Soteres, Brock, Bakay, & Luskin, 1996).

The phenotype and proliferative capacity in situ *of SVZa-derived cells* en route *to the olfactory bulb*

Earlier studies had concluded that cells *en route* to the postnatal olfactory bulb, which originate from the subventricular or subependymal zone, were postmitotic neurons (Altman, 1969; Kishi, 1987). This conclusion is consistent with the notion that migrating neurons do not undergo cell division (Purves & Lichtman, 1985). However, a number of studies have shown numerous mitotically active cells distributed throughout the pathway (Hinds, 1968; Altman, 1969; Bayer, 1983; Frazier-Cierpial & Brunjes, 1989; Menezes et al, 1995). Accordingly, we performed experiments to determine whether cells within the SVZa and SVZa-derived cells within the pathway are mitotically active. Neonatal rat pups were injected with the thymidine analog bromodeoxyuridine (BrdU) a few hours prior to perfusion, allowing us to detect the presence of dividing cells. We subsequently double-labeled sections obtained from these animals with an antibody to BrdU and with an antibody, TuJ1, that recognizes neuron-specific class III ß-tubulin (Lee, Tuttle, Rebhun, Cleveland, & Frankfurter, 1990a; Lee, Rebhun, & Frankfurter, 1990b). Our observation of numerous double-labeled TuJ1-positive/BrdU-positive cells distributed throughout the SVZa and pathway (Menezes et al, 1995) demonstrated that cells expressing a neuron-specific marker were capable of dividing. This contrasts with other regions of the CNS in which cells expressing neuron-specific markers are terminally postmitotic (Moody, Quigg, & Frankfurter, 1989; Lee et al, 1990a,b; Easter, Ross, & Frankfurter, 1993; Menezes & Luskin, 1994). Using a retroviral lineage tracer encoding E. coli ß-galactosidase, we also identified SVZa-derived cells within the pathway that were mitotically active (i.e., incorporate BrdU; see **Color Plate 4**, p. 187). Although dividing neuroblasts have been identified in the peripheral nervous system (LeDouarin, 1982; Memberg & Hall, 1995), the SVZa-derived cells are the only neurons in the central nervous system that have been demonstrated to possess the capacity for division. Our detection of triple-labeled (lacZ-positive, TuJ1-positive and BrdU-positive) cells further substantiates our assertion that SVZa-derived cells can be considered neuroblasts; i.e., cells with a neuronal phenotype that undergo cell division.

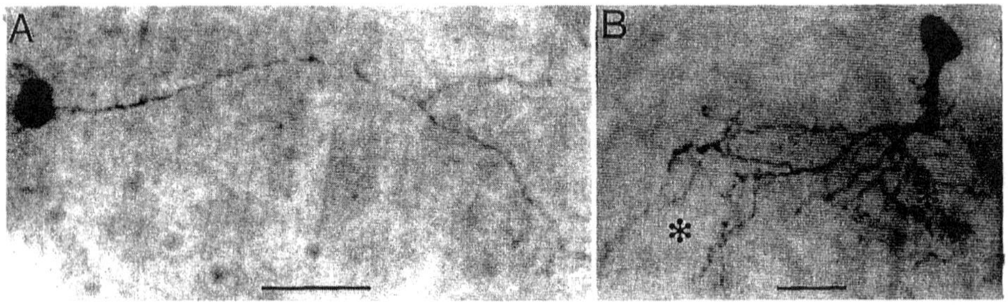

FIGURE 6. Morphological appearance of the two types of interneurons derived from SVZa progenitor cells of the neonatal forebrain. **A.** Photomicrograph of a representative immunohistochemically stained lacZ-positive granule cell with a cell body situated in the granule-cell layer from a rat injected with retrovirus at P0 and perfused at P28. The long, unbranched process extending into the external plexiform layer is characteristic of granule cells. **B.** Photomicrograph of a representative immunohistochemically stained lacZ-positive periglomerular cell from the glomerular layer of a P20 rat that received an injection of retrovirus at P2. The location of its cell body, situated next to the olfactory-nerve layer (top) and the extension of processes into either a single or adjacent glomeruli are typical of periglomerular cells; the processes of the labeled cell ramifies in two glomeruli (*asterisks*). Scale bars: A and B, 20 μm. (Modified from Luskin, 1993a with permission.)

The identity and neurotransmitter phenotype in vivo *of SVZa-derived cells in the olfactory bulb*

To characterize the properties of differentiated SVZa-derived cells, we examined their distribution, morphology and neurotransmitter expression in the olfactory bulb after they completed their migration and were no longer proliferative. The SVZa-derived cells were labeled by injections into the neonatal SVZa of the retroviral lineage tracer encoding lacZ or of PKH26. On the basis of morphology, SVZa-derived cells differentiated into granule cells or periglomerular cells, the types of olfactory-bulb interneurons situated in the granule cell and glomerular layer, respectively (Luskin, 1993a; Zigova et al, 1996). Most labeled granule cells had the typical bipolar morphology (**Figure 6A**), while the morphology of the periglomerular cells matched previous descriptions of their dendritic arborizations (**Figure 6B**; Price, 1970; Pinching & Powell, 1971). None of the labeled cells in the bulb resembled astrocytes (or stained with anti-glial fibrillary acidic protein (anti-GFAP) or oligodendrocytes, supporting the conclusion that the SVZa exclusively generates neurons.

Most of the olfactory-bulb interneurons contain the neurotransmitter GABA, but a subpopulation is known to contain dopamine, based on their expression of the enzyme tyrosine hydroxylase (TH; Halasz & Shepherd, 1983; Halasz, Ljungdahl, & Hökfelt, 1979; Gall, Hendry, Seroogy, Jones, & Haycock, 1987; McLean & Shipley, 1988). To determine the neurotransmitter phenotype of SVZa-derived cells, we injected BrdU into the neonatal SVZa and subsequently examined the neurotransmitters expressed by the BrdU-positive cells in the olfactory bulb. We observed many double-labeled BrdU-positive/GABA-positive granule and periglomerular cells, as well as double-labeled BrdU-positive/TH-positive periglomerular cells (Betarbet, Zigova, Bakay, & Luskin, 1995; in press). Our results demonstrate that the SVZa is a source of both GABAergic and dopaminergic interneurons of the olfactory bulb.

Conclusions and Future Directions

In summary, the interneurons of the olfactory bulb are generated postnatally from a specialized germinal zone surrounding the anterior part of the lateral ventricles, the SVZa, and its extension to the middle of the olfactory bulb. We currently have a limited understanding of the mechanisms that establish the SVZa as a distinct region. It is not known if the SVZa arises as a specialization of the underlying ventricular zone, which is the source of all the neurons of the cerebral cortex, or whether—prenatally or postnatally—the SVZa differentiates directly from the subventricular zone, which is predominantly a source of glia. The future identification of the factors and signals, both cellular and molecular, that influence the differentiation of the neuronal progenitor cells of the SVZa will provide insight into what determines cell fate as well as cell numbers in the central nervous system. By studying the migration of SVZa-derived cells to the olfactory bulb, we also hope to elucidate the mechanisms controlling neuronal migration, as well as neuronal differentiation. Ongoing studies comparing the ontogeny of the cerebral cortex to that of the olfactory bulb seek to identify the differential role(s) of cell-cell interactions and neurotrophic factors in the proliferation, migration and differentiation of components of each of these structures.

ACKNOWLEDGMENTS

The work described from this laboratory has been supported in part by a Biomedical Scholars Award from the Pew Charitable Trusts, the March of Dimes Birth Defects Foundation, and NIH grant NS28380 to M.B. Luskin. This chapter was adapted from: Luskin, M.B. (1993). Mammalian forebrain development: How cells find a permanent home and establish their identity. *The Journal of NIH Research*, **5**, 60-64.

Neural-Networks Models of Cognition
J.W. Donahoe and V. Packard Dorsel (Editors)

CHAPTER 3

A STATISTICAL FRAMEWORK FOR PRESENTING DEVELOPMENTAL NEUROANATOMY

Stephen L. Senft
Neuroengineering and Neuroscience Center (NNC)
and Department of Psychology
Yale University

ABSTRACT

Close analysis of the developing structure of mammalian brains can generate understanding and amazement. It is worthwhile to communicate this to a larger audience and to pursue it in greater depth. To do so we need dramatically more capable tools for coordinating and presenting the existing and incoming knowledge of neuroscience.

Neuroscientists have created a wide variety of methods for analyzing brain organization. This has resulted in vast amounts of highly detailed information, gleaned from many studies and expressed in many formats. At the same time, collective scientific efforts are deducing a growing number of *principles* that govern brain structure and development. Increasingly, these guidelines are capable of interrelating the great profusion of published findings.

These experimental and theoretical advances bring the opportunity and need for synthetic methods to systematically manage this consolidating knowledge, to make it more vivid and widely accessible. The technologies of simulation and computer graphics also are maturing, and are now at a level capable of animating and displaying biological principles and their anatomical consequences with high resolution and fidelity.

With these trends in mind, I have been designing an interactive computer-based framework for depicting neural organization and ontogeny. It incorporates a set of developmental rules, along with specific and statistical neuroanatomical information, to visualize brain development with network and synaptic detail. This tool is intended ultimately to facilitate study of any brain region. Its current features are illustrated for some simple aspects of cerebral cortex development.

Introduction

Today the flood of new scientific details demands, as never before, more efficient ways to organize data derived from brains.

Perceived problem

Nervous systems are touted as the most complex objects in the universe. They certainly are more complex than we can measure, more intricate than we can represent. Notwithstanding, scientists have created many analytic methods for probing brain organization, and are drawing out overwhelming numbers of details about neuroanatomy. These many partial insights are difficult to integrate, due in large part to their diversity and their diffuse distribution in a pre-electronic literature. There is an increasingly pressing need for additional synthesizing tools to coherently organize and present these burgeoning facts. This chapter outlines a strategy for portraying key aspects of developmental neuroanatomy that I have been creating to help address this problem.

Available methods

Tools for measuring the architecture of the brain range from those capable of detailing minute parts of single neurons to those capable of scanning entire brains. Many methods highlight the locations and branching structure of individual neurons or populations (e.g., Golgi impregnation, dye injection, antibody staining, even genetically introduced fluorescence; Marshall, Molloy, Moss, Howe, & Hughes, 1995). Computer reconstructions of such well-labeled objects in 3D, by manual or semiautomatic optical or electron microscopic methods, have evolved steadily for decades (Wann, Woolsey, Dierker, & Cowan, 1973; Villanueva & Le Bars, 1993), but remain limited in throughput. At another extreme are 3D imaging devices that visualize entire brains or brain regions and the tracts that connect them. Such devices provide comprehensive measures of global brain organization and mental processes (see Roland & Ziles, 1994), and can show even cellular organization (e.g., Jacobs & Fraser, 1994), but they do not reveal connectivity or synaptic detail.

Sampling problem

All of these startlingly informative tools, as well as those of intermediate range, are overwhelmed by sampling limitations and can report over only a relatively small window of spatial, temporal, or spectral resolutions. Each analytic technique provides fragmentary direct views of an integrated process that, we reason, extends with rich subtlety beyond the bounds of detection—becoming at some point complex enough and coherent enough to manifest as our own awareness. This age-old chasm between what we experience and imagine and what we can measure and depict continues to challenge us to create progressively more investigative and integrative methods.

It is incidental to the brain itself that our tools for gathering data from it are so bounded in their range of application and so incommensurate in the information they transduce. It is a result of historical forces and constraints of physics that, for instance, resolvable magnetic resonance information vanishes into pixillated mist at 10 microns or so, and does not directly give way to volumetric and time-lapse microscopy (which differ from whole-brain scanners both in depth of penetration and the physical entities they report); or that light microscopy does not merge seamlessly with the techniques of electron microscopy. If the brain's operation likewise is balkanized by such limitations in physics or dialect it is not obvious, except in the periphery.

Integration problem

By contrast, within our minds we can shift attention more freely. To illustrate, we can trace an easy path from rhodopsin's molecular role in transduction, across photoreceptor synapses and along axons into the lateral geniculate nucleus; glance sideward at rhythmic modulations arising in the septum and hippocampus, follow a phalanx of thalamic afferents into cortex, visualize retinal scenes mapped phasically into separate cortical laminae, and imagine how these patterns change over development. We patch in a wide variety of appropriate information at differing spatial and temporal scales, as needed, to flesh out the representational needs of the moment. Even so, our short-term memory is limited compared to the enumerated biodiversity and richness of detail of known brain circuits. None of us can access enough facts to do justice to the mounting backlog of neuroscientific knowledge.

Today, as never before, we need more efficient ways to organize data derived from brains. We need new ways to motivate and teach comparative and developmental neuroscience, to highlight overlaps and holes in our collective knowledge, to generate and evaluate anatomical and behavioral hypotheses in greater detail, and to predict how parts of the nervous system will develop under normal, experimental and regenerative circumstances.

Weight of numbers

Historically, the primary obstacle hampering an integrative understanding of brains has been the Sampling Problem. Today, even the early success of our measuring devices has become immense enough to present us with an equally challenging difficulty: the Integration Problem.

The impasse of large numbers is common to both challenges. Brains contain billions of cells, each of which has a complex branching pattern containing even more complex biochemical machinery. Cumulatively, we have quantitative structural information on perhaps several thousand specific neurons, and many fewer interacting circuits. But even if all of the cells and circuits had been measured, simply storing the data would take much more archival media than is available to the neuroscience community. In addition, and as with the information currently archived in the literature, it would take a numbingly long time to review all of the elements, and the final display would be impenetrably dense. *We need a rational way to accurately and concretely represent brain structure and its interactions, without being inundated by huge numbers.*

Perceived solution

It seems plausible that a researcher or student should be able to use a computer workstation to obtain detailed representations of any biological neural network of interest, using published data and direct measurements, available on line or on CD. In practice, though, existing neuroscientific knowledge is severely vignetted. The gaps often are more personal than inherent in our collective knowledge base: Others *do* know in great detail about the optic tract, hypothalamus, red nucleus or substantia gelatinosa, even if I do not but instead know about development of somatosensory cortex. What seems to be lacking most is a flexible 3D language for communicating brain organization. We need a visual vocabulary and syntax that is satisfying to anatomists, that can grow to accommodate genetic, biochemical and physiological information and that, ultimately, can be made to *operate*, in the sense of simulating the biological process of information digestion.

The metaphor of *three-dimensional navigation* through an abstract space has proven to be an effective tool for managing complex information. It has long been used as a mnemonic device: Each visited region is polled for its associated information. But, for the intricate and peculiar structures in neuroscience, the most compact metaphor is provided by the anatomy itself. Fortunately, the nervous system appears redundantly organized; thus a simplified scaffolding may suffice whose nodes can be unpacked to access additional data. One useful abstraction might consist of a three-dimensional (or better, a developmental) depiction of the locations and interconnections of all the named "nuclei" and "tracts" in the brain. This is manageable: Such objects number only in the thousands. We could create this by combining the pictorial precision and flair of computer graphics with the insights and measurements of developmental neurobiology and cell biology. We should be able to make representations as beautiful and accurate as Ramon y Cajal's compelling drawings of brain circuitry, only now three-dimensional, dynamic, and interactive.

A deeper aim is to present brain development as an expression of bioenergetic *principles*. Critical to this effort would be a formalization of growth rules for the cellular elements of the brain: the abstract control systems and constraints governing how cells divide, migrate, regulate process outgrowth and communicate intercellularly. If we are able to embed in a simulator such principles of brain organization, experimenters and students could model tangible hypotheses about how a region of CNS is organized and how it operates. Such interactive environments should enable one to choose alternative developmental rules and starting points. By manipulating logical variables, users might steer the development of brain regions into many possible architectures. From this they could better understand the potential contributions of key variables, and could readily learn and improve comparative functional neuroanatomy.

This is a remarkably stiff challenge: One needs to create a framework general enough to encompass or apply to the entire brain, yet flexible enough to show the detailed workings of neuropil. It should include interactions as focal as gap junctions or synapses and as diffuse as extracellular ionic or metabolic exchanges. Events must occur over periods as extended as developmental time or as evanescent as an action potential. One would like, as a byproduct of simulating development, to be able to follow microscopically realistic paths of patterned stimuli traveling from receptor sheets to effector organs, by way of anatomically accurate, non-hidden, neuron intermediates. At present, one merely can imagine such waves of information: flickering over forests of intricate growing arbors, like displays of the Southern or Northern Lights.

Related work

A wide variety of pertinent theoretical tools are being developed worldwide, in response to the recognized difficulty in handling the many advances that inform the study of nervous systems. Most notable is the Human Brain Project, which has many component aims (Huerta, Koslow & Leshner, 1993). Numerous groups are growing neuronal and axonal shapes algorithmically (e.g., Burke, Marks & Ulfhake, 1992; Van Veen & Van Pelt, 1994). Other well-known efforts emulate cellular physiology and its experience-based modifications, when provided with structure: GENESIS simulates mature multicellular networks; NEURON manipulates ionic channel parameters (see review by De-Shutter, 1992). There are attempts to build detailed volumetric models (Gally, Montague, Reeke, & Edelman, 1990) or extensive silicon networks with cell-like components (Mahowald & Mead, 1991). Too, there are systems for logical navigation through relational brain databases (Wertheim & Sidman, 1991; Willis & Koppe, 1991), or for multi-scale perusal of image and volume data (e.g., Roland & Ziles, 1994; Funka-Lea & Schwaber, 1994). A full list of the relevant studies would be very long: This is a broadly converging shared vision. Many important new works remain unpublished, but accessible over the Internet. An important set of tools from informatics will provide methods for searching and converting large bodies of knowledge per se into digestible form, and will create agent-based simulators of very large abstract systems.

This approach

The approach presented here arose from detailed experimental experience with the cellular anatomy of developing cerebral cortex (Senft & Woolsey, 1991). It shares characteristics with some of these other research efforts in depicting network architecture as collections of neuronal automata and neurons as collections of multi-variable compartments. It combines other didactic efforts by elaborating the metaphor of browsing in an anatomically accurate three-dimensional data base to help explain and teach concepts in neuroscience.

Emphasizing that function is often constrained by form, it attempts first to establish flexible methods to replicate the peculiar cellular forms found in neuroscience. It views the anatomical processes of neuron arborization and connectivity as central, and treats them as intrinsically dynamic and developmental problems. It has deferred, thus far, the critical questions of which physiological patterns this simulated anatomy can engender and how ongoing activity might sculpt these networks, but an interchange with other efforts is underway to collectively achieve those goals.

Its primary focus is on the question of how *statistics* can enable diffuse developmental concepts to generate discrete anatomical structures, even in the absence of exact specifications. The approach has been designed to incorporate general knowledge about the organization of nervous systems, so that a wide range of particular structures may be generated from different initial conditions. It uses high-performance computer graphics to visualize the resulting complex simulated circuitry. To date, the effort has intentionally focused on global applicability at the expense of immediate accuracy for any given brain region.

Key Principles of Brain Organization

Networks

The most fundamental abstraction surviving critical study of the brain is the concept of *neurons*. The most important feature of these elemental components is their *interconnectivity*. Enormous numbers of wildly elaborate sets of branched neuronal processes weld together using several classes of adhesions to generate the logical circuitry of the brain. These three-dimensional networks form dense local tangles of neuropil as well as compact tracts that wed distant regions together. They are interlaced with irregular glial networks, are perforated by blood vessels and capillary systems, and are immersed in a tortuous extracellular space.

Development

This intricate filigree is thought to be relatively static in its architecture, when assessed on time scales that encompass our everyday adult behavior and awareness. Hence researchers have focused on patterns of physiologic change that flicker over this framework—the "enchanted loom" envisioned by Sherrington. At the same time it is clear that, throughout life, brain morphology changes continuously on a microscopic scale (Purves, Hadley, & Voyvodic, 1986). While the linkage is not fully characterized, this variability appears to contribute to the ebb and flow of behavioral propensities (Merzenich, Kaas, Wall, Nelson, Sur, & Felleman, 1983). Therefore, accurate representations of neuroanatomy *inherently* should be dynamic. This conclusion becomes mandatory when we consider nervous system development. Here we encounter the important additional concepts of stem-cell lineages regulated in germinal layers, cellular mass migrations into target zones, outgrowth, arborization and remodeling of neurites, and selective synapse formation and stabilization. Importantly, we also encounter simplified intermediate brain architectures that, as Ramon y Cajal (1911) so elegantly pointed out, allow us to understand better the otherwise bewildering adult configurations.

A basic need

Even introductory readings in the neuroscientific literature identify hundreds of additional factors that are known to influence brain development and function. Virtually all relate structurally to these simple anatomical foundations. Thus far, no interactive 3D representational system exists that can arrange these basic concepts in a well-choreographed way. The following sections outline a beginning strategy.

Programming Brain Principles

Systematic abstraction

To assemble any given area of brain, synthetically, one might make generative algorithms that know merely that neurons have a range of somatic shapes and sizes, that they divide and migrate, that they transform from spheroids to arbors whose particular branching structure often covaries with functional type, and that they can synapse selectively with large sets of target cells. External specifications might then fine-tune the system to specify the details of geometry and timing needed to produce particular cell types and circuits. The simulation engine itself need only incorporate general biological properties and a selection of plausible algorithms that can organize the structure of a nervous system, regardless of the underlying but partially known biochemical mechanisms.

This is a ubiquitous approach for dealing with complex systems in science and mathematics: Simplifying equations are identified that, with different boundary conditions, explain a wide variety of particular situations. For the case describing regional brain development, such equations may remain mostly procedural, and may never have closed form. Nevertheless, it is helpful to separate the problem of identifying the pertinent variables and their interrelationships from the problem of specifying values. Moreover, such a strategy does not preclude making menus of rules and elaborately detailed prototypes available to users, so that in practice it takes little effort to recapitulate subtlely complex networks.

Primary assumptions

An animal's genome may be viewed as the primary natural representational system that programs brain development. We do not yet understand gene regulation thoroughly enough to elaborate this analogy with detailed descriptions of DNA algorithms. But we are virtually certain that the genome specifies neither the detailed pattern of connectivity for each cell in the brain (Sutcliffe,

Milner, Gottesfeld, & Reynolds, 1984) nor the precise number of cells in all regions (Williams & Herrup, 1988).

Given this constraint, elements of the brain might be described more naturally as a collection of *groups*, with sets of *statistics* that encompass all members of each group. Such "group" abstractions may or may not eventually map onto discrete genetic control systems. After all, statistics are not biochemical mechanisms. In a computer program, however, statistics can be *resampled* to produce the observed range of structures or behaviors for a group even in the absence of knowledge about underlying mechanisms. They also can be replaced by more detailed statistics when deeper knowledge is obtained. Thus, group statistics provide a compact and generalizable methodology for this synthetic enterprise: Small numbers of parameters suffice to determine the features of entire populations. As elaborated below, this permits one to construct complex virtual brain circuits of variable size and refinement.

Cellular framework

The statistical formalism entertains as distinct biological "groups" the large number of brain nuclei, subnuclei, and cell types that are generated at highly specific developmental locations and times, and that migrate, mature and interact with the rest of the nervous system as ensembles. (A variety of additional rationales for factoring the CNS into groups or cell assemblies can be found in the literature, e.g., Edelman, 1993; Palm, 1993). Thus, an underpinning for the brain framework can be made by placing representatives of each group as 3D clouds of cells, located at known morphological sites, using adult or embryological coordinates (e.g., Altman & Bayer, 1995) obtained from atlases or segmented 3D scans. To specify such clouds, we measure the position (relative to some more global frame of reference) for a subset of cells, and derive a mean location and standard deviation (and other moments such as skew, for non-Gaussian distributions) for each axis, which we assign to the named group.

We then *invert* this procedure, by randomly resampling these statistics, to derive a space-filling number of exemplar cells having the same distribution as the cells in the nucleus. We assume that only the (possibly highly) constrained statistical distribution of cells is relevant to the brain, not their exact location. The structural redundancy of brain allows us to use data derived from only a small percentage of the population to drive fuller representations. In this way we can fill in seams between sampled vignettes without having to measure and access voluminous data.

As with galaxies, one recognizes obvious prototypic forms: spherical, globular, annular, sheet-like, banded. In general, such nuclear distributions may be quite complex, but many variants can be accommodated by an underlying statistical data. It is an open question what menu of statistical descriptors will be needed to describe the full taxonomy of plausibly identifiable cell groups. Some cell clusters become so distorted by morphogenetic movements imposed by other developmental events that their distributions are difficult to describe analytically. It even may be that recapitulating their ontogeny is the most compact way of describing such adult configurations.

Segmental primitives

The envisioned model for brain architecture aims to produce views resembling ones visible through microscopes, at scales of networks to neuropil. Thus neurons need to be given more personality than dots. Brain cells in this model are constructed from sets of *tubular segments*, linked together to form branching arbors. Tubules can taper and have rounded ends and their diameters can be irregular and asymmetric. This permits forms for somata, ragged neurites, even glia, to appear biologically plausible.

Every segment is an element of a group and possesses a set of *fundamental* variables, always describing the element's location, length, diameter, orientation, and parent and child connections. Each element also has *accessory* variables for variously describing its appropriate subcellular anatomical features (e.g., density of axonal varicosities, numbers of neurofilaments) or capable of holding physiological variables (e.g., for steady-state or transient electrical signals). The *specific value* assigned to any variable is obtained by randomly sampling the group statistics corresponding to and constraining that variable.

This design provides the advantage that the underlying data structures can be relatively uniform and implemented efficiently in computer code. Also because of uniformity, the strategy of resampling from measured statistics can be replicated easily for any number of attributes, such as soma diameter, nuclear size and eccentricity, number and orientation of primary neurites, their diameter, taper, density of dendritic spines, and so on. This strategy can be extended as needed, and can produce millions of resampled objects as detailed and tangible as if each were measured from life. Such extra variables could provide enough definition within each tubule to support ionic and metabolic engines, and to generate visuals akin to Golgi or Nissl preparations.

Arborization

The 3D locations of *branch points* are the essential characteristics that must be specified in order to represent any desired shape of neuron. Branching patterns are characteristic, even defining, for cell types: bipolar, stellate, chandelier, pyramidal, etc. (Ramon Molinar, 1962; Lindsay & Scheibel, 1974; Pearson, Norris & Phelps, 1985). Analyses of neuron morphology characteristically define their numerous measures as statistical variations about means, such as the number, length or diameter of primary neurites (e.g., Juraska & Fifkova, 1979). Since canonical shapes often are seen when neurons are grown in isolation, some aspects of neuron form appear programmed endogenously (e.g., Coates, Fermini, Strahlendorf, & Strahlendorf, 1992), although many features obviously arise through environmental interactions.

In the model, average measures are assigned for a wide variety of arbor descriptors that in principle are amenable to direct observation, including path tortuosity, length before branching, branch point locations, spacing and angles, probability of synapsing with other groups in the vicinity, and so on. Statistical descriptors for temporal sequencing also are used to recapitulate neurite development. Timing parameters are obtained from observations on immature neurons and from inferential analysis of adult arbors. (See also Carriquiry, Ireland, Kliemann, & Uemura, 1991, and Caserta, Eldred, Fernandez, Hausman, Stanford, Bulderev, Schwarzer, & Stanley, 1995, for other statistical and fractal analyses of neurites; and see, for instance, Burke, Marks & Ulfhake, 1992; Nowakowski, Hayes & Egger, 1992; Van Veen & Van Pelt, 1994; Li, Qin & Wang, 1995, for parallel efforts to recapitulate neuronal form algorithmically.)

Growth hypothesis

Some of the ways in which brain characteristics are represented using these primitives at first may appear stilted, from the point of view of cell biology. For instance, when attempting to replicate known structures using this approach, it proved useful to construct sets of neuron prototypes from even more primitive groups whose statistics govern the cells' somata, linked to other groups whose statistics govern subsets of the cells' overall neurites.

As a biological rationalization, though, it seems possible that there exists in the genome only a finite repertoire of potential neuritic forms (perhaps as coordinately expressible transcription regions), and that each cell group developmentally expresses differing subsets from this morphogenetic roster. During development neurons may shift between growth states (either endogenously or as a response to exogenous stimuli), and different sets of statistics would describe the type of neurite emitted locally during each "epoch."

As examples, consider the differences in shape or timing between apical, basal and oblique dendrites of pyramidal cells (Ramon Molinar, 1962; Lindsay & Scheibel, 1974; Juraska & Fifkova, 1979), or the behavior of axons, which emit periodic offshoots as right angles into overlying tissue (Senft & Woolsey, 1991) then change character again when entering a target zone, to arborize explosively. This simulation-inspired hypothesis could be evaluated by further experimental study, but it has proved efficient, even if only from a descriptive point of view, to treat these subcellular features as morphogenetic quanta shared across cell types.

Connectivity

The hallmark of brain complexity is represented by the connections between cells located in disparate regions of the CNS. In this framework, we specify connections at the *group* level. The program realizes such physical links as elongated branched tubules that orient towards or away from an appropriate subset of targets. Simulated axonal tubules can sprout as if stimulated by diffusible substances released nearby or at a distance (Sato, Lopez-Mascaraque, Heffner, & O'Leary, 1994). Another primary characteristic of CNS is the range of local and specific connectivity that constitutes neuropil. There is no fundamental difference in this model between tract and neuropil representation, except that the cues used to orient to distant targets may be specified separately from those responsible for local arborization.

Junctions

Synapsis can be spatially both promiscuous and highly selective (Keller & White, 1987; Haydon & Drapeau, 1995). Synapses also vary in type from highly punctate sites between only two neuritic processes, to diffuse signals reaching many targets. In nature their temporal signaling characteristics, means of modulation, and downstream effects on anatomy and learned behavior are myriad. In this framework individual tubule segments maintain a long list of other arbitrary segments that they approach or touch, and keep track of the type and strength and precise location of "junctions" established between the segments. They are classified anatomically as pre- or post-, or even gap-junctional, and subclassified by functional type.

3D blueprint

The fundamental vertebrate plan involves formation and subsequent modification of a *neural tube*. Everywhere along this continuous and deformable surface there is an elaboration of a common ontogenetic theme: highly regulated division of sets of stem cells in proliferative zones. Hence, to rationally position the developing cell groups for even a small region of brain, it will be appropriate to have a flexible method for surface generation and its deformation by biological forces. Moreover, neurites generally grow along *interfaces*, in preference to isotropic growth in space. Such paired surfaces can provide the constraints to limit neurite outgrowth to biologically meaningful locations. Computer graphics excels at freeform surfaces (see especially Fujita, 1990, for an application to neural tube development), and this program has begun to implement surface descriptors, using non-uniform rational B-splines.

Stem cells

A detailed representation of brain development also needs to monitor cell cycling and cell division, and should be able to control the migration of neuroblasts from germinal zones into the areas in which they mature. As a result, it should be able to depict lineage relationships among the resulting neurons and glia. To emulate these events effectively one needs to represent control by mitogenic and growth factors. These agents often are diffusible substances that establish broad waves of maturational gradients, and which may be emitted and metabolized by specific groups.

Thus the simulation environment should also be able to effectively handle simultaneous diffusion of multiple biochemical species. To this end the tubular elements are embedded in a voxel-based "extracellular" sea (of adaptively set voxel size). Tubule segments can emit diffusing messages that are attenuated from voxel to voxel and consumed by segments at distant locations. One could inspect the spatial state of this system dynamically using volume-rendering visualization techniques. Currently, message diffusion has only elementary effects on the simulated anatomy.

Supporting elements

Blood vessels may be represented, with more algorithmic complexity, using anastomosing tubules having non-tree topologies (e.g., Kiani & Hudetz, 1991). But the endothelial vascular components themselves are patch-like, not tubular (unless highly collapsed and involuted). This example shows that simple tube and voxel primitives will be an insufficient basis for emulating all brain components. The same conclusion is reached when considering, say, myelin sheaths or growth-cone lamellae. Moreover, numerous explanatory mechanisms in neurobiology are subcellular, hence one would like to incorporate information derived from electron microscopy and biochemistry. Other data structures would be needed to gracefully depict subcellular organelles, such as Golgi bodies, mitochondria, or ribosomes. In the present implementation there is no systematic provision for defining membrane patches or ameboid shapes, but segmented tubules provide a convenient intermediate level of abstraction for erecting both supracellular and intracellular anatomical scaffoldings.

Timing

The birth dates, migration schedules, and maturational time tables are known in great detail for a large number of these brain groups (e.g., Altman & Bayer, 1990). Clouds of arborized cells constitute a unifying format for animating this information. Consequently, each group has several temporal variables, also statistically specified, for timing the onset and offset of behaviors such as cell division, translocation, and neurite extension.

Restatement of Approach

Overview

I have presented in broad strokes the outline of a framework to nucleate information in neuroscience for teaching and research. It advocates beginning at a phenomenological level. A set of statistical constraints and ad hoc rules are made to interact with a set of well-established principles of neuron growth to mock up specific aspects of brain development as clouds of elements that arborize and link to one another using simple procedures, specifiable at a *group* level.

Generality

To ensure generality, the program is designed to segregate the range of possible specific details from the range of plausible principles. Because not all users have the same assumptions, both the principles (e.g., "European" vs. "American" plans of cell determination) and the specifics (e.g., whether a projection exists between groups A and B) are to be assigned by user selection. Too, since one does not always logically factor biology into immutable categories (e.g., pyramidal cells could be viewed as a special class or merely as stellate cells with elaborate apical dendrites), the system permits dynamic rebinning of elements into new groups (e.g., stem-cell progeny might become elements of many anatomical groups, or cells might be recruited into a variety of functional groups, depending on context).

Advantages

The foremost *advantage* of its statistical and segmental design is that it is generic and allows one to repeatedly finesse the sampling problem: Measurements are made on small samples, and the derived statistics are resampled to generate as large a population as is needed. The generated population can have as tangible a structure as if one had imported traced data. One also can mix specific *camera lucida* data with statistical descriptors to handle special cases, for example to represent "identified" neurons having inherently few instances. This framework is useful for teaching, and this will be greatly enhanced as, through experience and consensus, one increases the accuracy of the tissue surfaces, cell placements, neurite arborization and interconnections managed by this approach. Too, if features become generated by more coherent algorithms, users will be able to hone their experimental designs by fine-tuning fundamental growth parameters in the computer. As underlying controls for neurite extension, branching and retraction correspond more closely to biological entities (Kater & Mills, 1991; Van Veen & Van Pelt, 1994; Li, Qin & Wang, 1995) researchers may be better able to conceptualize pivotal developmental phenomena that are hard to observe, such as neuropil formation (see also Gally et al, 1990).

Validation

Scientists routinely reduce their data to statistics, and often derive conclusions from statistical summaries in preference to arguing from individual instances. As a result of that enduring discipline it should be possible to convert a great deal of existing information into a form digestible by this program.

The strategies presented here do not guarantee that results can be accurately computable in practical terms. Given the inherent complexity of brain, we should expect models to grow complicated and cumbersome. For instance, it is clear that, in nature, most of the components in the framework are exquisitely sensitive to highly local effects of ions, growth factors, extracellular molecules, and other influences that are difficult to measure comprehensively. The present model simply provides skeletal locations for siting a number of such agents and a communal milieu for their interaction. For this framework, particularly in its early stages, it is not critical that a phenomenon be known in full detail. Instead, a major point of this work is that collectively we have a great deal of hitherto implicit knowledge to reap, by consolidating information, before we reach the limits of even current data or computer resources.

On the other hand, it is important that the primary variables of a system be identified. With experience, the range specified for each abstract variable might be expected to shrink, eventually approximating that measurable (in principle) from nature. The foremost *disadvantage* of the present formalism is that the requisite sets of variables are not specified by an underlying theory. The framework neither generates the variables sufficient to characterize the ontogeny of a group of neurons, nor assures the logical independence of variables that have been identified.

Thus, the challenge remains to establish rigorously the combination of statistical parameters required to describe all possible or merely all used forms of arborization, and to determine whether such parameters can be derived systematically from the forms of raw data at our disposal. It is also an open question whether all of the important anatomical phenomenology of the brain can be cast in terms of an interaction of fundamental principles and statistical constraints. However, the fact that genetic machinery ultimately is driving the specification of many of these features gives us additional reason to expect that a complete set can be identified, eventually.

For a while, a kind of Turing test may help to assess whether the approach is on track: Does its output appear plausible to trained anatomists? Too, are its algorithms becoming more compact and flexible, and do they converge in formal structure with underlying biochemical mechanisms? One interesting validating mechanism also is inherent in the design: The program can analyze raw morphometric data to derive a set of statistics that help it replicate the underlying arbor population. Because of the uniformity in data formats, *simulated* anatomy also can be analyzed as though it were imported data. Hence statistics produced by a simulation (based on data measured for one age) can be compared with *independent* statistics measured from nature (for the final age simulated).

How It Operates

Data specification

General information about neuroscience is expressed in the program in the form of separate state transition rules for various classes of element (somata, axons, dendrites). For instance, if a segmental element is "somatic" then its options include to fatten, to divide, to migrate (if it has not emitted any neurites other than perhaps its axon), to emit signals, or to autolyze. If the segment is "neuritic" then it may bud, resorb, elongate, shorten, branch, or orient with respect to a nearby object or tropic gradient (or as a computational expedience, with respect to the initial locations of objects emitting tropic compounds). Additional rules distinguish between axons and dendrites.

Separate, externally specified sets of statistical probabilities constrain the range of variation for these actions, such as the positioning of somata, or the angles neurites express when emerging from cell bodies, when elongating, and when branching. This extrinsic information can be provided several ways: tailored by hand (based on educated intuition and direct observation of stained neurons), derived from analysis of 3D confocal scans, or inherited via files output from automatic analysis of structure generated by prior simulations.

Group files

Statistics files contain arbitrary numbers of records, one record per group. Each line in a record defines the range of values for a particular property of the group (such as average branch angle, segment tortuosity, density of spines). Six parameters are used, and are interpreted as a distribution "law" (e.g., Gaussian or uniform), "N" (number of observations generating the statistics), and "mean," "standard deviation," "low" and "high" constraining values.

Global constraints for certain variables (such as branch angle, interbranch distance, segment diameters, branch probabilities) can be derived automatically from medial axis (skeleton) representations that are extractable analytically from appropriate 3D confocal data (Senft, 1995). A remaining challenge is to algorithmically detect *epochal* changes in arborization strategy, using adult and developing skeletonized neurites: The aim is to automatically generate temporal descriptions of neuron sub-trees, cast in terms of group statistics, that in turn can be used to recapitulate the branching patterns of a wide range of neuron populations.

Group descriptor records are concatenated by hand in quasi-temporal developmental order to produce cohorts of cells that express structures reminiscent of a variety of known cell types (bipolar, pyramidal, stellate, etc.). For efficiency, data files containing these statistics can incorporate two kinds of inheritance: (1) successive groups need specify only those variables that differ from groups specified previously; and (2) statistics files can be nested hierarchically. These features add greatly to the flexibility and reutilization of the statistical information. For instance, this permits one to site subtypes of cell groups in multiple locations, while easily respecifying only features like their range of somatic size or preferred branching angles. Ideally, of course, most such variations would be generated as a consequence of antecedent developmental events.

Matrix files

In nature, neuron populations characteristically project to a variety of target areas, within which they arborize preferentially. In this work (and possibly also in nature) such relationships are specified at the *group* level. They are represented to the computer through files containing sparse interaction matrices, whose entries indicate attraction or repulsion between groups. Interactions can become active conditionally (e.g., at certain distances or developmental times) and hierarchically (e.g., attempt to project to target C only after target B is found). At present, simulated axonal connections between groups describe meandering paths through space and do not detect all collisions. More realistic paths would result if axonal growth were constrained additionally by sets of deformable limiting surfaces.

Synapses

Synapses are represented as separate pre- and postsynaptic components, located at specific locations on segments. Each segment may have any number of anatomical synapses of any type, and can identify which other segments to which it is pre- or postsynaptic. A synapse is allowed to form when an axon from one group grows sufficiently close to an element of a target group. The "type" of synapse can be a property of both groups. Synaptic triads or serial synapses may result from these primitives. Synapses play an extremely important functional role in shaping real brains, but currently simulated synapses are almost purely anatomical, with few maturational and physiological consequences.

Simulations

Simulations progress by stepping an absolute clock (roughly corresponding to many minutes to hours of developmental time). Each group has internal timing mechanisms that are compared to the master clock, to govern when its elements mature and when they interact with other components. To manage these mechanisms for each group, the program currently provides a choice of two growth protocols.

The simpler, ad hoc, method *deals out* segments to preexisting group elements at a modulatable rate from a common preset quota. This requires that the designer know how to initialize the circuit and how detailed the result will be (such as how many somata there are, how many primary neurites each has, and the approximate order in which neurites elongate and branch). It generally avoids the question of how to trigger and limit developmental processes.

The more profound method provides *growth on demand* for each object. This concept is more comprehensible and familiar to biologists, and it admits a wide variety of causal mechanisms to regulate almost every aspect of growth: soma enlargement, emergence of primary neurites, rates of elongation, probabilities of branching or pruning, and so on. Using this more highly regulated method, stem-cell groups can spawn daughter groups whose elements express various migratory paths and fates. This represents a potential means for exploring the architectonic consequences of a variety of possible methods of regulating cell commitment, and ultimately it should require less voluminous specification of dependent measurements. But also it requires that much more profound cell-biological knowledge be encapsulated within the program, including a variety of signals that start and stop growth. Only a handful of these causative relationships have been incorporated thus far.

Clearly one will not always want to simulate the entire ontogeny of the nervous system merely to view a piece of adult brain. Therefore the program has been designed to permit a mixture of protocols. One can start a simulation at a given time point by substituting measurements and ad hoc settings pertaining to that age in place of fuller calculations based on deeper principles. One can save the results of a simulation during a given epoch, in the form of a set of statistics, to prime simulation of a subsequent epoch (since one has a statistical representation each time, it does not matter much that the two simulations differ in detail). Additionally, one may read in *precise* information for a set of cells obtained at the second age, and mix these with the statistically constrained examples of other brain elements.

In general, either mode may be invoked, as appropriate, since there are vastly differing depths of knowledge about mechanism for each region in the nervous system. For instance, if one has focal information about pyramidal-cell neurite outgrowth (e.g., Juraska & Fifkova, 1979), one could place such cells in post-migratory positions and run the system forward using those initial conditions. But if one also has information about lineage and cell determination, one might initialize the system as a germinal zone to investigate the conditions needed to reflect stem-cell behavior and migration.

The nervous system is too complex to simulate accurately in any but a few respects. This multi-mode design allows one to make wide didactic use of the program to represent phenomenology and to bridge temporal gaps where lack of knowledge produces obviously erroneous predictions—all the while refining the logic needed to generate systems at a deeper level. Although simulations inherently will be inexact for the foreseeable future, their design need not always accumulate error as simulated time progresses, because one can correct in mid-course using specific data derived from later developmental stages.

User control

At almost any point, even during data input or during circuit generation, users are free to select those groups which they would like to have actively participate in the definition and evolution of the simulated system. Selection is effected through lists, name matching, or 3D picks using the mouse. In addition, subsets of these "active" groups can be highlighted for closer inspection by changing color assignments or by making all other groups invisible. Sets of groups can be selected in concert, based on clonal or physical linkage or synaptic connectivity. Synapses can be viewed, by subtype, as decorations on neurites or as cloud clusters in the absence of neurites. Axes and scale bars may be toggled on or off interactively, and annotative information, in the form of group number and names, can be displayed in three dimensions next to the appropriate structures.

Large groups of neurons may be viewed mutually entwined, *in situ*, or the anatomy can be reformatted into an exploded "gallery" of 3D and logical (Sholl or connectivity) diagrams, where arrays of individual cells can be isolated and highlighted. Alternatively, all members of the gallery can be composited in 3D, to produce a canonical depiction of the average directions and modes of arborization for a cell class. Individual cell architectures can be exported to NEURON for detailed electrical analysis. To simulate surgical and chemical manipulations, target groups can be dragged to new locations, and augmented or depleted in number. The rest of the system can continue to develop, adapting to these changes.

Statistical format data files can be imported into the program combinatorially, and mixed freely with *camera lucida* data describing specific cells. Networks created this way and modified by user interaction can be reduced to new sets of statistics and saved for re-entry. Alternatively, every variable of every segment of every (active) group in the simulation can be sent to disk and re-imported. The generative statistics for any active group may be inspected and edited on line, as can values for any variable of any element in the simulation.

Example output

There are two primary display modes: an interactive sketch that depicts cell processes using simple lines, and a more detailed presentation in which the segments are drawn as fractal paths decorated with varicosities, spines, and irregular edgings. Either mode may be rotated, panned, and zoomed, using the keyboard or mouse. Oblique cutting planes can be manipulated in 3D to emulate the truncation of components, as if the system were physically sectioned into brain slices (networks continue to develop, intact). A log may be made of the viewpoints used in rapid sketch mode, to be used to pace the fractal display for automated output in greater detail to film or video.

The images shown here are a set of stills or frames from movie clips produced over the two years by the simulation program "ArborVitae." (Considerably more information can be conveyed using other media: either directly on the monitor as dynamic developmental views presented three-dimensionally and in stereo, under interactive control, or as movies sent to videotape.)

Color Plate 2 (p. 186) shows a set of somatic groups arrayed as arching laminae, to represent a sector of cerebral cortex (neurites exist in this simulation, but have been rendered invisible). Color is used to differentiate the overlapping cell groups assigned to the various "cortical layers." The arrangement

FIGURE 1. A set of immature "pyramidal" neurons, corresponding to the band of yellow cells in **Plate 2**. Each has an apical dendrite, early basal dendrites and a varicose axon.

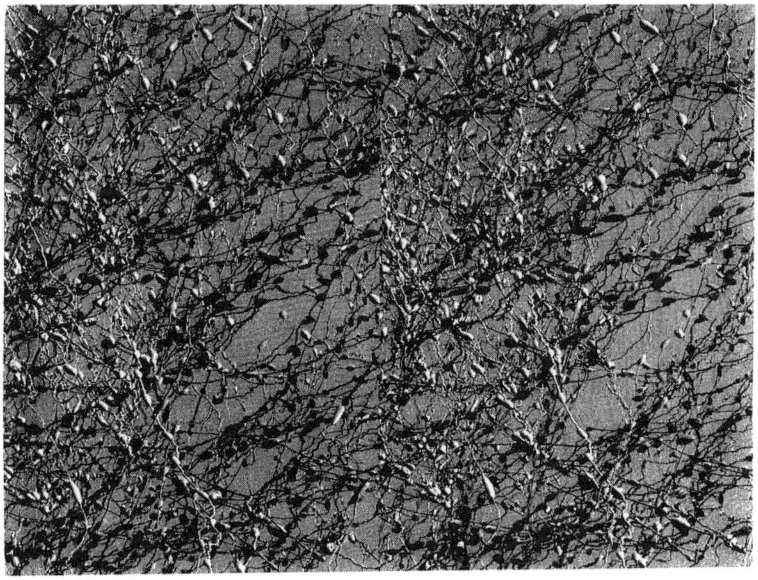

FIGURE 2. Stereo view of a field of highly varicose axonal processes, from a region like the thalamo-cortical projection path shown in **Plate 3**.

forms a laminated dome in three dimensions. It was generated by specifying that each group be distributed uniformly along two axes and normally along the third axis. The entire cloud was then warped morphologically in two axes to create the dome. (Ideally, each cell would be generated in a proliferative zone and made to migrate to its adult location.)

Figure 1 shows a set of immature "pyramidal" neurons, corresponding to the band of yellow cells in **Plate 2**. Each has an apical dendrite, early basal dendrites and a varicose axon. Cell nuclei are visible in variously eccentric positions within the somata, which exhibit a range of diameters.

Color Plate 3 (p. 187) shows a subset of elements from a larger simulation representing cell groups in the brainstem, thalamus and cortex. A set of fibers from the brainstem invade the thalamus from below. Groups of cells constituting a row of "barreloids" in the thalamus send their axonal processes into an overlying cortex containing arborized "stellate" cells. A set of unbranched "pial" cells delimits the top margin of the cortex.

Figure 2 shows a stereo view of a field of highly varicose axonal processes, from a region like the thalamo-cortical projection path shown above. The program created the varicosities by randomly sampling statistical specifications for average bead density, length and diameter. (Certain axonal processes remain visible in only the left or right image, having been clipped differently for each eye. This structural disparity requires more focused attention than usual to fuse the two fields.)

The program runs interactively on the Indigo class of Silicon Graphics workstation and has been written in C and GL (it may be redesigned for C++ and OpenGL). All variables have been cast as double precision. There currently are about 50 variables per segment (many are dormant but slated for physiology). There can be any number (often thousands) of segments per group, and any number (often hundreds) of groups in a simulation. The full system in **Plate 3** (of which about half is shown) has over 100 groups and over 130,000 elements, and took 20 minutes to generate, filling 40 megabytes (MB) of space. The program should scale relatively well, even on current machines, to a hundred variables per segment, and a thousand groups. Workstations now often have several hundred MB of fast memory, and gigabytes (GB) of local disk space. It is reasonable to anticipate that by the time details are added to this framework adequate to *begin* to satisfy both cell and molecular biologists and comparative neuroanatomists, affordable machines will exist with several GB of local memory, and terabytes of disk space. With such processing power, dramatically more complex and accurate systems would be generated. The precision of those future results will come as much from improvements in capturing the logic of biology in algorithmic form and from incorporating existing morphometry, as from the sheer increase in numbers of simulated elements.

Conclusion: A Call to Action

The purpose of this presentation is to encourage others to consider how to better achieve the overall goal of unifying neuroanatomical information. My hope is that this will enable more people to share the beauty and wonder that research in neuroscience engenders, that students can more facilely absorb an understanding of anatomy, and that researchers themselves will be able to frame more penetrating hypotheses.

I have presented a basic organizational framework that currently is relatively undetailed with respect to the daunting task at hand. But the emphasis in its design has been to provide a broad underpinning, in anticipation of later detail. It may be well enough founded, now, to mature dramatically in range and subtlety when supplied with greater numbers of measurement-based statistics. Only a small fraction of the effort to create this framework was expended on handcrafting statistical data to specify neuron classes, and only a very small number of the many published statistical measures of neurons have been incorporated. Imagine the richness of biological detail we might gather into usable form simply by entering precise numerical information for just the most prominent several dozens or hundreds of neuron types!

To approach this potential, this work must motivate a variety of researchers to adopt or create similar methods and to help cast knowledge of their specialties in exchangeable (statistical) formats. Some may be interested in extracting statistics of brain information from the literature and in establishing repositories on the Internet, or in verifying the input data and its transforms for accuracy. Others may be able to refine the statistical format itself. It will be necessary to derive theoretically a minimal set of features that adequately discriminate one group of neurons from another. In practice it appears that ad hoc sets of features can be provided relatively easily that give a plausible depiction of cell types and neural circuits. With more use, perhaps the natural feature sets will be approximated.

I believe that visual framing of even the phenomenology of developmental neuroscience will provide us with a dramatically clearer perspective on what has hitherto been accomplished collectively. Hopefully, though, a great deal more can be achieved by enriching this framework: The deeper developmental logic of both the program and its associated fields continue to evolve. The program has only preliminary internal representations for many central features: stem-cell regulation, constraint of cell migration, process outgrowth to spatial regions defined by deformable surfaces, electrical signaling and synaptic dynamics, message-passing by diffusion in voxellized intercellular space. The fields of cellular neurobiology and computer graphics have extensive information on each feature. It will be exciting to help tie these fields together to create a working user's menu of visible developmental principles for brain development. Here is one way to proceed.

Neural-Networks Models of Cognition
J.W. Donahoe and V. Packard Dorsel (Editors)

CHAPTER 4

EVOLVING ARTIFICIAL NEURAL NETWORKS IN PAVLOVIAN ENVIRONMENTS

José E. Burgos
Instituto de Psicología, Universidad Central de Venezuela
Escuela de Psicología, Universidad Católica Andrés Bello

ABSTRACT

This chapter introduces a computational approach to evolutionary interpretations of animal-learning phenomena, with an emphasis upon Pavlovian conditioning. The approach combines a neurocomputational algorithm (NCA) and a genetic algorithm (GA). The NCA determined the functioning of neural processing elements constituting artificial neural networks (ANNs). The GA consisted of a neurodevelopmentally informed genotype-to-phenotype transformation, and a scheme for simulating selection and reproduction. The interstimulus interval (ISI) function and the variation of optimal ISIs across preparations were chosen as motivating phenomena. On this basis, generations of ANNs were each trained in Pavlovian forward-delay procedures with different ISIs (either 2, 4, 8, or 16 time steps) and conditional stimuli (either CS1 or CS2). Individual fitness of a network architecture was identified with performance under these procedures. A major finding was that mean population fitness increased as a negatively accelerated function of generations. Although ANNs in different populations had comparable fitnesses at the end of evolution, larger ANNs emerged with selection for longer ISIs. Also, the magnitude of the Baldwin Effect (high innate responding to CS-alone preexposure trials) decreased as the selected ISI increased. Finally, when ANNs were trained under ISIs different from the ancestral ones, performances simulated ISI functions whose optimal values corresponded to the selected ISIs. These results are consistent with a synthesis of general-process and biological-constraint approaches to learning. *Key words*: Pavlovian conditioning, artificial neural networks, evolution, genetic algorithms, ISI functions.

Introduction

Recent applications of computers as research tools in empirical science involve simulations of biological phenomena. Such applications are collectively known as *Artificial Life* (ALife). ALife comprises all biological phenomena simulated through hardware or software (Langton, 1992). A premise of ALife research is that simulations of possible biological forms may provide useful research tools for achieving a more complete understanding of real biological forms (Cliff, 1991; Dawkins, 1987; Dennett, 1994; Emmeche, 1991; Langton, 1992, 1994).

One ALife research field combines *neurocomputational algorithms* (models of the structure and functioning of nervous systems) with *genetic algorithms* (models of evolution by natural selection). A main purpose of such combinations is to simulate the relation between learning and evolution (e.g., Ackley & Littman, 1992; Belew, McInerney, & Schraudolph, 1992; French & Messinger, 1994; Gruau & Whitley, 1993; Harp & Samad, 1991; Harvey, 1991; Hinton & Nowlan, 1987; Keesing & Stork, 1991; Miller, Todd, & Hegde, 1989; Paredis, 1991, 1994; Schaffer, Caruana, & Eshelman, 1990; Todd & Miller, 1991; Whitley, Starkweather, & Bogart, 1990).

In this chapter, I introduce a computational approach to the study of the relationships between learning and evolution. In contrast to other approaches, environment-behavior relationships are conceptualized in terms of the kinds of phenomena studied in that branch of experimental psychology known as *animal learning*, with an emphasis on Pavlovian conditioning. My main objective was to show how a computational approach can formulate and validate evolutionary interpretations of certain animal-learning phenomena. I also wished to show the relevance of such phenomena for formulating and validating computational approaches to the evolution of learning. Another distinguishing aspect of the present approach is that a neurodevelopmentally informed genetic algorithm was used. I wanted to show the implications of such a GA for the Baldwin Effect, a phenomenon that has received some attention from ALife researchers. In the first section, I describe two basic animal-learning phenomena (ISI functions and variations in the optimal ISI) from an evolutionary perspective. In the second section, I describe the approach in terms of the ontogeny-phylogeny dichotomy. In the third section, digital simulations are used to implement the approach. In the last section, I discuss some implications of the results for computational models of the biological bases of learning.

Biological Constraints on Pavlovian Conditioning

Pavlovian conditioning (Pavlov, 1927) comprises a class of learning phenomena observed under a particular experimental arrangement. With respect to the environment, the basic arrangement involves presenting an animal with instances of two kinds of events, a *conditional stimulus* (CS) and an *unconditional stimulus* (US), in a certain temporal relation. This relation is defined primarily in terms of the *interstimulus interval* (ISI), the time between CS onset and US onset, which represents a measure of the temporal contiguity between the CS and US. With respect to the organism, a response system is chosen in which the US evokes or elicits responses reliably and strongly before exposure to the arrangement.

The basic result is well known. Before training, the chosen response occurs only (or most strongly) in the presence of the US. This response is referred to as the *unconditional response* (UR). After training, the CS comes to evoke responses similar to those that were initially evoked only by the US. Responses evoked by the CS are referred to as *conditional responses* (CRs). An important property of CRs is that they tend to occur before US onset, for which reason Pavlovian conditioning is often said to involve learning to anticipate the occurrence of biologically critical stimuli.

The basic Pavlovian arrangement has been extremely fruitful for discovering a number of learning phenomena. Among the most fundamental and well established are the ISI function and the variation of optimal ISIs across preparations. An ISI function is a relation between the ISI and some behavioral measure, typically CR percentage (the percentage of CS presentations that evoke a response after training). ISI functions have an inverted U shape whose peak corresponds to the ISI value considered optimal for excitatory conditioning. As the ISI departs from the optimal value, the CR percentage decreases. This phenomenon has motivated a number of neurocomputational models of learning (e.g., Byrne, Gingrich, & Baxter, 1989; Grossberg, 1991; Hawkins, 1989; Sutton & Barto, 1981).

Optimal-ISI variation refers to the range of ISIs that produce maximal excitatory conditioning for different experimental preparations—such as the nictitating membrane response in rabbits (e.g., Frey & Ross, 1968; Schneiderman, 1966; Smith, Coleman, & Gormezano, 1969), autoshaping in pigeons (e.g., Gibbon, Baldock, Locurto, Gold, & Terrace, 1977), and conditioned suppression (Libby, 1951; Ross, 1961; Yeo, 1974) and taste aversion (e.g., Etscorn & Stephens, 1973; Garcia & Koelling, 1966) in rats. Optimal-ISI variation has not inspired any neurocomputational research. However, this phenomenon is intimately related to the ISI function when viewed from an evolutionary perspective.

Evolutionary interpretations of animal-learning phenomena arise from the notion of biological constraints. According to this notion, biological structure imposes restrictions upon the kinds of behavioral change an organism may show under certain environmental conditions (e.g., Bitterman, 1965; Bolles, 1970; Hinde, 1973; Seligman & Hager, 1972; Shettleworth, 1973). Certain biological structures are said to prepare organisms to learn under certain conditions and to contraprepare them to learn under others. In evolutionary biology, different structures permit organisms to satisfy the different adaptive demands posed by their specific environments. If an individual increases its reproductive success by changing its behavior in certain ways under certain environmental conditions, then the structure that mediates such a change, and whose blueprints are heritable through genetic transmission, will tend to predominate in the individual's progeny. And, if the environments of parents and progeny are relatively similar, then such a structure will mediate the same kind of behavioral change in the progeny, increase their reproductive success, and appear in subsequent progeny. Behavioral change may thus affect the course of phylogeny by providing a fitness advantage to organisms possessing the kind of structure that mediates such changes.

When the above interpretation is applied to ISI functions, we may say that high responding within a certain ISI range in present organisms is mediated by a certain kind of structure. Such a structure is a product of evolution insofar as it conferred a fitness advantage on the ancestors of those organisms by mediating learned responding within a similar ISI range. Because ISIs outside the range were not characteristic of the ancestral environment, learning under such ISIs was not selected for the ancestral organisms. The inability of present organisms to learn outside the ancestral ISI range, then, can be attributed to an absence of the necessary structure. Optimal-ISI variation arises from the fact that different organisms possess structures necessary to learn under different ISI ranges, insofar as those ranges were characteristic of their ancestral environments.

Nervous systems obviously represent critical biological structures for learning. Although all nervous systems are products of evolution by natural selection (insofar as their basic architectures contain elements that are heritable through genetic transmission), different nervous systems may result from different selection pressures. Thus, the kind of nervous system an organism possesses may mediate learning within a certain ISI range, but prevent learning outside that range, and thereby produce different ISI functions.

The critical effect of the architecture of the nervous system on learning has led to much experimental work on the biological bases of Pavlovian conditioning (e.g., Buonomano & Byrne, 1990; Kandel, 1985; Klopf, 1988; Klopf & Morgan, 1990; Moore, 1991; Moore, Berthier, & Blazis, 1990; Schmajuk & DiCarlo, 1991; Thompson, 1986; Thompson, Berger, & Madden, 1983). This work has served as a basis for constructing neurocomputational models that attempt to simulate certain properties of particular nervous systems. However, such research has not incorporated the *structure* of the neural network as an essential component of these models, algorithms, and simulations. A fundamental premise of the present paper is that an adequate computational account of the biological bases of Pavlovian conditioning (and of learning generally) must explicitly accommodate differences in network architecture within an evolutionary framework. In the next section, I introduce a computational approach motivated by this premise.

A Computational Approach

The present approach involves a combination of a neurocomputational algorithm (NCA) and a genetic algorithm (GA). The NCA and its neural-network implementation have been described elsewhere (Donahoe, this volume; Donahoe, Burgos, & Palmer, 1993; Donahoe & Palmer, 1994). Here, I describe the main components of the GA in terms of the ontogeny-phylogeny dichotomy, mentioning only those aspects of the NCA and its neural-network implementation that are pertinent to the simulations. I also describe how environment-behavior relations are conceptualized.

Ontogeny

I use the term *ontogeny* to refer to two processes that can be fully characterized at the level of the individual organism, namely development and learning. An individual organism is conceptualized as constituted by a genotype, a structural phenotype, and a functional or behavioral phenotype. Development involves the transformation of genotypes into structural phenotypes representing artificial neural networks (ANNs). Learning is conceptualized as changes in the behavior of ANNs under certain environmental conditions (see later).

Genotypes. Following canonical GAs, a genotype (an individual's genetic makeup) was conceptualized as a string of bits that represented genes. Each genotype was partitioned into fragments that encoded a neurodevelopmental parameter. All parameters were encoded by more than one gene (polygeny) and any given gene determined one and only one parameter (no pleiotropy). All genotypes within and across generations encoded for the same set of neurodevelopmental variables. (The **Appendix** shows the relation of genotype fragments to neurodevelopmental variables.)

Development. The transformation of genotypes into ANNs was determined by a neurodevelopmental algorithm (NDA) consisting of routines that simulated general aspects of the stages of proliferation, migration, differentiation, synaptogenesis, and elimination. This division corresponds roughly to the one used by neuroscientists to describe the development of nervous systems (e.g., Brown, Hopkins, & Keynes, 1991; Purves & Lichtman, 1985; see Luskin, this volume). As in other models, ANNs consisted of three kinds of layers of elements—input, hidden, and output.

The proliferation and migration routines determined the maximum number of hidden elements. These routines were based on a simplified version of the radial-unit model of neocortical histogenesis proposed by Rakic (1988a). In accordance with this model, developing nervous systems consist of three kinds of structures—proliferative units, glial fibers, and ontogenetic columns representing what, in ANN terminology, are referred to as *hidden layers*. Proliferative units and ontogenetic columns were related in a one-one manner through the glial fibers. Proliferative units thus constitute a kind of "protomap" of the future organization of hidden layers. The maximum number of proliferative units (and, hence, hidden layers) was encoded in the genotype. Proliferative units generated new elements in a manner functionally analogous to symmetric cell division. Elements were generated by two types of proliferative units. One type generated elements that eventually constituted the hidden layers of the sensory portion of an ANN, and the other type generated elements that eventually constituted the hidden layers of the motor portion. All proliferative units generated the same number of elements. Migration involved transferring elements from the proliferative units to ontogenetic columns. Elements from sensory and motor proliferative units could migrate at different rates, as determined by probabilities encoded in separate genotype fragments. Therefore, differences in the maximum number of elements per sensory and motor hidden layer in an ANN depended partially upon migration.

Element specification occurred during the differentiation routine, which simulated general functional aspects of neuronal differentiation in nervous systems (e.g., Austin & Cepko, 1989; Caviness & Sidman, 1973; McConnell, 1988, 1989, 1991; Miller, 1988; Rakic, 1988b). This routine determined which elements were excitatory and which were inhibitory using a probabilistic rule that assigned to each element parameters used in the NCA. Parameters for different kinds of elements were encoded separately and, hence, could be different within the same ANN.

Elements were connected through a synaptogenesis routine, whose basic parameter was the probability of a connection being formed between two elements. Synaptogenesis probabilities were encoded separately for different kinds of connections. As a simplifying restriction, only feedforward connections between adjacent layers were allowed (i.e., recurrent connections could not arise). Also, only lateral connections were allowed between excitatory and inhibitory neural-processing elements (NPEs) within a layer. This permitted simulation of a form of lateral inhibition. No lateral connections between inhibitory NPEs were allowed.

In contrast to other approaches, genotypes in the present approach did not directly encode the initial connection weights or the connections between specific units. Rather, connections were formed by conducting a Bernoulli trial for each pair of elements consistent with the connectivity restrictions mentioned above. The presynaptic element was connected to the postsynaptic element if and only if the Bernoulli trial was passed, which depended on the synaptogenesis probability encoded in the genotype. Once a new connection was formed, its initial strength was determined by the spontaneous activations of the pre- and postsynaptic elements. This strength represented the initial weight for that connection (i.e., the weight before any training). Spontaneous activations of NPEs were determined by genetically encoded parameters. The spontaneous activation parameters were encoded by separate genotype fragments for different kinds of NPEs. This strategy simulated a kind of activity-dependent synaptogenesis, which is known to occur extensively during cortical histogenesis (e.g., Cooper & Rakic, 1983; Corner & Ramakers, 1992; Eckenhoff & Rakic, 1991; Rakic, Bourgeois, Eckenhoff, Zecevic, & Goldman-Rakic, 1986).

After synaptogenesis, all input elements and hidden NPEs that did not send connections to other elements, as well as all hidden and output NPEs that did not receive any connections, were eliminated. Hence, the elimination of NPEs and connections depended only upon the availability of elements and connections (i.e., no parameters were encoded for this stage). Although different in detail, this routine is consistent with the idea of neuronal selection proposed by Edelman (1987). According to this idea, neurons in the developing brain (like organisms in a population) differ structurally and functionally from one another. Some neurons are more strongly activated by certain stimuli and, hence, more richly interconnected than others. Neurons that are not connected die out, leaving only neurons that have at least one presynaptic and one postsynaptic process.

Because most of the neurodevelopmental parameters were probabilistic, any given genotype could generate a class of ANNs that were similar in certain general aspects, but different in their specifics. Hence, strictly speaking, the present NDA implemented a one-many transformation of genotypes into structural phenotypes. In this manner, the present NDA permitted structural-phenotypic variation even with a fixed-length genotype.

Structural phenotypes. An ANN was a set of interconnected NPEs whose functioning was determined by the NCA. ANNs were made up of two subnetworks, sensory and motor. The sensory subnetwork consisted of one layer of input elements whose activations represented the occurrence of different environmental events, and one or more sensory-association (*sa*) layers, each associated with a set of *ca1* NPEs. The motor subnetwork consisted of one or more motor-association (*ma*) layers, each associated with a set of *vta* NPEs, and one layer of output NPEs whose collective activation determined the behavior of the ANN. (The nature of the *ca1* and *vta* NPEs is defined below.)

An input layer could consist of three kinds of elements, namely s_1, s_2, and *us*, whose activations were assigned according to a training protocol representing the environment of the ANN. During the simulations, the activations of s_1 and s_2 elements indicated the occurrence of different CSs, whereas the activation of the *us* element represented the occurrence of the US. The maximum numbers of s_1 and s_2 elements that a given ANN could have before elimination were directly encoded by separate genotype fragments. ANNs could have from zero to seven s_1 or s_2 elements, for a maximum total of 14 sensory input elements. All ANNs had only one *us* element.

Sensory-input elements were connected to NPEs in the first *sa* layer through plastic feedforward connections. The *us* element sent nonplastic connections directly to the *vta* NPEs, which in turn sent nonplastic connections directly to output NPEs. Therefore, output NPEs could be unconditionally activated by the *us* element through the *vta* NPEs. The *us-vta*-output pathways were analogous to ones that mediate URs in Pavlovian conditioning. Output NPEs could also be indirectly activated by s_1 or s_2 input elements by means of connections to *sa* and *ma* NPEs. The main difference between *sa* and *ma* NPEs concerned how synaptic plasticity was modulated by the activity of the *ca1* and *vta* NPEs (for a more detailed description, see Chapter 18, this volume, Donahoe et al, 1993, and Donahoe & Palmer, 1994).

The maximum numbers of *ca1* NPEs per *sa* layer and *vta* NPEs per *ma* layer were also genetically determined. This strategy was justified under the assumption that the corresponding parts of the mammalian brain (i.e., the CA1 output neurons of the hippocampus, and the ventral tegmental area—or VTA—respectively) are genetically conserved and, hence, show more intra-specific similarities relative to sensory- and motor-association cortex (Allman, 1990; Arbas, Meinertzhagen, & Shaw, 1991; Eccles, 1989). The number of output NPEs was genetically determined for similar reasons.

Functional phenotypes. Behavior of an ANN was defined as the mean activation of its output NPEs at a given time step (ts). The higher the mean activation of the output NPEs to a CS, the greater its chance was of being selected for reproduction. The present scheme thus simulated a form of directional selection, in which performance toward the upper limit of responding was favored. (As with other ALife approaches, the present approach adopted the propensity interpretation of individual fitness; Mills & Beatty, 1979. Under this interpretation, individual fitness does not refer to actual reproductive success but to an aspect of an individual's phenotype.) In the present case, the behavior after training was the phenotype of interest that affected and, hence, preceded reproduction.

Unlike other approaches, the environment was conceptualized in terms of temporal relations between environmental events as described by standard Pavlovian procedures. This conceptualization assumes that the same kinds of temporal relations defined by Pavlovian procedures also occur in natural environments and affect reproductive success. This is not to say, of course, that the specific nature of the events being related is unimportant, or that temporal relations are sufficient to understand learning phenomena at all levels of analysis. Clearly, the specific nature of the events represents a critical feature of an ecologically valid approach. But, temporal relations are undeniably important determinants of conditioning (cf. Gamzu & Williams, 1973; Rescorla, 1966, 1968). Focusing upon temporal relations among abstract events represents a momentary simplifying strategy rather than a theoretical proposition about the structure of natural environments.

The idea that natural environments consist of the same kinds of relations studied with Pavlovian procedures in laboratory environments has motivated investigations of the relation between Pavlovian conditioning and reproductive behavior in territorial species such as *Betta splendens* and *Trichogaster trichopterus* (e.g., Domjan & Hollis, 1988; Hollis, 1984, 1990; Hollis, Cadieux, & Colbert, 1989; Hollis, Dumas, Singh, & Fackelman, 1995; Hollis, Martin, Cadieux, & Colbert, 1984). These studies have shown that learning to anticipate territorial situations through a Pavlovian training allows such situations to be resolved more effectively. If trained individuals tend to be reproductively more successful, then aspects of their nervous systems that mediate such learning and are heritable would tend to appear in the individual's progeny (e.g., Hollis & Overmier, 1982). This kind of research demonstrates that Pavlovian conditioning may change reproductive success and, hence, the course of phylogeny.

Phylogeny

I use the term *phylogeny* to refer to lineages that result from implementing the selection and reproduction schemes of a GA, a kind of computational model of evolution by natural selection (Holland, 1975). In typical GAs, a performance measure is obtained for each individual in a population. Such a measure determines individual fitness. The fitness determines whether or not a given individual is selected for mating. In the present approach, the selection scheme was applied to a performance measure computed for each individual within a generation after training. Mating occurred between pairs of genotypes taken from a mating pool. Mating pools were formed for each generation by applying a rule known as tournament selection (e.g., Goldberg & Deb, 1991), which is more efficient and biologically plausible than standard rules such as roulette-wheel selection. Genetic variation was affected by crossover from combining the genetic makeup of the mating pair to form a new chromosome. In addition, the values of some of the genes constituting the new chromosomes had a low probability of undergoing mutation. The resulting chromosomes constituted a new population of genotypes forming a new generation.

The Baldwin Effect

When we combine ontogenetic and phylogenetic processes, individual learning (an nondevelopmental ontogenetic process) represents a guiding force in evolution (a phylogenetic process). This consideration would have a Lamarckian flavor if it were interpreted as the inheritance of acquired characters. However, as argued by Baldwin (1896), the relationship between learning and evolution can be interpreted in non-Lamarckian terms. This idea has come to be known as the Baldwin Effect (Simpson, 1953), and has received some attention from ALife researchers.

The key idea underlying the Baldwin Effect is that certain nondevelopmental ontogenetic changes, such as learning, increase individual reproductive success. Certain genetically transmissible traits (e.g., basic neural architecture) impose global constraints on how and under what circumstances learning occurs. The presence of these traits does not guarantee the occurrence of the changes in question but it does make it more probable, given the appropriate environmental conditions. If these conditions remain relatively constant over phylogenetic time, then the structures that permit the adaptive changes, as well as the changes themselves, will become progressively more frequent across generations.

In current ALife approaches, the Baldwin Effect has been observed as an increase in the mean population learning rate across generations to the point that, towards the end of evolution, individuals need virtually no training (e.g., Ackley & Littman, 1992; French & Messinger, 1994; Gruau & Whitley, 1993; Hinton & Nowlan, 1987; Keesing & Stork, 1991). This effect is largely due to the fact that, in these approaches, the initial connection weights are directly encoded in the genotype. Thus, evolution in these approaches involves searching for optimal initial weights. In these terms, the Baldwin Effect refers to the extent to which responding to a given stimulus is innate, that is, occurs after little or no training. In the present approach, however, initial connection weights were not directly encoded in the genotype. This strategy raises the question of whether the Baldwin Effect (defined here as the extent to which an ANN shows high responding to CSs without training) emerges in these simulations.

Simulations

Evolution simulations

An evolution simulation consisted of a lineage of nonoverlapping generations. A generation consisted of a sequence of development, training, and selection occurring for each member of a population of 100 individuals. The maximum possible number of generations produced by the GA was 100. Different lineages evolved independently of one another, and corresponded to different environmental conditions. Development was simulated through the NDA. Here, I focus upon the procedural aspects of training and selection.

Training consisted in the simulation of a number of CS-US pairings. Time was represented as a sequence of discrete time steps (ts). No assumptions were made about the specific real-time duration of a ts. At each ts, the activations and connection weights of all NPEs of an ANN were updated according to an asynchronous-random procedure, which is considered to be biologically more plausible than synchronous updating (e.g., Amit, 1989). Also, this update procedure has been found to induce stability in certain ALife models (e.g., Bersini & Detours, 1994). In the present simulations, a random sequence of all NPEs was generated at each ts, and then their activations and connection weights were updated in that order.

After development but before training, the parameter that determined the magnitude of spontaneous activations of NPEs was set to 0.1 for all NPEs. This strategy eliminated ANNs with NPEs that had high spontaneous activations as an outcome of evolution. Differences in the spontaneous-activation parameter played a role only during the synaptogenesis stage of the NDA by affecting the magnitude of the initial connection weights.

In the overall simulation experiment, eight environmental conditions were defined by a combination of four ISIs (2, 4, 8, or 16 ts) and two CSs (CS1 or CS2). A lineage of ANNs evolved independently for each condition. For ANNs in CS1 lineages, the CS was defined as the activation of all s_1 input elements with a value of 1.0, and zero activation of any s_2 input element. For ANNs in the CS2 lineages, the activation levels for s_1 and s_2 input elements were reversed. For all ANNs, the US was defined as the activation of the *us* element with a value of 1.0 at the last ts of a trial (i.e., CS and US were paired according to a forward-delay procedure). The ISI was defined as the number of ts from the first to the penultimate CS ts.

All ANNs were given 125 CS-US paired trials, the last 25 of which served as trials to evaluate individual fitness. Fitness was measured by the proportion of evaluation trials that elicited a CR. A CR for a given individual was defined as a mean activation across its output NPEs of 0.5 or more at the penultimate ts. The intertrial interval (the time between successive pairings) was assumed to be sufficiently long to allow for activations of all NPEs to decay to spontaneous levels.

After all individuals within a generation had been evaluated, a mating pool was formed through a series of tournaments in which five individuals were randomly sampled from the population and their fitness values compared. The individual with the highest fitness value was declared the winner of the tournament and selected for the mating pool. Then, all individuals were returned to the population. Thus, a genotype with a high fitness value could be represented many times in the mating pool. If the highest fitness value was shared by two or more individuals, then the tournament was dismissed and another tournament formed. Tournament selection ended after 100 winners had been selected. If tournament selection had not been completed after approximately one hour because almost all individuals had achieved the same high level of fitness, the evolution simulation was considered to have reached stability.

After the completion of a mating pool, reproduction took place on 50 mating trials. A mating trial involved randomly sampling two individuals from the mating pool, and recombining their respective genotypes with a probability of 0.8. Recombination consisted of selecting two loci randomly (the same in both parental genotypes) and translocating the middle segment of both parents. This strategy is known as *two-point crossover* (Cavicchio, 1970; DeJong, 1975; Holland, 1975; Schaffer, Eshelman, & Offutt, 1991). After mating, parents were returned to the mating pool, and alleles of the new genotypes were mutated with a probability of 0.001. The result of a mating trial was two genotypes consisting of genetic material (possibly mutated) from both parents. The end result of the reproduction process was a population of 100 genotypes constituting a new generation. Then, the development-training-selection sequence was repeated for the individuals of the new generation, and so on.

Before starting the simulations, the founder genotype population, referred to as G0, was generated. G0 was chosen from several randomly-generated populations, on the basis of the mean population fitness in a forward-delay procedure in which 200 pairings were randomly distributed between CS1-US and CS2-US, with an ISI of 4 ts. After this training, 50 more pairings randomly distributed between the two kinds of pairings were given to determine individual fitness. If the mean population fitness was greater than zero, then the genotype population was chosen as G0. This strategy was adopted after preliminary simulations showed that random founder populations in which few individuals had fitnesses greater than zero permitted very little or no evolution.

Lineages were generated as follows. First, eight ANNs were developed from each G0 to obtain 800 ANNs, which were randomly divided into groups of 100. Then, ANNs in each group were trained in a forward-delay Pavlovian procedure with one of the four ISIs and one of the two CSs. After evaluation, a new genotype population was created through the selection scheme described earlier. This new population constituted G1 (the next generation). Once the new genotypes were generated, new ANNs were developed and exposed to the same conditions as their ancestors. The same operation was repeated generation after generation, until the criterion for termination was satisfied. This strategy produced a total of eight lineages, labeled 2-CS1, 2-CS2, 4-CS1, 4-CS2, 8-CS1, 8-CS2, 16-CS1, 16-CS2, where the initial number denotes the ISI, and CS1 or CS2 denotes the kind of CS. I shall focus on the results for 4-CS1 and 8-CS1, since they faithfully represent the major points of interest here.

Effects on genotypes. The effects on genotypes were measured by changes in mean chromosomic overlap across generations, a measure of genotypic similarity. The overlap between any two genotypes could range from -1.0 (the two genotypes had no loci with the same genes) to 1.0 (the two genotypes had no differing loci). An overlap of 0.0 indicates that the two genotypes agreed (or differed) in 50% of their loci, which is the expected value for randomly generated genotype populations. The mean overlap for a generation was computed for all the possible combinations of two genotypes in a population of 100.

The left panel of **Figure 1** shows that mean chromosomic overlaps increased as a negatively accelerated function of generations. The overlap for G0 was close to 0.0, which is consistent with the expected value for a founder population. In general, individuals became genetically more similar as generations advanced. Overlaps increased within a few generations, and then stabilized at different values for ISIs of 4 and 8 ts. For 4-CS1, the mean overlap of the last generation was less than 0.5, which signifies that a considerable amount of genetic variation remained at the end of evolution. In contrast, the mean overlap at the end of evolution for 8-CS1 was substantially larger, indi-

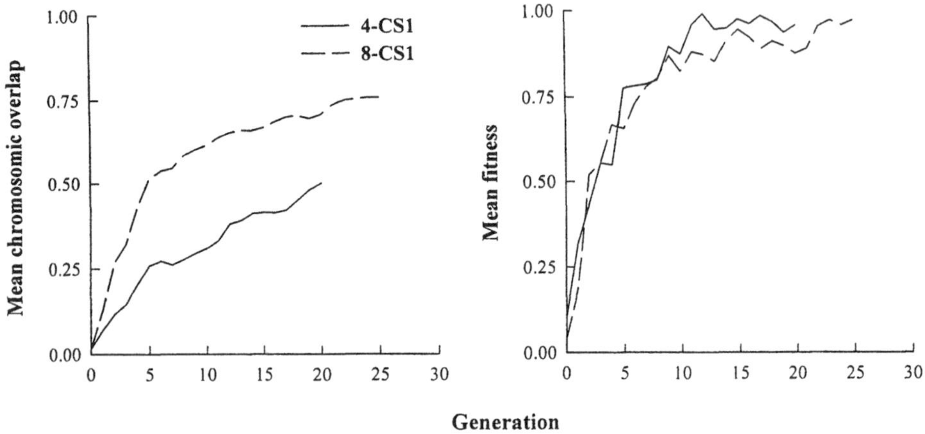

FIGURE 1. Evolution of mean population chromosomic overlap (left panel) and fitness (right panel) as a function of generations.

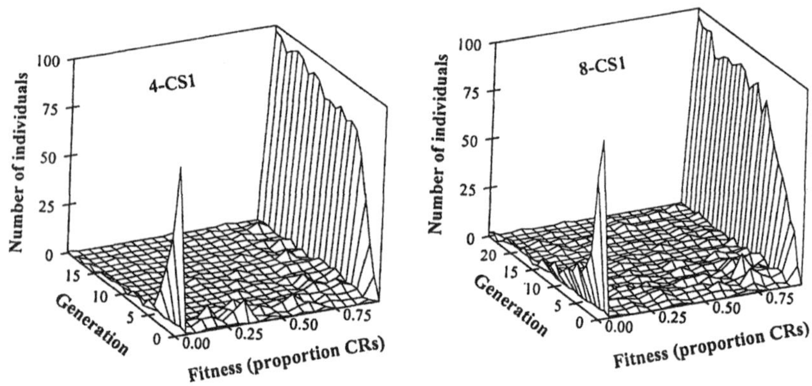

FIGURE 2. Frequency distributions of fitness values over generations for 4-CS1 (left panel) and 8-CS1 lineages (right panel). Shown are the number of individual networks having a fitness of a given value. Fitness was measured by the proportion of test trials with a CR.

cating that the longer ISI more tightly constrained the genetic variation needed to produce individual networks of high fitness.

Effects on fitness. The right panel of **Figure 1** shows that mean population fitness also tended to increase as a negatively accelerated function of generations. Lineages started with ANNs of low fitness (close to 0.0) and ended with high-fitness (close to 1.0) ANNs. ANNs in the last generation of both lineages had similar fitness. Hence, comparably fit ANNs emerged from selection for increased responding under different environmental conditions. Both lineages satisfied the fitness-stability termination criterion well before 100 generations.

Figure 2 shows 3D mesh plots depicting changes in the frequency distributions of fitness values across generations, for the same two ISIs—4-CS1 (left panel) and 8-CS1 (right panel). These plots give an idea of fitness variability within and between generations. Both plots show the same overall tendency, that is, a high number of low-fitness individuals at G0 followed by a sharp increase in the number of high-fitness individuals after several generations. Note that a few individuals showed low-fitness values even towards the end of evolution.

Effects on learning. A high fitness indicates only that a high proportion of evaluation trials satisfied the CR criterion of a mean output-NPE activation > 0.5. A second measure of the functional phenotypes was provided by the mean activation levels across trials for the first and last generation of each lineage.

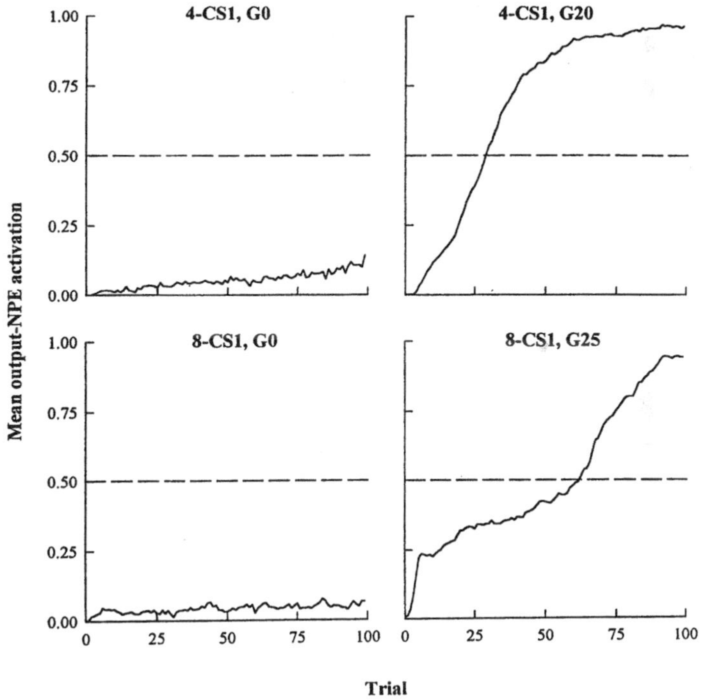

FIGURE 3. Mean output-NPE activations across trials for the first (left panels) and last generations (right panels) of 4-CS1 and 8- CS1.

Figure 3 shows average learning curves for the first (left panels) and last generation (right panels) with 4-CS1 (top panels) and 8-CS1 (bottom panels). The 0.5 criterion for CR occurrence is indicated on the X axes for reference. In the first generation, most output-NPE activations remained at near-zero values across training trials. In the last generation, output-NPE activations increased as a negatively accelerated function of trials for both lineages. At the beginning of evolution, only a few ANNs learned a little (but just enough to guarantee an output-NPE activation larger than 0.5 on at least one probe trial). However, by the end of evolution, most ANNs showed the maximal output-NPE activation towards the end of training.

Control simulations. The above results show that learning and behavior-dependent selection were sufficient for evolution to occur. However, it is unclear whether or not learning and behavior-dependent selection were necessary. To address this issue, two kinds of control simulations were run, No-Learning (NL) and Random-Fitness (RF). Both were run with the same ISIs and CSs as before, and using G0 as the founder population. In the NL simulation, selection was behavior-dependent, but the learning function was disabled so that ANNs could not learn. Thus, any changes in network architecture and performance were attributable to simulated natural selection alone. In the RF simulation, ANNs could learn, but individual fitness was a random number between 0

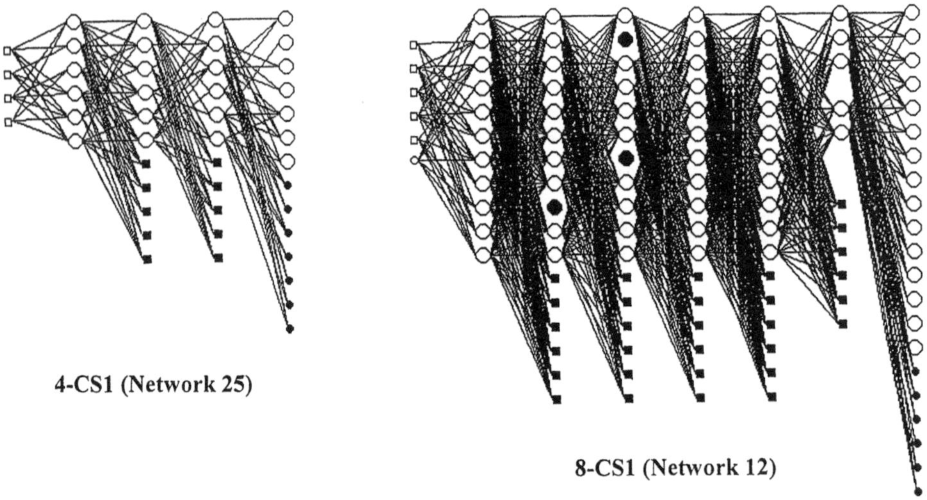

4-CS1 (Network 25)

8-CS1 (Network 12)

FIGURE 4. Representative ANNs from the last generation of 4-CS1 (left panel) and 8-CS1 (right panel).

and 1. Thus, changes were attributable to learning alone. In both control simulations, fitness remained close to 0.0 throughout 100 generations for all conditions. Hence, both simulated learning and natural selection were necessary for the evolution of increased responding.

Effects on structural phenotypes. The above results show that different selection pressures generated ANNs that were *functionally* equivalent (i.e., comparably fit) under the environmental conditions in which ancestral selection occurred. Were such ANNs also structurally equivalent? **Figure 4** shows a representative ANN from the last generation of 4-CS1 and 8-CS1. The open squares indicate s_1 input elements, the open small circles indicate s_2 input elements, filled squares indicate *cal* NPEs, and the small filled circles indicate *vta* NPEs. Open large circles represent excitatory NPEs and, in the rightmost layer, output elements; filled large circles represent inhibitory NPEs. The empty spaces within layers represent elements that were eliminated at the end of development. For simplicity, inhibitory NPEs are shown without connections. The presence of an inhibitory NPE, however, signifies that it received and sent at least one connection within the layer. The *us* input element was omitted for all ANNs. Although both ANNs had a fitness of 1.00, their architectures were clearly different.

The most salient difference is that selection under the longer ISI generated substantially larger ANNs than selection under the shorter ISI. Indeed, the 8-CS1 ANN has substantially more *sa* layers (connected to *cal* NPEs) as well as substantially more NPEs per input, hidden, and output layer. Also, the 8-CS1 ANN contained more inhibitory NPEs and a much richer connectivity than the 4-CS1 ANN. Note that these ANNs had few or no s_2 input elements, which was the case in practically all the members of the population. Similar results were observed for ANNs selected under other conditions. Overall, these results show that different selection pressures generated structurally different ANNs.

Test simulations

Different selection pressures generated ANNs that were functionally equivalent under the environmental conditions in which ancestral selection occurred. How would these ANNs behave under other environmental conditions? Various test simulations addressed this issue.

Procedure. In one test simulation, ANNs were given a CS-preexposure procedure to determine the extent of the Baldwin Effect (i.e., high responding to novel nonreinforced CS trials). ANNs from the last generation of each lineage were given 125 presentations of CS-alone trials, the last 25 of which were used as evaluation trials to determine CR proportion. All ANNs were exposed to the same CS duration experienced by their ancestors. In the other test simulations, ANNs from the last generation of each lineage were exposed to each of the environmental conditions used in the evolution simulations. All ANNs were given 125 trials of each condition, the last 25 of each were used as evaluation trials to compute CR proportion. The purpose of these simulations was to determine if the architecturally different ANNs from different populations were *functionally* different when exposed to environmental conditions that differed from the ancestral ones. The results of these simulations are expressed in terms of changes in the mean CR proportion for different ISIs.

CS preexposure: the Baldwin Effect. As the left panel of **Figure 5** shows, the Baldwin Effect decreased rapidly as a function of the ISI. ISI-2 ANNs showed the strongest effect, which indicates that they tended to respond with very high output-NPE activations from the very beginning of training. In a sense, then, evolution under ISI 2 resulted in ANNs that innately responded to the CS. ISI-4 ANNs also showed some degree of the Baldwin Effect, although to a substantially smaller extent. Finally, CR proportions for ISI-8 and ISI-16 ANNs remained at zero after 125 CS-alone preexposures. Hence, these ANNs did not show the Baldwin Effect.

The fact that initial connection weights were not directly encoded in the genotype undoubtedly contributed to these results. Indeed, because initial connection weights were determined only indirectly by the genotype, evolution in the present simulations seems to have involved primarily a search in architecture and parameter space, rather than in initial connection-weight space. One parameter used in the activation rule determined NPE spontaneous activations, which in turn affected initial weights. On this basis, one might argue that the ISI-2 ANNs showed the Baldwin Effect because they had larger initial

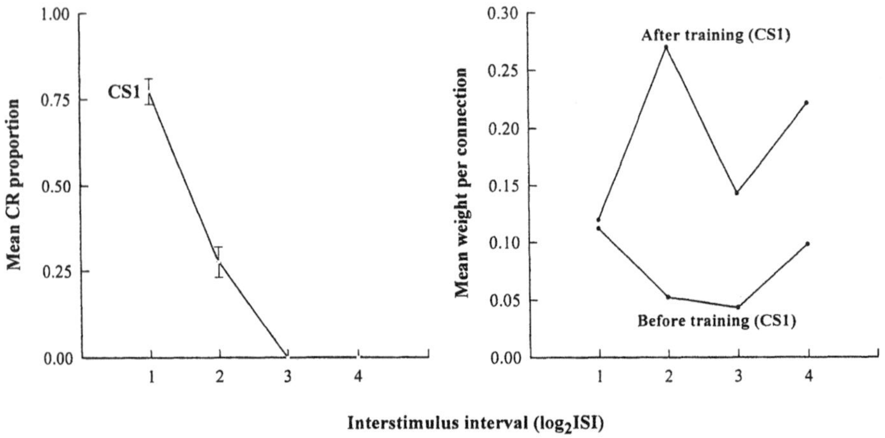

FIGURE 5. Mean CR proportions during CS-1 tests (left panel), and mean weight per connection before and after training (right panel) for ANNs during the last generation of each lineage.

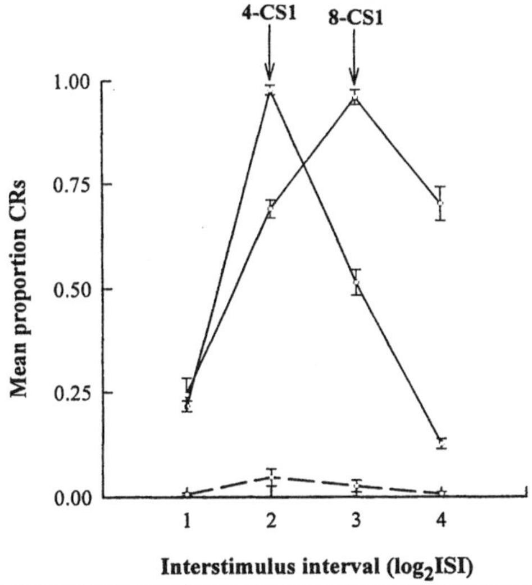

FIGURE 6. Mean CR proportions during tests of ANNs at different ISI values of the last generation of 4-CS1 and 8-CS1. The ISI throughout evolution is shown at the top of each function.

weights. As the right panel of **Figure 5** shows, the ISI-2 ANNs did not have markedly higher initial weights than the other ANNs. However, note that the difference between values of the connection weights before and after training for ISI-2 ANNs was substantially smaller than the difference for other populations. Considering that selection under an ISI of 2 generated the smallest ANNs of the simulations, a strong interaction between initial connection weights and the size of the ANN may underlie the Baldwin Effect observed in the present simulations.

ISI functions and optimal-ISI variation. Finally, **Figure 6** shows the effects of the ISI tests of ANNs in the last generation of 4-CS1 and 8-CS1. The arrows indicate the conditions under which the ancestors learned and were actually selected. Responding in both populations changed as an inverted U-shaped function of ISIs, with peaks that coincided with the ancestral ISI. Thus the ISI functions qualitatively resembled those observed experimentally. Optimal and ancestral ISI coincided for both populations of ANNs.

The broken line toward the bottom of **Figure 6** indicates the findings obtained with a test given to the 4-CS1 ANNs in which CS2 was now the reinforced stimulus. Responding to a CS different from the ancestral one was near zero. This finding is consistent with a Pavlovian-conditioning phenomenon known as *CS nonequipotentiality*: Different CSs are differentially effective in evoking a response after being paired with a given US, thus representing another form of biological constraint on learning (e.g., Bolles, 1970; Garcia & Koelling, 1966).

General Discussion

Genetic Algorithms and Network Architectures

This chapter illustrates how genetic algorithms (GAs) may be used to address the challenge of generating suitable architectures for neural networks. The input-output relations that networks are competent to mediate depend upon aspects of their structures, such as the number of neural-processing elements (NPEs). In normative neural-network research, the structure whereby the network *represents* the correlations between and among environmental and behavioral events is the product of handcrafting: The researcher designs (and redesigns) the architecture until it successfully mediates the input-output relations required by the task (see Barto and Sutton, this volume). GAs permit neural-network researchers to address the problem of neural architecture in a principled manner. In the present instance, the guiding principles of the GA were those of developmental neurobiology. Thus, unlike other applications to neural-network research, the GA used here did not determine the particulars of the architecture (e.g., the connection weights between NPEs). Instead, like the relation between the genome and the structure of the brain of living organisms, the simulated genes acted through a neurodevelopmental algorithm (NDA). The NDA, together with the GA, constrained the architecture such that the input-output relations favored by selection could be mediated. In this way, a single founder population of simulated genes subjected to the GA produced a *family* of architectures. This approach parallels the strategy employed by evolution in the design of the brain: The number of genes expressed in the nervous system—as large as it is—is much too small to specify precisely the structure of the brain. What phylo- and ontogenic selection produce is a range of variation, not a single variant, with most variants capable of mediating the favored relations (cf. Palmer & Donahoe, 1993).

The integration of a GA with an NDA permitted the founder population to respond not only to selection for one favored relation; e.g., Pavlovian conditioning with an ISI of 4 time steps (ts); but also to a number of other selection pressures; e.g., conditioning with other ISIs. Thus a single founder population acted upon by successive cycles of the same GA-NDA and neurocomputational, or learning, algorithm (NCA) produced different neural architectures. But, each member of the different families of networks was typically competent to mediate the input-output relations favored by selection. The fruitfulness of the approach is further documented by other simulations, not described here, in which this same founder population—subjected to still other selection pressures—produced other families of quite different network architectures, all of which responded suitably to the different demands (Burgos, 1996). The hybrid GA-NDA-NCA approach offers a promising general strategy for obtaining neural-network architectures approaching the competencies of living organisms and holding out the promise of significant engineering applications as well.

Genetic Algorithms and Learning

When an NDA is integrated with a GA, the same learning process—as defined by the NCA—is implemented in different neural architectures. That is, learning acts upon different simulated neural systems as occurs with organisms in the natural environment. The present findings may be used to make two broad points about the learning process. First, the same learning process is capable of yielding very different outcomes depending on the neural structures on which such processes act. The same NCA produced maximal responding at one ISI for one family of network architectures and at a different ISI for a different family. Thus, different outcomes of learning do not necessarily imply different learning processes. Such findings are consistent with *general-process learning theory*: A common learning process may have differing effects when acting upon different neural systems. Second, when a GA is expressed through an NDA, the appropriate functioning of the network continues to be dependent on the modification of connection weights by the NCA. Except for the very smallest networks produced in the present simulations (i.e., networks mediating conditioning with very short ISIs), appropriate responding required learning. That is, the Baldwin Effect did not occur since networks were not "born" with "innate" knowledge of how to respond. As in the natural environment, where learning is regarded as the "pacemaker" of evolution (Wilson, 1975), proper performance requires the concerted selecting effects of both the ancestral environment, as simulated by the GA, and the individual environment, as simulated by the NCA.

One specific aspect of the general-process approach merits special mention. When the different neural architectures were acted upon by the common learning process implemented by the NCA, different families of networks responded maximally to different ISIs *without the need to posit any explicit timing mechanism.* When the NCA modified the connection weights, the network as a whole performed as if there were a timing mechanism, although—in truth—there was none. There was timing but no timer. Timing *emerged* from interactions among the NPEs (see Buonomano & Merzenich, this volume). Moreover, an ordered temporal gradient of responding also emerged (see **Figure 6**) without an explicit timing mechanism.

Finally, the interdependences between GA, NDA, and NCA may be illustrated by the nature of the change in ANN size as the ISI changed. Note that performance with longer ISIs tended to be mediated by ANNs having larger numbers of NPEs (see **Figure 4**). This finding was replicated in a simulation using two handcrafted ANNs that differed only in the number of *sa* hidden layers (Burgos, 1996). Both ANNs were trained in a forward-delay procedure with either a short or a long ISI. As expected, the small ANN performed substantially better under the short ISI, whereas the large ANN performed substantially better under the long ISI. A value was computed for each ANN-ISI combination, representing the mean number of times an NPE was updated at a given updating position within a ts. The best performance yielded a mean number very close to 1.0, corresponding to the ANN-ISI combination in which the number of hidden NPEs was identical to the number of ts prior to reinforcement. Performance, then, improved as the mean number of times an NPE was updated at a given position within a ts approached 1.0. This result suggests, as a conjecture, that optimal conditions in this particular simulation may be approximated by a Latin square whose columns represent the possible update positions (as determined by the number of hidden NPEs) and whose rows represent the number of ts prior to reinforcement (as determined by the ISI). Indeed, the key feature of a Latin square of C x R items (where C is the number of columns and R is the number of rows) is that each item appears once and only once in each column and row (Dénes & Keedwell, 1974). In the present case, the items represent NPE updates. The above conjecture cannot be extended without modification to the ANNs evolved through the GA, for the smallest number of hidden NPEs in the smallest of these ANNs was substantially larger than the longest ISI. Therefore, optimal conditions in those ANNs may be better approximated by Latin rectangles. Perhaps the NCA parameters in these ANNs interacted in important ways with the architecture, for, in contrast to the handcrafted ANNs, such parameters differed across different classes of NPEs in the evolved ANNs. The update procedure may be a critical aspect here. In the present simulations, the order in which NPEs were updated was randomly and independently determined within each ts (i.e., an asynchronous, random update procedure was used). Such an update procedure is biologically plausible, as noted earlier, and may have determined the ways parameters and architectures interacted. In any case, a characterization of optimal conditions in terms of Latin squares (or rectangles) allows for a definition of optimal conditions that relates the temporal structure of the environment with the neural structure of organisms. The organism thus becomes a defining component of optimal conditions for learning.

ACKNOWLEDGMENTS

The present work is based on the author's doctoral dissertation, University of Massachusetts/Amherst. Preliminary versions were presented at the 18th and 19th Annual Conventions of the Association for Behavior Analysis (San Francisco, May 1992, and Chicago, May 1993), and the XVIII Annual Conference of the Society for Quantitative Analyses of Behavior, Washington, D.C., May 1995.

Fragment	*Parameter*
F_1	Maximum number of s_1 elements.
F_2	Maximum number of s_2 elements.
F_3	Number of *sa* proliferative units.
F_4	Maximum number of *cal* NPEs per *sa* layer.
F_5	Number of *ma* proliferative units.
F_6	Maximum number of *vta* NPEs per *ma* layer.
F_7	Maximum number of output NPEs.
F_8	Initial proliferation probability.
F_9	Migration probability for *sa* elements.
F_{10}	Migration probability for *ma* elements.
F_{11}	Probability that an *sa* NPE is excitatory.
F_{12}	Probability that an *ma* NPE is excitatory.
F_{13} to F_{17}	NCA parameters for *sa* excitatory NPEs.
F_{18} to F_{22}	NCA parameters for *sa* inhibitory NPEs.
F_{23} to F_{27}	NCA parameters for *cal* NPEs.
F_{28} to F_{32}	NCA parameters for *ma* excitatory NPEs.
F_{33} to F_{37}	NCA parameters for *ma* inhibitory NPEs.
F_{38} to F_{42}	NCA parameters for *vta* NPEs.
F_{43} to F_{47}	NCA parameters for output NPEs.
F_{48}	Probability of connecting an s_1 element to an NPE in the first *sa* layer.
F_{49}	Probability of connecting an s_2 element to an NPE in the first *sa* layer.
F_{50}	Probability of connecting two excitatory *sa* NPEs.
F_{51}	Probability of connecting an *sa* NPE to a *cal* NPE.
F_{52}	Probability of connecting an excitatory *sa* NPE to an inhibitory *sa* NPE.
F_{53}	Probability of connecting an inhibitory *sa* NPE to an excitatory *sa* NPE.
F_{54}	Probability of connecting an *sa* NPE to an *ma* NPE.
F_{55}	Probability of connecting two excitatory *ma* NPEs.
F_{56}	Probability of connecting an *ma* NPE to a *vta* NPE.
F_{57}	Probability of connecting an excitatory *ma* NPE to an inhibitory *ma* NPE.
F_{58}	Probability of connecting an inhibitory *ma* NPE to an excitatory *ma* NPE.
F_{59}	Probability of connecting an *ma* NPE to an output NPE.

APPENDIX. Relation of genotype fragments to developmental parameters.

PART TWO: NEURAL PLASTICITY

The simulations described in Part One offer promising beginnings to the principled determination of the architectures of artificial neural networks. Instead of handcrafting the architecture, which is the normative practice in cognitive psychology and artificial intelligence, network structure was the product of simulations constrained by experimental observations of neuroanatomy, developmental neurobiology, and evolutionary biology. The constraints employed were quite "gentle," however. That is, the simulations instantiated the *outcome* of basic neurobiological processes rather than the constituent processes themselves. For example, Senft's neuroanatomical simulations did not explicitly implement genetically determined processes, e.g., the molecular events governing cell-surface interactions, but the *effects* of those processes, e.g., the clustering of cells. Similarly, Burgos's genetic/developmental simulations did not implement basic neurodevelopmental processes, e.g., the role of NMDA receptors in the migration of neuroblasts along glial strands, but the consequences of those processes, e.g., the formation of columns of cortical neurons. The gentleness of these constraints prompts the question: What is the appropriate relation between simulation research and the biobehavioral science informing that research?

Two rules of thumb, or heuristics, seem useful: First, no aspect of a simulation may be inconsistent with experimental research. Simulating the outcomes of basic biobehavioral processes—and not the constituent processes themselves—is acceptable if those outcomes occur under the conditions being simulated. Thus, incomplete implementation of experiment-based knowledge is tolerable, but inconsistency is not. For example, the notion that a *single* unit gives rise to both excitatory and inhibitory efferents—which is often assumed within the parallel-distributed-processing approach of cognitive psychology—is not acceptable within a biobehavioral approach because the assumption is inconsistent with experimental observation: Axon terminals of a given neuron do not liberate both excitatory and inhibitory neurotransmitters. Second, biobehaviorally constrained simulations need to implement only those biobehavioral processes that are minimally necessary to mimic the phenomenon being simulated. Not all constituent processes must be explicitly represented within the simulation. The principle of parsimony applied to simulations means that there is no single "correct" level for carrying out biobehaviorally faithful simulations. Instead, the processes that are explicitly implemented in the simulation vary with the phenomena being simulated. For example, when the temporal and quantitative properties of transmitter release by a presynaptic neuron are simulated, at least some of the intracellular events leading to exocytosis must be implemented in the simulation. However, when the effects of transmitter release on the activation of a postsynaptic neuron are simulated, the intracellular events leading to release and activation may not need to be implemented. Of course, simulations carried out at different levels of analysis must be consistent with one another and, over time, phenomena will likely be encountered that require more global simulations to be replaced by those that implement additional processes. That level of analysis is appropriate at which orderly and valid functional relations are simulated between independent and dependent variables (cf. Skinner, 1950).

Because biobehavioral simulations must be faithful to experimental findings from all levels, researchers must be familiar with findings that do not explicitly inform the simulations if inadvertent inconsistencies are to be avoided. For that reason, two chapters in this section describe in some detail the processes involved in neurotransmission and long-term potentiation. Neurotransmission concerns the processes whereby compounds liberated by one neuron affect the activity of other neurons. Long-term potentiation concerns the enduring effects of neurotransmission (changes in synaptic efficacy), and provides the best current model of how learning and memory are mediated at the neural level. **Meyer**'s chapter on neurotransmission makes a number of points of general significance. Among these are that a substantial number of different neurotransmitters have been identified and that transmitters differ among themselves in the scope of their effects (so-called volume effects). Volume effects of the neurotransmitter dopamine will prove important in understanding how environmental feedback affects behavior. **Frey**'s chapter on long-term potentiation indicates the complexity of the cellular mechanisms by which long-lasting changes in synaptic efficacy occur. Of particular importance is the distinction between relatively short-duration effects (on the order of minutes or a few hours) and longer-duration effects (relatively permanent changes in synaptic efficacy that are dependent on the synthesis of new proteins; e.g., Frey & Morris, in press). Although most of the simulations in this volume are at the level of neural systems, all must honor what is known at the cellular and intracellular levels. Moreover, **Buonomano and Merzenich**'s simulations illustrate that even global phenomena—e.g., temporal and phonemic discriminations—can arise as emergent effects of interactions between cellular processes. Note especially that the characteristics of these processes are known from *independent* experimental findings, and are not mere inferences from the phenomena under investigation, a common strategy in psychological as contrasted with biobehavioral research (cf. Donahoe & Palmer, 1994).

Neural-Networks Models of Cognition
J.W. Donahoe and V. Packard Dorsel (Editors)

CHAPTER 5

PRINCIPLES OF NEUROTRANSMISSION AND IMPLICATIONS FOR NETWORK MODELING

Jerrold S. Meyer
Department of Psychology
Neuroscience and Behavior Program
University of Massachusetts, Amherst

ABSTRACT

Interneuronal communication in vertebrate nervous systems is carried out mainly by a process of chemical transmission using a variety of small molecules and peptides. After being released at synaptic junctions, neurotransmitters are usually assumed to act locally. However, diffusion is sometimes sufficient to permit these substances to reach more distant sites of action. Synapses were once considered to be structurally immutable in adult nervous systems, but they are now thought to exhibit considerable remodeling in response to normal experiences (including learning). The time course of neurotransmitter action varies enormously, ranging from rapid opening or closing of membrane ion channels to slower but longer-lasting alterations in neuronal gene expression. These slower processes, which are mediated by a cascade of second messengers and transcription factors, are likely to participate in the mechanisms of neural plasticity and information storage.

Introduction

Chemical communication by nerve cells was first demonstrated over 70 years ago by the physiologist Otto Loewi (1921). However, not until the 1950s and 1960s was chemical neurotransmission in the central nervous system (CNS) widely accepted. This was quickly followed by an explosion of research identifying new transmitters and peptides in the brain, visualizing the location of specific neurochemical pathways, and forging links between transmitter action and behavioral function. The current era is marked by yet another remarkable wave of activity aimed at understanding neurotransmission at the molecular biological level. As in other areas of science, progress over this long time span has been driven largely by technical innovations, which in turn have given rise to periodic reconceptualizations of the transmission process.

This chapter presents some of the features of neurotransmission that may be of interest in the construction of neural networks. I begin with a summary of major neurotransmitter systems and some examples of transmitter co-localization and interaction. This is followed by a discussion of pre- and postsynaptic aspects of the neurotransmission process. Topics covered in these sections include synaptic structure, transmitter disposition (synthesis, storage, release, and inactivation), mechanisms of signal transduction, novel messenger systems, and neurotransmitter regulation of gene expression. For simplicity, the focus of the chapter is on neurotransmitter release from vesicles ("packets" of neurotransmitter) within nerve cells, although extravesicular release has been well documented both under normal physiological conditions (i.e., leakage from the cytoplasmic transmitter pool) and under pathological or artificially induced conditions (i.e., "reverse transport" due to reversal of membrane ionic gradients or treatment with transporter substrates such as amphetamine in the case of dopamine and norepinephrine; Levi & Raiteri, 1993).

Neurotransmitter and Neuropeptide Systems

Classical transmitters

Here and in the sections to follow, it is useful to distinguish between the so-called "classical transmitters" and neuropeptides. The classical transmitters consist of a diverse group of small molecules that includes acetylcholine (ACh), the amino acids glutamate (Glu), gamma-aminobutyric acid (GABA), and glycine (Gly), the catecholamines dopamine (DA) and norepinephrine (NE), and the indoleamine serotonin (5-HT). Two additional classical transmitters are not treated in the present chapter, epinephrine (another catecholamine) and histamine.

Although space limitations preclude an exhaustive description of the localization and function of these transmitters, the main points may be offered. Readers interested in a more detailed coverage are referred to Meyer, Feldman, and Quenzer (1996). (1) ACh: Major clusters of cholinergic neurons are located in the basal forebrain (including the nucleus basalis and septum) and pons (pedunculopontine and dorsolateral tegmental nuclei). Ascending projections of these cells play significant roles in sensory processing, memory, and other aspects of cognitive functioning. Another important group of cholinergic neurons is found in the corpus striatum, where they participate in extrapyramidal motor control. (2) Glu and GABA: Glutamate-and GABA-containing cells are found throughout the brain. These amino-acid transmitters are the workhorses for fast excitatory and inhibitory transmission in the brain, respectively. From a functional perspective, Glu is of particular interest for its involvement in neural plasticity via the NMDA receptor (see below and Frey, this volume). (3) Gly: Although present in the brain, Gly is most heavily concentrated in the spinal cord where it functions as an important inhibitory transmitter. (4) DA: Dopaminergic neurons are clustered in just a few cell groups, with the most (behaviorally) significant clusters located in the substantia nigra and ventral tegmental area of the mesencephalon. These cells project rostrally to the striatum, limbic structures such as nucleus accumbens, amygdala, and septum, and the neocortex. Despite its restricted cellular localization, DA has been implicated in a variety of normal and pathological processes, including reward mechanisms, motor control, and schizophrenia. (5) NE: Norepinephrine-containing neurons are located in several cell groups in the pons (particularly the locus coeruleus) and medulla. Ascending projections are sent to almost all regions of the forebrain, which suggests that NE may regulate very basic aspects of neural functioning. One general theory of NE action has proposed that this transmitter mediates sensory attention or "vigilance" in response to salient sensory stimuli (Aston-Jones, 1985). (6) 5-HT: Like NE, the serotonergic system originates in multiple cell groups within the pons and medulla (mostly the raphe nuclei). Also like NE, serotonergic forebrain projections are very widespread throughout the cortex, diencephalon, and limbic system. In this case, however, the major global theory of 5-HT action posits that the transmitter participates in sensorimotor integration and in the modulation of central pattern generators responsible for rhythmic motor behaviors (Jacobs & Fornal, 1993).

Neuropeptides

Neuropeptides constitute a larger group of substances that differ from the classical transmitters in that they consist of chains of three to approximately 40 amino acids (see below for specific examples). The representative sampling of peptides presented here is organized into groupings based on function and/or localization. (1) Opioid peptides: This category, which includes some of the first peptides discovered in the brain, consists mainly of the enkephalins, ß-endorphin, and dynorphin. These substances are best known for their analgesic action, but they are also involved in reinforcement mechanisms and various vegetative functions. (2) Gut-brain peptides: These peptides are not only present in the brain, but also figure prominently in the gastrointestinal system. Examples are substance P, cholecystokinin (CCK), neuropeptide Y (NPY), neurotensin, and vasoactive intestinal peptide (VIP). (3) Neurohypophyseal peptides: There are two peptides in this group, namely vasopressin (also known as antidiuretic hormone) and oxytocin. These peptides were first discovered in the magnocellular neurons of the paraventricular and supraoptic nuclei, the axons of which project to the posterior lobe of the pituitary gland where both substances are released as hormones into the bloodstream. However, they have also been identified in other cell groups where they appear to serve a neurotransmitter function and exert significant behavioral actions (e.g., regulating various social behaviors). (4) Hypophysiotrophic peptides: The hypophysiotrophic peptides were first found in the hypothalamus and were first identified through their regulatory effects on the secretion of hormones from the anterior pituitary gland. However, like the neurohypophyseal peptides, these substances were later found in other brain areas where they serve additional functions. This group includes corticotropin-releasing hormone (CRH), gonadotropin-releasing hormone (GnRH), thyrotropin-releasing hormone (TRH), and somatostatin. CRH has received particular attention as a peptide involved in stress and anxiety.

Synaptic Structure and Sites of Neurotransmitter Action

The axons of neurons that contain ACh or amino-acid transmitters generally form tree-like branched structures that end in the kind of bulbous terminals illustrated in standard introductory textbooks. However, cells that synthesize catecholamines, 5-HT, or peptides give rise to axons bearing intermittent swellings termed varicosities. Like the more traditional axonal endings, these varicosities contain large numbers of synaptic vesicles and are the likely areas of transmitter release for those cells. For the sake of simplicity, I will henceforth use the terms "nerve terminal" or "nerve ending" to refer to all types of presynaptic structures, including varicosities.

Many synapses exhibit membrane specializations that are seen postsynaptically as electron-dense thickenings and sometimes presynaptically as so-called dense projections. The presynaptic specializations, called active zones, are considered to be sites of vesicle docking and release, whereas the underlying areas of membrane thickening on the postsynaptic side (called postsynaptic densities) are thought to possess a high concentration of transmitter receptors. However, these assumptions do not hold in all cases. First, some varicosities do not show any synaptic specializations and therefore may permit vesicle release at varying sites on the membrane (Descarries, Séguéla, & Watkins, 1991). Second, a recent study by Smiley, Levey, Ciliax, and Goldman-Rakic (1994) indicates that immunoreactivity for DA D_1 receptors on monkey cortical neuronal spines is usually *not* found at postsynaptic densities, but rather is displaced to other regions of the spine.

These and other findings (e.g., Garris, Ciolkowski, Pastore, & Wightman, 1994; Mitchell, Oke, & Adams, 1994) suggest that neurotransmitter molecules may sometimes diffuse some distance before reaching their sites of action. Agnati, Zoli, Pich, Benfenati, and Fuxe (1990) have called this type of signaling "volume transmission," in contrast to point-to-point "wiring transmission" involving transmitter action near the site of release. One of the first demonstrations of volume transmission was made by Jan and Jan (1983), who showed that a GnRH-like peptide in bullfrog sympathetic ganglia influenced cells that were not directly innervated by this substance. An intermediate form of transmission also occurs in which transmitter action is local, but in which each presynaptic cell innervates large numbers of postsynaptic cells either in the same anatomical target area or in multiple areas (e.g., when axonal branches from the same cell project to different brain regions). Indeed, the major monoaminergic systems (DA, NE, and 5-HT) exhibit some features of both volume transmission and the intermediate form just described.

Another traditional notion that has been challenged in recent years is the idea of synaptic immutability. The earliest demonstration of synaptic plasticity came from developmental neurobiologists, who showed that synaptic junctions are overproduced and subsequently pruned during normal ontogeny of the nervous system (see Luskin, Singer, this volume). Nevertheless, for many years researchers assumed that synaptic architecture was stable during adulthood. According to this view, plasticity in the mature organism would involve only alterations in the efficacy of existing synapses, not the growth of new synapses or loss of old ones.

Several lines of evidence argue against this view, however. For example, a number of studies suggest that increases occur in synaptic number or alterations in synaptic shape when animals learn new tasks (reviewed by Weiler, Hawrylak, & Greenough, 1995). Moreover, since the nervous system is con-

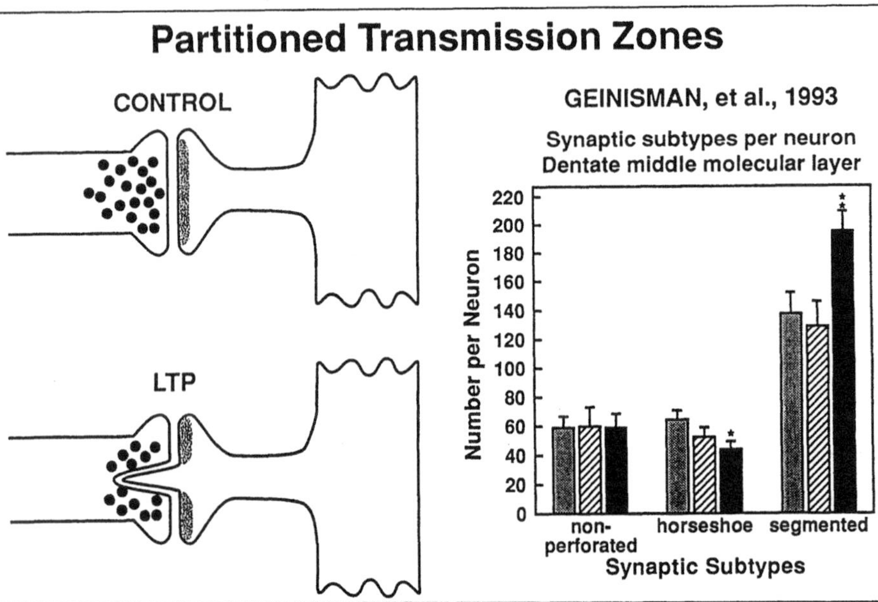

FIGURE 1. Influence of *in vivo* long-term potentiation (LTP) on synaptic subtypes in the hippocampal dentate gyrus. Male rats were subjected to four consecutive days of LTP in the dentate gyrus, after which the molecular layer of the dentate was studied by electron microscopy. LTP was associated with a significant increase in the number of segmented synapses (complete separation of postsynaptic densities [PSD]; see left panel) per neuron in the middle molecular layer. No change was found in horseshoe synapses (horseshoe-shaped PSD with a sectional partition) or atypical non-perforated synapses (synapses showing a morphology similar to perforated synapses but with a continuous PSD). Gray bars, unstimulated controls; hatched bars, Coulombic controls; black bars, potentiated subjects. From Weiler et al (1995). Original data from Geinisman, deToledo-Morrell, Morrell, Heller, Rossi, and Parshall (1993).

stantly engaged in information processing, synaptic turnover may well be an ongoing process that does not require explicit experimental intervention (Wolff, Laskawi, Spatz, & Missler, 1995). Even peripheral neurons that are not usually considered to be involved in learning may undergo synaptic remodeling over time. This has been documented in an elegant series of studies by Purves and his colleagues, who directly imaged changes in dendritic branching and synaptic connectivity of individual mouse autonomic ganglia cells (Purves & Hadley, 1985; Purves, Voyvodic, Magrassi, & Yawo, 1987).

Some investigators have theorized that changes in synaptic structure may involve perforated (sometimes also called segmented) synapses, that is, synapses characterized by one or more gaps (perforations) in the postsynaptic density (see Calverley & Jones, 1990). One type of change might involve conversion of nonperforated to perforated synapses as a consequence of appropriate afferent stimulation. **Figure 1** illustrates a possible example of this type of plasticity in rats undergoing *in vivo* long-term potentiation (LTP; see Frey, this volume). Alternatively, synapse numbers could be increased by the splitting of perforated synapses to create two "daughter" synapses in each case (Carlin & Siekevitz, 1983; Dyson & Jones, 1984). Whether this occurs in the brain remains to be proven.

Transmitter Synthesis, Storage, Release, and Inactivation

Classical transmitters

The synthesis of classical transmitters occurs by enzymatic catalysis. Transmitter synthesis rate thus may be controlled by several processes, including availability of the dietary precursor for the transmitter and regulation of the rate-limiting enzyme in the biosynthetic pathway. Such regulation often involves stimulation by neuronal activity and suppression by synthesis-inhibiting autoreceptors on the nerve terminal. The enzymes necessary for transmitter synthesis are made in the cell body of the neuron and then transported to the nerve terminals. Consequently, the enzymatic machinery for the synthesis of classical transmitters is concentrated near the sites of release, enabling relatively rapid replenishment of transmitter when demand is great. Transmitter molecules are usually synthesized in the cytoplasm of the nerve terminal and then transported into the vesicles for storage and eventual release (NE is a notable exception to this generalization; dopamine ß-hydroxylase, the final enzyme in the NE biosynthetic pathway, is found inside the vesicles). In typical electron micrographs, the vesicles used to store classical transmitters are relatively small and appear either clear (electron lucent) or with an electron-dense core.

The actual release process involves action potential-mediated opening of voltage-gated calcium (Ca^{2+}) channels, thereby initiating a biochemical cascade leading to fusion of the vesicle membrane with the plasma membrane of the nerve terminal (reviewed by Verhage, Ghijsen, & Da Silva, 1994). Large numbers of vesicles are present in each nerve ending. However, the release process is so rapid that only those vesicles that are already "docked" at specialized membrane release sites have sufficient time to engage in exocytosis when an action potential invades the terminal.

The relationship between nerve firing and transmitter release is more complicated than initially believed. For example, in some well-studied systems, a nerve impulse seems to stimulate release of no more than one vesicle from a given active zone (Korn, Sur, Charpier, Legendre, & Faber, 1994). Furthermore, even this seemingly meager result is not invariable; that is, vesicle fusion in such systems seems to be a stochastic process governed by various regulatory mechanisms that are just beginning to be understood (**Figure 2**). This feature of neuronal function is important for neural modelers to consider, as an increased probability of transmitter release has been reported to occur in hippocampal LTP (Bekkers & Stevens, 1990; Stevens & Wang, 1994). Another interesting feature of transmitter release is the absence of a linear relationship to firing rate. This has been documented in the DA system, where the amount of transmitter released *per impulse* is much greater when the cells are in burst-firing mode than when they are in single-spike mode (Gonon, Suaud-Chagny, Mermet, & Buda, 1991).

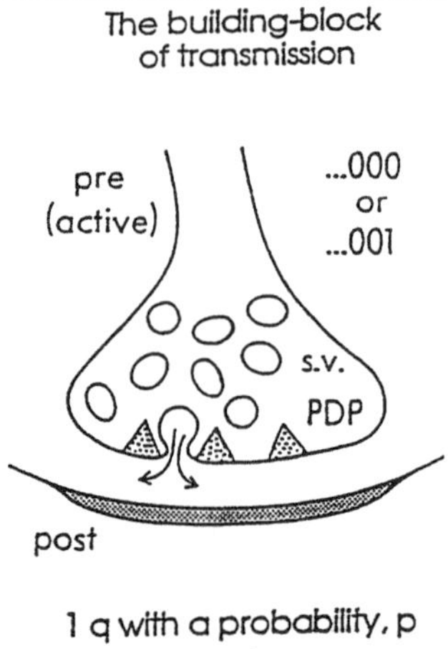

FIGURE 2. Hypothesized properties of vesicular release. A nerve terminal is illustrated with synaptic vesicles (s.v.) and a single active zone containing the typical presynaptic dense projections (PDP). Presynaptic activity is hypothesized to release the contents of either one vesicle (1 q: quantum) or none, with a probability p (hence the binary notation 000 or 001). P depends on the characteristics of the synapse and may be modifiable. From Korn et al (1994).

Regulation of transmitter release involves not only the firing rate of the cell, but also the action of terminal autoreceptors and heteroreceptors. Autoreceptors are receptors for the same transmitter released by that terminal. They usually function to inhibit further transmitter release (negative feedback), although a few instances of release-enhancing autoreceptors have also been reported. Heteroreceptors are terminal receptors for other transmitters that may act either to stimulate or inhibit release at that terminal (e.g., a cholinergic receptor on a DA nerve terminal). Many examples of heteroreceptors have now been identified (see Kalsner & Westfall, 1990), suggesting that local modulation of transmitter release (i.e., at the synapse) may, in some cases, be as important as modulation via traditional axodendritic and axosomatic synaptic connections.

The amino-acid and amine transmitters are removed from the synaptic cleft by a combination of cellular uptake and passive diffusion. Enzymatic metabolic processes lead to eventual degradation and clearance of transmitter molecules. On the other hand, ACh is inactivated directly by rapid enzymatic hydrolysis. Both methods permit efficient termination of synaptic signaling, although breakdown of ACh by acetylcholinesterase is particularly rapid.

Neuropeptides

In contrast to classical transmitters, a neuropeptide is formed as part of a larger propeptide, which is actually a protein that contains the active peptide within its linear structure. All proteins, including those associated with neuropeptides, are synthesized by the translation of messenger RNA (mRNA) on ribosomes in the cell body. Following certain chemical modifications, the protein product is packaged into a storage vesicle. These vesicles are noticeably larger than those used by classical transmitters, and they all possess an electron-dense core. The process of liberating the active peptide from the propeptide within the vesicle begins while the vesicle is en route to the nerve terminal. It should be clear from this description that every peptide molecule must originate in the cell body; no local synthesis can occur in the nerve terminals. This is an important difference from classical transmitter systems, because it means that peptidergic neurons take longer to replenish their transmitter stores when release rates are high. The exact amount of time needed for such replenishment ranges from many minutes to several hours or even days, depending on the rate of propeptide synthesis and packaging, and on the length of the axon (the velocity of fast axonal transport is approximately 100-200 mm per day).

Neuropeptide release is exocytotic and is probably mediated by mechanisms similar to those involved in the release of classical transmitters. Release-modulating autoreceptors and heteroreceptors have likewise been found in at least some peptide systems. One difference, however, is that the large peptide-containing vesicles do not seem to use the specialized release sites required by the smaller vesicles that contain classical transmitters.

The processes that terminate peptide transmission are not well understood in many cases. Nevertheless, in the best-characterized systems (e.g., the enkephalins), signaling is terminated by enzymatic degradation rather than by uptake.

Transmitter co-localization

Following the discovery and characterization of various neuropeptides, it soon became clear that peptides are co-localized in many instances with classical transmitters. Some well-known examples of this are the co-localization of DA with CCK or neurotensin in midbrain dopaminergic neurons, and the presence of GABA with substance P and with enkephalin or dynorphin in striatal output neurons. Storage of classical transmitters and peptides in different vesicular populations allows for the possibility of differential release of the two transmitters. Indeed, several model systems have indicated that higher rates of neuronal firing may be necessary for stimulation of peptide release than for the release of co-localized classical transmitter (Hökfelt, 1991; see **Figure 3**).

Co-release of two or more transmitters from the same nerve terminals allows for several types of interactions, some of which are illustrated in **Figure 4**. The situation shown in the second panel, which is commonly found in systems that co-release a peptide and a classical transmitter, may lead to either cooperative (i.e., additive or synergistic) or antagonistic effects postsynaptically. In most cases, the receptor populations stimulated by the two substances presumably function independently, even though an interaction may occur downstream at the level of post-receptor signaling mechanisms (i.e., second messengers). However, Fuxe and Agnati (1985) have cited examples in which one type of receptor may modulate the affinity and/or density of another receptor type within the postsynaptic cell membrane. Another kind of interaction, not shown in the figure, occurs when receptors on the nerve terminal modulate the release of a co-transmitter. For example, NPY and NE are co-localized in sympathetic nerve endings where they inhibit each other's release via terminal autoreceptors. Yet the two substances act together postsynaptically to produce constriction of vascular smooth muscle (Lundberg & Hökfelt, 1986).

Neurotransmitter Receptors and Signal Transduction Mechanisms

The concept of receptor subtypes

All transmitter systems that have been extensively studied have been found to possess more than one type of receptor. Often there are multiple levels of receptor diversity. For example, cholinergic receptors are broadly categorized as either nicotinic or muscarinic, however there are further subtypes within each of these general classes. It is highly likely that when closely related receptor subtypes are coded by distinct genes, the subtypes evolved from a common ancestral form through a process of gene duplication followed by mutation and/or recombination. Additionally, multiple receptor subtypes can be created from a single gene product by alternative mRNA splicing. For many years, knowledge of receptor diversity was based entirely on pharmacological analysis; i.e., the ability of certain drugs to selectively activate or inhibit some but not all of the actions of a neurotransmitter. In recent years, however, gene-

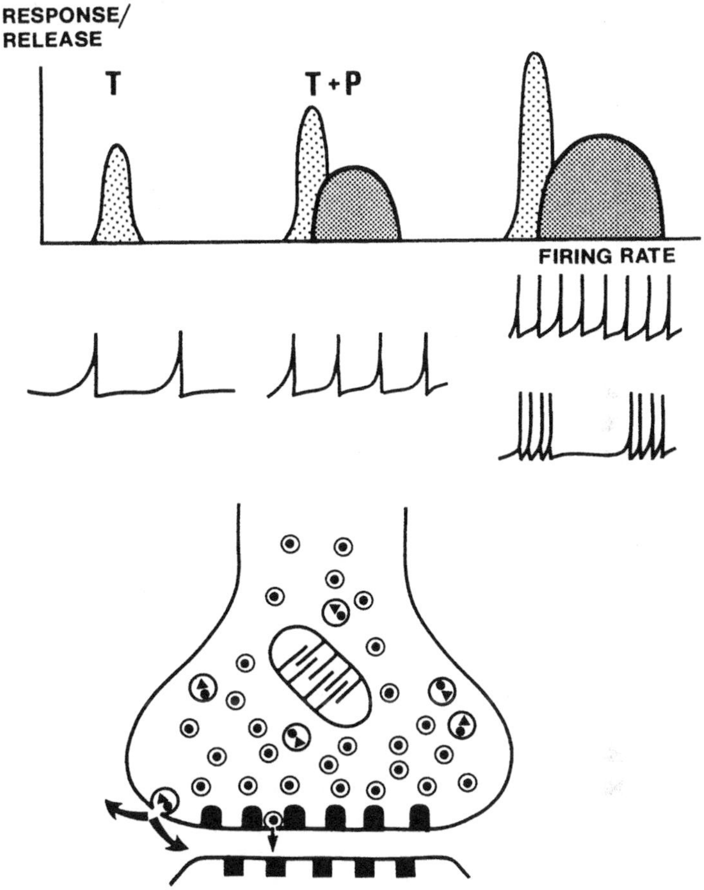

FIGURE 3. Differences in release between classical transmitters and neuropeptides. Top: When a classical transmitter (T) and a peptide (P) are co-localized in the same cell, only the transmitter is released at low firing rates. Higher rates of continuous firing or burst firing release both transmitter and peptide. Bottom: Classical transmitters and peptides are differentially stored and released. Small synaptic vesicles contain only the transmitter (dots), whereas large dense-core vesicles contain both transmitter and peptide (triangles). Small vesicles undergo exocytosis at active zones that face the synaptic cleft, whereas large vesicles frequently release their contents in extrajunctional areas. From Hökfelt (1991) and Lundberg and Hökfelt (1983).

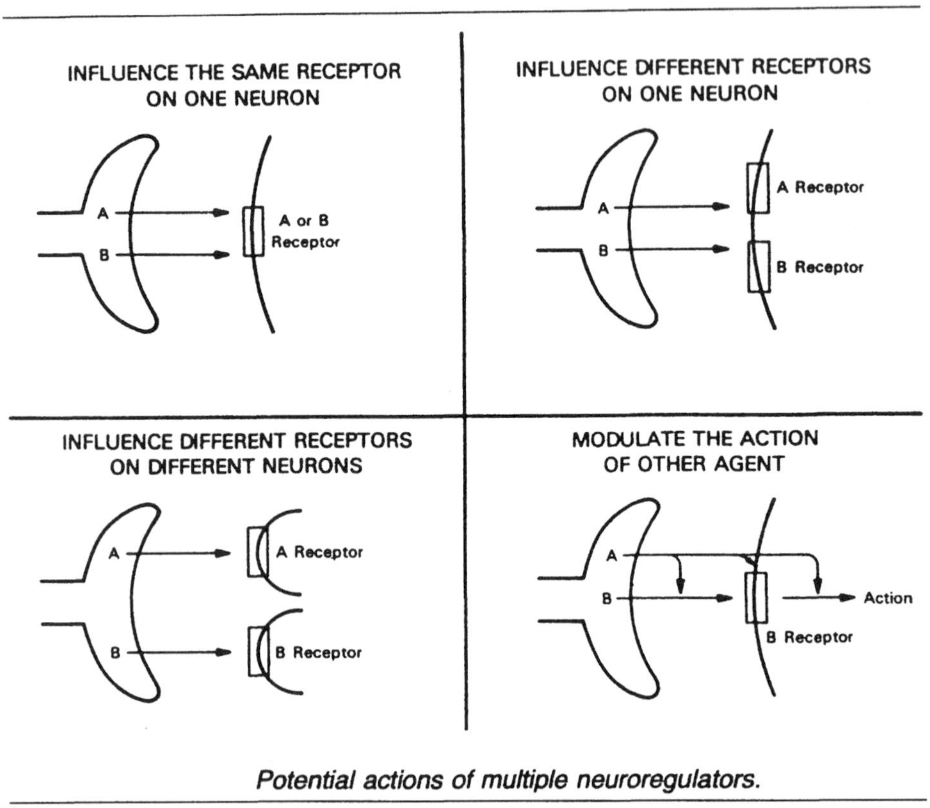

FIGURE 4. Co-localized neurotransmitters and peptides may exert a variety of different effects on postsynaptic target cells. From O'Donohue, Millington, Handelmann, Contreras, and Chronwall (1985).

cloning studies have verified the existence of many proposed receptor subtypes and have even demonstrated the presence of subtypes not previously detected by pharmacological investigation. We can only speculate as to the selection pressures underlying the evolution of receptor diversity; however, Schofield, Shivers, and Seeburg (1990) have proposed that one important advantage of such diversity is the increased information-handling capacity it confers on the nervous system.

The existence of multiple receptor subtypes for a given transmitter has as least two important implications. First, as we shall see, a variety of signal-transduction mechanisms are available by which receptors can influence post-synaptic activity. If various receptors for a transmitter use distinct transduction mechanisms, then the transmitter can exert quite different excitatory or inhibitory effects on its postsynaptic targets, depending on which subtype is present on each target cell. Second, although a neurotransmitter obviously recognizes and activates all of its receptor subtypes, the unique structure of each subtype allows for the development of synthetic compounds that exhibit selectivity for one (or some) subtype(s) vs. others. Selective agonists and antagonists are useful not only for identifying and characterizing different subtypes, but also for determining their involvement in behavioral and physiological functions.

Receptor superfamilies

The ligand-gated channel receptor superfamily. Transmitter receptors fall into two distinct superfamilies of membrane proteins: ligand-gated channels (LGCs) and G-protein-coupled receptors (GPCRs). LGCs are large proteins thought to be composed of five subunits that assemble in the membrane. Important LGCs include nicotinic cholinergic receptors (which are the prototype of this group); the three general subtypes of excitatory amino acid (i.e., glutamate) receptors, namely AMPA, kainate, and NMDA (each named for a relatively selective agonist); $GABA_A$ receptors; glycine receptors; and $5\text{-}HT_3$ receptors. LGCs tend to exhibit considerable heterogeneity in their subunit composition, thereby leading to variability in function.

As suggested by their name, LGCs contain a neurotransmitter binding site and an intrinsic ion channel that is gated by the transmitter. Whether receptor activation depolarizes or hyperpolarizes the postsynaptic membrane depends on the ionic permeability of the channel. Nicotinic, AMPA, kainate, and $5\text{-}HT_3$ receptors generally possess nonspecific univalent cation channels that cause depolarization due to sodium (Na^+) influx across the membrane. NMDA receptors are additionally permeable to Ca^{2+}, which confers second-messenger properties on this subtype (see below). Some AMPA/kainate receptors likewise show significant Ca^{2+} current depending on their subunit composition (Hollmann, Hartley, & Heinemann, 1991). In contrast, the channels associated with $GABA_A$ and glycine receptors are permeable to chloride (Cl^-) ions, the passage of which leads to membrane hyperpolarization and inhibition of cell firing.

Besides the recognition site for the neurotransmitter, some LGCs are known to possess additional binding sites capable of modulating receptor function. For example, $GABA_A$ receptor functioning is enhanced by benzodiazepine and barbiturate drugs acting at allosteric binding sites on the receptor complex (Macdonald & Olsen, 1994). NMDA receptors, on the other hand, are inhibited by phencyclidine (PCP) and related drugs (Wood, Tadimeta, Iyengar, Lanthorn, Monahan, Cordi, Sun, Vazquez, Gray, & Contreras, 1990). These compounds are believed to bind to a site within the channel, effectively blocking ion flow across the membrane.

Because of the close linkage between receptor activation and ion-channel opening, LGCs operate with a very short latency, i.e., a few msec. Dissociation of the transmitter from the receptor likewise causes rapid channel closing. Consequently, such receptors are well suited for fast signaling within the nervous system. One other noteworthy feature of many LGCs is their rapid desensitization (loss of activity) when continuously exposed to the transmitter or an agonist drug (e.g., see Ochoa, Chattopadhyay, & McNamee, 1989). This process may limit the extent of postsynaptic responding under conditions when presynaptic elements are highly active for a period of time.

Before leaving the topic of LGCs, it is important to discuss the unique voltage-sensing feature of the NMDA receptor. When the cell membrane is at the resting potential, the NMDA receptor channel is blocked by magnesium (Mg^{2+}) ions. Depolarization of the membrane removes this block, thereby permitting the channel to open in response to glutamate. The necessary depolarization may require repeated stimulation of the same synaptic input or simultaneous stimulation of multiple inputs, particularly if the inputs are individually weak. This mechanism underlies the ability of NMDA receptors to serve a critical role in activity-dependent plasticity (Malenka, 1995; also see Singer, Frey, this volume).

The G-protein-coupled receptor superfamily. GPCRs differ from LGCs both structurally and functionally. These receptors consist of only a single subunit that, in all known cases, is predicted to contain seven membrane-spanning (i.e., transmembrane) regions. Important examples of GPCRs are the muscarinic ACh receptors (five cloned subtypes thus far, designated m1-m5), all known DA receptors (five cloned subtypes, D_1-D_5), all known adrenergic receptors including the entire α- and ß-adrenoceptor families, all 5-HT receptors except for 5-HT_3 (this includes families of 5-HT_1 and 5-HT_2 receptors, along with 5-HT_4-5-HT_7), the so-called metabotropic Glu receptors (mGlu1-mGlu7), $GABA_B$ receptors, and all known neuropeptide receptors.

As implied by their name, GPCRs operate by coupling to members of a family of membrane proteins called G proteins. They are so-named because their functioning is regulated in part by the binding of guanine nucleotides. Each GPCR selectively interacts with some G proteins but not others, which partly accounts for the specificity of cellular actions produced by different members of the GPCR family.

G proteins contain three subunits, designated α, ß, and γ. When the molecule is activated by coupling to an agonist-stimulated receptor, the α subunit dissociates from the remaining ßγ dimer and then stimulates one or more types of membrane effectors that mediate the cellular response. In some cases, the effectors are ion channels such as for potassium (K^+; Nicoll, 1988). Direct modulation of ion channels by G proteins occurs with a latency of approximately 50-100 msec, which is an order of magnitude slower than channel opening of LGCs.

The second general group of membrane effectors are enzymes that participate in second-messenger synthesis or degradation (Hille, 1992). For example, the G protein G_s was named for its ability to stimulate adenylyl cyclase, thereby enhancing cyclic adenosine monophosphate (cAMP) formation. G_i, on the other hand, inhibits adenylyl cyclase and cAMP synthesis (second messengers are discussed further in the next section). It should be noted that the same type of G protein may be capable of participating in both ion-channel and enzyme modulation. This dual aspect of G protein action is shown in **Figure 5**. Cellular effects mediated by second-messenger changes have latencies measured in hundreds of milliseconds. However, these effects may also considerably outlast the initiating synaptic event. Second-messenger systems are thus well suited for slow but sustained signaling and for modulating the influence of fast neurotransmitters.

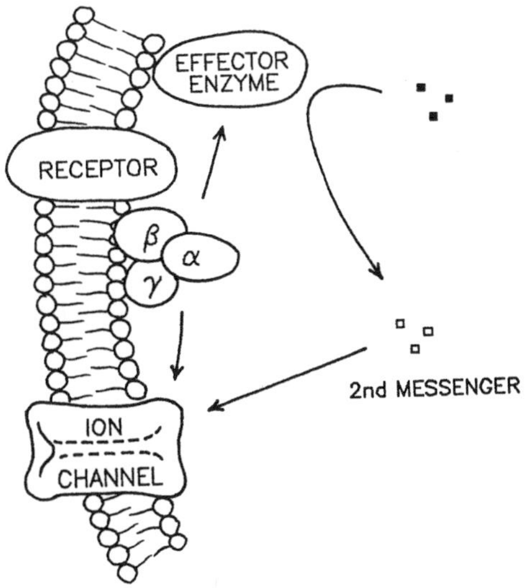

FIGURE 5. Mechanisms of action of G-protein-coupled receptors. An activated G protein (shown here as a complex of three subunits—α, ß, and γ) may directly modulate certain membrane ion channels or may stimulate effector enzymes that control intracellular second messengers. From Sternweis and Pang (1990).

If a cell possesses two or more receptor systems that function via the same G proteins and effectors, then synaptic inputs may converge to produce additive effects on cellular activity. This has been shown in hippocampal pyramidal neurons, where 5-HT_{1A} and $GABA_B$ receptors activate the same G protein to open K^+ channels and hyperpolarize the membrane. On the other hand, the multiplicity of receptor subtypes, G proteins, and effector systems allows the same transmitter to produce quite divergent effects in different cell types. For example, ACh produces a slow hyperpolarization in some cells by muscarinic-receptor-mediated enhancement of K^+ channel opening, but a slow depolarization in other cells by muscarinic receptor stimulation of the phosphoinositide second-messenger system. Finally, some cells possess both multiple transmitter inputs and multiple effector mechanisms for some of these inputs, thus exhibiting both convergence *and* divergence of signaling mechanisms (e.g., see McCormick & Williamson, 1989).

Second-messenger systems

Second messengers are small intracellular molecules that mediate the effects of first messengers, i.e., neurotransmitters and hormones. Some of the important second messengers in the nervous system are cAMP, cyclic guanosine monophosphate (cGMP), diacylglycerol (DAG), inositol trisphosphate (IP_3), and Ca^{2+} ions. Formation of cAMP and cGMP is catalyzed by the enzymes adenylyl cyclase and guanylyl cyclase, respectively. In retinal cells, cGMP levels are regulated by rhodopsin-induced stimulation of phosphodiesterase (PDE), the enzyme that degrades cGMP. DAG and IP_3 are second messengers that are jointly liberated from the membrane lipid phosphatidylinositol bisphosphate (PIP_2) by a PIP_2-specific phospholipase C. The final second messenger, Ca^{2+}, can enter the neuronal cytoplasm from several different sources. Thus, cytoplasmic Ca^{2+} concentrations can be elevated either by influx through NMDA receptor channels and/or voltage-gated Ca^{2+} channels, or by release from intracellular storage compartments such as the smooth endoplasmic reticulum.

Second messengers vary in their range of action within a cell. For example, Ca^{2+} has a very short range of action due to factors such as cytoplasmic buffering and sequestration by internal storage sites. In contrast, IP_3 and particularly cAMP seem to diffuse considerably longer distances before being metabolized. This is important in light of the fact that some mechanisms of plasticity may require the coordinated action of several second messengers. Therefore, if Ca^{2+} levels are locally elevated in a dendritic spine (perhaps due to the activation of NMDA receptor channels) but adenylyl cyclase-coupled receptors are not present in that spine, cAMP generated at another location in the dendrite might be able to reach the spine and interact with the Ca^{2+} to produce an appropriate postsynaptic response (Kasai & Petersen, 1994).

Second messengers generally operate through activation of protein kinases. These are enzymes that modify the functioning of various target proteins through the addition of phosphate groups to specific amino-acid residues (i.e., through phosphorylation). The electrically charged phosphate groups alter the conformation of the affected proteins, thereby influencing their biological activity. In the case of an enzyme, for example, its catalytic activity might be either increased or reduced in the phosphorylated state. Phosphorylated proteins are returned to their original state by enzymes called phosphatases. This process terminates the cellular effects of second messengers by reversing the phosphorylation-induced changes in protein function.

Each second messenger is associated with a particular type of protein kinase. For example, cAMP activates cAMP-dependent protein kinase (also called protein kinase A; PKA), whereas cGMP similarly functions via cGMP-dependent kinase. Ca^{2+} first binds to a receptor protein called calmodulin, and then this complex can activate Ca^{2+}/calmodulin-dependent protein kinase. DAG, one of the second messengers produced from PIP_2 hydrolysis, remains in the membrane and activates protein kinase C (PKC). The other messenger, IP_3, is liberated into the cytoplasm. It subsequently binds to a receptor on endoplasmic reticulum membranes and opens a channel for Ca^{2+} release from the endoplasmic reticulum into the cytoplasm (here we may consider Ca^{2+} to be a *third* messenger in the biochemical cascade). The phospholipase C/PIP_2 system (sometimes termed the phosphoinositide second-messenger system) can therefore activate a variety of Ca^{2+}-dependent mechanisms. One of these actually turns out to be protein kinase C, because this enzyme is activated by Ca^{2+} as well as DAG.

An interesting feature of the type II Ca^{2+}/calmodulin-dependent protein kinase (CaMK II) enables it to act as a kind of "molecular switch." When activated, this enzyme can engage in autophosphorylation, i.e., phosphorylation of itself. This "turns on the switch" by causing the kinase to become temporarily independent of Ca^{2+}. Eventually, the switch is turned off when the kinase becomes dephosphorylated. Some investigators have hypothesized that conversion of CaMK II and possibly other kinases to a stimulation-independent state may play a role in long-term potentiation or other forms of neuronal plasticity (Bliss & Collingridge, 1993; also see Frey, this volume).

Second-messenger-induced protein phosphorylation alters numerous functions related to synaptic transmission (**Figure 6**). Among the neuronal proteins subject to phosphorylation are the neurotransmitter-synthesizing enzymes tyrosine hydroxylase and tryptophan hydroxylase, a number of different transmitter receptors, voltage-gated ion channels, synaptic-vesicle proteins, cytoskeletal proteins, and nuclear proteins involved in gene regulation. Second-messenger-induced phosphorylation thus plays a critical role in virtually all aspects of neuronal signaling. Different populations of neurons may exhibit distinctive patterns of phosphorylated proteins, depending on the transmitter inputs and receptor subtypes found in each population, the G proteins and second messengers activated by those receptors, and the available substrates for the kinases that are stimulated.

Novel inter- and intra-cellular messengers

Gaseous messengers. An unusual development in the study of neurotransmitter mechanisms was the discovery of two gaseous messenger substances. The first and best characterized is nitric oxide (NO). NO is synthesized by the enzyme nitric oxide synthase (NOS), which has a widespread though heterogeneous

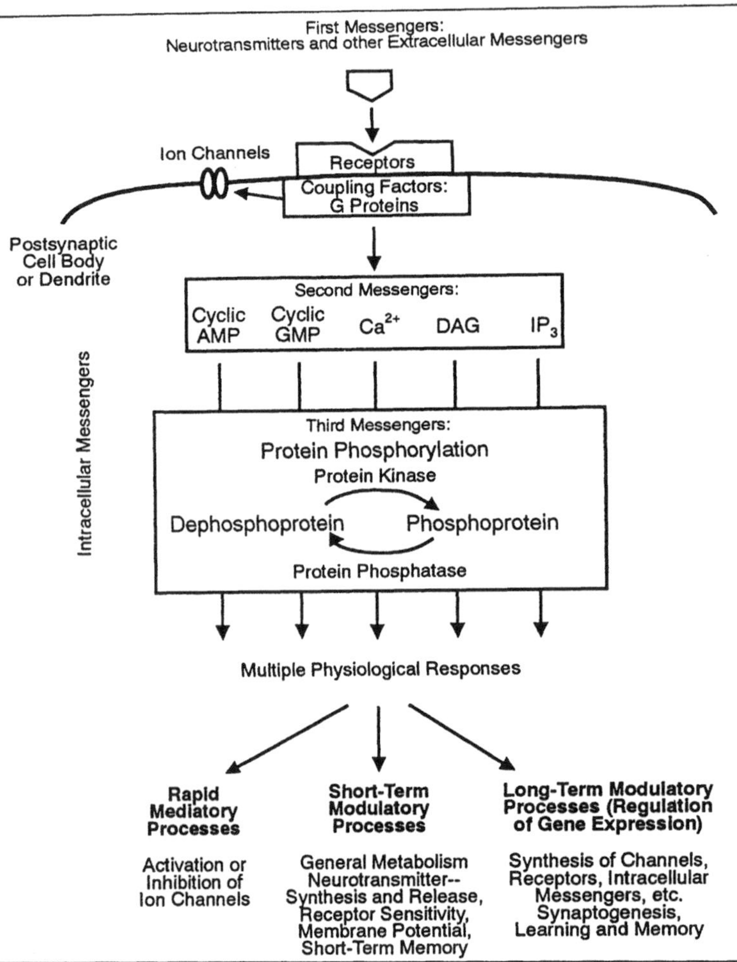

FIGURE 6. Role of second messengers and protein phosphorylation in nervous-system function. The figure illustrates the biochemical cascade triggered by neurotransmitter stimulation of G-protein-coupled receptors and culminating in rapid to long-term neuronal responses. Key roles in this process are played by second messengers and second messenger-activated protein kinases. From Hyman and Nestler (1993).

distribution in the brain. In a number of areas, NOS can be detected in various interneurons but not in long-axon projection pathways. This pattern has been observed in (1) cerebellar cortex, where the basket and granule cells but not the Purkinje cells are NOS-positive, (2) cerebral cortex and hippocampus, in which NOS is absent from pyramidal cells but is present in scattered interneurons in various layers, and (3) striatum, where the medium spiny projection neurons lack NOS but the enzyme is found in a population of aspiny interneurons (Vincent & Hope, 1992). The neuronal form of NOS is activated by increasing the intracellular Ca^{2+} concentration. In cerebellum and hippocampus, Garthwaite (1991) demonstrated that Ca^{2+} activation of NOS is likely mediated by excitatory amino-acid stimulation of NMDA receptors.

NO does not act on a traditional membrane-bound receptor. Rather, it binds to the heme iron present in the soluble form of guanylyl cyclase and activates the enzyme, thus enhancing cGMP formation. A close relationship between NO and guanylyl cyclase is supported by the similar regional distribution of NOS staining and staining for cGMP following stimulation of NO synthesis (Southam & Garthwaite, 1993). On the other hand, staining is not usually in the same cells within a given area. In most cases, either NOS is found in postsynaptic cells and cGMP in presynaptic elements, or the localization may be reversed. These and other findings strongly suggest that NO diffuses from its site of origin to other cells where it activates guanylyl cyclase.

Several features of transmission by NO are unique compared to other intercellular messengers. First, due to its high membrane permeability, it cannot be packaged or stored in any vesicular (or other) structure. It must instead be made and released upon demand. Second, rapid diffusion of NO along with the lack of any known uptake mechanism may permit the gas to diffuse for relatively (on a cellular scale) long distances and consequently to act on distant targets. Finally, as already mentioned, NO is able to directly stimulate second-messenger synthesis without the intervention of a membrane receptor.

The second putative gaseous messenger is carbon monoxide (CO), which may also serve an important signaling function via activation of cGMP. CO is synthesized by the enzyme heme oxygenase (HO), which has a somewhat different regional and cellular distribution than NOS. Unlike NOS, for example, HO is strongly expressed in Purkinje cells and pyramidal neurons (Maines, 1993; Verma, Hirsch, Glatt, Ronnett, & Snyder, 1993).

Figure 7 illustrates a proposed model of NO and CO function. According to this model, NO is mainly synthesized postsynaptically as a consequence of an NMDA receptor-mediated rise in intracellular Ca^{2+} (note that other mechanisms for elevating Ca^{2+} levels may also be important in some systems). NO then diffuses in a retrograde direction to produce various effects (including increased cGMP formation) in axon terminals. A number of studies have suggested that NO may be a critical retrograde messenger involved in the enhancement of presynaptic glutamate release in hippocampal long-term potentiation (Bliss & Collingridge, 1993; also see Frey, this volume). In contrast, CO may be formed either pre- or postsynaptically and thus can serve as either a retrograde or an intracellular messenger.

Lipid messengers. Certain lipids can also function as intracellular and/or intercellular messenger substances. One such compound is the unsaturated fatty acid arachidonic acid (AA), which is liberated from membrane phospholipids by the enzyme phospholipase A_2. Once released, AA undergoes a complex series of metabolic reactions that results in the formation of many biologically active compounds that are collectively called eicosanoids (these include various prostaglandins, thromboxanes, etc.). Eicosanoids can act either as second messengers by influencing the cell in which they are formed, or they can subserve a first-messenger type of function by diffusing to neighboring cells like NO. Various findings suggest that some of these eicosanoids may participate in both transmitter-release and signal-transduction mechanisms in some neuronal systems (reviewed by Shimizu & Wolfe, 1990).

Even more interesting, perhaps, is a recently discovered AA derivative named anandamide (Devane, Hanuš, Breuer, Pertwee, Stevenson, Griffin, Gibson, Mandelbaum, Etinger, & Mechoulam, 1992). This substance is the first compound to be given serious consideration as an endogenous ligand for cannabinoid (i.e., marijuana) receptors. One of the structures containing a high

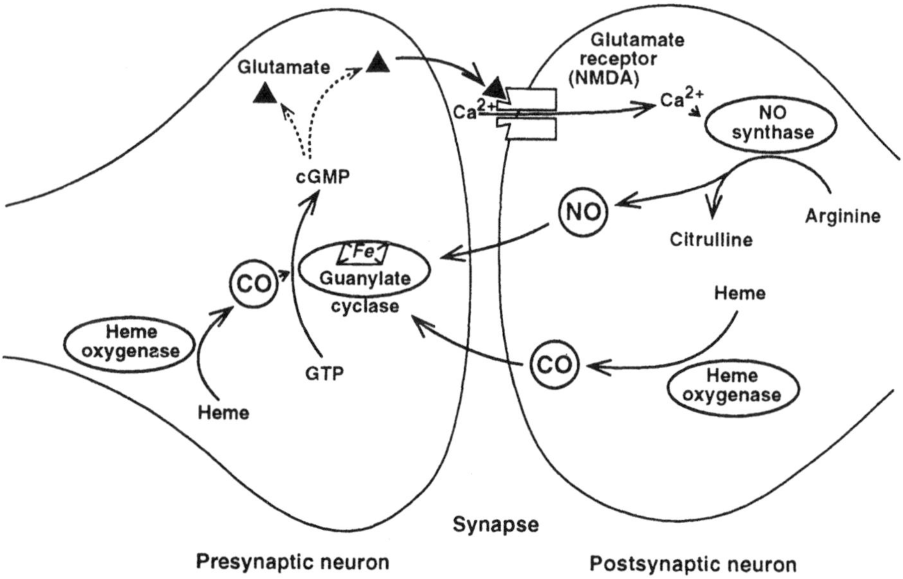

FIGURE 7. Proposed functioning of the gaseous messengers nitric oxide (NO) and carbon monoxide (CO). According to this model, NO is typically synthesized in postsynaptic cells, from which it diffuses into presynaptic terminals and activates the cGMP-forming enzyme guanylyl (guanylate) cyclase. CO, which also stimulates cGMP formation, may be synthesized either pre- or postsynaptically. One consequence of increasing presynaptic cGMP concentrations may be enhanced release of glutamate (▲) from the terminal. From Maines (1993).

concentration of cannabinoid receptors is the hippocampus (Herkenham, Lynn, Little, Johnson, Melvin, de Costa, & Rice, 1990), which corresponds well to the established effects of marijuana smoking on cognitive functioning. Assuming that anandamide is confirmed as an actual neurotransmitter or neuromodulator, it might eventually be found to play a significant role in normal learning and memory processes.

Neurotransmitters and Gene Regulation in the Nervous System

The final topic to be covered in this summary of neurotransmission concerns the regulation of gene activity. Endocrinologists have known for many years that certain hormones, for example gonadal and adrenal steroids, affect their target cells primarily by altering the transcription of various genes in those cells. Changes in gene transcription then lead to alterations in the rates of synthesis or (less frequently) degradation of various cellular proteins. Growing evidence now indicates that neurotransmitters can similarly regulate gene expression, a process that may well be important in learning and memory, as well as in long-lasting adaptive responses to drugs and other external agents.

There are two phases of gene activation triggered by synaptic input (Armstrong & Montminy, 1993). The initial phase is characterized by induction of "immediate-early genes" (IEGs; sometimes also called "early-response genes"). IEGs are usually expressed only at low levels in the absence of cellular excitation. When synaptic inputs are active, however, they are rapidly though transiently induced (e.g., mRNA levels may be significantly increased within 15 minutes but remain elevated for only another 15-30 minutes). Most IEGs code for nuclear proteins, several of which are discussed below.

The second phase of synaptic gene activation is the induction of "late-onset genes." As we shall see, these genes are slower to respond to stimulation because their induction is dependent on the action of IEG proteins. Although many important neuronal genes are undoubtedly subject to late-onset synaptic regulation, only a few examples have been well characterized. Among these are the genes for tyrosine hydroxylase and for several neuropeptides such as substance P (Armstrong & Montminy, 1993).

Transcription factors and gene regulation

Transmitters alter gene expression by means of a complex mechanism involving second messengers together with a family of proteins known as transcription factors. Molecular biologists have known for some time that most genes contain several distinct regions with differing functions. One of these obviously is the coding region specifying the nucleotide sequence of the RNA transcribed from that gene. However, "upstream" from the coding region is the so-called promoter region, within which are DNA sequences that serve as binding sites for these transcription factors. Several broad families of transcription factors have been identified, with exotic names such as leucine-zipper,

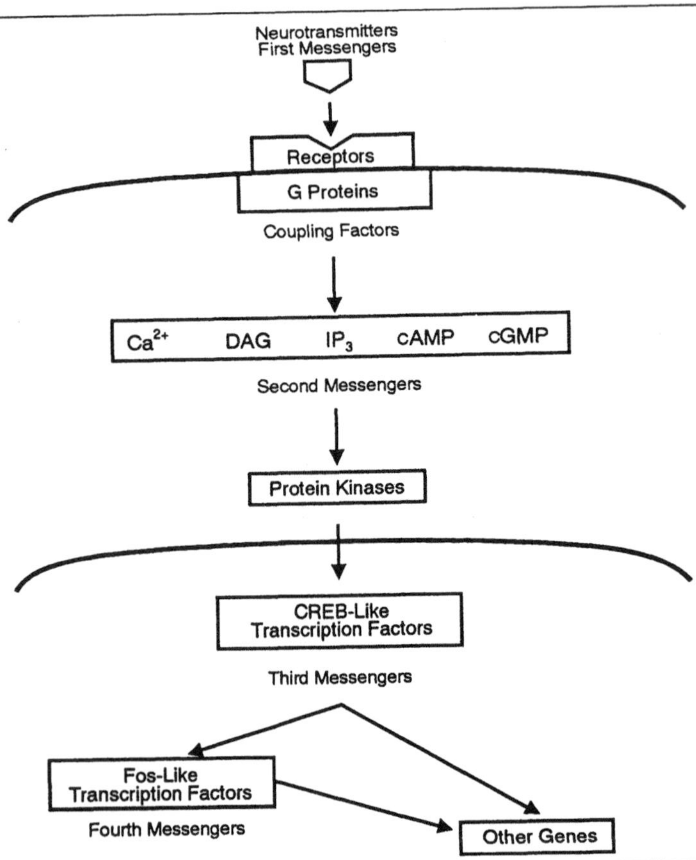

FIGURE 8. Neurotransmitter regulation of neuronal gene expression. Neurotransmitters influence the expression of a variety of neuronal genes through a mechanism involving second messengers, protein kinases, and several types of transcription factors. From Hyman and Nestler (1993).

zinc-finger, and helix-loop-helix proteins. These names refer to structural features that help mediate the interaction of the transcription factor with its DNA binding site. (For further details, see a contemporary biochemistry or molecular biology text such as Watson, Gilman, Witkowski, & Zoller, 1992.) The important point here is that binding of a transcription factor to a gene regulatory site can either enhance or suppress transcription of that gene.

As mentioned above, second messengers are linked with transcription factors in the neurotransmitter control of gene expression. Let us consider the example of transcriptional activation by cAMP. In this case, the cAMP-dependent protein kinase phosphorylates a transcription factor called the cAMP-response element binding protein (CREB). Once activated, CREB binds to a DNA regulatory site not surprisingly named the cAMP-response element (CRE). The interaction between CREB and CRE enhances transcription of the "downstream" gene (Vallejo, 1994). Interestingly, CREB can be phosphorylated by other kinases besides the cAMP-dependent protein kinase, which allows multiple extracellular signaling pathways to converge on the same target gene(s) in a given cell.

C-fos and other immediate-early genes

Transcription factors often function in a cascade. Thus, CREB and various other transcription factors induce a group of immediate-early genes called c-fos, fos B, c-jun, jun B, zif-268, and so forth (Morgan & Curran, 1989; Sagar & Sharp, 1993). The products of these IEGs are themselves transcription factors that play important roles in neuronal gene regulation. I will focus on the Fos and Jun families, which are among the best characterized IEGs.

C-fos is a proto-oncogene, which means that it is the normal cellular counterpart of a related gene that is present in certain cancer-causing viruses. The transcription factor encoded by the c-fos gene is called Fos. As with IEGs generally, levels of Fos protein and mRNA are low in most neurons in the absence of stimulation. However, stimuli that increase the appropriate second messengers (e.g., cAMP or Ca^{2+}) can rapidly induce Fos expression through the action of CREB and several other transcription factors known to participate in c-fos gene regulation. Depending on the neuronal population under investigation, such stimuli may be environmental or pharmacological. In this way, c-fos and other IEGs can be considered markers of neuronal activation (Morgan & Curran, 1989; Sagar & Sharp, 1993). Other research suggests that IEGs may play an important role in mechanisms of neural plasticity and learning (Abraham, Dragunow, & Tate, 1991).

Figure 8 presents a simplified summary of the overall sequence of events beginning with receptor activation and culminating in the activation of target genes in the cell. As shown in the figure, as many as four different levels of messengers may participate in this process. Through such complex regulatory mechanisms, synaptic activity can produce long-term adaptive changes in neuronal functioning.

Summary

At its simplest, the process of neurotransmission consists of invasion of a nerve terminal by an action potential, fusion of one or more vesicles with the presynaptic membrane, release of a few thousand molecules of one chemical substance into a synaptic cleft, stimulation of a single population of postsynaptic receptors located within that synapse, and the rapid elicitation of an excitatory or inhibitory postsynaptic potential due to the opening of ion channels on the postsynaptic membrane. However enticing this simplified view may be, we have seen that neurotransmission is often vastly more complicated. In particular, three themes emerge that are of special significance for the modeling of neural networks. First, several mechanisms are available for coincidence detection within a postsynaptic cell, including the voltage sensitivity of NMDA receptors and possibly also the long-range action of certain second messengers. Such mechanisms may underlie the type of synaptic strengthening envisioned by Donald Hebb almost 50 years ago (Hebb, 1949). Second, transmitters and other intercellular messengers sometimes act at sites distant from their synthesis and release. Incorporating this feature may lead to novel effects not seen in networks that use only point-to-point transmission. Finally, even in adult organisms, new synapses may form and (some) old ones disappear over time. Consequently, neural plasticity may involve not only strengthening and weakening of existing connections, but also changes in the "wiring diagram." Neurotransmitter-related alterations in gene expression play a pivotal role in both chemical and structural alterations in synaptic action.

ACKNOWLEDGMENT

Research conducted during the preparation of this chapter was supported by grant DA-06495 from the National Institute on Drug Abuse.

Neural-Networks Models of Cognition
J.W. Donahoe and V. Packard Dorsel (Editors)
1997 Elsevier Science B.V.

CHAPTER 6

CELLULAR MECHANISMS OF LONG-TERM POTENTIATION: LATE MAINTENANCE

Uwe Frey
Federal Institute for Neurobiology
Gene Regulation & Plasticity
Magdeburg, Germany

ABSTRACT

Hippocampal long-term potentiation (LTP) is the primary model for investigating mechanisms and processes involved in the establishment of certain forms of explicit memory in the mammalian brain. During the last decade much progress has been made in elucidating the cellular mechanisms underlying the induction and early expression of LTP. In contrast, little is known about the prolonged maintenance of this phenomenon. The sustained duration of LTP, however, is a prerequisite for its postulated role in information storage. In this chapter, I shall focus on results relevant to the prolonged maintenance of LTP, and show that the cellular processes involved in LTP are consistent with those believed to underlie learning and memory formation.

Introduction

At the beginning of the 20th century Cajal (1911) proposed that neuronal networks are not cytoplasmatically continuous, but communicate with each other at distinct junctions, which Sherrington termed *synapses*. External events are represented in the brain as spatio-temporal patterns of activity within preexisting neuronal circuits. Processes involved in learning and memory formation must therefore occur within these circuits, most likely at synaptic junctions. While Cajal favored the development of new connections during memory formation, Konorski (1948) and Hebb (1949) proposed that preexisting neuronal connections can be strengthened by simultaneous activation of pre- and postsynaptic sites. According to Konorski and Hebb, plastic changes must be located at synaptic junctions; consequently, research on modifications of neuronal plasticity has focused on the synapse rather than on other neuronal elements.

A number of theories were developed (e.g., Eccles, 1964; Griffith, 1966; John, 1967; Marr, 1969; Matthies, 1974) based on the coincidence-detection rule of Hebb and Konorski. In these earlier formulations, possible cellular mechanisms for the long-lasting functional change required for long-term memory were unknown. However, in the early seventies, Lomo and Bliss

(Bliss & Gardner-Medwin, 1973; Bliss & Lomo, 1973) first described a prolonged increase of synaptic efficacy after specific afferent stimulation in the dentate gyrus of the hippocampus, a structure required for the formation of certain forms of learning and memory (Penfield & Milner, 1958; Milner, 1972; O'Keefe & Nadel, 1978; Teyler & DiScenna, 1986; Thompson, 1986; Alkon, Amaral, Bear, Black, Carew, Cohen, Disterhoft, Eichenbaum, Golski, Gorman, Lynch, McNaughton, Mishkin, Moyer, Olds, Olton, Otto, Squire, Staubli, Thompson, & Wible, 1991). Brief high-frequency stimulation of excitatory connections between fibers of the perforant path and granule cells caused a dramatic and sustained increase in the efficacy of synaptic transmission. This phenomenon, which is called long-term potentiation (LTP), has been found with varying properties in all excitatory pathways of the hippocampus, as well as in several other regions of the brain (Racine, Milgram, & Hafner, 1983; Patrylo, Schweitzer, & Dudek, 1994). At present, however, only hippocampal LTP has been thoroughly investigated with respect to its duration and its association with certain forms of learning and memory formation. Hippocampal LTP is thought to serve as an elementary mechanism for the establishment of certain forms of explicit memory in the mammalian brain. This suggests that hippocampal LTP might be characterized by cellular properties similar to those of processes occurring during learning and memory formation in vertebrates.

Much progress has been made in elucidating the cellular mechanisms underlying the induction and early expression of LTP (early-LTP or E-LTP, which lasts 3 to 4 hours; for a review see Bliss & Collingridge, 1993). However, little is known about the prolonged maintenance of LTP. Due to the sustained duration of hippocampal LTP, which can last for several weeks in the intact animal (Bliss & Gardner-Medwin, 1973), this phenomenon is considered to employ a mechanism involved in the storage of distinct information. Here, I review results indicating that prolonged, late-LTP (L-LTP) in the hippocampus has cellular properties of the type needed for learning and memory formation.

The first published evidence that hippocampal LTP plays a role in memory was provided by Barnes (1979), who indicated that the speed of learning a spatial task was positively correlated with the persistence of LTP. It was hypothesized that persistent changes of synaptic efficacy in the hippocampus induced "naturally" during learning are responsible for the information storage involved in learning the task. Although other work supports these results (Skelton, Miller, & Phillips, 1985; Morris, Anderson, Lynch, & Baudry, 1986; Matthies, Ruethrich, Ott, Matthies, & Matthies, 1986; McNaughton & Morris, 1987; Morris, 1989; Castro, Silbert, McNaughton, & Barnes, 1989; Alkon et al, 1991; Korol, Abel, Church, Barnes, & McNaughton, 1993; Doyère, Burette, Rédini-Del Negro, & Laroche, 1993a; Doyère, Rédini-Del Negro, & Laroche, 1993b; Moser, Moser, & Andersen, 1993), the correlation of an increase in field potentials with processes underlying learning and information storage is very indirect. Field-potential recording *in vivo* has a number of disadvantages, such as the occurrence of temperature fluctuations (Moser et al, 1993) and the complexities associated with recording from a large neuronal population. Furthermore, little is known about the true function of the hippocampus during learning and memory formation. Nevertheless, the empirically obtained results suggest a role for LTP as one mechanism underlying information storage.

Mechanisms and phases of memory formation

In 1974 Matthies (1974) developed a hypothesis concerning the neuronal mechanisms of memory formation. The assumed phases of short-term, intermediate, and long-term memory (LTM)—with their different time courses, decay times, biological correlates, and sensitivities to interventions—reflect properties corresponding to the cellular mechanisms of a synaptic, synaptosomal, and nuclear regulation of memory formation (see **Figure 1**).

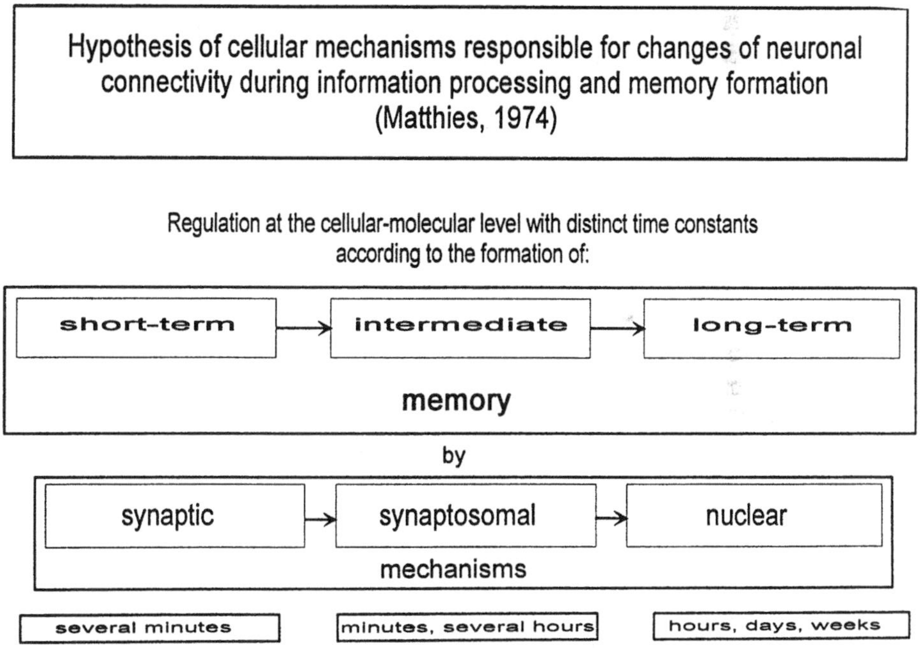

FIGURE 1. Mechanisms of permanent changes of neuronal connectivity. An illustration of the hypothesis of Matthies (1974) concerning the cellular mechanisms responsible for changes of neuronal connectivity during learning and memory formation.

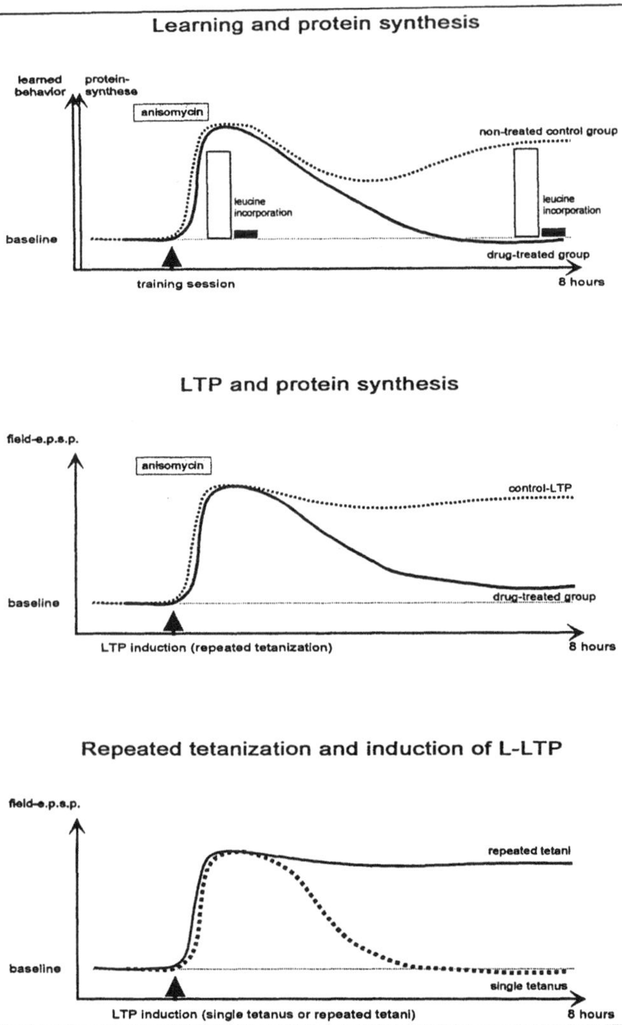

FIGURE 2. Common characteristics of processes during learning and hippocampal LTP. Distinct forms of learning and LTP depend on protein synthesis. The upper graph illustrates the amnestic effect of the reversible protein-synthesis inhibitor anisomycin (closed line) when it was administered during the training session. Experiments investigating protein synthesis during acquisition revealed a two-phase time course of leucine incorporation, which was blocked by anisomycin. The middle graph illustrates the action of anisomycin on the maintenance of hippocampal LTP, which was prevented after about 3-5 hours. The lower figure shows that repeated tetani are required for the induction of the protein-synthesis-dependent stage of LTP. Single tetanization (dotted line) only produces a short-term, protein-synthesis-independent phase.

It has been demonstrated that the formation of LTM requires the initiation of protein synthesis (Flexner, Flexner, & Stellar, 1963; Agranoff & Klinger, 1964; Barondes & Cohen, 1967; Pohle & Matthies, 1974; Shashoua, 1976; Grecksch, Ott, & Matthies, 1980; Grecksch & Matthies, 1980; Duffy, Teyler, & Shashoua, 1981; Flood, Smith, & Jarvik, 1980; Mizumori, Rosenzweig, & Bennett, 1985; Shashoua, 1985) and the subsequent post-translational addition of a fucose group (i.e., fucosylation) to newly synthesized proteins (Popov, Rüthrich, Pohle, Schulzeck, & Matthies, 1976; Popov, Schulzeck, Pohle, & Matthies, 1980; Shashoua, 1991). The synthesis and subsequent processing of the macromolecules seem to be controlled by the transsynaptic action of neuromodulators during learning-related convergence of neuronal signals on integrating neurons.

Learning experiments revealed a biphasic occurrence of protein synthesis (**Figure 2**). The early stage, during and immediately after the learning procedure, was characterized by the synthesis of mainly cytoplasmatic, soluble proteins. The later stage, 4 to 8 hours after acquisition, revealed an increase of membrane-bound proteins. These changes in the synthesis of macromolecules were most obvious in hippocampal regions, but also occurred in some neocortical structures (for review see Matthies, 1989b; Matthies, 1989a). It was suggested that the early phase of protein synthesis represents the synthesis of regulatory proteins. The regulatory proteins then control the formation of target proteins that, finally, remodel neuronal connectivity during late stages of memory formation. These experiments were the first step toward verifying the hypothesis of a phasic model of cellular mechanisms underlying memory formation in vertebrates.

The complexity of the learning process makes it very difficult to study these mechanisms in the intact animal. Therefore, LTP in hippocampal slices was used. This provided both an indirect tool to study processes occurring during learning and memory formation and a method to see whether the cellular mechanisms of learning and memory formation also underlie hippocampal LTP. Initially, LTP was regarded as a unitary phenomenon, so observations were mainly confined to the induction and maintenance of E-LTP. However, considering the findings with macromolecule synthesis during learning experiments, different stages and corresponding mechanisms might also occur with LTP, if it was directly involved in learning and memory formation. To verify this assumption, LTP must be investigated for at least 8 to 10 hours, the time required for the protein-synthesis-dependent formation of a permanent "memory trace" at the synaptic level.

Induction mechanisms of LTP

Hippocampal LTP in the CA1 region is commonly characterized by four basic properties: cooperativity, associativity, input-specificity and occlusion.

Cooperativity denotes that a distinct intensity threshold of afferent stimulation has to be achieved to obtain LTP. "Weak" high-frequency stimulation, which activates few afferent fibers, causes only a short-lasting post-tetanic potentiation, whereas higher intensities of tetanization can produce a longer-lasting potentiation (McNaughton, Douglas, & Goddard, 1978; Huang & Kandel, 1994). Our recent finer analyses demonstrate that even strong intensities of tetanization may be insufficient to produce L-LTP. Even "strong" tetanization must be repeated to initiate mechanisms responsible for the prolonged maintenance of LTP (**Figure 2**). These results are in accordance with the findings by Reymann, Malisch, Schulzeck, Brödemann, Ott, and Matthies (1985), and Huang and Kandel (1994).

Associativity means that, besides the activation of a certain number of afferent fibers, the postsynaptic cell needs to be active at the same time. Thus even a weak input can be potentiated if it occurs at the same time as a strong tetanus to a separate but convergent input (McNaughton et al, 1978; Levy & Steward, 1979). LTP is *input-specific*, i.e., only those inputs develop LTP which were active at the time of LTP induction. Finally, it is a widespread view that asymptotic LTP (maximal amount of potentiation at a given stimulus intensity) prevents further potentiation during its maintenance, once it has been established. This property is known as *occlusion* (de Jonge & Racine, 1985). Since most of the experiments were performed *in vitro* and observed for only minutes or a few hours, it remains to be determined whether cooperativity, associativity, input-specificity and occlusion are also characteristic of L-LTP.

For the induction of associative forms of LTP in the hippocampus, the activation of a distinct ionotropic glutamate receptor seems to be critical. The excitatory neurotransmitter glutamate acts as the main transmitter in the CA1 region and in the dentate gyrus. During strong and/or repeated tetanization, depolarization of the postsynaptic cell can be achieved via activation of different non-N-methyl-D-aspartate (non-NMDA) receptors. However, it is quite possible that depolarization can be produced by other than glutamatergic inputs (see Bliss & Collingridge, 1993). Depolarization is required for the removal of a Mg^{2+} block of NMDA receptor channels, which allows an influx of Ca^{2+} (for review see Coan & Collingridge, 1987; Collingridge & Bliss, 1987; Bliss & Collingridge, 1993). Thus, loading of the postsynaptic site with Ca^{2+} is one crucial step in LTP induction.

Increase in intracellular Ca^{2+} can be achieved through mechanisms other than activation of the NMDA receptor, i.e., by voltage-dependent calcium channels or the release of Ca^{2+} from intracellular stores. The rise in intracellular Ca^{2+} may fulfill a role as an intracellular activator of further processes that may lead to the initiation and expression of E-LTP and—in connection with other coincident factors—to the induction of L-LTP.

Experiments investigating postsynaptic Ca^{2+} elevation have demonstrated that a rise in intracellular Ca^{2+} appears insufficient to induce stable LTP

(Malenka, Kauer, Perkel, Mauk, Kelly, Nicoll, & Waxham, 1989a; Malenka, Kauer, Perkel, & Nicoll, 1989b; Harvey & Collingridge, 1992; Bliss & Collingridge, 1993). In addition, application of NMDA induces only a short-lasting (< 1 hour) increase in synaptic efficacy (Kauer, Malenka, & Nicoll, 1988; Bliss & Collingridge, 1993). In experiments in which a prolonged potentiation

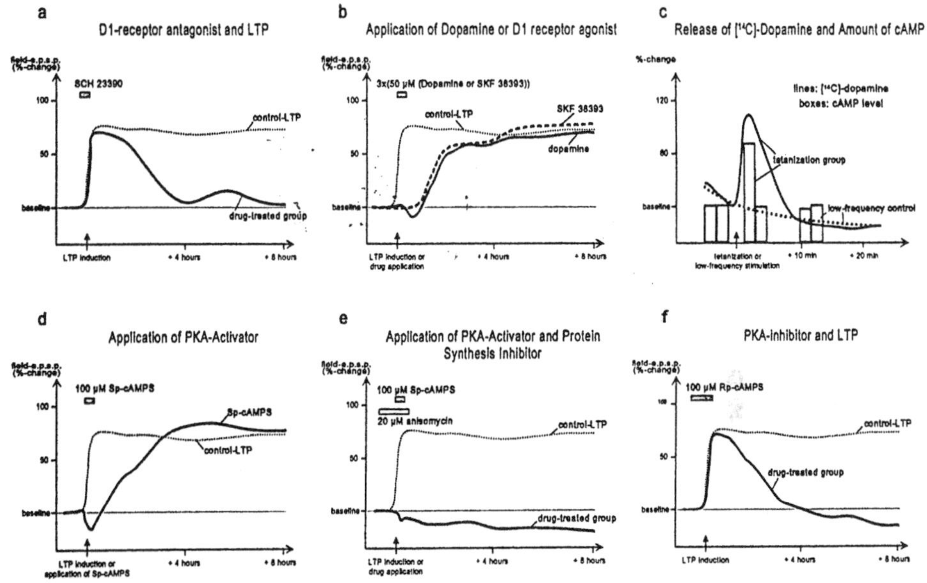

FIGURE 3. *In vitro* effects of dopamine and L-LTP in hippocampal CA1 neurons. This figure illustrates the requirement of mechanisms for L-LTP activated by dopaminergic receptors during and/or immediately after repeated tetanization. **a.** Application of a D_1-receptor antagonist during LTP using conventional repeated tetanizations prevents the late maintenance of LTP (L-LTP). **b.** Threefold application of dopamine or of the D_1-receptor agonist SKF 38393 reveals a late-onset potentiation. The dopaminergic D_1 receptor activates the cAMP/PKA complex. cAMP levels are shown in **c** during control stimulation (open boxes) and high-frequency stimulation (shaded boxes). Repeated tetanization produced a transient enhancement of cAMP levels (Frey et al, 1993a). Lines indicate the release of [^{14}C]-dopamine after low- (dotted line) and high-frequency (closed line) stimulation. Tetanization enhanced the release of [^{14}C]-dopamine in hippocampal slices, paralleling the increased cAMP levels after tetanization. **d.** The effect of transient application of the PKA activator Sp-cAMPS. The time course of Sp-cAMPS-induced potentiation resembles the time course of potentiation induced by dopamine or by the D_1-receptor agonist SKF 38393 in b. **e.** The potentiation induced by Sp-cAMPS can be prevented by the protein-synthesis inhibitor anisomycin, which parallels the blockade of L-LTP by anisomycin during conventionally induced LTP (see **Figure 2**). **f.** The action of the PKA inhibitor Rp-cAMPS on electrically induced LTP, illustrating the role of dopamine and the subsequent activation of the cAMP/PKA complex during hippocampal LTP in the CA1 region.

in the CA1 *in vitro* has been observed (Kamiya, Sawada, & Yamamoto, 1993), application of NMDA was paralleled by an elevated extracellular Ca^{2+} concentration, which is sufficient to induce LTP by itself (Reymann, Matthies, Frey, Vorobyev, & Matthies, 1986; Melchers, Pennartz, & Lopes Da Silva, 1987).

These results indicate the necessity of other coincident factors that are activated at the same time via monosynaptic or multisynaptic mechanisms. Broadly speaking, there is some evidence that a glutamatergic-metabotropic (mGlu) receptor must be active at the same time that intracellular Ca^{2+} rises (Reymann & Matthies, 1989; Izumi, Clifford, & Zorumski, 1991; Bashir, Bortolotto, Davies, Berretta, Irving, Seal, Henley, Jane, Watkins, & Collingridge, 1993). Application of mGlu receptor agonists produces a delayed increase in synaptic efficacy (Bortolotto & Collingridge, 1993). These results are still under investigation, and it is not yet known whether mGlu receptor-activated potentiation lasts longer than 4 hours (Manahan-Vaughan & Reymann, 1995). We have evidence that, for the induction of L-LTP, a second neurotransmitter must be present during LTP induction (see **Figure 3**).

Phases of LTP

LTP in the hippocampal CA1 region and the dentate gyrus consists of different phases (Krug, Lössner, & Ott, 1984; Frey, Krug, Reymann, & Matthies, 1988; Matthies, Reymann, Krug, Frey, Loessner, & Popov, 1989; Reymann, Frey, & Matthies, 1988b; Matthies, Frey, Reymann, Krug, Jork, & Schroeder, 1990; Frey, Huang, & Kandel, 1993a; Bliss & Collingridge, 1993). Long-term memory and L-LTP in the hippocampus—i.e., late maintenance of more than 4 hours—can be distinguished from short-term memory or E-LTP by inhibitors of protein and RNA synthesis (Krug et al, 1984; Stanton & Sarvey, 1984; Grecksch & Matthies, 1980; Frey et al, 1988; Otani, Marshall, Tate, Goddard, & Abraham, 1989; Frey, Seidenbacher, & Krug, 1993b; Nguyen, Abel, & Kandel, 1994; Frey, Frey, Schollmeier, & Krug, 1995a).

E-LTP consists of a short-term potentiation (STP)—lasting from 30 minutes to 1 hour—that is mediated by activation of the NMDA receptor and calcium-calmodulin-dependent kinase II, and an intermediate stage (LTP1)—lasting from 1 to 4 hours—that is dependent on activation of different protein kinases (for reviews see Reymann, 1993; Bliss & Collingridge, 1993). E-LTP is followed by the protein- and RNA-synthesis-dependent L-LTP (or LTP2); see **Figure 4** (Krug et al, 1984; Otani & Abraham, 1989; Otani et al, 1989; Frey et al, 1988; Frey et al, 1993b; Nguyen et al, 1994; Frey et al, 1995a). Furthermore, as shown recently, L-LTP can be functionally distinguished from E-LTP by its ability to undergo further plastic changes (Frey, Schollmeier, Reymann, & Seidenbacher, 1995b).

Early phase of LTP (E-LTP)

As already noted, the establishment of E-LTP depends on the postsynaptic elevation of intracellular Ca^{2+}. Several different Ca^{2+}-sensitive enzymes have been shown to participate in mechanisms responsible for modifications of

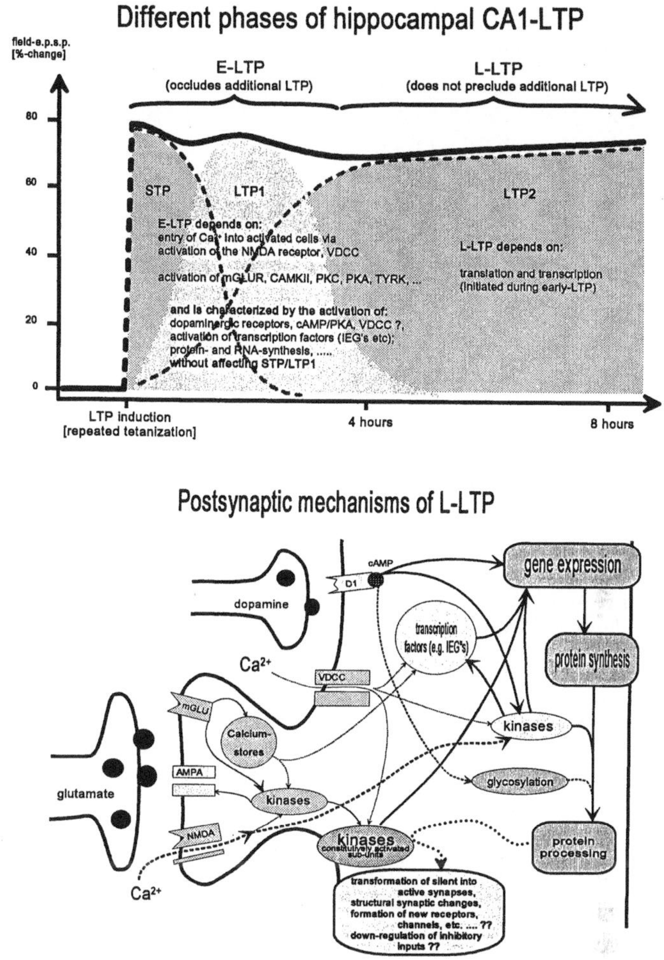

FIGURE 4. Hippocampal long-term potentiation occurs in phases. The upper graph illustrates the time course and underlying mechanisms of hippocampal LTP, which consists of two major stages: an early phase of LTP (E-LTP) lasting about 3-4 hours, followed by a late phase (L-LTP). In contrast to the early maintenance of potentiation, L-LTP depends on gene transcription and translation. The lower figure summarizes possible postsynaptic mechanisms required for the induction of L-LTP in the hippocampal CA1 region. Repeated tetanization of afferent fibers initiates a cascade of cellular events responsible for the induction of E-LTP and L-LTP. The prolonged maintenance of LTP depends on the activation of both glutamatergic and dopaminergic receptors. The convergent action of both transmitter systems and the subsequent activation and synergistic action of cellular second-messenger systems initiate gene transcription, translation, and post-translational processing of macromolecules required for the sustained duration of LTP.

neurin), proteases (e.g., calpain), as well as phospholipases and protein kinases synaptic efficacy. Recent results suggest a role of phosphatases (e.g., calci- (Baudry, 1986; Lovinger, Wong, Murakami, & Routtenberg, 1987; Malinow, Madison, & Tsien, 1988; Reymann, Brödemann, Kase, & Matthies, 1988a; Reymann, Frey, Jork, & Matthies, 1988c; O'Dell, Kandel, & Grant, 1991; Bliss & Collingridge, 1993; Qian, Gilbert, Colicos, Kandel, & Kuhl, 1993; Malinow, Schulman, & Tsien, 1989; Funauchi, Haruta, & Tsumoto, 1994; Shaw, Lanius, & Van den Doel, 1994; Wang & Stelzer, 1994; Fazeli, Breen, Errington, & Bliss, 1994; Muller, Molinari, Soldati, & Bianchi, 1995). Phosphorylation cascades have been most thoroughly investigated, particularly those involving protein kinases. Transient activations of different kinases appear at different stages of E-LTP. Thus, the activation of Ca^{2+}/calmodulin kinase II (Reymann et al, 1988a; Malinow et al, 1989; Ito, Hidaka, & Sugiyama, 1991; Silva, Stevens, Tonegawa, & Wang, 1992) and protein tyrosine kinase (O'Dell et al, 1991; Grant, O'Dell, Karl, Stein, Soriano, & Kandel, 1992) are known to be critical for the induction of LTP. Protein kinase C (PKC) plays an important role in the maintenance of E-LTP, or LTP1 (see Bliss & Collingridge, 1993). Since most of the experiments investigating the role of these enzymes studied only E-LTP, little is known about their contribution to L-LTP.

Transient activation of kinases would not necessarily produce a sustained potentiation of a few hours' duration. However, CAM kinase II, which is present in postsynaptic densities demarcating the active zone of excitatory synapses, might undergo a Ca^{2+}-dependent autophosphorylation, and then convert the enzyme into a constitutively active Ca^{2+}-independent kinase. There is also evidence that proteolytic cleavage of PKC can create a constitutively active molecule (for review see Bliss & Collingridge, 1993).

Whether the maintenance of E-LTP depends on the continued action of protein kinases is controversial. For example, the proteolytic cleavage of other proteins could also cause a persistent change in receptor and/or ion-channel function. Such mechanisms might be responsible for changes observed for the glutamatergic, ionotropic amino-3-hydroxy-5-methyl-4-isooxazole proprionic acid (AMPA) receptor. There is some evidence that the sensitivity of the postsynaptic AMPA receptor increases after tetanization. Stimuli may evoke an enhanced excitatory postsynaptic potential (i.e., EPSP), at least during E-LTP (Davies, Lester, Reymann, & Collingridge, 1989). This observation might be related to the recently described conversion of postsynaptically silent synapses into active synapses (Liao, Hessler, & Malinow, 1995; Isaac, Nicoll, & Malenka, 1995). These authors report that a high proportion of synapses in the hippocampus contain NMDA but not AMPA receptors. Such synapses are non-functional at normal resting potential, but acquire AMPA-type responses following LTP induction. A rearrangement of functional receptors might provide a postsynaptic mechanism for the induction and early expression of LTP. Further investigation is needed to determine if these modifications also take place during L-LTP.

Retrograde mechanisms in LTP

The most controversial issue concerning possible mechanisms for the early expression of LTP arose from findings that an increased transmitter release might also contribute to LTP expression (Lynch, Errington, & Bliss, 1985; Lynch, Errington, Clements, Bliss, Rédini-Del Negro, & Laroche, 1990; Feasey, Lynch, & Bliss, 1986; Errington, Lynch, & Bliss, 1987; Canevari, Richter-Levin, & Bliss, 1994; Richter-Levin, Canevari, & Bliss, 1995). If this were the case, then a retrograde messenger must go from the postsynaptic spine to the presynaptic terminal, since the probable trigger for the induction of LTP is the entry of Ca^{2+} through postsynaptic NMDA-gated channels. Recently, a number of candidates such as arachidonic acid, nitric oxide, carbon monoxide, and others have been proposed to function as the retrograde messenger (for a review see Bliss & Collingridge, 1993). Results describing an inhibition of E-LTP after blockade of these candidates are problematic and may be due, in part, to methodological difficulties such as the use of nonphysiological temperatures (Williams, Li, Nayak, Errington, Murphy, & Bliss, 1993). The time courses of the inhibitors of retrograde messengers also appear inconsistent. Increased transmitter release after LTP induction is greatest with STP, i.e., from a few seconds to an hour after LTP induction. However, the inhibitors of the candidates for retrograde signals spare STP.

A more realistic candidate to fulfill the function of a fast retrograde messenger could be K^+. K^+ is released from the postsynaptic cell during tetanization to a degree that partially reflects the level of NMDA-receptor activation (Collingridge, 1992). This signal could interact with presynaptic mGlu receptors activating phospholipase C. The activation of phospholipase C by these receptors is strongly potentiated by extracellular K^+ (Baskys & Malenka, 1991). Such a mechanism could elevate presynaptic Ca^{2+} and activate the ß isoform of PKC, which might enhance transmitter release (Colley & Routtenberg, 1993).

Late phase of LTP (L-LTP)

Mechanisms initiating L-LTP. Learning experiments demonstrate that protein-synthesis inhibitors prevent hippocampus-related learning after about 4 hours. Further, the formation of LTM requires the fucosylation of proteins, and the neurotransmitter dopamine is involved in the control of this process. Dopamine-induced improvement of LTM correlates with increased incorporation of fucose into hippocampal proteins (Flood et al, 1980; Jork, Grecksch, & Matthies, 1986). Similar to the processes occurring during learning, inhibition of protein synthesis during the induction of hippocampal LTP also prevents the prolonged maintenance of LTP (>3-6 hours). The requirement of fucosylation of macromolecules also seems to be critical for the establishment of L-LTP in the dentate gyrus and in the CA1 region (Krug, Jork, Reymann, Wagner, & Matthies, 1991).

Since activation of hippocampal glutamatergic receptors alone does not induce L-LTP, it has been suggested that another neurotransmitter, which is simultaneously present during tetanization, is involved in the establishment of L-LTP. The pattern and strength of stimulation used to induce LTP does not guarantee a clear monosynaptic glutamatergic activation of hippocampal target cells. In addition, inhibitory neurons and neurons releasing other transmitters are almost certainly activated as well. A number of transmitters, including norepinephrine, acetylcholine, and opiods, are known to modulate LTP (Bliss, Goddard, & Riives, 1983; Haas, Jefferys, Slater, & Carpenter, 1984; Stanton & Sarvey, 1987; Stanton & Sarvey, 1985b; Briggs & McAfee, 1988; Blitzer, Gil, & Landau, 1990; Stanton, Mody, & Heinemann, 1989; Bramham, Errington, & Bliss, 1988; Burgard & Sarvey, 1990; Bramham, Milgram, & Srebro, 1991; Bramham, 1992; Katsuki, Saito, & Satoh, 1992; Maeda, Kaneko, & Satoh, 1994; Sokolov & Kleschevnikov, 1995; Haas, Sergueeva, Vorobjev, & Sharonova, 1995).

Given the effects of dopamine on learning (Flood et al, 1980), we investigated whether dopamine is simultaneously released during LTP induction in the CA1 region of the hippocampus, and whether it is required for L-LTP. The hippocampus is innervated by dopaminergic fibers that course through the mesolimbic pathway (Baulac, Verney, & Berger, 1986; Grace, 1991), and there is evidence of the expression of the D_5 receptor in CA1 pyramidal cells. The D_5 receptor is related to the D_1 dopamine receptor that is coupled to adenylyl cyclase (Sunahara, Guan, O'Dowd, Seeman, Laurier, Ng, George, Torchia, Van Tol, & Niznik, 1991). We demonstrated that, in addition to glutamate and possibly other neurotransmitters, dopamine levels increase during conventional LTP induction (see **Figure 3c**; Frey, Schroeder, & Matthies, 1990). To determine whether dopamine might be the additional activator necessary for L-LTP, we showed that it plays a crucial role in the initiation of the mechanisms responsible for L-LTP in the hippocampal CA1 region (Frey, Hartmann, & Matthies, 1989a; Frey et al, 1990; Frey, Matthies, & Reymann, 1991). When specific inhibitors of the dopaminergic D_1 and D_2 receptors were administered during tetanization, L-LTP was prevented (**Figure 3a**).

The involvement of aminergic modulation of LTP has been proposed repeatedly (Dunwiddie, Roberson, & Worth, 1982; Stanton & Sarvey, 1985b; Gribkoff & Ashe, 1984; Bliss et al, 1983; Krug, Chepkova, Geyer, & Ott, 1983; Hopkins & Johnston, 1984; Robinson & Racine, 1985; Stanton & Sarvey, 1985c; Buzsaki & Gage, 1989). However, it has been suggested that the influences of these transmitters are only modulatory. We have demonstrated that activation of dopaminergic inputs to hippocampal CA1 neurons plays a critical role, perhaps acting as a cellular switch to establish the late phase of LTP. Furthermore, other observations have demonstrated that the transient application of dopamine alone initiates a delayed increase in both the population spike (Gribkoff & Ashe, 1984) and the synaptic response, i.e., the field-EPSP (Frey, unpublished data; see **Figure 3b**). These events may simulate the late phase of LTP. Similar results have been obtained by the application of norepinephrine in the dentate gyrus (Stanton & Sarvey, 1985b; Stanton & Sarvey, 1987).

In searching for the physiological significance of the simultaneous activation of two different inputs, it should be noted that dopaminergic neurons are assumed to mediate, in part, the effect of biologically significant reinforcing stimuli during learning (Stein, 1975; Bischoff, 1986; Donahoe, Burgos, & Palmer, 1993). Normal dopamine functioning also appears necessary for the establishment and maintenance of conditioned reinforcement, or incentive learning (Beninger, 1983). Thus prolonged enhancement of synaptic efficacy during the processing of specific sensory information becomes understandable. Dopaminergic facilitation of learning is consistent with observations from Packard and White (1989), who demonstrated a role of dopaminergic receptors in memory facilitation on learning tasks sensitive to both hippocampal and caudate lesions.

Recently it has been shown that repeated application (see **Figure 3b**) or transient application (Huang & Kandel, 1995) of a D_1 receptor agonist, SKF 38393, in CA1 synapses induces delayed potentiation whose time course is similar to that found after administration of dopamine. Because the D_1 receptor stimulates adenylyl cyclase and the D_1-receptor blocker efficiently prevents L-LTP (Frey, Schweigert, Krug, & Lössner, 1991), we explored whether the formation of cAMP is increased during LTP and whether L-LTP is dependent on cAMP-dependent protein kinase (PKA) or other cAMP-dependent processes. We found that the formation of cAMP was indeed transiently increased after and only after a threefold (i.e., repeated) tetanization was used for LTP induction. This is the stimulation protocol required for reliable L-LTP (Frey et al, 1993a; **Figure 3c**). This short-term elevation of cAMP was blocked by a D_1 antagonist, SCH 23330, and the NMDA-receptor blocker AP-5.

Because a large proportion of the adenylyl cyclase in the rat brain is sensitive to calmodulin in the presence of Ca^{2+}, the Ca^{2+} influx caused by activation of the NMDA receptor could stimulate adenylyl cyclase directly (Eliot, Dudai, Kandel, & Abrams, 1989; Chetkovich, Gray, Johnston, & Sweatt, 1991; Chetkovich & Sweatt, 1993). Since a transient increase in cAMP is required for the establishment of L-LTP, NMDA-receptor activation might be sufficient to induce a prolonged potentiation. However, this was not the case. Our experiments using the D_1-receptor blocker SCH 23330 suggest a synergistic activation of adenylyl cyclase by glutamate and dopamine receptors in the CA1. A similar increase in intracellular concentration of cAMP occurs after tetanization and after application of norepinephrine in the dentate gyrus, suggesting similar cellular mechanisms appear during L-LTP generation in different brain structures (Stanton & Sarvey, 1985a).

In contrast to the already discussed kinases, cAMP-dependent protein kinase appears not to play a role in E-LTP, although it is constitutively active in CA1 hippocampal neurons (Wang, Salter, & MacDonald, 1991; Greengard, Jen, Nairn, & Stevens, 1991). Recent observations reveal that application of specific PKA inhibitors are ineffective in blocking E-LTP but do block L-LTP (see **Figure 3f**; Frey et al, 1993a; Matthies & Reymann, 1993). Transient application of a membrane-permeable cAMP analog or of other PKA activators initiated a delayed potentiation that seemed to simulate L-LTP (see **Figure 3d,e**; Slack & Pockett, 1991; Frey et al, 1993a; Pockett, Slack, & Peacock, 1993). These results suggest that the cAMP-dependent activation of PKA is required for L-LTP.

Protein synthesis and LTP. In the early eighties it was demonstrated for the first time that hippocampal LTP is not a unitary phenomenon, but occurs in distinct phases. As already mentioned, intraventricular application of anisomycin, a reversible translational inhibitor, prevents the long-term maintenance of LTP in the dentate gyrus, an effect that parallels the block of LTM in several learning tasks (Krug et al, 1984).

These experiments were later reconfirmed for the dentate gyrus (Otani & Abraham, 1989; Otani et al, 1989), and a similar protein-synthesis dependency of L-LTP was shown for the CA1 region (Frey et al, 1988). The application of anisomycin before, during, or immediately after tetanization produced a gradual decrease of potentiation after 4-6 hours without affecting E-LTP. Application of anisomycin one hour after tetanization had no effect. A similar phenomenon was seen when LTP was induced in dendritic stumps of CA1 pyramidal cells of hippocampal slices *in vitro* (Frey, Krug, Brödemann, Reymann, & Matthies, 1989b). In these experiments the cell-body layer, the major site of protein synthesis, was surgically removed from the apical dendrites. The isolated dendrites revealed a pronounced E-LTP in the field-EPSP, as found in intact slices. However, the potentiation gradually decreased after about 4 hours, thus showing the same lack of L-LTP as observed in complete CA1 neurons after inhibition of protein synthesis with anisomycin. These results indicate, first, that there is a particular phase of L-LTP that depends on intact protein synthesis during and immediately after tetanization and, second, that the mechanisms responsible for L-LTP are located, at least partially, postsynaptically.

The foregoing results are consistent with the finding that the incorporation of radioactive-labeled amino acids into cytosomal proteins of hippocampal neurons is elevated for an hour immediately after tetanization (Duffy et al, 1981; Fazeli, Errington, Dolphin, & Bliss, 1988; Shashoua, 1988; Shashoua, 1990; Bullock, Lössner, Krug, Frey, Rose, & Matthies, 1990; Frey et al, 1991; Fazeli et al, 1993; Fazeli et al, 1994). This transient enhancement of protein synthesis coincides with the time window after tetanization during which the inhibition of protein synthesis prevents the initiation of L-LTP, the phase that depends on protein synthesis.

Learning experiments have shown a second peak in the elevation of leucine incorporation 8 hours after training (Lössner, Jork, Krug, & Matthies, 1982; Lössner, Schweigert, Pchalek, Krug, Frey, & Matthies, 1987). Recent experiments confirmed these results for hippocampal LTP, revealing an increase in radioactive-labeled proteins in synaptosomal membranes and postsynaptic densities 8 hours after tetanization (Bullock et al, 1990; Frey et al, 1991).

Parallel studies indicate that the formation of LTM exceeding 4 hours depends on fucosylation of proteins (Popov et al, 1980; Jork et al, 1986). Particular glycoproteins are completed by the post-translational formation of a fucose 1-2 galactose linkage, which can be prevented by pretreatment with 2-deoxy-galactose. The false sugar is incorporated into glycans instead of galactose. This incorporation of 2-deoxy-galactose cannot provide the fucose 1-2 linkage, thereby preventing LTM. Dopamine is involved in the fucosylation of hippocampal proteins during the formation of LTM (Jork, Lössner, & Matthies, 1979; Jork, Grecksch, & Matthies, 1982), and a similar mechanism may operate with hippocampal CA1 LTP.

Indeed, a similar effect has been described for hippocampal L-LTP (Krug et al, 1991; Angenstein, Matthies, Staeck, Reymann, & Staak, 1992). Pretreatment of rats with 2-deoxy-galactose prevented the formation of L-LTP without influencing transmission at the glutamatergic synapses or the induction of E-LTP. These results point to a role of fucosyl-glycoproteins during L-LTP. Based on these observations, the first stage of protein synthesis takes place immediately after tetanization. However, a further post-translational processing of proteins during E-LTP is required for the induction of L-LTP.

Which neurotransmitters and which intracellular messengers initiate the protein synthesis? For the hippocampal CA1 region, glutamate alone is not able to induce L-LTP. The repeated strong tetanization that induces L-LTP is characterized by the activation of cAMP/PKA cascade. Besides synergistically activating the glutamatergic NMDA receptor, dopamine may initiate processes required for the establishment of L-LTP. Indeed, recent experiments have found that the potentiation achieved by short-term activation of the D_1 receptor can be blocked by anisomycin (Huang & Kandel, 1995). The D_1 dopamine receptor activates adenylyl cyclase, and the consequent formation of cAMP may directly induce gene transcription (Montminy, Gonzalez, & Yamamoto, 1990) or activate PKA. The cAMP/PKA complex is primarily involved in mechanisms initiating L-LTP. Induction of potentiation by a membrane-permeable cAMP analog simulates L-LTP in its time course and is prevented by the protein-synthesis inhibitor anisomycin (see **Figure 3e**; Frey et al, 1993a). This finding suggests that the activation of dopamine receptors and the subsequent stimulation of cAMP/PKA might serve as an intracellular switch for the induction of L-LTP.

Different enzymes may be required for the processing of newly synthesized proteins. During E-LTP constitutively active, multifunctional kinases, such as PKC, could affect the rearrangement of macromolecules into functional proteins. It is not yet known which proteins are synthesized, although several macromolecules have been separated on two-dimensional gels (Fazeli, Corbet, Dunn, Dolphin, & Bliss, 1993) with some links to proteins involved in forming postsynaptic densities (Bullock et al, 1990).

RNA synthesis and LTP. The synthesis of mRNA is a necessary step in the establishment of LTM (for reviews, see Matthies et al, 1990; Alkon et al, 1991). This raises the question of whether mRNA synthesis is also involved in L-LTP. In contrast to experiments with protein-synthesis inhibitors, it has only very recently been demonstrated that synthesis of mRNA is probably required for L-LTP. Previous studies had shown no influence of the mRNA-synthesis inhibitor actinomycin D (Otani et al, 1989). It was therefore concluded that protein synthesis must be carried out by preexisting mRNA, suggesting an mRNA-independent (but protein-synthesis-dependent) phase of L-LTP.

In the foregoing experiments the authors only investigated E-LTP. Since the application of protein-synthesis inhibitors has a dramatic effect on the time course of L-LTP alone, studies of mRNA synthesis must examine LTP for longer time periods. Furthermore, prior experiments were carried out with anesthetized animals, and anesthesia can prevent gene expression (Dragunow, Abraham, Goulding, Mason, Robertson, & Faull, 1989a; Jeffery, Abraham, Dragunow, & Mason, 1990). Finally, a single tetanization was used to induce LTP, a stimulation pattern that may have led to a protein-synthesis-independent E-LTP, which may explain the failure to block LTP during the first 3 hours (Worley, Bhat, Baraban, Erickson, McNaughton, & Barnes, 1993; Huang & Kandel, 1994).

To overcome these possible shortcomings, two laboratories have reinvestigated the failure of mRNA-synthesis inhibitors to block LTP. Recent work from the laboratory of Kandel (Nguyen et al, 1994) has demonstrated that synaptic LTP in hippocampal slices *in vitro* was prevented after 1 to 3 hours when two structurally different RNA-synthesis inhibitors, ACD and 5,6-dichloro-1-ß-D-ribofuranosyl (DRB) were used. The drugs were only effective when administered during tetanization, i.e., during LTP generation. Thus a side effect of the drugs on mechanisms involved in LTP generation could not be excluded. A nonspecific action of these drugs on mechanisms required for E-LTP would also explain their early time point of action.

Since the question of whether mRNA synthesis takes place in LTP is crucial for understanding the mechanisms underlying L-LTP, we performed experiments not subject to the above-mentioned shortcomings (Frey et al, 1993b; Frey et al, 1995a). The effect of ACD was studied in two different systems, the CA1 *in vitro* and the dentate gyrus *in vivo*. First, we used hippocampal slices, thereby avoiding anesthesia, assuring a more controlled application of inhibitors, and allowing a better characterization of electrophysiological responses, i.e., the field-EPSP and the population spike. Second, we confirmed the experiments *in vivo* using freely moving rats. We also investigated the action of the structurally different transcriptional inhibitor DRB. Finally, we investigated LTP for a time course of 8 hours after its induction.

In contrast to the studies by Otani et al (1989) and Nguyen et al (1994), our experiments revealed that hippocampal L-LTP in the CA1 region and in the dentate gyrus can be partially prevented by ACD, when it is applied at an effective concentration and a sufficient time before LTP induction. The potentiation of the population spike amplitude decayed to baseline values after about 4 to 6 hours. The field-EPSP also showed a definite decay during the same time interval, but remained slightly potentiated. We can exclude the possibility that the prevention of L-LTP for the population spike and the decreased late phase of potentiation of the field-EPSP were due to toxic side effects of ACD. The drug did not influence the time course after low-frequency stimulation during either the *in vitro* or *in vivo* control experiments, and had no effect on E-LTP or L-LTP when administered immediately after tetanization. A second, structurally different, RNA-synthesis inhibitor, DRB, also decreased L-LTP.

It is still unclear whether the potentiated-EPSP component remains at a potentiated level compared to low-frequency controls, or if it is also fully inhibited by RNA-synthesis inhibition over time. Unfortunately, only the first 8 hours after LTP induction have been followed to avoid late-occurring nonspecific effects caused by the overall inhibition of RNA synthesis. Interestingly, our previous experiments investigating the influence of protein-synthesis inhibition on L-LTP showed a similar decay for both the population spike and the EPSP. (Frey et al, 1988).

Dendritically distributed mRNA (Steward & Wallace, 1995) might be responsible for the difference in the time course of potentiation of the population spike and EPSP after inhibition of mRNA synthesis. Preexisting dendritic mRNA may be responsible for whatever protein synthesis was observed, as well as for the effect of the protein-synthesis inhibitor anisomycin on the EPSP and the failure to influence the spike during E-LTP described by Otani et al (1989). However, if dendritically distributed LTP-specific protein synthesis occurs, it remains unclear why LTP can only be maintained for 3 hours when dendrites are separated from their somata (Frey et al, 1989b).

The assumed local synthesis of macromolecules in the dendrites may involve a yet-unknown soma-related factor, perhaps activated by a second convergent input to the activated neuron, such as provided by dopaminergic receptors. This input may have been cut off during the dissection procedure in dendritic-stump experiments (Jackson & Westlind-Danielsson, 1994). It is also possible that the vitality of dendritic stumps was diminished due to the microsurgical intervention and that, as a result, the stumps could produce E-LTP, but not L-LTP. Furthermore, dendritically distributed mRNA might be the result of previous activity-related restricted transport from the soma to the activated synapse, or molecules synthesized in synapses might stand for distinct factors initiating gene expression.

Finally, the time course of EPSP potentiation during L-LTP suggests that, besides mechanisms requiring RNA and protein synthesis, additional processes may be involved. These parallel processes include the prolonged activation of receptors or ion channels via phosphorylation, increases in receptor sensitivity, or even presynaptic mechanisms such as enhanced glutamate release. Further experiments must investigate such possible late-effector mechanisms.

Given that RNA-synthesis inhibitors were only effective when applied during and/or immediately after LTP induction, a conditioning stimulus that induces L-LTP must activate gene expression during and/or immediately after its application. During the last few years, attention has been focused on the identification of genes that might play a role in activity-dependent plasticity. Thus, in the hippocampus a variety of conditions, including the stimulation paradigms used to induce L-LTP, cause a rapid and transient activation of immediate early genes (IEGs), e.g., c-fos, zif/268 (Sheng & Greenberg, 1990; Abraham, Mason, Demmer, Williams, Richardson, Tate, Lawlor, & Dragunow, 1993; Dragunow, Currie, Faull, Robertson, & Jansen, 1989b; Nikolaev, Tischmeyer, Krug, Matthies, & Kaczmarek, 1991; Demmer, Dragunow, Lawlor, Mason, Leah, Abraham, & Tate, 1993; Jeffery et al, 1990; Cole, Saffen, Baraban, & Worley, 1989; Wisden, Errington, Williams, Dunnett, Waters, Hitchcock, Evan, Bliss, & Hunt, 1990; Worley et al, 1993; Link, Konietzko, Kauselmann, Krug, Schwanke, Frey, & Kuhl, 1995; Qian et al, 1993; Williams, Dragunow, Lawlor, Mason, Abraham, Leah, Bravo, Demmer, & Tate, 1995), and early-effector genes, such as tissue plasminogen activator (tPA), cyclooxygenase (Cox-2), and a ras-related gene, rheb (Qian et al, 1993; Yamagata, Andreasson, Kauafmann, Barnes, & Worley, 1993; Yamagata, Sanders, Kaufmann, Yee, Barnes, Nathans, & Worley, 1994).

As previously noted, E-LTP does not appear to depend on gene transcription. This conclusion does not preclude genes from being normally transcribed immediately after LTP induction but exerting their effects later, during L-LTP. Many of the IEGs encode transcription factors that may control the expression of downstream effector genes (Goelet, Castellucci, Schacher, & Kandel, 1986; Berridge, 1986; Sonnenberg, Rauscher, Morgan, & Curran, 1989; Morgan & Curran, 1989). More recent studies have identified genes that may themselves have effector function with the potential to quickly promote long-term alterations in neuronal phenotype during plastic processes, including LTP (Qian et al, 1993; Nedivi, Hevroni, Naot, Israeli, & Citri, 1993; Smirnova, Laroche, Errington, Hicks, Bliss, & Mallet, 1993; Yamagata et al, 1993; Yamagata et al, 1994). Further work is needed to characterize the role and function of newly synthesized gene products during LTP. Methodologically, the first step has been made in generating animals deficient in a particular gene; i.e., knockout animals. The known disadvantages, such as possible compensatory mechanisms during development, make it difficult to explain possible effects during LTP in these animals. Inducible gene knockouts would make a more specific characterization possible.

However, when all findings are considered, gene expression does appear to be required during LTP, and not only to refill stores of "housekeeping" proteins. That more than simply housekeeping functions are involved is shown by experiments in which the application of the RNA inhibitor ACD immediately after tetanization had no effect on LTP. Application of ACD at this time disrupts housekeeping functions, and yet L-LTP occurred.

Associativity, input-specificity, cooperativity, and occlusion during L-LTP. Are the four basic properties of LTP also valid for L-LTP? Most LTP studies have investigated only E-LTP. No experiments have been conducted to test whether *associativity* is required for the initiation of L-LTP. Further experiments are needed to determine whether L-LTP develops in a non-tetanized pathway with weak stimulation that occurs at the same time as a strong tetanus to a separate but convergent input or whether this pattern of stimulation causes only distinct phases of E-LTP.

E-LTP is *input-specific*, but this appears to be only partially the case with synaptic enhancement during L-LTP. Preliminary results indicate that the excitability of the entire stimulated cell may be changed during L-LTP: A second non-tetanized input potentiates the population spike during L-LTP (Frey et al, 1988; Reymann et al, 1988c). Since the activated-synapse population retains its enhanced synaptic efficacy as well as spike potentiation, these observations suggest a functional, plastic transformation of the entire cell. Synaptic input-specificity is still conserved since, in contrast to the elevated population spike, the synaptic response (i.e., the EPSP) seems to be unchanged for the non-tetanized input. However, the input-specificity may be only relatively conserved during L-LTP.

A potential central difficulty for the requirement of protein synthesis during L-LTP is raised by the assumption of synaptic input-specificity. How does the neuron achieve a restricted transport of newly synthesized proteins from the soma to the activated synapse? Preliminary results suggest a plausible account.

Recent findings permit a possible resolution of the apparent inconsistency between the input-specificity of L-LTP and the more general effects of protein synthesis. Commonly it is assumed that LTP is purely a local synaptic phenomenon, evidenced by its input-specificity. From this perspective, it was considered unlikely that a message is first sent to the nucleus of the cell, turning on or up-regulating a gene or genes, and then new macromolecules, targeted to modify a few stimulated spines, are sent to a distant dendritic compartment. Recently, an activity-regulated gene (arg3.1) was identified in response to synaptic plasticity (Link et al, 1995). Synaptic activity of the sort initiated by stimuli that induce L-LTP dramatically enriches mRNA levels of arg3.1 in the dendritic laminae. This enrichment, in conjunction with local protein synthesis at activated dendritic spines, provides a mechanism to modify stimulated synapses specifically. Other recent work demonstrates that protein synthesis also takes place within dendrites, where it is stimulated by NMDA-receptor-dependent synaptic activity (Feig & Lipton, 1993; Steward & Wallace, 1995). The differential localization of arg3.1 mRNA to dendrites provides the basis for local synthesis of the protein at activated postsynaptic sites. The restricted transport could elevate neuronal output, but in a manner that conserves synaptic input-specificity during L-LTP. In addition, the time course of potentials in a second non-tetanized pathway revealed a protein-synthesis-dependent elevation of neuronal excitability during L-LTP (Frey et al, 1988).

The third basic property of LTP is synaptic *cooperativity*. To initiate the protein synthesis upon which L-LTP depends, several prerequisites must be met. First, the LTP-inducing stimulus must reach a distinct intensity and be applied repeatedly. The cAMP/PKA cascade necessary for L-LTP must be activated by stimulation patterns that induce LTP. The transient up-regulation of the cascade might be achieved, as previously noted, by synergistic activation of NMDA and dopamine receptors in the CA1 region (Eliot et al, 1989; Chetkovich et al, 1991; Chetkovich & Sweatt, 1993; Frey et al, 1993a; Huang & Kandel, 1995) or norepinephrine receptors in the dentate gyrus (Stanton & Sarvey, 1985a). If the strong tetanization needed for L-LTP activates multiple neurotransmitter systems, then prolonged LTP might be characterized as a "multisynaptic" rather than a homosynaptic phenomenon. In that case, L-LTP would require two different forms of cooperativity for its induction: *cooperativity A*, involving strong depolarization of the postsynaptic cell to produce Ca^{2+} influx through NMDA-receptor-gated channels; and *cooperativity B*, involving activation of a different heterosynaptic input. One candidate for the heterosynaptic input might be dopamine for the CA1, which could lead to the activation of an additional second messenger, such as PKA, that is specifically required for processes involved in L-LTP (**Figure 4**). The two different cooperative properties of L-LTP might indicate a new principle of intercellular integration in the central nervous system. As is known from learning experiments, reinforcers are essential for the formation of prolonged memory traces.

Thus cooperativity B might implement the reinforcing action of aminergic systems at the cellular level.

We turn now to the fourth property of LTP—*occlusion*. Very recently it has been demonstrated that occlusion may not occur with L-LTP (Frey et al, 1995b). The widespread view was that asymptotic LTP, once established, prevented further potentiation, and that repeated tetanization did not prolong established L-LTP (de Jonge & Racine, 1985). The functional implication of this account was that potentiated neurons lose their capacity for further long-lasting facilitatory plastic changes in an activated input after the establishment of synaptic LTP. Thus synapses expressing LTP were not available for further processing of new incoming signals for a substantial period of time.

We have shown that hippocampal LTP precludes the induction of subsequent LTP only during E-LTP. Four hours after LTP induction, however, a newly delivered conditioning stimulus during maintained initial LTP produces a new potentiation of the earlier-activated synaptic input. Thus, established LTP is not a final state for the neuronal population, but preserves the capacity to react to incoming signals with further plastic changes. Since E-LTP is still maintained by the neuronal population, the neurons appear to reach a new level in responsiveness but are still available for further long-term processing of afferent signals. These findings suggest a new principle for signal processing during LTP. Although the cellular mechanisms responsible for occlusion during E-LTP and additional plastic changes during L-LTP are incompletely known, mechanisms specific to E-LTP could be responsible for the prevention of additional potentiation.

Perhaps intracellular cascades, which are necessary for the initiation of LTP, must become available again to generate new LTP. During E-LTP, a short-term potentiation was described that could reflect a NMDA-receptor calmodulin-dependent stage. Since the intracellular targets for Ca^{2+} and other agents may still be engaged with processes triggered by the initial potentiation, a second conditioned stimulus during this stage may fail to induce LTP. However, during L-LTP, when new macromolecules may be synthesized and processed, distinct mechanisms, such as transformed "silent" NMDA receptors that were activated during E-LTP, may return to pre-potentiation levels. These newly functional receptors would become available to new incoming signals, such as increased Ca^{2+} levels enabled by NMDA-receptor activation during a second conditioned stimulus.

Finally, very recent observations in my laboratory provide evidence for analog, long-lasting plastic events whose time course is similar to conventional, glutamatergic L-LTP. An increase in protease activity has been detected in perfusates from the dentate gyrus following potentiation (Fazeli, Errington, Dolphin, & Bliss, 1990). Also, the expression of the gene for the extracellular serine protease, tPA, is induced during LTP (Qian et al, 1993). Proteases may cleave the proteins required for the structural remodeling of dendrites observed during LTP (for reviews, see Baudry, 1986; Greenough, Armstrong, Comery, Hawrylak, Humphreys, Kleim, Swain, & Wang, 1994).

We have now investigated the establishment and maintenance of hippocampal LTP in mice genetically engineered to be deficient for tPA (Carmeliet, Schoonjans, Kieckens, Ream, Degen, Bronson, De Vos, Van den Oord, Collen, & Mulligan, 1994). Our results indicate that tPA-deficient mice develop a potentiation whose time course resembles that of conventional L-LTP. However, this increase in synaptic efficacy is distinct from conventional L-LTP and represents a novel form of *HE*terosynaptic *L*ong-lasting *P*otentiation (HELP; Frey, Müller, & Kuhl, 1996). Application of the $GABA_A$-receptor inhibitor bicuculline reveals a larger facilitation during normal synaptic transmission in mutant than in wild-type mice. LTP induction without GABA blockade and subsequent application of the $GABA_A$-receptor inhibitor prevented additional facilitation of synaptic transmission. This result indicates a reduced inhibition during LTP in mutant animals. Furthermore, continuous blockade of the inhibitory system by picrotoxin before and after induction of LTP inhibited the maintenance of potentiation in the mutant animal, whereas conventional L-LTP remained unaffected in the wild-type animal.

We therefore conclude that long-lasting potentiation in mutant mice in the absence of picrotoxin is maintained solely by GABAergic mechanisms. Our results indicate that tPA-deficient mice completely lack conventional L-LTP. HELP, however, does not require the expression of tPA and is mediated by NMDA-receptor-dependent modification of GABAergic transmission. This form of potentiation provides tPA-deficient mice with a CA1 output that is identical to that seen in wild-type mice during conventional L-LTP. Therefore, HELP might functionally compensate for L-LTP, and could explain why spatial-memory formation is unaffected in these mutant mice (Lipp, Wolfer, Bozizevic, Carmeliet, Collen, & Mulligan, 1993). This novel cellular plastic process might only occur in tPA-deficient mice. Alternatively, HELP may accompany conventional LTP and produce results that are indistinguishable from "normal" LTP. In this case, a neuronal population would employ two distinct mechanisms subserving a prolonged increase in synaptic efficacy. Heterosynaptic changes accompanying L-LTP but not E-LTP have also been detected in the dentate gyrus (Abraham, Bliss, & Goddard, 1985) and, as already noted, in a second non-tetanized pathway of the CA1 region (Frey et al, 1988; Reymann et al, 1988c). The use of tPA-deficient mice may provide a useful tool to further characterize the physiological consequences of, at least, HELP and conventional LTP.

Conclusion and Interpretations

LTP, at least in the hippocampus, is a non-unitary phenomenon in its time course and in its underlying cellular and intercellular processes. LTP is characterized by phases similar to those thought to be involved in the formation of memory. The early phase (E-LTP) lasts up to 4 hours and does not depend on intact protein and, probably, RNA synthesis. A later phase (L-LTP) requires the cooperative activation of two different neurotransmitter systems during LTP generation and the subsequent synthesis of mRNA and macromolecules. The cellular mechanisms responsible for the maintenance of L-LTP are not yet fully known, but dopamine-mediated processes appear to be involved. These processes control the post-translational fucosylation of proteins that are necessary for the formation of LTM and L-LTP. Newly synthesized macromolecules are transformed and constitutively active kinases target proteins to activated synaptic sites.

The induction of LTP requires a critical level of stimulation to activate the various mechanisms responsible for the determination of LTP duration. Thus, LTP requires an associative and cooperative induction, and—for the initiation of processes resulting in L-LTP—an additional, cellular induction signal. As shown, the induction signal may be the activation of cAMP/PKA through the dopaminergic D_1 receptor synergistically or cooperatively with the elevation of intracellular Ca^{2+} via the NMDA receptor.

New methods must be used to study cooperativity in L-LTP. Induction of L-LTP requires strong, repeated tetani but the electrical stimulation used in neurobiological experiments activates a relatively large number of afferent fibers. Therefore, fibers of different neurotransmitter systems, such as dopamine in the CA1, may be activated. It has not yet been verified if the coincident activation of multiple transmitters also occurs naturally when L-LTP is produced. Dopamine has been shown to play the role of a biological reinforcer during learning (Beninger, 1983). Since two different neurotransmitter systems are required for L-LTP, it was suggested that hippocampal LTP should be described as a "multisynaptic" phenomenon, at least with respect to the development of its late phases.

The observation that all of the processes required for the induction of L-LTP are initiated during or immediately after LTP generation indicates that E-LTP and L-LTP are not independent processes. They require similar, convergent intracellular mechanisms but the development of L-LTP involves the participation of additional processes. First, two different afferent-fiber systems must be active at the same time and, second, the activated postsynaptic neuron must be in an initial plastic state to react to incoming signals with prolonged plastic changes. The capacity of a neuron to react with further plastic changes returns only during L-LTP.

The characteristics of LTP and its underlying mechanisms are consistent with the view that hippocampal LTP may provide a primary model for the investigation of the processes underlying learning and memory formation. Nevertheless, L-LTP is a very complex phenomenon requiring additional clarification. The locus of plastic changes is, according to Konorski (1948) and Hebb (1949), the synaptic junction. However, recent results indicate that more than just the activation of a synaptic junction is involved in transforming and integrating incoming signals to a specific neuron into relatively permanent plastic changes at the synapse. Furthermore, as demonstrated in tPA-deficient mice, different interventions at the same neuron can produce the same neuronal outcome. Thus, the functional connectivity of neurons can be modulated not only by regulation of the synaptic efficacy of glutamatergic inputs, but also by a modification of inhibition. Therefore, conventional glutamatergic LTP may represent the cellular integration of input-specific afferent information. However, forms of potentiation such as HELP may provide a change in the output of the cell without a differentiated evaluation of the afferent signal. These processes may have very substantial implications for the functioning of a single neuron within a given neuronal network.

From an evolutionary perspective, LTP may represent a general neuronal property whose full expression emerges only gradually. LTP-like phenomena have been described in invertebrates (for reviews, see Castellucci, Frost, Goelet, Montarolo, Schacher, Morgan, Blumenfeld, & Kandel, 1986; Goelet et al, 1986) and for structures in the spinal cord (Pockett & Figurov, 1993). The full significance of long-lasting neuronal changes can only be understood within the anatomical context (network architecture) and environmental demands in which they appear. Nevertheless, the investigation of plastic events occurring in single neurons will sharpen our understanding of their contribution to the integrated action of the entire network.

ACKNOWLEDGMENTS

I am very grateful to Hansjürgen Matthies, Klaus G. Reymann, John Donahoe, Vivian Dorsel, and Ritchie Brown for their helpful comments on the manuscript. Portions of the recent experiments reported here were supported by the German BMBF, FKZ: 310258A.

Neural-Networks Models of Cognition
J.W. Donahoe and V. Packard Dorsel (Editors)

CHAPTER 7

TEMPORAL INFORMATION PROCESSING: A COMPUTATIONAL ROLE FOR PAIRED-PULSE FACILITATION AND SLOW INHIBITION

Dean V. Buonomano and Michael M. Merzenich
Keck Center for Integrative Neuroscience
University of California, San Francisco

ABSTRACT

The brain receives a continuous barrage of sensory input from the environment. These inputs are dynamic and constantly changing in space and time. The brain must make decisions based on both the spatial and temporal structure of sensory stimuli. The neural basis of most forms of temporal information processing are poorly understood. This is particularly true for the processing of complex temporal patterns in the range of tens to hundreds of milliseconds. We have proposed that a series of well characterized neuronal properties permit circuits of neurons to transform temporal information into a spatial code. Neurons exhibit a wide range of properties in addition to postsynaptic potential (PSP) summation and spike generation. Neuronal properties such as paired-pulse facilitation (PPF) and slow PSPs are well characterized, but their role in information processing remains unclear. We have developed a continuous-time neural-network model whose elements incorporate PPF and slow inhibitory postsynaptic potentials (IPSPs), that examines whether such networks can discriminate time-varying stimuli. The time constants of the PPF and IPSPs were estimated from empirical data, and were identical and constant for all elements in the circuit. By incorporating these elements into a circuit inspired by neocortical connectivity, we demonstrate that the network is able to discriminate different temporal patterns. Generalization emerges spontaneously. Our results demonstrate that known time-dependent neuronal properties enable a network to transform temporal information into a spatial code in a self-organizing manner, i.e., with no need to assume a spectrum of time delays or to custom design the circuit.

Introduction

The nervous system faces the challenging task of making sense of the "blooming, buzzing confusion" (James, 1890) of our sensory environment. The nervous system receives a continuous barrage of inputs and must make decisions in real time. Most real-world stimuli impinge on peripheral sensory layers and produce spatio-temporal patterns of activity. Depending on the nature of the stimuli, the nervous system uses information contained in the spatial, temporal or spatio-temporal pattern of activity of the sensory afferents.

Spatial information

Spatial information refers to information encoded in the pattern of sensory afferents activated independently of the temporal structure of the activation. In the sensory modality, deciding whether a stimulus was applied to the forefinger or ring finger is a simple spatial task. Much more sophisticated examples of spatial tasks are character recognition and vowel perception. Different letters activate different patterns of photoreceptors, and the neocortex must make decisions based on these patterns to interpret them. Similarly, there is relatively little temporal structure in vowels, which can be discriminated on the basis of the frequency of their formants and, more specifically, the relationship between their frequencies.

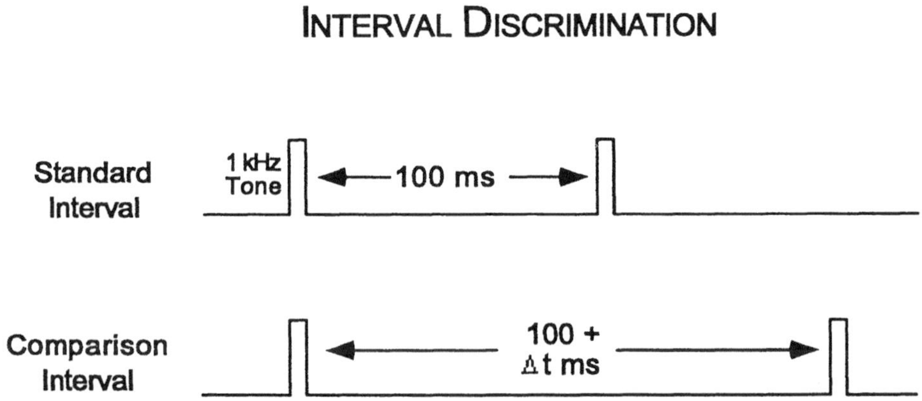

FIGURE 1. Interval discrimination. In a typical interval-discrimination task a subject listens to two tones (pips) separated by a standard interval. A second comparison stimulus is identical to the standard stimulus except that the interval between the pips differs from the standard interval by Δt. The subject makes a forced choice as to whether the first or second interval was the longest. The value of Δt may be varied to determine the psychophysical threshold for interval discrimination.

Temporal information

In a temporal task, information is encoded in the temporal structure of the stimulus. Psychophysical studies typically use interval- or duration-discrimination tasks to analyze temporal information processing. In an interval-discrimination task a subject listens to two brief auditory tones (pips) separated by a standard interval, and a second pair of pips separated by a comparison interval (**Figure 1**). The subject is required to decide which was the longer interval.

The analysis of this task is complicated by the fact that each pip should activate the same primary auditory afferents. That is, at the level of the primary sensory sheath, the short and long stimulus are identical in terms of the spatial pattern of activation of the primary afferent. To discriminate the short and long intervals, the nervous system must transform temporal information into a spatial code.

Most real-world stimuli are not purely spatial or temporal tasks, but spatio-temporal. The nervous system must use both spatial and temporal cues for complex sensory tasks such as speech perception, music perception, and motion processing (e.g., Gibbon & Allan, 1984; Tallal, Galaburda, Llinas, & von Euler, 1993).

In trying to understand how the nervous system performs complex signal processing, we believe that it is useful to discriminate between spatial and temporal processing. Although many stimuli contain information in both the spatial and temporal domains, the neural mechanisms underlying both forms of processing may be distinct, although not necessarily segregated into different parts or circuits in the nervous system. If we consider a position-discrimination task, it is clear that there is a peripheral representation of the stimuli; i.e., distinct groups of neurons are activated at different locations on the sensory surface. In the interval-discrimination task, however, the stimuli that define both the short and long intervals can only be discriminated on the basis of the time between the two pips. In order to solve this task, the nervous system must convert temporal information into a spatial code; i.e., at some level of the nervous system different populations of neurons must be activated during the different intervals. The neural mechanisms underlying temporal information processing are unknown.

Time-Dependent Neuronal Properties

To date, our understanding of how neurons perform computations is based on theoretical models in which interconnected units take the weighted sum of their inputs and generate their outputs via an activation function (cf. Anderson & Rosenfeld, 1988). These elements are meant to represent the summation of fast excitatory and inhibitory PSPs (EPSPs and IPSPs, respectively) to produce spike generation. Indeed, these models have been effective in performing complex computations and have provided many insights into how the nervous system processes information. However, neurons exhibit many additional properties such as paired-pulse facilitation (Zucker, 1993; Clark, Randall, & Collingridge, 1994), paired-pulse depression (Deisz & Prince, 1989; Nathan & Lambert, 1991; Fukuda, Mody, & Prince, 1993), slow $GABA_B$-mediated IPSPs (McCormick, 1989; Douglas & Martin, 1991), rebound facilitation (Landry, Wilson, & Kitai, 1984; Llinas & Muhlethaler, 1988; Buhl, Halasy, & Somogyi, 1994), spike accommodation (Madison & Nicoll, 1986), and voltage-dependent excitatory currents (Clark et al, 1994).

Paired-Pulse Facilitation

Paired-pulse facilitation (PPF) refers to a short-term form of homosynaptic facilitation in which the second of a pair of action potentials produces a larger EPSP as compared to the first. Homosynaptic PPF is thought to be due to residual Ca^{2+} in the presynaptic terminal (Zucker, 1993; Wu & Saggau, 1994). In general the magnitude of facilitation varies from 50 to 100%, with maximal facilitation occurring at approximately 50 ms. A second phenomenon known as paired-pulse depression (PPD) of IPSPs can also result in facilitation of EPSPs (Nathan and Lambert, 1991; Metherate & Ashe, 1994). PPD relies on the activation of presynaptic $GABA_B$ receptors and tends to peak at 100-300 ms (Davies et al, 1990; Fukuda et al, 1993).

Slow IPSPs

The inhibitory neurotransmitter GABA mediates most inhibition in the nervous system. The $GABA_A$ receptor mediates a ligand-gated fast IPSP; the $GABA_B$ receptor mediates a second messenger-gated slow IPSP. The slow IPSPs typically peak at 100-200 ms and last 500-1000 ms (Hablitz & Thalmann, 1987; McCormick, 1989).

To date few neural-network models have incorporated either of these properties, in part because their role in information processing is unclear. One possibility is that they contribute to the processing of temporal information in the range of tens to hundreds of milliseconds. Consider the interval-discrimination task: The first pip produces a pattern of activity in the network and triggers a series of time-dependent properties. When a second identical pip arrives 100 ms later it essentially arrives in a different cortical environment. Accordingly, as a result of PPF, some synapses will be facilitated and others will be inhibited. Thus the same input may produce a different pattern of activation in the network. These differences could be used to encode temporal information.

A Model of Temporal Information Processing

As a step towards addressing the question of whether time-dependent neuronal properties may underlie temporal processing, we used a neural network composed of integrate-and-fire elements that incorporated PPF and slow IPSPs in addition to fast EPSPs and IPSPs. We focused on PPF and slow IPSPs because they have been described in some detail in cortical neurons (Creager, Dunwiddie, & Lynch, 1980; McCormick, 1989; Nathan & Lambert, 1991) and can be incorporated efficiently into integrate-and-fire units. Furthermore, PPF may be particularly relevant to temporal processing because the amplitudes of EPSPs provide temporal information about recent spike occurrence.

Integrate-and-fire elements

In our simulation, we used integrate-and-fire elements similar to those used in some previous models (Wörgötter & Koch, 1991; Buonomano & Mauk,

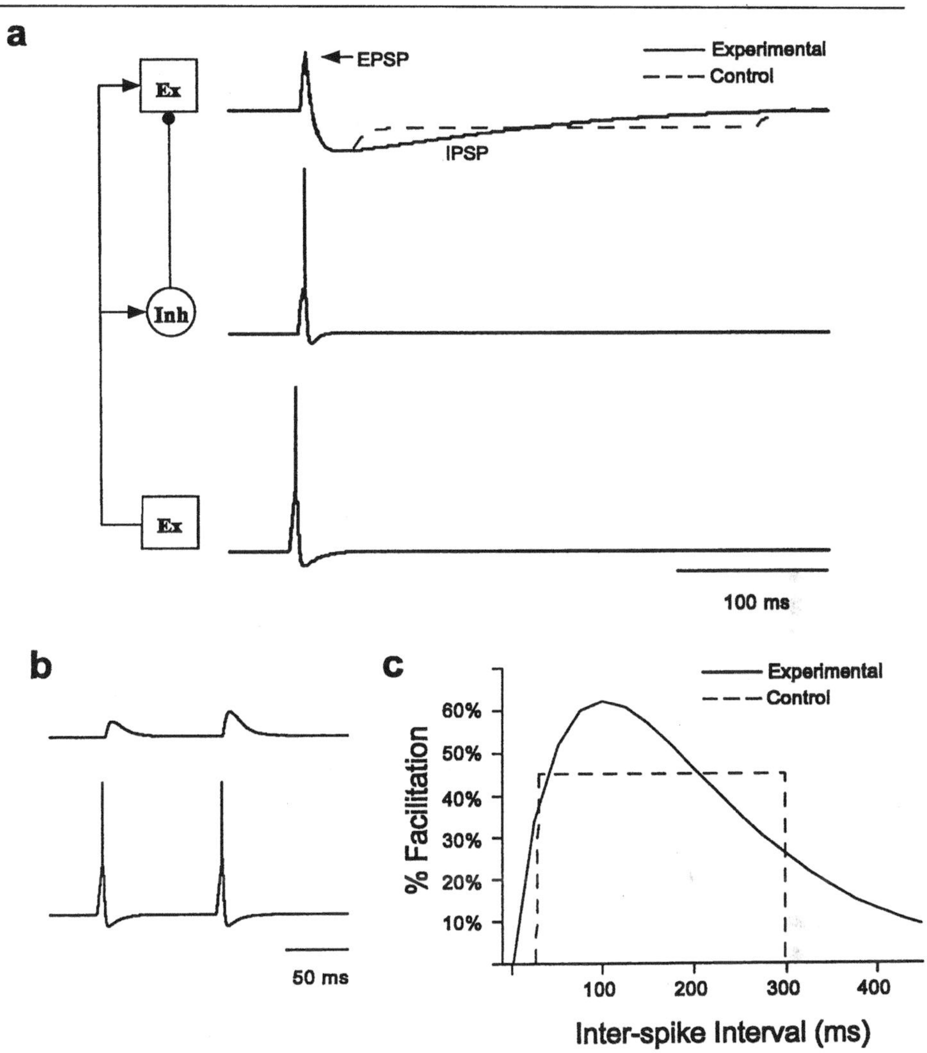

FIGURE 2. Integrate-and-fire elements that incorporate slow IPSPs and PPF. Traces represent the voltages of the simulated integrate-and-fire elements. **a.** Slow IPSP. By triggering a spike in the lower excitatory element (Ex) a suprathreshold EPSP is elicited in the inhibitory element (Inh), resulting in a fast EPSP followed by a slow IPSP in the upper Ex unit. The time constant of the slow IPSP is 80 ms. For illustrative purposes the connection strengths were increased for this figure. **b.** PPF. The second of two consecutive spikes in an Ex element produces a larger EPSP in the postsynaptic unit. **c.** PPF function. The PPF function was simulated with an alpha function peaking at 100 ms. For control experiments, the time-varying profile of the slow IPSP and PPF were transformed into step functions from 30-300 ms, dashed lines in a and c.

1994). Each element was composed of a single compartment with a membrane time constant. All synaptic currents were summed and, if the voltage of the compartment reached threshold, a spike was generated. Both excitatory (Ex) and inhibitory (Inh) cells produced postsynaptic currents that were described by an instantaneous rise and an exponential decay. The $GABA_A$ and $GABA_B$ fast and slow IPSP components were modeled as a single current with a slow decay (**Figure 2**). PPF of EPSPs was simulated with an alpha function (**Figure 2a**), with a peak facilitation of 60% occurring at 100 ms.

Network architecture

The excitatory and inhibitory elements were incorporated into a randomly connected circuit representing cortical layers IV and III (Douglas & Martin, 1989). The network was composed of 100 inputs, and 150 and 250 elements in layers 4 and 3, respectively. The Ex units projected forward to both the Ex and Inh units in the next layer (**Figure 3**). In keeping with experimental observations, 20% of the elements in each layer were inhibitory and 15-20% of the connections onto each element type were inhibitory.

Simulations of Temporal Processing

Interval discrimination

The simplest task studied was interval discrimination: Two pulses were presented on the same input channels with different intervals between them. Each pulse simulated a brief stimulus such as a tap, tone or flash. The first stimulus pulse initiated a set of excitatory and inhibitory interactions in the network. Due to time-dependent changes imposed by PPF and slow IPSPs, the network was in a different state upon the arrival of the second pulse. Thus, even if the second pulse was identical to the first, some units would have different probabilities of firing depending on the inter-pulse interval. These units could be used to encode temporal information. To demonstrate this, we added an output layer to the network and trained it to recognize interval-specific patterns produced in layer 3 by five different stimulus intervals (80, 130, 180, 230 and 280 ms). All excitatory units in layer 3 (Ex3 units) were connected to a number of output units equal to the number of stimuli being discriminated. A supervised-learning rule was used to train each output unit to respond to a given stimulus (Buonomano & Merzenich, 1995). The output layer and the supervised-learning rule are not meant to be part of a realistic simulation of temporal processing. However, they do provide a means for determining whether the activity pattern in the network is competent to discriminate between different intervals. With the exception of the changes of connection weights between the Ex3 and output units during training, there was no plasticity in the connection weights or time constants at any level of the network. If the output units were able to discriminate between the intervals, it could only be as an emergent product of interactions between the units. Indeed, after train-

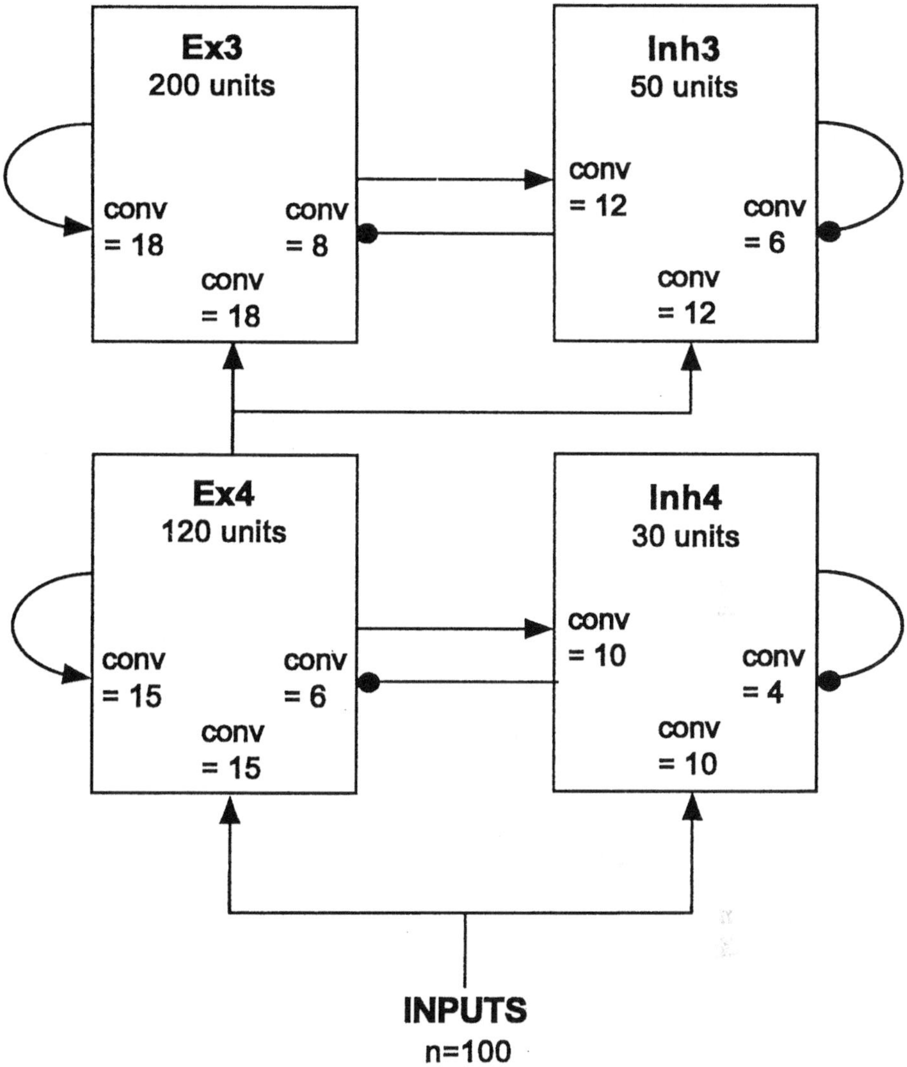

FIGURE 3. Network circuit and architecture. Integrate-and-fire elements were incorporated into a two-layer network representing neocortical layers IV and III. Each layer of the network contained excitatory (Ex4 and Ex3) and inhibitory (Inh4 and Inh3) units. Each box represents a population of elements of a given cell type. Each cell type received inputs from three other cell types. The convergence (conv) indicates the number of presynaptic inputs each element received. The connection probability between any two cell types was determined by the convergence value divided by the total number of presynaptic elements. (See text for details.)

ing, output units did spike in response to the appropriate interval, demonstrating that activity patterns produced in the Ex3 units contained sufficient information to code the temporal intervals (**Figure 4**).

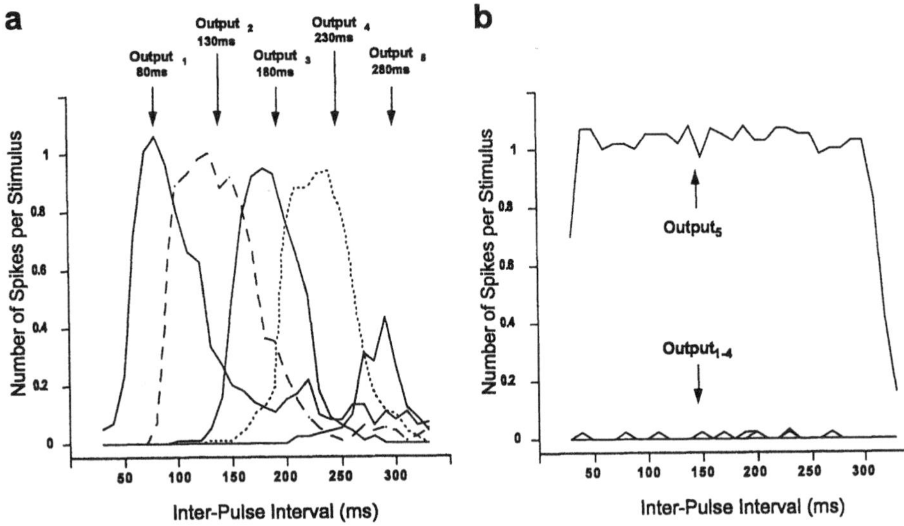

FIGURE 4. a. Interval tuning curves. A supervised-learning rule was used to train each of five output units to respond to one of five intervals (80, 130, 180, 230, 280). Each output element received inputs from all Ex3 units. In this manner, the response of each output element corresponded to the weighted average of a population of interval-sensitive Ex3 elements. Interval tuning curves were constructed from 100 presentations of each interval. Generalization was shown by the smooth tuning curves around each target interval. (See text for additional information about the tuning curve for output #5.) **b.** Control simulation. When the time course of the slow IPSPs and the PPF were described by step functions (see dashed lines in **Figure 2**), the network was unable to discriminate any of the intervals.

Temporal generalization

In addition to performing temporal discriminations, a biologically plausible model must exhibit generalization along a temporal dimension. The network was therefore tested with intervals varying from 30 to 330 ms, and interval tuning curves were constructed for the output units (**Figure 4a**). Although each output unit had been trained to respond to only one of five stimuli, each exhibited a tuning curve centered around its trained interval. Note that, with these tuning curves, the network forms a population code by which it can represent any interval between 30 and 300 ms. (Output unit #5, trained at the 280-ms interval, demonstrated a significantly worse tuning curve because a 280-ms interval approached the limit of the time constants of the network.) To demonstrate that the ability of the network to perform temporal discriminations was a result of the time dependency of the PPF and the slow IPSPs, we performed a control experiment in which PPF and slow IPSPs followed a step function (dashed lines in **Figure 2**). In these control simulations, output units were unable to discriminate among any of the trained intervals and exhibited flat tuning curves (**Figure 4b**).

Temporal-pattern discrimination

Interval discrimination is a simple temporal task that could be solved using delay lines. Discriminating sequences of inputs is a more complex task. In such a task, it is necessary to keep track of more than one interval, a difficult task for models that rely on delay lines. To examine the ability of the network to solve a more complex temporal task, we trained the network to discriminate among four different stimuli. Each stimulus consisted of four pulses with

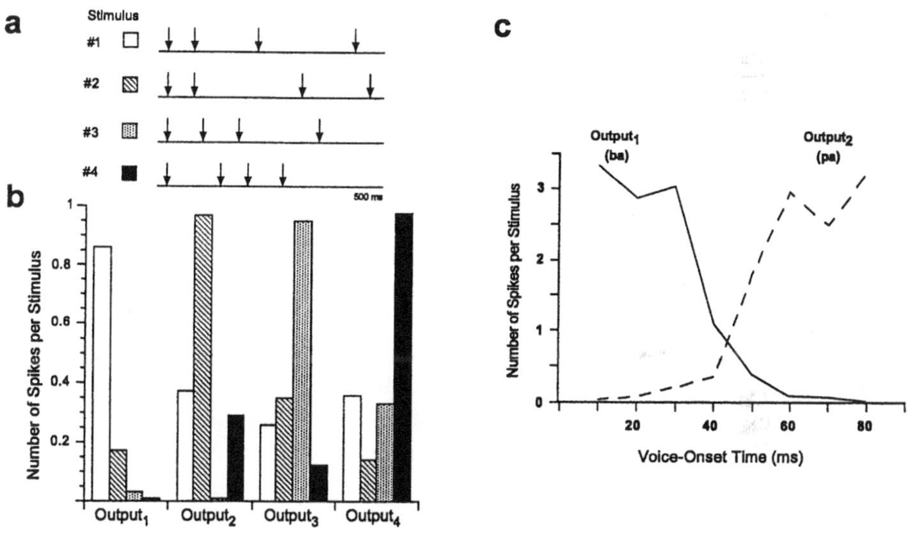

FIGURE 5. Random-sequence discrimination. **a.** Four different stimuli were used during training. Each stimulus consisted of four 5-ms pulses, with the intervals between pulses randomly selected between 50 and 250 ms. **b.** Each of four output units was trained to respond to one of the four stimulus sequences. After training, each output element responded preferentially, but not exclusively, to its training sequence. **c.** Phoneme discrimination. The two output elements were trained to discriminate 10- and 20-ms VOTs (/ba/) from 70- and 80-ms VOTs (/pa/). The network was then tested with intermediate VOTs. Tuning curves were constructed from 100 presentations of each stimulus.

randomly assigned inter-pulse intervals (**Figure 5a**). After training, we found that each output unit responded preferentially to the pattern of Ex3-unit activity that was elicited by the appropriate intervals (**Figure 5b**).

Phoneme discrimination

Speech perception is a preeminent example of a task that relies on temporal cues. An important cue for discriminating voiced and unvoiced phonemes (e.g., /ba/ and /pa/) is the voice-onset time (VOT). The VOT is the time between air release and vocal-cord vibration. "Ba" tends to have VOTs below 30 ms whereas "pa" has VOTs above 30 ms (Wood, 1976). A network with two outputs was first trained to discriminate /ba/ and /pa/ by training with the two shortest (10 and 20 ms) and the two longest (70 and 80 ms) VOTs. The network was then tested with intermediate values of simulated VOTs. After training, output units responded best to the VOT values on which they had been trained, and response magnitude decreased in a monotonic fashion as a function of the VOT (**Figure 5c**). Response curves were qualitatively similar to those observed in psychophysical studies of VOT discrimination (Wood, 1976).

Conclusions

Plasticity

We have shown that, using elements with realistic neuronal properties, temporal processing emerges as a result of state-dependent changes imposed on network dynamics. Without the need to change the parameters of any units, the network was able to perform interval, pattern, and phoneme discrimination. Further, generalization occurred to similar temporal patterns. We expect that increasing the complexity of the elements, through incorporating other neuronal properties and permitting synaptic plasticity, will further improve performance. A common form of synaptic plasticity, referred to as Hebbian association, assumes that synaptic strength increases if both the pre- and postsynaptic elements are coactive. However, our simulation experiments that incorporated Hebbian plasticity revealed difficulties in employing Hebb's rule with continuous-time networks having time-varying inputs. Hebbian association leads units to respond to the most frequent synchronous input patterns, which in the above tasks corresponded to the first pulse. And, of course, the first pulse was common to all of the intervals. Thus, the network became more responsive to the first pulse and not to the later pulses that contained the critical temporal information. Hebb's rule is well suited to reinforce simultaneous activity coming from spatially distinct inputs, but much remains to be learned about the temporal parameters that affect Hebbian plasticity. Furthermore, it is likely that more sophisticated learning rules that accommodate the role of both excitation and inhibition may be essential for learning temporal patterns.

Previous models

In work on temporal discrimination within psychology, the most prevalent accounts of temporal processing are internal-clock theories (Church, 1984; Macar, 1985). Internal clocks are hypothetical mechanisms in which a neural pacemaker generates pulses, with the number of pulses relating to a physical time interval recorded by some sort of counter. Internal-clock models have been used in a large body of research on human temporal psychophysics. To our knowledge, there are no physiological data that support the existence of pacemakers and counters processing temporal information on the order of hundreds of milliseconds.

More neurobiologically oriented models have often been based on the existence of delay lines or elements with a spectrum of different time constants (Braintenberg, 1967; Tank & Hopfield, 1987; Grossberg & Schmajuk, 1989), or thresholds (Antón, Lynch, & Granger, 1991). Indeed, the nervous system does use delay lines for the detection of interaural delays used for sound localization. However, these delays are on the order of a few hundred microseconds. For complex temporal stimuli on the order of tens to hundreds of milliseconds, delay-line models require an *ad hoc* architecture and lack biological support. Our simulations suggest that known time-dependent neuronal properties (not limited to PPF and slow IPSPs) with fixed and equal time constants permit a randomly connected network to transform temporal information into a spatial (place) code. This transformation occurs at each layer of the network, and may thus be amplified throughout the layers. The general conclusion that arises from our work is that temporal combination-sensitive neurons (e.g., Margoliash, 1983; Margoliash & Fortune, 1992) may emerge as a result of time-dependent changes in network state. That is, if stimulus *A* is presented to an animal followed by stimulus *B*, *A* produces a change in the state of the cortical network as a result of time-dependent neuronal properties. Then, stimulus *B* produces a pattern of activity that codes for *B* preceded by *A*, rather than for *B* alone. On this account, a place code develops for temporal relations between stimuli without the need to postulate any explicit timing mechanisms.

ACKNOWLEDGMENTS

This work was supported by National Institutes of Health Fellowship F32 MH-10431, Grant NS-10414, and Office of Naval Research Contract N00014-96-1-0206. Correspondence should be addressed to the senior author.

PART THREE: PERCEIVING

In the natural environment, complex *combinations* of stimuli guide behavior. Friends are recognizable through multiple stimuli—shape of eyes, color of hair, sound of voice, and so on. What biobehavioral processes mediate perceiving complex stimuli, and how may those processes be simulated?

Studies of the neural mechanisms of perceiving have identified a number of general characteristics of the perceptual systems of mammals: (1) Sensory processing begins with physical energies that stimulate appropriate receptors that then activate sensory pathways leading to thalamic nuclei that, in turn, are the origins of pathways leading to sensory areas of the cerebral cortex. Using the visual system as an example, photoreceptors activate ganglion cells in the retina that give rise to fibers that innervate the lateral geniculate nucleus (LGN). Fibers from the LGN then innervate cortical neurons in the primary visual area (V1) of the occipital cortex. (2) Neurons in primary sensory areas send efferents to neurons in sensory *association* cortex, with successive neurons responding most strongly to ever more complex combinations of stimuli originating from ever larger regions of the environment. Referring again to the visual system, cells in V1 respond to a particular orientation of line only when the line occurs in a small region of the visual field (i.e., V1 neurons have small receptive fields), whereas cells in visual association areas respond most strongly to certain combinations of lines and have large receptive fields (see **Tanaka**, this volume). (3) These feedforward pathways from receptor to sensory association cortex are complemented by recurrent pathways (see Chapter 1). For example, in the visual system, recurrent pathways extend from V1 to LGN, from deeper to more superficial sensory association areas, and from motor association to sensory association areas (see **Singer**, this volume). Of course, each sensory system has modality-specific characteristics as well. Again with reference to the visual system, V1 gives rise to two initially largely independent sensory association "streams"—a dorsal stream to the parietal lobes whose neurons are most strongly activated by movement and location, and a more ventral stream to the temporal lobes whose neurons are most strongly activated by form and color (Mishkin, Ungerleider, & Macko, 1983). (4) Lastly, cells in sensory association cortex that respond to complex combinations of stimuli are broadly tuned; i.e., they respond most strongly to one particular combination of stimuli but to other combinations as well (see **Singer** and **Tanaka**, this volume).

The overall structure of sensory systems (e.g., pathways from thalamic nuclei to cortex) is primarily determined by genetic factors. However, the specific connectivity between neurons within a brain region (e.g., within a thalamic nucleus or within a cortical region) is too detailed to be coded directly given the limited number of genes. Instead, the genes provide an extremely rich *potential* connectivity that is then winnowed to connections between neurons that are frequently coactive during prenatal and postnatal periods (see **Singer**, Burgos, and Luskin, this volume). Moreover, many of the same cellular processes underlie both prenatal and postnatal changes in connectivity; e.g., $Ca2+$ influx via NMDA receptors and second messengers instigated by monoaminergic neuromodulators (see **Singer** and Frey, this volume). Learning is truly postnatal development.

A central problem confronting the analysis of perceiving is how the diverse neural activity initiated by a complex stimulus can be integrated, or "tied together," so that behavior is guided by the concerted activity initiated by the stimulus and not by partial combinations or components alone. This is known as the *binding problem*, and a number of nonmutually exclusive mechanisms have been proposed whereby it may be addressed. Among these are the following: (1) Since all aspects of a stimulus initiate neural activity at approximately the same time (e.g., the neural activity produced by the form and location of the same object), synchronized neural activity may provide a basis upon which neural events in disparate brain regions may be integrated (see **Singer**, this volume; von der Marlsburg, 1985). (2) Neural activity in disparate brain regions may initiate activity in other regions that then activate recurrent pathways which simultaneously affect the disparate regions (see **Singer**, this volume, Edelman, 1987). (3) Biologically important events (e.g., reinforcers such as food) may occur after the diverse neural activity initiated by a complex stimulus, and such events activate neuromodulatory systems that project nonspecifically to large brain regions. The conjunction of the neuromodulator with coactivity between units in these diverse regions may modify the connections in a coordinated fashion (**Singer**, Frey, and Chapter 18, this volume). (4) The nonspecifically projecting output of the hippocampus to sensory-association cortex (Amaral, 1987) may simultaneously affect synaptic efficacies throughout those regions (**Gluck & Myers** and Chapter 18, this volume).

The simulations in Part Three (**Trehub**, **Hummel**, and **Gluck & Myers**) are sensitive to various of these characteristics of perceptual systems and address the binding problem in differing ways. What unites these differing efforts is their treatment of perceiving as the cumulative outcome of the concerted effect of fundamental biobehavioral processes.

Neural-Networks Models of Cognition
J.W. Donahoe and V. Packard Dorsel (Editors)
1997 Elsevier Science B.V.

CHAPTER 8

DEVELOPMENT AND PLASTICITY OF NEOCORTICAL PROCESSING ARCHITECTURES

Wolf Singer
Max-Planck-Institut für Hirnforschung
Frankfurt/Main, Germany

ABSTRACT

One of the basic functions of the cerebral cortex is the analysis and representation of relations among the components of sensory and motor patterns. It is proposed that cortex applies two complementary strategies to cope with the combinatorial problem posed by the astronomical number of possible relations: (1) Analysis and representation of frequently occurring, behaviorally relevant relations by groups of cells with fixed but broadly tuned response properties and (2) dynamic association of these cells into functionally coherent assemblies. Feed-forward connections are thought to be responsible for the specific response properties of neurons and reciprocal, associative connections for dynamic grouping. During development the architectures of both types of connections are susceptible to experience-dependent modifications, but become fixed in the adult. The feed-forward connections also appear to lose much of their functional plasticity, while the synapses of the associative connections retain a high susceptibility to use-dependent modifications. The reduced plasticity of feed-forward connections is probably responsible for the invariance of cognitive categories acquired early in development, while the persistent adaptivity of reciprocal connections is a likely substrate for the ability to generate representations for new perceptual objects and motor patterns throughout life.

Introduction

A fundamental feature of the cerebral cortex is the similarity of its organization across different areas (Szentágothai, 1979; Douglas, Martin, & Whitteridge, 1989). This suggests that the cortex performs computational operations of a general nature that support functions as diverse as perception, motor programming, remembering, planning, language processing, and reasoning.

What exactly these omnipotent processing algorithms are remains elusive, but the wealth of data gathered over the last few decades permits some educated guesses. Analysis of sensory systems suggests that it is one basic function of cortical modules to detect consistent relations among incoming signals, often referred to as features, and to represent such relations by responses of neurons. The iteration of this process is thought to lead eventually to descriptions of the consistent constellations of elementary features that characterize individual perceptual objects. The assumption is that cortical representations of motor programs have a similar format in which descriptions refer to the spatio-temporal relations among activated muscles. Because the number of possible feature constellations that are examined and eventually represented is astronomical, it is essential that cortical processing algorithms be capable of coping with combinatorial problems.

I propose that there are two main strategies. First, hard-wired neurons are used to detect and represent relations that are particularly frequent and important. Second, dynamic grouping mechanisms, allowing for a flexible recombination of responses from hard-wired neurons, enable different, higher-order relations to be analyzed and represented successively within the same hardware. Because most of the data relevant to this context are from the mammalian visual system, the two coding strategies and the associated adaptive mechanisms will be exemplified in this modality.

Two Strategies, Two Classes of Connections

Neurons in the primary visual cortex of mammals evaluate particular spatial and temporal relations among the responses of retinal ganglion cells and represent these relations by their feature-specific responses. Among the extracted features are the location, orientation and polarity of luminance gradients, their direction of motion, their spectral composition, and their interocular disparity, the last reflecting viewing distance. For the extraction of these features, signals of retinal ganglion cells have to be correlated with one another, and this appears to be achieved by selective recombination of inputs, as first proposed by Hubel and Wiesel (Hubel & Wiesel, 1962; Hubel, 1975) and supported by several recent studies (Chapman, Zahs, & Stryker, 1991; Jagadeesh, Wheat, & Ferster, 1993). Thus, in order to detect and to represent the joint firing of ganglion cells responding to the vertical outlines of an object, inputs from vertically oriented rows of ganglion cells are made to converge selectively on individual cortical cells (**Figure 1A**). This strategy of input recombination probably also is used for the evaluation of other relations that are analyzed at the level of primary visual cortex (V1). In all likelihood the same basic operations are iterated in prestriate cortical areas. As suggested by the substantial divergence of projections beyond V1 and by the functional specialization of neurons in prestriate areas, many of these operations appear to be performed in parallel, each of the areas evaluating particular subsets of higher-order relations in feature space (**Figure 1B**; Maunsell, 1995).

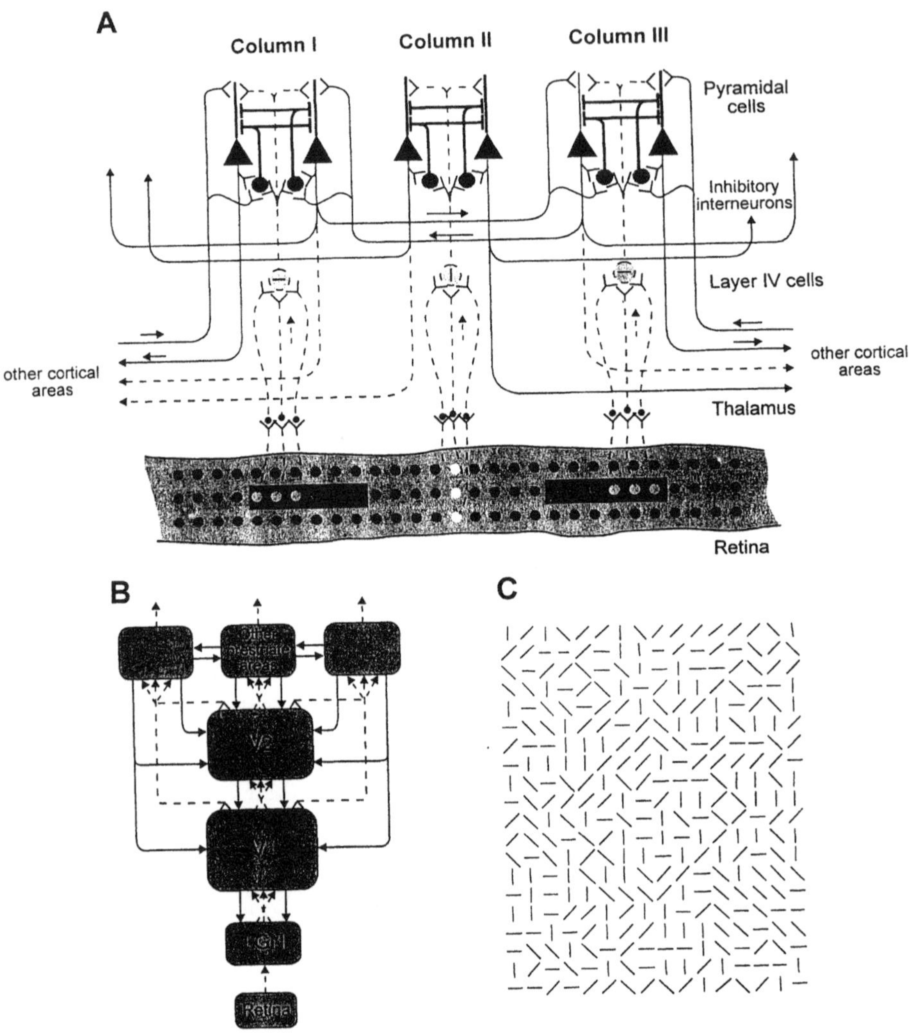

FIGURE 1. Schematic representation of feed-forward, RF-generating (broken lines) and reciprocal, assembly-forming connections (solid lines) within primary visual cortex (**A**) and between different areas of the visual cortex (**B**). Panel **C** shows an example of perceptual grouping on the basis of vicinity and colinearity. The colinearly arranged line segments defining the outlines of a diamond get grouped together and pop out from the randomly distributed line segments of the background. The figure can be segregated from ground due to the enhanced saliency of figure-defining contour elements.

The assumption is that the preference of cortical neurons for particular features results from the specific combination of converging feed-forward connections as exemplified in **Figure 1A** for cells tuned to horizontally (columns I and III) and vertically oriented contours (column II). The output of retinal ganglion cells (lightest circles in retina) that are aligned in horizontal (columns I and III) or vertical rows (column II), respectively, converges after relay in the thalamus onto cortical cells in layer IV. Because of this specific combination of inputs, layer IV cells acquire orientation-selective RFs tuned to vertical and horizontal orientations. The output of the orientation-selective layer IV cells is then relayed further onto pyramidal cells in other cortical layers and these in turn project with feed-forward connections to prestriate visual areas as indicated in **Figure 1B**. Note that this wiring diagram is highly simplified and omits most of the sophistication of intracortical circuitry. As indicated in **Figure 1B**, it is assumed that this strategy of evaluating and representing particular relations among input signals by selective recombination of feed-forward connections is iterated over the subsequent processing stages in prestriate visual areas. The assembly-forming connections (solid lines) are assumed to originate from and to terminate on pyramidal cells, thus assuring reciprocal excitatory interactions. In addition, they terminate on inhibitory interneurons which in turn synapse on pyramidal cells. As indicated in **Figure 1B**, grouping functions are also attributed to the reciprocal connections among cortical areas occupying the same level in the processing hierarchy, and to the back-projections from higher to lower processing stages. The latter are thought to bias grouping as a function of computational results obtained at the respective higher level. The general organization of these ensemble-forming interareal connections resembles that of the intra-areal grouping connections: They originate from pyramidal cells, are excitatory, and terminate on both pyramidal cells and interneurons in the target areas. In the present example, the tangential intra-areal grouping connections (**Figure 1A**) are proposed to link preferentially columns responsive to colinear contours, an architecture for which there is some experimental support (see arrangement of shaded RFs in the retina). Intrinsic connections span larger distances along trajectories corresponding to the location of orientation columns responding to colinear contours (Schmidt, Löwel, Goebel, & Singer, in preparation). The effect of the resulting grouping is that signals evoked by colinear contour borders get selected and bound preferentially for joint evaluation at subsequent processing stages.

Interestingly, however, this strategy of recombining inputs and generating cells with selective response properties is not pursued to exhaustive descriptions, neither of the elementary features represented in V1 nor of the immensely more complex constellations of features of natural objects. At all processing stages, neurons remain broadly tuned to variations of stimulus parameters along different feature dimensions. The responses of individual cells are ambiguous, and a full description of a particular feature or constellation of features can be obtained only by evaluating jointly the graded responses of a population of neurons. At first sight, such *coarse coding* appears uneconomical as it seems to require even more neurons to describe a particular feature or constellation of features than a strategy that uses the responses of individual, sharply tuned cells as descriptors. However, broadly tuned individual cells respond to many different features, so as a consequence, populations coding for different features overlap. Thus, a single cell can participate at different times in the analysis and representation of different features, and this can be exploited to reduce substantially the number of required representational units (Hebb, 1949; Braitenberg, 1978; Grossberg, 1980; von der Malsburg, 1985; Singer, 1985; Edelman, 1987, 1989; Palm, 1990; Abeles, 1991; Gerstein & Gochin, 1992; Young & Yamane, 1992).

The problem with overlapping population codes, however, is that natural visual scenes usually contain many image components that are adjacent or overlapping in Cartesian and in feature space, thus evoking simultaneous responses in overlapping populations of broadly tuned cells. In order to exploit the advantage of population coding, the responses related to a particular feature must be identified and labeled in a way that assures their joint evaluation at subsequent processing stages and prevents false conjunctions with responses evoked by unrelated features.

This requires a dynamic selection process that permits the grouping of distributed neuronal responses in ever-changing constellations. Due to dynamic grouping, signals are selected at one level and reassociated in a flexible and context-dependent way at the next level via the feed-forward connections. This allows for dynamic rerouting of signals within a fixed hardware configuration and circumvents the combinatorial explosion of representational units that would result if every possible feature or constellation of features had to be analyzed by selective recombination of feed-forward connections and represented by sharply tuned neurons. It can therefore be iterated over successive processing stages to analyze and represent in a versatile way relations of ever-increasing complexity, up to the level where the represented relations describe whole perceptual objects.

I propose that a system exploiting this strategy needs two classes of connections: first, feed-forward connections that are responsible for the generation of neurons with feature-selective receptive fields (RFs); and second, reciprocal connections among these neurons that serve to dynamically associate them into assemblies. To economize neurons, the former should generate cells preferring frequently occurring features that are suitable for the definition of perceptual objects. The latter, in contrast, should not contribute to the feature-specific RF structure but should allow for a maximum of combinatorial freedom in associating feature-coding cells. Thus, the constraints for the architecture, the development, and the use-dependent malleability of the two classes of connections are different. Before reviewing evidence on the development and adaptivity of "feed-forward, feature-extracting" and "reciprocal, assembly-forming" connections, the organization and putative mode of action of the latter requires brief discussion. The two classes of connections are depicted in **Figure 1A**.

Strategies for Response Selection

Dynamic selection and association of responses for further joint processing is accomplished best by enhancing their saliency. In principle there are two strategies to raise the saliency of distributed responses. The selected neurons can be made to discharge more vigorously, or they can be made to discharge in precise temporal synchrony. Both mechanisms enhance the impact of the selected responses, the first profiting from temporal and the second from spatial summation of synaptic potentials in the target cells. Grouping through synchronization has the additional advantage that it can operate at a fast time scale because no temporal integration is required and selection can occur at the level of individual action potentials. This permits multiplex grouping operations and may be beneficial when several groups need to be established simultaneously within the same cortical area.

Available evidence suggests that both strategies are used. The discharge rate of cells in V1 can be modified in a context-dependent way by concurrent stimuli presented remote from the classical RF (Nelson, 1985; Morrone, Burr, & Maffei, 1982; Blakemore & Tobin, 1972; Gilbert & Wiesel, 1992), and recently it has been demonstrated that cells in V1 responding to the component features of a perceptual figure respond more vigorously than cells which respond to similar features that are not part of a figure (Lamme, van Dijk, & Spekreijse, 1993; Lamme, 1995). This supports the hypothesis of response selection by modulation of discharge rate. Experiments on response selection by attentional mechanisms also show an enhancement of selected responses (Wurtz, Goldberg, & Robinson, 1980; Moran & Desimone, 1985).

Evidence for synchronization of distributed responses is more recent because it can be obtained only by recording simultaneously from more than one cell. But it is now firmly established that cortical cells can synchronize their discharges with a precision in the range of milliseconds (for reviews, see Singer, 1993; Singer & Gray, 1995). The analysis of correlated firing among simultaneously recorded neurons had initially been used as a tool to investigate neuronal connectivity. Hence, early cross-correlation studies did not consider stimulus- or context-dependent variations in correlation probability. Still, numerous studies revealed correlated firing among spatially distributed cortical neurons (Toyama, Kimura, & Tanaka, 1981a,b; Michalski, Gerstein, Czarkowska, & Tarnecki, 1983; Ts'o, Gilbert, & Wiesel, 1986; Ts'o & Gilbert, 1988; Aiple & Krüger, 1988; Hata, Tsumoto, Sato, Hagihara, & Tamura, 1988; Hata, Tsumoto, Sato, & Tamura, 1991; Gochin, Miller, Gross, & Gerstein, 1991; Schwarz & Bolz, 1991; Roe & Ts'o, 1992).

Later investigations have focused on stimulus-induced correlated firing, emphasizing the dynamic aspect and context dependence of synchronization phenomena. Stimulation-dependent synchronization always occurs with close to zero phase lag, is often associated with oscillatory firing patterns, and is not time locked to the stimulus, the latter indicating that synchrony is generated by neuronal interactions. Such dynamic synchronization has been found between (1) neurons distributed within the same cortical area (Gray & Singer, 1989; Engel, König, Gray, & Singer, 1990; Livingstone, 1991; Eckhorn, Schanze, Brosch, Salem, & Bauer, 1992; Gray & Viana Di Prisco, 1993; Eckhorn, Frien, Bauer, Woelbern, & Kehr, 1993; but see also Gawne & Richmond, 1993); (2) neurons distributed across different areas within the same hemisphere (Eckhorn, Bauer, Jordan, Brosch, Kruse, Munk, & Reitboek, 1988; Engel, Kreiter, König, & Singer, 1991; Bullier, Munk, & Nowak, 1992; Nowak, Munk, Chounlamountri, & Bullier, 1994); (3) neurons located in different hemispheres (Engel, König, Kreiter, & Singer, 1991; Nelson, Nowak, Chouvet, Munk, & Bullier, 1992); (4) between visual areas and multimodal association cortex (Bressler, Coppola, & Nakamura, 1993; Roelfsema, König, Engel, & Singer, 1994); and (5) between somatosensory and motor cortex (Murthy & Fetz, 1992).

Cells preferentially synchronize their responses if activated by contours of the same object, and can rapidly switch the partners with which they synchronize when stimulus configurations change (Gray, König, Engel, & Singer, 1989; Engel, König, & Singer, 1991). Evidence does suggest that synchronization probability is related to behavior. In strabismic cats, V1 neurons driven by different eyes no longer synchronize their responses. This may reflect the inability of strabismic subjects to fuse the images seen by the two eyes (König, Engel, Löwel, & Singer, 1993). When strabismus leads in addition to amblyopia, perceptual deficits are associated with disturbances in the synchronization patterns of cortical neurons rather than with abnormalities in the response properties of individual cells (Roelfsema, König, Engel, Sireteanu, & Singer, 1994). In animals trained to perform sensory-motor tasks, synchronicity was seen to increase both within (Abeles, Bergman, Margalit, & Vaadia, 1993; Vaadia, Haalman, Abeles, Bergman, Prut, Slovin, & Aertsen, 1995) and across areas (Bressler et al, 1993) in relation to problem solving. This dependence of synchronization patterns on stimulus configurations and performance supports the hypothesis that synchronization serves to select the responses of distributed neurons and associate them into coherent assemblies for joint processing. (For comprehensive reviews see Gray & Singer, 1989; Engel et al, 1990, 1991a,b; Livingstone, 1991; Bullier et al, 1992; Murthy & Fetz, 1992; Nelson et al, 1992; Eckhorn et al, 1988, 1992, 1993; Bressler et al, 1993; Gawne & Richmond, 1993; Gray & Viana Di Prisco, 1993; Nowak et al, 1994; Roelfsema et al, 1994.) Studies based on lesions and on selective manipulations of early experience have identified the tangential intra-areal (Löwel & Singer, 1992), the interhemispheric callosal (Engel et al, 1991; Nelson et al, 1992) and feedback connections (Nelson et al, 1992; Sillito, Jones, Gerstein, & West, 1994) as substrates of these synchronization phenomena.

Experience-dependent plasticity of feed-forward connections

The basic features of the feed-forward connections to V1 seem to require no experience for their expression, as many neurons develop their characteristic selectivity for elementary features before birth in monkeys, and prior to eye opening in other mammals (Hubel & Wiesel, 1963; Wiesel & Hubel, 1974). The same is true for the columnar arrangement of response properties and the layout of maps. The specification of these architectures is thus the result of evolutionary selection. Still, the expression of some of these features does depend on activity. Blockade of spontaneous retinal discharges prevents segregation of the afferents from the two eyes into ocular dominance columns (Stryker & Harris, 1986), suggesting the possibility that spontaneous activity promotes axon sorting. Ganglion cells in the developing retina engage in coherent oscillatory activity (Galli & Maffei, 1988; Meister, Wong, Baylor, & Shatz, 1991), providing the option to exploit synchronous activity for the identification of the origin and neighborhood relations of afferents.

However, a substantial fraction of neurons in V1, especially those in layers remote from thalamic input, develop feature-specific responses only if visual experience is available. Receptive-field properties and maps in these layers can be modified by manipulating visual experience during a critical period of early postnatal development (for reviews see Blakemore, van Sluyters, & Movshon, 1976; Frégnac & Imbert, 1984; Stryker, 1991; Rauschecker, 1991; Goodman & Shatz, 1993). Thus, there is room for epigenetic shaping of receptive field-generating feed-forward architectures.

This activity-dependent refinement of connections is based on a Hebbian correlation analysis. Synapses are strengthened if the probability is high that they are active in temporal contiguity with the postsynaptic target cell, and they destabilize if inactive while their target is driven by other inputs (Rauschecker & Singer, 1979; Miller, Keller, & Stryker, 1989). Neurons wire together if they fire together. Such a selection mechanism is ideally suited for generating architectures capable of extracting consistent, frequently occurring relations. Of the many afferents converging onto a particular target cell, only those that are frequently coactivated become consolidated. As a consequence, the cell becomes tuned to the stimulus configuration that produced this coherent input pattern. Accordingly, selective exposure to particular patterns increases the percentage of cortical cells tuned to these patterns, albeit within the limits of the genetically predetermined architecture. Thus, cells in V1 can be made to prefer certain orientations (Blakemore & Cooper, 1970; Hirsch &

Spinelli, 1970; Rauschecker & Singer, 1981) or directions of motion (Tretter, Cynader, & Singer, 1975) more than others, but they cannot be instructed to develop preferences for patterns that they would not respond to normally.

It is largely unknown to what extent the preferences for more complex constellations of features at higher processing stages are subject to experience-dependent specification. Cells tuned to feature constellations characteristic of faces are present in inferotemporal cortex of monkey babies (Rodman, O'Sca-

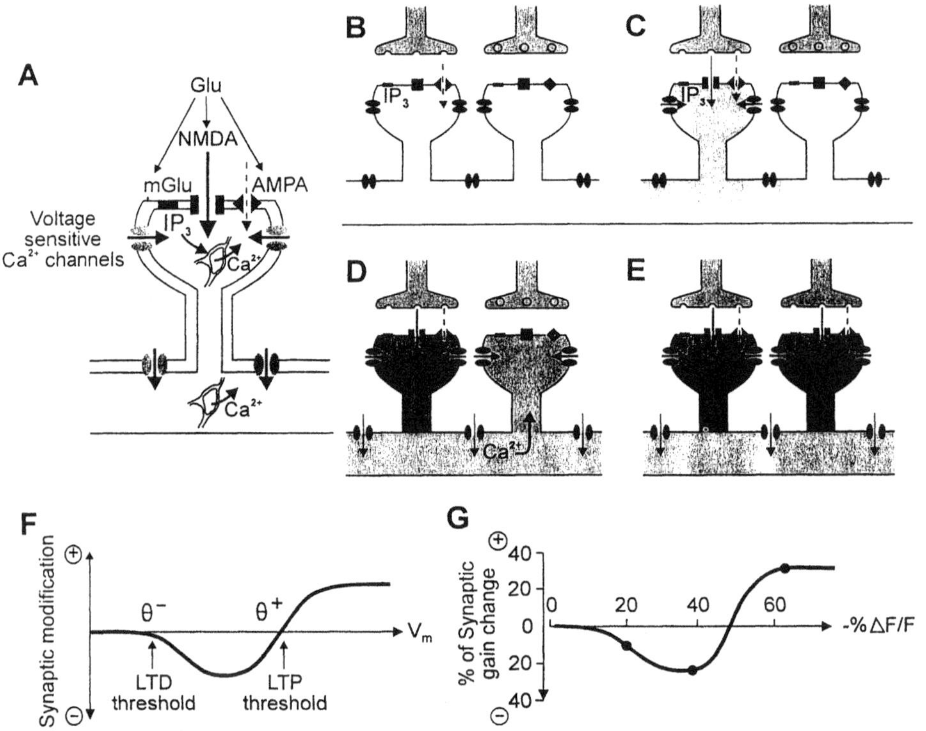

FIGURE 2. Putative synaptic processes likely to mediate the induction of both experience-dependent circuit selection during development and long-lasting synaptic gain changes in the adult. Here, only glutamatergic synapses are considered. **A.** Summary of ligand-gated and voltage-gated mechanisms that contribute to depolarization and modulate the concentration of Ca^{2+} ions ($[Ca^{2+}]i$) in the postsynaptic dendritic compartment. Glu: Glutamate; mGlu (rectangle, left): metabotropic glutamate receptor; NMDA (square, center): N-methyl-d-aspartate receptor; AMPA (diamond, right): α-amino-3-hydroxy-5-methyl-4-isoxazole propionic acid receptor; IP_3: Inositol-triphosphate. **B.-E.** Homosynaptic and heterosynaptic modifications of synaptic transmission for two inputs terminating on spines of the same dendritic segment. Mechanisms influencing $[Ca^{2+}]i$ are indicated by the same symbols as in A. Solid arrows indicate Ca^{2+} movements, and their thickness indicates the amplitude of the flux. The

laidhe, & Gross, 1993). This suggests that even complex relations get extracted and represented by genetically determined feed-forward architectures. Whether this is true also for patterns that are less stereotyped and significant than faces is unknown.

Synaptic mechanisms for selection of feed-forward connections

The selection process appears to be initiated by signals generated in the respective postsynaptic neurons, which evaluate synchrony in the activity of

---

density of the stippling reflects both the expected level of depolarisation and the increase of $[Ca^{2+}]i$. In B, C and D only the left-hand input is active, while in E both inputs are simultaneously active. The four conditions differ in the amplitude of the depolarizing responses of the postsynaptic dendrite. It is assumed that this amplitude is determined both by the activity of the modifiable synapses and by the state of other excitatory, inhibitory and modulatory inputs to the same dendritic compartment (not shown). **B.** The left input fiber discharges at low frequency. Only AMPA and mGlu receptors are activated; voltage-gated Ca^{2+} conductances are inactive. There is no substantial rise in $[Ca^{2+}]i$ and no lasting modification of synaptic transmission at the active synapse. **C.** The left input fiber discharges at higher frequency. Now both NMDA- and voltage-gated Ca^{2+} channels are moderately activated. $[Ca^{2+}]i$ rises to an intermediate level and leads to long-term depression (LTD) of the active synapse. There is only a small spread of depolarization to other spines. **D.** The depolarizing response is assumed to be stronger than in C, either because the left input fiber discharges at higher frequency or because it is active in conjunction with other excitatory inputs. Accordingly, NMDA receptor-gated and voltage-gated Ca^{2+} conductances are also more activated. The massive increase of $[Ca^{2+}]i$ in the activated spine leads to long-term potentiation (LTP). Moreover, depolarization spreads to other compartments of the cell and is thought to trigger action potentials. This spread of depolarization, aided perhaps by back-propagating Na^{+} spikes (Stuart & Sakmann, 1994; Yuste & Denk, 1995) activates voltage-gated conductances. This is assumed to lead to an intermediate rise in $[Ca^{2+}]i$ at the postsynaptic side of the inactive synapse, which as a result undergoes heterosynaptic depression. **E.** Conditions are as in D, except that the second input is now also active. This facilitates the recruitment of ligand-gated Ca^{2+} sources at the synapses of the second input and raises $[Ca^{2+}]i$ above LTP threshold so that the second input is no longer depressed, but undergoes LTP. Because the first input already causes substantial depolarization of the dendritic compartment, the second input can undergo LTP at activation levels well below those that would be required if the first input had not been activated. Most of the experience-dependent developmental-circuit changes can be accounted for by this scenario if one equates LTP and LTD with consolidation and disruption of synaptic connections, respectively. **F.** Illustration of the dependence of the polarity of synaptic-gain changes on the depolarization level of the dendritic compartment (Artola & Singer, 1993). Ordinate: direction of gain change. Abscissa: depolarization achieved during activation. Θ^- and Θ^+: thresholds for the induction of LTD and LTP. **G.** Experimentally determined (Hansel, Artola, & Singer, 1994) increases of dendritic Ca^{2+} concentration expressed as fluorescence change (%ΔF/F) of the Ca^{2+} indicator Fura2 (abscissa) after activation protocols leading to weak LTD (1st point), strong LTD (2nd point) and strong LTP (3rd point). Ordinate: Average amplitude of synaptic gain changes induced with the three stimulation protocols. As predicted in a model by Lisman (1989) there is a close correlation between changes in membrane potential (F), the increase of $[Ca^{2+}]i$, and the polarity and magnitude of synaptic-gain changes.

---

converging afferents (Rauschecker & Singer, 1979; Miller et al, 1989). The end result is morphological changes of axonal and dendritic arborizations (Friedländer, Martin, & Wassenhove-McCarthy, 1991; Antonini & Stryker, 1993; Kossel, Löwel, & Bolz, 1995). N-methyl-d-aspartate (NMDA) receptors have been assigned an important role in experience-dependent circuit selection. Their activation protects inputs firing in conjunction with the postsynaptic cell against destabilization, promotes heterosynaptic repression of other, inactive inputs, and is a necessary prerequisite for the reconnection of previously weakened connections (Kleinschmidt, Bear, & Singer, 1987; Gu, Bear, & Singer, 1989; Bear, Kleinschmidt, Gu, & Singer, 1990). This supports the hypothesis that NMDA receptors evaluate the coincidence between pre- and postsynaptic activation and that Ca^{2+} entry through the NMDA receptor serves as an early signal in the cascade leading to synapse stabilization (Bear, Cooper, & Ebner, 1987; Bear, Cooper, & Ebner, 1989; Fox & Daw, 1993). Synapses which are inactive while the postsynaptic cell is discharging weaken, and it has been proposed that this is because they cannot activate their NMDA receptors (Kleinschmidt et al, 1987; Gu et al, 1989; Bear et al, 1990). Synapses can also weaken if they are active while the postsynaptic cell is prevented from responding (Reiter & Stryker, 1988), another condition in which activation of NMDA receptors is unlikely.

These conditions are strikingly similar to those required for the induction of synaptic-gain changes in the adult such as homosynaptic long-term potentiation (LTP), heterosynaptic depression, and homosynaptic long-term depression (LTD), all of which are rapidly induceable and long-lasting changes of synaptic efficacy, first discovered in the hippocampus (**Figure 2**). The suggestion that these use-dependent changes in synaptic gain could also serve as a first step in experience-dependent circuit selection is receiving experimental support. *In vitro* studies on visual-cortex slices demonstrated LTP (Artola & Singer, 1987; Tsumoto, 1992; Kirkwood, Dudek, Gold, Aizenman, & Bear, 1993) and LTD (Artola, Bröcher, & Singer, 1990; Tsumoto, 1993; Kirkwood & Bear, 1994a,b), revealed a similar dependence of plasticity on NMDA-receptor activation as in developmental changes (Singer & Artola, 1995) and, most importantly, showed an age-dependent decline in the susceptibility to undergo LTP that paralleled the time course of the critical period (Kato, Artola, & Singer, 1991; Kato, Braun, Artola, & Singer, 1991; Crair & Malenka, 1995) for experience-dependent modifications. This decline is associated with a reduced contribution of NMDA receptor-mediated synaptic responses (Kato et al, 1991a,b; Crair & Malenka, 1995), whereby three factors seem to contribute: (1) a reduction of NMDA receptors (Tsumoto, Haghara, Sato, & Hata, 1987; Bode-Greuel & Singer, 1989; Fox, Sato, & Daw, 1989; Fox, Daw, Sato, & Czepita, 1992); (2) an increase of postsynaptic inhibition that prevents lifting of the Mg^{2+} block (Kato et al, 1991a,b); and (3) a developmental change in the gating characteristics of the NMDA receptor (Hestrin, 1992). Recently, it has been shown that rearing animals in the dark, which prolongs the critical period, also retards the decline in LTP susceptibility (Kirkwood, Lee, & Bear, 1995). Furthermore, there is evidence that manipulations which facilitate LTP induction, such as addition of the neuromodulators acetylcholine and norepinephrine (Bröcher, Artola, & Singer, 1992) or direct depolarization of postsynaptic cells, also favor the induction of experience-dependent modifications of receptive-field properties (Frégnac, Shulz, Thorpe, & Bienenstock, 1988; Greuel, Luhmann, & Singer, 1988). However, despite the attraction of these analogies, direct proof that LTP and LTD serve as first steps in developmental-circuit selection is still missing.

It is of particular interest that the experience-dependent selection of feedforward connections is not determined solely by local correlations of activity, but is supervised by attentional mechanisms. Sensory signals induce circuit changes only when the animals attend to these signals and use them for the control of behavior. Visual stimulation does not induce changes of ocular dominance and orientation selectivity of cortical neurons when applied during anesthesia. Likewise, monocular deprivation fails to induce circuit changes in awake animals when the signals conveyed by the open eye are inappropriate for visuo-motor coordination. This is the case, for example, when rotation of retinal coordinates interferes with visually guided eye, head, and body movements. Conversely, passive visual stimulation does produce changes in feedforward connections even under anesthesia when paired with electrical activation of central-core projections. (For a review of the literature on central gating of developmental plasticity, see Singer, 1990).

Experience-dependent modifications also fail to occur when the noradrenergic, the cholinergic, or the serotoninergic projections to the visual cortex are inactivated, the permissive effects of these modulatory projections being mediated by ß, M1 and S2 receptors, respectively (Kasamatsu & Pettigrew, 1979; Bear & Singer, 1986; Gu & Singer, 1993, 1995). No data are yet available on a putative gating function of the dopaminergic system. It is noteworthy that this blockade of experience-dependent circuit changes occurs without noticeable alterations of the neuron's responses to visual stimuli, suggesting that the control of long-lasting synaptic modifications by the modulatory systems is mediated by second-messenger interactions that do not directly interfere with the electrophysiological properties of the neurons. These results indicate that the developing brain has the option to shape its architecture not only as a function of frequently occurring input constellations, but also as a function of their behavioral relevance. This implies that there are evaluation systems in the brain which are able to judge the adequacy of sensory signals in the context of ongoing processing and to prevent sensory input from modifying circuits if it does not meet with the "expectancies" that are defined by the genetically constrained architecture of the brain. Disturbances of experience-dependent shaping of neuronal architectures during development can thus have several quite different causes. First, genetic predispositions could lead to the development of architectures that do not match well with the actual conditions of the environment to which the developing brain is exposed. If this mismatch is too large, experience-dependent adaptation of functional architectures may not occur or may remain incomplete. Second, there could be inborn malfunctions of the gating systems that evaluate the behavioral relevance of experience and permit use-dependent changes in circuitry. Such malfunctions could prevent experience-dependent shaping of architectures despite normal environmental conditions. Disturbances of these gating functions could become particularly relevant in the context of ontogenetic learning processes that require social interactions, such as filial imprinting in animals and the development of higher cognitive abilities in humans. Third, all cerebral functions could be normal but the environment could be impoverished, failing to provide the information which the developing brain requires to optimize its functional architecture. Thus there are multiple possibilities for disturbances of experience-dependent shaping of neuronal architectures, and it is to be expected that abnormalities in the development of neuronal connectivity are not infrequent. It may be worthwhile, therefore, to extend the search for pathophysiological causes of abnormal behavior to developmental disturbances in the wiring of the neuronal hardware, and to not confine clinical studies to the investigation of molecular and cellular disturbances only.

A large number of cellular mechanisms have been identified that change during early development in parallel with the decline of use-dependent plasticity. This suggests that numerous processes cooperate in the maintenance of use-dependent plasticity during the critical period. Modifications have been described for surface-recognition molecules at synaptic locations, for distributions of a variety of neurotransmitter receptors and voltage-gated Ca^{2+} channels, for the gating characteristics of NMDA receptors, for the laminar distribution of modulatory afferents, and for a large number of second-messenger systems. (For reviews see Singer, 1990, and Shaw, Needler, Wilkinson, Aoki, & Cynader, 1984). Recently, neurotrophins such as nerve-growth factor (NGF) and brain-derived neurotrophic factor (BDNF) have also been shown to play a role in experience-dependent selection of feed-forward connections, but the results are still inconclusive. The effects of adding NGF were found compatible with the hypothesis that input selection could be based on competition for neurotrophins released by the postsynaptic target in an activity-dependent way (Maffei, Berardi, Domenici, Parisi, & Pizzorusso, 1992; Carmignoto, Canella, Candeo, Comelli, & Maffei, 1993; Berardi, Cellerino, Comenici, Fagiolini, Pizzorusso, Cattaneo, & Maffei, 1994; Domenici, Cellerino, Berardi, Cattaneo, & Maffei, 1994). Application of BDNF but not of NGF was found to prevent sorting of thalamic afferents into ocular dominance columns (Cabelli, Hohn, & Shatz, 1995), and intracortical infusion of BDNF but not of NGF in monocularly deprived kittens quite unexpectedly promoted disconnection of non-deprived inputs, thus reversing the polarity of the selection mechanism (Galuske, Kim, Castren, Thoenen, & Singer, 1996; Singer, 1994).

Thus, while the rules governing activity-dependent circuit selection are reasonably well understood, much remains to be done to clarify the underlying molecular mechanisms.

Use-dependent plasticity of assembly-forming connections

Much less data are available on the developmental specification of feedback projections and reciprocal intra- and interareal cortico-cortical connections. Most of these pathways attain their final selectivity only during postnatal life, and their architecture is highly susceptible to activity-dependent modifications. In cat V1 the tangential intracortical connections already exhibit a crude periodic patterning prior to eye opening, suggesting some experience-independent selectivity in the organization of these connections. However, some authors (Lübke & Albus, 1992) have concluded that the patchy pattern of intrinsic connections at the time of eye opening closely resembles that seen in adults, requiring no further developmental refinement. But whether this selectivity is related to the columnar pattern of feature-specific cells that emerges at almost the same time is unknown. The tangential axons continue to grow beyond the time of eye opening, combining extension with refinement towards the highly selective mature pattern (Price & Blakemore, 1985; Luhmann, Singer, & Martinez-Millan, 1990; Callaway & Katz, 1990; Galuske & Singer, 1996; but see Lübke & Albus, 1992). Depriving kittens of vision delays this refinement (Callaway & Katz, 1991), and data from strabismic kittens indicate that vision-dependent selection of these intracortical connections follows a correlation rule in much the same way as has been established for the feed-forward connections, suggesting similar mechanisms of selection (Löwel & Singer, 1992). Columns exhibiting a low probability of coherent firing lose their reciprocal connections. One consequence is that cells in these columns also lose the ability to synchronize their discharges, even when activated conjointly with coherent stimuli (König et al, 1993), supporting the notion that cortico-cortical connections have a synchronizing action. Conversely, contiguous activation of spatially distant columns increases their mutual coupling to the extent that cells actually acquire two spatially separate receptive fields, the ectopic one reflecting the response properties of the remote columns (Singer & Tretter, 1976).

In V1 of the normally reared adult, the tangential intracortical connections selectively link columns with similar feature preferences and the density of connections decreases with distance (Ts'o & Gilbert, 1988; Gilbert & Wiesel, 1989; Malach, Amir, Harel, & Grinvald, 1993). This agrees with the postulate that the architecture of assembly-forming connections should reflect the *Gestalt* criteria for perceptual grouping (Gray et al, 1989; Engel et al, 1991). The organization of the tangential connections in V1 seems appropriate for the grouping of responses according to the criteria of vicinity and similarity (**Figure 1**). The fact that tangential connections are selected by experience according to a correlation rule makes the conclusion unavoidable that their mature architecture reflects, at least to some extent, the joint probabilities with which particular features co-occurred during early development. Such acquisition of knowledge about typical feature constellations by changes in architecture would be ideally suited to support figure-ground distinctions and perceptual grouping. Responses to feature constellations characteristic for perceptual objects would become grouped preferentially and routed together via feed-forward connections for further joint processing.

No data are available yet on the genetic constraints limiting epigenetic modifiability of these tangential cortical connections, nor is it known whether their use-dependent selection is gated by modulatory systems. Data are also lacking on the development and the epigenetic modifiability of reciprocal cortico-cortical long-range connections. Except for the callosal connections in the cat between contralateral areas 17, which seem to exhibit a dependence on experience similar to that of the intrinsic tangential connections (Innocenti & Frost, 1979), virtually nothing is known about the role of experience in the development of interareal and feedback projections. It is unlikely, though, that they would be less susceptible to epigenetic selection than the intra-areal connections.

Plasticity in the mature cortex

In the adult, use-dependent plasticity of feed-forward connections appears to be very limited, at least at lower levels of processing. The synapses of thalamic afferents become much less susceptible to undergo LTP (Crair & Malenka, 1995), and in V1 the structure of RFs can be altered only with invasive conditioning procedures (Frégnac et al, 1988; Greuel et al, 1988). But even at higher levels, such as the inferior temporal cortex of primates, extensive training is required in order to produce a statistically significant increase of neurons tuned to newly learned patterns (Miyashita, 1988; Sakai & Miyashita, 1991). The situation appears to be similar in other sensory cortices, although modifications of response properties have been described after conditioning and extensive stimulation in both auditory cortex (Weinberger, 1993; Recanzone, Schreiner, & Merzenich, 1993) and somatosensory cortex (Merzenich & Sameshima, 1993; Kaas, 1995; Merzenich & de Charms, in press). The changes of RF properties were confined to modifications of tuning width, size, and minor changes in preferred features. Drastic alterations in response properties of neurons can be observed in higher sensory areas and in motor centers as a function of changes in attention or the behavioral task. Since these modifications occur on a fast time scale, they are probably due to dynamic context-dependent rerouting of input signals rather than to synaptic modifications of feed-forward connections.

These findings seem to exclude that, in the adult, representations of new patterns are generated to any substantial extent by modifications of the RF-forming feed-forward connections, and shifts the focus to the assembly-forming associative circuits. Indeed, there is ample evidence for their malleability despite the fact that research on their function still has a short history. Most of the studies that have demonstrated LTP and LTD in neocortical slices of mature animals have actually investigated the malleability of reciprocal cortico-cortical connections, although this is rarely made explicit. Recordings typically are obtained from neurons in supragranular layers, and responses are investigated that are elicited either from white matter, layer IV, or adjacent regions within supragranular layers. If monosynaptic, and the claim is that they are, these responses are mainly if not exclusively due either to tangential or ascending intra-areal connections, or to long-range cortico-cortical projections. Thus, it is safe to conclude that, in the adult, cortico-cortical connections are highly susceptible to LTP and LTD and hence to use-dependent long-term modifications of their efficacy.

Evidence for use-dependent changes in the coupling strength of cortico-cortical connections is also available from *in vivo* recordings. Repeated coactivation of neuron pairs in the auditory cortex of monkeys led to enhanced synchronization of their discharges, and this effect occurred only when the animals paid attention to the tone used for activation (Ahissar, Vaadia, Ahissar, Bergman, Arieli, & Abeles, 1992). This is best explained by enhanced efficacy of connections that do not contribute to the RF proper but have synchronizing effects, a characteristic feature of tangential intracortical connections (König et al, 1993; Ts'o & Gilbert, 1988; Ts'o et al, 1986).

Several studies demonstrated striking rearrangements of retinotopic and somatotopic maps, both after prolonged stimulation of afferent pathways and denervation (Weinberger, 1993; Recanzone et al, 1993; Merzenich & Sameshima, 1993; Kaas, 1995; Merzenich & de Charms, in press; Chino, Smith, Kaas, Sasaki, & Cheng, 1995; Schmid, Rosa, & Calford, 1995; Das & Gilbert, 1995). Because these modifications occurred over large distances, they could not be accounted for by adaptive changes at the level of feed-forward connections, but had to be attributed to enhanced efficacy of tangential intracortical connections. It appears that under the extreme condition of deafferentation, the intracortical association connections can increase their efficacy so much that they can actually drive cells in remote columns and hence generate RFs. Recent evidence suggests that this increase in efficacy is associated with sprouting and the formation of new synaptic contacts (Darian-Smith & Gilbert, 1995).

Long-term changes in the efficacy of the ensemble-forming association connections at one level of processing are expected to alter the activity patterns conveyed by feed-forward connections to the next level. Hence, the response properties of neurons at this level should change. This seems to be in contradiction to the relative stability of RFs in the adult. One explanation is that the expected modifications may be revealed only if the effect of grouping operations is studied rather than the structure of classical RFs. This, however, will require application of more complex stimuli and the analysis of context-dependent response modifications. Context-dependent modifications of RFs have been shown in V1. Responses to stimuli presented in the classical RF are modulated by concomitant stimulation of neighboring regions in the visual field (Nelson, 1985; Morrone et al, 1982; Blakemore & Tobin, 1972; Gilbert & Wiesel, 1992), but also subsequent to intensive and repeated activation of these regions (Pettet & Gilbert, 1992).

Conclusions

Although data on use-dependent development of cortical circuits are still sparse, the following conclusions appear warranted. (1) During early postnatal development both the feed-forward, RF-generating connections and the reciprocal, assembly-forming connections are susceptible to experience-dependent modifications, and these use-dependent changes appear to obey a correlation rule emphasizing the role of coherent activity in circuit selection. Converging inputs conveying consistent messages become consolidated, the consistency criterion being repeated, correlated activation. Despite its substantial malleability, the architecture of the feed-forward connections appears to be constrained by genetic predisposition more than that of the assembly-forming connections, but more data are needed to substantiate this point. (2) After the end of morphogenesis the architectures of both connection systems crystallize, and in sensory cortices most of the RF-forming pathways also seem to lose the ability to undergo use-dependent gain changes, while this ability is retained by the assembly-forming connections. (3) This persistent functional malleability of assembly-forming connections is the likely basis for the generation of new representations, because it allows for a rapid and flexible association of feature-representing neurons into new constellations. It is also responsible for the short- and long-term changes in cortical maps observed after extensive stimulation or denervation. (4) The rules and induction mechanisms underlying use-dependent gain changes of assembly-forming connections in the adult are similar to those supporting circuit selection during development. It is therefore attractive to assume that LTP or LTD serves as an initial step in both processes. As LTP and LTD are phenomena that can rapidly be reversed, they alone cannot suffice to generate new representations, i.e., durable associations of feature-specific cells. They may, however, play an important role for the rapid and context-dependent association of neurons and hence for the flexible selection and routing of activity across subsequent processing stages. Permanent associations could form if synapses that often undergo LTP, or whose potentiation is not rapidly reset, would eventually strengthen irreversibly. Adult plasticity would thus appear as the continuation of developmental processes, with the only difference that it no longer leads to modifications of the blueprint of the architecture and operates mainly by regulating the efficacy of assembly-forming connections.

ACKNOWLEDGMENTS

I wish to thank A. Kreiter, P. König, and P. Roelfsema for valuable comments on the manuscript, R. Goebel for composing the pattern in Figure 1C, R. Ruhl for preparing the figures, and I. Pipacs for editing the text.

Neural-Networks Models of Cognition
J.W. Donahoe and V. Packard Dorsel (Editors)

CHAPTER 9

INFEROTEMPORAL CORTEX AND OBJECT RECOGNITION

Keiji Tanaka
The Institute of Physical and Chemical Research (RIKEN)
Saitama, Japan

ABSTRACT

Cells in the anterior part of the inferotemporal cortex (area TE) of the monkey brain selectively respond to various moderately complex object features, and those that respond to similar features cluster in a columnar region oriented vertically to the cortical surface. Cells within a column respond to similar but not identical features. Data from optical imaging in TE have suggested that the borders between neighboring columns are not discrete, but that there is a continuous mapping of complex-feature space within a larger region containing several partially overlapping columns. This continuous mapping may be used for various computations, such as production of the image of an object at different viewing angles, illumination conditions, and articulation poses.

Introduction

Visual object recognition, i.e., recognition of objects from their visual images, is a key function of the primate brain. This recognition is not a process of template matching between the input image and stored images, but a flexible process in which considerable change in an image—arising from differing illuminations, viewing angles, and articulations of the object—can be tolerated. In addition, our visual system can deal with novel objects based on previous visual experience with similar objects. Generalization may be an intrinsic property of the primate visual system. This chapter searches for the neural organization essential for these flexible aspects of visual object recognition in the anterior part of the inferotemporal cortex.

The anterior part of the inferotemporal cortex (area TE) represents the final stage of the ventral visual cortical pathway, which is thought to be essential for object vision in the monkey. This pathway starts at the primary visual cortex (V1) and leads to area TE via relays in several non-primary visual cortical areas (V1-V2-V4-TEO-TE). A bilateral lesion of TE (Dean, 1976; Gross, 1973) or deafferentation of TE by a complete bilateral lesion of the middle stage (Yaginuma, Osawa, Yamaguchi, & Iwai, 1993) results in severe deficits in the visual recognition or discrimination of objects. TE, in turn, projects to several polymodal brain sites outside the visual cortex, including the perirhinal cortex (area 35/36), the prefrontal cortex, the amygdala, and the striatum of the basal ganglia. The perirhinal cortex projects to the hippocampus via the entorhinal cortex (area 28).

I do not assume that visual object recognition is completed within the circuits leading to and including area TE, but rather that it depends on neural activities distributed over the brain beyond the visual cortex. However, because area TE is located at the final common stage in the processing of visual images of objects, I assume that area TE and its afferent circuits are essential for much of the flexible properties of visual object recognition. The perirhinal cortex may also be a key structure for visual object recognition, because recent lesion studies have shown its importance for the execution of visual delayed nonmatching-to-sample independent of the hippocampus (Eacott, Gaffan, & Murray, 1994; Gaffan, 1994; Gaffan & Murray, 1992; Meunier, Bachevalier, Mishkin, & Murray, 1993; Murray & Mishkin, 1986; Suzuki, Zola-Morgan, Squire, & Amaral, 1993; Zola-Morgan, Squire, Amaral, & Suzuki, 1989; Zola-Morgan, Squire, Clower, & Rempel, 1993). However, too little is known about the perirhinal cortex to identify the nature of its contribution to visual object recognition.

This review article is eccentric in that our own data are emphasized and the citation of other references is selective. The selection is not based on the value of the studies, but on their relevance to the subject at hand. For an overview of studies on the inferotemporal cortex, readers should consult other reviews (e.g., Rolls, 1991; Miyashita, 1993; Gross, 1994; Desimone, Miller, & Chelazzi, 1994), especially for discussions of the mechanisms of short-term memory of object images. I first summarize the data from unit-recording experiments to show that cells in TE respond to moderately complex object features, and that those responding to similar features cluster in a columnar region. The process by which the selectivity occurs is then looked for in the afferent pathways to TE. Data on optical imaging of TE is next introduced in the discussion of the function of TE columns. This discussion is further elaborated to include how the concept of an object emerges in the brain.

The recordings introduced in this article were from the part of TE lateral to the anterior middle temporal sulcus (AMTS), which is often referred to as TEd (dorsal part of TE), and were obtained from anesthetized preparations.

Stimulus Selectivity of Cells in TE

An obstacle to the study of neuronal mechanisms of object vision has been the difficulty in determining the stimulus selectivity of individual cells. Great variety exists among objects in the natural world, and we do not know how the brain partitions this diversity into its constituent features.

Single-unit recordings from TE were initiated by C.G. Gross and his colleagues (Gross, Bender, & Rocha-Miranda, 1969; Gross, Rocha-Miranda, & Bender, 1972). They found that cells in TE had large receptive fields, most of

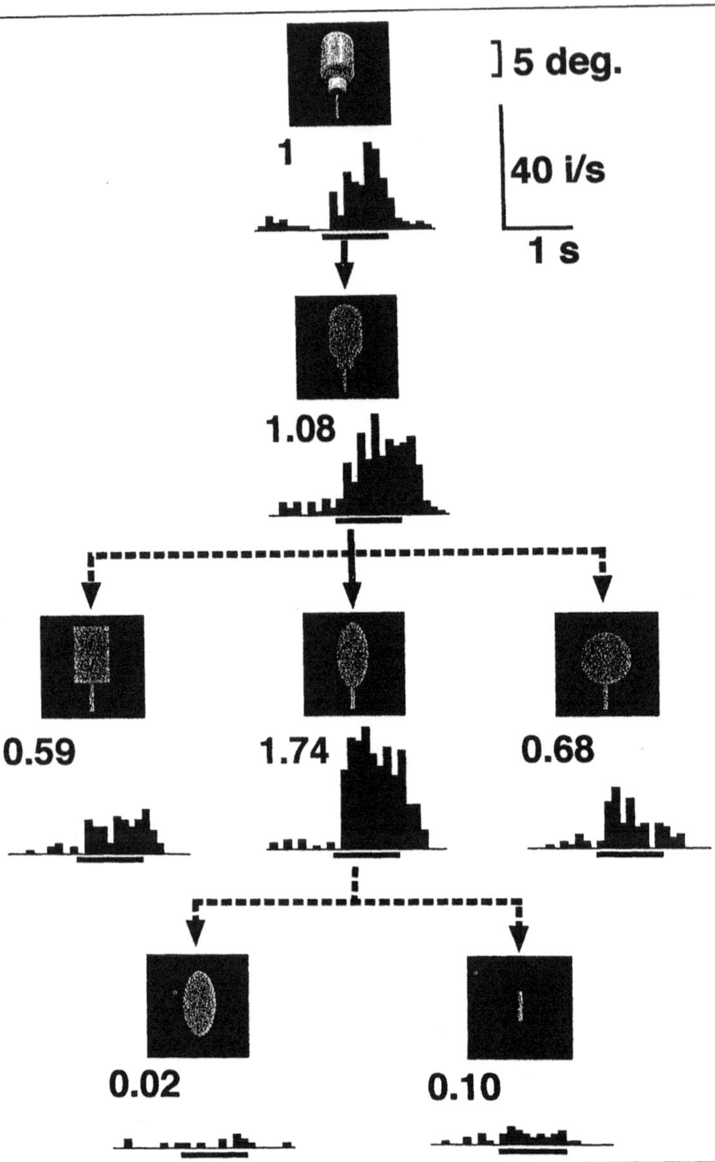

FIGURE 1. An example of the reduction process used to determine the feature critical for the activation of cells in the ventral visual pathway. The responses were averaged over ten repetitions of the stimuli. The underlines indicate the period of stimulus presentation, and the numbers above histograms indicate the magnitude of the responses normalized by the response to the image of a water bottle. This cell was recorded from TEd.

which included the fovea, and that some cells responded specifically to a brush-like shape with many protrusions, or to the silhouette of a hand. They extended the study of stimulus selectivity by using two different methods, one constructive and the second reductive. In the first method, they used Fourier descriptors that were defined by the number (frequency) and amplitude of periodic protrusions from a circle. A contour of any shape can be reconstructed by linearly combining elementary Fourier descriptors of frequencies and amplitudes. Some cells responded specifically to Fourier descriptors of a particular range of frequencies with considerable invariance for the overall size of the stimulus (Schwartz, Desimone, Albright, & Gross, 1983). However, this method was not very promising, since the same group of authors found that the response of a TE cell to a composite contour was far from the linear combination of its responses to the elementary component contours (Albright & Gross, 1990). Fourier descriptors do not appear to provide the basis by which the inferotemporal cortex represents objects.

The second method used by Gross was the reductive method. Diverse objects were initially presented within the receptive fields of individual cells. After an object had been found that effectively stimulated a cell, paper cutouts simulating various features of the object were presented to characterize more precisely the nature of the critical stimuli (Desimone, Albright, Gross, & Bruce, 1984). Our work extended the reductive method by devising a procedure in which the test features were systematically varied with the aid of a specially designed image-processing computer system (Fujita, Tanaka, Ito, & Cheng, 1992; Ito, Fujita, Tamura, & Tanaka, 1994; Ito, Tamura, Fujita, & Tanaka, 1995; Kobatake & Tanaka, 1994; Tanaka, Saito, Fukada, & Moriya, 1991).

In the new method, spike activities were first isolated from a single cell. Then, many three-dimensional imitations of animals and plants were presented to determine which stimuli were effective. Next, images of the effective stimuli were taken by a video camera and presented by the computer to a TV monitor to determine the most effective stimulus. Finally, the image of the most effective stimulus was simplified, while maximal activation persisted, by sequentially removing parts of the features contained in the image. The minimal stimulus that continued to activate the cell maximally was determined to be the critical feature for that cell. **Figure 1** exemplifies the process for a cell recorded from TEd, in which the effective stimulus was reduced from the image of a water bottle to the combination of a vertical ellipse and a downward projection from the ellipse.

After the reduction process had been completed, the image was modified to examine the selectivity further. **Figure 2** exemplifies this latter process for a second TE cell, whose domain of selectivity was most clearly determined. The cell responded maximally to a pear model within the routine set of object

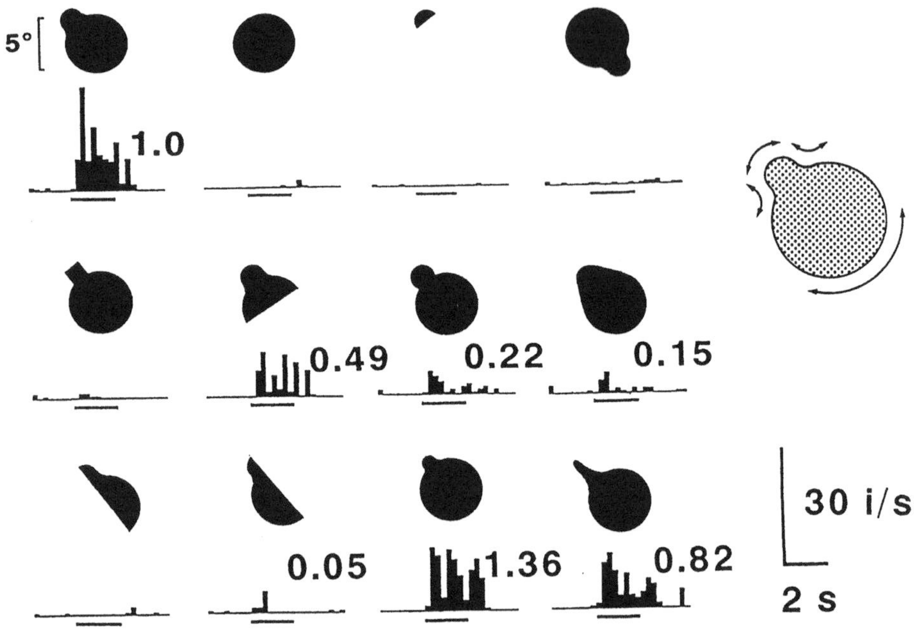

FIGURE 2. An example of further study of selectivity after the reduction process was completed. This second cell is different from the cell shown in **Figure 1**, but was also recorded from TEd.

stimuli, and the critical feature was determined to be "a rounded protrusion from a rounded body with a concave smooth neck." The body or the head by itself did not evoke any responses (**Figure 2**, first line). The head had to be rounded because the response disappeared when it was replaced by a square, and the body had to be rounded because the response decreased by 51% when it was cut in half (second line, left). The neck had to be smooth and concave because the response decreased by 78% or 85% when the neck was replaced by one that was either straight or had sharp corners, respectively (second line, right). The critical feature was neither the right upper contour nor the left lower contour because neither stimulus by itself evoked responses (third line, left). The width and length of the projection were also not critical (third line, right).

A third example of a TE cell is shown in **Figure 3**. The cell responded strongly to the face of a monkey toy (top left), and the critical feature was determined to be a configuration in which two black spots and one horizontal black bar were arranged in a gray disk (top line, second from left). Both the bar and the spots were indispensable (top line, right two) and the circular outline was also essential (bottom left). The contrast between the inside and outside of the circular contour was not critical (bottom line, second from right). However, the spots and bar had to be darker than the background within the outline (bottom right).

By determining the critical features for hundreds of cells in TE, we reached the conclusion that most cells in TE require moderately complex features for their activation, such as the 16 examples in **Color Plate 1** (p. 185). The critical features were more complex than orientation, size, color, and simple textures, which are known to be extracted and represented by cells in V1. Some of the features were shapes that were moderately complex, while others were combinations of such shapes with color or texture. Responses were selective for the contrast polarity of the shapes: That is, contrast reversal of the critical feature reduced the response by more than half in 60% of tested cells, and replacement of the solid critical features by line drawings of the contour reduced the response by more than half in 70% of tested cells (Ito et al, 1994).

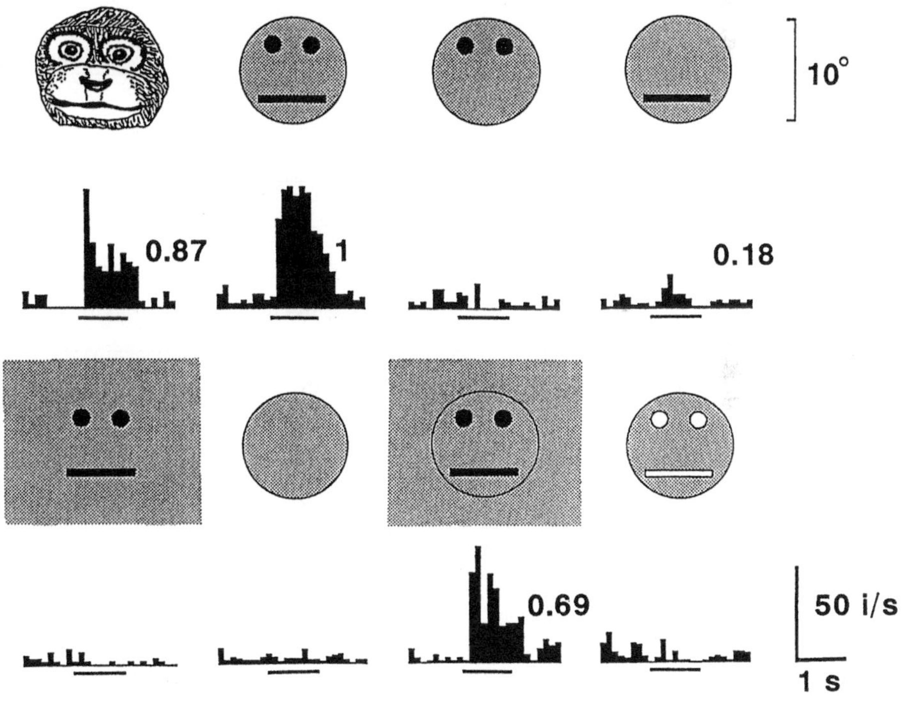

FIGURE 3. A third example of a TEd cell.

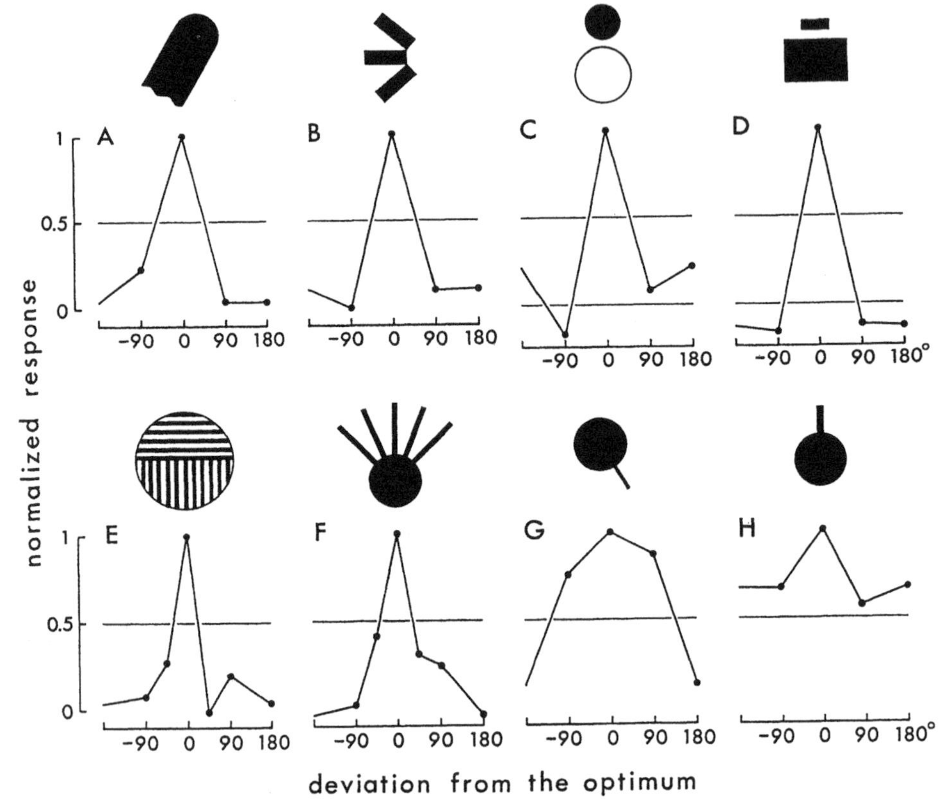

FIGURE 4. Tuning of responses of eight TEd cells for the orientation of a stimulus.

Invariance or Selectivity for Position, Orientation, and Size

Gross et al (1972), in their pioneering experiments, found that cells in TE had large receptive fields within which the critical stimulus continued to evoke responses. Using sets of shapes whose critical features were determined by the reduction method, and other shapes in which the critical features were somewhat modified, we demonstrated that selectivity for shape was generally preserved over large receptive fields (Ito et al, 1995). Although selectivity was generally preserved for the critical features, responding was also affected by their orientation in the frontoparallel plane (Tanaka et al, 1991) and by their size (Ito et al, 1995).

Figure 4 shows the data from eight cells for the tuning of responses to orientation in the frontoparallel plane. Rotation of the critical feature by 90° decreased the response by more than half for most cells (**A-F**). The tuning of the remaining cells was broader: The response was reduced by more than half only when the feature was rotated 180° (**G**), and some cells showed only smaller decreases even with rotation of 90° and 180° (**H**).

The effects of stimulus size on the response varied more among cells. Twenty-one percent of tested TE cells responded to ranges of size of more than four octaves of the critical features, whereas 43% responded to size ranges of less than two octaves. Tuning curves for four TE cells, two from each group, are shown in **Figure 5**. The tuned cells may be in the process of forming size-invariant responses or, alternatively, both size-dependent and -independent processing of features occurs in TE.

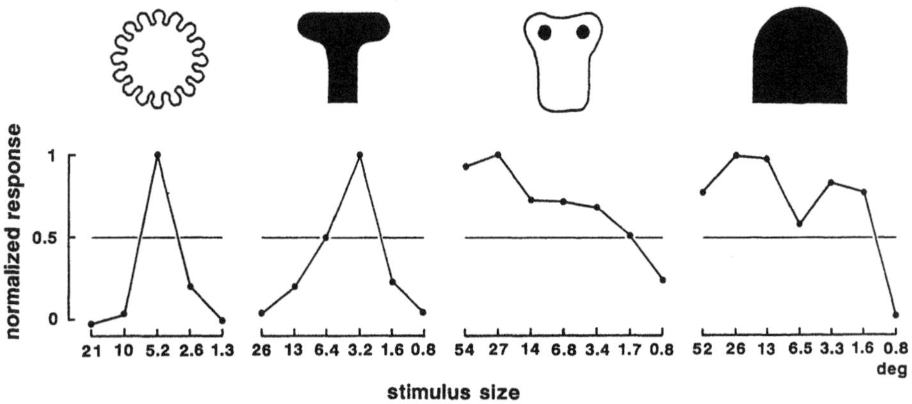

FIGURE 5. Tuning of responses of four TEd cells for the size of a stimulus.

Columnar Organization in TE

How are cells with various critical features distributed in TE? By recording from two TE cells simultaneously with a single electrode, we have found that cells located near one another in the cortex have similar stimulus selectivity (Fujita et al, 1992). The critical feature of one isolated cell was determined by using the procedure described above, while—at the same time—responses of another isolated cell, or of non-isolated multi-units, were recorded. In most cases, the second cell responded to the optimal and suboptimal stimuli of the first cell. The selectivity of the two cells varied slightly, however, in that the maximal response was evoked by slightly different stimuli, or the mode of the decrease in response differed when the stimulus was changed from the optimal stimulus. **Figure 6** shows an example of the latter cases.

To determine the spatial extent of the clustering of cells with similar selectivity, we examined the responses of cells recorded successively along long penetrations made either vertically or obliquely to the cortical surface (Fujita et

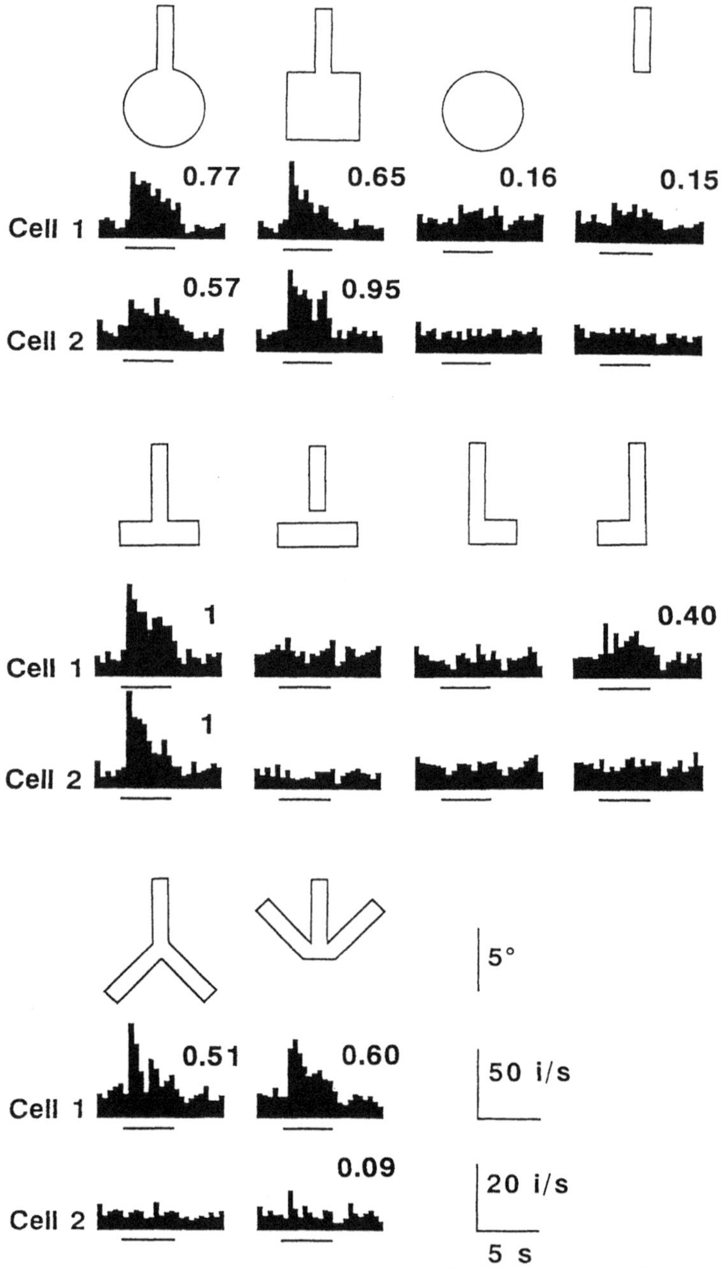

FIGURE 6. An example of simultaneous recording from two nearby neurons in TEd.

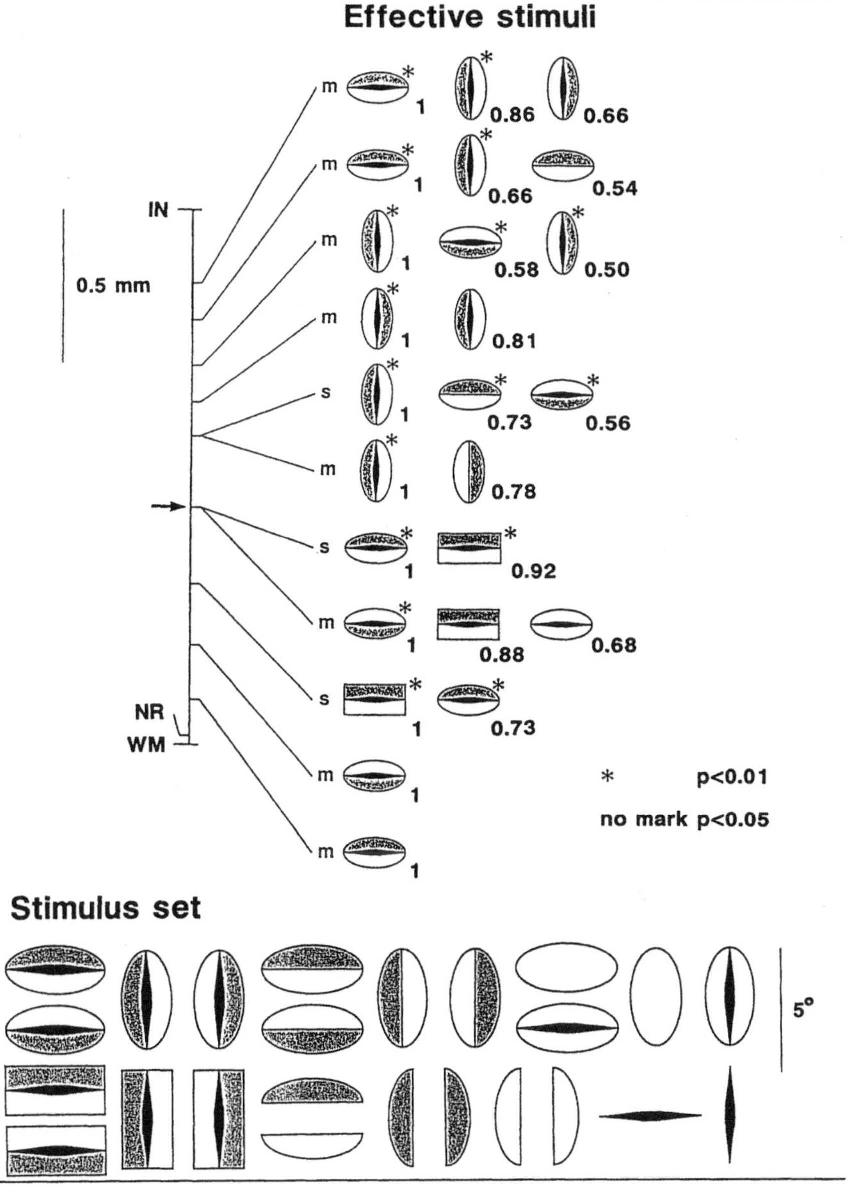

FIGURE 7. Responses of cells recorded along a vertical penetration in TEd. The responsiveness of the cells was tested with the set of stimuli shown at the bottom, which was made in reference to the critical feature of the first cell indicated by the arrow. Effective stimuli are listed separately for individual recording sites in their order of effectiveness (m indicates multi-unit and s single-unit recording).

al, 1992). The critical feature of a cell at the middle of the penetration was first determined. Then, we presented a set of stimuli including the optimal feature for the first cell, its rotated versions, and ineffective control stimuli. Cells recorded at different positions along the penetration were tested only with the fixed set of stimuli. As shown in **Figure 7**, cells recorded along the vertical penetrations generally responded to the same critical feature as the first cell or to some closely related stimulus. The cells responding to common stimuli spanned nearly the entire thickness from layer 2 to layer 6. In the penetrations made obliquely to the cortical surface, the situation was different. The cells that were responsive to the critical feature of the first cell, or to related stimuli, were restricted to a small region neighboring the first cell. The horizontal extent of the region averaged 400 μm. The cells outside this region either did not respond to any of the stimuli in the set or responded to control stimuli that were ineffective in activating the first cell.

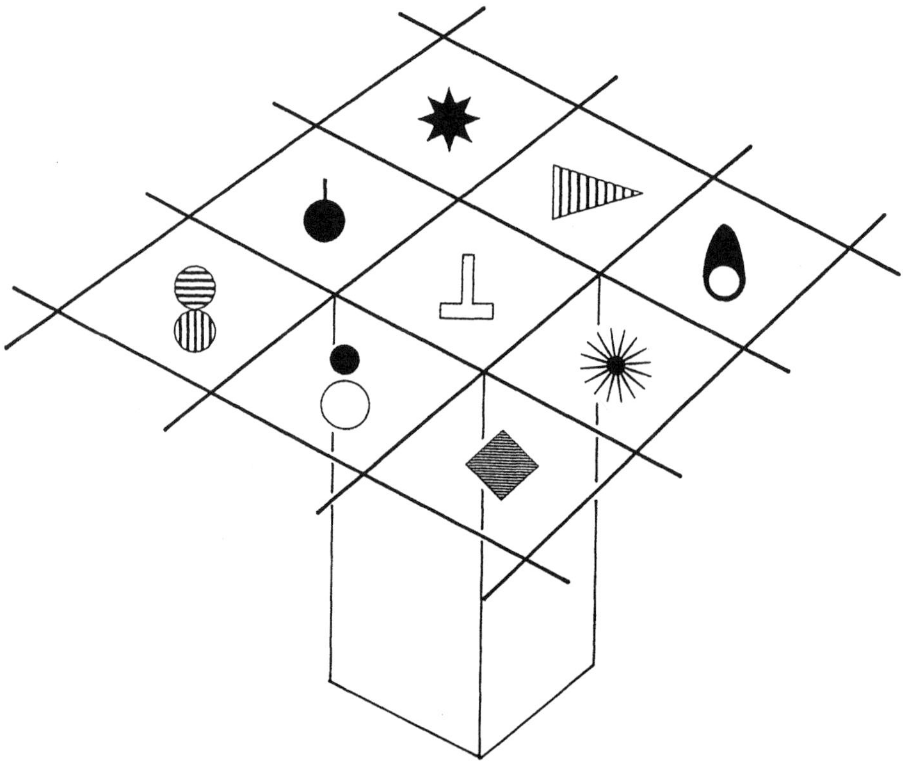

FIGURE 8. Schematic drawing of the columnar organization in TE.

In summary, the TE region is composed of columnar modules in which cells within a column respond to similar but not identical features (**Figure 8**). The width of a columnar module across the cortical surface may be slightly greater than 400 μm. The span of a column along an oblique penetration is smaller than the true width of the column when the penetration crosses its periphery. The number of modules, which was estimated by a division of the whole surface area of TE into 500-by-500-μm squares, was 1300 modules. Although the columns are drawn as discrete units in **Figure 8**, the overlap and continuity between neighboring columns is estimated later in the chapter using optical imaging.

Organization of afferents to TE

Selective responses to complex features, which were first observed in TE cells, can be traced to earlier stages in the afferent pathway to TE. We have found that cells requiring such complex features for their maximal activation were already present in TEO and V4 (Kobatake & Tanaka, 1994), although their proportion was small.

To compare such cells in TEO and V4 with cells in TE, we compared the responses of individual cells to a fixed set of simple features versus individually determined critical features (Kobatake & Tanaka, 1994). The fixed set of stimuli consisted of: (1) 16 bars at four orientations differing by 45° and two sizes (either 0.5° x 2° or 0.5° x 10°) and (2) 16 squares of four different colors and two sizes (0.5° x 0.5° or 2.5° x 2.5°). Stimuli both darker than and lighter than the background were included. This set was sufficient to evoke good, although submaximal, responses in cells in V2 and V4 that showed selectivity only in the domain of orientation or color, and size. Most cells in TE either did not respond to the simple stimuli included in the set, or their responses were minimal compared with their responses to the individually determined complex critical features (as shown in **Figure 9**, left).

TEO and V4 contained cells with various levels of stimulus selectivity. Some TEO and V4 cells showed no or negligible responses to any of the simple stimuli, as had cells in TE, while others showed moderately strong responses to some of the simple stimuli in addition to the maximal response to the complex critical features (as shown in **Figure 9**, right). The remaining cells even responded maximally to some of the simple stimuli. **Figure 10** shows the proportion of occurrence of three groups of cells classified by the magnitude of their maximum response to the simple stimuli normalized for the overall maximum response of the cell.

TEO and V4 were characterized by a mixture of cells with various levels of selectivity. We may take this mixture as evidence that selectivity is constructed through local networks in these regions. If we randomly sample cells from a local network in which selective responses to complex features are constructed by integrating simple features, the sample should include cells having various levels of selectivity. Cells located close to the input end of the network should be maximally activated by simple features, those close to the output end should respond only to complex features, and those at intermediate stages should show intermediate properties.

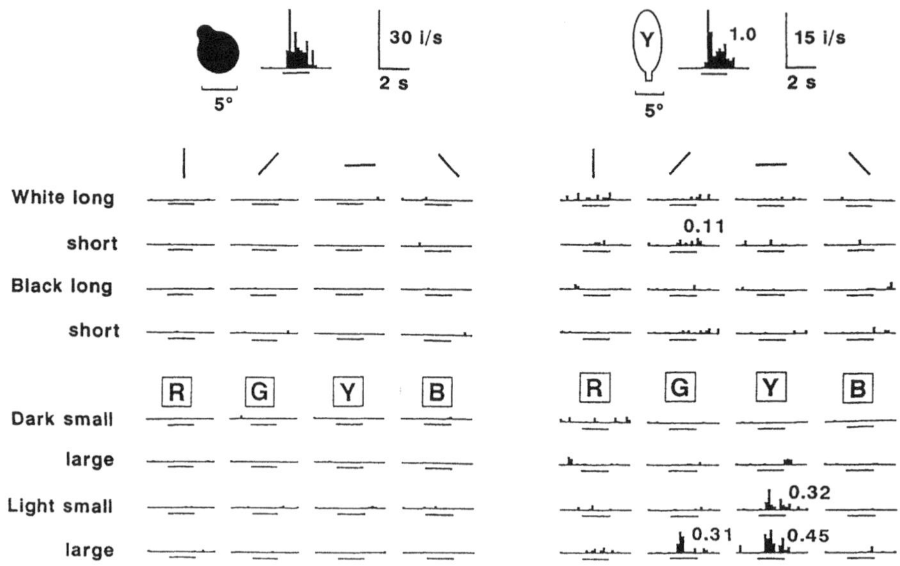

FIGURE 9. Responses of a TEd cell (left) and TEO cell (right) to a set of simple stimuli. Their responses to individually determined critical features are shown at the top of the figure.

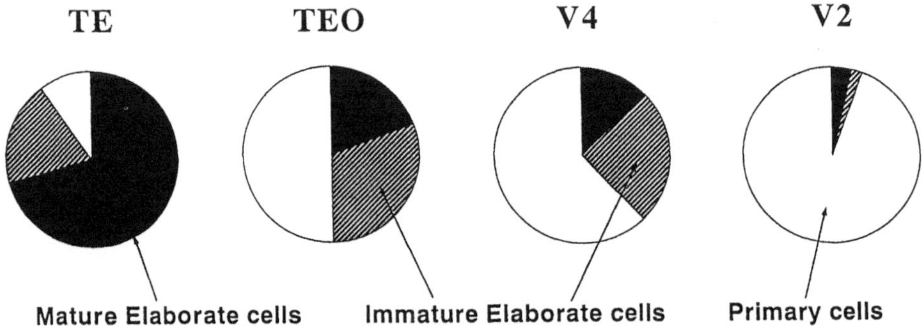

FIGURE 10. Proportion of cells with different levels of selectivity to complex features. The maximum response to the simple stimuli was less than 0.25 of the response to the complex critical feature for Mature Elaborate cells, between 0.25 and 0.75 for Immature Elaborate cells, and greater than 0.75 for Primary cells.

Although we did not find clear evidence of selectivity to complex features in V2, slightly stronger responses to complex patterns than to simpler stimuli—such as bars and gratings—were not unusual in this area (Lehky, Sejnowski, & Desimone, 1992). Selectivity to moderately complex features may develop gradually throughout the lower stages and become apparent in V4 and TEO.

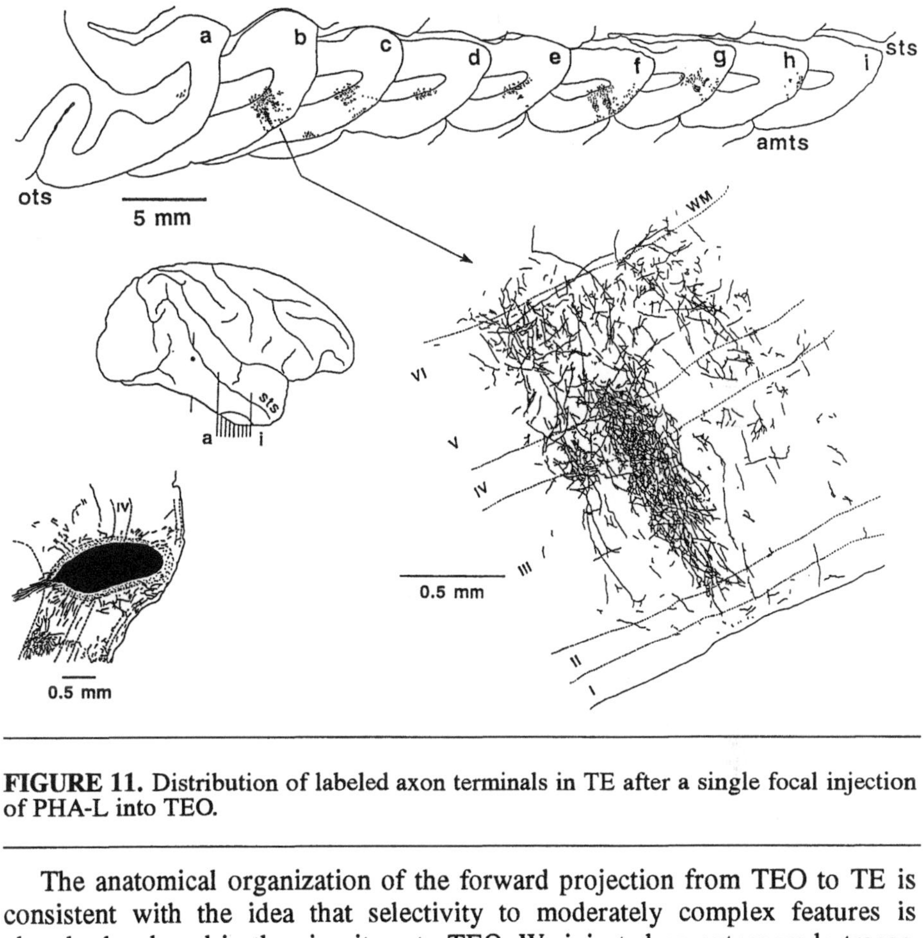

FIGURE 11. Distribution of labeled axon terminals in TE after a single focal injection of PHA-L into TEO.

The anatomical organization of the forward projection from TEO to TE is consistent with the idea that selectivity to moderately complex features is already developed in the circuit up to TEO. We injected an anterograde tracer, PHA-L, into a single small region (the horizontal width of the injection sites was 330-600 μm) of the part of TEO representing the central visual field, and observed labeled axon terminals in TE (Saleem, Tanaka, & Rockland, 1993). Labeled terminals were largely limited to three to five focal regions in TE (**Figure 11**). In each of the projection foci, the labeled terminals were not restricted to the middle layers, but were distributed to columnar regions extending from layer 1 to layer 6. The horizontal width of the columnar foci was

200-380 μm, which was slightly smaller than the electrophysiologically determined width of columns in TE. As noted above, the receptive fields of cells in TE are large, usually including the fovea, and no retinotopical organization has been found in TE. Thus, the specificity of connections from TEO to TE should be defined in the feature space, not in the retinotopical space. It is suggested that outputs from a single site of TEO may carry information about a particular complex feature, and they are sent, therefore, only to a limited number of foci in TE.

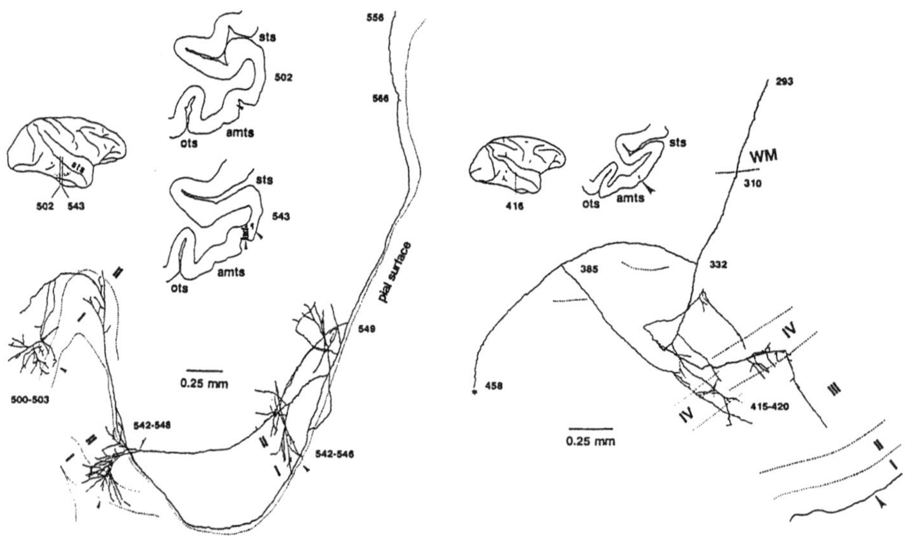

FIGURE 12. Two single axons projecting from TEO to TE. The axon shown in the left half made terminal arbors only in layers 1 and 2, whereas the axon on the right made arbors only around layer 4.

Although the overall distribution of labeled terminals was elongated to form a columnar region, this does not necessarily mean that individual axons make arbors in columnar regions. Indeed, single axons reconstructed from serial sections were heterogenous in shape. Some axons terminated exclusively in layer 4 and the bottom of layer 3, as the one shown in **Figure 12**, right. Some other axons terminated only in layers 1 and 2, as the one shown in **Figure 12**, left. There were also single axons that terminated in elongated regions, including both middle and superficial layers. It is possible that a single site in TEO sends many different kinds of information about a complex feature to a column in TE, and that they interact with each other through the local network within the TE column.

In summary, two characteristics may emerge first in TE: One is the columnar arrangement of cells with overlapping yet slightly different selectivity in local regions, and the other is the invariance of responding for stimuli at different positions in the visual field. The receptive fields of cells in TE are large, including the fovea, and the selectivity of responses is essentially constant throughout these large receptive fields. A significant fraction of the cells in TEO and V4 respond to moderately complex stimuli as in TE. However, the receptive fields of cells in TEO and V4 are still much smaller than those of cells in TE and are retinotopically organized (Boussaoud, Desimone, & Ungerleider, 1991; Kobatake & Tanaka, 1994). This indicates that there are two steps in the formation of cells that respond to integrated features and are invariant with respect to changes in stimulus position. First, the selectivity is constructed for stimuli at a particular retinal position in TEO and V4, and then the invariance is achieved in TE by obtaining inputs of the same selectivity but having receptive fields at different retinal positions.

Optical Imaging of the Columnar Organization

To characterize further the spatial properties of columnar organization in TE, we used the technique of optical imaging (Wang, Tanaka, & Tanifuji, 1994). The intrinsic signals, which are thought to originate mainly in the increase of deoxidized hemoglobin in capillaries around regions of elevated neuronal activity (Frostig, Lieke, Ts'o, & Grinvald, 1990), were measured. The cortical surface was exposed and illuminated with red light tuned to 605 nm. Activated neuronal tissue utilizes oxygen from hemoglobin, so the density of deoxidized hemoglobin increases in nearby capillaries. Because deoxidized hemoglobin absorbs much more light at 605 nm than oxidized hemoglobin, the region of cortex with elevated neuronal activities becomes darker in the reflected image.

Method

To find visual stimuli that would be effective for the area of TE exposed for optical imaging, and to establish the relation between the optical changes and elevated neuronal activities in TE, we combined single-cell electrophysiological recordings with optical imaging. Unlike previous studies of V1 and V2, the single-cell recordings were conducted in separate sessions prior to the optical imaging session. Activities of single cells were recorded with a microelectrode, and the critical features were determined for 15 to 25 cells recorded in six to eight penetrations at different positions. The optical imaging was performed while the critical stimuli and various control stimuli were presented. Five to 25 different stimuli were used, each of which was presented 24 or 40 times for four seconds.

To determine the change in activity due to the effect of the critical stimuli, the intensity of the image obtained when a critical stimulus was presented was divided by the intensity obtained under baseline conditions. Two baseline conditions were used, one in which no visual stimulus was presented and a second in which stimuli were presented but not containing the critical features. The first baseline condition measured the activity generated by the critical features against the level of background activity, and the second against the level of activity induced by visual stimuli that did not contain the critical features.

Findings

Each of the critical features determined in the preceding unit-recording sessions activated two to 12 dark spots within the imaged region of TE (3.3 by 6.1 mm). The locations of the spots were different for different features, and one of the spots covered the position of the electrode penetration from which the critical feature was determined. The average diameter of individual spots was 490 μm, which approximately coincided with the width of columns in TE inferred from unit-recording experiments (Fujita et al, 1992). Note that observable changes in the optical signals are obtained only if a large proportion of the cells within a column were activated by the critical feature. Thus, clustering of cells that responded to a moderately complex feature was confirmed.

The set of visual stimuli used in one block of optical imaging included three critical features, which were determined for three different cells recorded in the same penetration. Two of them were combinations of two colors of different luminosity, and the third one included a gradation of color from light blue to dark blue. All three stimuli evoked dark spots around the electrode penetration. The spots all covered the position of penetration, but they overlapped only partially. Each of the spots was about 500 μm in size, and the size of the overall region was 1100 μm. The stimuli were similar in that each contained a change in luminosity.

Similar stimuli also produced partially overlapping activations in another case. The stimuli were a series of faces presented from different viewing angles. All of the five cells recorded in an electrode penetration selectively responded to the sight of a face. For one of the cells, the image of a face could be simplified to a combination of eyes and nose, but we did not find simplified stimuli for the remaining four cells. Three of the five cells maximally responded to frontal images, and the other two maximally responded to profiles. Thus, we included five different views of one face in the stimulus set used for optical imaging.

In the optical-imaging phase of the experiment, all of the face stimuli evoked activation spots around the penetration (we failed to recover the exact location of the electrode penetration in this case). Each activation spot partially overlapped the other four spots, and, moreover, the center positions of the spots systematically moved in one direction as the face turned from the left profile to the right profile through the frontal and 45-degree images. Individual spots were 300-400 μm wide, and the overall region they covered was 800 μm in the axis along which the centers of the spots moved. This systematic shift of activation was also obtained with the face of another person: At the same viewing angle, the two faces activated the same region. The possibility that the systematic shift was caused by movement of a part of the face (e.g., the nose) was excluded because, even at different horizontal positions, the frontal image always activated the region at the center.

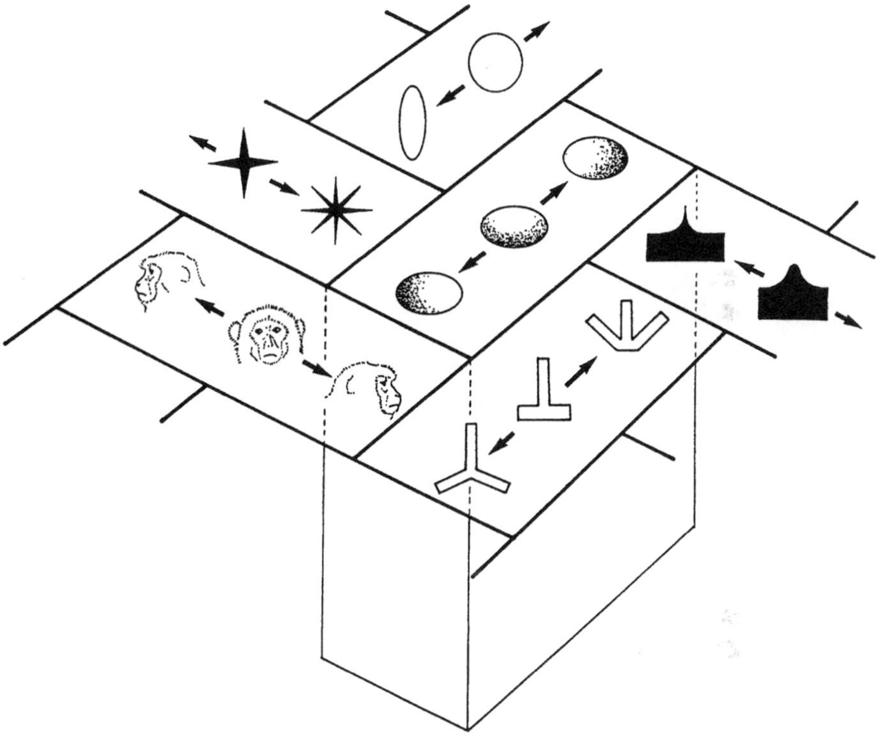

FIGURE 13. Revised schematic diagram of columnar organization in TE.

These facts suggest that several columns representing different but related features overlap one another and together compose a larger-scale unit in TE. The findings obtained when the viewing angle of a face was varied also suggest that some complex features are continuously mapped within larger scale groups of cells (**Figure 13**). Whether the mapping is continuous throughout a large part of TE, or whether discontinuities exist between groups of cells whose diameters are approximately one mm, is yet unknown. Considering the large dimensions of the feature space that TE should represent, the latter is more likely.

FIGURE 14. Twenty-eight shape stimuli used in delayed matching-to-sample training with adult monkeys.

Changeability of Selectivity in the Adult

The stimulus selectivity of cells in TE changes in the adult according to changes in the visual environment. To demonstrate such changes, we trained two adult monkeys to discriminate 28 moderately complex shapes, shown in **Figure 14**, using a behavioral test apparatus containing a visual display equipped with a touch screen to monitor responding. We used a delayed matching-to-sample task (Kobatake, Tanaka, & Tamori, 1992; Kobatake, Tanaka, Wang, & Tamori, 1993) in which one of the shapes served as the sample stimulus, and changed randomly from trial to trial. After the screen for the sample display had been touched, a 16-second period elapsed (the delay period) during which the sample stimulus was not present. Then, five shapes (the comparison stimuli) were presented at different display positions. One of the comparison stimuli was the same as the sample stimulus, and if the screen for this stimulus was touched the monkey received a drop of juice. After a year of training, the monkeys were placed under anesthesia for repeated recordings from cells in TE. For individual cells, we determined the best stimulus from the set of animal and plant models we had used to investigate critical features in naive monkeys. The response to the best object stimulus was then compared with the responses of the same cell to the shape stimuli used in training. Because of time constraints, we did not perform the reduction process in this experiment.

In TE of the trained monkeys, about one-fourth of the cells gave a maximum response to some of the stimuli used in the training. Conversely, only 5% of TE cells in untrained animals responded maximally to these stimuli. These results indicate that the number of cells responding to the training stimuli increased due to the year-long discrimination training. However, the spatial organization of the modified cells in the cortex is yet to be studied. We do not know whether new columns were formed for the discrimination of training stimuli, or whether some of the cells in the columns present before training became tuned to the training stimuli. Whether similar changes happened in TEO and V4 is also unknown.

Earlier work by Sakai and Miyashita (1991) had demonstrated the effects of discrimination training on stimulus selectivity in adult monkeys, but only indirectly. Using a delayed matching-to-sample procedure with Fourier descriptors as stimuli, they reinforced responding directed toward one of two comparison stimuli. Unlike the prior study, the sample and comparison stimuli were not identical, but were arbitrarily related to one another by the procedure. After training for about a month, during which the arbitrary sample-comparison correspondence was acquired, recordings from TE cells were made using the same task paradigm. Some cells responded to two stimuli composing a pair, and these responses were significantly more frequent than would be expected by chance. Considering that the pairing of stimuli was arbitrary, the dual responsiveness of the cells to the paired stimuli could have been formed only through the adult training.

In two studies, responses of cells in TE and surrounding regions changed within the course of recording from the same single cell. Miller and his colleagues (Miller, Li, & Desimone, 1991; Li, Miller, & Desimone, 1993) found that, as the newly introduced stimulus became familiar, responses to the stimulus decreased in cells at the border region between TE and the perirhinal cortex. This effect is to be distinguished from habituation of responses to successive presentation of the same stimulus, because the decrease occurred even after several intervening presentations of different stimuli. However, because the changes that Miller et al (1991) observed were opposite in direction to those we observed after long training, the two phenomena are not likely to be related. Rolls et al (1989) found that responses of cells in TE and the ventral bank of the superior temporal sulcus to a set of faces changed rapidly after the introduction of a new set of faces. These changes included both increased and decreased responses. In our own experiments, the changes observed in the responses of cells likely occurred over a longer time period because the performance of the monkeys improved throughout the five months of training.

Functions of the TE Columns

Columnar organization suggests that an object feature is not represented by the activity of a single cell, but by the activity of many cells within a single columnar module. Representation of features by columnar modules whose constituent cells have overlapping selectivity to the effective stimuli satisfies two apparently conflicting requirements in visual recognition: robustness to subtle changes in input images, and precision of representation. While the image of an object projected to the retina changes due to changes in illumination, viewing angle, and articulation of the object, the global organization of outputs from TE changes very little. The clustering of cells with overlapping and slightly differing selectivity buffers the columnar module against these changes.

Representation by multiple cells with overlapping selectivity can be more precise than a mere summation of representation by individual cells. A similar argument has been made for hyperacuity (Erickson, 1968; Snippe & Koenderink, 1992). The position of the receptive fields changes gradually in the retina with a large overlap among nearby cells. By taking the difference between the activity of nearby cells, an acuity much smaller than the size of the receptive fields is produced. A similar mechanism to that in retinal space may work in feature space with largely overlapping and gradually changing selectivity, as discussed by Trehub in this volume. A subtle change in a particular feature that does not markedly change the activity of individual cells can be coded by the differences in the activity of cells with overlapping and slightly different selectivity.

The function of the columnar organization in TE may go beyond the discrimination of input images. The optical-imaging experiments suggest that there is a continuous mapping of features within cortical units about one mm in size across the cortical surface. This may have two functional consequences. One effect is to make an evenly distributed variety of cell properties along the feature axis. Continuous mapping may be a tool to make the full divergence without omission (Malach, 1994; Purves, Riddle, & LaMantia, 1992). A second effect is to permit computations that involve variations on the feature to be carried out by local neuronal circuits. These computations may transfer the image of an object for three-dimensional rotations and for production of the image under different illumination conditions and articulation poses.

A series of studies recently performed with slices of rat motor cortex suggest that there are two kinds of connections between pyramidal neurons through their axon collaterals (Thomson & Deuchars, 1994). Pyramidal cells within a narrow columnar region 50 to 100 μm wide are tightly connected by synapses on the basal dendrites or the proximal part of the apical dendrites. Activation by these connections produces rapidly rising, large excitatory postsynaptic potentials (EPSPs) that cause the pyramidal cells to tend to fire synchronously. The second anatomical arrangement also favors the synchronous firing of pyramidal cells within a narrow column. Input axons appear to make synapses indiscriminately on the apical dendrites of these pyramidal cells, which are gathered together to form bundles (Peters & Yilmaz, 1993). The resulting narrow column corresponds to the "minicolumn" of Mountcastle (1978). In contrast, pyramidal cells with a longer horizontal distance are connected by synapses at the distal part of the apical dendrites. The EPSPs are small and slowly rising, but long lasting: They may contain the N-methyl-D-aspartate (NMDA) type of glutamate receptors.

Taken together, we may draw a schema for area TE in which cells within a minicolumn make up a functional unit, receiving common inputs and mutually exciting one another. Nearby minicolumns exert weak, but long-lasting, effects on each other. After a minicolumn is activated by the retinal visual input, subthreshold activation propagates from it to nearby minicolumns, forming a pattern of activation with a focus. The focus of activation may move from one minicolumn to another, through interaction with distant activation foci in TE or with other brain sites. This mechanism may be used for various kinds of computation that the visual system must conduct to realize the flexibility of visual recognition, such as transfer of the image of an object for three-dimensional rotations, production of the image under different illumination conditions, and, in the case of faces, recognition of the image with different expressions. In this way, the columnar organization of TE may provide an overlapping and continuous representation of object features upon which various kinds of calculations can be performed.

Binding Activities in Distant Columns

Because object features to which individual TE cells responded were only moderately complex and cells within a single column responded to similar features, the computation performed within a column can provide only information on partial (but not necessarily local) features of object images. To represent the whole image of an object, computations in several or several tens of different columns must be combined. This raises the problem of "binding," that is, how are different sets of activity discriminated when there are more than two objects in nearby retinal positions? The receptive fields of TE cells are too large to discriminate different objects according to their retinal positions. The problem of binding exists regardless of the presence or absence of so-called "concept units" in brain sites beyond TE. Concept units, if present, have to discriminate different sets of TE activity originating in different objects.

One possible mechanism to solve the problem is the synchronization of firing of different modules (Engel, Konig, Kreiter, Schillen, & Singer, 1992; Singer, 1993). If the image of an object synchronizes the firing of some modules, and the activity of those modules is not coordinated with the firing of other modules activated by simultaneously present different objects, then this synchrony can provide a basis by which the features of the same object may be bound together. Firing synchronized with oscillations has been found between cells in the cat visual cortex, and some context dependency of the synchronization has also been reported. Although oscillating firing has not been found in TE (Young, Tanaka, & Yamane, 1992; Tovee & Rolls, 1992), non-periodic synchronization may be present in TE.

Another possible mechanism of binding in TE is selection by attention (Crick, 1984). We can pay attention to only one object at a time, or a few at most. If representation of the features of an attended object is enhanced and that of other objects is suppressed, the binding problem is resolved. This is a likely mechanism, because strong effects of attention have been found on responses of TE cells (Richmond & Sato, 1987; Moran & Desimone, 1985; Spitzer, Desimone, & Moran, 1988; Chelazzi, Miller, Duncan, & Desimone, 1993).

A third possibility is that a single object initiates various separate sites of activity in TE that are then integrated through interactions involving loops of activity with retinotopically organized activity in earlier stages of the ventral pathway. TE projects back to TEO, V4, V2, and even V1 (Rockland, Saleem, & Tanaka, 1994; Rockland & Van Hoesen, 1994). There are also step-by-step feedback projections. Kawato and his colleagues (Kawato, Inui, Hongo, & Hayakawa, 1991; Kawato, Hayakawa, & Inui, 1993) have made a similar suggestion concerning the potential importance of feedback projections in integrating the disparate neural activity generated by a single object. (See also Singer, this volume.)

"Face Neurons" in the Anterior Part of the Superior Temporal Sulcus (STPa)

In the early 1980s, cells were found in STPa that selectively responded to the sight of a face (Bruce, Desimone, & Gross, 1981; Perrett, Rolls, & Caan, 1982; Perrett, Mistlin, & Chitty, 1987; Rolls, 1992; Yamane, Kaji, & Kawano, 1988; Young & Yamane, 1992). These so-called face neurons have been extensively studied. There are reports that such cells are also present in TE itself (Baylis, Rolls, & Leonard, 1987; Tanaka et al, 1991), in area TG (Nakamura, Matsumoto, Mikami, & Kubota, 1994), and in the amygdala (Leonard, Rolls, Wilson, & Baylis, 1985; Nakamura, Mikami, & Kubota, 1992). The meaning of "selectivity" varied among these studies: Some presented only a few non-face stimuli, and most did not test partial features of the image of a face. However, a few used an image made with scrambled patches of a picture of a face, and this scrambled face was not effective (Bruce et al, 1981). A few other studies found cells that were not activated by a face without the eye or by the eye only (Perrett et al, 1982; Rolls et al, 1985). We found systematically arranged columns in TE that responded to different views of faces. These data indicate that, in order to activate some face cells, all of the essential features that make up a face must be present. Thus the image of a face may activate more complex features than some other objects represented by cells in TE.

The presence of face neurons cannot be generalized to the representation of other objects. Faces are important media for social communication between individuals. Faces of monkeys have special significance for monkeys and, for laboratory monkeys, so do those of human beings. There are reports that the activity of face neurons in STPa represents individual differences and expressions by means of a population coding (Baylis, Rolls, & Leonard, 1985; Young & Yamane, 1992; Rolls & Tovee, 1995). Discriminating faces from other objects is only a preliminary stage in representing expressions or features of individual faces. The units prepared for population coding of individual differences and expressions of faces may appear as "cognitive units" for faces. This view is supported by the finding that some STPa cells respond to the view of body poses or movements or hand actions (Perrett, Smith, Mistlin, Chitty, Head, Potter, Broennimann, Milner, & Jeeves, 1985; Perrett, Harries, Bevan, Thomas, Benson, Mistlin, Chitty, Hietanen, & Ortega, 1989; Perrett, Hietanen, Oram, & Benson, 1992; Oram & Perrett, 1994). Body movements and hand actions often express important information about the relation between individuals in the scene, or between the individual in the scene and the observer. These groups of cells in STPa may be specially prepared for social communication.

Object Recognition by Activities Distributed over the Brain

Sakata and his colleagues have recently found shape-selective cell activity in areas of the intraparietal sulcus (Sakata & Kusunoki, 1992; Taira, Mine, Georgopoulos, Murata, & Sakata, 1990). Many cells in the lateral bank of the sulcus selectively responded to visual images of switches that the monkey had been trained to manipulate. The switches varied in shape, so the monkey positioned its hand differently when it reached toward different switches. The discharges began when the monkey saw the switches, and—because the discharges decreased when the task was performed in a dark room—the activity was evoked, in part at least, by visual inputs.

In addition, some cells in the more posterior part of the sulcus responded to more primitive features of stimuli, including the 3D orientation of a pole or the 3D tilt of a plane. Using a computer graphic system that permitted the presentation of stimuli with or without retinal disparity, they found that the neural activity initiated by the images declined when binocular disparity was eliminated (Kusunoki, Tanaka, Ohtsuka, Ishiyama, & Sakata, 1993; Tanaka, Kusunoki, Ohtsuka, Takiura, & Sakata, 1992; Tanaka, Murata, Taira, Shikata, & Sakata, 1994). Thus, binocular disparity was a critical cue for the responsiveness of these cells.

Taken together, these findings seem to indicate a flow of information concerning the 3D shape of objects that the monkey may manipulate. This coincides with the proposition by Goodale and his colleagues (Goodale, Milner, Jakobson, & Carey, 1991; Goodale & Milner, 1992), based on human clinical data, that the dorsal pathway leading to the parietal cortex is responsible for "visuo-motor control," but not for "spatial vision" (Mishkin, Ungerleider, & Macko, 1983). For an object to be manipulated, its 3D structure should be perceived. There may be a representation of object shape in the dorsal pathway that is independent of the representation of objects in the ventral pathway and, probably, that represents the shape only coarsely. This dorsal representation may influence the representation of objects in the ventral pathway through the indirect connection via the parahippocampal structures (Van Hoesen, 1982; Suzuki & Amaral, 1994) or via regions in the superior temporal sulcus (Seltzer & Pandya, 1978, 1984, 1989, 1994).

The accumulated findings favor the idea that there are no cognitive units representing the concept of objects, but that instead the concept is found in activities distributed over various regions in the brain. When the visual image of an object is presented, it is processed in the ventral visual pathways and a representation, including its similarity to other objects and different views of the same object under different conditions, is reconstructed there. This representation in the ventral visual pathway utilizes population coding on two levels. First, the image of the object is represented by a combination of multiple partial (including both holistic and local) features represented by different columns in TE. Because the partial features are represented in an analog manner, the combinatorial representation can be understood as a combination of similarities to different "prototypes" of object images (Edelman, 1995). The second level of population coding occurs in the representation of partial features. The features are represented by multiple cells within a TE column that have overlapping selectivity. Triggered by inputs from the inferotemporal cortex, emotional information about the object is read out in the amygdala, associations with other objects are read out through the perirhinal cortex, and the behavioral significance emerges from the prefrontal cortex. The visual image of the object is also processed in the dorsal pathway, and information necessary for the monkey to manipulate the object is read out in the parietal cortex. All this recovered information, distributed over the brain, may constitute the neural basis for the concept of an object.

PLATE 1 (Tanaka, Chapter 9). Sixteen examples of the critical features of cells in TEd. They are moderately complex.

PLATE 2 (Senft, Chapter 3). A set of somatic groups arrayed as arching laminae, to represent a sector of cerebral cortex (neurites exist in this simulation, but have been rendered invisible). Color is used to differentiate the overlapping cell groups assigned to the various "cortical layers."

PLATE 3 (Senft, Chapter 3). A subset of elements from a larger simulation representing cell groups in the brainstem, thalamus and cortex. A set of fibers from the brainstem invade the thalamus from below.

PLATE 4 (Luskin, Chapter 2). A representative SVZa-derived cell that recently synthesized DNA in the migratory pathway of a P3 animal. The SVZa was injected with retrovirus (encoding lacZ) on P0, and BrdU was administered 4 and 2 hours prior to perfusion. The anti-BrdU staining was visualized immunohistochemically using the ABC-immunoperoxidase method and VIP as a chromogen. The chromatin of the BrdU-positive cells appears dark purple; it is surrounded by the blue ß-galactosidase histochemical reaction product in the cytoplasm of the lacZ-positive cells. Note that although the double-labeled cell in the middle of the photomicrograph recently replicated, it has an elongated cell body and a leading process, characteristic of a migrating neuron. Scale bar: 20 μm. Dorsal is up, anterior is to the right. (Modified from Menezes et al, 1995 with permission.)

PLATE 5 (Georgopoulos, Chapter 13). Top: Preferred directions of 475 directionally tuned cells recorded during a 3D reaching task. Lines are vectors of unit length. (From Schwartz et al, 1988. Reproduced with permission; copyright by the Society for Neuroscience.) Middle: The same population of cells (light blue lines) shaped for a movement in the direction indicated by the yellow line. Their length is proportional to the changes in cell activity associated with the particular movement direction illustrated. The direction of the population vector (orange) is close to that of the movement. (From Georgopoulos et al, 1988. Reproduced with permission; copyright by the Society for Neuroscience.) Bottom: 95% confidence cone for the direction of the population vector (line in the center of the cone); the movement direction (yellow line) is within the cone. (From Georgopoulos et al, 1988. Reproduced with permission; copyright by the Society for Neuroscience.)

PLATE 1 (Tanaka, Chapter 9).

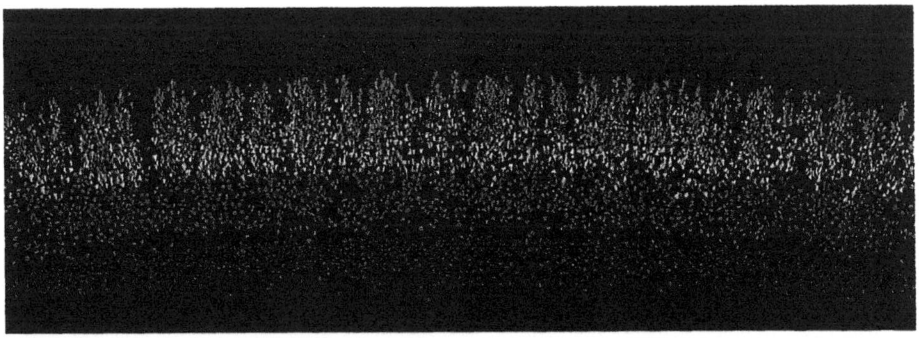

PLATE 2 (Senft, Chapter 3).

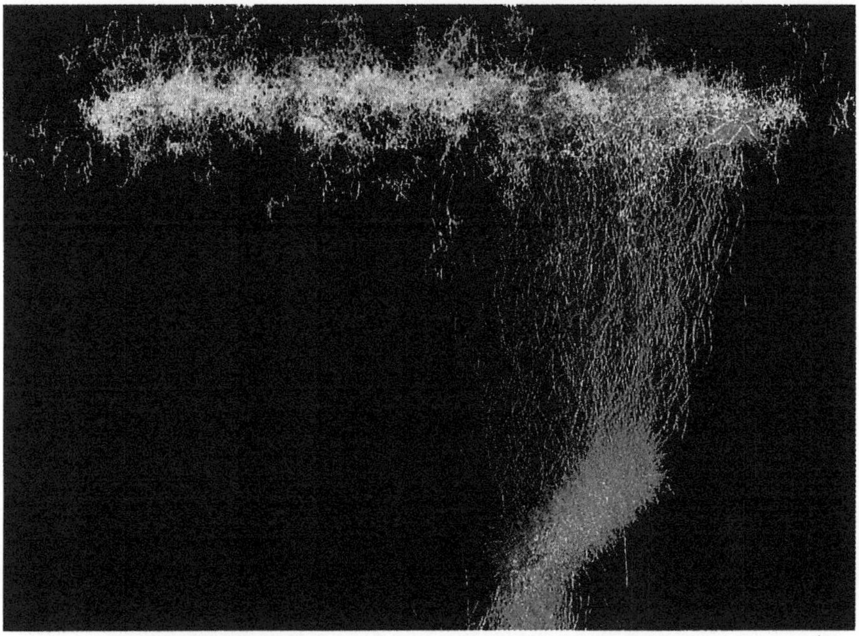

PLATE 3 (Senft, Chapter 3).

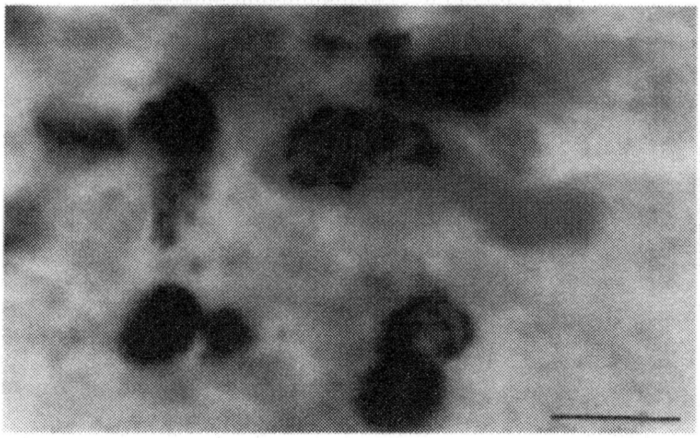

PLATE 4 (Luskin, Chapter 2).

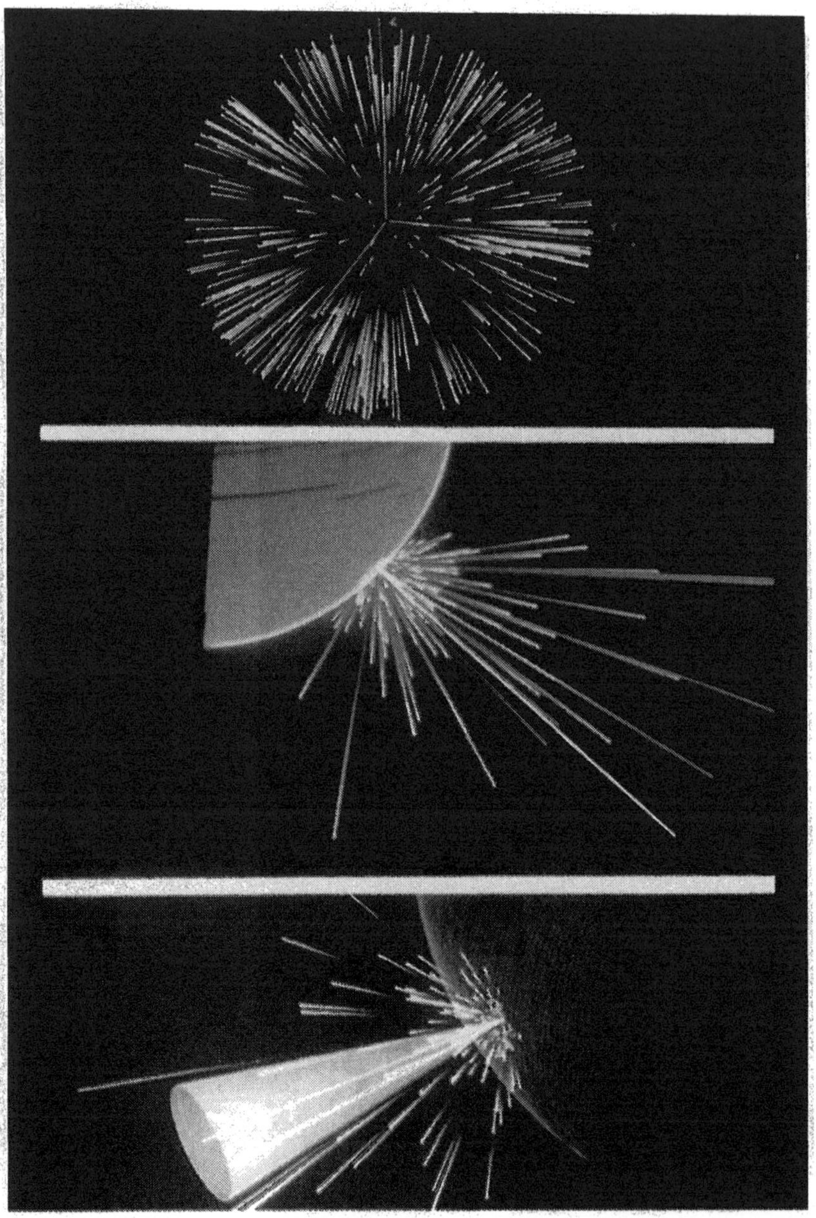

PLATE 5 (Georgopoulos, Chapter 13).

J.W. Donahoe and V. Packard Dorsel (Editors)

CHAPTER 10

SPARSE CODING OF FACES IN A NEURONAL MODEL: INTERPRETING CELL POPULATION RESPONSE IN OBJECT RECOGNITION

Arnold Trehub
Department of Psychology
Neuroscience and Behavior Program
University of Massachusetts, Amherst

ABSTRACT

Response to faces as measured by cell discharge in the temporal cortex of monkeys suggests a sparse cell-population coding of complex visual stimuli. The prevailing view assumes that a sparse population code requires the joint contribution of a relatively small group of cells (a neuronal ensemble) for effective coding and recognition. This assumption is based primarily on the consistent observation that single cells in the temporal cortex are broadly tuned rather than narrowly tuned to individual faces. It has been argued that the joint activity of a relatively small number of broadly tuned cells, each responsive to a different constituent feature of a face, could form an ensemble code selective enough to distinguish individual faces. In the present study, schematic faces were presented as stimuli to a model neuronal system for visual pattern learning and recognition. This model effectively codes individual faces by means of competitive activity among single cells during recognition instead of by ensemble coding. The computer simulation permitted an analysis of the activity profiles of all tuned cells during learning and recognition of the faces. All cells were found to be broadly tuned even though coding was mediated by the discrete output of single cells on a competitive basis in a sparse neuronal population rather than by the joint activity of a group of cells. The results show that the observation of broad tuning of cells in temporal cortex under typical experimental conditions does not warrant the conclusion that neuronal ensembles are required for the coding of individual faces. Suggestions are made for changes in the design of experiments to better test hypotheses about the coding of faces (or any other complex visual patterns).

Introduction

A central question for our understanding of visual pattern recognition in the brain is how neurons in the visual system code perceived objects. Face recognition is a particularly important aspect of complex pattern recognition and, following the early reports of face-selective cells in the temporal cortex of monkeys (Gross, Rocha-Miranda, & Bender, 1972; Desimone, Albright, Gross, & Bruce, 1984), there has been a major effort to understand the neuronal coding of faces (see, for example, Kosslyn & Mumford, 1991; Bruce, Cowey, Ellis, & Perrett, 1992).

In the typical experimental procedure, the spike discharge of single cells in the inferior temporal cortex of the monkey is recorded while the animal is presented with pictures or drawings of faces. It has been observed that cells which selectively discharge in response to faces as a stimulus class exhibit broad tuning curves in response to the faces of particular individuals. Mainly on the basis of this observation, the prevailing view is that single cells cannot adequately account for selective recognition of individual faces. Instead, it has been proposed that the neuronal processing is in the form of a sparse population code wherein face recognition requires the joint contribution of a small population of cells, each selectively responsive to the presence of a different facial feature (Baylis, Rolls, & Leonard, 1985; Young & Yamane, 1992). In this formulation, it is the *pattern* of activity over an ensemble of cells (a joint activity vector) that constitutes the recognition code (Gross, 1992; Gross & Sergent, 1992).

An unresolved issue is how a neuronal population code, sparse though it may be, can selectively evoke a correct recognition response to a particular member of a stimulus category. This paper examines the activity levels of individual cells in a simulated neuronal model of visual object recognition when the system is required to learn and recognize each face in a group of line-drawn faces. Analysis of cell response profiles suggests an alternative interpretation against the common view that sparse coding of a complex visual pattern such as an individual face implies a neuronal ensemble of separately coded features. The results indicate that instead of a coding scheme based upon an ensemble of separate features, a sparse group of cells where each is holistically tuned to a different exemplar of a particular face provides effective face recognition.

Brief Description of Model

The neuronal model simulated here (Trehub, 1991, chapters 2, 3, 4, 5, and 7) consisted of five key integrated mechanisms: (1) a 16x16-cell foveal retina; (2) a mechanism for triggering saccadic excursions to regions of high edge density in the visual field; (3) a putative post-retinal mechanism for positioning the centroids of retinotopic excitation patterns close to a standard internal axis (stimulus capture); (4) a learning mechanism for tuning synaptic transfer weights on individual adaptive cells (filter cells) in a detection set to patterns of retinal stimulation; (5) a competitive (winner-take-all) mechanism that selects a recognition response contingent on the relative activation levels of cells in the detection set in the context of each stimulus. The spike frequency of each cell can be considered as a positive monotonic function of its activation level.

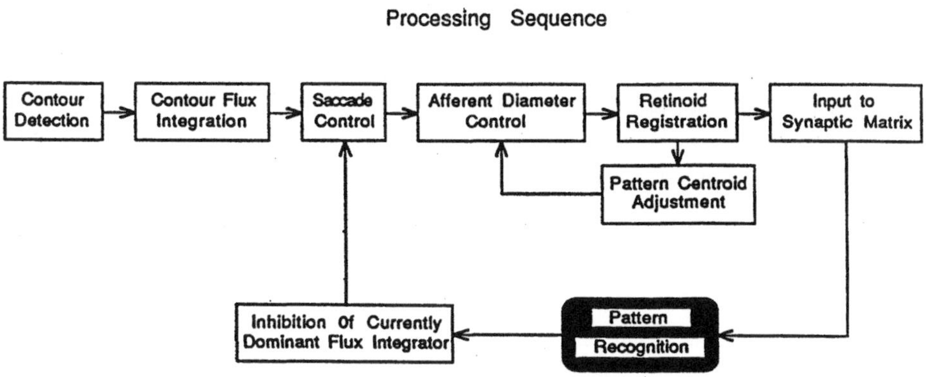

FIGURE 1. Processing sequence for parsing an object in a complex visual environment.

Figure 1 illustrates the processing sequence for parsing an individual face out of the set of 10 faces presented in the model's visual field. The total field is analyzed by an array of retinotopically indexed cells (flux detectors), each of which receives input from a relatively small region of the complete retinal field. Each flux detector integrates the amount of visual contour excitation in its particular retinal region and discharges with a frequency proportional to its total excitation. The contour flux detectors feed a matched array of cells that control visual saccades. The flux detector with the highest discharge frequency captures control of the saccadic apparatus and directs a saccade to the circumscribed visual region that provides its input. Thus, the region of visual space with the highest contour density will be fixated first.

There is a field constriction mechanism that limits the fovea-centered aperture of retinotopic input to a short-term memory module called a retinoid (Trehub, 1977). The retinoid can translate retinotopic excitation patterns over cells in an egocentric coordinate space. It is quadrantally organized, and automatically locates and positions pattern centroids on a standard reference axis within the visual system by shifting its visual pattern so that excitation is balanced within a threshold of tolerance over all quadrants. At the start of the parsing process, the visual field aperture is constricted to a small window on the stimulus field and an initial tolerance level is set for hemifield mismatch in the retinoid system. The centroid of the current effective visual pattern is then shifted to the standard egocentric reference axis. The visual aperture is progressively enlarged in a stepwise fashion and, at the same time, the system relaxes its tolerance for quadrantal-excitation imbalances. At each step, the system seeks to adjust the current centroid of the stimulus component within the afferent aperture so that it lies approximately on the reference axis. When the visual aperture reaches a limiting size, the pattern of retinoid excitation in its standardized position within the aperture is projected to a neuronal mechanism for learning and recognition called a synaptic matrix (Trehub, 1991).

A schematic of the synaptic matrix is shown in **Figure 2**. Its structural properties and the learning rule can be briefly summarized as follows. Retinotopic afferents S_{ij} are in discrete point-to-point synapse with a following set of

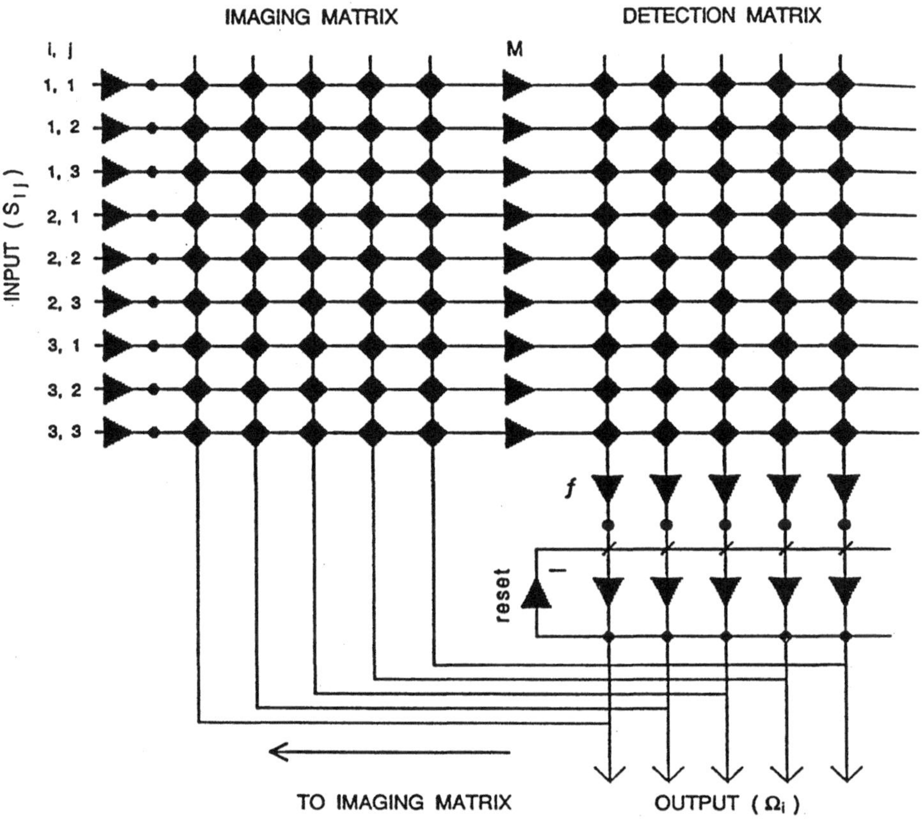

FIGURE 2. Schematic of a synaptic matrix. Afferent inputs are designated S_{ij}. Mosaic cells are designated M. Dots represent fixed excitatory synapses; short oblique slashes represent fixed inhibitory synapses; filled lozenges represent adaptive excitatory synapses. Reset neuron (-) generates an inhibitory postsynaptic potential to reset all class cells when discharged. Given any arbitrary input to the synaptic matrix, the class cell coupled with the filter cell having the highest product sum of afferent axon activity and corresponding transfer weights will fire first and inhibit all competing class cells.

neurons, called mosaic cells (*M*). The axon of each mosaic cell is in parallel adaptive synapse with all members of a set of cells in the detection matrix, which are called filter cells (*f*). Each filter cell is in discrete synapse with an output neuron called a class cell (Ω). Each class cell integrates the activation input from its coupled filter cell. The axon of each class cell bifurcates and sends a collateral back in adaptive synapse with the dendrites of all mosaic cells (*M*) in the imaging matrix. Finally, a reset neuron (marked -) receives excitatory input from the axons of all class cells (Ω) and sends its own inhibitory input back in parallel synapse with all class cells. Integration of filter-cell input to paired class cells, together with the reset mechanism, ensures that the class cell that receives the highest activation from its coupled filter cell will fire first and inhibit all competing class cells.

One-trial learning of a visual stimulus pattern takes place by modification of adaptive synapses on filter cells in the detection matrix and mosaic cells in the imaging matrix. (In this simulation, processes taking place in the imaging matrix will not be discussed.) The magnitude of learning-related changes in synaptic transfer weight (ϕ) are determined according to the following expression:

$$\underset{b \rightarrow Lim}{\phi_{im}} = b + S_{im}\,(c + kN^{-1}) \tag{1}$$

where $\underset{b \rightarrow Lim}{\phi_{im}}$ is the transfer weight of synapse ϕ_{im}, from the basal value (b) to the saturation limit (*Lim*), on an adaptive filter cell m; b is the initial transfer weight of the unmodified synapse; c is a fixed synaptic contribution from the active axonal contact on ϕ_{im}; kN^{-1} is a proportional synaptic contribution taking account of N coactive axons on the cell m at the time of learning, and a synaptic modification constant k; and S_{im} is the activity level of axonal input at ϕ_{im}. The product sum of afferent axon activity over the mosaic cell array (*M*) and the corresponding synaptic transfer weights (ϕ) on each filter cell determine its activation level.

The parameter values used in the present study for stimulus capture, learning, and recognition of faces were the same as used in a previous simulation of self-directed learning in a complex environment (Trehub, 1991, chapter 12).

Procedure

The stimuli that were presented to the model consisted of schematic faces (in pixel display) that had been used in previous experiments to explore perceptual classification in humans. A subset of 10 faces taken from the original line-drawn stimuli used in studies by Reed & Friedman (1973) and Nosofsky (1991) were digitally scanned and reduced in size so that each face was approximately 18 pixels in height. All 10 faces were presented together throughout the simulation.

FIGURE 3. Bit-mapped face stimuli. Faces 1-5 are in category A; faces 6-10 are in category B.

The 10 faces could be separated into 2 different categories with 5 faces in each category on the basis of a multidimensional (MDS) analysis of eye height, eye separation, nose length, and mouth height (**Figure 3**; Reed & Friedman, 1973; Nosofsky, 1991). In the current study, each of the faces was assigned an identifying name and a letter designation indicating that it belonged to category A or B (e.g., Tim-A, Ned-B). Before the start of the recognition procedure, synaptic transfer weights on one filter cell (f_1) in the detection set were tuned (weights selectively increased by the learning mechanism) to a random pattern of retinal excitation. This cell evoked the response "RANDOM" whenever it was the most active filter cell in the detection set. On all subsequent trials the neuronal model was presented with all 10 faces in a single display. On each trial, the model retina automatically fixated on an individual face in a quasi-random fashion. The task was to capture a face, report its name (face recognition), and give its category designation. If the response was correct, the operator typed in "YES" and another face was captured and the procedure repeated. If the response was wrong, the operator typed in "NO" and a previously unmodified filter cell (e.g., f_2) in the detection set was synaptically tuned to the retinal pattern of the captured face (the current exemplar) by the intrinsic properties of the learning mechanism. Then the operator typed in the appropriate name and category designation which would be evoked by the model whenever f_2 was the most active filter cell. Again, the system captured another face and the same procedure was repeated. Notice that on the first recognition trial, the only possible response that the system could make was "RANDOM", since it had nothing else in its response repertoire. Each response to a captured face was counted as a trial whether the response was correct and followed immediately by a new capture, or whether it was incorrect and resulted in the exemplar-tuned synaptic modification of another filter cell in the detection set (learning). The simulation proceeded until 400 trials were completed. Performance was examined for face recognition and category designation in each of 16 sequential blocks of 25 trials for each block.

Results

Face recognition and categorization

Figure 4 shows the learning curves over all blocks. The percentage of correct responses for both face recognition and categorization was characterized by a curve with an initial rapid rise over the first 50 trials followed by deceleration of improvement. The categorization response improved more rapidly than did the recognition of individual faces. At the end of the 400 trials, correct performance for both recognition and categorization was at the 96% level. The conclusion that categorization of faces improved more rapidly than the recognition of individual faces was based on the following considerations. In the simulation, a correct identification of a face also evokes its correct

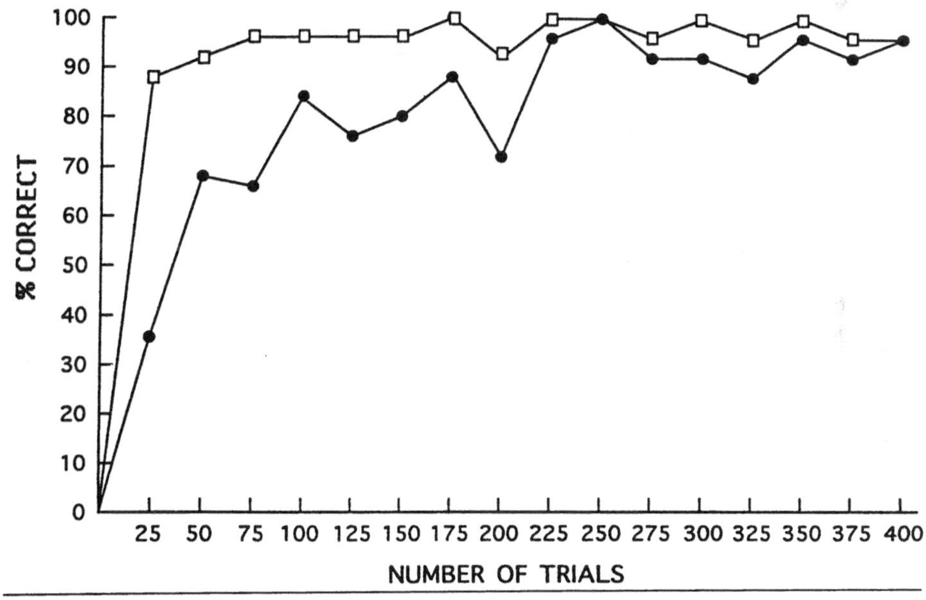

FIGURE 4. Plot of performance. Points plotted show the percent of correct responses over 16 blocks of 25 trials each. Filled circles indicate level of performance for face recognition in each of the successive trial blocks. Open squares indicate level of performance for category designation on corresponding blocks.

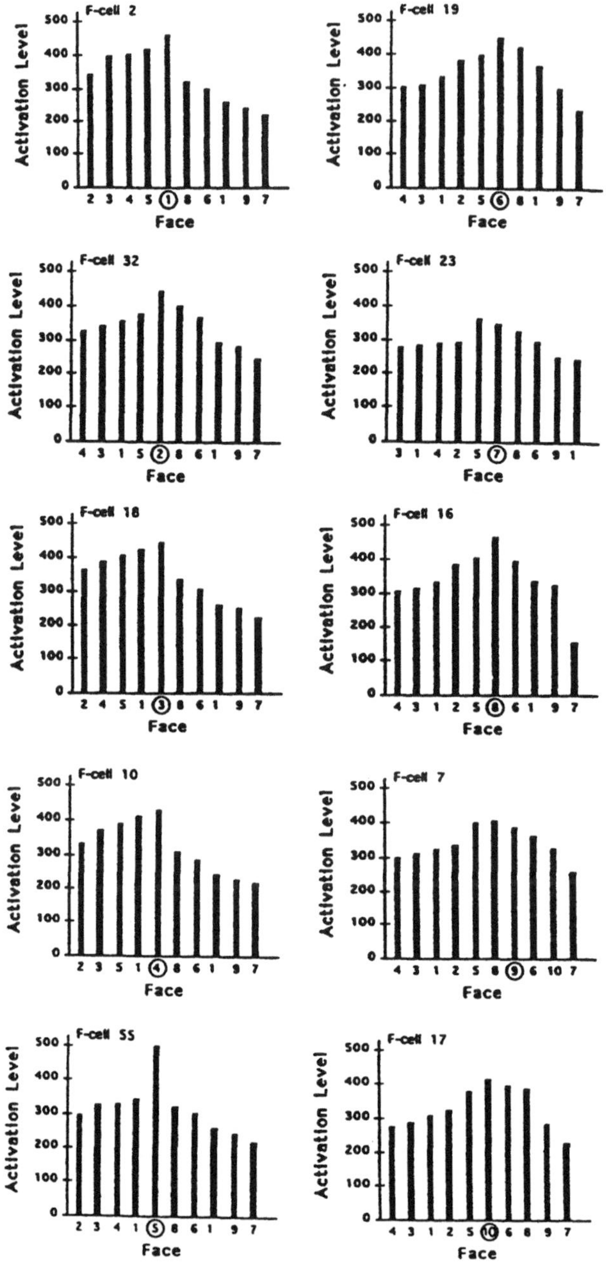
F-cell 2
F-cell 19
F-cell 32
F-cell 23
F-cell 18
F-cell 16
F-cell 10
F-cell 7
F-cell 55
F-cell 17
Activation Level
0
100
200
300
400
500
Face

category. Only those trials where the identification response was wrong could provide information about the rate of category learning (cat) relative to the rate of recognition learning (rec). Since two categories of faces (A and B) were presented, there was a 50% chance that a wrong identification response would nevertheless name a face in the correct category. Hence on each block of trials we would expect a relative advantage for category performance on the basis of chance alone. Only if the magnitude of the advantage were greater than expected by chance could we conclude that categorization improved more rapidly than recognition. Thus in order to determine if the rate advantage for categorization was significantly greater than chance expectation, the following formula was applied on each block of trials:

$$\text{Exp Adv [cat]} = \text{\% Correct [rec]} + (100 - \text{\% Correct [rec]})/2 \qquad (2)$$

The differences between the observed percentage of correct categorization and the Expected Advantage [cat] over all 16 blocks of 25 trials provided the data on which to assess the rate of improvement in categorization. There was an unbiased advantage for categorization ranging from +20% on block 1 to +6% on block 8. Over the last eight blocks, the categorization advantage ranged from +4% to 0%. A total of 70 filter cells in the detection set had been synaptically tuned by the learning mechanism to exemplars of the captured faces. All filter cells exhibited broad tuning curves over the faces that were captured. This is illustrated in **Figures 5** and **6** where the activation levels of 10 different filter cells are shown in response to each of the 10 faces. These were randomly selected from the cells that signaled the correct response in a sample drawn from the last 50 trials in which 10 different faces were captured.

Cell response profiles

On each trial, the neuronal model selects a discrete recognition response on the basis of competitive activity among cells. The filter cell with the highest activation level evokes its associated name for the face that has been captured while inhibiting the output effects of all other cells. It is important to notice that a filter cell that has been selectively tuned to a particular face can exhibit a stronger response to other faces. This will not degrade the effectiveness of the

---

FIGURE 5 (left). Activation levels of each of 10 sampled filter cells (F-cells) in response to exemplars of each of the 10 face stimuli. Activation response of each F-cell plotted against each face. Each F-cell in the left column had been synaptically tuned during learning to one of the 5 faces in category A. Each F-cell in the right column had been tuned to one of the 5 faces in category B. For each cell, the face that it had learned is indicated by being circled. In each of the 10 plots, faces in category A are ordered so that the level of F-cell activation evoked by each face grades down to the left of the distribution; faces in category B are ordered so that the activation they evoked grades down to the right of the distribution.

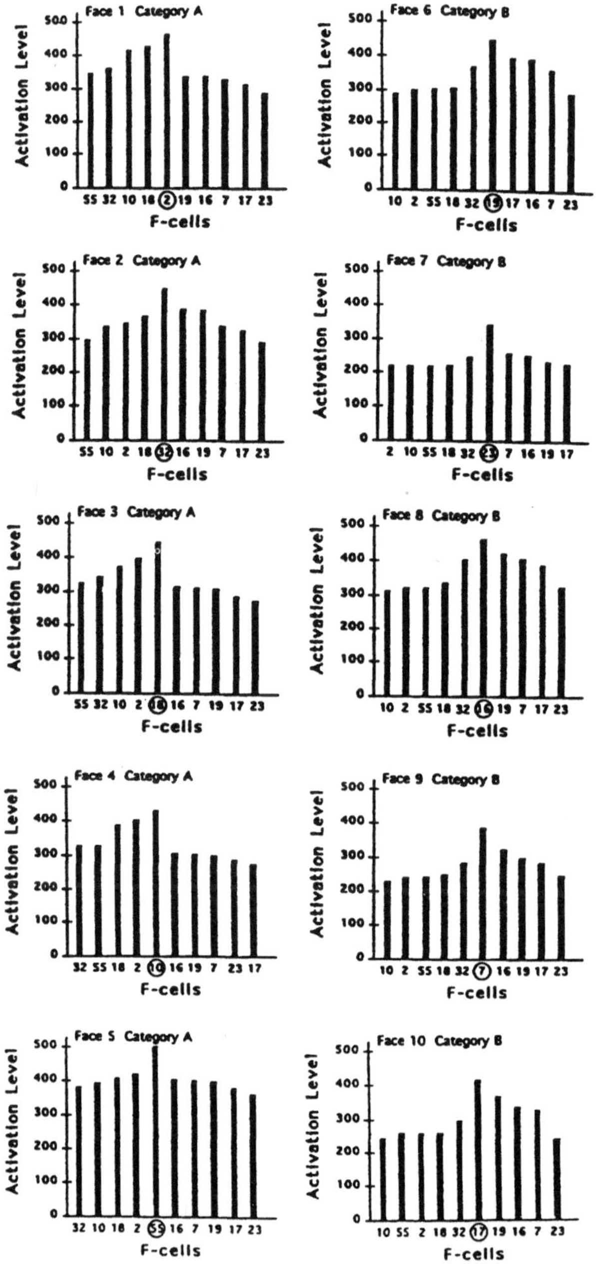
Face 1 Category A
Activation Level
0 100 200 300 400 500
55 32 10 18 2 19 16 7 17 23
F-cells
Face 2 Category A
Activation Level
0 100 200 300 400 500
55 10 2 18 32 16 19 7 17 23
F-cells
Face 3 Category A
Activation Level
0 100 200 300 400 500
55 32 10 2 18 16 7 19 17 23
F-cells
Face 4 Category A
Activation Level
0 100 200 300 400 500
32 55 18 2 10 16 19 7 23 17
F-cells
Face 5 Category A
Activation Level
0 100 200 300 400 500
32 10 18 2 55 16 7 19 17 23
F-cells
Face 6 Category B
Activation Level
0 100 200 300 400 500
10 2 55 18 32 19 17 16 7 23
F-cells
Face 7 Category B
Activation Level
0 100 200 300 400 500
2 10 55 18 32 23 7 16 19 17
F-cells
Face 8 Category B
Activation Level
0 100 200 300 400 500
10 2 55 18 32 16 19 7 17 23
F-cells
Face 9 Category B
Activation Level
0 100 200 300 400 500
10 2 55 18 32 7 16 19 17 23
F-cells
Face 10 Category B
Activation Level
0 100 200 300 400 500
10 55 2 18 32 17 19 16 7 23
F-cells

recognition system as long as the response of the correct cell is higher than any other cell in the detection set at the time that its learned exemplar or a pattern most similar to it is captured. For example, it can be seen in **Figure 5** that F-cell 7 gives a stronger response on the trials in which faces 5 and 8 were captured than on the trial in which face 9 (the face it had learned) was captured. Yet, as **Figure 6** shows, the response of F-cell 7 to the capture of face 9 is stronger than any of its competing filter cells when face 9 is the effective stimulus.

The overall selectivity of the recognition system can be characterized by the number of competing filter cells which approach the peak activation level on each trial in which there is a correct response. This is illustrated in **Figure 7** which shows the distribution of the number of competing cells with activation levels within 10% of the peak on all correct trials. It was found that on 31% of the trials there was no competing cellular activity within 10% of the peak response. On 41% of the trials there was only one competing cell within this range. The general shape of the selectivity distribution is similar to the reported distribution of discrepancies between population vectors (ensembles) of unit responses in cells of the macaque inferotemporal cortex and corresponding stimulus (face) vectors (Young & Yamane, 1992).

Implications

It is clear from these results that filter cells exhibit a graded response to all faces and therefore do not exhibit a punctate code. This finding is consistent with some kind of sparse coding mechanism for face recognition. However, the question of how sparse coding is used to ensure reliable recognition is more problematic. It is commonly proposed that a sparse population code entails the *joint* activity of a relatively small number of cells (an ensemble), each making its own necessary contribution to the set of encoded features which, taken together, characterize an individual face (Baylis, Rolls, & Leonard, 1985; Gross, 1992; Gross & Sergent, 1992; Young & Yamane, 1992). Similarly, in the more general context of object recognition, it has been suggested that objects are coded by sparse combinations of active cells where each cell represents the presence of a particular complex partial feature of the object (Tanaka, 1993; Tanaka, Saito, Fukada, & Moriya, 1991). Let us call this kind of code *sparse-code 1*. This approach to the problem of object recognition postulates a structure of overlapping feature detectors (mini-templates) in the visual system

---

FIGURE 6 (left). Activation evoked by each face plotted against each F-cell. For each face, the cell that had learned it is indicated by being circled. Cells that had learned faces in category A are ordered so that their response levels grade down to the left; cells that had learned faces in category B are ordered so that their response levels grade down to the right.

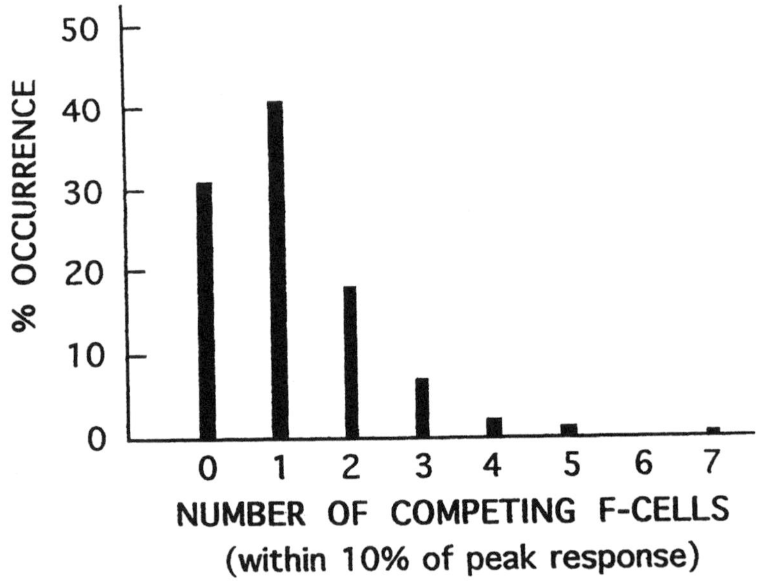

FIGURE 7. Bar graph showing the percent of trials in which differing numbers of competing F-cells had an activation response within 10% of the observed peak response.

that are assumed to be distributed in replicated fashion over the visual field. The critical notion is that each stack of feature detectors (putatively in columnar organization) is tagged by its retinotopic location so that a complete object can be uniquely defined by the whole concurrent activation pattern of a *set* of particular detectors at their particular locations.

However, a caution must be raised here. If, at any given moment, there is only one object in the visual field, then an activated set of spatially indexed features might provide a unique definition of that single object. But what if there is more than one object in the visual field, as is normally the case in the natural world? Under the normal circumstance, we would need a biologically plausible mechanism that is able to map our complex retinal activation patterns onto just those discrete groups of spatially-indexed feature detectors that correspond to each of the separate objects in view (the binding problem). This is not a trivial problem. Indeed, it remains one of the serious obstacles for the general class of pattern recognizers based upon the principle of detecting and combining partial features.

An alternative interpretation of sparse coding is suggested by the operating characteristics of the neuronal model (Trehub, 1991) that generated the results obtained in the present simulation study. In this model, when a face is captured within the visual afferent aperture, all filter cells show increased activation, but the cell with the highest activation level in response to the current retinal (proximal) stimulus evokes the appropriate discrete output. Hence, for each proximal stimulus a single cell can generate a code precise enough for effective recognition (Barlow, 1972, 1985; Konorski, 1967; see also Konishi, 1991). The joint contribution of other coding cells is not required.

If this is the case, why did effective performance in the present simulation require that more than 10 filter cells be synaptically tuned to learn 10 faces? The answer is revealed in the difference between the distal (environmental) and the proximal (retinal) stimulus. When a face is in the visual field, we do not know exactly where its features will be registered on the retina. At one time, fixation might be centered on the upper part of a given face; at another time on a lower part of the same face. The features of a constant distal stimulus may excite differing proximal patterns on the retina at different times. Variations in fixation of no more than 1 degree in visual angle can result in significant changes in the distribution of foveal excitation. Each retinal pattern represents only an exemplar of a given stimulus. Some exemplars may vary from previously learned patterns to the extent that they exceed the capacity of the recognition system to generalize correctly . Thus effective recognition of a face (or any other complex pattern) requires that different cells be tuned to at least a few different exemplars of the face in order to facilitate proper generalization and compensate for fortuitous shifts in exemplar capture (Trehub, 1991). In this sense, the sub-population of exemplar-tuned cells that *individually* signal a particular face also constitute a sparse code for that face. Notice, however, that this code is significantly different from sparse-code 1 in that it does not require the *joint* activation of an ensemble of exemplar-tuned cells to achieve effective recognition. Let us call this kind of neuronal code *sparse-code 2*.

The characteristic strategy for investigating selective coding of faces (or other objects) in neurophysiological experiments has depended on finding cells in which the peak spike rate is systematically evoked by the presentation of particular faces in an arbitrary set of stimuli (Desimone et al, 1984; Perrett, Mistlin, & Chitty, 1987; Young & Yamane, 1992; see also Tanaka et al, 1991; Gallant, Braun, & Van Essen, 1993). Implicit in this strategy is the general assumption that if the output of a cell is to be a reliable indicator of a particular object, the cell must respond more vigorously when that object is seen than when any other object is seen. This investigatory approach precludes the possibility of uncovering a neuronal recognition mechanism based upon competitive discrimination by sparse-code 2. For example, under the usual paradigm, F-cell 7 in the present simulation (**Figures 5** and **6**) would be thought to more likely code for face 5 or face 8 than for face 9, which it actually learned and correctly recognized within the competitive recall model (Trehub, 1977, 1991).

If the neuronal brain mechanism for face recognition in the monkey is organized on the principle of sparse-code 2 then several implications for the interpretation of single-cell recordings follow: (1) broad tuning of many cells in response to a particular face (or any other complex pattern) does not straightforwardly imply an ensemble code; (2) discovering a set of exemplar-tuned cells requires that we record the *concurrent* responses of a large number of cells to many presentations of each face (the distal stimulus) in the stimulus set because the retinal pattern (the proximal stimulus) that is captured is likely to vary over time even for identical faces as a result of shifts in fixation; (3) given the effect of variation in fixation, it would be helpful for the interpretation of results to monitor fixation throughout an experiment; (4) the critical indicator of selective coding is not the relative spike rate of a cell in response to different stimuli, but rather the rate of its output relative to other cells responding at the same time.

Neural-Networks Models of Cognition
J.W. Donahoe and V. Packard Dorsel (Editors)

CHAPTER 11

STRUCTURE AND BINDING IN OBJECT PERCEPTION

John E. Hummel
University of California, Los Angeles

ABSTRACT

This chapter discusses the problem of structure in shape perception and object recognition. This problem, and the characteristics of its solution, provide a powerful set of tools for understanding many properties of human object recognition, including the strengths and limitations of our ability to recognize objects in novel viewpoints, and the role of time and attention in object recognition. Many of the problems surrounding the representation of structure center on the *binding problem*—the problem of representing what goes with what. Many properties of human object recognition can be understood in terms of the ways the visual system solves the binding problem. The chapter uses two recent models of object recognition to illustrate these points: Hummel and Biederman's (1992) JIM, and Hummel and Stankiewicz's (1996) JIM.2. This chapter is written from a behavioral and computational perspective rather than a neuroscientific one. The emphasis is on the abstract nature of the representations and processes the visual system must bring to bear on the problem of structure, and on the behavioral implications of those representations and processes.

Constancy and Structure in Object Perception

The great problem in the study of human object recognition is to understand *object constancy*, our capacity to recognize objects despite variations in the image presented to the retina. This capacity takes two forms. The most commonly studied is recognition despite variations in viewpoint. We can recognize objects in a wide variety of views even though different views can present radically different images to the retina. This capacity is particularly challenging to understand because human object recognition is robust to some, but not all, variations in viewpoint. Recognition is indifferent to the location of the image on the retina, left-right reflection (Biederman & Cooper, 1991a; see also Tanaka, this volume), scale (Biederman & Cooper, 1992), and some rotations in depth (Biederman & Gerhardstein, 1993, 1995; but see Tarr & Bülthoff, 1995). However, it is sensitive to rotation in the picture plane (as when an object is upside down; Jolicoeur, 1985, 1990; Tarr & Pinker, 1989, 1990). The second property of human object constancy is our capacity to generalize over variations in an object's exact 3D shape. This capacity has at least two familiar and important manifestations. First, we are remarkably good at recognizing objects as members of a class, such as "chair" or "car," rather than just as specific instances, such as "my office chair" or "Toyota Camry." And second, we can easily recognize novel members of known object classes: The first time we see a Dodge Viper, it is easy to recognize it as a car, even if we have never before seen a car with exactly that shape.

Together, these properties make human object recognition challenging to understand because they defy explanation in terms of simple geometric laws. The mathematics of projective geometry is well understood, and if human recognition performance were predictable in terms of it, then models based on these laws would provide an adequate and intuitive account of human object recognition. But these laws do not explain our capacity for object recognition. A system based strictly on the laws of projective geometry (e.g., one that somehow converted retinal images into object-centered representations of 3D shape) would be equally able to accommodate all variations in viewpoint (which the human is not), but would not tolerate variations in an object's shape (which the human does).

These and other properties of human object constancy have led some researchers to postulate that we recognize objects on the basis of *structural descriptions* specifying the object's parts (or features) in terms of their relations to one another (Biederman, 1987; Clowes, 1967; Marr & Nishihara, 1978; Palmer, 1978a; Sutherland, 1968; Winston, 1975). The most explicit such theory to date is Biederman's (1987) *recognition by components*, and its variants (Bergevin & Levine, 1993; Dickenson, Pentland, & Rosenfeld, 1992; Hummel & Biederman, 1990; 1992; Hummel & Stankiewicz, 1996). According to this theory, objects are represented as collections of *geons* in particular categorical relations. (Geons are simple volumes such as cylinders, bricks, and cones, distinguished by categorical contrasts in the properties of image edges; Biederman, 1987.) For example, a coffee mug would be represented as a curved cylinder (the handle) side-attached to a straight vertical cylinder (the body). The relations are critical: If the curved cylinder were attached to the top of the straight cylinder, then the object would be a bucket rather than a mug (Biederman, 1987). This type of representation provides a natural account of many properties of human object recognition. Note that it will not change if the mug is translated across the visual field, moved closer to or farther from the viewer, or rotated in depth (provided the handle does not disappear behind the body). But rotating the mug 90° about the line of sight (so that the body is horizontal and the handle is on top) will change the description. Like human object recognition, this description is sensitive to rotations about the line of sight, but insensitive to translation, scale, left-right reflection, and some rotations in depth. It is also insensitive to things such as the exact length of the handle or the exact width of the body, making it suitable as a basis for class recognition (Biederman, 1987).

There is behavioral evidence for parts-based structural descriptions in human object recognition. Biederman and Cooper (1991b) have shown that one object image will prime another to the extent that they depict the same geons in the same relations. (If a person views and names an object at one time, and then views and names the same object, say, five minutes later, then his naming response is likely to be faster the second time than it was the first time—and faster than the time required to name a new image. This increase in naming speed is referred to as *priming*. A component of this priming is mediated by specifically visual representations of shape; Biederman & Cooper, 1991a,b, 1992.) There is also substantial evidence that the relations between an object's features or parts play an important role in the representation of object shape (Goldmeier, 1972; Hummel & Stankiewicz, 1995; Palmer, 1977, 1978b; Saiki & Hummel, 1996a,b; in press). For example, Hummel and Stankiewicz (1995) showed that human object perception is more sensitive to the relations between an object's parts than to the parts' positions relative to any single reference point (i.e., coordinates of the type proposed in normalization-based theories of object recognition; see, e.g., Olshausen, Anderson, & Van Essen, 1993; Poggio & Edelman, 1990; Ullman, 1989; Ullman & Basri, 1991). These and other findings suggest that we represent objects in memory as structural descriptions specifying the interrelations among their parts (see Quinlan, 1991, for a review). Of course, this is not to say that objects are represented exclusively as structural descriptions (Farah, 1992; Hummel & Stankiewicz, 1996; Tarr & Pinker, 1990). We shall return to this point shortly.

The defining property of a structural description is that it specifies the *relational structure* of an object's shape. Rather than representing an object's parts in terms of their coordinates relative to the origin of a reference frame, a structural description represents an object's parts in terms of their positions relative to one another (see Hummel, 1994; Palmer, 1978a). How might the visual system generate a structural description from an object's image? This question is particularly interesting, because relational structures are difficult to represent in systems of discrete processing elements, such as nervous systems and artificial neural networks (see Fodor & Pylyshyn, 1988).

This chapter discusses the problem of structure in shape perception and object recognition: Given a representation of the local features in an object's image, how can the visual system generate a description of the object's parts and their interrelations, and then use that description to recognize the object? This problem, and the characteristics of its solution, provide a powerful set of tools for understanding many properties of human object recognition. Many of the problems surrounding the representation of structure center on the *binding problem*—the problem of representing what goes with what—and many properties of human object recognition can be understood in terms of the ways the visual system solves the binding problem in the context of shape perception.

The chapter will use two models—Hummel and Biederman's (1992) JIM, and Hummel and Stankiewicz's (1996) JIM.2—to illustrate these points. Although the chapter emphasizes the representation and processing of relational information in vision, the issue of structure is very general. Relational structures play an important role in virtually all aspects of cognition, including language, categorization (Barsalou, 1993; Saiki & Hummel, 1996a,b), similarity judgment (Medin, Goldstone, & Gentner, 1993), and reasoning (Gentner, 1983; Gick & Holyoak, 1980; Holyoak & Thagard, 1989). The role of structure in these domains bears important relationships to the role of structure in vision (see Jackendoff & Landau, 1991; Logan, 1994). Indeed, many of the routines my colleagues and I have proposed for processing structure in the context of vision also apply to the representation and processing of structure in the domain of reasoning (Hummel & Holyoak, 1992, 1993; Hummel, Meltz, Thompson & Holyoak, 1994; see also Shastri & Ajjenagadde, 1993).

This chapter is written from a behavioral and computational perspective rather than a neuroscientific one: The focus is more on the nature of the problem the visual system is solving, and on the abstract nature of the representations and processes it brings to bear on its solution, than on the neurophysiological details of that solution.

Structural Description and Dynamic Binding

Generating a structural description from an object's image entails solving three related problems (Hummel & Biederman, 1992). Consider generating the description *cone on top of brick* from the image of a cone on top of a brick. First, the local features (e.g., contours and vertices) of the cone and brick must be segmented into groups so that the features of one do not interfere with the interpretation of the other. (This problem is a variant of the familiar figure-ground problem in vision.) Likewise, the parts' attributes (e.g., the shapes of their cross sections and axes, their orientation, aspect ratio, etc.) must also be bound into sets. And finally, the representation of the cone must be bound to the relation *above*, while the brick is bound to *below*.

The representation of these bindings is critically important. Bindings can be either *dynamic* or *static*. A dynamic binding is one in which a single representational unit (e.g., symbol, neuron, or collection of neurons) is used in many different combinations without sacrificing its independence. For example, one unit might represent cones and another might represent the *above* relation; a cone above another part would be represented by explicitly tagging these units as bound together. A static binding is one in which a separate unit is pre-dedicated for each conjunction. For example, one unit might respond to cones above other parts, another might respond to cones below other parts, and so forth. Structural description requires dynamic binding (Hummel & Biederman, 1992). The number of units required to pre-code all possible part-relation conjunctions would be enormous (growing exponentially with the number of relations). More importantly, static binding sacrifices the independence—and therefore the similarity structure—of the bound entities: The similarity of a cone above something to a cone below something is completely lost in a representation where each part-relation binding is coded by a separate unit. This loss of similarity structure is a fundamental property of static binding that cannot be overcome even with sophisticated static codes, such as Smolensky's (1990) tensor products (Hummel & Biederman, 1992; Hummel & Holyoak, 1993). Dynamic binding is thus a prerequisite to structural description. Understanding how the visual system represents structure means understanding how it solves this dynamic binding problem.

There is some evidence that synchrony of firing is one basis for dynamic binding in biological visual systems (Engel, König, Kreiter, & Singer, 1991; Gray, König, Engel, & Singer, 1989; Gray & Singer, 1989). The basic idea is that neurons belonging to the same perceptual group (e.g., because they are responding to different parts of the same contour or surface) generate spike trains roughly in synchrony with one another; those same neurons tend to fire out of synchrony when they belong to different groups. Albeit controversial (see, e.g., Tovee & Rolls, 1992), this idea has attracted a great deal of attention in the neuroscientific and computational vision communities. (See Singer, this volume.)

From a computational standpoint, the question of how dynamic binding is *represented* (e.g., by synchrony or some other means) is less important than the questions of how it is *established* (e.g., how do neurons know *whether* they should be grouped) and how it can be *used* to perform useful work. This chapter is concerned primarily with the latter issues: How can dynamic binding be established and used for structural description and object recognition?—and, in addition, with the related question: What are the benefits and costs of dynamic binding in object perception? Both models reviewed here use synchrony for dynamic binding, but neither is committed to this convention. What is more important is the manner in which the binding is used and the properties that result from it. The next section reviews Hummel and Biederman's (1992) JIM (for *John and Irv's Model*) model of object recognition. JIM serves to illustrate how dynamic binding can be used for structural description, and some of the behavioral properties that result from this approach to object recognition. The following section presents some important limitations of JIM—and of dynamic binding in general—and reviews Hummel and Stankiewicz's (1996) JIM.2 model of how the visual system might recognize objects given these limitations.

The advantages of dynamic binding: JIM

JIM is a seven-layer artificial neural network (**Figure 1**) that uses synchrony of firing for dynamic binding and structural description. As input, it takes a line drawing of an object. Units in its first two layers represent local features

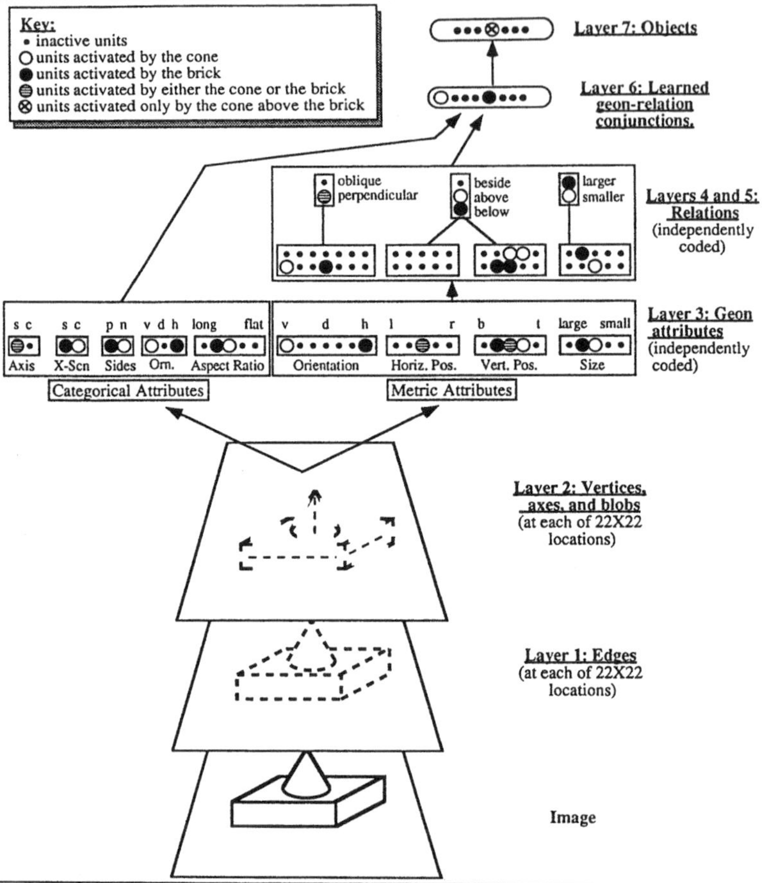

FIGURE 1. The architecture of JIM (Hummel & Biederman, 1992). Units are coded to depict their responses to the image in the Key. In Layers 3 and above, large circles indicate units activated in response to the image and dots indicate inactive units. Units in Layer 1 represent image edges (discontinuities in surface orientation and depth). Layer 2 represents the vertices, axes, and "blobs" (elongated, oriented regions of activity) defined by conjunctions of edges in Layer 1. Layer 3 represents the geons in an image in terms of their properties: Axis shape (*Axis*)—straight or curved; Cross section shape (*X-Scn*)—straight or curved; whether the *Sides* are parallel (*p*) or non-parallel (*n*); Coarse orientation (*Orn.*)—vertical (*v*), diagonal (*d*), or horizontal (*h*); *Aspect Ratio*—elongated (*long*) to flattened (*flat*); Precise orientation (*Orientation*); Position along the horizontal dimension of the visual field (*Horiz. Pos.*)—left (*l*) to right (*r*); Position along the vertical dimension of the visual field (*Vert. Pos.*)—bottom (*b*) to top (*t*); and *Size*—large or small. Layers 4 and 5 represent the relative orientations, locations, and sizes of the geons in an image. Units in Layer 6 respond to specific conjunctions of units in Layers 3 and 5, and units in Layer 7 sum the outputs of Layer 6 units over time to respond to complete objects.

(such as contours, vertices, and axes) at each of 22 x 22 locations in the image. The model's intermediate layers represent a structural description of the object's shape. Units in Layer 3 respond to the shape attributes of individual geons (e.g., whether a given geon has a straight or curved axis, etc.), and units in Layers 4 and 5 code the categorical relations between geons (e.g., whether one geon is above or beside another, etc.). Units in Layers 6 and 7 use the outputs of Layers 3 and 5 as the basis for object recognition. The model's input is completely viewpoint-specific. (For example, a given object will activate one set of local feature units when it is in the upper left of the visual field, and a completely different set of units when it is in the lower right.) However, recognition is largely viewpoint-invariant. As with human object recognition, JIM's ability to recognize an object is completely unaffected by the location of the image in the visual field, the size of the image, some rotations in depth, and even left-right reflection. Also like that of the human, JIM's performance suffers when an object image is misoriented in the picture plane. Importantly, the model achieves these invariances on the basis of a single exposure to each object: Trained on just one view of an object, JIM (like the human) can subsequently recognize that object in a wide range of new viewpoints (even new orientations in depth). This capacity distinguishes JIM from most other models of object recognition, and from all models based on normalization or transformation procedures (Olshausen et al, 1993; Poggio & Edelman, 1990; Ullman, 1989; Ullman & Basri, 1991).

Dynamic binding plays a central role in JIM's capacity for structural description and object recognition. When an image is presented for recognition, edge and vertex units in Layers 1 and 2 group themselves into sets corresponding to geons by synchronizing their outputs. Units representing features of the same geon fire in synchrony with one another, and out of synchrony with units representing the features of other geons. Synchrony is established on the basis of strictly local interactions between contour and vertex units, making it possible to group features into geons without any prior knowledge of the geons' identities. Because of these interactions, the outputs of local feature units arrive at Layer 3 in packages corresponding to geons (i.e., one geon per unit of time). This fact is critically important because it makes it possible for the units in Layer 3 to code the attributes of geons in a completely independent fashion. In turn, this independence gives rise to all of the model's other properties.

This point bears elaborating. Each unit in Layer 3 responds to one geon attribute. For example, one unit will respond to *any* geon with a *curved cross section* (such as cones and cylinders), another will respond to any geon with a straight cross section (such as wedges and bricks); other units code whether a geon's sides are parallel or non-parallel, whether its axis is straight or curved, its aspect ratio, and still others code its orientation, size, and location in the image. A given unit will respond to any geon with its target property, regardless of the geon's other properties. For example, the curved-cross-section unit will respond to any geon with a curved cross section, regardless of the geon's location, size or orientation in the image. Similarly, the units that code a geon's location in the visual field will respond to any geon in their target location; the same is true of the units that code orientation and size. That is, every attribute is represented in a form that is completely independent of every other attribute.

This independence gives rise to two critical properties as a natural consequence. One is invariance with viewpoint. The same units will respond to, say, a cone (*curved cross section*, *non-parallel sides*, and *straight axis*) regardless of where the cone is located in the visual field, its size or its orientation (barring "accidental" views, such as when the cone is viewed end-on, projecting a circle to the retina; see Lowe, 1987). The second critical property of the independence is that it automatically captures the similarity structure of different geons, and different geon-relation combinations (i.e., objects). For example, the representation of a cone is more similar to that of a cylinder (*curved cross section*, *parallel sides*, and *straight axis*) than to that of a brick (*straight cross section*, *parallel sides*, and *straight axis*). Units in Layers 4 and 5 use a geon's metric properties (size, location, and orientation) to compute the relations between separate geons. These units are independent of those that code geon shape attributes. For example, the same unit codes the *above* relation regardless of what is above what. As a result, if two objects are similar in the relations among their parts but differ in the parts' shapes, then their representations will overlap on the relation units relations, but not on the part attribute units; and if two objects are similar in their parts but differ in their relations, then their representations will share geon attribute units but not relation units. In general, the independence of the shape and relation units makes it possible for the representation to completely capture an object's attribute structure (see Hummel & Biederman, 1992).

In turn, this independence is possible only because of the dynamic binding established in Layer 2. This binding is carried forward in Layers 3 through 6. In Layer 3, it binds shape attributes into packages corresponding to geons, and in Layer 5, it binds relations to geons. For example, a cone above a brick would be represented as two sets of synchronized units: the cone attribute units in Layer 3 will fire in synchrony with the *above* unit in Layer 5 while the brick attributes fire in synchrony with the *below* unit. The cone-above set must fire out of synchrony with the brick-below set. If the features of the cone "accidentally" fire in synchrony with those of the brick, then the resulting pattern in Layer 3 would specify *curved cross section*, *straight cross section*, *parallel sides*, and *non-parallel sides* without specifying how they go together. Such a pattern cannot distinguish a brick and a cone (*curved cross section* with *non-parallel sides* and *straight cross section* with *parallel sides*) from a cylinder and a wedge (*curved cross section* with *parallel sides* and *straight cross section* with *non-parallel sides*), or even from a single geon with a combination of these properties.

In summary, representing shape attributes and relations independently gives rise to some very desirable properties, but it depends critically on dynamic binding.

Behavioral implications of JIM. One line of support for this type of representation is the ease with which it can account for the properties of human object recognition. As noted previously, human object recognition is completely invariant with translation, scale, and left-right reflection; it is largely invariant with orientation in depth, but sensitive to orientation in the picture plane. JIM's recognition performance captures all these aspects of human object recognition. These properties are detailed in Hummel and Biederman (1992), so I shall describe only two of them (arguably the most interesting) here: invariance with reflection, and sensitivity to orientation in the picture plane.

Human object recognition is invariant with left-right reflection in the sense that an object image will visually prime its left-right reflection *just as much as it primes itself* (Biederman & Cooper, 1991a). (As elaborated shortly, this complete invariance obtains only with prime-probe delays on the order of minutes; with delays of about 3 seconds, images prime themselves more than they prime their reflections; Stankiewicz, Hummel, & Cooper, 1995.) This finding strongly suggests that at least a component of the visual representation of shape is completely indifferent to left-right reflection. This is especially interesting because not all objects are even the same 3D shape under left-right reflection. A left-right reflection of a left shoe is not a left shoe; it is a right shoe. Similarly, there exists no object in the 3D world that would project to the retina as the left-right reflection of a grand piano. (Grand pianos are not bilaterally symmetrical.) Nonetheless, we happily recognize these fictional objects as grand pianos. In general, our propensity to ignore left-right reflection in recognition speaks to a deep disrespect for the laws of projective geometry. Although invariance with reflection is impossible to explain in terms of a 3D geometric approach to object recognition, it can be explained straightforwardly in terms of geon attributes and relations. In JIM, "left-of" and "right-of" are both coded as the single relation "beside," making the representation indifferent to left-right orientation. As a result, JIM's recognition performance is completely invariant with left-right reflection: Trained to recognize a grand piano as a piano, JIM would just as easily recognize the reflection of a grand piano as a piano. (For the same reason, JIM could not distinguish a piano from its reflection, although JIM.2 could.) This property derives directly from the nature of JIM's structural descriptions.

One of the most robust findings in the human object-recognition literature is sensitivity to orientation in the picture plane. We are faster to recognize an upright image of an object than we are to recognize an upside-down image. Moreover, the time to name an object increases almost linearly with the degree of misorientation (for reviews, see Jolicoeur, 1990; Quinlan, 1991; Tarr, 1995). This finding has led numerous researchers to postulate that we recognize misoriented objects by "mentally rotating" them to an upright orientation: The greater the misorientation, the further we must rotate the image, hence the linear function. However, the story is not quite so simple. In some studies (e.g., Jolicoeur, 1985), responses are slightly *faster* when an object is completely upside down (i.e., 180° off upright) than when it is almost upside down (slightly less than 180°). It is impossible to account for this effect in terms of simple linear "mental rotation." However, it follows naturally from JIM's structural descriptions. Tested with images rotated in the picture plane, JIM shows a monotonic decline in performance to 135°, and a slight increase in performance at 180° (Hummel & Biederman, 1992). Importantly, it does so without performing rotations of any kind. Rather, performance declines because rotations in the picture plane perturb the geons' relations and categorical orientations. For example, consider a lamp whose shade is above its base when the lamp is upright. The shade will appear both above and beside the base when the lamp is rotated 45°; it will only be beside the base at 90°, and it will be beside and below the base at 135°. The mismatch between the relations in the rotated image and the relations in the upright lamp continues to increase through 135° of rotation, so performance gets worse. Performance improves at 180° because the mismatch in the relations (and the geons' categorical orientations) actually decreases. Although the above/below relations are still reversed at 180°, the spurious "beside" relation disappears.

In summary, JIM simulates the major invariances and view sensitivities of human object recognition, and its ability to do so derives directly from the structural descriptions it uses for recognition. In turn, these descriptions depend critically on the independent coding and dynamic binding of shape attributes and relations. As discussed in the next section, another line of support for the role of such representations in human shape recognition comes from some of their limitations, and what those limitations imply about human object recognition.

The limitations of dynamic binding: JIM.2

Although an independent coding of part attributes and relations affords several advantages, it is also subject to a number of important limitations. As noted previously, independent coding makes a representation heavily dependent on dynamic binding. JIM requires all an object's geons to fire cleanly out of synchrony with one another. If the features of two or more geons happen to fire in synchrony "accidentally," then the binding of geon attributes to one another and to their relations will be completely lost. The resulting representation will be virtually useless for recognition (see Hummel & Biederman, 1992;

Hummel & Stankiewicz, 1996). JIM is therefore premised on strong assumptions about the speed and reliability with which dynamic binding can be established: It must be error-free and it must be established prior to recognition. Unfortunately, both these assumptions are almost certainly wrong. Dynamic binding, by synchrony or any other means, is likely to be error-prone because it requires a process to actively assign units to groups. Perceptual grouping (e.g., binding image features into geons) is especially likely to be error-prone because it requires the simultaneous consideration of multiple, often contradictory cues. For the same reason, it is virtually guaranteed to be time consuming (see Hummel & Stankiewicz, 1996).

By contrast, object recognition is both reliable and fast. Indeed, it is arguably too reliable and too fast to depend as critically on dynamic binding as JIM does (see Hummel & Stankiewicz, 1996). Apparently, human object recognition is not subject to the kind of strong binding constraints demanded by a completely independent coding of attributes and relations. But this does not imply that structural descriptions play no role in human object recognition. Rather, it suggests that structural descriptions—or, more generally, any representation based on a fully independent attribute code—is not, by itself, sufficient to explain human object recognition.

In the attempt to understand how human object recognition might evidence the strengths of a structural description without the limitations of a fully independent attribute code, Hummel and Stankiewicz (1996) have proposed a hybrid model of object recognition. This model, JIM.2, augments dynamic binding for structural description with static binding for speed and reliability. The model's architecture is illustrated in **Figure 2**. As with JIM, the first two layers of JIM.2 code local image features (contours, vertices, and axes) separately at each location in the visual field; Layers 3 and 4 use the outputs of these feature detectors to generate a structural description of an object's shape; and Layers 5 and 6 use those descriptions as the basis for object recognition. Synchrony of firing, established by local interactions in Layers 1 and 2, groups image features into geons. This model contains a set of units—like JIM's Layers 3 and 5—that represent geon shape attributes and relations in a completely independent fashion (the *Independent Geon Array*, or *IGA*, in JIM.2's fourth layer). The IGA has all the strengths and limitations of JIM's independent geon and relation code: complete preservation of object similarity structure and limited invariance with viewpoint, but strong dependence on dynamic binding. This part of JIM.2 gives the model the properties of a structural description when dynamic binding is established correctly.

But dynamic binding takes time. When an image is first presented to the model, all the local feature units (contours, vertices, and axes) tend to fire at once, even if they belong to separate geons. During this period, the representation generated on the IGA is virtually useless for recognition because it does

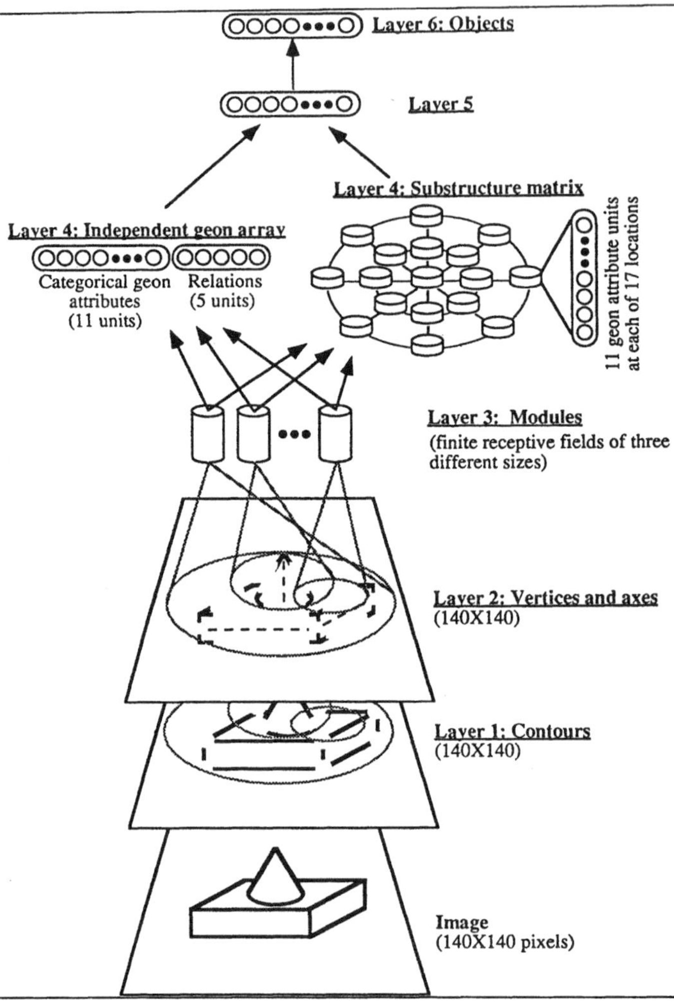

FIGURE 2. The architecture of JIM.2 (Hummel & Stankiewicz, 1995a). Units in Layer 1 represent the contours in an object's image. Units in Layer 2 respond to vertices (heavy solid lines) and axes of parallel and non-parallel symmetry (dashed lines). Layer 3 consists of gated modules (cylinders) with finite circular receptive fields (light ellipses over Layers 1 and 2). Receptive fields of three sizes are distributed in a hexagonal lattice over the visual field. Layer 4 has two parts: The *Independent Geon Array* (*IGA*) contains 11 units that code the properties of geons (*Axis*, *Cross Section*, *Aspect Ratio*, and *Sides*, as in JIM) and 5 units that code a geon's categorical relations to other geons; the *Substructure Matrix* (*SSM*) contains 11 shape-attribute units at each of 17 positions in a circular reference frame. Units in Layer 5 respond to specific patterns of activation on Layer 4, and units in Layer 6 sum the outputs of Layer 5 units over time to respond to complete objects.

not specify how shape attributes go together into geons. However, JIM.2's fourth layer contains an additional collection of units—the *Substructure Matrix* (*SSM*)—that preserve the separation of geon attributes, even when multiple geons fire in synchrony. The SSM is a coordinate-like representation in which separate collections of units respond to geon attributes at each of 17 positions in a circular reference frame: Each unit in the SSM codes a static binding of one geon attribute to one location in the reference frame. Although geons that fire in synchrony with one another are superimposed on the IGA, they remain separate on the SSM.

The mapping of feature outputs (in Layer 2) to the IGA and SSM is controlled by a collection of gated modules in Layer 3. Modules have finite receptive fields at a variety of locations and sizes over the image (see **Figure 2**). Whenever a collection of feature units fires—whether that collection belongs to a single geon or to multiple geons—the modules map their outputs simultaneously to both the IGA and the SSM. Each feature is mapped into the SSM in a

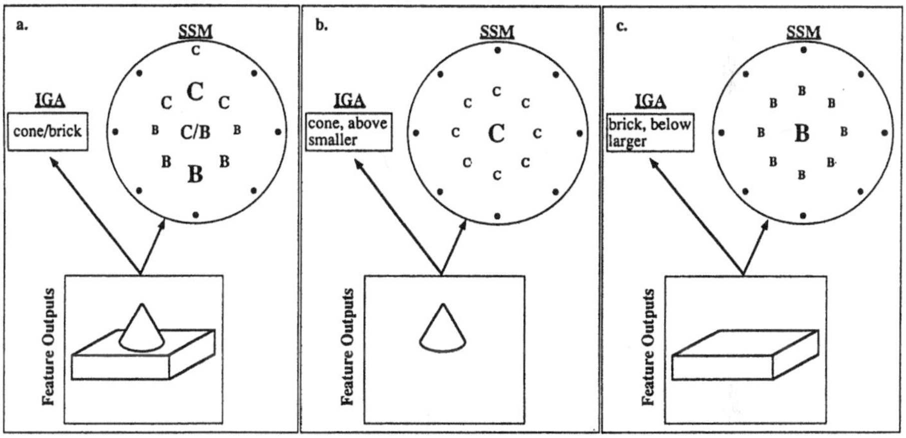

FIGURE 3. Illustration of the mapping of feature outputs (Layer 2) to Layer 4's Independent Geon Array (IGA) and Substructure Matrix (SSM). **a.** When all an object's features fire at once, the properties of the object's geons (here, a brick and a cone) are superimposed on the IGA but are kept separate on the SSM. SSM locations responding to the cone are marked with Cs, those responding to the brick are marked with Bs, and those responding to neither are indicated with a dot. Image locations are coded coarsely by SSM locations, so a given geon will activate a range of locations in the SSM. The strength of a location's response to a given geon is caricatured in the size of the corresponding letter (e.g., small Cs depict weak responses to the cone and large Cs depict strong responses). **b.** When the cone fires by itself, it will be represented by itself on both the IGA (along with its relations to the brick) and the SSM. **c.** When the brick fires by itself, it will be represented by itself on both the IGA (along with its relations to the cone) and the SSM.

location corresponding to its location relative to the whole feature collection. Consider **Figure 3**. The modules will map the features of the cone (which are in the upper region of the feature collection) to the upper region of the SSM, and map the features of the brick (which are in the lower region of the feature collection) to the lower region of the SSM. Although they will be superimposed on the IGA, the cone and brick will remain separate on the SSM. Once the features of the cone and brick start to fire out of synchrony with one another (**Figure 3b** and **3c**), the modules will map their outputs to the IGA and SSM in separate packages (one package per unit time).

The result of these operations is the following. When an image is first presented to the model, all its features will fire at once, producing a view-sensitive representation on the SSM. (This representation is invariant with the location and size of the image in the visual field, but sensitive to the orientation in which the object is depicted.) If the view is familiar, then the model will recognize the object immediately based on this representation. But if the view is unfamiliar (e.g., it is a rotation or a left-right reflection of a familiar view), then the representation on the SSM may be insufficient for recognition. Thus, the advantage of the static attribute-location binding in the SSM is that it frees the model from dependence on dynamic binding, permitting rapid recognition of familiar object views; the disadvantage is that it is view-sensitive. But given enough time, the local feature units in Layers 1 and 2 will segment themselves into geon-based sets. As each geon fires, the IGA represents it in terms of its shape attributes and relations to the other geons in the object: Once dynamic binding is established, JIM.2 represents the object as a structural description. Like the structural descriptions in JIM, the structural descriptions in JIM.2 are invariant with left-right reflection and some rotations in depth.

Behavioral implications of JIM.2. The most immediate prediction of this model is that recognition will be fast and initially view-sensitive (by virtue of the SSM), but given time to generate a structural description, recognition will be relatively viewpoint-invariant (by virtue of the IGA). To turn these observations into specific predictions, it is necessary to estimate (1) how long it takes to generate a structural description on the IGA, and (2) how long the view-sensitive representation lasts on the SSM. (As long the SSM remains active, recognition will be somewhat view-sensitive, even after a view-invariant representation is generated on the IGA.) As discussed in Hummel and Stankiewicz, a reasonable estimate of the time required to generate a structural description from an image is on the order of 150 to 300 ms. Recent behavioral results from my laboratory (discussed below) suggest that the activation trace on the SSM may last a few seconds. Naturally, both these estimates depend on several factors (such as the viewing conditions, the nature of the task, and so forth), but they provide a ballpark estimate of the time course of view-sensitive and view-invariant recognition. Given these estimates, JIM.2 predicts that object recognition will be viewpoint-sensitive for a few seconds, and then later become more robust to variations in viewpoint.

Ellis and Allport (1986) report findings consistent with this prediction. They had subjects view two pictures in succession and say whether they depicted objects with the same name. Subjects were faster to respond "yes" if the pictures depicted the same object (e.g., two pictures of the same car) than if they depicted different objects with the same name (two different cars). At inter-stimulus intervals (ISIs) under two seconds, they found that response times were faster still if the objects were depicted at the same orientation in depth: Recognition was initially sensitive to viewpoint. But at longer ISIs, this same-view benefit disappeared: Eventually, recognition was robust to changes in viewpoint. These data suggest that recognition is mediated by two representations, one fast but viewpoint-sensitive, the other slow but viewpoint-robust (Ellis & Allport, 1986). Using a similar paradigm, Ellis, Allport, Humphreys, and Collis (1989) later showed that the same-view benefit is insensitive to translation and, under some circumstances, scale. Although it is view-sensitive, even the fast representation is not strictly retinotopic (Ellis at al, 1989). These data are strikingly consistent with the predictions of JIM.2: The SSM supports rapid recognition that is robust to translation and scale, but sensitive to orientation in depth (and left-right reflection). By contrast, the IGA is slow but robust to variations in viewpoint.

A related line of support for JIM.2 comes from the literature on information integration across eye movements. Irwin and his colleagues (e.g., Carlson-Radvanski & Irwin, in press; Irwin, 1992; Irwin & Andrews, 1995) have shown that the information preserved across saccadic eye movements has the properties of a structural description rather than a literal representation of the positions of local features in the image. (Among other things, it is capacity-limited in a manner consistent with the capacity limits of dynamic binding.) Interestingly, it takes approximately 300 ms to initiate a saccade. It is tempting to speculate that part of this time reflects the visual system's generating a structural description prior to the saccade: The description takes time to generate, so the visual system postpones the saccade until it has been generated. Other factors undoubtedly contribute to the delay as well, but the relationship is suggestive.

The relationship between binding and view-invariance in JIM.2 also predicts that there will be a relationship between visual attention and view-invariance. Dynamic binding of the type required for structural description requires visual attention (Enns, 1992; Logan, 1994; Treisman, 1993; Treisman & Gelade, 1980; Treisman & Schmidt, 1982; see Stankiewicz et al, 1995, for a review). However, attention is not necessary for object recognition (Tipper, 1985; Tipper & Driver, 1988; Treisman & DeShepper, 1995). If attention is necessary for dynamic binding, then whatever representation mediates the recognition of unattended objects must be based on static binding. It is possible that something like the SSM serves this function.

Support for this conjecture comes from some recent findings regarding the role of attention in priming for left-right reflections of object images (Stankiewicz et al, 1995). Recall that Biederman and Cooper (1991a) showed that visual priming is invariant with left-right reflection. JIM and JIM.2 account for this invariance in terms of the structural descriptions generated in response to attended images. If we assume that the SSM mediates recognition in the absence of visual attention, then JIM.2 predicts that: (1) ignored objects will be visually primed, (2) this priming will obtain even if the first and second images fall in different parts of the visual field (i.e., like the SSM, the priming should be indifferent to translation), and (3) the priming will be sensitive to left-right reflection in that an ignored image will prime itself in the same view, but not in the left-right reflected view. This is exactly the pattern observed by Stankiewicz et al (1995). When subjects attended to an object on one trial, both the same view and its reflection were visually primed on the next trial. Ignored objects were primed only in the same view, but this priming obtained even though the first and second images never fell in the same part of the visual field. Moreover, the effects of view (same vs. reflected) and attention (attended vs. ignored) were strictly additive: Attended objects in the same view enjoyed the same advantage in priming over attended objects in a reflected view as ignored objects in the same view enjoyed over ignored objects in a reflected view. This additivity is consistent with the suggestion that the representation mediating the same-view priming (e.g., the SSM) is independent of the representation mediating the reflected-view priming (e.g., the IGA; see Stankiewicz et al, 1995).

The fact that ignored objects were primed in the same view despite the images' falling in different parts of the visual field is more interesting than it appears. Priming for ignored images was no more sensitive to translation across the visual field than priming for attended images. This finding suggests that attention plays no role whatsoever in the visual system's capacity for translation invariance. Because the SSM is just as indifferent to translation as the IGA, this is exactly the pattern predicted by JIM.2.

The priming observed by Biederman and Cooper was completely invariant with reflection, in that an image primed its reflection just as much as it primed itself. By contrast, the priming observed by Stankiewicz et al was at least partially view-sensitive: Even in the attended condition, priming was greater for the same view than for the reflected view. The difference between the findings of Biederman and Cooper and those of Stankiewicz et al is due to a difference in the time between prime and probe trials. Biederman and Cooper used a long-term priming paradigm in which the time between successive presentations of an image was on the order of minutes (with several other images presented during the intervening period). By contrast, Stankiewicz et al observed their same-view advantage using a short-term priming paradigm in which successive presentations of an image occurred within about 3 seconds of one another (with no other images in between). When Stankiewicz et al used a long-term paradigm like that of Biederman and Cooper, the priming they observed was completely invariant with left-right reflection (like that observed by Biederman and Cooper). Ignored objects showed no long-term priming at all. Apparently, priming in the SSM is short-lived compared to priming in the IGA (Stankiewicz et al, 1995).

Summary and Extensions

Representing object structure entails dynamically binding independent shape attributes into relational structures. The benefit of this approach is that it completely preserves an object's attribute structure, giving view-robust recognition as a natural consequence. The cost is that dynamic binding is expensive, requiring both time to establish (and, in the case of synchrony, even to represent), and finite attentional resources to maintain. The strengths of this approach to object representation provide a very natural account of the strengths of human object recognition (Hummel & Biederman, 1992). And, perhaps more interestingly, the limitations provide a natural account of many of our limitations (Hummel & Stankiewicz, 1996).

Naturally, there remain many open questions about the visual representation of object structure. For example, although there is reason to believe that dynamic binding of some form is a prerequisite to the representation of relational structures (see Fodor & Pylyshyn, 1988), it is still unclear how nervous systems accomplish dynamic binding. Synchrony is a candidate mechanism, but the role of synchrony remains controversial. And even if synchrony proves to be one means for dynamic binding, it is certainly possible that there are others. (See Singer, Trehub, this volume.)

Other aspects of the representation of structure remain a mystery as well. Cognitive scientists and neuroscientists are almost completely in the dark about the nature of the relations represented in biological visual systems. JIM and JIM.2 are based on highly simplified categorical relations, such as "above," "below," and "beside." Although this representation is almost certainly insufficient, virtually nothing is currently known about what the "right" set of relations is. Similarly, it is unknown whether one set of relations supports visual recognition and a different set supports, say, visual reasoning. These questions remain virtually unexplored in both the behavioral and neuroscientific communities. But whatever their answers turn out to be, it seems certain that the problem of structure is one that the visual system has adapted to solve, and that the nature of its solution has profound implications for the properties of human object recognition.

Neural-Networks Models of Cognition
J.W. Donahoe and V. Packard Dorsel (Editors)

CHAPTER 12

A NEURAL-NETWORK APPROACH TO ADAPTIVE SIMILARITY AND STIMULUS REPRESENTATIONS IN CORTICO-HIPPOCAMPAL FUNCTION

Mark A. Gluck and Catherine E. Myers
Center for Molecular and Behavioral Neuroscience
Rutgers University

ABSTRACT

We previously presented a top-down, connectionist model of cortico-hippocampal interaction during classical conditioning (Gluck & Myers, 1993) which argues that the hippocampal region adapts stimulus similarity according to two basic principles: a bias to differentiate—or reduce similarity—between stimuli that are predictive of different salient future events, and a bias to increase similarity between stimuli that co-occur and/or predict similar future events. The original cortico-hippocampal model of Gluck and Myers (1993) addressed a wide range of intact and hippocampal-lesion data. It also generated a number of novel and testable predictions about behavioral phenomena that should be hippocampal dependent. In more recent work (Myers, Gluck, & Granger, 1995) we have shown how at least one aspect of the proposed hippocampal-region function could arise from the substrate of the entorhinal cortex, the major sensory input to the hippocampal formation. Other work has shown how the role of cholinergic modulation from the medial septum can be incorporated into the model and relevant behavioral data explained (Myers, Ermita, Harris, Hasselmo, Solomon, & Gluck, 1996). This modeling suggests instead that the hippocampus is always active during normal learning. What should differ consistently between intact and hippocampal-lesioned animals are the strategies or methods by which they master tasks. This viewpoint suggests that attempts to differentiate and identify "hippocampal-dependent" or "hippocampal-independent" tasks may not be the most useful way to proceed. Rather, empirical emphasis should focus on tasks that can be solved by both intact and hippocampal-lesioned animals, and concentrate on ways in which generalization or transfer performance differs between the two groups.

Introduction

The hippocampal region (**Figure 1**) is composed of a group of structures located deep within the brain, and includes the hippocampus itself as well as the nearby dentate gyrus, subiculum and entorhinal cortex. The entorhinal cortex receives highly processed information from the multimodal cortical-

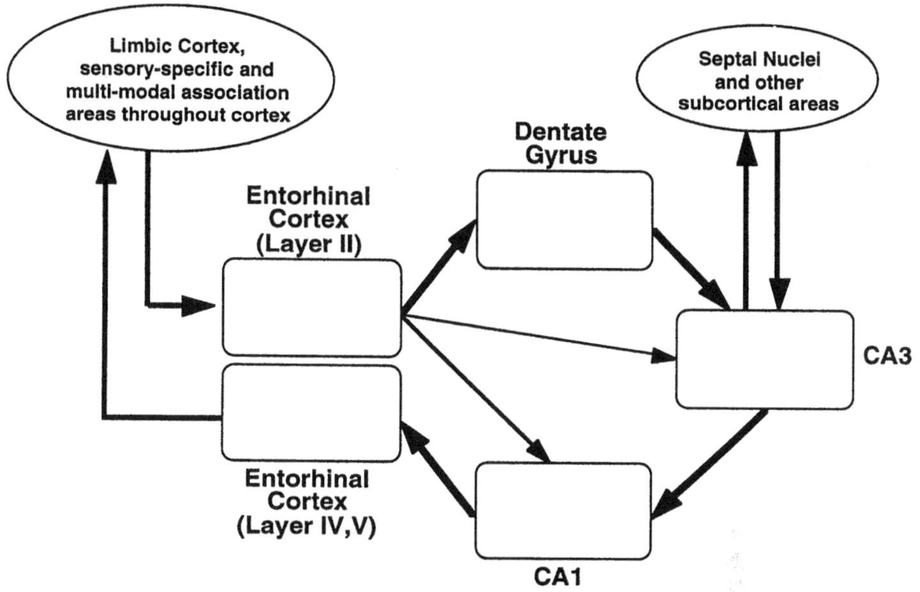

FIGURE 1. Schematic of major information-flow pathways in the hippocampal region. Highly processed, multimodal inputs enter entorhinal cortex and proceed in a largely unidirectional pathway through the hippocampal formation (dentate gyrus, hippocampal fields CA3 and CA1 and subiculum—not shown) before returning to entorhinal cortex and thence back to the same cortical areas where they arose. There is also a bidirectional pathway through the fornix connecting the hippocampus with subcortical areas such as medial septum. Many other connections exist in addition to the major ones shown here.

association areas. Information flows in a roughly unidirectional fashion from the entorhinal cortex to the dentate gyrus, to hippocampus, to the subiculum and back to the entorhinal cortex before returning to the same sensory areas where it originally arose (Amaral & Witter, 1989). In addition to this basic pathway, there are a large number of direct connections between the structures of the region and to other structures such as medial septum.

Damage to the hippocampal region in humans produces a characteristic anterograde-amnesia syndrome, strongly impairing the learning of new "episodic" or "declarative" information, the kind of information about individual events and experiences that is generally accessible to conscious control (Squire, 1987). In contrast to episodic or declarative memories, which are often acquired in a single exposure, other kinds of memory are acquired incrementally over many exposures. These "procedural" memories often survive hippocampal-region damage relatively intact. Thus, animals with lesion to the hippocampal region can still learn to choose the novel item from a pair of objects where one object was seen immediately before (Zola-Morgan & Squire, 1992), acquire simple discriminations of singly presented odor stimuli in an operant task (Eichenbaum, Fagan, Mathews, & Cohen, 1988), and learn to navigate to an escape platform when starting from a constant location in a pool (Eichenbaum, Cohen, Otto, & Wible, 1991). Similarly, human hippocampal-damaged amnesics are not impaired in learning simple classification tasks (Knowlton, Squire, & Gluck, 1994), or elementary motor skills or responses (Cohen, 1984; Gabrieli, McGlinchey-Berroth, Carillo, Gluck, Cermack, & Disterhoft, 1995). All of these spared tasks can be solved by incremental formation of habits or tendencies, without requiring episodic memories of any individual learning trial.

Interestingly, the extent of impairment often depends critically on the precise extent of the lesion. One example is the latent-inhibition effect, in which prior unreinforced exposure to a stimulus retards later learning to respond to that stimulus (Lubow, 1973). Latent inhibition is attenuated by broad hippocampal-region damage (Solomon & Moore, 1975; Kaye & Pearce, 1987) but *not* by damage strictly limited to the hippocampus and sparing entorhinal cortex (Honey & Good, 1993; Reilly, Harley, & Revusky, 1993). Similarly, odor-discrimination reversal is impaired by hippocampal lesion but actually *facilitated* after entorhinal lesion (Otto, Schottler, Staubli, Eichenbaum, & Lynch, 1991). These and other results (e.g., Jarrard, 1993; Zola-Morgan & Squire, 1992) suggest that the different substructures of the hippocampal region make differentiable contributions to the processing of the region as a whole; however, the specific assignment of function to substructure, and knowledge of the ways in which they interact, are as yet unclear.

Stimulus Representation and the Hippocampus

A top-down model of adaptive representations in the hippocampal region

We previously presented a top-down, connectionist model of cortico-hippocampal interaction during classical conditioning (Gluck & Myers, 1993). The fundamental explanatory mechanism of this model was **similarity**. Similarity can be defined operationally as the degree to which an organism generalizes what it knows about one stimulus to another stimulus (James, 1896; Shepard, 1958; Tversky, 1977). There are many ways to conceptualize the psychological notion of similarity. One approach is to view stimulus items as represented by points in a high-dimensional **psychological space**; within this geometric model of similarity, small inter-item distances are representative of high similarity, while large inter-item distances are representative of low similarity (Shepard, 1957). Initially, the distance between points in this psychological space may reflect physical similarity of stimuli. For example, a 1000-Hz tone is physically more similar to a 1200-Hz tone than to a 600-Hz tone, and the arrangement of these stimuli in psychological space may reflect this ordering (**Figure 2A**).

If two stimuli have high similarity, what is learned about one will generalize strongly to the other; this will facilitate transfer between them, but impede learning to discriminate them. Thus, similarity can have important consequences for learning. In Gluck and Myers (1993), we reviewed evidence that similarity in psychological space is not fixed, but changes during learning in response to correlations between stimuli. For example, if an animal learns that food is signaled by the 600- and 1000-Hz tones but not by the 1200-Hz tone, then the similarity in psychological space may be altered to reflect this (**Figure 2B**). As a result, there may come to be high similarity, and hence high generalization, between stimuli that are similar in meaning. Related accounts of adaptive similarity have been proposed within cognitive psychology for theories of concept learning (Kruschke, 1992; Nosofsky, 1984) and categorical perception (Harnad, Hanson, & Lubin, 1994).

In several recent papers we described comparative behavioral analyses that, we argued, suggest that animals (and possibly people) with damage to the hippocampal region are unable to alter stimulus similarity based on experience. While these hippocampal-damaged subjects can still learn whether or not to

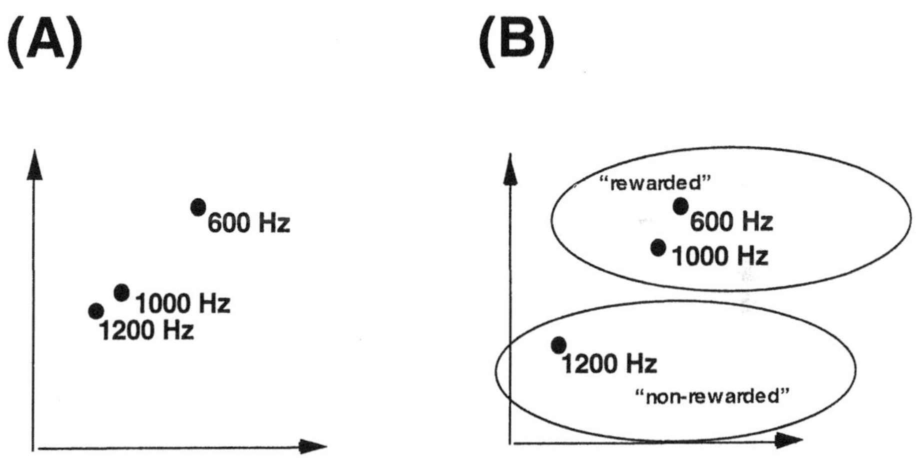

FIGURE 2. If stimuli are represented as points in a high-dimensional "psychological space," then the similarity, or tendency to generalize between them, is illustrated by the distance between them in this space. **A.** Initially, stimulus locations may reflect physical similarity: The representation of a 1000-Hz tone may be closer to that of a 1200-Hz than a 600-Hz tone. **B.** Similarity is assumed to alter as a function of experience; if both the 1000-Hz and 600-Hz tones predict reward but the 1200-Hz tone does not, similarity (and distances in psychological space) may change to reflect this.

respond to an individual stimulus, their performance is notably impaired on many tasks involving learning relationships between stimuli—especially in the absence of explicit reinforcement. This observation led to our proposal that one function of the hippocampal region is to adapt stimulus similarity to facilitate learning (Gluck & Myers, 1993). More specifically, we proposed that the hippocampal region adapts stimulus similarity according to two basic principles. The first is a tendency to differentiate—or reduce similarity—between stimuli that are predictive of different salient future events. We called this principle **predictive differentiation**. For example, if stimulus A predicts reward but B does not, similarity between them should decrease, minimizing generalization between them and facilitating their discrimination. The second principle is a tendency to increase similarity between stimuli that co-occur and/or predict similar future events. We called this principle **redundancy compression**. Thus, if A and B reliably co-occur (and are therefore likely to make similar predictions about future reward), the similarity between A and B is increased, enhancing generalization and decreasing discriminability between A and B.

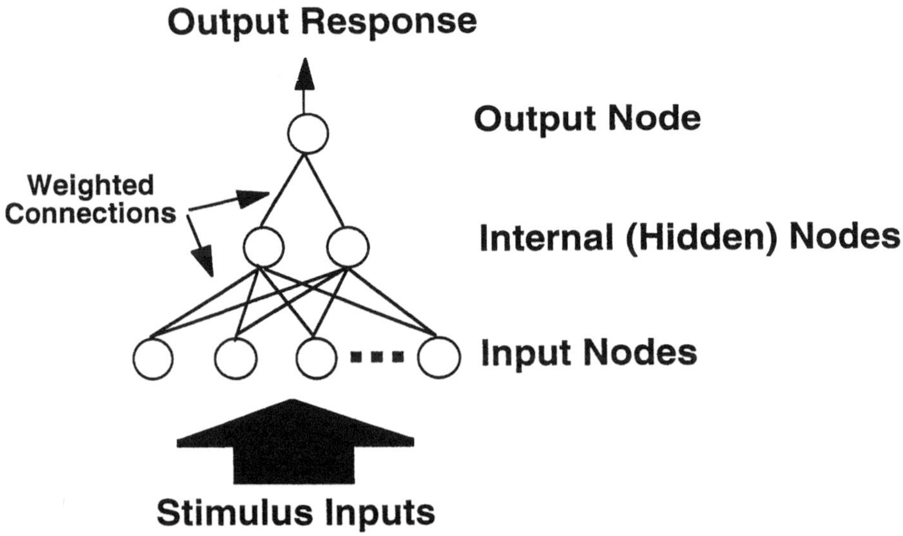

FIGURE 3. An example of a connectionist network. Input activations representing the presence or absence of inputs feed through weighted connections to activate an internal or hidden layer; these activations in turn pass through weighted connections to activate an output-layer node. The activation of this output node is interpreted as the strength (or probability) of a response. The activations of the hidden-layer nodes represent a remapping of the inputs, and representational similarity here leads to generalization at the output node. All connection weights are assumed to be modifiable.

For most simple training paradigms in which there is a simple set of stimulus-stimulus correlations, our theory makes clear predictions for learning and generalization that can be deduced at a verbal or qualitative level, without computer models or simulations. In brief, our theory predicts that these changes in similarity will be evident in normal intact animals, but missing or altered in animals with damage to the hippocampal region. This provides us with a clear and formal theory for how and where we expect hippocampal-lesioned animals to differ behaviorally from normal animals. Although the theory can be applied usefully at just this verbal-qualitative level, a formal instantiation of the theory allows the specification of exact predictions, as well as providing a way to explore more subtle interactions among multiple stimuli in complex training paradigms.

A connectionist model of adaptive similarity in cortico-hippocampal function. Classically conditioned learning can be implemented within a connectionist network model (**Figure 3**). Inputs representing the presence or absence of stimuli (including contextual stimuli) activate an internal or "hidden" layer of nodes through weighted connections; these internal nodes, in turn, activate an output node through weighted connections. The activity level of the output node defines the strength (or probability) of a conditioned response. The remapping of stimuli to activations that occurs at the hidden layer is a representation of stimuli similar to that conceptualized in **Figure 2**: Namely, inputs that activate similar (or dissimilar) representations in the hidden layer will be more (or less) likely to activate similar output responses.

All the connection weights in the network of **Figure 3** are assumed to be adaptive—that is, changeable through experience—and learning consists of modifying these weights until the correct conditioned response is given to each pattern of input stimuli. Algorithms are known for training the upper layer of weights between the hidden- and output-layer nodes, so long as the desired activation of the output-layer node is known. For example, the least-mean-square (LMS) algorithm (Widrow & Hoff, 1960) is a correlational rule directly related to psychological learning theories (Rescorla & Wagner, 1972) as well as putative mechanisms of synaptic plasticity such as long-term potentiation (LTP; Donegan, Gluck, & Thompson, 1989; Levy, Brassel, & Moore, 1983; Brown, Kariss, & Keenan, 1990; see Frey, this volume).

The desired activations at hidden-layer nodes are not known a priori, however, and thus more complex mathematical procedures are needed to allow training the lower layer of weights. Although algorithms do exist that can train these multi-layer weights—notably the error backpropagation algorithm (Rumelhart, Hinton, & Williams, 1986)—their mathematical complexity makes it difficult to imagine how such algorithms might be implemented in the cerebellar and cerebral cortices (although see Schmajuk & DiCarlo, 1990). However, if the desired response of the hidden nodes to an input pattern were somehow known, then the lower layer of weights could be trained using a second application of LMS. In Gluck and Myers (1993) we proposed that the hippocampal region can provide this kind of information regarding desired new stimulus representations in long-term memory, formalized here as the activations of the hidden-layer nodes.

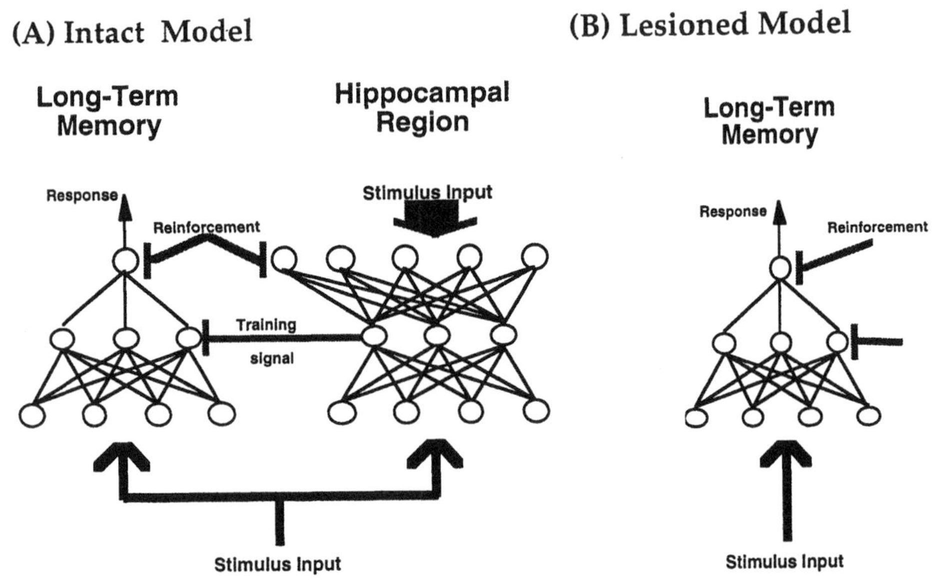

FIGURE 4. The cortico-hippocampal model (Gluck & Myers, 1993). **A.** The intact model. A hippocampal-region network learns to reconstruct its inputs, plus a prediction of US arrival, while forming new stimulus representations in its internal layer that compress redundant information but differentiate predictive information. These new representations are acquired by a cortical network, which learns to map from them to a prediction of classification or behavioral response, and which is the site of long-term memory. **B.** The lesioned model. Disabling the hippocampal network is assumed to result in the cortical network's no longer being able to acquire new representations, although it can still learn to map from existing representations to new behavioral responses.

A model of the hippocampal region is shown in the rightmost portion of **Figure 4A**, and is termed an autoencoder (Hinton, 1989). This network is broadly similar to the basic network shown in **Figure 3**, except for the addition of enough output nodes to reconstruct the entire input pattern as well as predict the conditioned response. This network may be trained by an algorithm such as error backpropagation. The hidden-layer activations, determined in the hippocampal-region network, are transferred incrementally to a cortical network (in the leftmost portion of **Figure 4A**), which is assumed to be the site of long-term memory storage. The cortical model can then use one application of LMS to modify the input-hidden-layer weights, and a second independent application of LMS to modify the hidden-output-layer weights.

Within this model framework, hippocampal-region damage is simulated by disabling the hippocampal-region network, as shown in **Figure 4B**. In this case, the cortical network is assumed to be unable to modify its lower-layer weights, although it can still modify its upper-layer weights. Thus, even without hippocampal input, the cortical network can continue to learn new behavioral responses based on preexisting (and now fixed) hidden-layer representations. The computational processing remaining in **Figure 4B** following removal of the hippocampal-region network can, therefore, be compared to behavior after a broad hippocampal-region lesion, including the hippocampal formation and entorhinal cortex (the so-called H+EC lesion). These intact and lesioned models can be applied to simple associative-learning preparations such as classical conditioning, and accurately capture the behavior of intact and lesioned animals in a variety of paradigms (Gluck & Myers, 1993; Myers & Gluck, 1994). We discuss a few examples below.

Evidence for redundancy compression. The first putative hippocampal function, redundancy compression, increases similarity between co-occurring or redundant stimuli. One of the simplest paradigms in which redundancy compression is expected is **sensory preconditioning**. Consider two distinct stimulus cues, A and B, perhaps two tones. The representations they evoke should be highly distinct—like the 1000-Hz and 600-Hz tones shown in **Figure 2A**. As such, there should be very little generalization between A and B. If A is subsequently paired with the US (A+ training), a test presentation of B should evoke very little response. However, if prior to the A+ training there are repeated non-reinforced trials pairing A and B, redundancy compression should make the representations of co-occurring cues A and B more similar, as in the 1000-Hz and 600-Hz tones in **Figure 2B**. This will increase generalization between A and B, so that subsequent A+ training will transfer partially to B, with the result that presentation of B should evoke conditioned responding. This sensory-preconditioning effect is seen in intact animals (Thompson, 1972) and also in the intact model (**Figure 5A**). Because our model assumes that sensory preconditioning arises from hippocampal-dependent representational compression during the pretraining phase, the effect is not present in the lesioned model (**Figure 5A**). Similarly, hippocampal damage through fimbrial lesion attenuates sensory preconditioning in rabbits (Port & Patterson, 1984).

Conversely, in compound preconditioning, preexposure to AB retards later learning to discriminate them in both normal animals and children (Lubow, Rifkin, & Alek, 1976). The intact model shows this behavior (**Figure 5B**) since preexposure increases generalization and retards discrimination of A and

B; this hippocampal-dependent effect does not occur in the lesioned model (Gluck & Myers, 1993). Thus, our model makes the *novel prediction* that there should be a seemingly paradoxical facilitation of learning after hippocampal-region damage in compound preconditioning; this prediction remains to be experimentally tested.

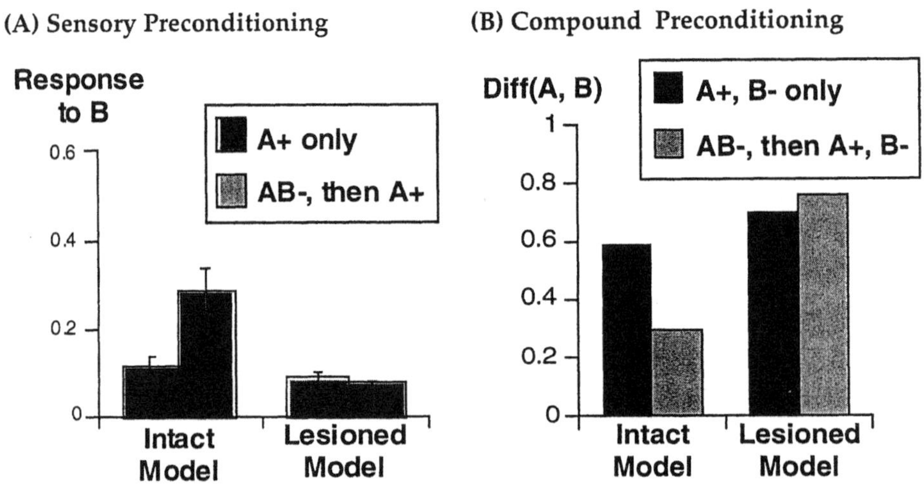

FIGURE 5. Simulations with intact and lesioned models. **A.** Sensory preconditioning: Unreinforced preexposure to a compound AB followed by training to respond to A produces stronger responding to B alone than a control condition with no preexposure. The intact, but not the lesioned, model shows this effect. **B.** Compound preconditioning: Unreinforced preexposure to AB in the intact model slows later training to discriminate A and B, as shown by less relative difference in responding to A and B (Diff(A,B)). The model predicts hippocampal-region lesion should eliminate the effect.

Evidence for predictive differentiation. In our theory, the hippocampal region is also assumed to mediate predictive differentiation, decreasing the similarity between stimuli that are to be mapped to different outputs. The simplest paradigm in which differentiation is expected is a discrimination task in which two stimuli A and B are associated with different responses (e.g., A predicts the US but B does not). In the intact model, the hippocampal-region network constructs new internal representations that decrease the similarity between A and B. These new differentiated representations are acquired by the cortical network's hidden layer, which can then map these representations to different responses as the task requires. Interestingly, the lesioned model shows no particular deficit on this task, and learns as quickly as the intact model (**Figure 6A**; Gluck & Myers, 1993). This is because the preexisting (fixed) hidden-layer representations in the lesioned model's cortical network are likely to partially distinguish A and B for this simple task, and thus all the network must do is map these distinguishable representations to the correct responses.

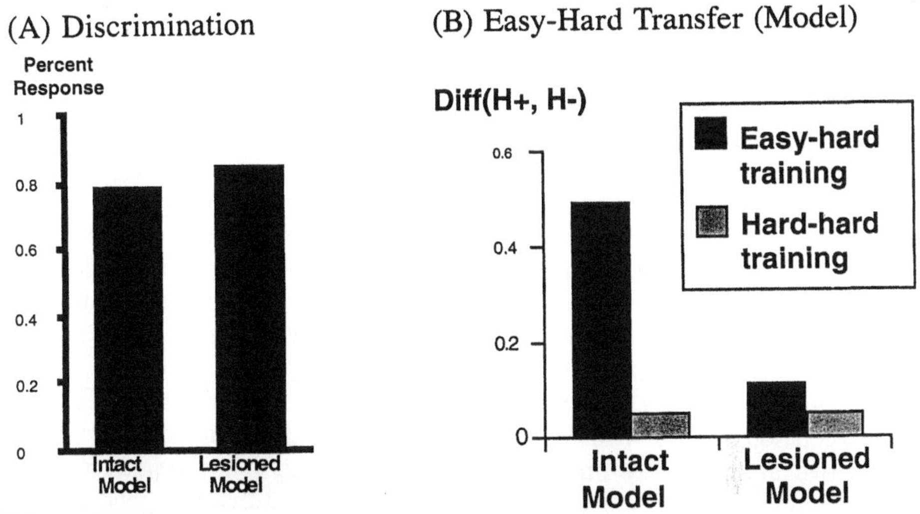

FIGURE 6. Simulation results. **A.** Stimulus discrimination: Training to respond to A but not to B. The intact model forms new stimulus representations that differentiate A and B, and then maps them to opposite responses. The lesioned model simply maps from preexisting (fixed) representations in the cortical network to the correct responses. There is no impairment in conditioned-discrimination learning in the lesioned model. **B.** Easy-hard transfer: Learning a hard discrimination between A and B is facilitated by prior training on an easier discrimination along the same stimulus continuum in the intact, but not the lesioned, model. This is shown by a smaller relative difference in responding to A and B (Diff(A,B)) after a fixed amount of training in the lesioned model than in the intact model. This prediction of hippocampal-dependence remains to be tested. Adapted from Myers et al (1995).

Consistent with the model's implications, hippocampal lesions do not impair learning a simple discrimination across a variety of preparations (e.g., Good & Honey, 1991; Jones & Mishkin, 1972; Port, Romano, & Patterson, 1986; Zola-Morgan, Squire, Rempel, Clower, & Amaral, 1992; Zola-Morgan & Squire, 1986; Jarrard, 1993; Ross, Orr, Holland, & Berger, 1984; Silveira & Kimble, 1968). In some cases, hippocampal-region damage has even been shown to facilitate learning (e.g., Eichenbaum, Fagan, & Cohen, 1986; Eichenbaum et al, 1988; Eichenbaum, Otto, Wible, & Piper, 1991; Port, Mikhail, & Patterson, 1985; Schmaltz & Theios, 1972). Our model provides an interpretation of this seemingly paradoxical facilitation of learning after hippocampal lesion. **Figure 6A** shows that the lesioned model learns somewhat faster than the intact model, which must spend time constructing new representations that are not strictly needed for the training task. (But, as described later, these new representations may be relevant for future generalizations of this learning).

The additional time and effort spent by the intact model in constructing these new and differentiated stimulus representations can be very helpful if the task changes so that the same cues are relevant but their meaning has changed. A simple example of this occurs in the easy-hard transfer paradigm, in which animals are first trained on an "easy" discrimination (e.g., black vs. white) and then transferred to a "hard" discrimination along the same stimulus continuum (e.g., dark gray vs. light gray). This transfer facilitates learning more than an equivalent amount of pretraining on the hard discrimination itself (e.g., Lawrence, 1952; Riley, 1968; Terrace, 1963). The intact model correctly shows this effect (**Figure 6B**; Gluck & Myers, 1993): The difference in output activations to the two hard stimuli, Diff(H+,H-), was greater following pretraining on the easy task than following pretraining on the hard task. During pretraining on the easy task, the hippocampal-region network differentiates the representations of the two stimuli, which predict different outcomes. Since these two stimuli differ on only a single dimension (e.g., brightness), that dimension will be differentiated, with the result of decreased generalization between stimuli with differing values of this dimension. This differentiation will help in the subsequent hard task, involving the same dimension. In the control condition, with pretraining on the hard task, the same mechanisms operate, but they are slower because the stimuli are harder to distinguish. Therefore, the pretraining is not as effective. In the lesioned model, with no differentiation mechanisms, the easy-hard transfer effect is not obtained. This leads to the novel prediction that hippocampal-lesioned animals will not show easy-hard transfer, another prediction that remains to be tested.

Application to contextual sensitivity. The model also applies to the effects of hippocampal lesion on contextual processing. Many of the learning deficits associated with hippocampal damage can be described as contextual deficits, since they suggest an inability to incorporate information about the environmental conditions under which an event occurs (Hirsh, 1974). It should be noted that this does not reflect a general inability to perceive contextual cues, since lesioned animals can still learn to discriminate contexts (e.g., Good & Honey, 1991; Phillips & LeDoux, 1994). What seems to be disrupted in the lesioned animal is the ability to use context to interpret the *meaning* of conditioned cues (Myers & Gluck, 1994). For example, in the eyeblink preparation, if a normal animal is trained to respond to a cue in a specific context, and that cue is then presented in a novel context, the response to that cue will often be weaker in the new context; this effect is eliminated in hippocampal-lesioned animals (**Figure 7A**; Penick & Solomon, 1991; Honey & Good, 1993).

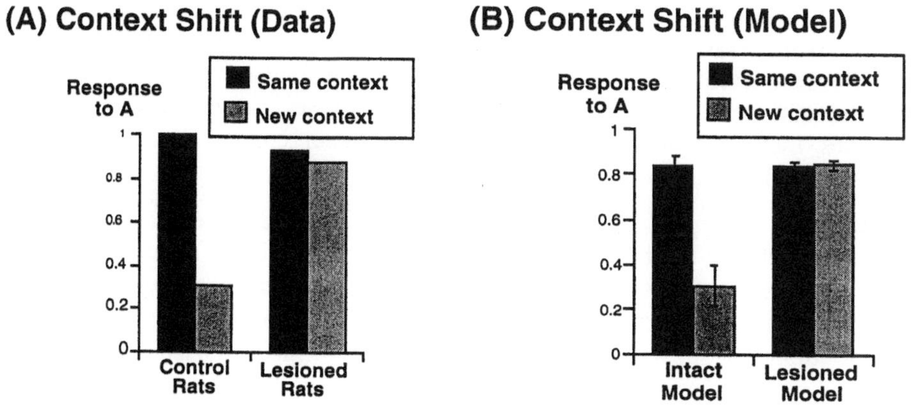

FIGURE 7. Simulation of context effects. **A.** In normal animals, a conditioned response to A may decrease if A is presented in a new context (Hall & Honey, 1990); hippocampal-lesioned animals do not show this response decrement after a context shift (Penick & Solomon, 1991; Honey & Good, 1993; figure replotted from data presented in Penick & Solomon, 1991.) B. The intact, but not the lesioned, model correctly shows this response decrement with context shift (Myers & Gluck, 1994).

[Although this context-shift phenomenon is robust in the rabbit eyeblink preparation, it does not occur in all preparations, cf. Bouton & Peck, 1989. In Myers and Gluck (1994), we discuss how our model provides an interpretation of this variance across species and preparations.]

As the hippocampal-region autoencoder learns to generate CRs, it also learns to reconstruct its inputs, including contextual cues. Thus, contextual information is included in the representations formed in the autoencoder's internal layer. As a result, the context influences the internal-layer representations in such a way that, if the CS is later presented in a new context, the representation of that CS will be more weakly activated than it was in the training context. **Figure 7B** shows the simulated response decrement after context shift in the intact version of the model, which parallels observations in the intact animal shown in **Figure 7A**. The model also correctly predicts that with extended training the hippocampal representation will exclude irrelevant contextual information and become more and more context-dependent (Myers & Gluck, 1994). There is some evidence of this kind of time dependence of contextual sensitivity in animals (Hall & Honey, 1990; see also Myers & Gluck, 1994, for review). In the lesioned model, however, no new representations are formed, so there is no means for incorporating contextual information into the representation of a CS. As a result, the lesioned model is relatively insensitive to context (**Figure 7B**). This is consistent with the occurrence of the strong responding generally shown by hippocampal-lesioned animals in a new context (**Figure 7A**). The cortico-hippocampal model can similarly account for results from a range of context studies (Myers & Gluck, 1994), including the release from latent inhibition with context shift. Moreover, the model provides a computational instantiation and elaboration of several existing qualitative theories that have implicated the hippocampus in context learning (Hirsh, 1974; Penick & Solomon, 1991).

Limitations and open issues. Although the Gluck and Myers (1993) model accounts well for the behavioral data from lesion studies of animal conditioning, it makes no substantive contact with the anatomical and physiological substrates that could give rise to this behavior. In Myers et al (1995), we made a first attempt at showing how at least one aspect of the proposed hippocampal-region function could arise from the substrate of the entorhinal cortex. Much of our more recent work focuses on extending this approach, integrating bottom-up and top-down modeling. An important theme of this research has been an endeavor to use *existing* bottom-up models developed by other researchers to implement the top-down processes suggested by our own previous work.

A bottom-up model of redundancy compression in entorhinal cortex

Although the intact model of **Figure 4A** adopts the simplifying assumption that the hippocampal region functions in a unitary fashion, the hippocampal region is, in fact, anatomically divided into several distinct structures, as shown in **Figure 1**. We have hypothesized that the representational function computed in our intact model's hippocampal region may be subdivided, and the subfunctions localized in different anatomical sites around the region. In particular, we proposed that stimulus-stimulus redundancy compression could emerge from the anatomy and physiology of superficial entorhinal cortex (Myers et al, 1995).

This suggestion is based on a prior bottom-up model of superficial piriform (olfactory) cortex, which argued that the anatomy and physiology of the structure are sufficient to implement hierarchical clustering of odor inputs (Ambros-Ingerson, Granger & Lynch, 1990). In brief, they proposed a competitive network model in which local recurrent inhibition silences all but the most strongly responding pyramidal cells; these "winning" cells then respond to a family or cluster of inputs with similar features. Recurrent feedback from the piriform cortex to olfactory bulb is also assumed to allow iterative responses to odors, from which successively finer-grained (hierarchical) classifications can be constructed. One aspect of this model is that, since similar inputs tend to be clustered to similar output responses, the network performs redundancy compression of exactly the sort we have previously proposed to occur in the hippocampal region (Myers et al, 1995). In particular, if two inputs co-occur, they will be treated as a single compound input. Later, if one of the inputs occurs alone, the network will tend to treat this as a degraded version of the compound input, and assign it to the same cluster as the compound.

Interestingly, the piriform cortex and entorhinal cortex elide in rat, and their superficial layers are closely related anatomically and physiologically, suggesting the possibility of related functionality (Woodhams, Celio, Ulfig, & Witter, 1993; van Hoesen & Pandya, 1975; Price, 1973). Specifically, superficial entorhinal cortex contains pyramidal cells with sparse nontopographic connections with afferents in layer I (van Hoesen & Pandya, 1975) with denser feedback connections to local inhibitory cells (Kohler, 1986). And, entorhinal cortex shows NMDA-dependent, theta-induced long-term potentiation (LTP) (deCurtis & Llinas, 1993). Noting this similarity, Gluck and Granger (1993) suggested that entorhinal cortex performs a similarity-based clustering operation similar to that proposed to occur in piriform cortex.

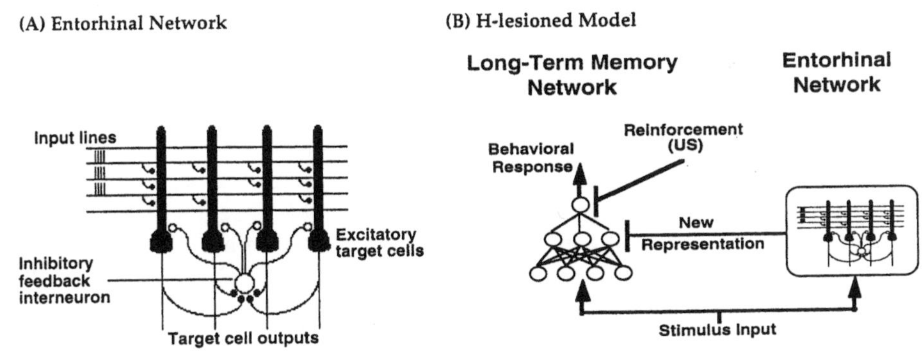

FIGURE 8. A. In the entorhinal model, target cells are excited by sparse afferents, and in turn activate local inhibitory-feedback interneurons. Feedback silences all but the most strongly activated target cells. Synaptic plasticity makes these "winning" target cells more likely to "win" in response to similar inputs in the future. The resulting network activity is constrained by stimulus-stimulus redundancy compression. **B.** The H-lesioned model, in which an entorhinal-cortex network provides new compressed representations to the internal layer of a long-term memory network.

In sum, then, we have proposed that the entorhinal cortex is sufficient to implement the redundancy-compression aspect of the representational changes we ascribed to the hippocampal region as a whole (Myers et al, 1995). A model implementing these proposed processes, and based on the physiologically and anatomically motivated model of Ambros-Ingerson et al (1990), is shown in **Figure 8A**. One difference between the piriform and entorhinal models is that the piriform model assumes repetitive sampling and input masking, based on recurrent connections from piriform cortex to olfactory bulb. We have not assumed this in the entorhinal model, and so it only performs a single-stage, similarity-based clustering or compression of its inputs. The resulting network is similar to the unsupervised competitive-learning systems developed by Kohonen (1984), Rumelhart and Zipser (1985), Grossberg (1976), and others. A second important difference between the piriform and entorhinal cortices is that, while the piriform cortex is primarily an olfactory area, the entorhinal cortex receives input from a broad spectrum of polymodal cortices, as well as from the piriform cortex. Thus, we have suggested that, while the piriform cortex might be sufficient to implement redundancy compression within the olfactory domain, the entorhinal cortex is required to implement redundancy compression between stimuli from different modalities, or across the polymodal features of a single stimulus (Myers et al, 1995).

Evaluating the entorhinal model. The present hypothesis assumes that a selective hippocampal lesion (the "H" lesion) that does not otherwise damage entorhinal cortex might allow redundancy compression processes to survive. Behaviors that depend mainly on these processes should continue to be exhibited after H lesion, while behaviors that require other representational processes such as predictive differentiation should be disrupted. We can evaluate this situation by constructing an H-lesioned model, as shown in **Figure 8B**, in which the hippocampal-region network of **Figure 4A** is replaced by an entorhinal network. [Full details of this model implementation are given in Myers et al, 1995.] The long-term memory network continues to operate as in the intact model, except that the new representations provided by the entorhinal network are biased only by stimulus-stimulus redundancy compression, not by the other representational biases attributed to the hippocampal region as a whole.

Once again, consider latent inhibition, in which unreinforced preexposure to a cue slows subsequent learning to respond to that cue (Lubow, 1973). In our intact model, latent inhibition is caused by compression of the preexposed cue with co-occurring and equally non-reinforced contextual cues. The subsequent increase in learning time results because the model must first redifferentiate the cue from the context before a response can be selectively associated with the cue (**Figure 9A**); the effect is correctly absent in the lesioned model (Myers & Gluck, 1994). Because this effect is assumed to depend primarily on redundancy compression, the entorhinal network in the H-lesioned model is sufficient to produce latent inhibition (**Figure 9B**). Consistent with the model, rats with selective (ibotenate) hippocampal lesions but no entorhinal damage show normal or even enhanced latent inhibition (Honey & Good, 1993; Reilly et al, 1993).

More important, however, is how the model interprets subtle differences in transfer behavior following latent inhibition. When shifted to a new context

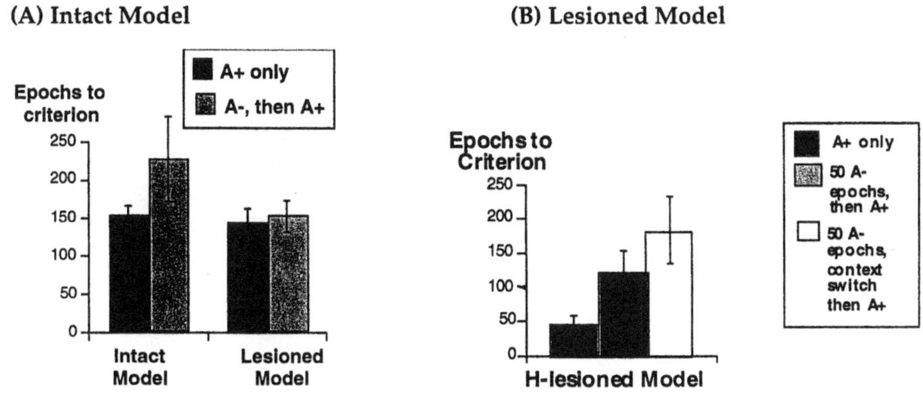

FIGURE 9. Latent inhibition in the model. **A.** In the intact model, prior exposure to A slows subsequent learning to respond to A, compared with a non-exposed condition; this effect is correctly eliminated in the H+EC-lesioned model (Myers & Gluck, 1994). **B.** In the H-lesioned model, however, latent inhibition is preserved, consistent with animal data on selective H-lesion (Myers et al, 1995). Additionally, the H-lesioned model correctly exhibits a release from latent inhibition if a context shift occurs between exposure and training phases.

between exposure and acquisition phases, the intact animals show release from latent inhibition, but the H-lesioned animals do not (Honey & Good, 1993). This implies that the superficially similar latent-inhibition phenomenon observed in intact and H-lesioned animals may result from different underlying processes. The intact, but not the H-lesioned, model shows a similar release from latent inhibition after a context shift, consistent with the animal results (Myers et al, 1995). The difference in context sensitivity in the models results from different hippocampal-region processing. The intact model compresses information during the preexposure phase by incorporating contextual information with the representation of the cue. Then, in the subsequent acquisition phase, it must differentiate these. But, if there is a context shift, the cue's representation is already differentiated from the new context, resulting in a release from latent inhibition. The H-lesioned model compresses information during the preexposure phase by reducing the amount of representational resources used to encode the cue. This compression cannot be undone without hippocampal-mediated differentiation, and the only way to accomplish this during the acquisition phase is to "hunt" for features of the compressed representation that happen to distinguish cue and context. This is made no easier by a context shift and, therefore, the H-lesioned model shows no release from latent inhibition (**Figure 9B**).

This work represents a first attempt to map a behaviorally driven "top-down" model onto actual physiological substrates. This is done by showing that at least one portion of the proposed function of the hippocampal region could emerge from the circuitry of the entorhinal cortex. Open questions for future work are where the other remaining components of the proposed function could be sited, and how these modules and the presumed entorhinal clustering module interact—both during learning and after various kinds of selective lesion. Other open questions involve how other brain structures modulate hippocampal-region processing in physiologically realistic ways, and how the hippocampal region actually interacts with cortex during learning. These issues are partially addressed in the next two sections.

Incorporating models of septohippocampal cholinergic modulation

The original cortico-hippocampal model concentrated on the pathways carrying information in and out of hippocampus but, as mentioned above, there is also an important bidirectional pathway through the fornix connecting the hippocampus with subcortical structures (see **Figure 1**). One important input through the fornix is a modulatory cholinergic input from the medial septum. Hasselmo and Schnell (1994) have suggested that this cholinergic input can be used to switch the hippocampus between two processing states: In the presence of acetylcholine (ACh), the hippocampus stores new information and suppresses pattern reconstruction via the recurrent collaterals in CA3. In the absence of ACh, storage is suppressed, and pattern reconstruction is allowed to occur to retrieve stored patterns for transfer to long-term cortical or cerebellar storage. This hypothesis is consistent with physiological studies (Hasselmo, Schnell & Barkai, 1995) showing that ACh both enhances synaptic plasticity and suppresses activity in hippocampal stratum radiatum (the site of synapses from recurrent collaterals) more than activity in stratum lacunosum-moleculare (the site of synapses from extrinsic inputs). It is also consistent with behavioral data showing that classical conditioning is retarded after interruption of septohippocampal cholinergic inputs by medial septal lesion (Berry & Thompson, 1979) or the anticholinergic drug scopolamine (**Figure 10A**; Solomon, Solomon, Vander-Schaff, & Perry, 1983).

We have approximated Hasselmo's storage-recall switching hypothesis within Gluck and Myers's original cortico-hippocampal model by noting that the tendency of the hippocampal-region network to store new information, as opposed to simply processing it and recalling old information, is determined by the hippocampal network's learning rate (Myers et al, 1996). Disrupting septal input can therefore be approximated by lowering this learning rate—although *not* the rate at which this information is transferred to the cortical network, *nor* the rate at which cortical associations develop. The consequence of this depression of hippocampal learning rates is to strongly retard classical conditioning in the model in a manner that is proportional to the amount of depression (**Figure**

10B). The anticholinergic drug scopolamine causes a similar dose-dependent effect in both rabbits (Solomon et al, 1983) and humans (Solomon, Groccia-Ellison, Flynn, Mirak, Edwards, Dunehuew, & Stanton, 1993).

Consistent with the model's behavior, the empirical data suggest that the effect of disrupting septohippocampal cholinergic pathways is not to abolish learning, but rather to delay its onset: More trials are needed before learning occurs, but once it begins, learning proceeds at approximately normal rates and eventually reaches the same asymptotic level as under normal conditions (Myers et al, 1996). It is interesting to note the relationship between our model of septohippocampal modulation during conditioning and the two-stage conditioning model proposed by Prokasy (1972). Prokasy proposed that the period during which the behavioral response was acquired was preceded by a prior period in which the response remained relatively constant at its baseline level. Both the absolute baseline level and the temporal duration of this first stage were assumed to vary with individuals. Prokasy's model therefore predicts the S-shape acquisition curve seen in intact subjects (**Figure 10A**). The two-stage model is consistent with the cortico-hippocampal model. The cortico-hippocampal model assumes that the hippocampal region forms stimulus representations that are then acquired by the cortical network. The cortical network cannot solve the task until these hippocampal-mediated representations are formed and transferred. Therefore, the intact model shows an initial period of baseline responding until the hippocampal representations evolve (**Figure**

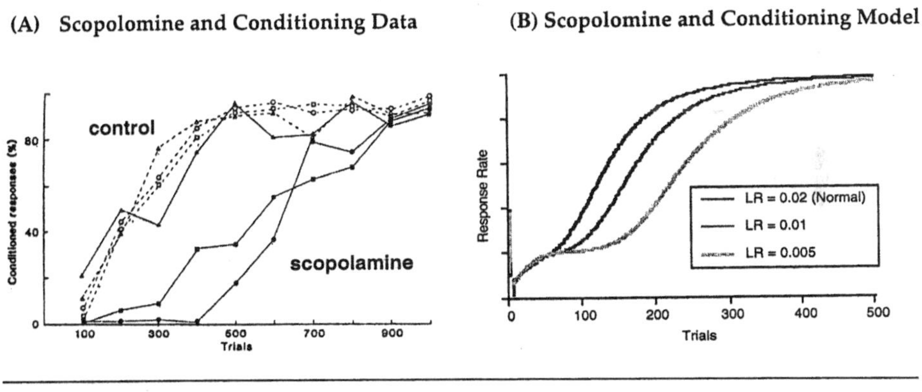

FIGURE 10. A. Classical eyeblink conditioning in rabbits is disrupted by systemic administration of the anticholinergic scopolamine. The onset of learning is delayed but, once begun, learning proceeds at normal rates (Solomon et al, 1983). **B.** A similar effect is obtained in the model by lowering the learning rate of the hippocampal region (Myers et al, 1996). This is an approximation of Hasselmo and Schnell's (1994) hypothesis that septohippocampal ACh determines the rate at which the hippocampus stores new information.

10B). The effect of decreasing the hippocampal learning rate to simulate anticholinergic drugs can, therefore, be interpreted as an extension of the initial phase within Prokasy's two-stage model. This is also consistent with the interpretation by Thompson, Berger, Berry, Hoehler, Kettner, and Weisz (1980) of the functional consequences of medial septal lesions.

Dose dependence of cholinergic and anticholinergic drugs. An important aspect of the modeling work is its generation of predictions regarding the effects of cholinergic and anticholinergic drugs. In our pilot studies, raising the learning rate of the hippocampal network improved learning in the model to some extent (Myers et al, 1996). This is consistent with data showing that patients with basal forebrain damage may show temporary memory improvements given the cholinergic agonist physostigmine (Chatterjee, Morris, Bowers, Williamson, Doty, & Heilman, 1993). Similarly, patients with Alzheimer's disease, who typically show basal forebrain damage and reduced brain acetylcholine levels (Whitehouse, Price, Struble, Clark, Coyle, & DeLong, 1982), show some cognitive and memory improvements given physostigmine (Thal, Fuld, Masur, & Sharpless, 1983; Davis & Mohs, 1982) or the cholinergic agonist Tacrine (Knapp, Knopman, Solomon, Pendlebury, Davis, & Gracon, 1994; Manning, 1994).

Beyond some optimal level, however, further increases in the hippocampal learning rate actually retards learning in the model (Myers et al, 1996). **Figure**

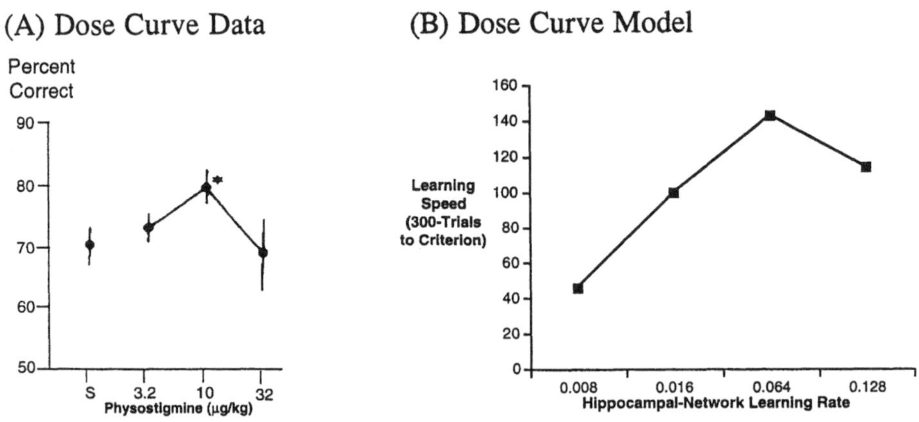

FIGURE 11. A. Classical eyeblink conditioning can be facilitated by moderate doses of physostigmine; but beyond some optimal dose, there is no facilitation or even impairment. Adapted from Ogura & Aigner (1993). **B.** A similar "dose dependency" results in the model from increased hippocampal learning rates. After some initial improvement, further increases in learning rate do not facilitate learning. Adapted from Myers et al (1996).

11B shows this dose-dependent effect of increased hippocampal learning rate in the intact model. This retardation with very high learning rates is a general property of connectionist networks (cf. Jacobs, 1988).

Figure 11A shows that the cholinergic-esterase inhibitor physostigmine can, at moderate doses, improve learning in normals, but at dosages exceeding this optimum level, either no facilitation (e.g., Santucci, Kanof, & Haroutunian, 1989; Ogura & Aigner, 1993; Sweeney, Bachman & Coyle, 1990) or actual impairment may occur (e.g., Ennaceur & Meliani, 1992; Miyamoto, Narumi, Nagaoka, & Coyle, 1989; Dumery, Derer, & Blozovski, 1988). Similarly, although moderate doses of the cholinergic agonist oxotremorine can improve learning in aged rats, higher doses actually impair learning (Markowska, Olton, & Givens, 1995). The model therefore provides an account for these empirical phenomena, which have been problematic in the clinical pharmacology literature. The ability of cholinergic drugs to produce a robust memory facilitation in normal subjects is important given the recent pharmacological interest in producing drugs that facilitate memory. One important aspect of future modeling work is to more accurately investigate expected dose-response curves in normal subjects and subjects with reduced brain acetylcholine levels.

Discussion

The original cortico-hippocampal model of Gluck and Myers (1993) addressed a wide range of data from intact and hippocampal-lesioned organisms. It also generated a number of novel and testable predictions about behavioral phenomena that should be hippocampal dependent. The theory underlying this model also makes contact with a wide range of preexisting, albeit more qualitative, theories of hippocampal-region function.

For example, Eichenbaum, Cohen, and their colleagues (Eichenbaum, Otto, & Cohen, 1992a; Eichenbaum et al, 1992b) proposed that the hippocampus is needed to form flexible representations during learning. This proposal is related to our demonstration that the absence of appropriate stimulus representations during learning can result in altered transfer performance in hippocampal-lesioned animals (Gluck, Myers, & Goebel, 1994). We have already noted how our model instantiates qualitative suggestions that the hippocampal region is necessary for contextual processing—even though our model does not assume context per se is the hippocampus's chief domain (cf. Hirsh, 1974; Penick & Solomon, 1991).

Similarly, our model holds that configural learning may often require hippocampal-dependent, stimulus-stimulus learning and may, therefore, be especially susceptible to hippocampal damage (Myers & Gluck, 1994). This is consistent with Rudy and Sutherland's (1989) view that configural learning is especially hippocampal dependent. This relationship was highlighted in a recent article by Alvarado and Rudy (1995), in which these authors wrote "...our suggestion that it is the degree of associative conflict between the configural representation and the representation of the elements that determines when the hippocampal formation is essential can be considered to be an implication of Gluck and Myers's principle of predictive differentiation" (p. 1062).

Our model stands in stark contrast to qualitative theories that the hippocampus is involved primarily in spatial learning (e.g., O'Keefe & Nadel, 1978). Clearly, spatial learning is extremely disrupted in animals with hippocampal-region damage (e.g., Morris, 1983; Jarrard, 1993); it is also true that "place cells" form in the hippocampus that respond preferentially when the animal is in a particular region of space (e.g., O'Keefe, 1979; McNaughton, Chen, & Marcus, 1991). However, our model and theory suggest that the hippocampus is involved in all kinds of learning that depends heavily on the formation of new representations. For this reason, our theory predicts that spatial learning, which presumably involves associating arbitrary views and proprioceptive information into concepts of "place," might be especially sensitive to hippocampal damage—even though spatial learning per se is not the function of the hippocampus. This argument is similar to that advanced by many others who have considered possible information-processing roles for the hippocampus (e.g., Donahoe & Palmer, 1994; Eichenbaum et al, 1988; Taube, 1991).

One of the most important goals of this modeling work was to produce a simple description of the information-processing role of the hippocampal region in associative learning. The specific proposal was that the hippocampal region adapts stimulus representations to produce generalization and thereby facilitate learning. The model demonstrated that this simple mechanism was, in fact, sufficient to generate a range of behaviors observed in intact animals and animals with hippocampal-region damage. The model also demonstrated that it was not necessary to posit specialized stimulus-specific or task-specific mechanisms to explain this range of behaviors. We know that at least some of this proposed hippocampal-region function can emerge from the biological substrate—and we hope that future work will show that the entire function can be so mapped onto brain regions. We also know that the basic story is consistent with subcortical modulation by the medial septum and with stimulus processing in sensory cortex.

The modeling also provides a framework for understanding why several previously problematic phenomena might occur—including the facilitation of discrimination learning after hippocampal-region damage, the elimination of latent inhibition with a broad hippocampal-region lesion but not a smaller hippocampal-only lesion, and the U-shaped dose-response curve for cholinergic drugs in normal subjects. Additionally, the modeling provides several novel predictions about which tasks ought to be hippocampal dependent, such as compound preconditioning and easy-hard transfer. These predictions provide means for testing the model, but they also can be used to guide research toward productive paradigms.

The model also demonstrates the difference between a structure being active during a behavior versus being necessary for that behavior. The fact that discrimination learning might be unaffected—or even facilitated—by hippocampal damage has been interpreted to suggest that the hippocampus is inactive—or even inhibitory—during conditioned learning. This suggestion is at odds with neurophysiological data showing that the hippocampus is not only active during learning, but performs very specific functions. For example, hippocampal activity is strongest during initial learning and decreases as a task becomes fully mastered (Sears & Steinmetz, 1990).

Our model suggests, instead, that the hippocampus is *always* active during normal learning. The fact that some tasks do not strictly *require* mediation by the hippocampal region does not contradict this. What should consistently be different between intact and hippocampal-lesioned animals are the strategies or methods by which they master these tasks. A simple example comes from the reversal paradigm: Although both intact and lesioned animals are assumed to solve the initial discrimination, the model predicts that only intact animals construct new representations that differentiate the stimuli. Thus, the intact animals perform differently from the lesioned animals on the reversal phase. Therefore, statements that some behaviors are "hippocampal dependent" while others are "hippocampal independent" may not be the most useful conceptualizations. Instead, experimental tasks that can be solved by both intact and hippocampal-lesioned animals should be used with the focus on the different ways in which generalization or transfer performance proceeds.

PART FOUR: BEHAVING

This section is concerned with behavioral processes. We treat these processes separately from those mediating perception only because of the limitations of language; the sequential nature of language is ill-suited to depict the parallel, interacting nature of biobehavioral processes. Neural-network simulations, as imperfect as they may be in a given instance, are conceptually better equipped to describe such processes and to explore their implications.

Before introducing the chapters on behaving, we identify two general points arising from simulations of perceiving. First, as the phenomena being simulated grew more complex, the simulations tended to become less tightly constrained by the relevant biobehavioral science. For example, the simulations of object perception (Hummel) assumed a hierarchical network structure consisting, in part, of elements called *geons*. Geons are perceptual primitives from which more complex objects could be synthesized. Although the neuroanatomy of perceiving is consistent with a hierarchical network structure, neurophysiological work (see Tanaka) indicates that the effective stimuli for activating neurons in visual-association cortex are not so neatly defined as the concept of geon suggests. Moreover, the stimuli that most effectively drive neurons undoubtedly vary depending on the history of the individual with respect to particular combinations of stimuli and the consequences for behaving in their presence (e.g., Sakai & Miyashita, 1991). In spite of differences between the simulated primitives and those identified through direct neurophysiological observations, the simulations usefully illustrate that, given a biological mechanism whereby the activity of disparate units may be bound together, the concerted effects of perceptual primitives can specify complex perceptions. A second general point is that simulation research not only provides a means for exploring the implications of basic biobehavioral processes, but also suggests potentially promising avenues for new experimental research. For example, neurophysiological research indicates that a complex stimulus activates a substantial population of broadly tuned neurons. However, simulations indicate that the activity of populations of broadly tuned units can guide behavior by means of different mechanisms, e.g., through the concerted output of a substantial number of units (Gluck & Myers) or through the output of a single unit whose activity survives inhibitory interactions with other, less strongly activated, units (Trehub). Electrophysiological research must distinguish between these alternative mechanisms for achieving specificity of output from a population of broadly tuned units.

In this section, **Georgopoulos**'s electrophysiological work indicates that behaving as well as perceiving arises from the activity of a population of broadly tuned neurons. Each neuron fires most rapidly when executing a movement in a particular direction, and less strongly as the movement departs from the preferred direction. (Note that broad tuning is not an inherent property of neurons, but the net effect of interactions between that neuron and other neurons synapsing upon it.) The direction and vigor of movement can be described by a population vector that is the resultant of the activity of the individual neurons within the population. Of course, the vector itself has no reality status; it is simply a descriptive tool to summarize the concerted effect of a population of individual neurons. Thus, when Georgopoulos states that "one of the possible operations of the network may be described as the computation of the neuronal population vector," a claim is being made about the *effect* of the neuronal population and not about the *cause* of the movement. In short, locutions that characterize movements as produced by population vectors, or changes in direction of movement as produced by rotating population vectors are employing convenient metaphors to summarize the neural processes that produce behavior.

Palmer's chapter draws upon behavioral research to better define the task that confronts the effort to simulate biobehavioral processes through neural networks: Any of a wide range of stimuli must be able to guide any of a wide range of responses. Neurophysiological research indicates that this is a formidable task because both perceiving and behaving involve the activity of *populations* of neurons (i.e., the binding problem arises). Further, behavioral research indicates that environment-behavior relations emerge as the cumulative effect of a selection process (selection by reinforcement), and that complex relations are products of prolonged *histories* of selection. The net effect depends on the precise nature of the selection history (see the Training Algorithm, Chapter 1) as well as the structure of the nervous system simulated by the neural network (see the Genetic Algorithm, Chapter 1). The chapters by **Berthier** and **Gullapalli** are directed at different aspects of the selection history. **Berthier** emphasizes the developmental history. For example, reaching in adults is the product of the coordinated activity of three motor systems controlling the trunk, arms, and fingers, respectively. These systems myelinate successively, which permits selection to modify the synaptic efficacies of these three populations of neurons during different developmental periods. **Gullapalli** emphasizes the individual history. Specifically, he examines the effects on behavior of changes in the criteria for reinforcement as training progresses from simple to more complex behavior. Procedures that reinforce successively closer approximations to complex behavior—*shaping* procedures—permit reinforcers to select behavior of a complexity that would otherwise be unattainable.

Neural-Networks Models of Cognition
J.W. Donahoe and V. Packard Dorsel (Editors)
1997 Elsevier Science B.V.

CHAPTER 13

MOTOR CORTEX: NEURAL AND COMPUTATIONAL STUDIES

Apostolos P. Georgopoulos
Brain Sciences Center
Veterans Affairs Medical Center, Minneapolis
and Departments of Physiology and Neurology
University of Minnesota Medical School, Minneapolis

ABSTRACT

The motor cortex can be regarded as a network of neurons processing, *inter alia*, spatial motor information. A basic component of this information is the direction of movement in space. Experimental studies in behaving monkeys have shown that the impulse activity of single motor-cortical cells relates to this component in an orderly fashion, such that the frequency of cell discharge is a sinusoidal function of the direction of movement, with the direction for which cell discharge is highest denoting the "preferred direction" of the cell. The neuronal ensemble of such directionally tuned cells can be regarded as a network in which each cell is represented as a vector pointing in the cell's preferred direction. The network operates to generate a signal in the direction of a desired movement. We regard this operation as the vectorial summation of the cell vectors, weighted by a scalar measure of the intensity of cell activation. The resulting vector sum is called the "neuronal population vector." Analysis of experimental data has shown that the population vector points in the direction of the movement. In addition, the population vector can be calculated as a time-varying signal and, as such, is a robust predictor of the direction of the upcoming movement during the reaction time and during instructed and memorized delays. Finally, it has proven a good tool for monitoring and deciphering directional information when more complex directional operations are performed.

Introduction

The motor cortex is located on the lateral and medial surface of the cerebral hemisphere, just in front of the central sulcus, and is the major precentral motor area. Several premotor areas have been identified anterior to the motor cortex. They are interconnected with the motor cortex and, taken together, constitute a highly interactive group of motor areas in the cerebral cortex. This chapter focuses on the coding of directional information by single cells and neuronal populations in the motor cortex. Applications of these analyses are discussed, together with neural-network modeling studies to which the coding scheme has led.

Directional Tuning of Single Cells

When an organism is reaching in space, cell activity during the reaction time relates primarily to the direction of the movement and less to its extent (Georgopoulos, 1990; Fu, Flament, Coltz, & Ebner, 1995). Specifically, cells in the motor cortex (Georgopoulos, Kalaska, Caminiti, & Massey, 1982; Georgopoulos, Schwartz, & Kettner, 1986; Kalaska, Cohen, Hyde, & Prud'-homme, 1989; Caminiti, Johnson, & Urbano, 1990; Schwartz, Kettner, & Georgopoulos, 1988; Schwartz, 1992), as well as in other motor structures (Kalaska, Caminiti, & Georgopoulos, 1983; Fortier, Kalaska, & Smith, 1989; Caminiti, Johnson, Galli, Ferraina, Burnod, & Urbano, 1991), are directionally selective and broadly tuned with respect to the direction of movement. Cell activity is highest for a movement in a particular direction (the cell's "preferred direction") and decreases progressively with movements farther and farther away from this direction. These changes in cell activity relate to the direction, not the endpoint of the reaching movement (Georgopoulos, Kalaska, & Caminiti, 1985). Quantitatively, cell activity is a linear function of the cosine of the angle between the preferred direction of the cell and the direction of a particular movement (Georgopoulos et al, 1982; Schwartz et al, 1988), as follows:

$$d_i(\boldsymbol{M}_l) = b_i + a_i \cos\Theta_{C(i)M(l)} \qquad (1)$$

where $d_i(\boldsymbol{M}_l)$ is the discharge rate of the ith cell with movement in direction $\boldsymbol{M}_l$, b_i and a_i are regression coefficients, and Θ is the angle formed between the cell's preferred direction $\boldsymbol{C(i)}$ and the direction of movement $\boldsymbol{M(l)}$. Equation 1 holds both for 2D reaching movements performed on a plane (Georgopoulos et al, 1982) and for free 3D reaching movements (Georgopoulos et al, 1986; Schwartz et al, 1988). An example is shown in **Figure 1**. The preferred directions differ for different cells and are distributed in the whole 3D directional continuum (**Color Plate 5**, p. 188, top panel; Schwartz et al, 1988). Finally, pairs of cells with similar preferred directions tend to show excitatory synaptic interactions whereas pairs of cells with opposite preferred directions tend to show inhibitory synaptic interactions (Georgopoulos, Taira, & Lukashin, 1993).

Directional Coding by Neuronal Populations

The broad directional tuning indicates that a given cell participates in movements of various directions, and that, conversely, a movement in a particular direction will involve the engagement of a whole population of cells. Therefore, a unique signal for the direction of a movement could reside in the activity of the neuronal ensemble. We proposed a vectorial code for the recovery of this signal from the neuronal ensemble (Georgopoulos, Caminiti, Kalaska, & Massey, 1983; Georgopoulos et al, 1986; Georgopoulos, Kettner, & Schwartz, 1988), as follows: (1) A cell is represented as a vector that points in

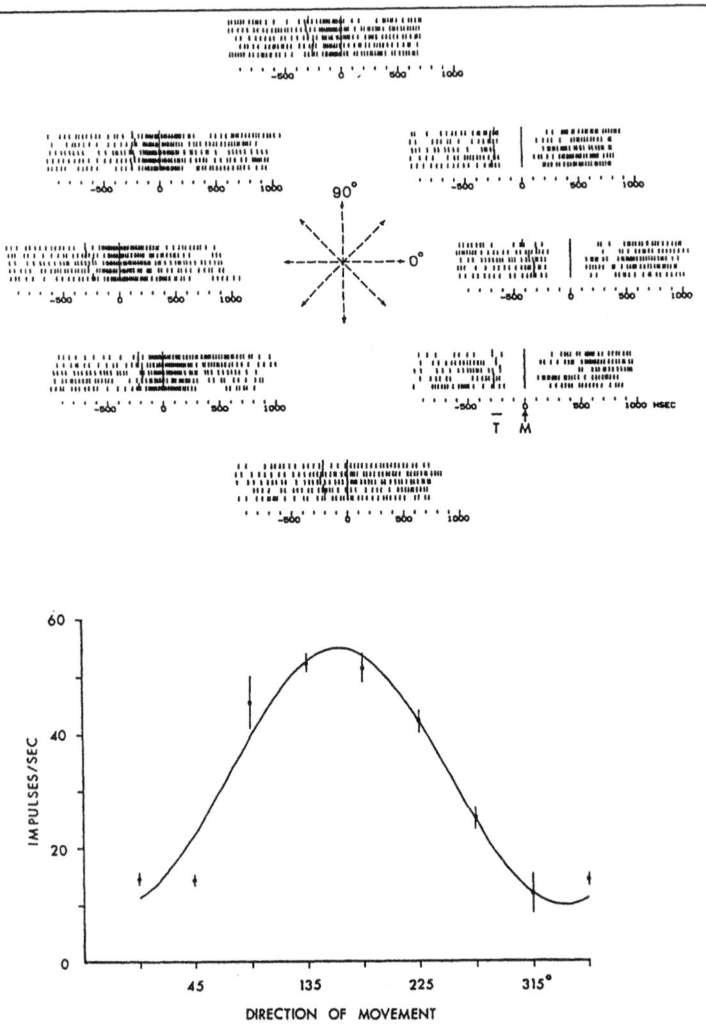

FIGURE 1. Directional tuning of a cell recorded in the arm area of the motor cortex during 2D reaching. Top: Impulse activity during five trials of reaching in the directions indicated in the drawing at the center. Short vertical bars indicate the occurrence of an action potential. Rasters are aligned to the onset of movement (M). Longer vertical bars preceding the onset of movement indicate the onset of the target (T); those following the movement indicate, successively, the entrance to the target zone and the delivery of reward. Bottom: Average frequency of discharge (± SEM) from the onset of the stimulus until the entry to the target zone are plotted against the direction of movement. Continuous curve is a cosine function fitted to the data using multiple-regression analysis. (From Georgopoulos et al, 1982; reproduced with permission. Copyright by Society for Neuroscience.)

the cell's preferred direction. (2) Cell vectors are weighted by the (scalar) change in cell activity during a particular movement. (3) The sum of these vectors (i.e., the population vector) provides the unique outcome of the ensemble coding operation. This operation can be expressed as follows:

$$\boldsymbol{P}_l = \sum_{i}^{N} w_{il} \boldsymbol{C}_i \qquad (2)$$

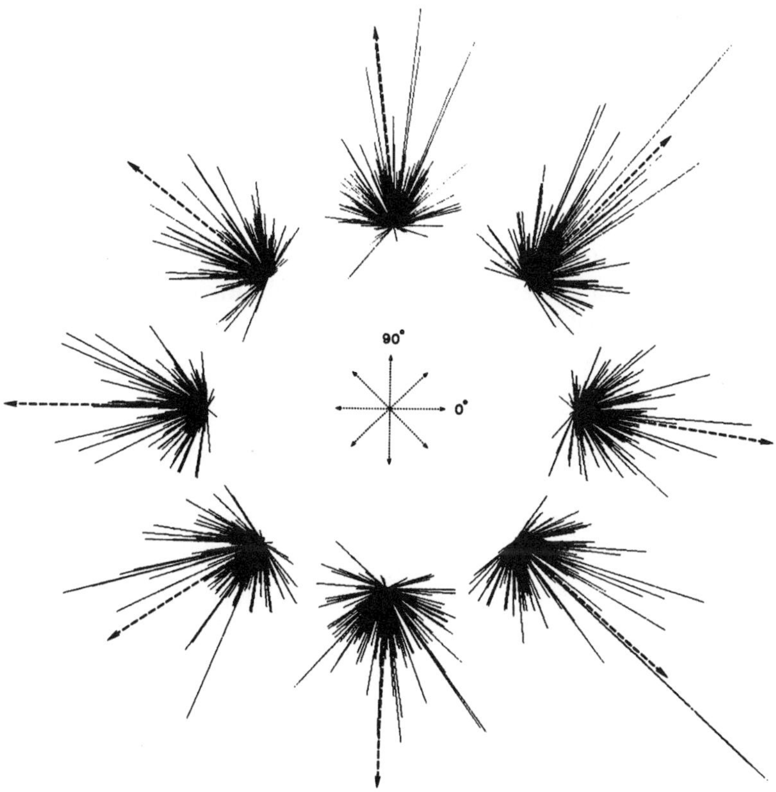

FIGURE 2. Neuronal-population coding of the direction of reaching illustrated for a motor-cortical population (N = 241 cells) and 8 movement directions on a 2D working surface. Vectorial contributions of single cells (continuous lines) add to yield the population vector (dotted arrow). Each cluster represents the same population; the movement directions are shown in the diagram at the center. The population vector points in or near the direction of the movement. (From Georgopoulos et al, 1983; reproduced with permission.)

It was found that the population vector points in the direction of the movement (Georgopoulos et al, 1983; Georgopoulos et al, 1986; Georgopoulos et al, 1988; **Figure 2** and **Color Plate 5**, p. 188, middle and bottom panels). The population-vector approach has proved useful not only in studies of motor cortex (Georgopoulos et al, 1983; Georgopoulos et al, 1986; Georgopoulos et al, 1988; Kalaska et al, 1989; Caminiti et al, 1990) but also in studies of other brain areas, including the cerebellum (Fortier et al, 1989), the premotor cortex (Caminiti et al, 1991), area 5 (Kalaska et al, 1983), and area 7 (Steinmetz, Motter, Duffy, & Mountcastle, 1987).

Three aspects of the population-vector analysis are remarkable: its simplicity, its robustness, and its spatial outcome. The ongoing calculation of the population vector is a *simple* procedure, for it (1) assumes the directional selectivity of single cells, which is apparent, (2) weights vectorial contributions by single cells on the basis of the change in cell activity, which is reasonable, and (3) relies on the vectorial summation of these contributions, which is practically the simplest procedure to obtain a unique outcome. The population vector is a *robust* measure, for it can convey a good directional signal with only a small number of cells (Georgopoulos & Massey, 1988). Finally, the population vector is a *directional* measure, isomorphic to the direction of movement in space. Indeed, the population analysis transforms aggregates of purely temporal spike trains into a directional signal.

Some general properties of the neuronal population vector

The neuronal population vector predicts the direction of reaching for movements of differing origins. When monkeys made movements that started from different points, the population vector predicted well the direction of the reaching movement (Kettner, Schwartz, & Georgopoulos, 1988; Caminiti et al, 1991), even when the preferred directions of individual cells shifted somewhat with different movement origins (Caminiti et al, 1990).

The direction of reaching is predicted well by neuronal population vectors in different cortical layers. In these studies, the population vector was calculated from two separate sets of cells recorded in the upper (II and III) and lower (V and VI) layers of the motor cortex (Georgopoulos, 1990). The average absolute angle between the population vector calculated from cells in the upper layers and the direction of movement was 4.31° (SD = 2.98°) for eight different movement directions, compared to 2.32° (SD = 2.06°) for the lower layers (Georgopoulos, 1990). This finding suggests that the ensemble operation of the population vector can be realized separately in the upper and lower layers. This is important because information can then be distributed to different structures according to the differential projections from the upper and lower layers (Jones & Wise, 1977).

The neuronal population coding of the direction of reaching is resistant to loss of cells. The population coding described above is a distributed code and as such does not depend exclusively on any particular cell. This robustness was evaluated by calculating the population vector from progressively smaller samples of cells randomly selected from the original population (Georgopoulos et al, 1988). It was found that the direction of the population vector can be reliably estimated from as few as 100-150 cells (**Figure 3**), and from many fewer if optimal algorithms are used (Salinas & Abbott, 1994).

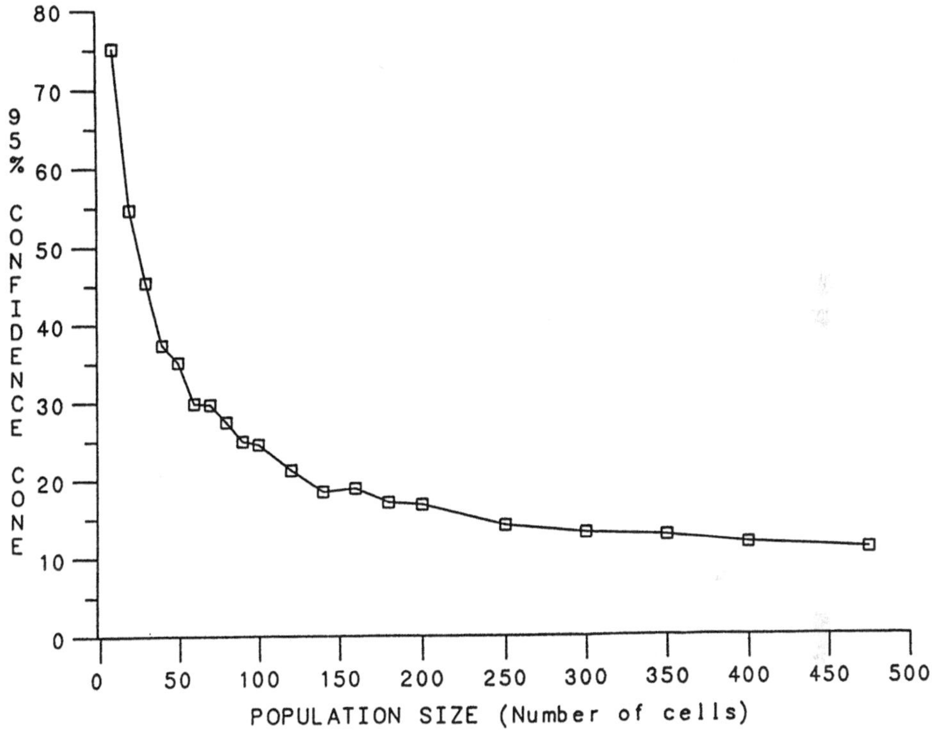

FIGURE 3. Directional variability of the population vector (ordinate) plotted against population size (abscissa). Points in the graph are means of eight half-angles of 95% variability cones, one for each of the eight movement directions used. Units on the ordinate are in degrees. (From Georgopoulos et al, 1988; reproduced with permission by the publisher.)

The neuronal population vector transmits directional information comparable to that transmitted by the direction of movement. In the standard two-dimensional movement task used in our studies, monkeys (Georgopoulos et al, 1982) and human subjects (Georgopoulos & Massey, 1988) moved a manipulandum from the center of a planar working surface to a target on the circumference of a circle. In this case, movement directly toward the target is the ideal direction. If the subject's movements were directly toward the target, performance would be perfect, and the movement could be said to transmit the maximum possible information. However, movements are rarely dead on target, and, therefore, the information transmitted is rarely maximal. The greater the dispersion of the movement endpoint around the target, the less the information transmitted. This dispersion may be parceled into errors in the amplitude of movement and in the direction of movement, and these two types of errors may be studied separately. Since we were primarily interested in the control of the direction of movement, subjects were instructed to "move in the direction of the target" with no restrictions on the amplitude or endpoint of the movement. This provided a purely directional task. The information transmitted by the direction of movement was calculated from a "performance matrix" in which the ideal and actual directions were tabulated (Georgopoulos & Massey, 1988).

Essentially the same technique was used to calculate the information transmitted by the direction of the population vector. Since the population vector is the vectorial sum of weighted contributions of individual cells, and since these weights change from trial to trial due to inter-trial variability in neuronal discharge, the direction of the population vector varies somewhat from trial to trial. This variation in the direction of the population vector was treated in exactly the same way as the direction of movement, and the information transmitted calculated. Based on the information transmitted by the direction of movement and the direction of the population vector (Georgopoulos & Massey, 1988), we found the following. First, the information transmitted by both of these measures increased as the input information increased, but more slowly than the maximum possible, tending to saturate at high levels. This loss of information was probably due to noise generated during the initial (perceptual) and successive (perceptual-motor) processing stages. Second, the information transmitted by the population vector was consistently greater than that transmitted by the movement vector by approximately 0.5 bits. Thus, some information is lost between the motor cortex and the movement. However, this loss differed from that due to noise, for it did not increase with increasing stimulus information but remained constant at about 0.5 bits at all levels of input information. The second form of loss could occur during processing in other motor structures or during the biomechanical implementation of movement.

The neuronal population vector predicts the direction of dynamic isometric force. The dynamic relations of cell activity in the motor cortex to the direction of 2D isometric force has been investigated more recently (Georgopoulos, Ashe, Smyrnis, & Taira, 1992). The following experimental arrangement allowed dissociation of the dynamic and static components of the force.

Monkeys produced pure force pulses on an isometric handle in the presence of a constant force bias such that the net force (i.e., the vector sum of the monkey's force and the bias force) was in a visually specified direction. The net force developed over time had to maintain the specified direction and to increase in magnitude until it exceeded a criterion intensity threshold.

The most interesting case occurred when the directions of the net and bias forces differed by being, for example, orthogonal. In order for the task to be performed successfully under these conditions, the animal's force has to change continuously in direction and magnitude. At every moment during force development, the vector sum of the movement force and the bias force had to be in the visually specified direction. This experimental arrangement effectively dissociated the animal's force vector, the direction of which changed continuously in a trial, from the net-force vector, the direction of which remained invariant. Eight net-force directions and eight bias-force directions were employed.

Recordings of neuronal activity in the motor cortex revealed that the activity of single cells was directionally tuned in the absence of bias force, and that this tuning remained invariant when the same net forces were produced in the presence of different directions of bias force. These results demonstrated that cell activity does not relate to the direction of the animal's force, since the net force was equivalent to the dynamic component of the force exerted by the animal after a static component vector (equal and opposite to the force bias) was subtracted. These findings suggest that the motor cortex provides a dynamic force signal during force development with other, possibly subcortical, structures providing the static compensatory signal. This latter signal could be furnished by antigravity neural systems, given that most static loads encountered are gravitational in nature. According to this general view, the force exerted by the subject consists of dynamic and static components, each of which is controlled by different neural systems. These signals converge in the spinal cord and provide an ongoing integrated signal to the motoneuronal pools.

The foregoing results establish that coding of directional information applies to the motor output in general, even in the absence of joint motion. Moreover, the direction specified by the motor cortex is not that of the total force exerted by the animal but of the dynamic component of the force; i.e., the component of the force remaining after subtraction of a constant, static force.

Time-varying properties of the neuronal population vector

The population vector is a robust predictor of the direction of movement, as shown by the following analyses in which the population vector was calculated as a time-varying signal at short successive intervals (e.g., every 10 or 20 ms). The relation of this signal to an ongoing or a planned movement was assessed.

The neuronal population vector predicts the movement trajectory in continuous tracing movements (Schwartz 1993, 1994). In this experiment monkeys smoothly traced sinusoids with their index finger from one end of the display screen to the other. The direction of the population vectors, calculated successively along the trajectory, changed throughout the sinusoidal movement, closely matching the smoothly changing direction of the finger path. Moreover, a neural "image" of the sinusoidal trajectory of the movement was obtained by connecting successive population vectors tip-to-tail (Georgopoulos et al, 1988). This finding suggests that the length of the population vector carries information about the instantaneous velocity of the movement.

The neuronal population vector predicts the direction of reaching during the reaction time. This is the simplest case of predicting *in time* the direction of an upcoming movement. Given that changes in cellular activity in the motor cortex precede the onset of movement by 160-180 ms, on the average (Georgopoulos et al, 1982), it is an important finding that the population vector predicts the direction of the upcoming movement during the period when the movement is being planned (Georgopoulos, Kalaska, Crutcher, Caminiti, & Massey, 1984; Georgopoulos et al, 1988). An example is shown in **Figure 4**.

The neuronal population vector predicts the direction of reaching during an instructed-delay period. In these experiments monkeys were trained to withhold the movement for a period of time after the onset of a visual cue signal, and then to move in response to a "go" signal. During this instructed delay period, the population vector in the motor cortex (computed every 20 ms) gave a reliable signal specifying the direction of movement that was later triggered for execution (Georgopoulos, Crutcher, & Schwartz, 1989b).

The Neuronal Population Vector Deciphers Complex Directional Processing

The results summarized above underscore the operational usefulness of the neuronal population vector for monitoring in time the directional tendency of the neuronal ensemble. We took advantage of this property and used the population vector *as a probe* to decipher the neural processing of directional information during various cognitive operations.

Memory holding

In these experiments (Smyrnis, Taira, Ashe, & Georgopoulos, 1992) monkeys were trained to move a handle on a 2D working surface in directions specified by a light on the plane. They first moved the handle to "capture" a light on the center of the plane, and then moved the handle in the direction indicated by a peripheral light (the *cue* signal). A signal to move (the *go* signal) was given by turning off the center light. The following tasks were used. In the *non-delay* task, the peripheral light was turned on at the same time as the center light went off. In the *memorized delay* task, the peripheral light stayed on for 300 ms (cue period) and the center light was turned off 450-750 ms later

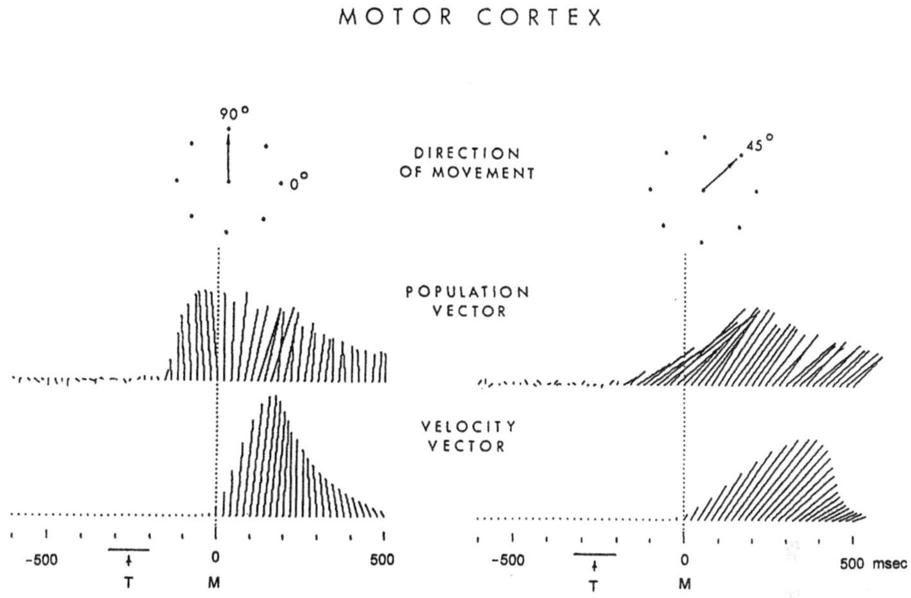

FIGURE 4. The population vector points in the direction of movement well before the movement begins. Top: The results for two movement directions in 2D space are illustrated. Middle: The population vector was calculated every 20 ms. Bottom: The average instantaneous (20 ms bin) velocity of the movement is also shown. Before target onset (T), the population vector is very small in length and its direction varies from moment to moment. Well before the onset of movement (M), the vector increases in length and points in the direction of the upcoming movement. (From Georgopoulos et al, 1984; reproduced with permission by the publisher.)

(delay period). Finally, in the *non-memorized delay* task the peripheral light stayed on continuously, whereas the center light went off 750-1050 ms after the peripheral light came on.

Recordings in the arm area of the motor cortex allowed the population vector to be calculated every 20 ms, following the onset of the peripheral light. We were interested in two aspects of the information carried by the population vector—its direction, which can be interpreted as the directional information carried by the population signal, and its length, which can be regarded as the strength of the directional signal. We found that the direction of the population vector during the memorized delay period was close to the direction of the target. The length of the population vector was similar in the cue period but longer during the memorized than the non-memorized part of the delay. Three phases were distinguishable. First, the population vector showed an initial increase in length that started approximately 100 ms following the cue onset and peaked at 250 ms. We interpret this initial peak as reflecting an *encoding* process. This increase was very similar for both memorized and non-memorized delay tasks. The second phase differed for memorized and non-memorized tasks, having a stronger, sustained signal during the memorized delay than the non-memorized delay period. We refer to this as a *holding-in-memory* process. Finally, following the onset of the go signal, the population vector length increased similarly for all tasks, which reflects a *movement-generating* process.

The directional information carried by the population vector in the memorized task identified the memorized information in a direct fashion. Moreover, this analysis provided some insight into the time course of encoding and holding in memory of directional information. An interesting aspect of these findings is that the increase in the signal during the memorized delay period was observed *in the absence* of the target, although one might have expected the signal to be stronger in the presence rather than in the absence of the visual stimulus. This finding strengthens our interpretation that the increased signal was a memory rather than a sensory signal. It also raises the more general possibility that the motor cortex may be particularly involved in memorial processes when only part of the sensory information about an upcoming movement is present.

Memorized complex trajectory

The studies summarized above dealt with motor-cortical activity during a task that required memorization of the direction of a straight movement. A related question concerns the neural mechanisms subserving memorized, *complex* movement trajectories (Hocherman & Wise, 1991). We investigated this problem in a recent study (Ashe, Taira, Smyrnis, Pelizzer, Georgakopoulos, Lurito, & Georgopoulos, 1993) in which monkeys were trained to perform from memory an arm movement requiring an orthogonal bend, up and to the left, following a waiting period. They held a 2D manipulandum over a spot of light at the center of a planar working surface. When this light went off, the animals were required to hold the manipulandum there for 600-700 ms and then move the handle up and to the left to receive a liquid reward. At the time of movement, there were no external signals specifying the time to initiate or the trajectory of the movement. Following 20 trials of the memorized movement trajectory, 40 trials were performed with visually triggered movements in radially arranged directions. The activity of 137 single cells in the motor cortex was recorded extracellularly during performance. A substantial percentage (62.8%) of cells changed activity during the waiting period prior to the beginning of movement. Other cells did not change activity until after the minimum waiting time was over, and, occasionally, cell activity changed almost exactly at 600 ms after the center light was turned off. However, the most interesting observation was that a few cells changed activity *exclusively*

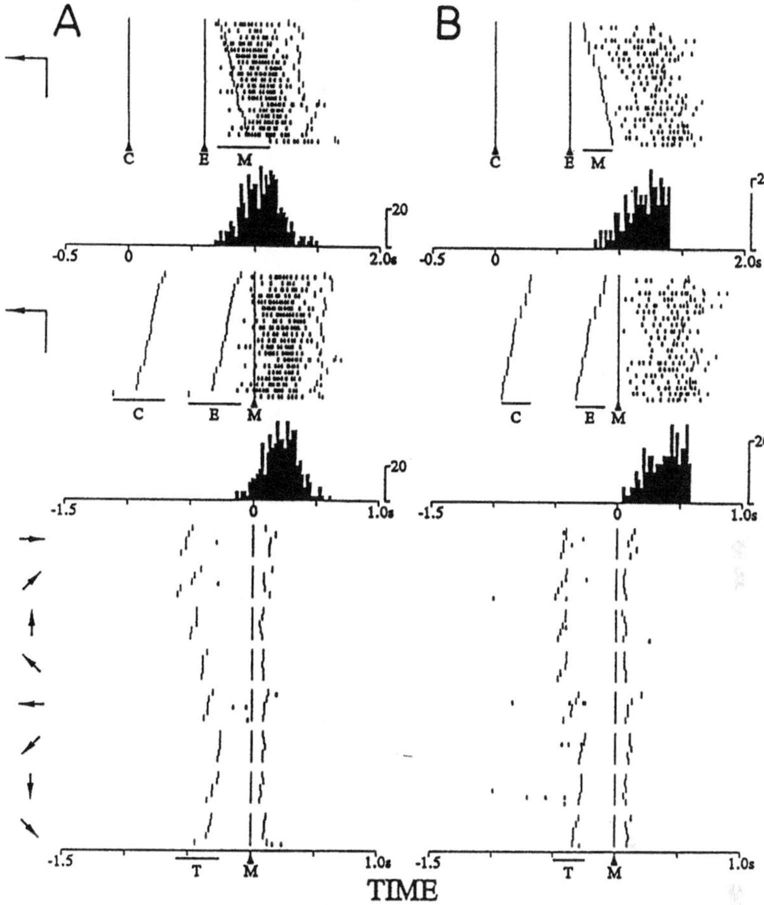

FIGURE 5. Examples of two cells (**A** and **B**) that showed changes in activity during the execution of the memorized movement trajectory (top) but not during movements of the visually instructed task (bottom). For each cell, the following are plotted. Top: Twenty trials of impulse activity (short vertical bars) are shown aligned to the onset (C) of the beginning of the delay (i.e., the time that the center light went off). Subsequent longer vertical bars indicate the end (E) of the waiting period, the exit of the handle from the center (M), and the entrance into the endpoint window. The vertical scale in the histogram indicates impulses. Middle: The same data are plotted aligned to the exit of the handle from the center window. Bottom: Five trials of visually triggered movements in the direction shown to the left of the rasters are shown aligned to the exit (M) of the handle from the center window. The longer vertical bar preceding M indicates the onset of the visual target (T). Filled triangles below the horizontal axes indicate fixed events; horizontal bars indicate the range of time for the event indicated below them. (**A**, cell Pi054u/6; **B**, cell Pi062u/4). (From Ashe et al, 1993; reproduced with permission by the publisher.)

during the execution of the memorized movement (**Figure 5**). These cells were completely inactive during performance of similar movements in a visually guided control task. Together, these findings suggest that *performance of a movement trajectory from memory may involve a specific set of cells*, other than those activated during visually guided or memorized movements. This conclusion is consistent with the results of recent modeling studies (Lukashin, Wilcox, & Georgopoulos, 1994).

Mental rotation

The mental-rotation task required a transformation of an intended movement direction. We first carried out psychological experiments in human subjects. Then we trained monkeys to perform the same task and recorded the activity of single cells during performance of the task. Finally, we tried to relate the neural results with the behavioral observations from the human studies, and to interpret the latter in terms of the former. The objective was to coordinate neurophysiology with cognitive psychology as closely as possible.

The task required subjects to move a handle at an angle from a reference direction defined by a visual stimulus on a plane. Since the reference direction changed from trial to trial, the task required that, for a given trial, the direction of movement be specified according to this reference direction. In the psychological studies (Georgopoulos & Massey, 1987) human subjects performed blocks of twenty trials in which the angle above and its departure (counterclockwise or clockwise) were fixed, although the reference direction varied. Seven angles (spaced between 5° and 140°) were used. The basic finding was that the time to initiate a movement (reaction time) increased in a linear fashion with the angle. The most parsimonious hypothesis to explain these results is that subjects arrive at the correct direction of movement by shifting their motor intention from the reference direction to the movement direction, traveling through intermediate angular space. This idea is very similar to the mental-rotation hypothesis of Shepard and Cooper (1982), which accommodates the monotonic increase of reaction time with orientation angle during a judgment task. The task is to decide whether a given line drawing is the same as, or a mirror image of, a comparison drawing. The subjects are said to rotate a visual image when responding. In fact, the mean rates of rotation and their range among subjects were very similar in the perceptual (Shepard & Cooper, 1982) and motor (Georgopoulos & Massey, 1987) studies. Moreover, when the same subjects performed both perceptual and motor rotation tasks, their processing rates were positively correlated (Pellizzer & Georgopoulos, 1993), which implies similar processing constraints for both tasks.

In the neurophysiological studies (Georgopoulos, Lurito, Petrides, Schwartz, & Massey, 1989a; Lurito, Georgakopoulos, & Georgopoulos, 1991), two rhesus monkeys were trained to move the handle 90° and counterclockwise from the direction of the reference stimulus. These trials were intermixed with others in which the animals moved in the direction of the target. The time-varying neuronal population vector was calculated during the reaction time, with the following results: When the animal's arm moved toward the reference stimulus, the vector pointed in the direction of the stimulus; when the arm moved away from the stimulus, the vector rotated from the direction of the stimulus to the movement direction through a counterclockwise angle. This is illustrated in **Figure 6**. It is remarkable that the population vector rotated at all, especially through the smaller, 90° counterclockwise angle.

These results showed clearly that whatever the cognitive processes involved in the task, they were accompanied by a rotation of an analog signal. The occurrence of a true rotation was further documented by a transient increase, during the middle of the reaction time, in the recruitment of cells whose preferred directions were between the stimulus and movement directions. This indicated that rotation of the population vector was not the result of varying activations of only two cell groups, one whose preferred directions centered on the stimulus and the other on the movement direction. This rotation process, "sweeping" through a directionally tuned ensemble, provides for the first time a direct visualization of the neural processes accompanying a dynamic cognitive process.

In summary, the results of these studies reveal the neural correlates of a dynamic cognitive representation (Freyd, 1987)—a time-varying, dynamic representation of direction in the motor cortex showing a transformation of direction when required and achieved. On the behavioral level, the mean rotation rate and range of rates observed with monkeys were very similar to those obtained with humans.

Context-recall memory scanning

In the preceding studies, we identified the neural correlates of a mental-rotation process as an orderly rotation of the neuronal population vector from the direction of the stimulus to that of the movement, through successive directions within a specified angle. This rotation paralleled the spatial rule operating in the mental-rotation task, which required the production of a movement at an angle from a stimulus direction. In the present study (Pellizzer, Sargent, & Georgopoulos, 1995) we sought, instead, to determine the neural correlates of a cognitive process that was not based on a spatial constraint but on the serial position of stimuli within a sequence. Given an arbitrary sequence of stimuli on a circle, with one identified as the test stimulus, the criterion response was movement toward the stimulus that followed the test stimulus in the sequence. This task is a visuomotor version (Georgopoulos & Lurito, 1991; Pellizzer & Georgopoulos, 1993) of a context-recall memory scanning task (Sternberg 1969). Previous psychophysical studies (Pellizzer & Georgopoulos, 1993) suggested that the processing mechanisms differed between mental-rotation and

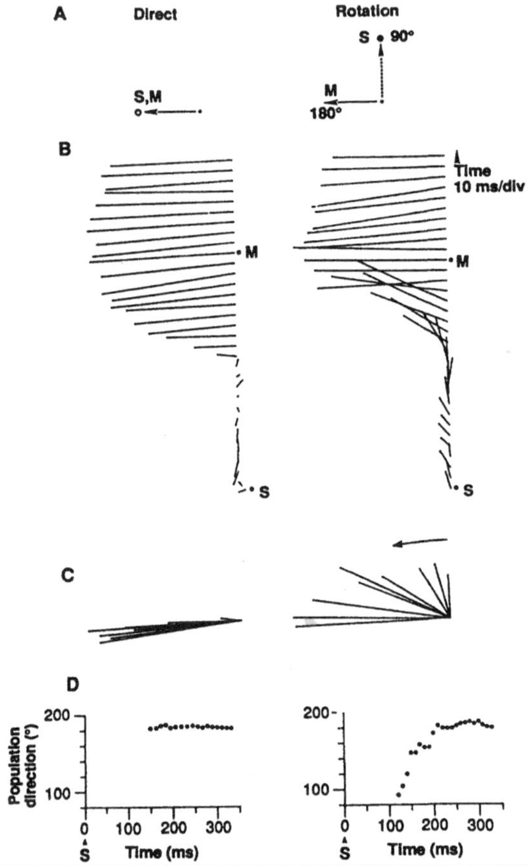

FIGURE 6. Results from a direct (left) and a rotation (right) case in a mental-rotation task. **A.** Task description. Unfilled and filled circles indicate the dim and bright lights, respectively. Interrupted and continuous lines with arrows indicate the stimulus (S) and movement (M) directions, respectively. **B.** Neuronal population vectors calculated every 10 ms from the onset of the stimulus (S) at positions shown in A until after the onset of the movement (M). When the population vector lengthens, for the direct case it points in the direction of the movement, whereas for the rotation case it points initially in the direction of the stimulus and then rotates counterclockwise and points in the direction of the movement. **C.** Ten successive population vectors from B are shown in a spatial plot, starting from the first population vector that increased significantly in length. Notice the counterclockwise rotation of the population vector (right). **D.** Scatterplots of the direction of the population vector as a function of time, starting from the first population vector that increased significantly in length following stimulus onset (S). For the direct case, the direction of the population vector is in the direction of the movement (~180°); for the rotation case, the population vector rotates counterclockwise from the direction of the stimulus (~90°) to the direction of the movement (~180°).

context-recall tasks. In order to determine the neural mechanisms in the latter task, we recorded single-cell activity in the motor cortex of a monkey trained to perform a context-recall and, as a control, an instructed-delay task. We also re-analyzed the neural data from the mental-rotation study (Lurito et al, 1991) to provide a comparison to the present context-recall study.

In the control task, a yellow stimulus was presented in one of eight directions and stayed on for 400 ms, after which it turned blue. This provided the go signal for the monkey to exert a force pulse such that a force feedback cursor exceeded a certain threshold. In the context-recall task, three yellow stimuli (list stimuli) were presented successively (every 400 ms) at different positions on the circle, and stayed on the screen. After an additional 400 ms, one of these stimuli (excepting the last) turned blue. This identified the test stimulus, and also served as the go signal. The monkey then moved the cursor in the direction of the stimulus that followed the test stimulus in the sequence. During the response time, the patterns of neural activity in the motor cortex initially resembled those associated with the direction of the second stimulus. When the test stimulus was the first in the sequence, cell activity continued to reflect the direction of the second stimulus which in this case was the appropriate motor response. However, when the test stimulus was the second in the sequence, neural activity changed to reflect the pattern associated with the direction of the third stimulus which was now the appropriate motor response. This switch was abrupt, occurring ~100-150 ms after the go signal, and was evident both in the activity of single cells and in the time-varying neuronal population vector, which changed direction within ~50-60 ms.

These findings reveal neural correlates of a switching process that is different from mental rotation (Lurito et al, 1991). Additional evidence for the differing nature of the two neural processes was provided by an analysis of the directional selectivity of cells that changed activity during the response time. In a rotation process, the set of cells that change activity during the response time should include cells whose preferred directions are intermediate between the stimulus and response directions. Indeed, this is what was observed (see Figure 13 in Lurito et al, 1991). In contrast, a switching process, such as postulated for the context-recall task, should not involve the activation of cells in directions intermediate between the test stimulus (*S*2) and motor response (*S*3). And, this was precisely what was observed (see Figure 4 in Pellizzer et al, 1995). The time taken to derive the direction of movement in the mental-rotation task reflects a transformation, whereas time in the context-recall task reflects a selection process. Finally, these studies provide an insight into the neural mechanisms of these processes in a particular brain area, namely the motor cortex. But, it is obvious that other brain areas are likely to be involved. Additional experiments are needed to delineate such areas and elucidate their contributions to performance.

Neural-Network Modeling of Motor-Cortical Directional Operations

The motor cortex can be essentially regarded as a neural network whose elements are directionally tuned. One of the possible operations of the network may be described as the computation of the neuronal population vector. The broad directional tuning of single cells seems to be a general property of the population operation, and broad tuning has emerged for units in the hidden layer of a three-layer network trained to calculate the population vector (Lukashin, 1990).

The contribution of interactions among cells to the computation of the population vector also requires examination. It is known that there are extensive local interconnections (Huntley & Jones, 1991) among cells in motor cortex and these promote functional neuronal interactions (Stefanis & Jasper, 1964; Asanuma & Brooks, 1965). Our objective was to (1) identify the nature of these interactions among directionally tuned cells in the motor cortex, (2) to study interactions among units in an artificial neural network made of directionally tuned elements with massive interconnections, and (3) to compare the findings obtained from the motor cortex with those from artificial neural networks. We found the following (Georgopoulos et al, 1993). First, in the motor cortex, interactions between cells were more than twice as frequent when they were tuned than when they were not. The interaction between pairs of cells ranged from strongly positive (i.e., excitatory) to strongly negative (i.e., inhibitory) as the angle between the preferred directions of the cells varied from 0° (i.e., same preferred direction) to 180° (i.e., opposite preferred directions). Second, the same trend was found between the directionally tuned elements of a massively interconnected, dynamic artificial network during the computation of the population vector. Third, when computation of the population vector was stable, the strength of the synaptic interactions was low. In the best (i.e., most stable) case, the mean synaptic strength tended toward $2/N$, where N is the number of elements in the network. This is consistent with the finding that cortical cells in an area are extensively but weakly interconnected (Martin, 1988). Such findings tend to validate the correspondence between the motor-cortical and the artificial neural network and open the possibility of using such networks to interpret the cognitive operations involved in mental-rotation and context-recall tasks.

The time-varying directional operations discussed in the preceding sections have recently been modeled using a massively interconnected artificial neural network consisting of directionally tuned neurons. The outcome of this simulation has reproduced the neuronal population vector (Lukashin & Georgopoulos, 1994a,b) and many of the experimental findings. This work has led to a novel hypothesis concerning how the memorized trajectories of complex movements could be stored in the synaptic connections of overlapping neural networks (Lukashin et al, 1994). In brief, there is a general-purpose network that is involved in all movement, memorized or not, that carries no information about trajectories of specific paths of movement (e.g., circles, ellipses, scribbles, etc.), and that, if activated in isolation, would produce straight-line trajectories. There are also networks that are highly specific for a particular trajectory (e.g., clockwise circle) and that are interconnected with the general-purpose network. When a specific trajectory is to be performed, the appropriate specific network coordinates with the general-purpose network to produce the desired trajectory. Simulations demonstrate that the size of the specific network need be less than 5% of the size of the general-purpose network for the desired trajectory to be stored and reproduced (Lukashin et al, 1994). It is noteworthy that such very specific cells have, in fact, been observed at low proportions in neurophysiological recordings during the performance of memorized trajectories (Hocherman & Wise, 1991; Ashe et al, 1993).

How specialized are these small networks of cells and how do such networks come about in the first place? As yet, we can only speculate on these issues. With respect to the general-purpose network, it is reasonable to assume that it is present at birth, since it is assumed to subserve all movement. There are several possibilities concerning the specialized networks. One is that there are a number of small-size networks, specific for basic paths of movement (e.g., straight lines, curves, and combinations thereof—"motor-shape primitives"), present at birth. Then, learning other, complex motor acts would consist of adjusting the connection strengths between the general-purpose and specific networks. This idea implies that all of the specialized networks are used routinely, although not as frequently as the general-purpose network. Another possibility is that innate specific networks code for more complicated shapes and are large in number. The mechanism of motor learning would then be similar to that described above, but only a small number of the specialized networks would be required. Under these circumstances, a number of the complex, specialized networks might never be used. Such a situation would be parallel to that encountered in the immune system, in which a potential exists for making a large number of antibodies but only some are actually made, depending on the exposure of the organism to specific antigens. In both cases there is a selection—of a specialized trajectory or of an antibody—from a large ensemble available. Finally, an intermediate possibility would be that we begin with motor-path primitives, but end with more complex trajectories by combining these primitives with the general network to become very specialized and to form other trajectories in novel associations.

Concluding Remarks

The topics reviewed above demonstrate the richness of neurophysiological, behavioral, and neural-network modeling studies of the direction of movement in space. They underline the heuristic value and power of the neuronal population-vector analysis in deciphering directional neuronal operations. And, a scheme for the possible translation of motor-cortical directional commands to muscle activations via spinal interneuronal systems has been proposed (Georgopoulos, 1988, 1996) and modeled (Lukashin, Amerikian, & Georgopoulos, 1996).

The population-vector analysis provides quite a general scheme. It need not be confined to physical space but may be generalized to arbitrary spaces and dimensions. Indeed, applications have been made successfully to the coding of faces in the inferotemporal cortex of the monkey (Young & Yamane, 1992) and have been suggested for coding combinations of finger movements (Georgopoulos et al, 1993).

ACKNOWLEDGMENT
This work was supported by PHS grant NS17413.

Neural-Networks Models of Cognition
J.W. Donahoe and V. Packard Dorsel (Editors)

CHAPTER 14

SELECTIONIST CONSTRAINTS ON NEURAL NETWORKS

David C. Palmer
Department of Psychology
Smith College

ABSTRACT

Repeated cycles of selection among variable units of behavior is offered as a model of adaptive complexity for neural-network researchers because of its simplicity, power, and consistency with behavioral research. How such elementary processes can yield complex behavior is discussed, and aspects of variability, reinforcement, and replication of units are explored in turn. It is concluded that a network that simulates the reinforcement of behavior can be powerful, but that training programs of evolving contingencies are necessary as well. Cogent simulations require an understanding of the behavior to be simulated as well as an appropriate network model.

Introduction

The following observations are offered to those neural-network researchers whose primary motivation is to exploit the methodology not primarily to solve engineering problems but to model adaptive behavior in nature. To these researchers a network that, say, sorts mail is of interest only to the extent that it does so in a behaviorally plausible way, that is, as a human or even another animal might do the task. Such a network might be demonstrably inferior to another that had been specifically crafted to do the job but might, nevertheless, be chosen for study because it helps us evaluate a theory of behavior. Thus, what is already known about principles of behavior constrains our models. It is the goal of this paper to discuss these constraints.

What counts as a constraint? Unfortunately, there is no overarching consensus among behavioral scientists on "what is already known about principles of behavior" generally, nor, perhaps, is there any single phenomenon the interpretation of which everyone endorses. In what follows I make no attempt to survey impartially different viewpoints; rather, I will present only the position that to me is most cogent, that of selectionism. Selectionism is the view that complexity in both the structure and behavior of organisms can be explained as the cumulative product of repeated cycles of differential selection on variable substrates. Selectionist accounts are appealing because of their parsimony and power: Only relatively simple processes are invoked, there is no appeal to control by processes or agencies that themselves defy explanation, and the space, or universe, of potential products of selection is virtually unlimited. In the few pages available to me I intend to discuss only the most general implications of a selectionist approach to behavior for neural-network simulations; the reader who wishes more detail is referred to Donahoe and Palmer (1994).

Selectionism

Contingencies of selection operate at different levels, are mediated by different mechanisms, and have been analyzed and codified by different research traditions. Stable, adaptive changes in structure and behavior that emerge over generations are typically explained by appealing to the principle of natural selection and are studied by the sciences of genetics and evolutionary biology. The processes are well known and widely accepted as adequate explanations of complexity in the structure of organisms: Genetic variation arising from breeding or mutation ensures variability among individual organisms; some variants are better suited to survive to breeding age or to have more offspring than others in the prevailing circumstances; consequently the relevant variations tend to be represented in higher proportions in subsequent generations. Changing contingencies can select new, increasingly complex classes of organisms. Thus organisms can be as admirably adapted to a unique set of environmental demands as if they were designed by an omnipotent engineer.

Adaptive changes in behavior in an individual organism are typically explained by appealing to the principle of reinforcement and are studied by the sciences of neurobiology and the experimental analysis of behavior. The relevant processes are not yet thoroughly understood, nor universally accepted as explanations of all adaptive behavior, but they are analogous to those of natural selection and are potentially as powerful: Behavior varies from moment to moment; some units of behavior are more likely than others to lead to important consequences (reinforcers); these behaviors tend to be repeated under similar conditions in the future. Changing contingencies can select new and increasingly complex classes of behavior. Thus complex behavior can be as admirably suited to a unique set of environmental demands as if it were willed by an omniscient mind.

Selection has been invoked at other levels of analysis as well, from the evolution of cultures (Campbell, 1975) to the locomotion of protozoans (Baldwin, 1895), the operation of the immune system (Jerne, 1955), and problem solving in humans (Popper, 1972). With respect to modeling behavior with neural networks, both selection over "generations" and selection within the "lifetime" of a network are relevant. Genetic algorithms can be used to select neural architectures that most effectively solve a particular problem or class of problems by simulating the selective breeding process (cf. Burgos, this volume). Alternatively, the connectivity of networks of a given architecture can be modified by simulating the selective effects of reinforcement of complex repertoires in organisms; as a consequence, any output of the system that meets some criterion will tend to be repeated (replicated) under similar conditions. These approaches are endorsed here not only because of the potential power of selection to produce complex adaptations, but because they appear to be analogous to the way nature has solved problems.

Common to all selectionist systems is the replication of units over time with blind variability in the form, structure, or behavior of the unit. By blind variability, I mean simply that the variability is not guided by future exigencies. The finch does not know whether climatic changes in the next few years will favor birds with short, stout beaks or long, slender ones. The size and shape of the beaks of its offspring will vary, but not in order to satisfy a particular future contingency. The hungry rat exploring a Skinner Box does not know that a particular response will lead to food delivery. Its behavior varies but it does not do so in order to meet the particular contingency in effect. However, variability in form or behavior need not be random. We are more likely to be born with an extra finger than an extra liver; past contingencies of survival have ensured that vital organs and functions tend to survive minor genetic variation, e.g., through redundant genetic coding and repair mechanisms. In the field of behavior, if one response is no longer reinforced we are more likely to execute other responses in our repertoire than a completely novel response. Thus, in some cases, variation in nature can be guided by the past, but it is always blind to the future. If future changes are similar to past changes, variability might seem to be goal directed. For example, in the Galapagos Islands, the types of seeds available to finches vary with rainfall, and different types of seeds favor different types of beak. Since trends in weather can fluctuate widely over relatively short intervals, variability in beak structure is adaptive. However, eye structure should be relatively invariant since the optics of vision are presumably essentially constant from generation to generation. Thus, because of long exposure to such evolutionary contingencies, we might expect finch offspring to vary more widely in beak size and shape than in the form or function of the eye. Although this appears to be an example of variation with respect to a future contingency it is so only to the extent that the future is like the past.

Clearly, the ability of a selectionist system to "solve a problem" posed by a future contingency depends on the gap between the current state of the system, the required future state of the system, and the variability of the system. If contingencies are stable, variability is detrimental; wide variations are more likely to reduce one's fitness than to enhance it. On the other hand, if contingencies vary widely, large variations might be adaptive. To the neural-network researcher attempting to model selectionist systems, there is no single answer to the amount of variability to build into a network, for nature herself has no single answer. One can either model particular systems of particular organisms using as a guide our current understanding of the mechanisms underlying such variability, or one can let the contingencies in the network's "world" determine the appropriate variability of the system.

The Universe of Products of Selection

The universe of potential products of selection is that set of every conceivable variant over unlimited cycles of replication. The size of this set is not entirely unconstrained—it is constrained by the mechanisms that generate variability, for example—but one can see that it can be enormous. It will be helpful here to consider an artificial selectionist system that is fully constrained by our assumptions. For instance, slightly modifying an example invented by Dawkins (1986), we can imagine a system that begins with a string of typewriter keyboard characters and replicates it (generates "offspring") with minor random variations in length and in the choice of characters at each position in the string. These second-generation strings are then replicated with variations, and the process continues over many generations. It is clear that the potential space of this system is infinitely large and, although most of it is gibberish, it includes not only the complete works of Shakespeare, but everything else that has ever been written or ever could be written with the characters from our keyboard. Thus, such a system is potentially very powerful. If we want a Keats sonnet or an Ogden Nash limerick, it is there. If we want to write ravishing poetry, we need not be inspired by the muse; we merely have to sift through our universe of variations until we find something that suits our purpose.

The universe of potential body forms would presumably be the products of every permutation of DNA bases. Again, most of these products would be biological gibberish, but tucked away in this enormous universe is every organism that ever lived, in addition to every variety of hopeful monster that might be selected by future contingencies. The universe of behavior is less easily imagined, as the units of behavior and the origins of variability in behavior are not completely understood, and in any case would be specific to individual organisms. However, a perusal of the Guinness Book of World Records stretches our notions of the potential behavior of members of our own species. The permutations of human behavior are extraordinary, not to say ridiculous.

These imaginary universes of potential products of selection illustrate the adequacy of a system of replication with variation to satisfy contingencies in the relevant system. Of course, to find a particular item in one of these universes can be a formidable task. To find a particular string of letters in Dawkins's universe by searching randomly would take forever. In order for such a system to be useful there must be a strategy for sifting out the candidates that suit our purposes. Fortunately, for every item in the set, there is a family tree that we can trace to bring us straight to that string. If there were some sort of algorithm for making choices at every branch of a family tree, we could arrive at our string in relatively short order. For example, Dawkins chose to select strings that more and more closely approximated Hamlet's observation about the shape of a cloud,

Methinks it is like a weasel.

In his selectionist system each generation was winnowed down to a single, best string which served as the parent of the next generation of variants. Although the space of variations was enormous, his simulation converged on the correct string in a few tens of generations. The actual number of generations in which a solution is found depends on arbitrary parameters such as the fecundity and variability of the strings. The feature of interest is not the actual number of generations required, but the evident superiority of a systematic search over a random search through the universe of possible variations.

It is contingencies of selection that choose paths through the universe of variations and determine the composition of the population of units at any generation. Systematic contingencies can select any element in the universe and can do so relatively efficiently. In Dawkins's simulation the contingencies were systematic because the topography of the desired variant was specified by his selection algorithm; in every generation it was clear which variants were most suitable. In the evolution of species, selection is systematic because contingencies of survival typically change slowly relative to the reproduction rate of organisms. As a consequence, evolution can keep pace with changing contingencies, and we observe that species are typically wonderfully suited to the local contingencies of survival. We marvel at the uncanny camouflage of the walking stick, shaped over countless generations, presumably, by the increased chance of escaping predation that accrued to any individual that looked more like a twig than its cousins. If contingencies change abruptly, however, biological evolution may be unable to keep pace, and the observer will no longer marvel at the match between the organism and the requirements of its world; species become extinct. In general, then, replication with variation can generate virtually limitless candidates for the solution to a problem; however, finding that solution requires a systematic search carried out by contingencies of selection.

The relevance of these observations to neural-network researchers is twofold: First, a network that "replicates" behavior with variation would not only simulate the behavior of many organisms, it would be, in principle, extremely powerful in generating potentially adaptive behavior. Second, demonstrating the potential of such a system requires a program of selection contingencies that pick out the correct branches in the family tree of the ultimate target behavior. Unfortunately, although the problem of designing a selectionist network is relatively straightforward, the problem of developing a program of contingencies necessary to demonstrate its power can be formidable. The former problem requires technical expertise, the latter requires behavioral expertise. In particular, we must understand something about the history of the behavior we are trying to shape. It is easy to shape more forceful leverpresses in a rat or more accurate letter formation in a child because it is clear what is required at every step. However, the kinds of complex behavior that are usually of greatest interest are often the product of sequences of other behavior or are built up of smaller units of behavior, and the lineage of the target behavior may be unclear. What do we do when we solve a math problem in our head, for example? How do we answer questions about past events? What is the difference between hearing something and listening to something? How do we compose sentences? Designing an appropriate network architecture is only a part of answering such questions. Moreover, there are many ways in which to design a selectionist system. Dawkins's technique for generating poetic strings was presumably not the same as Shakespeare's. If our networks are to cast light on the behavior of organisms, they must be guided by what is known about the selection of behavior. We begin with a consideration of the selection contingencies themselves.

Reinforcement

Reinforcement, as studied in most experimental procedures, is, to a first approximation, binary; that is, it either occurs or does not occur. As far as information for the use of the network is concerned, it merely indicates that a response has met the minimum requirements of a contingency in a particular setting. This is not to say that the effect of reinforcement is invariant; to the contrary, the effectiveness of a reinforcer on a target response depends on the nature of the response, various motivational variables, the magnitude, quality, and delay of the reinforcer, the strength of incompatible responses, and the history of the organism under similar conditions. However, under appropriate motivational conditions—such as those in effect in most conditioning experiments—the variability in behavior owing to variability in the effectiveness of the reinforcer is swamped by the variability in behavior owing to the contingency of reinforcement. Ultimately both sources of variability need to be accounted for, but complexity in both behavior and structure can arise from contingencies in which the selection contingency is all-or-none. Thus, to the extent that reinforcement in organisms is insensitive to variations in behavior, it is an example of unsupervised learning. Backpropagation algorithms may be computationally useful but do not simulate reinforcement. Nothing in the reinforcement signal tells the network which dimension of the response was critical in meeting the contingency, and the omission of reinforcement tells the network nothing about the deficiencies of the response. The cases, in which, say, the amount of reinforcement varies in proportion to the intensity or duration of the response, are too exceptional to vitiate this conclusion.

Variability is fundamental to products of selection

A binary reinforcement signal merely indicates that the minimum requirements of a particular reinforcement contingency have been satisfied. It can not specify a blueprint for responding. Any variation that meets the contingency is eligible for reinforcement. We notice that behavior varies from instance to instance, and over repeated trials many unique responses will be strengthened. These will all have had a common effect in meeting the contingency but need not be identical or even similar in other respects. A sieve sifts out particles exceeding a minimum size, but it does not specify the shape, composition, precise size, or any other feature of the particles. Thus, just as one can escape predation by hiding, by playing dead, by looking fierce, by tasting bad, and so on, a rat in an operant chamber can press a bar with its left paw, its right paw, both paws, its nose, and so on, and the barpressing contingency will still be met. It is true that we can insist that the rat press the bar with its left paw and withhold reinforcement for any nonconforming topographies, but a change of focus reveals the same fundamental variability: The rat can press the bar forcefully or gently, with its toes or its heel, for a short or long duration. Regardless of the specificity of our contingency, behavior is free to vary in any respect that is not specified. It is, of course, logically possible that a contingency could completely specify the value of every parameter—this would be analogous to specifying the architecture, activity, connection weights, and so on, of every node in a neural network—but a system of such contingencies would be satisfied only rarely. If Dawkins had simply waited for a single mutation to match his target string, he would still be waiting; in fact, using modern computers to generate string variations, he could expect to wait at his terminal for many billions of years before a variation met the requirements of his contingency! In nature, contingencies necessarily tolerate variability.

When contingencies of reinforcement are stable over the course of one's observations, responses can be seen to divide into two classes, those that meet the prevailing contingency and those that do not. It is customary to refer to those responses that meet an enduring contingency as a functional response class (Catania, 1973), but whether a class, so defined, "hangs together" as a useful unit of analysis depends on its orderliness with respect to changing contingencies (Skinner, 1935). For example, rats occasionally press the lever when rearing and exploring the upper corners of the front of an operant chamber, as we might grab a shelf when climbing a wobbly ladder, but presumably these adventitious leverpresses do not vary systematically with food deprivation or schedules of reinforcement. In network simulations a target behavior may be emitted because our algorithms have built variability into the network. Such responses are not cause for celebration until they vary in an orderly way with contingencies, that is, until we can produce them on demand.

Thus selection is a blunt tool. The unsettling implication of this prosaic truth is that what we are trying to model with our neural networks is a fuzzy target. Variability is fundamental to the classes of objects, attributes, or behaviors that emerge from programs of selection, yet the classes are orderly. Fortunately this variability is naturally simulated by neural networks; even when the output vector of a network is the same, networks are unlikely to be replicated in every detail. In some cases, this may be seen as a weakness. Some of the behavioral domains that cognitive scientists hope to understand are superficially amenable to formal modeling. Language, for example, has been modeled with the formalisms of linguistics. However, the units of analysis in formal systems are typically fully constrained and are thus incommensurate with the units of analysis arising from a selectionist system. (For example, "grammatical sentence" has a precise meaning in a formal linguistic system, but among native speakers the term is fuzzy.) Of course, it is a strength of network research, not a limitation, that it simulates actual behavior rather than a formal model of behavior. (Cf. Palmer & Donahoe, 1992, for further discussion of this point.)

Replication: What is selected?

As noted above, in a static environment, behavior can be seen to vary in an orderly way with motivational procedures, called establishing operations (Keller & Schoenfeld, 1950; Michael, 1982), and with contingencies of reinforcement. Thus, in a stable context, under typical motivational conditions, it is customary to say that behavior is a function of its consequences, or, simply, that reinforcement strengthens or maintains behavior upon which it is contingent. However, a more general statement is that reinforcement alters the stimulus control of behavior. That is to say, reinforcement alters the strength of antecedent-behavior relations. When a response is followed closely in time by a reinforcer, the response becomes more likely to occur *under the same conditions* in the future. When the context is constant, response rate, or probability, changes as a function of its consequences, because the environment is not a variable. When the context changes, however, we notice a decrement in responding, revealing the evocative role of antecedent conditions. These antecedent conditions include prior behavior of the organism as well as, we presume, proprioceptive and interoceptive stimulation. Of course, we demonstrate functional control by antecedents with variables that can be manipulated; thus we commonly say that reinforcement alters environment-behavior relations.

Clearly this is not a claim that every response is elicited by a specific antecedent stimulus (i.e., as in S-R psychology). As already mentioned, responding is a function of many variables acting together. Rather, it is a claim that reinforcement alters the strength of behavior through a change in the control of that behavior by the conditions that prevailed at the time the behavior was emitted. (See Donahoe, Palmer, & Burgos, in press; Dinsmoor, 1995; for more detailed discussions of this issue and related topics.)

With respect to neural networks, then, the appropriate unit of selection appears to be the connection strength between elements. This, of course, is typical in network research; it is perhaps for this reason that neural-network simulations appear to be so promising as tools for interpreting complex behavior. However, some network researchers may feel that their work is constrained by a different interpretation of operant behavior, specifically, one in which reinforcement alters the strength of spontaneously emitted behavior (e.g., Stein, Xue, & Belluzzi, 1994). This view has been supported by the science of behavior (e.g., Skinner, 1938), but only in stable environments and only because a science of behavior, as such, does not require modeling the physiological substrate of behavioral regularities.

Behavior is fluid and parallel

If the behavior of organisms were serial and punctate, so that only a single, discrete unit of behavior occurred at any moment, reinforcement could efficiently select environment-behavior relations that satisfied a contingency. However, organisms are continually engaged in a variety of overlapping, fluid behaviors from postural adjustments, orienting responses, exploratory responses, and locomotor behavior to manipulations of objects. In addition, one may engage in a variety of behaviors within the skin, such as subvocal speech, conditioned and unconditioned perceptual behavior, and autonomic activity. Some subset of these responses might be relevant to a particular contingency; for example, grasping and eating food requires the coordination of orienting, perceptual, and motor behavior. Nevertheless, part of the organism's activity is typically irrelevant to any given contingency. To be maximally efficient, reinforcement should alter the probability of only those responses required by the contingency. However, there is nothing in the reinforcement signal itself that can differentiate its effects on the organism. Thus there is a "credit-assignment" problem: How are the effects of reinforcement distributed in a network so that target behaviors consistently emerge in a particular contingency?

Heterogeneity of response systems

The organism is not a *tabula rasa*, and the architecture of biological networks is not homogeneous. Some environment-behavior relations are more easily conditioned than others. Some response systems can be recruited for a wide variety of functions, while other response systems are narrowly committed to a particular function and modulation by environmental events is limited. We use our hands for everything from scratching a mosquito bite to waving to the Queen, and hand movements can be quickly conditioned to an arbitrary stimulus, including stimulation arising from our own overt and covert behavior. Pupil dilation, heart-rate and blood-pressure changes, sweating, and other homeostatic activity serve to maintain the internal economy of the organism and are not so easily recruited by arbitrary stimuli. We can get a friend to move his hand in a novel way just by asking him to do so; by asking, we can also get him to stop breathing, but only briefly, and a request for heart-rate changes will meet little success at all. Such differences in controlling relations are perhaps the main reason that the operant-respondent distinction is so widely honored.

To a limited extent, this offers a structural solution to the credit-assignment problem: Different response systems may be differentially susceptible to reinforcement. Some responses, such as those that regulate circulation, respiration, blood pressure, digestion, and other dimensions of the internal economy of the organism, are narrowly committed to a single function; it would not be adaptive for these autonomic responses to vary widely with arbitrary contingencies of reinforcement. Other responses, such as those executed by the limbs, hands, and fingers of humans can be recruited for virtually any arbitrary task. This functional distinction corresponds roughly to that between behaviors mediated by smooth muscles and those mediated by skeletal muscles. Some theorists have argued that the effects of operant and classical conditioning are restricted to skeletal and smooth muscles respectively (e.g., Konorski & Miller, 1937), but the issue remains unresolved. Although attempts to demonstrate the sensitivity of visceral responses to operant conditioning have been equivocal and controversial (Miller & Dworkin, 1974), changes in the control of skeletal behavior with reinforcement can be much more consistently demonstrated. Nevertheless, this observation does not suggest a general solution to the credit-assignment problem; it merely reflects the fact that some of the activity of organisms, committed to one function, is not free to vary widely in the service of other functions. However, even if the effects of reinforcement were restricted to skeletal responses, there are still an enormous number of responses that inevitably occur in parallel. How can the effects of reinforcement be restricted further to those responses that are actually required by a contingency of reinforcement? In freely moving organisms the environment is heterogeneous and unpredictable; much of the activity of such organisms must be sensitive to modification. Moreover, it is the ability of networks, biological or artificial, to discover adaptive solutions to novel problems that excites our interest.

The domain of the reinforcement signal

In response systems that are typically studied with the operant procedure, both the potential response and the potential discriminative stimulus, if not completely arbitrary, can at least vary over a tremendous range. We can arrange a reinforcement contingency so that an experimental subject will press the left key when he receives a light shock to his finger, or an elbow in his ribs, or a touch on the shoulder, or when he smells lilacs, or when he sees one of indefinitely many visual stimuli, or hears one of indefinitely many auditory stimuli, or when he senses that the entire room is tipping or speeding up, or when he tastes ginger, vinegar, or anise. Moreover, we can get him to press it with any one of ten fingers, his nose, his palms, his elbows, and so on. Under the appropriate motivational conditions our subject can acquire a virtually unlimited number of discriminated operants. If reinforcement is to be explained by the strengthening of neural pathways it is evident that the responsible physiological mechanisms must be diffuse. That is, the reinforcement signal must be poised to alter an enormous range of pathways at every moment.

Temporal contiguity

Reinforcement modifies the strength of responses that immediately precede it in the same context. The effectiveness of a reinforcer is reduced if it is delayed following a target response, with greatly reduced reinforcing effects found with delays of about half a minute (Grice, 1948; Lattal & Metzger, 1994; Lattal & Gleeson, 1990; Logan, 1960; Schlinger & Blakely, 1994; Wilkenfield, Nickel, & Blakely, 1992). However, observation of a laboratory animal, such as a rat, in even an impoverished environment, reveals that it can engage in dozens of conspicuous behaviors in a 30-second interval. If reinforcement has even a modest effect on all of them it would appear to reduce the effectiveness of a reinforcement contingency: The target behavior would be competing with many irrelevant responses. However, the research on delay of reinforcement reveals a gradient of effectiveness. The more closely reinforcement follows behavior, the greater the effect on the behavior. This gradient of effectiveness provides one constraint on the selection of behavior: Those responses that precede the reinforcer most closely in time are strengthened the most.

The power of temporal contiguity is revealed by a common classroom demonstration known as the "superstition experiment" (Skinner, 1948). In this demonstration a hungry pigeon, previously trained to eat readily from the food magazine, is given brief access to food every 15 seconds. It is commonly observed, following such a procedure, that the bird is engaging in some stereotypic behavior such as pacing back and forth, pecking, or turning—behavior apparently captured by adventitious contingencies of reinforcement.

> The conditioning process is usually obvious. The bird happens to be executing some response as the hopper appears; as a result it tends to repeat this response. If the interval before the next presentation is not so great that extinction takes place, a second "contingency" is probable. This strengthens the response still further. (Skinner, 1948, p. 168)

Thus, under some conditions, temporal contiguity between reinforcer and behavior is sufficient to strengthen a target behavior. There need be no causal relationship between the behavior and its consequences, and "superstitious" behavior can be supported. What then distinguishes adaptive behavior from superstitious behavior? In a sense, all behavior maintained by reinforcement is superstitious. The causal relationship between behavior and its consequences is not available as a stimulus to the organism; it sometimes plays a role in human behavior, but even so the relationship cannot be determined with certainty for any individual future contingency. That temporal contiguity should be adequate to select adaptive behavior is a reflection of contingencies in nature. Actions typically have immediate consequences that, in an everyday sense, are caused by the action. For some of these consequences, it is appropriate that the behavior should occur again under similar conditions. It is true that our actions may initiate a cascade of events with delayed consequences, and that these may also be important. However, it appears that delayed consequences are not sufficiently reliable for organisms to have evolved sensitivity to them. There is only one event, or constellation of events, that immediately follows a response, but there are an infinite number of delayed consequences.

It may appear that humans are exceptional in being sensitive to long-term consequences. We plant seed in the spring and harvest in the fall; we toil in school and qualify for a job only after we graduate some years later; we hang the umbrella on the doorknob to remind us to bring it to work the next day. In every case of control by delayed consequences there appears to be some sort of mediating behavior; we may analyze and describe the contingencies, in which case our behavior may be under the immediate control of our own verbal behavior rather than the long-term consequence itself. If a ten-dollar bill were to drop into your lap right now, as a would-be reinforcer for some long past behavior, the credit-assignment problem would be conspicuous. Without collateral data and mediating analysis, there would be no effect on the target behavior.

We seem to have arrived at contradictory conclusions: At every moment, reinforcement must be poised to strengthen an enormous number of potential environment-behavior relationships, but its effects are actually restricted to the immediately preceding events. A possible resolution is that a uniform, diffuse reinforcement signal is broadcast to all elements of the network, but only those connections between recently active elements are strengthened. With this design, the network would be simple but versatile; it could learn to emit reliably any arbitrary behavior in any circumstance, provided that its behavior were sufficiently variable that the target response were emitted at least once. Immediately after the response was emitted, a reinforcement signal would be sent to all parts of the network. But only connections mediating the target response would be recently active, and only those would be strengthened.

Behavioral discrepancy

Temporal contiguity between behavior and its consequences is not sufficient to alter the stimulus control of that behavior. The consequence must be, speaking loosely, "surprising" to the organism. If the current context has been reliably paired with a reinforcer, then a full range of conditioned behavior will be evoked by the context even before the reinforcer is delivered. If there is no difference between this constellation of behavior and the behavior actually evoked by the reinforcer itself, then there is no additional change in the stimulus control of behavior. If the dog salivates as much to the sound of the can opener as to the food itself, the food will be a less effective reinforcer once the can opener has operated. There is relatively little discrepancy between the conditioned behavior to the can opener and the unconditioned behavior to the food itself. However, the sound of the can opener will be an effective reinforcer unless it, too, is highly correlated with some antecedent event.

If the dog is thoroughly satiated with food there will be relatively little salivation to the food, even in the absence of the sound of the can opener. In this case, too, there will be relatively little effect on the stimulus control of behavior, since there is little discrepancy between behavior evoked by the reinforcer and that evoked by other events. Thus, temporal contiguity between behavior and a consequence is not sufficient to strengthen behavior; there must be a behavioral discrepancy as well (Donahoe, Crowley, Millard, & Stickney, 1982; Kamin, 1968; 1969; Rescorla & Wagner, 1972; vom Saal & Jenkins, 1970).

What is the effect of a single reinforcement?

Although reinforcement procedures are typically applied repetitively throughout experimental sessions and perhaps for many sessions, it is not the case that repetition of contingencies is necessary for the selection of behavior. A single trial is often sufficient to establish a new discriminated operant in strength. The behavior of rats in mazes and cats in puzzle boxes led early researchers to view acquisition as a gradual process and to speak of learning curves, but Skinner (1938) showed that a single reinforcer could effect a considerable increase in the probability of a barpressing response in a hungry rat. Casual interpretation of human behavior reveals that behavior is commonly conditioned in a single trial. We usually need to be told only once where a meeting will be held, or the price of an item in a store, or the grade we received on an exam. Thus, the physiological changes underlying learning—those changes simulated, however abstractly, by our network models—must happen rapidly.

Skinner was able to observe single-trial learning in the laboratory only when the organism had been thoroughly adapted to the chamber and to the operation of the food magazine, that is, only when competing behavior was weak. A naive rat commonly startles and freezes at novel sounds such as the operation of the food magazine, and, even when hungry, a naive rat may spend more time sniffing and exploring the experimental chamber than in eating freely available food. Thus, the behavior of an organism is an integration of the effect of many concurrent contingencies. Although the effect of a single reinforcement can be substantial, this can be reliably demonstrated with a novel response only under tightly controlled conditions.

If the effect of a single reinforcement is apparent only with tight experimental constraints, what are we to make of our everyday experience, in which we learn names, dates, places, anecdotes, weather forecasts, and many other things with only a fleeting exposure to them, often with little obvious reinforcement? If a famished rat in a Skinner Box represents the ideal preparation in which to observe one-trial learning, we should hardly expect to observe it in the complex world of an adult human. Nevertheless, it would be quite anomalous for an adult human to take more than one trial to learn that flipping a switch had a conspicuous effect such as turning on a heater, opening a panel, operating a CD player—or delivering a morsel of food.

One is tempted to assert that rapid learning reflects the difference between rats and humans, that we have much larger brains, and, simply, that we are smarter than they are. Indeed, it would be absurd to argue that species differences are irrelevant, but are we to conclude that humans are more easily conditioned than rats or, abstracting, that larger networks are more easily conditioned than smaller ones? It is not obvious that this should be the case, and one might plausibly argue that, because of the greater number of potential competing responses at any moment, large networks would tend, all other things being equal, to be more variable than smaller networks. Greater variability would no doubt be adaptive if novel behavior were required but might tend to retard the emergence of a single, dominant response. As an analogy, the person who weighs every imaginable alternative may solve some formidable and obscure problems but may take a long time to solve everyday problems with obvious solutions. In any case, explaining an anomaly by resorting to claims of species differences, in the absence of either relevant experimental evidence or theoretical justification, is empty and has no explanatory force. Fortunately other explanations are possible.

One-trial learning is commonplace in human behavior only when the target behavior is already a strong response in the subject's repertoire but just happens to be weaker than other behavior under the prevailing circumstances. The behavior to be conditioned when an adult human learns that the weatherman predicts rain, or that a wall switch turns on the overhead fan, or that the soup-of-the-day is clam chowder are all responses that have been thoroughly conditioned under similar circumstances. When we listen to a weather report we are already inclined to assert that it will rain, or, as it may be, that it will be fair. The same variables that make us tune in to the weather report in the first place increase the likelihood that we will make statements about the weather. The weather report itself merely selects some subset of those statements. When we discover a switch, we are already confident that it will turn something on. In a restaurant, "clam chowder" is one of some dozens of responses that are already weakly evoked by the setting long before the waiter recites the daily specials. The effect of the contingency of reinforcement is to slightly modify the control of the target response by the current setting so that it becomes stronger than the myriad other responses that tend to be evoked in that setting. At any moment, the potential behavior of an experienced organism can be thought of as a panorama. Reinforcement of a well-practiced response is analogous to increasing the illumination on a figure in the panorama to make it stand out from the background.

In contrast, some responses have little or no baseline strength. They have to be drawn into the panorama from scratch, as it were. When we hear someone speak in an unfamiliar tongue we usually find ourselves wholly unable to repeat what was said. Even repeating a single word may be difficult. We may have to try many times before we get it even approximately right, particularly if it is composed of unfamiliar phonemes. When we first learn to ride a bicycle, to operate a clutch, to swim, to juggle oranges, to "walk the dog" with a yo-yo, the relevant behavior is shaped for the first time, and it commonly takes many trials before successful behavior consistently emerges. The target behavior has no baseline strength; our behavior is highly erratic at first and smooths out over repeated trials.

To a naive rat, an operant chamber is a feast of unfamiliar smells, textures, and other stimulation. The target response, pressing the bar, often does not appear as a smooth unit during baseline conditions. That is, either the rat does not press the bar at all or does so with a topography that differs considerably from the form of the response after prolonged training. For example, the rat may rest a paw on the bar as it rears its body, or it may sniff and nibble at the bar, operating the microswitch in either case. Thus, there are many irrelevant features of the behavior of the rat, irrelevant, that is, with respect to the contingency of reinforcement. Over the course of continued exposure to the contingency, the relevant elements of the response tend to be selected and many of the irrelevant ones tend to extinguish. The behavior appears to be a smooth and efficient unit. Although this fitful emergence of optimal performance might appear to be inefficient, it is adaptive. It is important to recall that another contingency of reinforcement might select nibbling or rearing rather than barpressing as the target response; nothing in the phylogenetic or ontogenetic history of the rat has made barpressing a preferred unit of response. Thus the emergence of units of behavior often takes multiple trials because of the weeding out of variability, not because reinforcement is ineffective on any one trial.

However, once barpressing has become differentiated as a smooth unit of behavior, the effect of a single instance of reinforcement will be more evident, particularly in transitions between schedules of reinforcement. For example, when reinforcement is withheld, barpressing will extinguish: It will virtually disappear from the observable repertoire of the rat. However, extinction is not symmetrical with acquisition; the effect of prior training has not been erased. If the contingency is reinstituted, the response returns to full strength almost immediately. To some extent this is because the response does not have to be differentiated from a variable substrate again; it appears as a smooth unit immediately. A second reason is particularly relevant to neural-network models: When a response is put on extinction, the organism does not return to its baseline state. The effects of reinforcement are not eliminated by extinction trials; rather the response loses strength relative to competing behavior just to the point that it is seldom emitted in the relevant context. Once it stops being emitted, there are, of course, no more extinction trials, and further weakening does not occur. Moreover, if competing behavior is strong, the loss of strength of the target response may be quite small. When the contingency of reinforcement is renewed the strengthening effect need only be sufficient to nudge the response into prominence for it to appear in apparently substantial strength.

Winner takes all

Thus we can view behavior as a cauldron of competing responses. The overt behavior of the organism usually seems to be unitary, smooth, and organized, but the strength of those alternative behaviors that typically emerge during extinction reveals that the context potentiates a variety of behaviors, presumably simultaneously. We occasionally see blends of incompatible behavior: In speech we may stutter and halt, or begin one metaphor and end with another, or utter a Spoonerism. When avoiding an oncoming pedestrian we may falter left and right in a clumsy pas-de-deux. But more commonly our behavior has a fluid unity. At the behavioral level of analysis this poses no problems; indeed it is a blessing. The problem emerges at the physiological level of analysis—and for those who would simulate behavior with neural networks. How is the typically smooth performance of the organism to be squared with the claim that, at any moment, many responses have appreciable strength?

One possibility is that there is a "winner-take-all" relationship among competing responses. That is, the strongest response at any moment gets emitted. Responses of substantial, but lesser, strength are suppressed. Physiologically, this can presumably be effected by inhibitory interconnections among motor neurons, at least within response systems. This would permit the organism to emit compatible responses simultaneously but would suppress behavior that competes with the dominant response within any response system. We might walk left around a tree, or right around a tree, while solving anagrams in our head, but we would not, as a consequence of some sort of behavioral averaging, walk into the tree.

Experimental chambers are typically simple environments, permitting only a relatively narrow range of behavior; when running wheels, chains, string, bits of wood, and multiple levers are available, the behavior of a rat will be more variable than in a chamber with a single manipulandum, and the effect of an explicit contingency of reinforcement will be less apparent. When explicit contingencies of reinforcement are arranged for two behaviors concurrently, it is commonly observed that an organism will appropriately allocate its time between the two behaviors, but it is more accurate to say that it allocates its time with respect to all concurrent contingencies, natural as well as contrived. The proportion of time allocated to a particular activity depends, not just on the schedule of reinforcement for that particular response, but on the richness of that schedule relative to all concurrently available schedules. A robot in an industrial assembly line will weld joints in sheet metal over and over again, relentlessly, until it wears out; its human counterpart will stop for coffee breaks, for lunch, when the foreman is out, when the whistle sounds, when a pretty woman walks by the window, when any one of myriad other distractions occur, and he may go on strike at last. When simulating behavior with neural networks it is tempting to regard the performance of the robot as an objective, and, indeed, for instrumental purposes that may be appropriate; for the purposes of demonstrating the relevance of the workings of a neural network to the behavior of organisms, however, the factory worker is a better model.

Acquired reinforcement

Some stimuli, such as food, water, warmth, and sexual contact, are innately reinforcing, presumably because organisms that devote much of their behavior to getting these things are more likely to produce viable offspring than those who are not so reinforced. Other stimuli, such as money, hearing one's self praised, good grades, and so on, acquire their reinforcing properties in the lifetime of the individual, presumably through being paired with other reinforcers. Moreover, sometimes reinforcement can be mediated by our own overt behavior, or even our covert behavior—that is, behavior below the threshold of observability to another. (Note that the observability of a response is not some essential property of that response; rather, it depends on the observer and his tools. Galvanic skin response, tongue movements, and salivation are all undetectable to the casual observer. There is no reason in principle to limit the scope of our interpretations to behavior that is typically observable without special apparatus.) For example, we may plan a chess maneuver "in our head," or rehearse a debating point covertly before springing it on our spouse. We may struggle covertly with a math or logic problem, or engage in various recall strategies as we attempt to remember a name. If these covert exercises are "successful" we will emit the behavior overtly at a later time. Thus, not only can reinforcement mechanisms alter environment-behavior relations, both elements—environment and behavior—may recursively recruit reinforcement mechanisms if they have been established as acquired, or conditioned, reinforcers.

Acquired reinforcers can be quite idiosyncratic: One child glows when praised by a teacher, another is embarrassed, and a third, hoping to impress the tough set, is surly. Since there is nothing that predicts what arbitrary events will be of value to an organism in its particular circumstances, natural selection could not have arranged a separate mechanism for acquired reinforcement, one activated by money, one by praise, etc. It must be the case that acquired reinforcers recruit the same reinforcement mechanisms as primary reinforcers.

It is beyond the scope of the present paper to discuss the variables contributing to the establishment of acquired reinforcers. However, to a first approximation, any event that is highly correlated with an unconditioned reinforcer will itself become a reinforcer. Once established, an acquired reinforcer functions like a primary reinforcer and can in turn alter the function of antecedent stimuli. Thus, in chains of behavior or events, early elements can function as acquired reinforcers even though they may long precede an unconditioned reinforcer. Thus we come under the control of "signs of success." Our behavior is controlled by the immediate consequences that have been correlated with other reinforcers in the past. Acquired reinforcement is important in understanding the selection of complex human behavior such as problem solving, behavior chains, and, in general, any behavior that is not maintained by promptly delivered, biologically important consequences. Since this includes most complex human behavior, it is clear that acquired reinforcement must be represented in our simulations.

Sources of Variability

Central to any selectionist account of complexity is variation in the substrate that is being replicated. Genetic variability is understood to arise from sexual reproduction and from mutations produced by defects in transcription of the genetic code, radiation, toxins, and other insults to the organism. In behavior, the units of analysis are not so well understood at the physiological level. Even so, it is possible to identify several sources of variability in behavior.

When we flip a light switch, swat a mosquito, or laugh at a joke, our behavior has a conspicuous unity. No observer would dispute that the behavior occurred. Much of our behavior is of this sort, with simple, discrete responses being particularly clear examples. However, the human brain contains billions of richly interconnected neurons, most of which fire at some appreciable baseline rate. That rate increases when the rate of input to the cell from other neurons increases, and the rate increases still further if synaptic efficacies have been modified by reinforcement. Thus reinforcement is just one of the variables affecting the probability with which a neuron fires. The activity of the populations of neurons that participate in any behavior will be modified by reinforcement but will not be fully determined by it. Thus, even under conditions of low stimulation and when the organism is at rest, the nervous system is not an array of dormant switches but a boiling stew of activity. Moreover, even the simplest of behaviors—raising a finger toward a letter key, turning a head toward a voice—are mediated by many millions of muscle fibers which are activated in turn by as many neurons. What is a simple, orderly event at the behavioral level is a highly complex confluence of events at the neuromuscular level. There are indefinitely many ways that a population of muscle fibers can execute a response. No matter how small the variance in the physiological processes that control each molecular event, the number of permutations of possible variations will ensure that no two responses, however interchangeable functionally or similar in form at the behavioral level, are ever identical in their neural substrates. However, the integration of the molecular events yields great order at the level of the contingency: The lever still gets pressed. In this light, it is not response variability that needs to be explained but response unity.

Although no neural-network models remotely approach the complexity of the nervous system, they model this variability well—or at least much better than other approaches in normative cognitive science. Variability is produced in part by the sheer number of variables that contribute to the control of an output vector. The response is usually evaluated by a very coarse sieve: Did the target response occur, or did it not? In complex networks it is typically possible for a target response to be mediated by an enormous number of network states. In addition, it is common to include a random process in one or more of the parameters of networks, reflecting, in part, our ignorance of the molecular operation of the nervous system we are trying to model. Thus, although the complexity of the target response itself is grossly simplified in simulations, the complexity of control of the response is modeled relatively well.

A second source of variability is in the environment, or in the stimulation of the organism by the environment. Owing to practical constraints on network size, this variability is usually modeled poorly in network simulations. It is common to represent an environmental event by the activation of one or only a few input nodes, a coarse model of the effect of environmental stimulation. Because of the enormous number of sensory receptors and because of the dynamic relationship between the environment and the organism, it is clear that no two events ever affect the organism in precisely the same way. It is necessarily the case that contingencies of reinforcement tolerate variability not only in response topography but in the controlling stimuli as well.

Variability in contingencies of reinforcement is another source of variability in behavior. Generally, selection contingencies tend to reduce variability: A hungry rat pressing a lever on a schedule of intermittent reinforcement is, at least at a molar level, doing the same thing over and over. (Analogously, contingencies of survival tend to weed out variations that do not meet some minimum requirement.) In contrast, an extinction procedure tends to lead to an increase in behavioral variability (e.g., Schwartz, 1980, 1982). This can be simulated in a network in a relatively straightforward way: If extinction trials reduce the control of a response by a particular context, the response will eventually become weaker than competing behavior. In an organism, or network, with a rich repertoire, variability will be a natural consequence. Extinction procedures may also have a second, more direct effect on variability; the abrupt termination of a reinforcement contingency can have a potentiating effect on other behavior. The organism becomes more active, is more likely to attack conspecifics or to engage in other available activities (Azrin, Hutchinson, & McLaughlin, 1965). Instances of the operant response are likely to vary more in force, duration, and length of runs than during the reinforcement condition (Mechner, 1958; Notterman, 1959). Thus extinction can serve a motivational function with widespread effects on behavioral variability.

Conclusion

As noted above, designing a network that exemplifies selection is only one step toward simulating behavior; we also need to determine the "family tree" of selecting contingencies. When the behavior to be modeled is complex, designing an adequate selection program can be formidable. For example, we might design a model that selected variations in chains of DNA bases, but although such a model would, in principle, be able to produce the genetic code for an amoeba, a cat, or a human being, we do not know enough about the required genetic codes to guide the selection process. Similarly, in models of complex behavior we need a thorough understanding of the behavior to be simulated. We must distinguish between basic units of behavior and higher-order behavior which is built up of these units. Selection from a variable substrate is an appropriate explanation for the units, but only indirectly for the higher-order behavior. A child's early attempts to use a can opener, to unlock a cabinet, to pronounce a new word all appear to be examples of variable behavior shaped by its consequences. But solving a novel multiplication problem, say, 342 x 259, does not. Of course, as a parlor game we could just begin generating "ballpark" answers that are shaped and winnowed by our audience; the answer would be preceded by a closely related candidate answer, and so on. But usually when faced with such a problem we just get down to work and calculate the precise answer; the answer would not be selected from closely related candidates. However, the sequence of behavior leading to the answer appears to defy explanation in selectionist terms only because it is not a behavioral unit. A change of focus would reveal that the whole performance was composed of smaller units that could indeed be so understood. Simulating such a performance with a selectionist network would require identifying appropriate units of analysis.

Thus, the adequacy of our simulations depends on appropriate network architecture, appropriate scheduling of contingencies of selection, and a fair understanding of the behavior we are trying to simulate. An important advantage of neural-network research is that it helps us to evaluate our understanding of these three domains and may help us bootstrap our way to a better understanding. Because of the complexity of the subject matter, there may be no better alternative.

Neural-Networks Models of Cognition
J.W. Donahoe and V. Packard Dorsel (Editors)

CHAPTER 15

ANALYSIS OF REACHING FOR STATIONARY AND MOVING OBJECTS IN THE HUMAN INFANT

Neil E. Berthier
Department of Psychology
Neuroscience and Behavior Program
University of Massachusetts, Amherst

ABSTRACT

The psychological literature emphasizes the roles of neural maturation, the actor-environment relationship, dynamics of the body, and learning of appropriate motor actions in determining the course of motor development. This chapter relates these factors to the model of Berthier (1996) and extends that model to the case of catching a moving object.

Introduction

In the early decades of this century the study of motor development was viewed as a central topic in developmental psychology. Leaders in the field investigated how motor control develops and how neural and environmental factors interact to guide development. These early theories of motor development were part of an intense nativist-environmentalist debate, and the theoretical view of motor development emerging from this period has often been distorted and misunderstood. A goal of this chapter is to reconsider some of these early views and relate them to a more modern view of how infants might gain control of their motor systems. A mathematical model of the development of reaching is presented that views infants as exploring possibilities for action and selecting those actions that are the most effective and efficient. The last section of the chapter extends the model to the catching of moving objects by infants and investigates how infants "predict" future positions of a moving object.

The Development of Motor Control

Early theoretical analysis

Although John B. Watson was not a developmental psychologist, his behaviorist viewpoint had clear implications for the study of development. In contrast to most other areas of psychology, his influence in developmental psychology provoked a strong reaction against the environmental determination of behavior (Horowitz, 1992). The two leading figures in the area of motor development, Arnold Gesell and Myrtle McGraw, took seemingly nativist positions in opposition to Watson. Partly because of the nativist-environmentalist debate, Gesell's and McGraw's statements were interpreted as arguing for the nativist position that motor development was almost entirely determined by the unfolding of a genetic plan. However, neither Gesell nor McGraw was an extreme nativist, and their theoretical contributions provide for a significant role of the environment in determining development. They discovered many important facts about motor development, but their work led to a decades-long hiatus in research on the development of motor control. I will consider these earlier views more fully because many are still relevant to current problems and because they directly influenced later views of the development of motor control.

Gesell and McGraw are considered maturationists. That term loosely means that the critical determiner of development is the underlying maturation of the nervous system. A strong maturationist position contends that development is simply the unfolding of one's genetic inheritance, with the environment playing only a minor role in determining the developmental outcome.

Arnold Gesell. Arnold Gesell was a leader in developmental psychology in the early part of this century. From 1911 to 1948 he was the director of the Yale Clinic of Child Development, where he and his colleagues studied the development of both normal children and children from special populations (Thelen & Adolph, 1992). Gesell argued that development was directed and determined by maturation of the central nervous system and that this maturation was controlled by the genes. He concentrated on careful observation and description of the development of behavior (Thelen & Adolph, 1992), and believed that the mechanisms underlying development were largely biological and should be studied by biologists. A large part of his research legacy was the generation of developmental norms—scales that describe when the average child first shows particular behaviors, and time ranges for the appearance of these behaviors. Gesell and his coworkers used motion-picture film to record infant behavior, and provided detailed and extensive descriptions of children's development.

Typical of Gesell and his colleagues' work are the studies performed by Halverson (1931, 1933) tracing the development of prehension. In these studies infants were studied longitudinally from 4 to 60 weeks of age. During experimental sessions they were presented with a set of various graspable objects while their behavior was recorded on motion-picture film for later frame-by-frame analysis. Direct, unfilmed observations were also conducted through a one-way window. In Halverson's 176-page monograph (Halverson, 1931), the development of reaching was described in excruciating detail. He described ten types of grasps, at least seven types of approach, and four stages in the approach pattern. He concluded that the increase in reaching proficiency was due to anatomical growth of the digits, maturation of the neuromusculature, training, and increased sensibility of the fingertips. If the role of psychology in the study of motor development is simply to describe behavior, Gesell and his coworkers largely met this goal.

Myrtle McGraw. Myrtle McGraw performed most of her work in motor development at Columbia University during the 1920s, '30s, and '40s. Unlike Gesell's, a large part of McGraw's work was aimed at determining how change occurred. A good example is her work on the development of creeping and crawling (McGraw, 1941). This study was based on 82 infants and used both motion-picture film and written descriptions. McGraw identified nine stages in the development of crawling but, unlike in Gesell's work, the stages were informed by neuroanatomical findings about the development of neuromuscular systems. Conel (1939) had shown that the neuroanatomy of the infant brain was very different from that of the adult brain: Infant cerebral cortices lack the internal connectivity and dendritic complexity seen in adults. The descending cortical connections are also immature at birth and largely unmyelinated. Conel's (1939) results led McGraw to conclude that the cortex of the human infant was largely non-functional at birth and that brainstem and spinal systems controlled behavior in neonates. These assumptions about the functional capability of the infant nervous system led McGraw to describe crawling as, first, a reflexive activity controlled by brainstem and spinal areas, then an inhibited and disorganized activity during the period when the motor cortex gained functional connections with spinal areas and, finally, a voluntary and adaptive behavior when the cerebral cortex became fully involved.

For McGraw, the role of learning in the control of movement was minimal until the cortical systems become almost fully functional. McGraw's classic study (1935) of the development of a pair of twins, Johnny and Jimmy, led to the conclusion that training has little effect on behavior until the neural system is "ready." In this study, McGraw provided one of the twins with special training on a task, such as roller skating, and then compared how well the specially trained twin performed relative to his untrained sibling. She concluded that early training had little effect on the long-term development of the skill.

The legacy of the maturationists. For several decades after Gesell's and McGraw's work, psychologists showed little interest in pursuing research in motor development. These fallow years were primarily the result of the strong maturationist views of developmental psychologists and the belief that Gesell and McGraw had largely "solved" the problem of motor development. The view that neonates were reflexive creatures whose behavior is controlled by subcortical areas persisted well into the 1970s. Then, other work—such as that of Haith (1980) on neonatal eye movements and Zelazo, Zelazo, and Kolb (1972) on the adaptability of walking—indicated that infant behavior was purposeful, directed, and cognitively interesting.

Modern views of motor development

The psychological study of motor control and motor development re-emerged in more recent decades largely as a result of the growth of cognitive psychology and the development of dynamic-systems theory. Cognitive psychology suggested that movements were cognitively planned and programmed, and thus of interest to psychologists. Dynamic-systems theory provided a set of tools and ideas within which movement could be better understood.

Dynamic-systems theory. Kugler, Kelso, and Turvey (1980) used recent progress in the mathematical study of nonlinear systems to devise a new way of viewing motor behavior. They suggested that the mechanical properties of the organism's motor systems and of the environmental situation are the main determiners of actions. In the context of motor development, Thelen (1995) has argued that behavior can be modeled as a set of stable states or stable cycles, and that changes in behavior reflect phase transitions between those stable states.

Dynamic-systems theory has been most useful in understanding cyclic or rhythmic behavior. Thelen (1995) argues that rhythmic movement is the result of interactions between underlying component oscillators, each with its own physical characteristics. For example, Thelen, Kelso, and Fogel (1987) studied kicking movements of infants and concluded that kicks were best understood as spring-like oscillatory movements. With each kick, energy was injected into the limb and the mass-spring properties of the limb determined the kinematics of the kick. Goldfield, Kay, and Warren (1995) examined the activity of infants in a "Jolly Jumper" and found that they began with a few tentative bounces, but quickly settled on a pattern that resulted in stable oscillations of the Jumper.

In the field of motor development, the dynamic-systems approach has been greatly influenced by the work of Eleanor Gibson. Gibson argued against a cognitive-representation theory of mental development, and for a theory based on direct perceptions of the affordances of the environment (Gibson & Spelke, 1983). A key assumption of her approach is that infants actively explore their environment and learn the affordances available to them (Gibson, 1988). In the field of motor development, these ideas have led dynamic-systems theorists to emphasize the role of the infant's exploration of the dynamics of the situation when learning appropriate and adaptive ways of moving. The Jolly Jumper study (Goldfield et al, 1995) shows how infants learn the dynamics of the system in order to behave in a way that leads to stable bouncing. The dynamic-systems theorist talks about exploration by the infant of the environment-infant dynamic system and the "soft-assembly" of behavior that fits that system (Thelen, 1995).

A second major idea from the Gibsonian theory is that the actor cannot be considered in isolation from the situation in which the action occurs. This leads dynamic-systems theorists to argue that one must understand the loads imposed on the body, and the kinematics and dynamics of the body-environment system, to understand motor control. For example, Thelen, Fisher, and Ridley-Johnson (1984) showed that stepping could be elicited at an age when it did not normally occur if the infant was placed upright in water, thereby reducing the load on the leg musculature.

Dynamic-systems theory has been very successful in the domain of rhythmic activities such as kicking, stepping, and bouncing, but it was not clear how the theory could be extended to inherently non-rhythmic behaviors such as reaching. More recently, Thelen and coworkers have examined the development of reaching using the dynamic-systems approach, with some success (Thelen, Corbetta, Kamm, Spencer, Schneider, & Zernicke, 1993). Studying four babies longitudinally, they found that young infants initially have individual reaching styles but that, later in development, all infants come to reach similarly. This convergence of motor patterns presumably arises because all babies ultimately face the same dynamic constraints in reaching (Thelen et al, 1993).

Cognitive approach. The cognitive approach to the study of behavior is currently the predominant approach in both developmental psychology and the domain of motor control. While cognitive psychologists come in all types, common assumptions are that mental representations and their manipulations are keys to understanding cognition, and that the flow of information through the brain and its processing are the most important topics for study. In the domain of motor control, cognitive and applied psychologists usually assume that motor activity is the result of a planning process, and that a major problem in planning is the generation and parameterization of motor programs (Rosenbaum, 1991; Schmidt, 1982). Unlike other areas of cognitive psychology, however, motor-control researchers have investigated extensively how learning occurs, and what type of information is important in the learning process (Schmidt, 1982).

E.J. Gibson has also had considerable impact on cognitive developmentalists, even though the Gibsonian framework emphatically rejects the notion of mental representation. Among Gibsonians studying motor development, von Hofsten has used cognitive constructs to describe the mechanisms underlying infant behavior. For example, von Hofsten and Lindhagen (1979), in discussing the success of infants in catching moving objects, conclude that the behavior "reflect[s] a basic human capacity to time-coordinate one's behavior with external events, and to foresee in one's actions future positions of moving objects." Similarly, more recent work by von Hofsten, Spelke, and Feng (1993) argues that infants predict the future positions of moving objects by reasoning about the physics of the world.

Summary

The above work provides some of the key ideas that should be incorporated into a general theory of motor development. As argued by the maturationists, one must acknowledge the role of neural development in both guiding and constraining what motor actions are possible and likely. However, the maturationists can be faulted when they contend that experiential factors are ineffective before a certain stage of development. It seems possible that experience plays a role in early motor development (e.g., von Hofsten, 1993). The arguments of dynamic-systems theory that actor-environment dynamics are a key constraint on movement and that motor development is a soft-assembly of behavior determined by the infant's exploration are compelling. It is difficult to see how an infant could track changes in the dynamics of arm movement using a readout of the genetic code. There are also cognitive aspects to the problem of motor development. How do infants predict the future position of a moving object in order to intercept it? How does an infant determine how to preshape hand configuration for different-shaped objects? Clearly, these are challenging problems for the infant to solve within the context of development. However, it is not clear that these problems are solved through manipulations of mental representations.

A Mathematical Model of Infant Reaching

Kinematic features of infant reaching

Reaching is an appropriate problem for the study of motor development. Its development takes place over an extended period and is influenced by the maturation of spinal and brain systems. Mechanically, the arm is a complex, dynamic system that is continually changing in strength and size. The infant must also deal with the challenges of accommodating to different external loads on the arm. Finally, the manner in which infants reach is also indicative of their perceptual and cognitive capabilities.

Early work on infant reaching by Halverson (1931) and White, Castle, and Held (1964) was primarily descriptive and relied on photographic methods. This work focused on the way infants approached a graspable object, and described the various grasps that infants employed during development.

The use of video and, later, modern kinematic motion systems in the domain of infant reaching was pioneered by von Hofsten (von Hofsten, 1979; von Hofsten & Ronnqvist, 1988). These systems allowed hand speed during the reach to be estimated. von Hofsten found that the infant's hand was constantly accelerating and decelerating during reaching. This pattern of movement is dramatically different from the adult pattern, where simple reaches are accomplished with a single acceleration and deceleration of the hand. von Hofsten (1979) termed each acceleration and deceleration of the hand a movement unit. He hypothesized that, because each unit was defined by an acceleration of the hand, movement units were actually the underlying submovements of the reach. To test whether movement units represent the functional submovements of the reach, the curvature of the hand path has been compared with the timing of the movement units. If movement units represent a corrective sequence of submovements, then one would expect that each submovement would result in a change in hand direction and that peaks of curvature would be associated with minima of the hand-speed profile. Fetters and Todd (1987) found in a longitudinal study of 10 infants at 5, 7, and 9 months that, indeed, 413 out of 425 curvature peaks (i.e., 97%) were associated with speed minima. Other studies, while generally supportive, have not found such a tight association of curvature peaks with speed minima. von Hofsten (1991) analyzed reaches from 19- to 31-week-old infants and found that 644 out of 862 curvature peaks (i.e., 79%) coincided with speed minima. Moreover, Matthew and Cook (1990), analyzing data from young infants, found that some directional corrections occurred *within* movement units.

While the results showing that changes in curvature occur *between* movement units are consistent with movement units being the components or action units of the reach, the results showing direction changes *within* movement units appear—at first glance—inconsistent with this view. There are two possible explanations for the within-movement unit changes in direction. First, in the cited studies, directions and angles were measured using coordinates of Cartesian space. Thus, an impulse of force generated by the elbow musculature would result in simple rotation at the elbow. Such a movement would be elementary, but would result in curvature of the hand path. Second, it is likely that current algorithms for decomposing reaches into movement units do not identify all of the submovements. Most algorithms use an approach of peak-and-valley finding and define movement units as a region of the hand-speed profile in which a peak of criterion height is demarcated by valleys of criterion depth. These algorithms usually treat a region of the reach that has several small "bumps" in the hand-speed profile as a single movement unit. Each of these bumps, however, might reflect an underlying submovement.

If infant reaches are composed of a sequence of submovements, what might the underlying submovements look like, and how might we fully decompose reaches into submovements? There is consensus on the kinematics of adult submovements. Georgopoulos (1986) observed that, when adults reach for positions or objects without an emphasis on accuracy, hand-speed profiles show a dome or bell shape. Others have shown that these hand-speed profiles are well fit by "minimum-jerk" polynomials (e.g., Flash & Hogan, 1986). Minimum-jerk speed-profile polynomials are fourth order and result from the assumption that the reacher is minimizing the total jerk (third time derivative of position) of the hand movement. These results have been taken to support the hypothesis that adults plan movements that minimize jerk, and these types of polynomials provide very good fits to the empirical data.

An attractive feature of minimum-jerk speed polynomials is that they can be specified by three parameters—a height, a width, and a position in time. If we assume that infants have elementary minimum-jerk submovements, is there a way we can determine the number and parameters of the submovements? Berthier (1996) has proposed an algorithm for accomplishing this task based on the results of Flash and Henis (1991). Flash and Henis forced adults to make movements in two steps by switching targets during the first movement. They found that these two-step movements could be decomposed into two elementary submovements, each of which was minimum-jerk. In regions of overlap (i.e., when both submovements were activated), the total movement profile was simply the sum of the submovements.

The Berthier (1996) algorithm used a stripping procedure that sequentially decomposed the reach into minimum-jerk submovements from the largest to the smallest submovement. Because Flash and Henis (1991) found that the submovements summed, one can work backwards and find and fit the largest submovement and subtract it from the data. One can then find the next largest submovement, estimate its parameters, and so on. At each step, the fitting of the three parameters of the polynomial is done by a gradient-descent method.

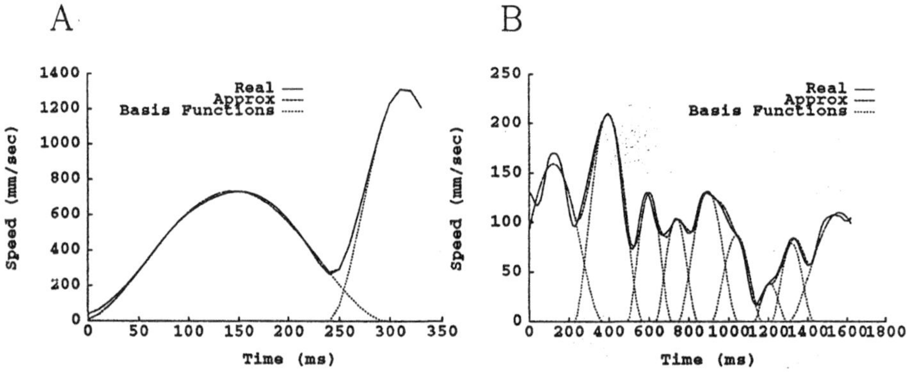

FIGURE 1. Batting- and reaching-speed profiles for movements by a 6-month-old infant. The end of the time series is the time of contact with the target object for each of the movements. **A.** An out-and-back batting movement which resulted in the infant's batting the toy out of the experimenter's hand. **B.** A more typical reach by the same infant that occurred just before the bat. Both movements have been decomposed into an underlying set of submovements (dashed lines) with the experimentally recorded data given by the solid line.

Figure 1A shows the result of the decomposition of an out-and-back "batting" movement by a 6.5-month-old infant. As can be seen, the algorithm decomposed the movement into two submovements. The sum of the two minimum-jerk submovements closely fits the data. **Figure 1B** shows the decomposition of a more typical infant reach. I found that the algorithm worked very well to decompose most infant reaching movements. Given the high quality of the fit in **Figure 1**, it seems likely that infant reaches are composed of a sequence of submovements that are simple in form and similar to adult movements. von Hofsten's (1991) description of movement units as varied and irregular in shape led him to conclude that the action units of the reach were generated by a relatively complex process. However, the current data suggest that a relatively simple process could underlie the sequence of submovements in infant reaching.

The finding that infants reach in a sequence of submovements does not force one to conclude that the sequence is correcting (i.e., changes the course of movement). Matthew and Cook (1990) provide evidence that the infant's hand moves in a correcting path during a reach, but Ashmead, McCarty, Lucas, and Belvedere (1993) could find no evidence that infants corrected their reaches when a target was shifted. The current state of the literature leaves open the question of corrections during reaches.

Infant reaching as the time-optimal solution to a stochastic control problem

Is there a reason that infants use multiple submovements in reaching for objects, or are multiple-submovement reaches simply a reflection of neural and biomechanical limitations of the infant? Berthier (1996) proposed that infants adopt these reaching kinematics, not because they are the only types of movements that infants can make, but because moving this way is both time-efficient and reliable. Instead, infant reaching was modeled as a stochastic optimal-control problem, where the amount of stochasticity was related to the level of motor control the infant possessed at particular times in development. Berthier used a reinforcement-learning algorithm known as *Q*-learning to compute optimal reaching strategies, but hypothesized that infants solve the problem by trial-and-evaluation learning.

Figure 1 indicates that it is not a limitation of strength that causes infants to employ multi-submovement reaches. The figure shows an out-and-back batting motion in which the infant retracted her hand from a toy held by an experimenter and then rapidly accelerated her hand to knock the toy out of the experimenter's hand. The rapid forward motion covered the same distance and roughly the same path as a typical reach, but the distance was covered in a single high-speed movement. The difference between this case and a typical reach is that at contact with the target, the infant's hand is not in position to effect a grasp of the toy. This result suggests that some characteristic of the reaching-to-grasp task caused the infant to adopt the multi-submovement kinematics. Berthier (1996) suggested that this characteristic was the demands of accurately positioning the hand for grasping.

The Berthier model builds on the work of Meyer, Abrams, Kornblum, Wright, and Smith (1988), who studied the speed-accuracy trade-off captured by Fitts's Law (Fitts, 1954). Meyer et al assumed that adults could reach for a target with any number of submovements but that faster submovements come at the cost of accuracy. The task for the adult is then to plan a movement that maximizes the probability that the hand arrives in the target zone in the minimal time. Meyer et al (1988) analytically identified reaches that met this criterion and found a good fit to actual reaches by adults.

The same trade-off between speed and accuracy might be faced by infants when reaching, but—because of the state of their development—the problem might be more severe. It seems likely that reaching with high average speeds might lead infants to entirely miss the target. It might be more time efficient and reliable for infants to reach in several small movements when they have poor control over their hands.

Movement as transitions between submovements. Berthier modeled infant reaching as a discrete-step Markovian decision task. The infant is assumed to have the hand at a particular position in space and is faced with a decision of which movement to generate next. The submovement selected by the infant then causes a state transition to a new state. The current level of motor control the infant possesses is captured by a stochasticity parameter that determines the variability of the state-transition mapping. If the stochasticity is low, it is likely that a particular submovement would cause a state transition to a particular next state. If the stochasticity is high, a particular submovement could cause transitions to one of many different next states.

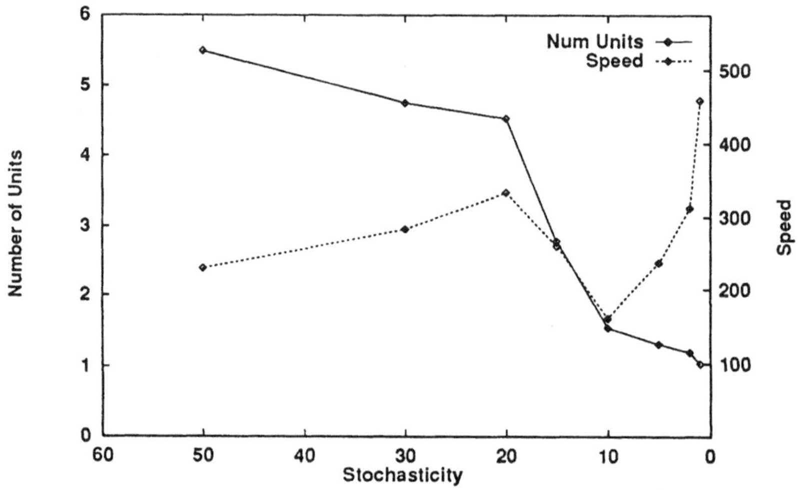

FIGURE 2. The simulated number (solid line) and speed (dashed line) of submovements as a function of age. Increasing age was modeled as decreasing stochasticity in the arm model. The values of stochasticity are x10^{-5}.

Berthier simulated a planar reaching task and showed that with low levels of stochasticity (high levels of motor control) the time-optimal solution is to reach for the target using a single submovement. With high levels of stochasticity (low levels of motor control) the time-optimal solution is to reach for the target with several submovements. When the results of simulations with various levels of stochasticity are plotted, the kinematics match the data obtained from infants. **Figure 2** shows how increasing age (decreasing stochasticity) affects the number and speed of submovements. As stochasticity decreases, the number of submovements decreases and their distance increases. Speed of the first submovement at first increases with age, but when stochasticity decreases to a critical point, the speed of the first submovement decreases and then increases. The latter prediction that speed will decrease and then increase with age has been supported by work of Berthier and McCarty (1996). In a simple reaching task, they found that five-month-olds reached significantly faster than seven-month-olds, and that seven-month-olds reached significantly more slowly than nine-month-olds. As predicted by the model, when the infants showed slower and then faster hand speeds, they also showed steadily decreasing numbers of submovements.

Berthier (1996) directly assessed the level of motor control of six 6.5-month-old infants by measuring the angular error during reaching. For each submovement, the angle between the direction to the target and the direction the hand actually went during that submovement was computed. **Figure 3** shows these angular errors in both azimuth and elevation. As can be seen, the errors are distributed as a Gaussian with a mean of zero. The 95% confidence intervals on the means were plus and minus a few degrees and included zero. These data show that, on the average, infants are heading directly for the target. If the model correctly captures the process and the directional error of movement reflects a limitation on the infant's ability to control the arm, then these data show that infants are reaching as effectively as possible under the circumstances.

Berthier (1996) then compared the measured error with the error assumed in the simulations of the model. The error from the simulations is plotted as the third curve in **Figure 3**, and is comparable in shape and magnitude to the measured error. Because the simulation error was taken from simulations that showed kinematics similar to 6.5-month-old infants, the comparison of simulation and actual error provides strong support for the underlying assumptions of the model.

Learning of submovements. So far, we have addressed how well the model actually predicts the behavior of infants and examined whether the critical assumption of error in movement is supported by actual infant data. The model also hypothesizes that the observed strategies of movement come about via a learning mechanism. It is assumed that particular strategies arise because the

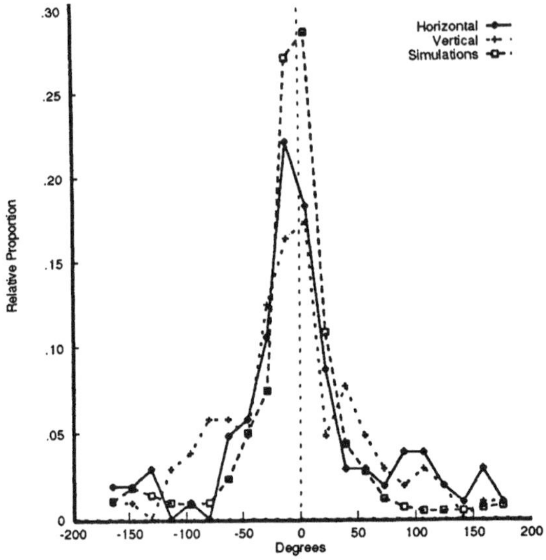

FIGURE 3. The distributions of the angles between the direction of the target and the direction of the hand movement. One angle (degrees vertical) indicates directional error above or below the target, and the other (degrees horizontal) indicates directional error to the left or right. Angular error is also shown from the simulated reaches of **Figure 2** (Simulations).

infant actively explores the effects of particular movements and selects movements in the future that have been shown to be efficient in the past. In this way infants determine which movements are most effective for the state of development of their neural control systems. As the infant grows, new strategies of movement replace older strategies that are no longer efficient.

The simulations above used *Q*-learning (Watkins & Dayan, 1992) to compute the optimal reaching strategies for particular levels of stochasticity. *Q*-learning is a reinforcement-learning procedure where a controller learns to control movement in a plane by trial and evaluation, and *Q* refers to the *quality* of the action. In *Q*-learning, the system computes a *Q*-function that maps states and actions to an evaluation function that informs the controller of the expected total cost of executing that action from that state. *Q*-learning proceeds with the system's choosing actions, noting the consequences of actions, and then updating the *Q*-function. At the end of training, the *Q*-function contains all the information needed for the controller to execute optimal decisions; the *Q*-function implicitly defines optimal strategies of reaching. The general procedure is memory based in that during reaching little "planning" is performed and movements are selected by reference to the *Q*-function.

Although Q-learning effectively computed time-optimal reaching strategies in the above simulations, and provides a congenial metaphor for how infants might learn efficient reaching strategies, the above implementation of Q-learning is not a realistic model of how infants might learn to reach. The first deficiency is that the simulations start with a random strategy of reaching. It is clear from McGraw's work (McGraw, 1941) and from more recent work by von Hofsten (1982) that young, pre-reaching infants do not randomly flail their arms about. McGraw (1941) believed that subcortical systems initially control infant movement. She concluded that subcortical systems provide an initial set of behaviors that are later supplanted by more flexible and powerful control as the cortical systems mature.

McGraw (1941) ruled out any role for experience and learning until the descending cortical systems mature, but more recent work by von Hofsten (1982) and Ennouri, Dubon, Notides, and Bloch (1994) shows that early arm movements of human neonates are goal and task oriented and somewhat flexible. In a theoretical paper, von Hofsten (1993) argued against the idea that early arm movements are fixed and reflexive and proposed instead that pre-reaching movements are action oriented from very soon after birth. von Hofsten (1993) suggested that infants discover successful ways of reaching by searching a "task space" for efficient solutions.

The existence of early, baseline reaching strategies would be helpful for two reasons. First, these strategies would put bounds on the infant's motor actions. It seems likely that if infants simply executed random arm movements early in learning they would strain or damage their muscles or arms. Early crude strategies would protect the integrity of the arm during early learning.

Second, baseline reaching strategies would hasten learning. In many control situations, reinforcement-learning algorithms are very slow because of the large number of possible control actions. Barto (1990) has suggested that the use of a crude controller in early learning would bootstrap learning because (1) it would focus learning on regions of the state space that were likely to hold the solution, and (2) the learning system would not have learn to control the "easy" part of the system, but could focus on learning to control the more difficult and nonlinear parts of the problem.

Extension of the model to catching a moving object

A more realistic and powerful knowledge structure. The previous model assumed that infants were attempting to reach for objects in minimal time and that the determining factor in which strategies fulfilled this goal was the infant's ability to control its movements. Infants were assumed to determine which strategies matched their current level of motor control through exploratory learning, the results of which were memorized, i.e., stored in a look-up table. A major failing of the above simulations is that look-up tables are both psychologically unrealistic and computationally limiting.

Using look-up tables to store information leads to an unacceptable increase in the size of the data structure for many real-life problems. In the simulations of arm movement, even with a simple model of the arm and a limited work-space, 32,400 states were required. And, each of these states was in turn associated with 486 Q-values, one for each action. Thus, 15,746,400 floating point variables would be required to store all the Q-values! Obviously, a more cognitively interesting problem with a more complicated state space and set of actions would require prohibitively large look-up tables.

In the above simulations, the look-up-table approach also leads to slow learning because of generalization problems. Because of the storage requirements for a complete Q-function table, we combined the Q-values for neighboring states in a 10-by-10 mm space into a single Q-value. This significantly reduced the required storage but often led to errors because the stored Q-value was not correct for all of the states within the 10-by-10 space. The Q-values might be approximately correct for most of the states in the space, but incorrect for at least one of the states. This inappropriate generalization sometimes led to execution of clearly inappropriate actions. In addition, a look-up table limits the amount of beneficial generalization that can occur. For example, there may be a group of states that share the same Q-values and within which it would be appropriate to generalize. In this case, one would only have to learn the correct Q-values for a single state and then generalize it to the appropriate neighboring states. Finally, look-up tables do not make sense psychologically. It is difficult to imagine an infant using a large database that stores all the appropriate actions for each of the possible states of the arm. For these reasons the use of artificial neural networks to approximate the Q-function was explored in the current research.

Neural-network knowledge structures. Many types of artificial neural networks have been used to store control information. Among the most useful methods for on-line learning are the radial-basis-function networks. These networks use a relatively small number of basis functions to represent and interpolate the desired function (Poggio & Gerossi, 1990), statistical methods (Geman et al, 1992; Gullapalli, 1992), and random representations that use an expansion recoding of the input to represent information (Albus, 1981). A psychological model of reaching, the Knowledge Model (Rosenbaum et al, 1995), uses related methods to represent the knowledge underlying reaching.

We used Cerebellar Model Arithmetic Computers (CMACs) to represent the Q-function because they were inspired by the anatomy and physiology of the cerebellum, an area that is believed to be important in adaptive control and in controlling reaching (Albus, 1981; Berthier, Singh, Barto, & Houk, 1993) and because CMACs have been shown to work very well in control problems (Miller, Kraft, & Glanz, 1992; Sutton & Whitehead, 1994).

Catching a moving object. Several studies have examined the ability of infants to catch moving objects (von Hofsten & Lindhagen, 1979; von Hofsten, 1980; von Hofsten, 1983; Robin, Berthier, & Clifton, 1996). This task requires infants to perceive the direction and speed of the target object and to anticipate its future position in order to effect a successful grasp. Infants as young as 5 months are successful at least some of the time in grasping a moving target. The task is interesting from a cognitive perspective because, while older infants readily catch moving objects, it is not clear if they succeed by "predicting" the future position of the moving object or if they employ some less cognitive, more lower-level process.

von Hofsten (von Hofsten & Lindhagen, 1979; von Hofsten, 1983) was the first to study infants reaching for moving objects. In his studies, objects moved at various speeds in a horizontal arc in front of the infant. It was found that infants as young as 5 months were able to catch objects moving at a speed of as much as 120 cm/sec. Analysis of the kinematics of reaching led von Hofsten to conclude that infants predict the future position of a moving object in order to obtain it (von Hofsten, & Lindhagen, 1979; von Hofsten, 1983). Recently, von Hofsten, Spelke, and Feng (1994) found that infants could catch a moving object that was moving diagonally in front of them in a frontal plane. von Hofsten et al (1994) concluded that infants use their knowledge of Newton's Second Law to predict the future position of a moving object.

Robin et al (1996) have not emphasized the physical reasoning of infants in predicting the future locations of moving objects. They found that infants aimed their hands in front of the moving object and gradually closed the distance between the two. Such a movement strategy might simply be the result of some type of visuo-motor coupling.

While contacting a moving object is *prima facie* evidence that infants are able to predict the future position of that object, it is not clear from the existing data if infants truly predict movement of objects in the cognitive sense of the word. Full-blown prediction would imply an ability to sense the current position, heading, and speed of an object, and then to compute the position of that object at some future time. Given the difficulty that children and adults have catching thrown objects, it seems unlikely that infants possess the ability to predict the motion of arbitrarily moving objects. Instead, it seems likely that an infant's ability to catch a moving object is based on some simpler strategy. The major goal of the current simulations was to test whether a simple mechanism might be sufficient to explain the ability of infants to catch moving objects.

The current simulations used *Q*-learning to compute the optimal reaching strategies for catching a moving object. The workspace and arm model are identical to Berthier (1996). In the current simulations, the target was moving at various speeds in an arc in front of the infant and the *Q*-function was stored in an artificial neural network instead of a look-up table.

The artificial neural network that stored the *Q*-function was a set of CMACs, one for each action. The input to each CMAC was the *x* and *y* positions of the hand and the angular position of the moving object at the current time. In the current simulations the object moved along an arc in a 120-by-120 mm workspace. The object entered the workspace 350 ms after the start of the trial at (120,80), moved through (100,100), and exited at (80,120). The simulated path of the target is identical to that of an object moving on the end of a 100 cm pole, as was used in the studies with infants. The simulated speed of the object was either 10 or 20 cm/sec.

Because of the difficulty of the problem and because *Q*-learning with a large number of actions requires extensive training, the number of actions in the current simulations was reduced from the number used in the stationary-target simulations. This was done by noting that in many reaching experiments the speed of a reach is linearly dependent on the distance of the reach. This assumption allowed us to use action vectors of two dimensions, a speed and a direction, with the distance being computed from the speed element of the action vector. Sixty actions were used, with directions going from -10 to 125 degrees in steps of 15 degrees, and speed going from 80 to 580 mm/sec in steps of 100 mm/sec. The distance was computed by dividing the speed by four. The temporal duration of a submovement was given by dividing the distance covered during the submovement by the speed and adding 100 ms for initiating the movement.

The use of a relatively small number of discrete actions limits the positions to which the hand can be moved. Because we wanted to keep the number of possible actions to a minimum, we allowed a larger target zone than with the stationary-target-object simulations. In the current simulations, we used a target zone of 20 by 20 mm centered on the target, and increased the stochasticity accordingly. Simulations with this arrangement using the stationary target indicated that a stochasticity of .0002 with the stationary target mapped to a stochasticity of about .0008 with the moving target.

The input to each CMAC was the state of the system, the *x* and *y* positions of the hand and the current angular position of the target. At each step the connection weights of the CMAC were updated so that the output was brought closer to the target *Q*-value for the executed action. The CMACs used 28 subdivisions (tilings) of the state space. There were four three-dimensional, 12 two-dimensional and 12 one-dimensional tilings. Each tiling divided the dimension into eight equal intervals, so that each of the one-dimensional tilings was composed of eight stripes, each of the two-dimensional tilings was composed of 64 rectangles, and the three-dimensional tiling was composed of 512 cubes.

Table 1 shows the average number of submovements in a reach, the average position of a grasp for successful reaches, and the percent of reaches that resulted in a successful grasp for both speeds of the moving object. Because

Stochasticity	N Submovements	Position of Grasp	Percent Grasp
	10cm/sec Target Speed		
.000001	1.0	(117.5,82.4)	100.0
.00001	1.0	(117.5,82.4)	99.7
.0001	1.22	(113.7,86.1)	91.8
.0002	1.41	(109.7,89.4)	79.5
.0005	2.31	(87.3,110.0)	76.9
.0007	2.31	(87.3,110.3)	72.3
	20 cm/sec Target Speed		
.000001	1.0	(117.5,82.5)	100.0
.00001	1.0	(117.5,82.4)	99.5
.0001	1.09	(114.5,85.8)	62.9
.0002	1.19	(110.5,89.6)	71.0
.0005	1.90	(73.1,121.6)	9.1
.0007	1.47	(98.4,101.0)	32.7

TABLE 1. The average number of submovements in a reach, the average position of a grasp for successful reaches, and the percent of reaches that resulted in a successful grasp for both speeds of the moving object.

the object was moving at constant velocity across the workspace from (140,62) to (62,140) passing through (100,100), the position of the object at contact implicitly gives the movement time and average speed of the hand. With low levels of stochasticity, reaches invariably resulted in grasp of the target at both the fast and slow target speeds. Higher stochasticities resulted in an increase in the number of misses and increases in the latency to grasp.

The current simulations match experimentally observed behavior: The simulations predict that the speed of the object will affect the probability of grasp. This prediction is consistent with findings from von Hofsten and Lindhagen (1979), in which infants were tested with objects moving at 3.4 to 30 cm/sec. As predicted by the simulation, infants had greater difficulty with faster-moving objects.

The model assumes that decreasing stochasticity reflects increasing motor control on the part of the developing infant. In the current simulations, decreasing stochasticity led both to increased probability of grasping and to grasping the target at an earlier point in its path. While no data are available on the point of grasp as a function of age, substantial data exist showing improving success over the first year (von Hofsten & Lindhagen, 1979; von Hofsten, 1983; Robin et al, 1996). As with the current simulations, this improved performance was dependent on the speed of the target object; infants learn to catch slower-moving objects before they learn to catch faster-moving objects (von Hofsten, 1983).

The result that performance is greatly dependent on the speed of the object and the age of the infant is not surprising, but infants have at least one strategy at their disposal that would be relatively unaffected by the speed of the target or their level of motor control. A possible solution to the problem would be to extend their hand into the path of the target object and to wait for the object to approach their hand. This strategy might even work better with faster-moving objects than slower-moving objects because the infants would not have to extend their hands for as long a period of time. For the range of speeds modeled here and used by von Hofsten (von Hofsten & Lindhagen, 1979; von Hofsten, 1983) and Robin et al (1996), infants do not adopt this strategy but instead "track" the moving object. This "tracking" allows the infant to grasp the target as soon as possible and essentially makes the problem an optimal-time problem as modeled here. For faster target-object speeds this does not result in as high a success rate as the "wait" strategy described above, but may be a carry-over from the infant's experience with stationary objects.

The current simulations demonstrate that it is possible to catch moving objects without any knowledge of physics and without reasoning about the path of the moving object. In the current simulations, there is no knowledge of physics nor is there any module that predicts the future position of the moving object. The current simulations work because the network learns a relationship between the object's position and velocity at a particular time, and which reaches are successful for those values.

It remains to be seen how powerful this simple mechanism is. Clearly, the network used here does not generate predictions for arbitrarily moving objects, but learns to solve a particular problem. If the current model captures the essentials of how infants contact moving objects, one would predict that infants would not succeed with arbitrarily moving objects, and that the likelihood of success would depend on the infant's previous experience with moving objects. Some movement trajectories should be substantially easier for the infant to cope with than others.

The current simulations address one of the major shortcomings of the earlier Berthier (1996) model. The moving-object simulations used an artificial neural network to store the information that guides reaching by the infant. The use of the neural network is not only more psychologically realistic, but allows faster learning because it provides for beneficial generalization between neighboring states. A major problem in the current simulations is that individual

CMACs were used for each of the actions. While multiple CMACs worked acceptably for the current task, learning would be accelerated if generalization could occur across actions. Further, the number of CMACs (actions) determines how long it takes to solve a particular problem, with simulation time being prohibitive for a few hundred actions. This severely limits the usefulness of this neural-network architecture for larger problems.

The current computational limitations arise mainly because the control problem is formulated with discrete actions. However, the problem actually involves control actions that are real valued. In the context of connectionist reinforcement-learning algorithms, discrete-action problems are much better understood than real-valued action problems. Gullapalli (1992) has, however, explored the use of reinforcement learning with real-valued actions and developed a controller for a robot insertion task that uses an artificial neural network with real-valued outputs. He showed that such an approach could be very successful if a function was implemented to "shape" learning (see Gullapalli, this volume). Extension of the present model to such a neural network is currently being explored.

Conclusions

The present work provides insight into the mechanisms underlying the development of infant reaching. The current simulations and those of Berthier (1996) easily and gracefully predict the basic kinematics of reaching and the time course of development. The model also relates to theoretical approaches to motor development and can accommodate the constraints and mechanisms of neural development. In the simulation of contact of moving objects, the model provides a parsimonious hypothesis of how infants might catch moving objects and makes predictions about how well infants might solve particular problems. Further experimentation and simulation is required to determine how well the modeled mechanisms can account for behavior in a wider range of situations.

Throughout the present chapter, there has been a tension between the theories and hypotheses we might want to investigate, and the theories and hypotheses that are implementable and understandable in a mathematical model. As the computational understanding of the current type of control problems increases, the class of implementable theories will increase. Until then, it will be a continuing struggle between the theories that are most interesting and those that are good candidates for investigation.

ACKNOWLEDGMENTS

The current research was supported by NSF grant SBR-9410160. The author thanks Phillip Lee for helping with the initial moving-object simulations, and Michael McCarty and Rachel Clifton for their many helpful comments.

Neural-Networks Models of Cognition
J.W. Donahoe and V. Packard Dorsel (Editors)

CHAPTER 16

REINFORCEMENT LEARNING OF COMPLEX BEHAVIOR THROUGH SHAPING

Vijaykumar Gullapalli
Princeton University

ABSTRACT

This chapter describes the utility of *shaping*, a procedure used in experimental psychology for the reinforcement learning of complex behavior. The principle underlying shaping is that complex behavior can be acquired by first learning related simpler behavior. Simulation results illustrating this approach are presented for a control task in which a neural network is trained using reinforcement learning for a five-degree-of-freedom robot hand.

Introduction

An important topic occupying the attention of researchers is the process underlying the learning of complex behavior via reinforcement feedback. When complex behavior is learned, the problem of "strengthening" behavior, which is the focus of most learning experiments and theories, is preceded by the problem of getting the desired complex behavior *to occur*. For reinforcement to occur, the target response must occur first. Indeed, spontaneous occurrence of the desired behavior becomes extremely unlikely as the complexity of the behavior increases. Nevertheless, it is possible to train humans and other animals to produce very complex behavior by suitably manipulating the reinforcement contingencies and the environment.

One approach to developing complex behavior, well known to animal learning researchers, is the method of shaping by successive approximations (e.g., Skinner, 1938; Staddon & Ettinger, 1989; Donahoe & Palmer, 1994). The principle underlying shaping is that learning complex behavior can be facilitated by first learning related simpler behavior. Although defining related simpler behavior might be difficult when covert cognitive behavior such as language or mathematics is to be learned, it is relatively easy to determine a sequence of approximations that will lead to mastery of a target *overt* behavior. Indeed, shaping has been most useful for teaching motor skills to animals, and, for the same reason, shaping can also prove useful for training artificial learning systems to perform as skilled controllers.

In this chapter, we present simulation results illustrating the utility of shaping in training robot controllers via reinforcement learning. In these simulations, we trained a neural network to control a dynamic simulation of a five-degree-of-freedom robot hand on a key-pressing task. Before describing the task and the implementation in detail, we present a brief overview of the shaping procedure.

Shaping

Shaping is an ancient animal-training procedure that has been studied by experimental psychologists interested in animal learning (Honig & Staddon, 1977). The term "shaping" itself has been attributed to the psychologist Skinner (1938), who used the technique to train animals, such as rats and pigeons, to perform complicated sequences of actions for rewards. Skinner describes how the technique is used to train pigeons to peck an illuminated spot on a pecking disk:

> We first give the bird food when it turns slightly in the direction of the spot from any part of the cage. This increases the frequency of such behavior. We then withhold reinforcement until a slight movement is made toward the spot. ... We continue by reinforcing positions successively closer to the spot, then by reinforcing only when the head is moved slightly forward, and finally only when the beak actually makes contact with the spot...
>
> The original probability of the response in its final form is very low; in some cases it may even be zero. ... By reinforcing a series of successive approximations, we bring a rare response to a very high probability in a short time. ... The total act of turning toward the spot from any point in the box, walking toward it, raising the head, and striking the spot may seem to be a functionally coherent unit of behavior; but it is constructed by a continual process of differential reinforcement from undifferentiated behavior, just as the sculptor shapes his figure from a lump of clay. (Skinner, 1953, pp. 92-93)

The phrase "reinforcing a series of successive approximations" expresses the essence of shaping. Given the task of training an animal to produce complex behavior, the trainer must (1) judge what constitutes an approximation to, or a component of, the target behavior, and (2) determine how to differentially reinforce successive approximations so that the animal easily learns the target behavior.

Unfortunately, neither of these two components of shaping has been formalized rigorously in the psychology literature, even though shaping is widely used both in psychological studies and to train pets and circus animals. Staddon (1983), for example, observes that the trainer often has to rely on an intuitive understanding of the way the animal's behavior is generated when determining which behavioral variations are precursors to the target behavior and how to reinforce these precursors. Variations in the behavior of individual animals also must be taken into account when making these judgments.

The limitations of relying on intuition when judging approximations to the target behavior are especially apparent when the behavior under consideration is cognitive (covert) in nature. However, when the overt behavior of the animal is being shaped, behavioral approximations become equivalent to physical distances, and it is therefore easier to determine a sequence of approximations that will lead to mastery of the target behavior. It is therefore not surprising that shaping has been used most often for teaching motor skills to animals. For the same reason, shaping can also prove useful for training artificial learning systems to perform as controllers of motor behavior.

Several neural-network researchers have noted that training a controller to perform one task can facilitate its learning a related second task (e.g., Selfridge, Sutton, & Barto, 1985; Gullapalli, 1990; Wieland, 1991). Selfridge, Sutton, and Barto (1985) studied the effect of shaping a controller to balance a pole mounted on a cart. They observed that overall learning times were typically shorter when a previously trained controller was retrained on a modification of the cart-pole system than when an untrained controller was trained from scratch. This was demonstrated for several types of modifications including increasing the mass of the pole, shortening the pole, and shortening the track. (See Barto & Sutton, this volume.)

Wieland (1991) illustrated the utility of shaping using a different version of the cart-pole task in which the controller had to simultaneously balance *two* poles mounted on a cart. Because it is easier to solve the two-pole balancing problem when the pole lengths are very different than when the pole lengths are almost equal, Wieland trained a controller by starting with poles of lengths 1.0m and 0.1m and gradually increasing the length of the shorter pole to 0.9m. Although it is very difficult to balance poles with lengths as close as 1.0m and 0.9m, the shaping process resulted in a controller that was able to do so. Wieland and Leighton (1988) also studied the utility of shaping schedules for accelerating other learning methods based on gradient-descent procedures.

Other applications of shaping in neural-network research have been in the area of training recurrent nets. Allen (1989) trained recurrent nets to generate long sequences of outputs using a shaping procedure that involved initially training the nets with short target sequences and introducing longer sequences gradually over training. Another related form of shaping was studied by Nowlan (1988). In this case, a robust attractor state for a recurrent network was developed by first training initial states near the attractor, and then gradually increasing the distance of the initial states from the attractor.

In this chapter, we focus on the utility of shaping for training controllers via direct reinforcement-learning methods. The behavior of a controller was shaped over time by gradually increasing the complexity of the control task as the controller learned. At the same time, the evaluation function used to compute the reinforcement delivered to the controller was also changed to reflect the increasing complexity of the task. This procedure is analogous to the manner in which shaping is used to train animals, and to the manner in which shaping was used in some of the studies cited previously.

Shaping Control Behavior

To shape the behavior of the controller, one has to determine (1) a series of approximations to the target behavior and (2) how to differentially reinforce successive approximations to the target behavior. The task under consideration involved controlling a robot hand as it pressed keys on a calculator keypad. A detailed description of the task follows.

Key pressing using a robot hand

The control task used to examine the utility of shaping involved pressing keys on a simulated calculator keypad using the index finger of a simulated dynamic model of the Stanford/JPL hand. The finger had three degrees of freedom (joints) and the motion of the hand was restricted to a plane parallel to the x-y plane of the calculator face. Thus, in all there are five degrees of freedom to be controlled. The task is depicted in **Figure 1**. The axes of rotation of the finger joints of the Stanford/JPL hand are as follows. The first joint (linking the finger to the palm) permits rotation about an axis parallel to the z-axis, thereby permitting abduction/adduction movements of the finger. The other two joints have axes of rotation that are perpendicular to both the first link of the finger and the z-axis, permitting motion similar to that of the phalanges of a human finger.

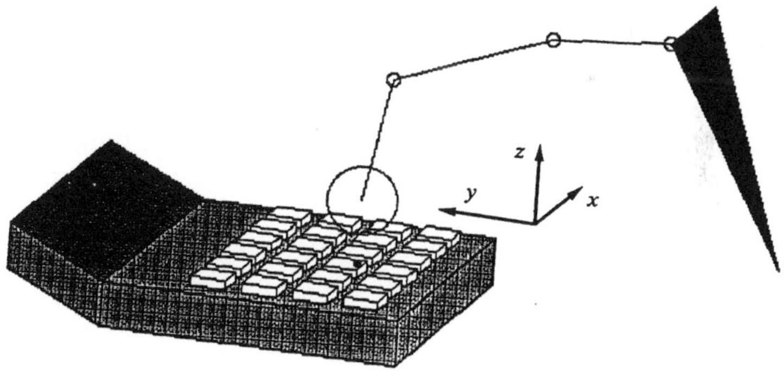

FIGURE 1. The arrangement for the key-pressing task. The simulated Stanford/JPL hand and the calculator are shown to scale. Only the index finger of the hand is indicated because only that finger was used in the task. The triangle represents the palm of the hand; the large circle represents the fingertip. The small dark circle on the calculator face is the "footprint" of the center of the fingertip.

The control actions are positioning commands that locate the hand in its plane of motion and position the three joints of the index finger. The key to be pressed is specified by setting a single bit in a 24-bit command input vector supplied to the controller. Additional inputs to the controller include proprioceptive feedback of the positions and velocities of the finger joints and the hand, a fingertip force sensation, and a binary "key-pressed" sensation that is set whenever a key is pressed and reset whenever a new target key is specified. To successfully press a key, the fingertip must depress the key to the level of the face of the calculator ($z = 0$). Provision of the hand position and velocity feedback permits the controller to learn to press any key starting from any initial hand configuration.

Because pressing a key involves positioning the fingertip over the key, then pressing, and then releasing it, the task is fairly complicated. The controller has to learn a sequence of actions to execute a single key-press operation. As one might imagine, the probability of a reinforcement-learning controller's generating such a complex sequence of behavior by chance (i.e., through stochastic search) is infinitesimal. A simple evaluation criterion that only signals successful key presses is therefore not useful for training the controller. Fortunately, we can shape the target behavior by defining successive approximations to the key-pressing operation, and providing more informative differential evaluations that facilitate learning.

The first problem in implementing shaping was to determine a series of approximations to the target behavior. This was not very difficult for the key-pressing task. A fairly intuitive series of approximations to the key-press operation is the following:

(1) Raise the fingertip so that it is not in contact with the keypad surface (thereby preventing accidental key strikes).

(2) Move the fingertip towards the target key, keeping the fingertip raised.

(3) Position the fingertip over the target key, keeping the fingertip raised.

(4) Position the raised fingertip over the target key, then press down with the fingertip.

(5) Position the raised fingertip over the target key and press down until the key is fully depressed.

(6) Position the raised fingertip over the target key, press down until the key is fully depressed, and then release the key by raising the fingertip.

The second problem in implementing shaping is deciding how to differentially reinforce approximations to the target behavior. This is more complicated. To differentially reinforce the controller as it learns a series of approximations, the training agent (or critic) has to maintain a behavioral history of the controller and infer from that history how well the controller has learned each approximation. Based on this inference, the critic must determine if the controller requires further training on a particular approximation or if it is ready to be trained on the next, more sophisticated, approximation. While it is wasteful to continue training the controller on an approximation that it has already mastered, switching to a more sophisticated approximation too quickly can also be detrimental to learning the ultimate target behavior. To best realize the benefits of shaping, accurate judgment of the controller's ability at every stage in training is therefore necessary. Clearly, the time frame over which the behavioral history is maintained and evaluated is a factor in making these judgments.

Our approach to the problem of differentially reinforcing the controller's behavior is (1) to restrict the time frame over which the controller's behavior is evaluated to individual training runs, i.e., individual attempts at the target behavior, and (2) to require that the behavior progressively satisfies the criteria for all the approximations to the target behavior, starting with the simplest, *in each attempt* at the target behavior. This approach allows the reinforcement criteria to be switched to increasingly sophisticated approximations quickly over training, while at the same time ensuring that the controller is trained on an approximation only if it has successfully met the criteria for all simpler approximations. Moreover, this approach sidesteps the question of how to infer the controller's ability at a given stage of training from its behavioral history over a longer time frame.

In the key-pressing task, for example, the critic maintains the behavioral history of the controller only over individual training runs. Each run begins with a new target key being assigned and ends after a fixed number of time steps have elapsed. During the course of each training run, the performance of the controller is evaluated using a series of criteria, each attuned to a corresponding approximation to the criteria given in the list above. The criterion used to determine the evaluation was selected at each time step based on the state of the hand and the portion of the above series of approximations that had already been accomplished. For example, if the fingertip was raised but not located over the target key, an evaluation criterion that rewarded motion towards the target key while keeping the fingertip raised was selected. If the fingertip had already been positioned over the target key, downward motion of the fingertip while keeping it over the target key was rewarded, and so on.

Initially, the controller might spend the entire duration of a training run learning to satisfy the criterion for the simplest approximation. With time, the controller learned to consistently satisfy the criteria for the simpler approximations, and the frontier of learning shifted to approximations closer to the target complex behavior. Thus, most of a training run was spent in learning the approximation to the target behavior at the current frontier of learning.

Implementation details

This section provides additional implementation details on the simulations. We describe the controlled process, the architecture of the neural network used to implement the learning controller, and the evaluation function(s) used. The networks used in these simulations were all standard fully connected feedforward networks (e.g., Rumelhart et al, 1986; Hinton, 1987).

The robot. As described above, the robot used in the key-pressing task was a dynamically simulated Stanford/JPL hand (see **Figure 1**). Only the index finger of the hand was used. Furthermore, the motion of the hand base was restricted to a plane parallel to the x-y plane. The control actions were positioning commands that moved the hand base in its plane and positioned the three joints of the index finger.

Proprioceptive feedback of the position and velocity of each joint was provided to the controller. Additionally, a simulated fingertip-force sensation was used to provide feedback during key presses. This force was a function of the height of the fingertip above the keypad, as shown in **Figure 2**.

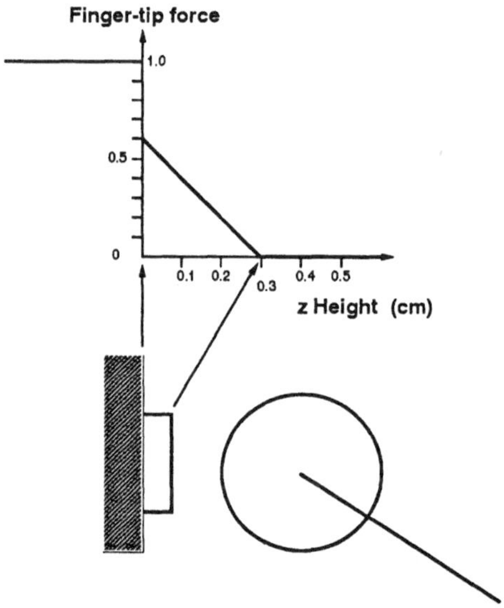

FIGURE 2. The output of the simulated fingertip force sensor as a function of the height of the fingertip above the keypad.

The controller. The overall architecture of the controller is shown in **Figure 3**. The inputs to the network included a 24-bit command input specifying the target key, feedback of the position and velocity of the hand and of the force on the fingertip, and finally, a representation (an efference copy) of the network's output on the previous time step. The network had 30 backpropagation units in the hidden layer and five Stochastic Real-Valued (SRV) units in the

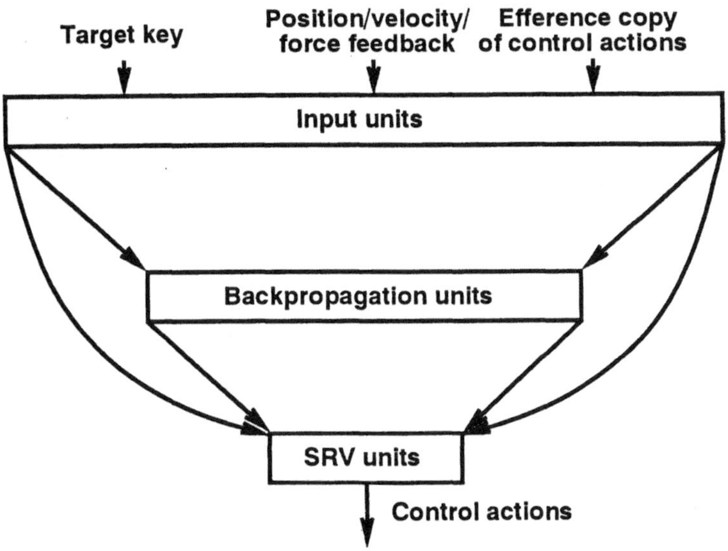

FIGURE 3. The general architecture of the controller used in the key-pressing task.

output layer. As shown in the figure, the output units are connected to both the input and the hidden units.

The most popular description of backpropagation units is given in Rumelhart, Hinton, and Williams, 1986. SRV units are described in detail in Gullapalli, 1990. These latter units are designed to learn real-valued outputs through reinforcement feedback. The procedure for training networks with SRV output units and backpropagation hidden units is also described in Gullapalli, 1990.

The backpropagation units used the logistic function ($f(x) = 1/(1+e^{-x})$) to compute their outputs. The initial weights of all the units were set to random values selected from a uniform distribution over the interval [-0.5, 0.5]. The learning rate *alpha* was set to 0.01 for the hidden (backpropagation) units and 0.001 for the SRV units. The learning rate on the variance, *rho*, was also set to 0.001 for the SRV units.

The evaluation. In each training run, the evaluation was computed using a sequence of criteria based on the portion of the key-press operation already accomplished. (See earlier description of the criteria.) For the purposes of computing the evaluation, the fingertip was considered to be positioned over the target key when the *x-y* distance to the key (denoted *d*) became zero and the hand velocities became very small. The key was considered to be pressed when the fingertip height (denoted z) became 0. In the following, the height of the fingertip at the previous time step is denoted z_{t-1} . Also note that each key on the keypad is 0.3 cm high (see **Figure 2**).

The procedure used for computing the evaluation was the following. If the fingertip was not yet positioned over the target key,

$$r = \begin{cases} 0.5(1 - |1-z|/6) & \text{if } z < 0.3, \\ 0.5(e^{-d} - |1-z|/6) & \text{otherwise.} \end{cases}$$

Once the fingertip was positioned over the target key,

$$r = \begin{cases} 0.7r_{\text{PRESS}} + 0.3d & \text{if key has NOT been pressed,} \\ r_{\text{RELEASE}} & \text{otherwise,} \end{cases}$$

where

$$r_{\text{PRESS}} = \begin{cases} \begin{cases} 0.5(1 - z/3) + 0.5 & \text{if } z < z_{t-1}, \\ 0.5(1 - z/3) & \text{otherwise} \end{cases} & \text{if } z > 0.3 \\ (1 - |z|/3) & \text{otherwise,} \end{cases}$$

and

$$r_{\text{RELEASE}} = \begin{cases} \begin{cases} 0.5(1 - |1-z|/3) + 0.5 & \text{if } z_{t-1}, \\ 0.5(1 - |1-z|/3) & \text{otherwise} \end{cases} & \text{if } z < 0.3 \\ 1 - |1-z|/3 & \text{otherwise.} \end{cases}$$

If the evaluation computed under any of the above conditions was less than zero, it was set equal to 0.

Training methodology

In order to keep computer simulation time reasonable while retaining all the essential aspects of the key-pressing task, we restricted the choice of the target keys to three keys, which are highlighted in **Figure 4**. These keys were chosen to require a broad range of motion of the fingertip. The controller was trained in a series of training runs, each of which began with a new target key being picked randomly. The probability of picking the key used in the previous training run was 0.1, while that of picking either of the other two keys was 0.45.

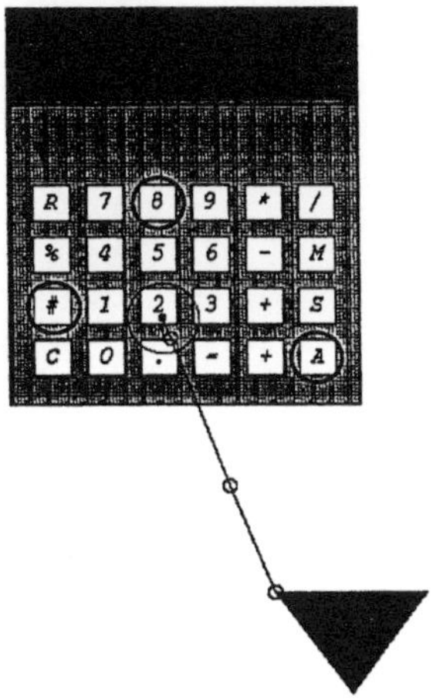

FIGURE 4. The set of target keys used in the key-pressing task, shown encircled by bold circles.

The initial hand configuration used in a training run was the configuration of the hand at the end of the previous training run. However, if the previous training run left the fingertip touching the keypad, the last two finger joints were repositioned so that the fingertip was no longer in contact with the keypad. In either case, the initial velocities were set to zero. Each training run lasted 15 time steps, during which the controller was trained using the appropriate evaluation criterion at each time step as described above. The sensory feedback to the controller was also updated at each time step.

For the purpose of comparison, we also attempted to train a controller on the key-pressing task without resorting to shaping. An identical training procedure was followed in this case, with the sole modification being the use of a single evaluation criterion that only rewarded pressing the target key.

Results

The performance of the controller on the key-pressing task after 25,000 training runs is shown in **Figures 5** through 7. Each figure contains four panels that show the motion of the hand over twelve time steps when pressing each of the three target keys. Panel **a** contains three strip charts that show the values of three quantities at each time step over the course of a key-press operation. These quantities are the distance from the fingertip to the target key (D, ranging from 0 to 9 centimeters), the height of the fingertip above the calculator face (Z, ranging from 0 to 1 centimeter), and the payoff, or evaluation (P, ranging from 0 to 1). Panels **b**, **c**, and **d** show the motion of the hand during the key-press operation from three different viewpoints.

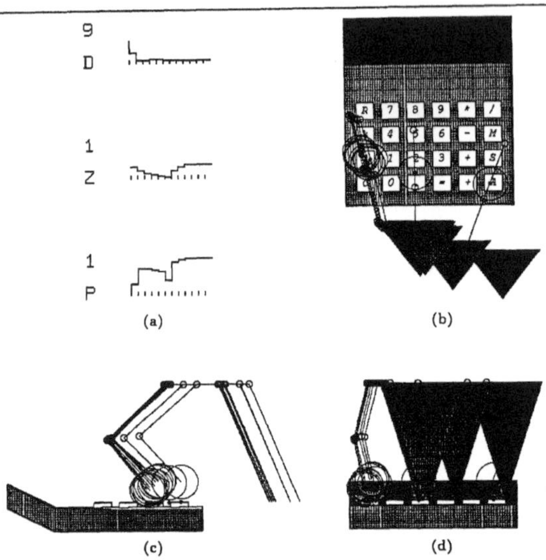

FIGURE 5. Pressing the # key on the calculator keypad. The initial position of the fingertip was the final position after pressing the A key. The dark triangle denotes the initial location of the hand. See text for details.

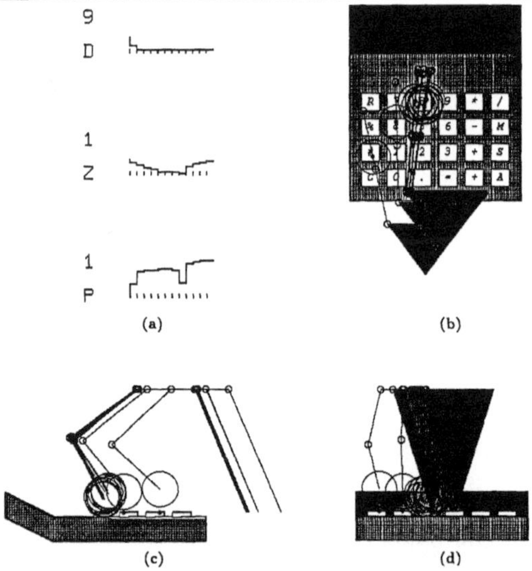

FIGURE 6. Pressing the 8 key on the calculator keypad. The initial position of the fingertip was the final position after pressing the # key. The dark triangle denotes the initial location of the hand. See text for details.

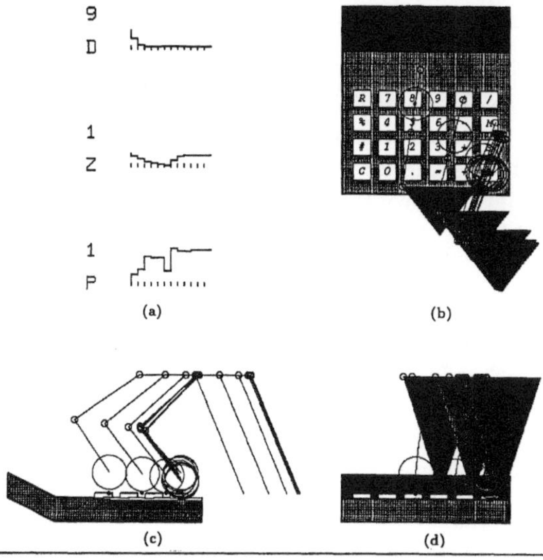

FIGURE 7. Pressing the A key on the calculator keypad. The initial position of the fingertip was the final position after pressing the 8 key. The dark triangle denotes the initial location of the hand. See text for details.

The strip charts show that the fingertip was lowered as it approached the target key until it made contact with the key and began to depress it. Then, once the key was fully depressed, the fingertip was raised to release the key. The sharp drop in the evaluation on the time step when the key was fully depressed (i.e., $z = 0$) was due to the switching of the evaluation criteria from one that rewarded downward movement to one that rewarded upward movement.

As evidenced by these figures, the controller learned to successfully execute the key-press operation for all three keys. Note that due to the dynamic nature of the hand model, the motion of the fingertip depended on the initial state of the hand, thus the figures presented here are merely representative samples. We tested the controller's performance with the hand starting in 10,000 random initial states and, in all the test runs, the target key was pressed successfully.

By comparison, *without shaping*, the controller did not learn the key-press task even after 500,000 training runs. Moreover, this was true even when the controller was trained to press just a single target key (the 8 key). These results support the views expressed at the beginning of the chapter regarding the difficulty of using reinforcement learning to generate complex behavior without the benefit of shaping. Without shaping, the probability of the complex behavior is so low that the target response never occurs, and therefore the reinforcer is never delivered.

Summary

In this chapter, we used a control problem to demonstrate how learning to solve complex problems can be facilitated by the shaping process of first learning to solve simpler subproblems. In this example, the shaping procedure proved indispensable for training the controller. Shaping is a natural way of introducing the trainer's knowledge into reinforcement-learning systems. In the key-pressing task, for example, the trainer's knowledge helped determine the series of approximations used in the shaping procedure and the inputs (sensations) to provide the controller. This example attests to the viability of reinforcement-learning approaches to learning complex control behavior when suitable shaping techniques are used.

PART FIVE: REINFORCEMENT LEARNING

Biobehavioral research indicates that reinforcement learning is the fundamental means by which experience changes responding. In reinforcement learning, change is dependent on a time-varying, global feedback signal that contains no information about the response other than some measure of its overall characteristics at that moment (see Chapter 1). Based on research at the behavioral level, the likelihood of a response recurring in an environment is a function of the prior consequences of the response in that environment (Donahoe & Palmer, 1994; Palmer, this volume). At the cellular level, synaptic efficacies change when pre- and postsynaptic neural coactivity is accompanied by a neuromodulator (Frey, this volume). Neuromodulators, such as dopamine, are liberated by neural systems that project *nonspecifically* from midbrain nuclei to cortical and subcortical regions (see **Donahoe** and **Montague**, this section). Stimuli that strengthen (reinforce) behavior—such as food for a hungry animal—activate neurons in these midbrain nuclei, and cause the volumetric release of the neuromodulator. Ultimately, the synapses whose efficacies are most greatly modified are those between neurons that are most reliably coactive when the neuromodulator occurs. And, these are the synapses along pathways mediating the environment-behavior relation preceding the reinforcer. On any one occasion, the efficacies of many irrelevant synapses may be changed. However, over time, the *cumulative* effect of the reinforcer is to modify the relevant synapses. Through this process, a nonspecific reinforcing system produces specific effects.

The emergence of specific effects as a cumulative product of global processes appears to be a general selection strategy exploited by evolution. On the level of species, death comes to *all* organisms, but—over time—those *particular* genotypes having greater reproductive fitness tend to predominate. On the level of the individual organism, a reinforcing stimulus affects *all* immediately prior environment-behavior relations, but—over time—those *particular* relations that most reliably precede the reinforcer predominate. On the cellular level, a nonspecific neuromodulator modifies synaptic efficacies between *all* concurrently active pre- and postsynaptic neurons, but—over time—those *particular* synapses that are most reliably activated in conjunction with the neuromodulator are most affected. Finally, on the intracellular level, the proteins synthesized in the nucleus (due to second messengers initiated by the conjunction of neurotransmitter and neuromodulator) migrate diffusely down the dendrite where they affect *all* previously stimulated receptors, but—over time—those *particular* receptors most reliably stimulated are most altered

(Frey & Morris, in press). Moment-to-moment processes at these differing levels appear chaotic, but their cumulative effects are orderly.

Two prerequisites for reinforcement have been identified: temporal contiguity and discrepancy (Palmer, this volume; see also Chapter 1). The role of temporal contiguity between the critical events has been documented on the behavioral (e.g., Gormezano & Kehoe, 1981), neural (e.g., Stein & Belluzzi, 1989), and cellular levels (e.g., Magee & Johnston, 1997; Markham, Lübke, Frotscher, & Sakmann, 1997). The role of a reinforcer-instigated discrepancy was first identified at the behavioral level (e.g., Kamin, 1968, 1969; Rescorla & Wagner, 1972). Now, findings reported in **Schultz**'s chapter provide insights into the processes associated with discrepancy at the neural level. Information at the cellular level is also becoming available (e.g., Abbott, Varela, Sen, & Nelson, 1997), but much remains to be uncovered concerning the neural systems and cellular processes involved in the biological computation of discrepancy.

Research at all levels of analysis points to selection by reinforcement as the basis of learning in living organisms, and workers in other fields have also appreciated its potential power; e.g., in applied mathematics (Werbos, 1974), artificial intelligence (Minsky, 1961), engineering (Widrow & Hoff, 1960, and philosophy (Dennett, 1981). The chapter by **Barto and Sutton** provides an abstract analysis of reinforcement learning in artificial intelligence that identifies its special strengths and characterizes some of the key conceptual issues. In reading this chapter, it is important to appreciate the nature of their distinction between *agent* and *environment* because the concept of agent includes only a part of what is included in the biobehavioral concept of organism. In artificial intelligence, the essence of the discrepancy requirement is captured by *temporal-difference models*, which are used by **Montague** and by **Moore and Choi** to simulate reinforcement learning in a variety of situations (see also chapters by Berthier and Gullapalli, this volume).

Among the major challenges to reinforcement learning as a general approach to complex behavior are these: (1) How can selection processes, which operate on a moment-to-moment basis, account for learning when environmentally mediated reinforcers are often delayed beyond the limits of the contiguity requirement? (2) How can reinforcers select behavior that entails complex response topographies and extended sequences of movements? Efforts to address the temporal challenges are described in chapters by **Donahoe, Barto and Sutton**, and **Moore and Choi** (see also Berthier, this volume). The complexity challenge is addressed in **Clouse**'s chapter through the implementation of *shaping* (see also Gullapalli, this volume).

Neural-Networks Models of Cognition
J.W. Donahoe and V. Packard Dorsel (Editors)

CHAPTER 17

ADAPTIVE DOPAMINERGIC NEURONS REPORT THE APPETITIVE VALUE OF ENVIRONMENTAL STIMULI

Wolfram Schultz
Institute of Physiology
University of Fribourg, Switzerland

ABSTRACT

This chapter describes how midbrain dopaminergic neurons in behaving monkeys detect and process environmental stimuli having appetitive value for the behavior of the subject. The majority of dopaminergic neurons show rather uniform, brief activations following presentation of appetitive stimuli almost exclusively. These stimuli include both primary rewards and conditioned stimuli predicting such rewards. Dopaminergic neurons are not activated by aversive or innocuous stimuli, or show quantitatively smaller activations following neutral and aversive stimuli that are physically very similar to and that occur in close temporal and contextual proximity to the appetitive stimuli. Stimuli that activate dopaminergic neurons must occur unpredictably. Primary rewards are only effective when they are not preceded by a conditioned, reward-predicting stimulus. Otherwise, the response is transferred to the earliest reward-predicting stimulus and no longer occurs to the primary reward. This transfer is reminiscent of Pavlovian stimulus substitution, suggesting that single dopaminergic neurons react to the most salient reward-related stimulus in a manner that parallels the behavior of the subject. When a predicted reward fails to occur, dopaminergic neurons are depressed in their activity at exactly the time when the reward would have occurred. Taken together, these results suggest that dopaminergic neurons code a deviation or error in the prediction of reward in close correspondence with the Rescorla-Wagner model of animal learning, according to which the gain in associative strength depends on the degree of stimulus unpredictability. Further, the transfer of responding from the primary reward to the conditioned, reward-predicting stimulus demonstrates the adaptive capacities of dopaminergic neurons: They are capable of responding to a wide variety of intrinsically neutral environmental stimuli associated with rewards. The message of dopaminergic neurons is broadcast as a global reinforcement signal to the large majority of neurons in the striatum (caudate nucleus and putamen) and to many neurons in the frontal cortex, where it is able to influence neuronal and synaptic activity occurring around the time of the dopaminergic signal. The responses of dopaminergic neurons and the associated architecture strongly resemble the adaptive-critic module of temporal-difference models of reinforcement learning developed on the basis of behavioral theories. These models, after having learned the predicted outcome of behavior, modify synaptic strength of a target actor according to the prediction. They have been used for constructing a variety of biologically plausible artificial networks that efficiently learn even complex tasks. These arguments suggest that the dopaminergic message can be effectively used as global reinforcement signal for adapting behavior according to the value of environmental stimuli.

Introduction

Two decades of psychopharmacological work have firmly established the crucial involvement of ascending midbrain dopaminergic systems in prime motivational functions. Experiments have studied dopaminergic functions by selective lesions and by administration of direct and indirect dopamine agonists and antagonists while assessing concurrent changes in various aspects of behavioral reactivity—including incentive learning, electrical self-stimulation, and drug self-administration (Wise, Spindler, de Wit, & Gerber, 1978; Beninger & Hahn, 1983; Fibiger & Phillips, 1986; Robbins & Everitt, 1992; Wise & Hoffman, 1992; Robinson & Berridge, 1993). Based on this extensive body of evidence, electrophysiological studies in behaving animals were undertaken to explore the detailed mechanisms of the motivational role of dopaminergic systems. These studies revealed that dopaminergic neurons were activated by specific environmental stimuli eliciting overt behavioral reactions (Miller, Sanghera, & German, 1981; Steinfels, Heym, Strecker, & Jacobs, 1983; Schultz, 1986). More recent studies have demonstrated that dopaminergic neurons are rather specifically activated by rewarding components of environmental stimuli occurring in specific situations (Romo & Schultz, 1990; Ljungberg, Apicella, & Schultz, 1992; Mirenowicz & Schultz, 1994). In particular, it appeared that two stimulus characteristics were necessary and sufficient for the responsiveness of dopaminergic neurons to external stimuli. First, the stimulus should be a primary reward, a conditioned stimulus predicting such a reward, or a new, potentially rewarding stimulus. Second, the stimulus itself should not be predicted by another phasic stimulus. This chapter will discuss the possible function of these responses by assessing the behavioral role of stimuli that are effective in driving dopaminergic neurons, and by exploring how the dopaminergic message sent to neuronal circuits in the striatum and frontal cortex could act to control the subject's behavior.

Values, Rewards and Reinforcers

Certain stimuli in the environment possess special value for the welfare of the individual subject by determining its survival and the survival of the species. Arbitrary, intrinsically neutral stimuli can acquire conditioned value by being associated with such stimuli. Stimuli having value serve three basic behavioral functions. Through their attention-grabbing or alerting function, they interrupt ongoing behavior, elicit immediate behavioral reactions and change the priorities of behavioral action. Through their reinforcing function, they increase behavioral reactivity to repeated presentations of these stimuli. Through their emotional function, they set up internal states of subjective feeling. The character of such values can be appetitive or aversive. Appetitive stimuli give rise to approach behavior following orienting reactions, they increase approach behavior with repeated stimulus presentation and thereby enhance the impact of the stimulus (rewarding or positive-reinforcing value),

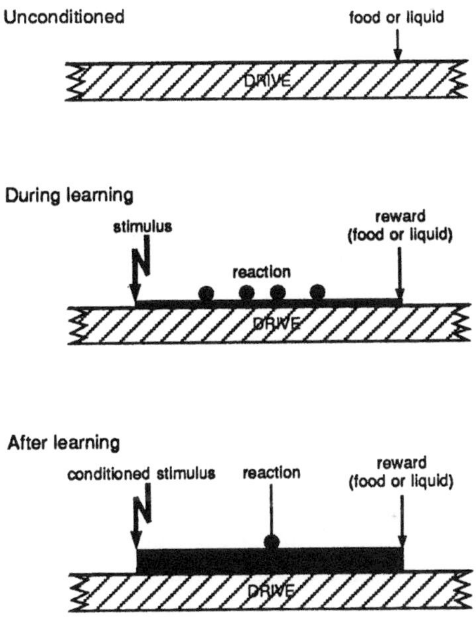

FIGURE 1. Processing of appetitive stimuli during learning. Top: Independent of learning, the occurrence of a primary food or liquid object will, often on the basis of a corresponding hunger or thirst drive state, induce approach behavior by the organism. Middle: During learning, the repeated pairing of an arbitrary, intrinsically neutral stimulus with a primary reward induces the organism to begin reacting to the stimulus. Bottom: Following sufficient stimulus-reward pairings, the intrinsically neutral stimulus becomes a conditioned, reward-predicting stimulus. It induces an internal motivational state by evoking an expectation of the reward, often on the basis of an underlying drive, and elicits the behavioral reaction. The scheme applies to both classical conditioning, where reward delivery automatically follows the conditioned stimulus, and instrumental (operant) conditioning, where reward delivery requires the active participation of the organism following the conditioned stimulus.

and they induce internal states of pleasure and hedonia. Aversive stimuli give rise to withdrawal behavior following orienting reactions, increase withdrawal behavior with repeated stimulus presentation, and thereby serve to reduce or escape the impact of the stimulus (punishing value) and induce internal states of fear and anger.

These processes can be illustrated in a basic paradigm of motivated reactions, which will be limited to appetitive reactions for reasons of application to dopaminergic neurons. The presentation of a food or liquid elicits approach behavior (**Figure 1**, top). According to different theories, the efficacy of these stimuli is determined to varying degrees by an internal hunger or thirst drive based on a corresponding homeostatic challenge (e.g., reduced blood glucose or increased blood electrolyte concentrations in the hypothalamus). Repeated pairing of such a "primary" reward with an arbitrary, intrinsically neutral stimulus will result in very similar and increasingly frequent approach behavior to this arbitrary stimulus (**Figure 1**, middle) until this now "conditioned" stimulus reliably elicits the behavioral response (**Figure 1**, bottom). It has been suggested that a central representation of the conditioned stimulus is associated with a central representation of the reward, such that the occurrence of the conditioned stimulus evokes a representation of the primary reward and predicts its occurrence (Dickinson, 1980). On the possible basis of a corresponding drive, the conditioned incentive stimulus would elicit an expectation of the primary reward and set an internal motivational state leading to the behavioral reaction (Bindra, 1968). It is important to note that the conditioned stimulus appears to elicit behavioral reactions that are very similar to those elicited by the primary reward itself, suggesting that the conditioned stimulus becomes a substitute for the primary reward in many respects (Dickinson, 1980). Important components of the behavioral response are transferred from the primary reward to the conditioned, reward-predicting stimulus. This stimulus substitution occurs in all forms of appetitive and aversive learning and constitutes a basic Pavlovian process.

These concepts explain why primary rewards and conditioned, reward-predicting stimuli may elicit very similar approach behavior. As primary rewards are usually referred to as unconditioned stimuli, the following questions arise: (1) To what extent are these primary rewards really unconditioned? (2) What would be gained by allowing arbitrary stimuli to become conditioned stimuli? Although food objects are usually referred to as primary rewards, every visitor to foreign countries immediately understands that the appetitive value of food objects is not necessarily intrinsically determined or innate, leaving the conclusion that so-called "unconditioned" primary rewards are also affected by experience. What, then, is unconditioned with primary rewards? It might be the taste experienced when the object activates the gustatory receptors, but that again may very well be conditioned. It appears that the final variable determining the primary, unconditioned biological value of nutrient objects may be the vegetative reactions following its ingestion, which may require up to several minutes. Such vegetative reactions are determined by innate constraints and constitute hard-wired biological constants serving as an internal reference determining the ultimate value of external objects. If approach behavior in early life is initially determined by instincts, then the majority of primary rewards become conditioned during life experience, thereby acquiring all the behavioral characteristics of unconditioned stimuli. Thus, the value of nutrient objects is determined by the vegetative reactions following its ingestion, and the value of environmental stimuli associated with these objects is expressed by the behavioral reactions leading to their acquisition. The advantage of an organism's capacity for learning to react both to primary rewards and to arbitrary, conditioned stimuli lies in the enormous variety of stimuli that can be potentially processed by an individual. For obtaining objects of prime biological importance, the individual becomes capable of adapting to the variable number of stimuli that actually occur in an environment with limited resources (Friston, Tononi, Reeke, Sporns, & Edelman, 1994).

Given the central behavioral importance of primary rewards and conditioned, reward-predicting stimuli, one might assume that the brain processes reward information by employing robust neuronal systems with high internal redundancy that continue to maintain functioning even after partial destruction. However, the largely conditioned nature of reward-related stimuli also requires a highly adaptive system with a minimum of hard-wired connections, which is nevertheless internally redundant. The physically inhomogeneous nature of rewards does not allow them to affect the brain through specific peripheral receptor systems tuned to a limited range of physical stimuli. Rather, the brain needs to employ neuronal mechanisms with a high degree of adaptive capacity for extracting the reward information from inhomogeneous and inconstant stimuli.

Appetitive Value of Stimuli Activating Dopaminergic Neurons

Primary rewards

The majority of dopaminergic neurons (75-85%) are activated by phasically occurring primary appetitive stimuli, such as solid or liquid food, whereas the remaining neurons are not influenced by any stimuli that we have tested. Responses are very phasic, occurring with latencies of 50-110 ms and lasting less than 300 ms. Neural responses are observed when the animal touches a small morsel of hidden food during exploratory movements in the absence of other phasic stimuli or receives a drop of liquid at the mouth outside of any behavioral task or while learning a task (Romo & Schultz, 1990; Ljungberg et al, 1992; Mirenowicz & Schultz, 1994). The responses do not discriminate between different rewards, but do discriminate between reward and nonreward. Neurons activated by primary rewards do not respond or are occasionally depressed in their activity when non-food objects are touched—even when their shape is similar to that of food objects—or when a fluid valve is audibly operated without actually delivering liquid. Only very few dopaminergic neurons are activated by innocuous primary aversive stimuli, such as an air puff to the hand or a drop of hypertonic saline to the mouth. However, these responses are not strong enough to result in an average population response (Mirenowicz & Schultz, 1996).

Conditioned stimuli

The majority of dopaminergic neurons (50-75%) are activated by conditioned visual or auditory stimuli that have become valid predictors of reward (Schultz, 1986; Schultz & Romo, 1990; Ljungberg et al, 1992; Mirenowicz & Schultz, 1994). Responses do not discriminate between visual and auditory stimuli; the same neurons respond to both modalities. Conditioned stimuli are generally slightly less effective than primary rewards, both in terms of response magnitude in individual neurons and in terms of the fraction of neurons activated.

In most situations, dopaminergic responses show an all-or-none discrimination between appetitive and neutral or aversive stimuli (Ljungberg et al, 1992; Mirenowicz & Schultz, 1996). Only very few neurons are also activated by conditioned aversive light or sound stimuli (Mirenowicz & Schultz, 1996). In situations in which neutral or aversive stimuli are presented in close temporal proximity and random alternation with physically very similar appetitive stimuli, the discriminative capacity of dopaminergic neurons becomes quantitative and the all-or-none discrimination is lost. For example, a small box that opens rapidly in front of the animal does not by itself activate dopaminergic neurons. However, responses occur to every opening if the box on some trials contains a visible morsel of food (Ljungberg et al, 1992) or when it opens without food in random alternation with an identical, food-containing box next to it (Schultz & Romo, 1990). In both experiments, responses to opening of the baited boxes are stronger than to the empty boxes. Animals perform an indiscriminate ocular orienting response to each opening, but approach only the baited box with the hand.

Novel stimuli

Dopaminergic neurons are activated by novel stimuli as long as they elicit behavioral orienting reactions (e.g., ocular saccades). Neuronal responses as well as orienting reactions subside after several tens of stimulus repetitions (Ljungberg et al, 1992), although particularly salient stimuli may remain somewhat effective for over 1,000 trials (Hollerman & Schultz, unpublished).

Conclusion

Given their responses to a limited range of stimuli—primary rewards, conditioned, reward-predicting stimuli, and novel, potentially rewarding stimu-

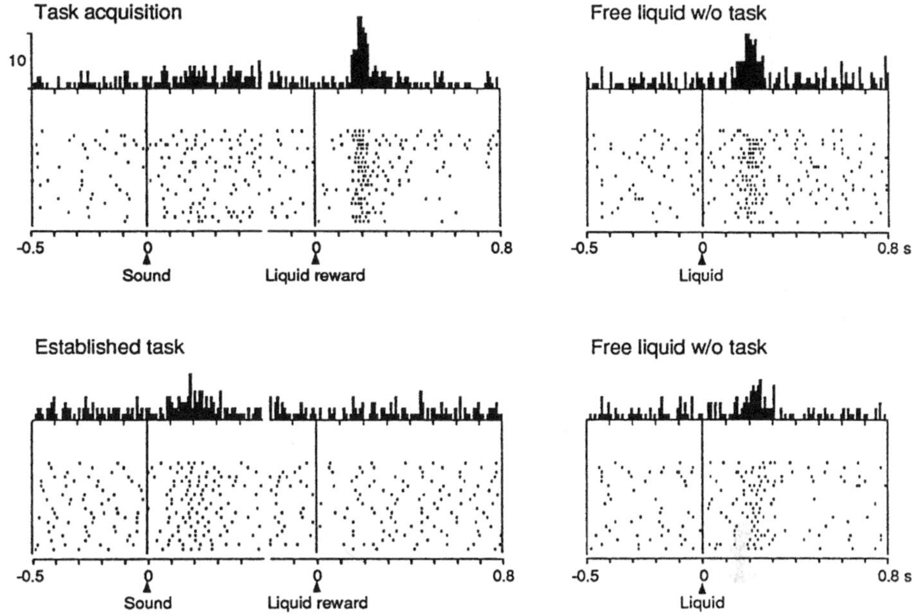

FIGURE 2. Responses of dopaminergic neurons at different phases of learning a reaction-time task. Top: Response to a largely unpredicted liquid reward during initial sound-reward pairings, but with very little response to the sound stimulus (left). The same neuron also responds to the liquid delivered unpredictably outside of any task (right). Bottom: Several days later during established task performance, responding of another dopaminergic neuron to the now-conditioned sound but the absence of responding to the liquid reward (left). It is as if the response had been transferred from the primary reward to the conditioned stimulus predicting it. The response of the same neuron to the liquid delivered unpredictably outside of any task is maintained, suggesting unchanged efficacy of the liquid (right). Note, different time bases are used because of the sound-reward intervals. Histograms are composed of neuronal impulses shown as dots below. Each dot denotes the time of a neuronal impulse, and its distance from the stimulus indicates real-time intervals. Each line of dots shows one trial, the sequence of trials being from top to bottom. Reproduced with permission from Mirenowicz and Schultz (1994).

li—dopaminergic neurons report the motivational value of environmental stimuli. As far as nutrient objects are concerned, the response is not related to obtaining or ingesting these objects but to the presentation of the associated stimuli. The dopaminergic response is evoked by the mere presentation of the stimulus and does not depend on a specific behavioral response being rewarded. Dopaminergic neurons respond almost exclusively to stimuli of appetitive value, but they do not appear to discriminate in a simple manner between different appetitive stimuli. They signal the presence of stimuli of appetitive value without indicating further details of the stimuli. Thus, dopaminergic neurons report a basic motivational value assigned to environmental stimuli, as opposed to other brain systems that might be more concerned with the explicit properties of individual appetitive stimuli. The response occurs in the majority of dopaminergic neurons and appears to be rather homogeneous—both in terms of similarity of responses to different stimuli and different degrees of conditioning, and in terms of similarity of responses by different neurons. This population response, together with the profuse innervation of target structures, constitutes a high degree of internal redundancy which allows the system to maintain its function in reporting the appetitive value of environmental stimuli despite partial destruction that may occur with aging or following inadvertent exposure to neurotoxic substances.

Adaptive Properties of Dopaminergic Neurons

The response of dopaminergic neurons is transferred systematically from primary rewards to conditioned, reward-predicting stimuli, and the response to the conditioned stimulus induces a loss of responding to the primary reward (**Figure 2**). The response to the conditioned stimulus itself can be further transferred to an earlier predictive stimulus with loss of response to the now-predicted conditioned stimulus. Thus, the dopaminergic response systematically occurs to the earliest reward-predicting stimulus, whereas subsequent predicted stimuli lose their efficacy for activating dopaminergic neurons (**Figure 3**; Schultz, Apicella, & Ljungberg, 1993). The responses to the conditioned stimuli become progressively smaller as the stimulus occurs farther away from the primary reward, the maximum interval between effective conditioned stimulus and primary reward being about 10-15 s. The response transfer occurs both during initial learning and in well-learned behavioral situations. During learning of an instrumental leverpressing task, the response is transferred from the liquid delivered at the animal's mouth to reward-predicting conditioned visual or auditory stimuli (Ljungberg et al, 1992; Mirenowicz & Schultz, 1994). In well-learned behavioral situations, the response transfers from a morsel of food encountered during self-initiated arm movements to conditioned visual or auditory stimuli (Romo & Schultz, 1990), and from a free drop of liquid occurring at an unpredicted moment to a conditioned auditory stimulus predicting such liquid (Mirenowicz & Schultz, 1994). Responses to both the earliest conditioned stimulus and the subsequent stimuli may coexist for a limited period during learning, whereas the transfer is immediate when it occurs between established behavioral situations. This suggests that response transfer is a basic neurobiological phenomenon with dopaminergic responses that is not the result of a general reduction of sensitivity to primary rewards and is independent of the learning context itself.

The transfer of the response of dopaminergic neurons from primary rewards to conditioned, reward-predicting stimuli strongly resembles Pavlovian stimulus substitution of behavioral reactions. The neuronal response transfer indicates that the conditioned stimulus becomes a substitute for the primary reward in eliciting a neuronal response. This reveals three important characteristics of the response of dopaminergic neurons. (1) The principal effective stimulus for dopaminergic neurons is the earliest reward predictor. In parallel with behavioral conditioning, dopaminergic neurons react to exactly the same stimulus as does the organism's behavior. Dopaminergic neurons thus track the central stimulus for approach behavior. However, it should be noted that the dopaminergic response to conditioned stimuli can be independent of the organism's

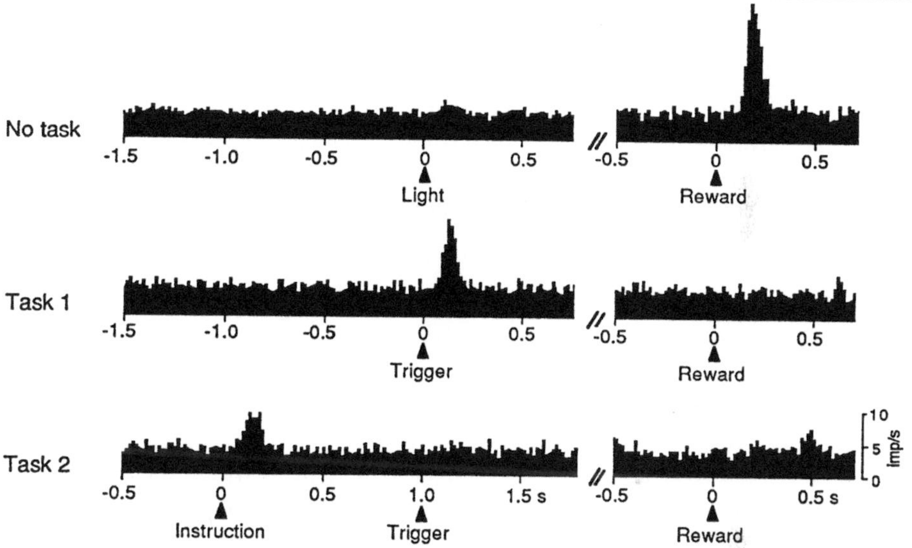

FIGURE 3. Responses of dopaminergic neurons to unpredicted primary reward showing response transfer to progressively earlier reward-predicting stimuli. All displays are population histograms obtained by averaging normalized peri-event time histograms for all dopaminergic neurons recorded in the behavioral situations indicated, independent of the presence or absence of a response. Top: In the absence of any behavioral task, no population response was observed in 44 neurons tested with a small light (data from Ljungberg et al, 1992), but there was an average response in 35 neurons to a drop of liquid delivered at a spout in front of the animal's mouth (Mirenowicz & Schultz, 1994). Middle: Response to a reward-predicting (trigger) stimulus in a spatial-choice reaching task, but absence of response to reward delivered during established task performance (23 neurons; Schultz et al, 1993). Bottom: Response to an instruction cue preceding by a fixed time of 1 s the reward-predicting trigger stimulus in an instructed spatial-reaching task (19 neurons; Schultz et al, 1993). Time bases are different because of varying intervals between conditioned stimulus and reward.

behavior; it also occurs when the animal erroneously fails to react. (2) Somewhat surprisingly, the neuronal responses engage the large majority of neurons in a functionally rather homogeneous system. This indicates a rather holistic influence of the reward signal on a wide variety of neuronal activities underlying behavioral output. (3) In parallel with behavioral conditioning, the response of dopaminergic neurons changes from one stimulus to another, suggesting that the neuronal response undergoes conditioning in a way similar to the rewarded behavioral response. This allows a reward-predicting function to be acquired by a large number of arbitrary stimuli to which the dopaminergic neurons then respond. This enormously increases the chance of the neurons' detecting the limited number of appetitive stimuli present in a given environment.

Relations to Associative Learning Theories

Learning theories

Contemporary learning theories characterize learning as acquisition of associative strength by a conditioned stimulus. The increment in associative strength during each learning episode in which the conditioned stimulus is paired with a reward is determined by the equation

$$\Delta V = \alpha \beta \; (\lambda - V). \qquad (1)$$

The gain in associative strength ΔV is determined by $(\lambda\text{-}V)$, where V is the current associative strength of the conditioned stimulus on that episode and λ is the maximum associative strength that could be sustained by the reward. The $\alpha > 0$ and $\beta > 0$ parameters represent the conditioned and unconditioned stimulus salience, respectively (Rescorla & Wagner, 1972; Mackintosh, 1975; Pearce & Hall, 1980). Thus, $(\lambda\text{-}V)$ reflects the extent to which the conditioned stimulus has already been established as a predictor of reward or, in other words, the extent to which the reward is currently predicted by the conditioned stimulus. When $V = \lambda$, the associative strength of the conditioned stimulus is sufficient to fully predict the occurrence of the reward and no further learning will occur (i.e., ΔV becomes zero). By contrast, when $\lambda > V$, the associative strength of the stimulus does not fully predict the reward and the ΔV term is positive and leads to further increments in associative strength. In this sense, the $(\lambda\text{-}V)$ term represents the extent to which the animal has failed as yet to learn the full predictive relation between the conditioned stimulus and the reward. For this reason, $(\lambda\text{-}V)$ could be said to represent an error in the prediction of reward. Learning algorithms that serve to minimize the $(\lambda\text{-}V)$ term can be considered as error correcting. Recent work has suggested that this learning rule is formally equivalent to the delta rule (Widrow & Hoff, 1960) of artificial neuronal networks (Sutton & Barto, 1981), thus allowing single neuronal elements, at least theoretically, to implement this error-driven learning rule.

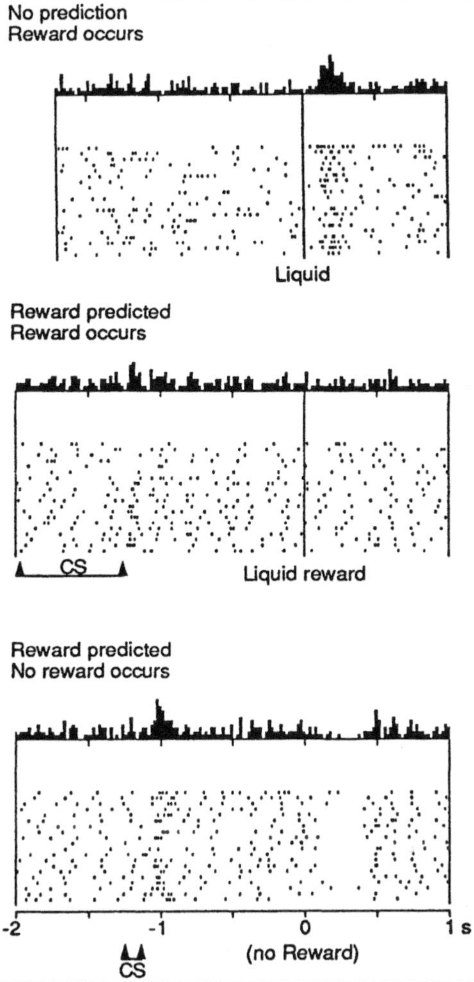

FIGURE 4. Dopaminergic neurons code an error in reward prediction. Top: A drop of liquid reward occurs in the absence of prediction (an error in the prediction of reward). The dopaminergic neuron was activated by this unpredicted occurrence of the liquid. Middle: A conditioned stimulus predicts a reward, and the reward occurs according to the prediction (no error in the prediction of reward). The dopaminergic neuron is activated by the reward-predicting stimulus (response scattered in the raster to the left) but fails to be activated by the predicted reward (right). Bottom: A conditioned stimulus predicts a reward, but the reward fails to occur because of lack of response by the animal. The activity of the dopaminergic neuron is depressed precisely at the time when the reward would have occurred. Note the depression occurring >1 s after the conditioned stimulus without any intervening stimuli (an internal process of reward expectation).

Apart from their role in learning, rewards have a separate, equally crucial function of maintaining established behavior. An established behavioral response will undergo extinction in the absence of the continued occurrence of rewards. However, whereas the efficacy of rewards during learning depends on their unpredictable occurrence, their occurrence is fully predictable during established behavioral performance. This distinction might help to delineate the potential involvement of neurons whose responses to reward show different characteristics in the two phases of learned behavior.

Do dopaminergic neurons report an error in the prediction of reward?

A central tenet of associative learning theories is the dependence of learning on the unpredictability of reward (Rescorla & Wagner, 1972; Mackintosh, 1975; Pearce & Hall, 1980; Donahoe, Crowley, Millard, & Stickney, 1982). The responses of dopaminergic neurons to reward-related stimuli occur under very similar conditions. This suggests that dopaminergic responses follow the (λ-V) term and signal an error in the prediction of reward. The occurrence of a reward without being predicted could be considered an error in reward prediction. Accordingly, dopaminergic neurons are activated by primary rewards only when the time of reward delivery is not predicted by any conditioned phasic stimulus (**Figure 4**, top). By contrast, the error signal should be small when a predicted reward actually occurs. Dopaminergic neurons are activated by the reward-predicting stimulus, but little or no activations are recorded in response to the predicted reward (**Figure 4**, middle). The transfer of neuronal responding from primary rewards to conditioned, reward-predicting stimuli apparently abolishes the primary reward response. Thus, neurons responding to reward encountered during spontaneous movements in the absence of phasic reward-predicting stimuli lose this response when a conditioned stimulus elicits the movement (Romo & Schultz, 1990). Neuronal responses activated by drops of liquid outside of any task disappear when the same reward is delivered at regular intervals of 4-5 s, each drop of liquid serving as a predictor for the next drop (Ljungberg et al, 1992). In learning situations, neurons respond to reward before the task is acquired but lose the response when a stimulus has become a valid reward predictor (**Figure 2**; Ljungberg et al, 1992; Mirenowicz & Schultz, 1994). Accordingly, when a more complicated task, such as a spatial delayed-response task, is acquired through a series of intermediate tasks, the learning phase of each intermediate task is accompanied by dopaminergic responses that disappear when the learning curve of each intermediate task reaches its asymptote (Schultz et al, 1993).

An inverse error in reward prediction does occur when a predicted reward fails to be delivered. Dopaminergic neurons are activated by the reward-predicting stimulus but are depressed in their activity at exactly the time at which the reward would have occurred (**Figure 4**, bottom). This was observed in all

22 neurons tested in various behavioral situations in which reward was withheld deliberately by the experimenter or failed to occur following an error of task performance by the animal (Ljungberg, Apicella, & Schultz, 1991; Schultz et al, 1993; Hollerman & Schultz, unpublished). Interestingly, the sharply delimited neuronal depression did not occur to an immediately preceding phasic stimulus. Rather, the last external stimulus occurred when the animal had touched the wrong response lever more than 1 s earlier, suggesting that dopaminergic neurons have access to an internal clock monitoring the time of expected reward delivery.

Reward responses are not conditioned to the general learning context; they occur whenever the reward occurs unpredictably in time. A reward occurring 500 ms earlier than predicted elicits a dopaminergic response for a few transitory trials, after which the reward becomes ineffective again (Hollerman & Schultz, unpublished). By contrast, the reward prediction provided by the general behavioral context alone does not abolish a response, as shown by the fact that responses to unpredicted rewards persist for several months of experimentation in the laboratory. This occurs even though the situation is highly associated with reward. Thus, the response to primary reward is only abolished when a conditioned stimulus predicts the time of the reward.

Responses to a conditioned stimulus also depend on whether that stimulus is unpredicted itself, and are reduced when the stimulus is signaled by a preceding cue (**Figure 3**; Schultz et al, 1993). Extensive overtraining attenuates responses to conditioned stimuli (Ljungberg et al, 1992), probably because the stimuli become predicted by the events in the preceding trial during highly stereotyped and automated task performance.

The responses of dopaminergic neurons show two prominent characteristics that have so far not been found together in other brain structures: temporal reward unpredictability, together with a rather uniform response of nearly the whole neuronal population. Reward responses and other forms of reward-related activity are found in several brain structures—such as the striatum, amygdala, orbitofrontal cortex and anterior cingulate cortex—but these responses usually occur in well-established behavioral tasks where reward is fully predicted by the conditioned stimuli (Niki & Watanabe, 1979; Thorpe, Rolls, & Maddison, 1983; Nishijo, Ono, & Nishino, 1988; Apicella, Ljungberg, Scarnati, & Schultz, 1991; Schultz, Apicella, Scarnati, & Ljungberg, 1992). In addition, these structures show several forms of specific task-related activity of which the reward responses constitute only a fraction. Also, this reward signal is not sent in a global, divergent fashion to a large number of postsynaptic neurons. These characteristics suggest that a function of such signals is to maintain established behavior rather than bring about learning. Thus, reward information may be treated differently by the brain in certain conditions, may have functions that vary according to the learning or established phase of behavior, and may occur in different neuronal systems with architectures corresponding to the possible function of the reward signal.

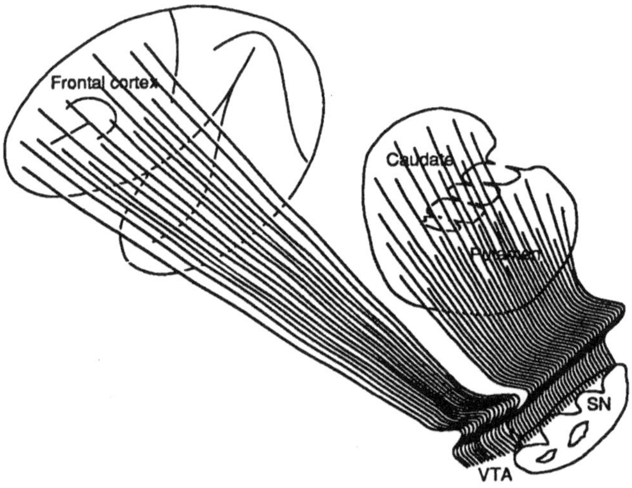

FIGURE 5. Dopaminergic message sent to striatum and cortex. The rather homogeneous population response of dopaminergic impulses to appetitive stimuli and its progression from the substantia nigra to the postsynaptic structures can be schematically viewed as a wave of synchronous, parallel activity. Our results show that the responses to appetitive stimuli of dopaminergic neurons in group A9 of the substantia nigra projecting to the striatum are indistinguishable from the responses in the dorsomedial substantia nigra and adjoining group A10 projecting to the frontal cortex.

Postsynaptic Influence of the Dopaminergic Message

The characteristic response of dopaminergic neurons, combined with the particular anatomical organization, provides clues to the way the dopaminergic signal may be used by the brain. The rather homogeneous response of dopaminergic neurons and the divergent nature of dopaminergic terminal arborizations suggest that the dopaminergic message is broadcast as a parallel population signal to postsynaptic structures (**Figure 5**). Dopaminergic neurons innervate virtually every neuron in the dorsal striatum (caudate nucleus and putamen) and the ventral striatum (the nucleus accumbens in rodents), and a considerable fraction of neurons in the frontal cortex. The striatum is engaged in closed loops involving the frontal cortex and, in addition, receives input from postcentral sensory and association cortex and from limbic cortical and subcortical structures (**Figure 6**). Neurons in the striatum and frontal cortex display very specific, behavior-related activities coding stimulus meaning, short-term

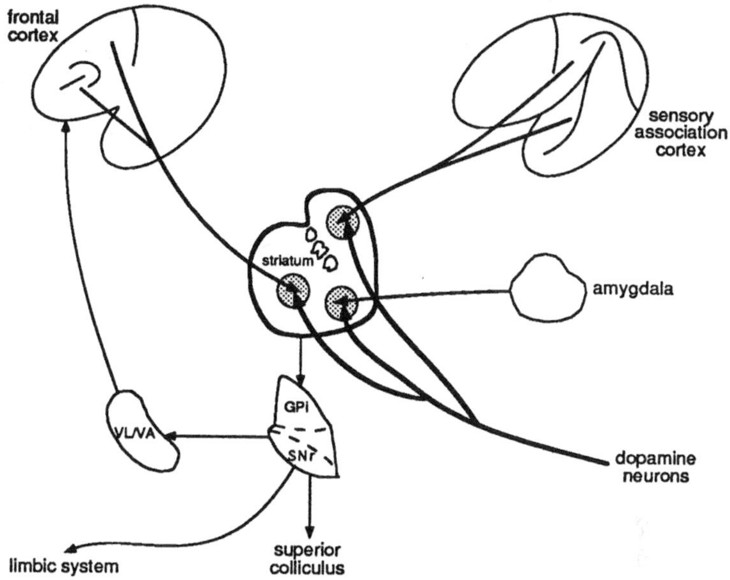

FIGURE 6. Influences of the dopaminergic system on basal ganglia circuitry. The highly simplified scheme shows the three major inputs to the striatum from (1) the frontal lobe, (2) the postcentral primary sensory and association cortex, and (3) cortical and subcortical limbic structures. The message from the basal ganglia is sent back to the frontal lobe. Dopaminergic neurons act on these circuits in the striatum and thus are in a position to influence important forebrain centers that organize and control behavioral output. The striatum denotes both the dorsal caudate and putamen and the ventral striatum, including nucleus accumbens. For reasons of simplicity, the dopaminergic projection to the frontal lobe has been omitted (upper left: dorsolateral, orbital and medial prefrontal cortex, medial and lateral premotor cortex, primary motor cortex).

memory, preparation of action, expectation of reward and execution of movements, thereby participating in transferring internal intentional states into overt actions for obtaining rewarding goals (Schultz, 1995; Schultz, Romo, Ljungberg, Mirenowicz, Hollerman, & Dickinson, 1995). Dopaminergic varicosities in striatum and cortex form synapses on dendritic spines which are equally contacted by cortical afferents (Freund, Powell, & Smith, 1984; Goldman-Rakic, Leranth, Williams, Mons, & Geffard, 1989; Smith, Bennett, Bolam, Parent, & Sadikot, 1993). An estimated 10,000 cortical terminals and 1,000 dopaminergic varicosities contact each striatal neuron (Doucet, Descarries, & Garcia, 1986; Wilson, 1995; Groves, Garcia-Munoz, Linder, Manley, Martone, & Young, 1995), thus allowing the dopaminergic inputs to play a crucial role in determining the efficacy of cortical influences on striatal neurons. Although inputs from heterogeneous cortical areas may remain largely segregated in the striatum (Alexander, DeLong, & Strick, 1986), the estimated number of 10,000 cortical inputs to a given striatal neuron indicates some degree of integration of information. It is quite possible that afferents from different but functionally related cortical areas converge on single striatal neurons, as suggested by convergent projections from somatotopically related areas of primary somatosensory and motor cortex into discrete striatal regions (Flaherty & Graybiel, 1993).

As pointed out by Flaherty and Graybiel (1994), the patterns of divergence and convergence are particularly suitable for allowing new associations to form. The wave of parallel dopaminergic population activity arriving at postsynaptic structures could lead to simultaneous release of dopamine from a large number of sites onto the majority of postsynaptic neurons. The observed phasic responses with high instantaneous frequencies seem particularly suited for efficiently releasing dopamine at synaptic sites, as higher impulse frequencies lead to disproportionally larger dopamine release following rapid saturation of dopamine reuptake (Gonon, 1988). Dopamine concentrations inside the synaptic cleft are at least 15-20 times higher than a few micrometers away (Kawagoe, Garris, Wiedemann, & Wightman, 1992), allowing a localized influence of dopamine release on specific parts of postsynaptic neurons. In this way, the nondifferential global population signal of dopaminergic neurons would be able to influence the control of behavioral action by widespread projections to structures involved in highly differentiated information processing.

A more formal assessment of this interaction may provide some indications on how the nondifferential dopaminergic activity could be applied as a teaching signal for specifically influencing processing in the striatum (**Figure 7**; see also Frey, this volume). In a reduced model, let A and B be two sets of inputs converging on a single striatal neuron I, each individual input contacting a dendritic spine of that neuron. The stems of the same spines are indiscriminately contacted by dopaminergic input X. Let us assume that inputs from dopaminergic neuron X are activated by a reward-related stimulus, and that only cortical input A is activated at the same time by some specific aspect of the same stimulus. Dopaminergic neuron X transmits the message that a reward-related stimulus has occurred without giving further details, whereas cortical input A carries detailed information about specific aspects of the same reward-related stimulus—such as modality, body side, color, texture, position, surrounding, or whether it is food, fluid or a conditioned sound or light—and may also code the details of an approach movement (different components of the same stimulus being coded by specific activity in different inputs A). Input B remains inactive as it codes different stimuli or different components of stimuli that are not now occurring. Through the simultaneity or near simultaneity of activity in A and X, the activity of neuron X could modify synaptic transmission between A and I but leave the transmission at the inactive B→I synapse unchanged. Thus, the message about a reward-related stimulus coming from dopaminergic neuron X specifically modifies the A→I neurotransmission. The key function of dopaminergic neuron X would be to signal the stimulus (primary reward or reward-predicting stimulus) that is particularly important for behavior, acting as a kind of gate for the highly structured activity occurring in cortico-striatal and limbic-striatal connections.

As an example, corticostriatal synaptic activity related to a particular sensory signal or movement during which reward was encountered would be modified when dopaminergic neurons were nearly simultaneously activated by the reward event. The fact that primary reward occurs after the sensory signals or movements that produce it rules out a strict simultaneity in real time between dopaminergic responses and activity in postsynaptic structures. Dopamine released after reward delivery needs to influence synaptic efficacies to

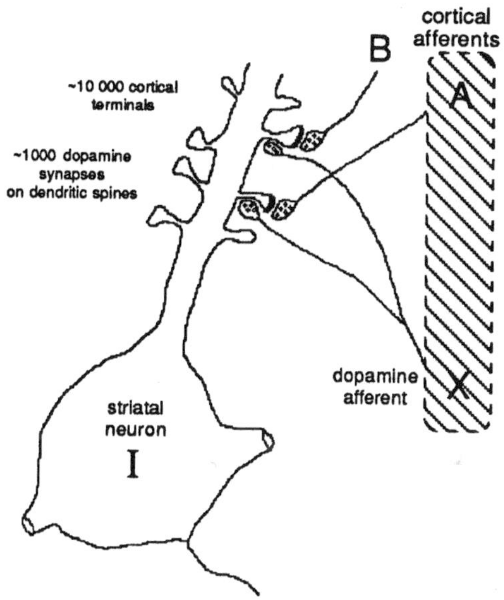

FIGURE 7. Synaptic arrangement of inputs from cortical and dopaminergic neurons to medium spiny striatal neurons. The dendritic spine is contacted by a cortical terminal and *en passant* by a dopaminergic axon. In the basic design of hypothetical dopamine-dependent heterosynaptic plasticity induced by a reward-related stimulus, cortical neurons A and B converge at the tip of different dendritic spines on a single striatal neuron I. The synaptic weights of these connections are modifiable. The modification occurs only when dopaminergic input X, coming indiscriminately to the stems of the same dendritic spines, is active at about the same time. In the present example, cortical input A, but not B, is active when dopaminergic neuron X is activated by a reward-related stimulus. This leads to a modification of the A→I transmission, but leaves the B→I transmission unaltered. Anatomical data from Freund et al (1984), drawing modified from Smith and Bolam (1990).

postsynaptic structures whose activity has recently peaked. This assumes some briefly enduring events, such as a hypothetical eligibility trace (Sutton & Barto, 1981; Houk, Adams, & Barto, 1995) that lasts for a short lapse of time in neuronal or synaptic activity of postsynaptic structures, upon which the dopaminergic signal could act to modify synaptic processing.

Neuronal-Network Models Using Dopamine-like Reinforcement Signals

The response of dopaminergic neurons to primary rewards provides a suitable basis for constructing neuronal-network models that use a biologically derived reinforcement signal for modifying synaptic weights implementing learning. Several characteristics appear to be particularly well suited for such reinforcement models. (1) Dopaminergic neurons respond to rewards. However, responses occur only when the reward appears unpredictably in time. Thus they signal an error in the prediction of reward in correspondence with the Rescorla-Wagner model of behavioral learning, and provide a biologically plausible means whereby rewards could be processed by single neuronal elements. The importance of temporal aspects of the predictability of stimuli is suggested by the fact that the temporal variation in reinforcer occurrence affects learning (Dickinson, Hall, & Mackintosh, 1976). (2) The acquisition of responsiveness to arbitrary stimuli demonstrates that dopaminergic neurons not only respond to primary rewards but that their responses are also influenced and modified by the same reward signal they are reporting. At the same time as they send a reward message to postsynaptic structures, they are themselves influenced by that message. (3) The way the dopaminergic message is sent to postsynaptic structures is very appropriate for influencing a large number of neurons irrespective of their specific function or current behavioral relationship. The majority of dopaminergic neurons respond rather homogeneously to physically different reward-related stimuli and produce a global reward message that is synchronously broadcast to a large number of neurons engaged in different activities underlying the current behavior. This influence would specifically affect those neurons or synapses that are selectively activated by the same environmental signal that also activated the dopaminergic neurons within a short time span. In this way, the global, nondifferential dopaminergic signal has a differential effect on active postsynaptic elements.

What is known about dopaminergic responses to reward-related stimuli has been utilized to varying degrees in neuronal models. (See Chapter 18.) A straightforward reinforcement system reports the correct execution of a desired behavioral response by homogeneously broadcasting an all-or-none reward signal independent of predictability to a large number of postsynaptic elements where it exclusively reinforces the weights of those synapses that are active around this time (Wickens & Kötter, 1995). Processing units in such model networks acquire properties very similar to those of biological single neurons in primate parietal association cortex (Mazzoni, Andersen, & Jordan, 1991). Networks based on the architecture of the basal ganglia are able to acquire conditional oculomotor responses (Arbib & Dominey, 1995). All of these models take account of the global nature of the reinforcement signal but do not exploit the particular response properties of dopaminergic neurons.

On the basis of the Rescorla-Wagner model, a class of very efficient reinforcement models termed temporal-difference models has been developed independently which, amazingly, reproduce virtually all of the salient characteristics of dopaminergic neurons outlined above (Sutton & Barto, 1981; cf. Donahoe & Palmer, 1989; see also Barto & Sutton, this volume). The adaptive-critic module of these networks initially reacts to unpredicted positive outcome (the critic mimicking responses to unpredicted primary reward) and, upon learning, transfers the response to a predicting stimulus (the adaptive component learning to respond to the conditioned, reward-predicting stimulus). By analogy to higher-order conditioned stimuli, signals that predict progressively earlier outcomes have proven to be particularly helpful in learning rather long sequences of behavior directed at remote outcomes. The general architecture of temporal-difference models remarkably resembles the architecture of the basal ganglia. Notable similarities are (1) the global influence of the critic on the actor elements, which resembles the nigrostriatal dopaminergic projection, and (2) the way in which the critic learns to react to the conditioned stimulus by being influenced by the reinforcement signal, possibly resembling the striatonigral connection.

In particular, by generating outcome-predicting signals, temporal-difference models have been able to learn a wide variety of different behavioral tasks, from balancing a pole on a wheel to playing backgammon and chess. (See Clouse, this volume.) As their strong correspondence with the activity of dopaminergic neurons has become understood, these models have been applied to the architecture and functions of the basal ganglia (Barto, 1995; Houk et al, 1995). A very similar model strongly resembling dopaminergic activities trains neuronal networks to perform ocular reactions on the basis of the acquired value of stimuli, the value-signaling module functioning like the adaptive critic of temporal-difference models (Friston et al, 1994). A related form of predictive learning has recently been used for modeling the foraging behavior of honeybees in unknown territory (Montague, Dayan, Person, & Sejnowski, 1995; see also Montague, this volume).

ACKNOWLEDGMENTS

The experiments in our laboratory were conducted and parts of the concepts were developed together with P. Apicella, J.R. Hollerman, T. Ljungberg, J. Mirenowicz and R. Romo. Several theoretical consequences were stimulated by discussions with A. Dickinson and R. Suri. The work was supported by the Swiss National Science Foundation, the Fyssen Foundation (Paris), the Fondation pour la Recherche Médicale (Paris), the United Parkinson Foundation (Chicago), the Roche Research Foundation (Basel), the National Institutes of Mental Health (Bethesda) and the British Council.

Neural-Networks Models of Cognition
J.W. Donahoe and V. Packard Dorsel (Editors)

CHAPTER 18

SELECTION NETWORKS: SIMULATION OF PLASTICITY THROUGH REINFORCEMENT LEARNING

John W. Donahoe
Department of Psychology
Neuroscience and Behavior Program
University of Massachusetts, Amherst

ABSTRACT

Evolution through natural selection has addressed the problem of modifying synapses throughout large networks of neurons by exploiting diffusely projecting neuromodulatory systems. When pre- and postsynaptic neurons are coactive, synaptic efficacies increase or decrease dependent upon whether the neuromodulator dopamine is simultaneously present or absent. Salient characteristics of this process can be simulated with selection networks, artificial neural networks whose architecture instantiates a processing system whose connection weights are modified by a scalar reinforcing signal. This arrangement resolves both the temporal paradox and the binding paradox, twin challenges to any attempt to interpret complex behavior by means of neural networks. Further, by exploiting an emergent property of selection networks—acquired reinforcement—critical aspects of imagining, thinking, and language acquisition can also be interpreted.

Introduction

This chapter has four primary goals: (1) to identify critical experimental findings that inform and constrain the search for a biobehaviorally plausible account of plasticity (learning), (2) to describe a general approach to simulating the mechanisms whereby behavior is acquired, (3) to describe a general approach to simulating the mechanisms whereby configurations of stimuli come to guide behavior, and (4) to expose some of the implications of this approach for complex behavior. A deep and comprehensive natural-science-based understanding of human behavior lies in the future, but its broad outlines are emerging in the present.

Biobehavioral constraints

The preceding chapters have identified a very substantial body of research that must be honored by any attempt to achieve a biobehaviorally faithful neural-network interpretation of complex behavior (i.e., cognition). Let me highlight a few of the central findings.

Temporal constraints. From experimental work at the behavioral level, we know that evolutionarily important events—such as food for a food-deprived organism—change the way in which the environment guides behavior only when the environment and behavior occur in close temporal proximity to such events (see Palmer, this volume). This conclusion is complemented by findings at the cellular level in which synaptic efficacies change only when certain neurotransmitters, and the intracellular processes to which they give rise, occur *concurrently*—as with glutamate and dopamine during long-lasting hippocampal long-term potentiation (LTP) (see Frey, and Meyer, this volume; Stein & Belluzzi, 1989; Stein, Xue, & Belluzzi, 1993). The consequences of these findings for neural-network simulations are clear: The algorithm(s) simulating the modification of synaptic efficacies (i.e., the updating of connection weights) must be temporally constrained, perhaps to something on the order of 100 ms. Moreover, the simulated processes must operate locally at the individual synapse if the exquisite specificity of the environmental guidance of behavior is to be achieved. (To say that the critical processes operate locally does not preclude the influence of events originating at more remote neural sites if those events have local effects at the time synaptic efficacies are changed.)

The temporal constraints imposed by both behavioral and cellular findings raise special problems for efforts to interpret complex behavior: Behavior often changes when events of direct biological significance occur after considerable delay or are absent altogether. For example, the scientist spends long hours in the laboratory, but the interval between that behavior and food on the table is great indeed. Somehow, neural networks intended to simulate the behavior of experienced organisms must function *as if* they were exempt from severe temporal constraints while remaining consistent with research imposing a short temporal horizon on the mechanisms of plasticity. The approach developed in this chapter attempts to accomplish these seemingly incompatible tasks. The apparent conflict between the moment-to-moment restrictions enforced by biobehavioral research and the longer-term sequences of events to which experienced organisms are sensitive is designated the *temporal paradox*.

Discrepancy constraint. Behavioral and neuroscientific research have led to a second general constraint: The biologically important events that foster learning do so only when those events are unexpected. Or, more technically, biologically important events change the environmental guidance of behavior only when the behavior they evoke is not otherwise occurring at that moment (see Palmer, this volume; see also Kamin, 1969; Donahoe, Crowley, Millard, & Stickney, 1982; Rescorla, 1968; Rescorla & Wagner, 1972). Cellular research is consistent with this behaviorally based conclusion. A biologically significant event, such as apple juice introduced into the mouth of a food-deprived monkey, no longer activates dopamine-producing neurons if the juice is preceded by another stimulus, such as the onset of a light, with which it has been reliably paired during the animal's past (see Schultz, this volume). And, the work of Frey (this volume) and others (summarized momentarily) also implicates dopamine in synaptic plasticity. The specific biological mechanisms implementing the discrepancy are not fully characterized, but a number of non-mutually exclusive candidates are available. These include negative-feedback mechanisms operating at the levels of both neural-systems and intracellular processes. As examples, (1) the activation of dopaminergic neurons, which is initially produced by inputs from biologically important events, may be attenuated by other inputs activated by stimuli that reliably precede the biologically important event due to "collision" between the two inputs (Shizgal, Bielajew, & Rompre, 1988; cf. Schultz, Romo, Ljungberg, Mirenowicz, Hollerman, & Dickinson, 1995) or through inhibitory circuits recruited by inputs to the dopaminergic neurons (Houk, Adams, & Barto, 1995) and (2) the response of dopaminergic neurons may be attenuated when the release of dopamine acts on inhibitory autoreceptors located on the dopaminergic neurons themselves (Kalsner & Westfall, 1990; Meyer, this volume; cf. Groves, Garcia-Munoz, Linder, Manley, Martone, & Young, 1995).

Binding constraint. Behavioral and neuroscientific findings are the source of a third general constraint on neural-network simulations of complex behavior—the binding constraint. At the behavioral level, any of a great number of stimuli may guide any of a great number of responses. Therefore, the biological mechanisms of synaptic plasticity, whatever they are, must be competent to affect many synapses throughout the brain simultaneously (see Palmer, this volume). Findings at the level of neuroscience point in the same direction. Stimuli—especially the complex combinations of environmental events typical of the real world—activate *populations* of neurons in sensory and sensory-association areas (see Tanaka, this volume). The conclusion that behavior is guided by the activity of populations of neurons is shared both by those who view the size of the neuronal population as sharply restricted by inhibitory interactions (see Trehub, this volume) and by those who envision larger, hierarchically organized populations of elements (see Hummel, this volume). Somehow, the neural mechanisms of perceptual learning must bind together a population of spatially diverse neurons into coherent patterns of firing.

A similar need for binding is required for the neural activity underlying responding. At the behavioral level what is regarded as a single response, such as elevating the arm, is at the neural level the concerted expression of the contraction of thousands of muscle fibers brought about by the activity of thousands of neurons (see Georgopoulos, this volume). As with perceptual processes, motor processes require the coordinated activity of populations of neurons. Accordingly, the biological mechanisms of plasticity must change the efficacies of many synapses throughout the nervous system while, at the same time, confining the changes to the "correct" synapses; i.e., those needed to mediate the selected environment-behavior relation.

The full task confronted by the mechanisms of plasticity is even more formidable: In order for the environment to guide behavior appropriately, not only must synapses be modified to form the two populations of neurons mediating the relevant perceptual and motor processes, but synapses of the many neurons interconnecting these two populations must be modified. In short, one population of neurons mediating sensory processes must appropriately activate a second population of neurons mediating motor processes. (Ignored for now is a further complication: Sensory and motor processes are interdependent; i.e., the activity of one may affect the other.) The need for the mechanisms of plasticity to be able to modify the synaptic efficacies of large populations of neurons while restricting the modified synapses to those needed to mediate specific environment-behavior relations constitutes the *binding paradox*.

Biological mechanisms of neural plasticity

How has evolution through natural selection addressed the dual challenges of the temporal and binding paradoxes? The first clue comes from the work of James Olds who, together with his then graduate student Peter Milner, discovered that electrical stimulation of certain regions of the brain strengthened the responses that preceded them (Olds & Milner, 1954). That is, such stimulation functioned as a *reinforcer* of behavior.

Nonspecific projection systems. Subsequent neuroanatomical work indicated that the various neural sites for which electrical stimulation was reinforcing had the common effect of stimulating neurons in a region of the midbrain known as the ventral tegmental area (VTA). Other work showed that axons from cells in the VTA *projected widely* within the brain, including (among others) to the motor-association areas of the frontal and prefrontal cortex (Fallon & Laughlin, 1987; Lindvall, Bjorklund, & Divac, 1978; see **Figure 1** for a schematic diagram of these diffuse dopaminergic projections.) Furthermore, stimulation of VTA neurons caused dopamine to be liberated in these areas (Hoebel, 1988; Stellar & Stellar, 1985). Because of their widespread distribution, dopaminergic neurons have the potential to affect the functioning of many other neurons. Consistent with this possibility, dopaminergic neurons have large projection fields and numerous synapses located along their axon shafts by which they can affect substantial populations of neurons (Groves, Lindner, & Young, 1994). Because synapses on a single dopaminergic axon can affect the responses of many neurons to other transmitters, dopamine is often designated a *neuromodulator* rather than a classic neurotransmitter.

Dopaminergic neurons commonly make synapse at the base of many of the numerous spines formed on the dendrites of target neurons. And, these spines often contain at their tips glutaminergic synapses receiving inputs from other neurons. Because glutamate is the primary excitatory neurotransmitter in the brain, and thus plays a central role in neural processing, the juxtaposition of dopaminergic and glutaminergic synapses suggests an interaction between their effects. Further, the joint effects of glutamate and dopamine may be spatially restricted to the relatively isolated intracellular compartments provided by the spines. Thus, all levels of observation—neuroanatomical, ultrastructural, and biochemical—indicate that the reinforcing effects of dopamine can be simultaneously widespread and specific to particular synapses. In short, evolution through natural selection appears to have "solved" the binding paradox by exploiting diffusely projecting neuromodulatory systems. This strategy confers yet another benefit: As the brain grew larger over evolutionary time, a dopamine-dependent mechanism of synaptic plasticity could accommodate growth by simply extending its projections further (Donahoe & Palmer, 1994). Point-to-point specification of the connectivity of the brain by the genome was not required, and would have been impossible in any case because the complexity of the human brain exceeds the capacity of the genome to specify it.

Given that synaptic plasticity is dependent on interactions between the dopaminergic neuromodulatory system and the glutamate transmitter system, what is the nature of that interaction? Here, the focus is upon the *functional* effect of glutamate-dopamine interactions on synaptic plasticity. Experimental studies (e.g., Frey, this volume; Chioda & Berger, 1986; see Wickens & Kotter, 1995, p. 195-198 for a review) indicate that the conditions necessary for long-term modifications in synaptic efficacy are the following: (1) Efficacy increases if activation of the presynaptic neuron causes the release of glutamate that, in turn, activates the postsynaptic neuron and if their coactivity is followed within less than a few hundred ms by a pulse of dopamine. That is, synaptic efficacies are modified by a heterosynaptic mechanism. (2) Efficacy decreases when glutamate from the presynaptic neuron activates the postsynaptic neuron and their coactivity is *not* followed by a pulse of dopamine. The first effect is known as LTP and the second as long-term depression (LTD). Other combinations of pre- and postsynaptic activity with or without the presence of dopamine have not been shown to have any consistent long-term effect on synaptic efficacy.

In keeping with experimental findings concerning the conditions that produce LTP and LTD, the present computer simulations incorporated a learning algorithm in which coactivity in the pre- and postsynaptic units increased the connection weight between the units if a reinforcer occurred, but decreased it otherwise (Donahoe, Burgos, & Palmer, 1993). For LTP, or acquisition, the change (Δ)—here, an increase—in the connection weight w between the ith presynaptic unit and the jth postsynaptic unit from time-step t-1 to time-step t is given by

$$\Delta w_{ij}(t) = w_{ij}(t) - w_{ij}(t-1) = \alpha_j a_i(t-1) a_j(t) r(t) \qquad (1)$$

where α $(0<\alpha<1)$ is the learning-rate parameter, a $(0<a<1)$ is the activation level of a unit, and r is the strength of the dopaminergic reinforcing signal (where r is the momentary increase in the activation of VTA units from time-step t-1 to t. According to Equation 1, the higher the activation levels of the pre- and postsynaptic neurons and the stronger the dopaminergic reinforcing signal, the greater the increase in connection weight. (In the actual simulations, other factors also affected the magnitude of the change in connection weight; e.g., competition was implemented among the multiple presynaptic inputs for control over potential receptor sites on the postsynaptic unit as described in the appendix of Donahoe et al, 1993.)

For LTD, or extinction, the change—now, a decrease—in connection weight between a pre- and postsynaptic unit is given by

$$- \Delta w_{ij}(t) = \beta_j a_i(t\text{-}1) a_j(t) \qquad (2)$$

where ß ($0 < \beta < 1$, $\beta < a$) is the extinction-rate parameter. Thus, the greater the coactivations of the pre- and postsynaptic units when the dopaminergic reinforcer is absent, the greater the decrease in connection weight.

Simulation of the Selection of Behavior

Figure 1 provides a schematic representation of the major dopaminergic projections originating from the VTA. Our present concern is with projections to the frontal cortex and basal ganglia, for these structures are the most directly implicated in the control of behavior. The significance of projections to the hippocampal region is considered later in the chapter. **Figure 2** depicts a neural-network architecture that *functionally* represents the sensory- and motor-processing units and their relation to the simulated nonspecific reinforcing system. A network of this general architecture is called a *selection network* because, as shown below, the reinforcing system has the cumulative effect of selecting those pathways that mediate environment-behavior relations.

For now, consider only the motor subnetwork of **Figure 2**. Units in the motor subnetwork become activated as the result of environmental stimulation and endogenous intranetwork events that produce "spontaneous" unit activity, (Donahoe, Palmer, & Burgos, in press). The behavior of the network is simulated by two classes of output units—R units and CR/UR units. R units simulate the operant behavior of the network; i.e., the great majority of behavior commonly described as voluntary (cf. Skinner, 1938). CR/UR units simulate the reflexive behavior of the network; i.e., behavior elicitable by environmental stimuli as a consequence of natural selection. Elicited behaviors are conventionally designated as unconditioned responses (URs) when reflexively elicited and as conditioned responses (CRs) when elicited by arbitrary stimuli that have previously been paired with reinforcing stimuli. The lines within the network represent pathways between units, with heavier lines indicating pathways having strong initial connection weights. Heavier lines are present for connections within the dopaminergic reinforcing system. These include the class of units stimulated by the reinforcing stimulus (S^R), the pathways from the VTA

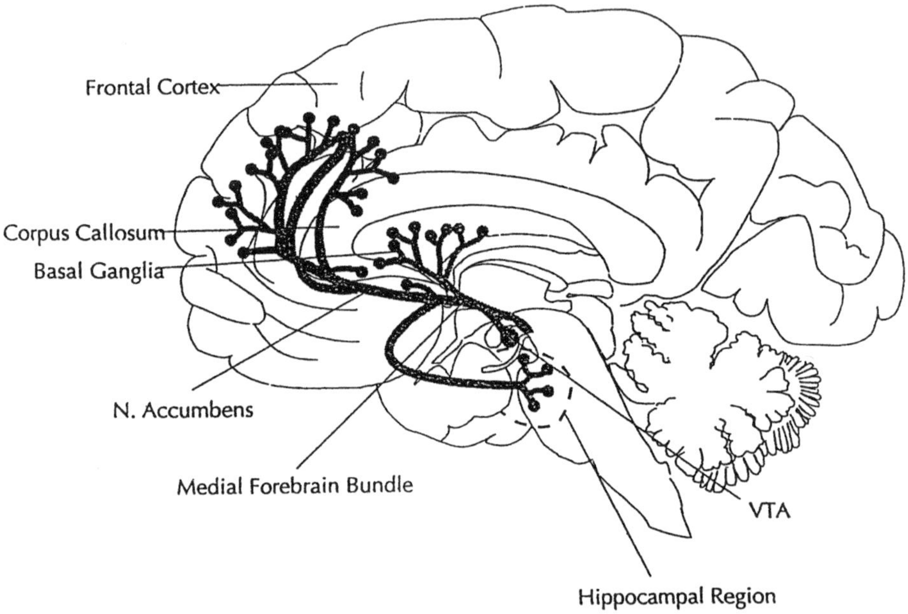

FIGURE 1. Schematic diagram of major dopaminergic systems of the brain. Of special concern are projections from the ventral tegmental area (VTA) to the prefrontal cortex, basal ganglia of the neostriatum, and the hippocampal region (including CA1 and subiculum).

(which innervate all units within the motor subnetwork), and the reflexive CR/UR system. Lighter lines indicate pathways between the largely glutaminergic neural-processing units, whose connection weights are small at the beginning of training and are modifiable by the nonspecific reinforcing system.

Two procedures are used for the behavioral study of plasticity, and salient outcomes of both can be simulated with networks of the type shown in **Figure 2**. In the first procedure—the classical or Pavlovian procedure—some stimulus (e.g., S1) activates a class of input units and is followed shortly thereafter by the reinforcing stimulus, S^R, which activates its own class of input units. A standard laboratory example of the classical procedure is provided by Pavlov's dog given food after the sound of a metronome. In the second procedure—the operant or instrumental procedure—input units are activated unsystematically by the environment and, if the target behavior (R) occurs, then S^R units are activated. A standard laboratory example of the operant procedure is Skinner's rat pressing a lever for food. Whichever procedure is simulated by the neural network, the environment activates input units and these, in turn, probabilisti-

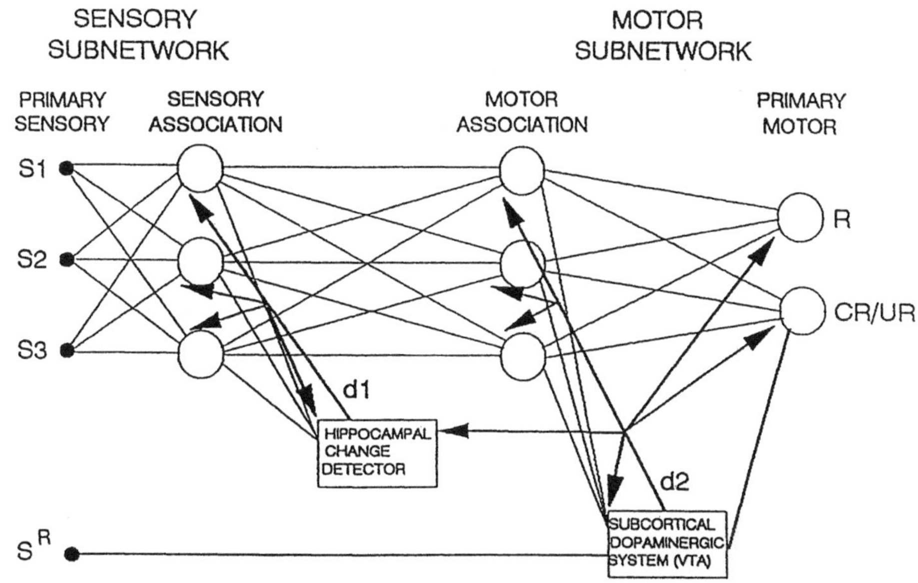

FIGURE 2. Basic architecture of a selection network. The reinforcing stimulus (S^R) activates the simulated ventral tegmental area (VTA), which then nonspecifically projects a reinforcing signal (heavy lines with arrows) to connections in the motor subnetwork. The strength of the VTA signal is proportional to the positive difference (d2) between the activation of the VTA at the current moment (time step t) and the previous moment (time step t-1). The VTA signal also modulates the strength of a second nonspecific reinforcing signal from the simulated hippocampus to sensory-association units in the sensory subnetwork. The strength of the hippocampal signal is proportional to the difference (d1) between the activation levels of inputs to the hippocampus at times t and t-1. (See text for a description of the functioning of a selection network.)

cally activate sensory-association, motor-association, and ultimately output units. In the classical procedure, S1 input units and S^R units are activated within a short time interval of one another with the result that these two classes of units and CR/UR units, which are activated by the S^R units, are coactive during training. Since S1 units and CR/UR units are reliably and strongly activated in concert with VTA units, connection weights along pathways leading away from S1 units and toward CR/UR units are most rapidly strengthened by the simulated dopaminergic system. Over time, the action of the nonspecific projection system increases connection weights along pathways extending from the class of S1 input units to the class of CR/UR output units. As shown in the left panel of **Figure 3**, the cumulative effect of this process simulates the acquisition of CRs. A similar process of selection of pathways by the reinforcer occurs with the operant procedure. Now, however, in addition to the activation by the environment of some input units (although not necessarily any *particular* input unit) and of S^R input units by the reinforcing stimulus, other units—R units—are also activated in temporal proximity to activation of the dopaminergic system. Activity in the class of R units is enforced by the procedural arrangement in which S^R units are activated only *after* the R unit has been activated. Acquisition in the operant procedure as measured by the levels of activation of an R and CR/UR unit is shown in the right panel of **Figure 3**.

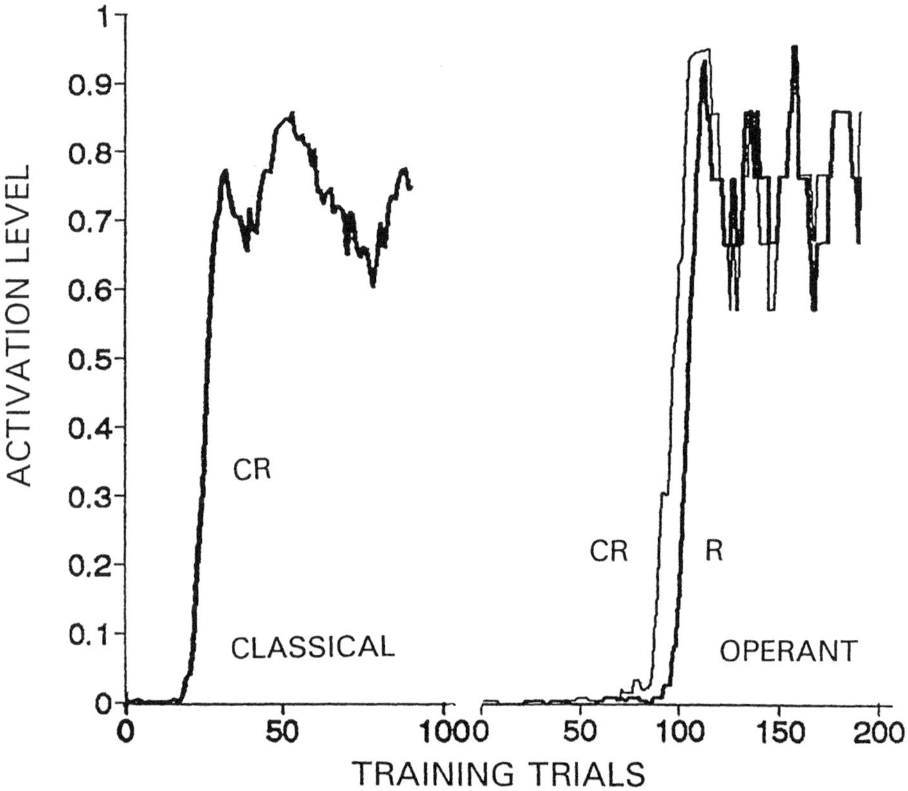

FIGURE 3. Simulations of conditioning with the classical (left panel) and operant (right panel) procedures. The activation levels of the output units for the reflexive, or conditioned, response (CR) and the operant (R) are shown. Conditioning begins more slowly with the operant procedure because the reinforcing stimulus occurs only on trials in which the activation level of R is greater than zero. Within the operant procedure, conditioning of the CR generally proceeds more rapidly than the R (see Donahoe and Palmer, 1994).

To summarize, synaptic efficacies increase when coactivity of pre- and postsynaptic neurons is accompanied by dopamine from the nonspecific projection system. Neural networks embodying this arrangement—i.e., selection networks—are competent to simulate acquisition with both the classical and operant procedures. (For simulation of other conditioning phenomena, see Donahoe et al, 1993; Donahoe & Palmer, 1994; Donahoe, Palmer, & Burgos, in press). Although classical and operant procedures are quite distinct in terms of the behavioral contingencies they impose—an S1-S^R contingency in the first case and an R-S^R contingency in the second—the outcome of both procedures may be simulated by a common mechanism for modifying synaptic efficacies. (The simulation of the results of both procedures assumes, of course, that the units mediating the relevant environment-behavior relations can be affected by the nonspecific dopaminergic system. If the relevant units can*not* be affected by that system, as is likely the case with autonomically mediated behavior, then the outcome of the procedure cannot be simulated without incorporating additional processes.) Of the two procedural arrangements, the operant is by far the more important for the emergence of complex behavior because new environment-behavior relations are not restricted to reflexive behavior but include the full behavioral potential of the organism. The entire behavioral repertoire of the learner is within reach of a diffusely projecting dopaminergic system: Whatever units are coactive prior to the occurrence of the reinforcing stimulus—no matter how many or how dispersed within the motor subnetwork—their connection weights can be modified by the reinforcer.

Acquired reinforcement

If the temporal interval between unit coactivity and the reinforcing signal is sufficiently short, then modifying synaptic efficacies by means of a nonspecifically projecting neuromodulatory system resolves the binding paradox. Within the motor subnetwork, an indefinitely large number of motor-association and motor units can have their connection weights changed in a coordinated fashion, resulting in the formation of a population of units whose concerted activity generates the target behavior. But what of the temporal paradox? How can a suitable temporal interval occur given the relatively long interval (on the neural time scale) that inevitably elapses between the occurrence of the target behavior (e.g., lever pressing by a rat) and the occurrence of the reinforcing stimulus (e.g., consumption of a food pellet)? Or, in terms of a neural network, how can binding occur when the activity of units in the motor subnetwork—particularly a multisynaptic subnetwork—may occur many time steps *before* the VTA is activated by S^R input units?

Environmentally mediated acquired reinforcement. The resolution of the temporal paradox draws upon work at both the behavioral and neural levels. At the behavioral level, research indicates that acquisition of even relatively simple responses, such as leverpressing, does not readily occur when the target response is followed by a putative reinforcer with delays typical of behavioral procedures. The interval between the execution of the target response and direct contact with the reinforcer is generally too great for the response to be easily acquired. To illustrate, in order for leverpressing to be rapidly acquired, the rat must first receive pairings of the sound of the feeder with food. Only after the rat avidly approaches the food tray upon hearing the sound of the feeder is leverpressing easily acquired (Skinner, 1938; cf. Spence, 1947). The sound of the feeder allows leverpressing to be more immediately reinforced than it would be by food alone.

As shown in the simulation of the classical procedure (see the left panel of **Figure 3**), a stimulus that has been paired with a reinforcer can itself activate the nonspecifically projecting reinforcing system. After pairing with the reinforcer, the stimulus—such as the sound of the feeder—activates VTA units via connections from motor-association units (see **Figure 2**). Thus, the sound of the feeder provides a more immediate *acquired reinforcer* for the target response, with the result that connection weights are modified with shorter delays than if activation of the VTA waited upon the stimulation of S^R units by food.

Internally mediated acquired reinforcement. With acquired reinforcement, activation of the nonspecific reinforcing system is mediated by environmental stimuli that intervene between the target response and the occurrence of the ultimate reinforcer. However, the activation of the nonspecific reinforcing system via connections from motor-association units will occur even if such intervening environmental events are not present. Within the motor subnetwork shown in **Figure 2**, activation of the nonspecific reinforcing system has two effects during operant procedures: (1) Connections from motor-association units to R units are strengthened (i.e, the target response is acquired) and (2) connections from motor-association units to the VTA are strengthened. Because of this second effect, selection networks are able not only to emit the target response but also to change their own connection weights before the occurrence of either the target response or the external reinforcer. If a network has a relevant history of selection by reinforcement, then the environment can activate the VTA via connections from activated motor-association units. In short, an experienced network acquires the capacity to strengthen its own connections. The process whereby this comes about is called *internal reinforcement* (Donahoe et al, 1993; Donahoe & Palmer, 1994, pp. 97-99; Houk et al, 1995; Wickens & Kotter, 1995; cf. Goldman-Rakic, 1987; Goldman-Rakic, Chafee, & Friedman, 1993). Indeed, internal reinforcement is required when the motor subnetwork contains more than one layer of motor association units, which simulates behavior mediated by multisynaptic pathways. In a motor subnetwork containing two layers of motor-association units, acquisition was retarded when only one layer could implement internal reinforcement via connections to the VTA, and was prevented when the internal reinforcement circuits were eliminated altogether and the VTA could be activated only by direct pathways from S^R units (Donahoe & Palmer, 1994, pp. 97-101). Thus, natural selection has resolved the temporal paradox by implementing internal-feedback circuits that, together with intervening environmental stimulation, activate the nonspecific reinforcing system earlier than would be the case if changes in synaptic efficacy were dependent on the occurrence of innate biologically significant events (see also Wickens & Kotter, 1995).

Simulation of the Selection of Stimulus Configurations

Through exploiting a nonspecifically projecting reinforcing system augmented by feedback circuits that internally activate the reinforcing system, natural selection has met the twin challenges posed by the binding and temporal paradoxes. However, those challenges have thus far been addressed only as they appear in motor systems in the brain; i.e., in frontal cortex and basal ganglia. But, behavior is mediated by populations of neurons in sensory as well as motor systems. How are coherent populations of neurons formed in parietal-temporal-occipital cortex? Such populations must also be formed in these areas if behavior is to be guided by complex configurations of stimuli.

The resolution of the binding and temporal paradoxes in sensory-association cortex, in particular, is less clear than in frontal cortex and basal ganglia—as imperfect as is our understanding of the latter. Nevertheless, the general outlines of the answer seem functionally similar: Neuromodulators released by nonspecific projection systems play a central role.

Formation of primary sensory neural populations

The primary focus here is upon the formation of neuronal populations in sensory-*association* cortex, but a few words about primary sensory cortex are in order. In primary sensory regions, the temporal and spatial structure of the environment acts upon the somatotopically organized structure of the brain to produce coactivity in neighboring neurons. For example, coactivity between neurons in primary sensory cortex shapes the connectivity of the visual system through activity-dependent mechanisms operating during postnatal development (e.g., Engel, Kreiter, Konig, & Singer, 1991). The effect of coactivity is dependent on the presence of neuromodulators from nonspecifically projecting subcortical nuclei. For example, during cortical development, changes depend not only upon coactivity between glutaminergic neurons (Bear, Kleinschmidt, Gu, & Singer, 1990), but also upon the neuromodulators norepinephrine and acetylcholine (see Singer, this volume; also Bear & Singer, 1986). Axons liberating norepinephrine arise from cells in the locus coeruleus and those liberating acetylcholine from substantia innominata, septal nuclei, diagonal band, and dorsolateral pons. The conjunction of coactivity and neuromodulator produces populations of neurons that respond coherently to similar stimuli arising from nearby regions of the environment, as in the sensitivity of cells within a visual cortical column to similar spatial frequencies and orientations within its receptive field (e.g., De Valois & De Valois, 1988; Hubel & Wiesel, 1968).

Formation of sensory-association neural populations

In experienced organisms, behavior is typically guided by complex combinations of stimuli (see Tanaka, this volume). As an illustration, in space perception, one's location within a familiar room is specified by the particular combination of stimuli arising from the walls and objects within the room. This combination guides walking. For another, in reading, the complex visual stimuli arising from words must be integrated with each other and with the auditory and articulatory processes that reflect the grapheme-phoneme correspondences of English. This combination guides reading. How are neuronal populations formed whose concerted activity specifies such complex combinations of stimuli?

The formation of functional populations of neurons depends critically upon interactions between the hippocampus and sensory-association cortex. Axons from polysensory neurons in sensory-association cortex are the origins of multisynaptic pathways that serve as inputs to the hippocampus (Amaral, 1989). Processing within the hippocampus converges on CA1 hippocampal neurons whose axons give rise to multisynaptic pathways that *diffusely* project back to sensory-association cortex. This reciprocal neuroanatomical arrangement allows the output of hippocampal CA1 cells to affect the functioning of polysensory cells throughout sensory-association cortex. More specifically, CA1-mediated feedback from hippocampus is hypothesized to alter synaptic efficacies between coactive pre- and postsynaptic neurons in sensory-association cortex. The synaptic and intracellular mechanisms mediating the proposed effect of nonspecific CA1-hippocampal feedback are not clear. As one possibility, glutamate, which is known to be liberated by some CA1 efferents, may potentiate the response of polysensory cells to their initially weak inputs from several sensory systems, thereby increasing the likelihood of activating the postsynaptic neuron (cf. Jay, Burette, & Laroche, 1995). Given that coactivity is thereby made more likely between presynaptic sensory neurons and postsynaptic polysensory neurons, the eligibility of these synapses for modification is increased. This increased eligibility for modification can then be exploited by nonspecific dopaminergic projections from the VTA that are known to innervate the sensory-association cortex of primates, but not rats (Berger, Gaspar, & Verney, 1991). Thus, the reinforcing system that modifies synaptic efficacies in the frontal lobes can serve a functionally equivalent role in the sensory-association cortex of the parieto-temporal-occipital lobes. Through the coordinating action of the reinforcing dopaminergic system, populations of sensory-association neurons form and activate, in turn, appropriate populations of neurons in the motor system. On this view, the hippocampus detects higher-order conjunctions of activity in multiple sensory channels and then "teaches" the sensory-association cortex to represent these conjunctions by strengthening synaptic efficacies among the relevant postsynaptic neurons; i.e., those that are activated by their multisensory inputs at times when the reinforcing system is activated. (Interactions between neurons in hippocampus and sensory-association cortex play important roles in other complex behavior, including memory, but they are not considered here. For an overview, see Donahoe & Palmer, 1994).

Simulation of formation of polysensory cells

The sensory subnetwork (see **Figure 2**) can be used to simulate the formation of polysensory cells with the proposed mechanism. Suppose that some behavior is reinforced in the presence of two stimuli (e.g., S1 and S2) but not

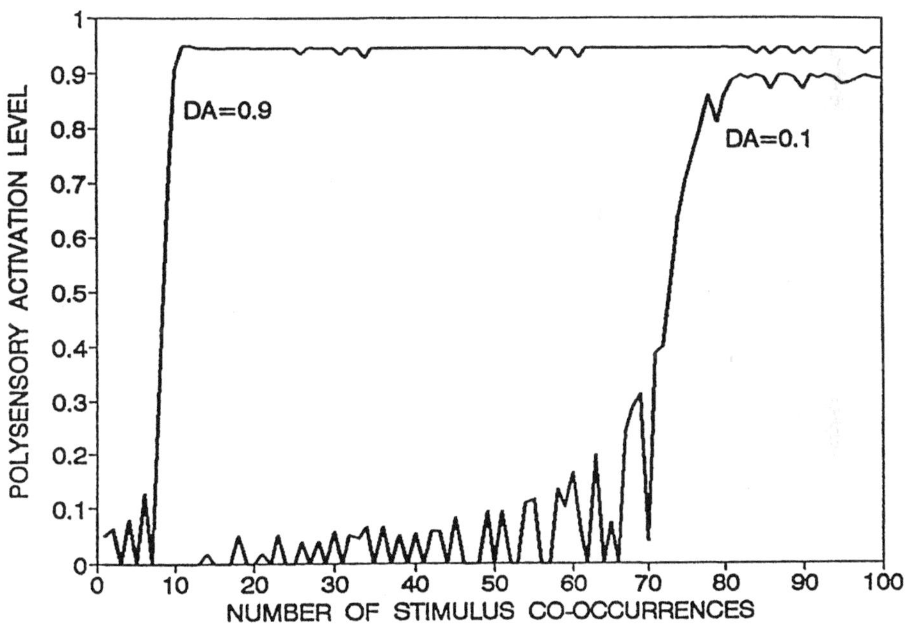

FIGURE 4. Simulations of the formation of S1-S2 polysensory units. Shown are the activation levels of S1-S2 units produced by the co-occurrence of inputs to the polysensory units from S1 and S2 units (see text). Initially, the S1 and S2 inputs only weakly activate the polysensory S1-S2 unit. Under the control of hippocampal output, S1-S2 units become polysensory, either rapidly with a strong DA signal (DA = 0.9) or more slowly with a weaker DA signal (DA = 0.1).

otherwise. As a consequence, behavior eventually occurs upon the joint occurrence of S1-S2, but not when either S1 or S2 is presented singly. **Figure 4** presents the results of a simulation of the formation of a polysensory S1-S2 unit that is strongly activated by the co-occurrence of the two stimuli. When the dopaminergic signal is strong (DA = 0.9), the cell rapidly becomes polysensory; when the dopaminergic signal is weak (DA = 0.1), the cell develops more slowly. (See Gluck and Myers, this volume, for a detailed development of a hippocampal simulation that is broadly consistent with the approach taken here.)

The net result of this process is that a nonspecifically projecting reinforcing system augmented by internal-feedback circuits has the potential to resolve the temporal and binding paradoxes in both the motor and sensory systems. Through the action of reinforcers—innate and acquired, environmentally and internally mediated—populations of neurons are selected in both sensory and motor systems that, together, implement complex environment-behavior relations in a coordinated fashion.

Role of inhibitory units. Although nonspecific neuromodulatory systems address the binding problem very effectively, they encounter a potential difficulty: Because a nonspecific reinforcing signal modifies synaptic efficacies between *all* coactive pre- and postsynaptic neurons, many of the modified synapses would likely involve neurons that were coactive for reasons causally unrelated to the reinforced environment-behavior relation. This would lead, if unchecked, to the modification of many irrelevant synapses with the ultimate consequence of impairing efficient neural processing.

Natural selection has addressed this potential undesirable side effect of nonspecific reinforcing systems through the evolution of local inhibitory circuits. Units implementing inhibitory circuits have been successfully simulated in selection networks. Consider a network whose layers of excitatory units are richly interconnected to excitatory units in other layers. When inhibitory units are not present between excitatory units within a layer, units in that layer compete strongly with one another for control over units in downstream layers on which their outputs converge. The result is that many upstream units must be active in order to strongly activate a downstream unit. This uneconomical use of units also arises when inhibitory units are not present between excitatory units within downstream layers. Now, a unit in an upstream layer weakly activates many units in the downstream layer. This produces interference among the upstream units in their ability to produce nonoverlapping patterns of activation in downstream units. Competition and interference among units are greatly reduced when local inhibitory units are implemented. The addition of inhibitory circuits permits a network to mediate multiple input-output relations by forming a series of local "winner-take-all" circuits. Experimental work at both the behavioral (Blough, 1975) and neural levels (Hartline & Ratliffe, 1957) demonstrates the effect and existence of such circuits.

Implications for Complex Behavior

In the concluding section of the chapter, some implications of the approach for complex behavior are outlined. Three sets of phenomena are touched upon: (1) conditioned perceiving, (2) problem solving, and (3) language acquisition. Before considering these implications, some general comments about the present approach are in order.

Biobehavioral and cognitive-science approaches to neural networks

The foregoing treatment of neural networks has been constrained and informed by experimental work at the behavioral and neural levels of analysis. However, much potentially relevant biobehavioral research has not been incorporated. For instance, nothing in the simulations directly reflects the intracellular second-messenger cascade that produces changes in synaptic efficacies. Only the presumed net effect of those processes—the change in synaptic efficacy—has been simulated. For another, although the general form of the network architecture is faithful to the neuroanatomy of a nonspecific dopaminergic reinforcing system, many aspects of that system have not been simulated. Only one feedback pathway has been implemented—e.g., from prefrontal cortex to VTA—although other feedback pathways are present in the brain—e.g., from the basal ganglia of the neostriatum to the VTA and substantia nigra. It is unlikely—and even undesirable—for any simulation to instantiate *all* biobehavioral knowledge because the simulation would then be as complex (and as difficult to understand) as the processes whose outcomes it simulates. Instead, it is sufficient for the simulation to capture the phenomena to the desired particularity, employing those processes and structures minimally necessary to accomplish that goal, while not violating any known experimental finding. To the extent that the outcome of the simulation is consistent with the phenomenon, then—to that extent—are the simulated processes competent to produce the phenomena. In short, the simulation demonstrates that those processes "explain" the phenomena. Of course, as the range of phenomena encompassed by a simulation increases, the need to incorporate additional findings may also increase (see Chapter 1).

The approach to neural networks pursued here differs fundamentally from parallel distributed processing in normative cognitive science. In cognitive science, the structure of networks and the algorithms modifying their connection weights are typically constrained only by knowledge of the behavioral output of the organism and by logico-mathematical considerations, not by direct information about the internal processes and structures mediating that output. Internal events are merely *inferred* from their external behavioral effects, a strategy that is fraught with perils of two sorts: (1) Behavioral observations do not sufficiently constrain inferences about underlying events to make any particular inference compelling. Who would claim that a reasonable way to understand the internal workings of a computer is to look at only its inputs and outputs? (2) Internal events that are solely the product of inferences from behavior invite circular reasoning. That is, behavioral observations provide the basis for the inferences, but then the validity of the inferences is judged by their consistency with the behavior that led to the inferences in the first place (Donahoe & Palmer, 1994, p. 9-10).

The treatment of neural networks nicely illustrates the differences between the biobehavioral and cognitive-science approaches. In biobehavioral science, the great bulk of evidence points toward reinforcement learning as the means by which synaptic efficacies (connection weights) are modified—although, of course, much remains to be known about the workings of the process. A quick survey of ten recent cognitive-science texts on my bookshelf revealed that only three mentioned reinforcement learning at all, and one of these was dismissive. Indeed, the most commonly used learning algorithm employs a technique known as backpropagation (Rumelhart, Hinton, & Williams, 1986), which lacks biological plausibility. Neural networks were mentioned in more of the texts—six of the ten—but the units and architectures of the networks were unrelated to the relevant neuroscience except in one instance. Others have commented on the lack of correspondence between the neural networks in cognitive science and in neuroscience (e.g., Crick & Asanuma, 1986). In cognitive science, it is sufficient for the outcome of the simulation to mimic the observed behavior. Such an approach may be quite reasonable for many engineering purposes (see Barto and Sutton, this volume), but it differs from the function of simulation in biobehavioral science. In biobehavioral science, simulation explores the implications of processes and structures that are the fruits of independent experimental analyses. (For further discussion of this issue, see Van Orden, Bosman, Goldinger, & Farrar, and Barnes & Hampson, this volume; cf. Donahoe & Palmer, 1989; Rumelhart, McClelland, & The PDP Group, 1986.)

Conditioned perceiving

One of the most important omissions from the neural networks thus far considered are *recurrent* connections. Recurrent connections extend from units within the network to other units that are more "superficially" placed; e.g., from motor-association units to sensory-association units. Until this point in the presentation, only *feedforward* connections have been considered; e.g., from sensory- to motor-association units. Through the operation of recurrent connections, the activity of a unit at one time can affect its own activity at some later time. And, in the brain, recurrent connections between regions are often as plentiful as feedforward connections: The feedforward pathways from thalamus to cortex are complemented by equally rich recurrent connections from cortex back to thalamus. Recurrent connections have not been considered for simplicity's sake alone; a nonspecifically projecting reinforcement system is as competent to modify connection weights along recurrent pathways as feedforward pathways, and with the same ability to resolve the temporal and binding paradoxes.

Only one effect of recurrent connections is considered here—the establishment of conditioned perceiving as a normal accompaniment of learning (Donahoe & Palmer, 1994, pp. 275-277, 333-340). Assume that recurrent connections extend from motor-association back to sensory-association units (see **Figure 2**). Intracortical and cortico-thamalic connections of this sort are well documented (e.g., Fuster, 1989). Consider now the events that would transpire during the acquisition of *any* environment-behavior relation. Environmental stimuli activate units first in the sensory and then the motor subnetwork. The former include units whose concerted activity mediates perceiving. When the target behavior occurs, the reinforcing system is engaged and increases the connection weights of all recently coactive synapses. This process strengthens synapses along pathways leading from activated input units to the target output units. When recurrent connections are present, an additional effect occurs: Units in motor-association areas activate pathways that connect to sensory-association units. If active recurrent connections make synapse with recently activated sensory-association units, then the reinforcing signal strengthens such recurrent connections from motor-association to sensory-association units. Which sensory-association units are most likely to have been recently activated and, hence, have their connections from recurrent pathways strengthened? The most recently activated sensory-association units are precisely those that were activated by the stimulating environment prior to the reinforcer. Thus, an *automatic* accompaniment of the strengthening of feedforward connections to units in the motor subnetwork is the strengthening of recurrent connections from motor-association units to sensory-association units. The functional significance of strengthening these recurrent connections is that whenever those motor-association units are activated in the future—by pathways from whatever other units—some of the units in sensory-association cortex that mediated the original perception will be reactivated. Thus, some of the neural activity underlying perceiving is automatically conditioned when learning occurs. This conditioned activity in sensory-association units is the physical counterpart of conditioned perceiving; i.e., imagining and reminiscing. The unit activity underlying conditioned perceiving can then reafferent units in the motor subnetwork thereby affecting the likelihood of previously reinforced behavior.

Problem solving

The ability to solve problems—to "think"—is one of the hallmarks of our species. Such behavior is largely covert and the environmental reinforcers (if any) follow a temporally extended sequence of many responses. What maintains this covert sequence of behavior that is reinforced by the environment after such long delays? The processes of conditioned perceiving and internal reinforcement provide potential routes to an answer. First, the present environment triggers perceptual processes that generate covert motor activity previously conditioned to those processes. Then—via recurrent connections—conditioned perceiving occurs that, in turn, evokes further covert motor activity previously conditioned to those conditioned perceptions. Repeated cycles of these covert processes are maintained by feedback circuits mediating internal reinforcement. Internal reinforcement occurs when motor-association units activate the nonspecific reinforcing system as a result of reafference. (For a more complete biobehavioral treatment of problem solving, see Donahoe & Palmer, 1994; pp. 277-293.)

In its simplest form, the operation of the internal reinforcing system can be illustrated by the following simulation. A stimulus (S2) was paired with an environmental reinforcing stimulus (S^R) while, on other trials, another stimulus (S1) was paired with S2 alone. After S2-S^R pairings had established functional internal reinforcing circuits that could be activated by S2, conditioning to S1 occurred even though S1 had never been paired with an environmental reinforcer. These training conditions were simulated using a network architecture that had been produced by a genetic algorithm (see Burgos, this volume). This particular genetic algorithm favored the evolution of an architecture that could learn to respond to either of two stimuli after 8 time steps. **Figure 5** shows the results of the simulation: Responding was acquired to S1 through the action of the internal reinforcement mechanism alone. Note that the genetic algorithm that guided the development of the architecture did not select for the capacity to learn under these conditions, but simply for learning to respond to either of two stimuli. Learning via the internal reinforcement mechanism alone was an emergent product of the evolutionary process. (This simulation exemplifies a phenomenon known as higher-order conditioning; Pavlov, 1927).

The internal reinforcement system is the origin of another emergent process that contributes to problem solving—*short-circuiting*. Assume that there are two parallel neural processes, each of which is potentially competent to produce the target response. The first is a cortical process requiring a number of cycles of reafferent relations between units in the motor- and sensory-association subnetworks before motor units are activated that produce the target response. The second process involves a more direct pathway whereby units in sensory-association cortex can activate the motor units of the target response, but by means of a pathway that initially provides a pattern of activation that is less discriminable than the pattern produced by the less direct, reafferent pathways. As learning progresses, the target response would first be mediated by the slower, but more distinctive, reafferent process. However, over time, the target response would come to be mediated by the faster, more direct perceptual process. For example, a shift might occur between cortical control of the motor system to largely subcortical control involving the basal ganglia of the neostriatum (see **Figure 1**). This shift to the more rapid process (i.e., the one that more quickly mediates the target response) would occur *automatically* because the delay in both internally and environmentally mediated reinforcement would be less than for the longer process. And, to the extent that verbal behavior is correlated with cortical involvement, a shift of the behavior from conscious to automatic processing would be anticipated by this account. Measures of neural imaging (see Raichle, this volume) and of retention (e.g., Knowlton, Mangels, & Squire, 1996) document a shift in neural processing of the type described. The central point is that short-circuiting is an emergent product of the neural mechanisms of internal reinforcement. Although there is more to problem solving and thinking than conditioned perceiving and internal reinforcement, important aspects of this complex behavior appear understandable in these terms.

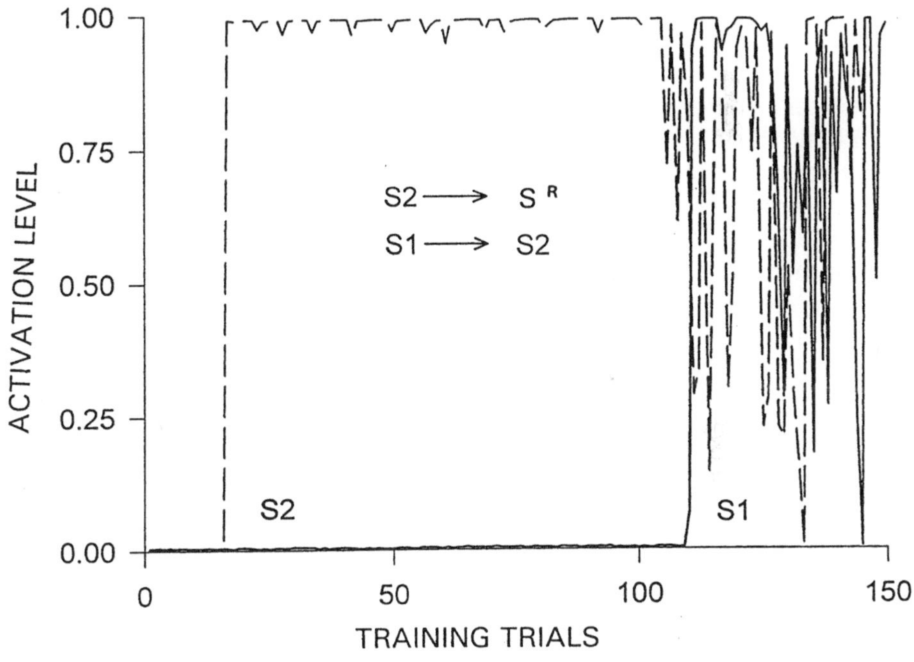

FIGURE 5. Conditioning produced by circuits mediating internal reinforcement. A stimulus (S2) was paired with a reinforcing stimulus (S^R) that directly activated the nonspecific reinforcing system. Another stimulus (S1) was never paired with S^R, but was paired with S2, which acquired the ability to activate the reinforcing system though feedback pathways extending from units in the motor-association subnetwork to the reinforcing system.

Language acquisition

The final complex phenomenon for which the biobehavioral mechanisms of reinforcement provide at least a partial solution is language acquisition. Our species uniquely acquires verbal behavior, and does so without continuous reinforcing stimulation from the environment. Indeed, the apparent absence of reinforcement has been thought by some to constitute an insurmountable impediment to a biobehavioral account of language acquisition (e.g., Brown & Hanlon, 1970; for critical reviews, see Donahoe & Palmer, 1994, pp. 317-319; Palmer, 1986). Verbal behavior and its acquisition pose the most formidable challenges to a biobehavioral approach, but the process of acquired reinforcement—both environmentally and internally mediated—suggest one path to its resolution.

Consider a child acquiring the ability to have its behavior guided by verbal stimuli (i.e., to comprehend speech). Although the auditory stimuli of speech are complex, the ability of such stimuli to control behavior does not present special problems for the reinforcement process itself. For example, a child "follows directions" and its behavior is reinforced by the environmental consequences of that behavior (the sought-after candy is obtained) and by acquired reinforcers from caretakers (the child is told that it has done well).

Recall, however, that whenever any environmental stimulus acquires the ability to guide motor responses it also acquires the ability to activate the internal reinforcing system. Of all behavior, human vocal behavior is uniquely able to capitalize on this latter process. To a greater degree than any other behavior, human vocal behavior produces stimuli (vocal speech) that are similar to the stimuli that, though prior conditioning, already guide behavior. That is, if a child vocalizes "baby" upon seeing a doll, that vocalization produces a stimulus—the speech sound /baby/—that is physically similar to the auditory stimulus /baby/ that already controls behavior and, therefore, activates the internal reinforcing system of the speaker. In the vernacular, /baby/ is already a meaningful stimulus. Thus, when the child says "baby," that response *immediately* produces a stimulus that activates the internal reinforcing system. There is an immediate, environmentally instigated and internally mediated acquired reinforcer for speaking. In addition, the more closely the child's vocal responding produces stimuli that approximate the ones produced by adult vocalizations, the greater the activation of the internal reinforcing system. In short, vocal behavior—and vocal behavior uniquely—immediately and appropriately engages the neural mechanisms of internal reinforcement. According to this interpretation, language acquisition is *not* directly dependent on receiving reinforcing feedback from others but on the prior comprehension of speech by the listener himself. That is, for language to be acquired, comprehension must precede production—as it does. Only then are the stimuli produced by the child's vocalizations able to activate the internal reinforcing system.

Biobehavioral accounts of verbal behavior require further experimental analyses to identify the relevant processes and further simulation research to explore the full implications of those processes. Nevertheless, no complex phenomenon seems beyond the reach of the approach and many phenomena appear within its grasp.

ACKNOWLEDGMENTS

I wish to express my indebtedness to Dr. José E. Burgos, with whom these simulations were conducted and, especially, to Dr. David C. Palmer, with whom these ideas were discussed over some years. Both are former students and present colleagues. Portions of the work were supported by the National Science Foundation and by a Faculty Research Grant from the University of Massachusetts at Amherst.

Neural-Networks Models of Cognition
J.W. Donahoe and V. Packard Dorsel (Editors)

CHAPTER 19

REINFORCEMENT LEARNING IN ARTIFICIAL INTELLIGENCE

Andrew G. Barto and Richard S. Sutton
Department of Computer Science
University of Massachusetts, Amherst

ABSTRACT

This chapter provides an overview of an approach to the study of learning that, in broad terms, has developed as a part of the field of Artificial Intelligence (AI), where it is called *reinforcement learning* due to its roots in reinforcement theories of animal learning. We introduce the field from the perspective of AI and engineering, describing some of its key features, providing a formal model of the reinforcement-learning problem, and defining basic concepts that are exploited by solution methods. Detailed discussion of solution methods themselves and their history are very broad topics that we do not attempt to cover here.

Introduction

This chapter describes an approach to the study of learning that has developed largely as a part of the field of Artificial Intelligence (AI), where it is called *reinforcement learning* due to its roots in reinforcement theories that arose during the first half of this century. Reinforcement learning in AI consists of a collection of computational methods that, although inspired by animal-learning principles, are primarily motivated by their potential for solving practical problems.

Although the ideas of reinforcement learning have been present in AI since its earliest days (e.g., Minsky, 1954, 1961; Samuel, 1959), several factors limited their influence. Chief among them is that AI research in the 1960s followed the allied areas of psychology in shifting from approaches based in animal behavior toward more cognitive approaches. This shift left little room for reinforcement theories. Although critics have argued convincingly that one cannot understand or generate *all* intelligent behavior on the basis of reinforcement principles alone, reinforcement-learning theorists believe that AI systems and cognitive theories that steer clear of these basic learning principles are handicapped as well.

A related factor that limited the influence of reinforcement-learning principles in AI is the belief that they were too computationally weak to be of much use. However, there is now ample evidence that reinforcement learning can be very powerful. Some of the most impressive accomplishments of artificial learning systems have been achieved using reinforcement learning. For example, Tesauro (1994, 1995) designed a system that used reinforcement learning to learn how to play backgammon at a very strong masters level; Zhang and Dietterich (1995) used reinforcement learning to improve over the state of the art in a job-shop scheduling problem; and Crites and Barto (1996) obtained strong results on the problem of dispatching elevators in a multi-story building with the aim of minimizing a measure of passenger waiting time. These are all very large-scale problems that present formidable difficulties for conventional solution methods.

In addition to these successes, the growing interest in reinforcement learning among current AI researchers is fueled by the challenge of designing intelligent systems that must operate in dynamic real-world environments. For example, making robots, or robotic "agents," more autonomous (that is, less reliant on carefully controlled, fully anticipated conditions) requires decision-making methods that are effective in the presence of uncertainty and that can meet time constraints. Under these conditions, learning seems essential for achieving skilled behavior, and it is under these conditions that reinforcement learning can have significant advantages over other types of learning.

Despite much recent progress in machine learning, including new learning methods for artificial neural networks, most machine-learning research has focused on learning under the tutelage of a knowledgeable "teacher" that can explicitly tell the system how it should respond to a set of training examples. Although supervised learning, or learning from examples, as this type of learning is called, is an important component of more complete systems, it is not by itself adequate for the kind of learning that an autonomous agent must accomplish. It is often very costly, or even impossible, to obtain instructions that are both correct and representative of the situations in which the agent will have to act. In uncharted territory—where one would expect learning to be most beneficial—an agent has to learn from its own experiences rather than from a knowledgeable teacher. The primary source of information and feedback in reinforcement learning is this interaction with an environment. Of course, an agent should also be able to take advantage of the knowledge and experience of other agents to the extent that it can, but it should not subordinate its own intrinsic goals, determined by its definition of what events are intrinsically reinforcing, to the more superficial goal of meeting the specifications set of another, possibly fallible, agent.

Reinforcement learning has developed into an unusually multidisciplinary research area. Researchers from AI, artificial neural networks, robotics, control theory, operations research, and psychology are actively involved. In this chapter we introduce the field largely from the perspective of AI and engineering. We describe some of the key features of reinforcement learning, provide a formal model of the reinforcement-learning problem, and define basic concepts that are exploited by solution methods. Reinforcement-learning methods themselves and their histories are very broad topics that we do not attempt to cover here. The reader should consult Barto (1992); Barto, Bradtke, & Singh (1995); Kaelbling (1993); and Sutton (1992) for some of these details and extensive bibliographies. We also do not discuss how this model of reinforcement learning relates to details of animal-learning theory or to neuroscience. The reader should consult Barto (1992, 1994) for some references to this literature.

Some Key Features

A good way to introduce some of the key features of reinforcement learning is to consider a few of the examples and possible applications that have motivated and guided its development:

(1) A master chess player makes a move. The choice is informed both by planning—anticipating possible replies and counter-replies—and by immediate, intuitive judgments of the desirability of particular positions and moves.

(2) An adaptive controller adjusts parameters of a petroleum refinery's operation in real time. The controller optimizes the yield/cost/quality tradeoff based on specified marginal costs without sticking strictly to the set points originally suggested by human engineers.

(3) Phil prepares his breakfast. When closely examined, even this apparently mundane activity reveals itself as a complex web of conditional behavior and interlocking goal-subgoal relationships: Walking to the cupboard, opening it, selecting a cereal box, then reaching for, grasping, and retrieving it. Other complex, tuned, interactive sequences of behavior are required to obtain a bowl, spoon, and milk jug. Each step involves a series of eye movements to obtain information and to guide reaching and locomotion. Rapid judgments are continually made about how to carry the objects or whether it is better to ferry some of them to the dining table before obtaining others. Each step is guided by goals, such as grasping a spoon, or getting to the refrigerator, and is in service of other goals, such as having the spoon to eat with once the cereal is prepared and of ultimately obtaining nourishment.

(4) A mobile robot decides whether it should enter a new room in search of more trash to collect or start trying to find its way back to its battery-recharging station. It makes its decision based on how quickly and easily it has been able to find the recharger in the past.

These examples share features that are so basic that they are often overlooked. All involve an *interaction* between an active decision-making agent and its environment in which the agent seeks to achieve a *goal* despite variations or *uncertainties* in the environment. The agent's actions are permitted to affect the future state of the environment (e.g., the next chess position, the level of reservoirs of the refinery, the next location of the robot), thereby affecting the options and opportunities available to the agent at later times. Correct choice requires taking into account indirect, delayed consequences of action, and thus may require foresight or planning. At the same time, the effects of actions cannot be fully predicted, so the agent must frequently monitor its environment and react appropriately. These three features—interactivity, uncertainty, and explicit goals—are key features of problems requiring intelligent adaptive behavior. Reinforcement learning is centered on such problems.

Another key feature of the reinforcement-learning approach is that it explicitly considers the *whole* problem of a goal-directed agent interacting with an uncertain environment. This is in contrast to many approaches that address a putative subproblem without addressing how it fits within a larger picture. We have already mentioned, for example, that much of machine-learning research is concerned with supervised learning without explicitly specifying how such an ability would finally be useful. Other AI researchers have developed theories of planning without considering its role in real-time decision making or the question of where the predictive models necessary for planning would come from. Whether or not these approaches are yielding useful results, it is clear that their focus on isolated subproblems has now become an important limitation.

Reinforcement learning takes the opposite tack by starting with a complete, interactive, goal-seeking system. All reinforcement-learning systems have an explicit goal, can sense aspects of their environments, and can choose actions to influence their environments and goals. Goals that involve planning address its interplay with real-time action selection and the question of how environmental models are acquired. Goals that involve supervised learning do so informed by a very specific role that specifies which capabilities and features are critical, and which are not.

Being complete in this sense does not, of course, mean that the reinforcement-learning approach currently fills in all the details, or even suggests how they should be filled in. Reinforcement learning is developing in an *abstract* framework that, while very broad in scope, requires imposing additional structure to address certain kinds of questions. There are many directions in which the reinforcement-learning model we describe here can be profitably specialized and extended.

An Example

The familiar children's game of naughts and crosses (Tic-Tac-Toe) provides a very simple example of reinforcement learning. Two players take turns playing on a three-by-three board. One player plays crosses (X's) and the other naughts (O's) until one player wins by placing three marks in a row—horizontally, vertically, or diagonally—as the "X" player has in **Figure 1**.

If the board fills up with neither player getting three in a row, the game is a draw. Because a skilled player can play so as never to lose, let us assume that we are playing against an imperfect player whose play is sometimes incorrect,

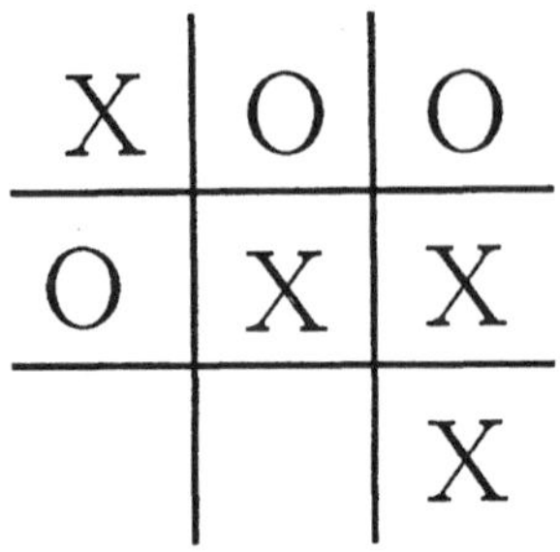

FIGURE 1. Naughts and Crosses (Tic-Tac-Toe). Two players take turns until one of them wins by placing three of his marks (X or O) in a row in any direction.

thereby allowing us to win occasionally. How might one construct a player that will find the imperfections in its opponent's actions and learn to maximize its chances of winning?

Although this is a very simple problem, it cannot readily be solved in a fully satisfactory way by classical techniques. For example, the classical "minimax" solution from game theory is not correct here because it assumes perfect play by the opponent. A minimax solution would never reach a game state from which it could lose, even if in fact it always won from that state because of incorrect play by the opponent. Classical optimization methods for sequential decision problems, such as dynamic programming (e.g., Bertsekas, 1987), can *compute* the optimal solution for any opponent, but require a complete specification of that opponent, including the probabilities with which the opponent would make each move in each board state. We assume this information is not available a priori for this problem, as it is not for the vast majority of problems of practical interest. On the other hand, such information can be estimated from experience, in this case by playing many games against the opponent. About the best one can do with classical methods is to first build from experience a model of the opponent's behavior up to some level of confidence, and then apply dynamic programming to compute an optimal solution given the approximate opponent model. Functionally, this is not that different from some reinforcement-learning methods.

Here is how the naughts-and-crosses problem could be solved most easily using a simple reinforcement-learning approach. First we set up a table of numbers, one for each possible state of the game—i.e., one for each possible configuration of X's and O's on the three-by-three board. Each number provides an estimate of the probability of our winning from that state. Assuming we always play X's, then for all states with three X's in a row the probability of winning is 1, because we have already won. Similarly, for all states with three O's in a row, the correct probability is 0 as we cannot win from them. All the other states, the nonterminals, we set initially to the same value, say 0.5, representing a 50% chance of winning.

Now we play a great many games against the opponent. To select our moves we examine the states that would result from each of our possible moves (one for each blank space on the board) and look up their estimated probabilities of winning. Most of the time we select as our move the one that leads to the state with the highest estimated probability of winning. Occasionally, however, we select randomly from one of the other moves instead; these are called *exploratory* moves because they cause us to experience states that might otherwise never occur. A sequence of moves made and considered during a game can be diagrammed as in **Figure 2**.

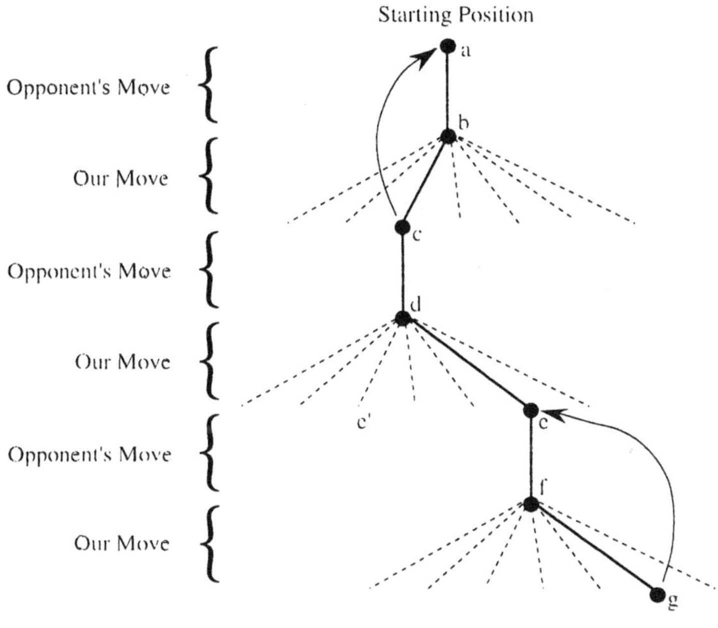

FIGURE 2. Moves in Naughts and Crosses. The bold lines represent the moves taken during a game. The dashed lines represent moves that we (our algorithm) considered but did not make. Our second move was an exploratory move, meaning that it was taken even though some other alternative move, that leading to e', was more highly ranked. Exploratory moves do not result in learning, but each of our other moves does, causing backups as suggested by the curved arrows and detailed in the text.

Now, while we are playing, we change the probability estimates for the states in which we find ourselves during the game. We attempt to make them more accurate estimates of the probabilities of winning from those states.

Informally, we say that the probability estimate for the state after each regular move is "backed up" to the estimate for the state after our preceding move, as suggested by the arrows in **Figure 2**. More precisely, the probability estimate for the earlier state is moved a fraction of the way from its current value to the value of the later state. Letting x_k denote the state after our kth move, and letting $V(x_k)$ denote the estimated probability of winning from that state (the *value* of state x_k), the update rule can be written:

$$V(x_k) := V(x_k) + \alpha[V(x_{k+1}) - V(x_k)],$$

where α is a small positive fraction called the *step-size parameter*.

This update rule performs quite well with this task. For example, if the step-size parameter is reduced properly over time, this method will converge for any fixed opponent to the true probabilities of winning from each state given optimal play by the algorithm (Singh, Jaakkola, & Jordan, 1994). Furthermore, the moves then taken (except on exploratory moves) will, in fact, be the optimal moves against the opponent. If the step-size parameter is not reduced to zero over time, then a player using this rule will also play well against opponents that change their play slowly over time. This update rule is closely related to the method Samuel used in his 1959 program for learning how to play the game of checkers (Samuel, 1959). Sutton (1988), who refined and analyzed algorithms like this, called them *temporal-difference methods*.

This example is very simple, but it illustrates some of the key features of reinforcement-learning methods. First, there is the emphasis on learning while interacting with an environment, in this case with an opponent player. Second, there is a clear goal, and correct behavior requires planning or foresight that takes into account delayed effects of one's choices. The simple reinforcement-learning player of naughts and crosses will, for example, learn to set up multi-move traps for a short-sighted opponent. It is a striking feature of reinforcement learning that it can achieve the effects of planning and lookahead *without* using a world model or carrying out an explicit search over sequences of choices. To be sure, planning using world models can be useful, but it is not always worth the effort.

On the other hand, the naughts-and-crosses example is so simple that it might give the false impression that reinforcement learning is restricted to such tasks. Although naughts and crosses is a two-person game, reinforcement learning also applies in the more natural context in which there is no explicit external adversary. Naughts and crosses involves a relatively small, finite-state set, whereas reinforcement learning can be applied to very large or even infinite-state sets. For example, Tesauro (1994, 1995) combined the algorithm described above with an artificial neural network to acquire impressive skill in playing backgammon, which has a huge number of states—approximately 10^{20}. The neural network provides this program with the ability to *generalize* from its past experiences, so that in new situations it selects moves based on information saved from similar situations faced in the past, as determined by its network. Thus, how well a reinforcement-learning system can work with problems having very large state sets is intimately tied to how appropriately it can generalize from past experience. Methods for supervised learning, which focus almost exclusively on the problem of forming appropriate generalizations, are most relevant to this aspect of reinforcement learning. A neural network is clearly not the only, or necessarily the best, way to do this.

Other features of the naughts-and-crosses example are not essential to reinforcement learning, for example, learning with no prior knowledge beyond the rules of the game. However, although many other reinforcement-learning examples begin similarly devoid of knowledge, reinforcement learning by no means entails a *tabula rasa* view of learning and intelligence. On the contrary, prior information can be incorporated into a reinforcement-learning system in a variety of ways that can be critical for efficient performance (Clause & Utgoff, 1992; Lin, 1992; Maclin & Shavlik, 1994; Mitchell & Thrun, 1993).

The naughts-and-crosses player also had to look ahead one step in order to evaluate the possible immediate results of a move. To be able to do this, it had to have a model of the game that allows it to "think about" how its environment will change in response to moves that it may never make. However, the naughts-and-crosses player used its model in only a very simple way, whereas other reinforcement-learning systems make much more extensive use of environmental models (e.g., Barto et al, 1995; Moore & Atkeson, 1993; Sutton, 1990, 1991). They can generate hypothetical experiences from which they can learn in the same way that the naughts-and-crosses player learns from real experience, or they can "reason" about the consequences of possible behavior and make various kinds of plans. These more complicated model-based reinforcement-learning systems can include a full range of high-level, symbolic processing, and an important aspect of reinforcement learning is the improvement of environmental models through learning. Thus, although reinforcement learning is often associated only with very low-level processing, this is by no means an essential aspect of the approach.

On the other hand, there are reinforcement-learning methods that do not need any kind of environmental model at all. Watkins (1989) called these *primitive methods*. Systems using only primitive methods cannot even think about how their environments will change in response to a single action. Because models have to be reasonably accurate to be useful, primitive methods can have advantages over more complex methods when the crucial bottleneck in solving a problem is difficulty in constructing a sufficiently accurate environmental model. Primitive methods are also important building blocks for model-based methods.

The naughts-and-crosses player had access to the complete state of the game, but reinforcement learning can also be applied when part of the state is hidden, or when different states appear to the learner to be the same (e.g., Whitehead & Ballard, 1990; Jaakkola, Singh, & Jordan, 1995). Finally, the naughts-and-crosses player is a reinforcement-learning system on just one level. The decisions refined by learning are about the primitive moves of the game. Recalling our comments about the abstract nature of the reinforcement-learning framework, nothing prevents reinforcement learning from working at higher levels, for example, where each of the "actions" is itself the application of a possibly elaborate problem-solving method (e.g., Maes & Brooks, 1990; Mahadevan & Connell, 1991; Singh, Barto, Grupen, & Connolly, 1994). In hierarchical learning systems, reinforcement learning can work simultaneously on several levels (Dayan & Hinton, 1993; Singh, 1991, 1992).

Fully satisfactory solutions are of course not yet available in all cases. Most of the theoretical results that exist so far, in fact, apply only to problems that share with the naughts-and-crosses problem the use of a tabular representation of a finite set of state values and access to complete environmental states (Barto et al, 1995). However, many reinforcement-learning researchers, like many other AI researchers, are willing to forge ahead when theoretical guarantees are lacking, and many applications of reinforcement-learning methods have been realized in ways that go considerably beyond available theory. Moreover, some of these applications have been very successful, as in the examples mentioned above by Tesauro (1994, 1995), Zhang and Dietterich (1995), and Crites and Barto (1996).

The Credit-Assignment Problem

In his famous paper "Steps Toward Artificial Intelligence," Minsky (1961) presented the basic ideas of "success-reinforced decision models" and discussed the major computational problem that complex reinforcement-learning systems would have to solve to be successful. He called this the *credit-assignment problem*:

> In applying such methods to complex problems, one encounters a serious difficulty—in distributing credit for success of a complex strategy among the many decisions that were involved (p. 17).

Later researchers distinguished between *temporal* and *spatial* aspects of the problem. Temporal credit assignment concerns determining which actions in the sequence of preceding actions were responsible for an eventual success (or failure). For example, if you win a chess game, how should you apportion credit among all the moves you made? The spatial aspect of the problem, on the other hand, concerns allocating credit to the many, possibly simultaneous, decisions that finally yielded an overt action. For example, if in winning the chess game your temporal credit-assignment mechanism assigned a certain amount of credit to a particular move, how should you further apportion this credit among the various decisions that caused you to select it? Both aspects of the credit-assignment problem remain central problems for modern reinforcement-learning systems.

The approach to temporal credit assignment used by many reinforcement-learning systems, including the naughts-and-crosses player described above, was introduced to AI in Samuel's program for learning how to play checkers (Samuel, 1959). The idea is that a reinforcement-learning system should not have to wait to learn until an externally supplied reinforcement signal occurs. The checkers player, for example, should not have to wait until the end of a game to receive reinforcement. The player should be able to produce for itself internal reinforcement when it achieves important subgoals during a game. Moreover, the player should be able to *learn* to recognize when important subgoals are achieved. Samuel's method for doing this is related, as previously noted by Minsky (1961), to the phenomenon of conditioned reinforcement in animal learning. An event that regularly precedes a reinforcing event can itself acquire the ability to reinforce still earlier activity; i.e., the event becomes a conditioned reinforcer. Conditioned reinforcers can, in turn, confer reinforcing qualities upon earlier events, making them into conditioned reinforcers as well. This suggests a recursive mechanism by which a system can learn long sequences of actions that ultimately bring about real success, that is, success as determined by the ultimate primary reinforcer. Much of modern reinforcement learning exploits this process.

The spatial aspects of credit assignment are not unique to reinforcement learning. For example, the error backpropagation method for adjusting the weights of a multi-layer artificial neural network is a spatial credit-assignment method widely used in supervised learning (Rumelhart, Hinton, & Williams, 1986). Although this algorithm can be adapted to address temporal aspects of credit assignment, it ordinarily only addresses the spatial aspects by apportioning the credit (in this case, the blame) among the weights of a complex network for the errors made by the network as a whole. Similarly, reinforcement-learning systems have to adjust their decision rules even if some other mechanism produces timely reinforcement. Reinforcement-learning systems can use a variety of methods developed for supervised learning, although some methods are better suited than others due to the different demands of reinforcement learning.

The general approach to credit assignment taken by reinforcement-learning systems is the major feature distinguishing them from methods based more directly on evolutionary metaphors, such as genetic algorithms (Goldberg, 1989; Holland, 1975). Like reinforcement learning, evolutionary methods can be used to adapt the interactive behavior of an agent to achieve an explicit goal, but they do so without assigning credit on an intra-individual basis. For example, if an agent does well, credit is assigned to *all* of its behavior, independently of how specific components of this behavior were related to success; full credit will even be given to behavior that was not expressed during the agent's lifetime. As a consequence, evolutionary methods, when used alone, may be inherently less efficient than methods that assign credit by taking into account intra-individual details about an agent's decision mechanisms and how they are marshaled over time. On the other hand, by not attempting intra-individual credit assignment, evolutionary methods are not misled by credit's being incorrectly assigned. In any event, we do not consider evolutionary methods to be especially well adapted to the reinforcement-learning problem. Although evolution and learning, especially reinforcement learning, share many features and can naturally work together, as they do in nature, they do not have equal access to the same credit-assignment mechanisms.

The Reinforcement-Learning Problem

Although certain learning algorithms are commonly associated with reinforcement learning, it is more useful to define reinforcement learning in terms of learning *problems*, or collections of problems, rather than as a collection of algorithms. Here we present a model of the problem that many AI researchers have adopted in their approaches to reinforcement learning. This model is based on the *Markov Decision Process* formalism that has been widely studied by decision theorists (e.g., Bertsekas, 1987; Ross, 1983).

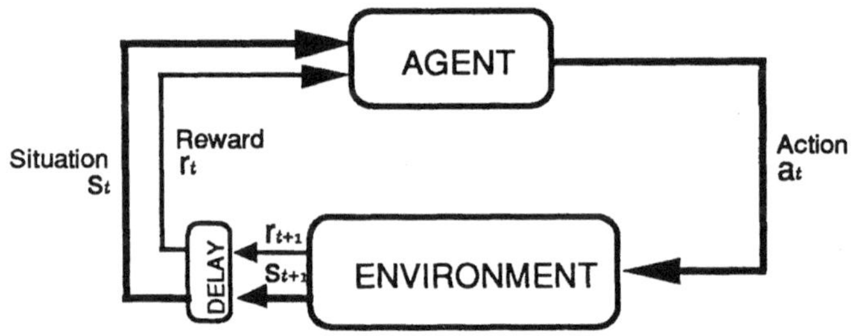

FIGURE 3. A Reinforcement-Learning Model. A reinforcement-learning agent and its environment interact over a sequence of discrete time steps. The *actions* are the choices made by the agent; the *situations* provide the agent's basis for making the choices; and the *rewards* are the basis for evaluating these choices.

The agent-environment interface

Reinforcement learning is about learning how to act to achieve a goal. A fruitful way of modeling such learning is based on viewing a decision maker, or *agent*, as a control system that is trying to develop a strategy by which it can make its environment behave in a favorable way (where "favorable" has a precise meaning). A simple type of strategy maps each situation to a probability distribution over the actions that are possible for that situation. Upon determining that it is in a new situation, the agent selects an action according to the probability distribution for that situation. As the agent learns, it changes this mapping, called its *policy*, based on its accumulating experience.

To make the model more specific, think of the agent and its environment as interacting over a potentially infinite sequence of discrete time steps t = 1,2,3,... At each time step t, the reinforcement-learning agent finds itself in a *situation*, $s_t \in S$, and on that basis uses its current policy to choose an *action*, $a_t \in A(s_t)$, where $A(s_t)$ is the set of actions available for situation s_t. One time step later, in part as a consequence of its action, the agent receives a numerical *reward*, $r_{t+1} \in \Re$, and finds itself in a new situation, $s_{t+1} \in S$ (**Figure 3**). Reinforcement-learning methods specify how such experiences produce changes in the agent's policy, which tells it how to select an action in any situation. Roughly speaking, the agent's objective is to find a policy that maximizes the amount of reward it receives over the long run.

It is important to understand the degree of abstraction this model involves. It is a very abstract and flexible model that can be applied at many different levels to many different problems. The actions, for example, could be low-level controls such as the voltages applied to the motors of a robot arm, or high-level decisions such as whether or not to have lunch or go to graduate school. Similarly, the situations can take a wide variety of forms. They could be low-level situations, such as direct sensor readings, or high-level ones, such as symbolic descriptions of the objects in a room. Some of the things making up a situation could even be entirely mental or subjective. For example, the agent could be in the situation of not being sure where an object is, or of having just been "surprised" in some clearly defined sense. Similarly, some actions could also be totally mental or computational, e.g., they may control what the agent chooses to think about, or where it focuses its attention. In general, actions can be the results of any decisions we learn how to make, and the situations can be anything we can sense that might be useful in making the decisions.

In particular, it is a mistake to think of the interface between a reinforcement-learning agent and its environment as the physical boundary between a robot's or an animal's body and the external environment. Usually the boundary is drawn closer to the agent. For example, the motors and mechanical linkages of a robot and its sensing hardware should usually be considered parts of the environment rather than parts of the learning agent, even though these parts were probably designed to make the learning agent's task easier. Similarly, if we apply the model to a person or animal, the skeleton, muscles, and sensory organs should all be considered part of the learning agent's environment. Reinforcers, too, may presumably be computed inside the physical bodies of natural and artificial learning systems, but are considered external to the reinforcement-learning agent.

The general rule we follow is that anything that cannot be changed arbitrarily by the learning agent is considered external to it and, thus, part of its environment. Note that we do not assume that events in the environment are unknown to the agent, only that they are incompletely controllable. For example, the agent will often know quite a bit about how its reinforcers are computed as a function of its actions and the situations in which they occur. But we always consider the reward computation to be external to the agent because it defines the problem facing the agent and, thus, is beyond its ability to change arbitrarily. In some cases, in fact, the agent may know *everything* about its environment and still face a difficult reinforcement-learning problem, just as we may know exactly how a puzzle like Rubik's cube works but still be unable to solve it. The agent-environment boundary represents the limit of the agent's *control*, not of its knowledge.

The agent-environment boundary can even be located at different places for different purposes. In a complicated robot, many separate reinforcement-learning agents may be operating at once, each with its own boundary. For example, one agent may make high-level decisions that form part of the situations faced by a lower-level agent that implements the high-level decisions. In practice, the agent-environment boundary is determined once one has selected particular sensations, actions, and reinforcers, and thus identified a particular decision-making problem of interest.

The reinforcement-learning model is a considerable abstraction of the problem of learning to make decisions based on their consequences. It proposes that whatever the details of the sensory and control apparatus, and whatever objective one is trying to achieve, any problem of learning goal-directed behavior can be reduced to three signals passing back and forth between an agent and its environment: One signal represents the choices made by the agent (the actions); a second signal represents the basis on which the choices are made (the situations); and a third signal defines the goal of learning (the rewards). We do not claim that this framework is adequate to usefully model *all* decision-learning problems, but it has proven to be widely applicable. Of course, the situation and action representations will vary greatly from application to application, and will strongly affect performance. In reinforcement learning, as in other kinds of learning, such representational choices are at present more art than science.

Goals, rewards, and returns

In reinforcement learning, the concept of goal is modeled by a special scalar signal called the *reward* that passes from the environment to the agent. Informally, the agent's goal is to maximize the total reward it receives. This means not just immediate reward, but reward over the long run.

The use of a scalar reward signal to formalize the idea of a goal is one of the most distinctive features of reinforcement learning. Although this way of formulating goals might at first appear limiting, in practice it has proven to be very flexible and very widely applicable. The best way to see this is to consider examples of how it may be used. For example, to train a robot to walk, researchers have provided reward on each time step proportional to the robot's forward motion. In learning to run a maze, the reward is often zero except upon reaching the goal, when it becomes +1. Another common approach in maze learning is to give a reward of -1 for every time step that passes prior to reaching the goal; this encourages the agent to reach the goal as quickly as possible. To train a robot to find and collect empty soda cans for recycling, one might give it a reward of +1 for each empty can collected. One might also give the robot punishers when it bumps into things, or when people yell at it. For an agent learning to play backgammon or chess, the natural rewards for winning, losing, and drawing are +1, -1, and 0, respectively.

It is important to remember that rewards define the ultimate goal of the learning process. The rewards delivered to a reinforcement-learning agent should represent what you really want the agent to do. In particular, the reward signal is not the place to impart to the agent prior knowledge about *how* to achieve what you want it to do. For example, a chess-playing agent should be rewarded only for actually winning, not for achieving subgoals such as taking its opponent's pieces or gaining control of the center of the board. If these kinds of subgoals are rewarded, the agent might find a way to achieve the subgoals without achieving the real goal, e.g., taking the opponent's pieces even at the cost of losing the game. The reward signal is a way of communicating to the robot *what* it should achieve, not *how* it should be achieved.

Newcomers to this model of reinforcement learning are sometimes surprised that the rewards—the definition of the goal of learning—are computed in the environment rather than in the agent. Certainly, most ultimate goals for animals are recognized by computations occurring inside their bodies, e.g., by sensors for recognizing food and hunger, pain and pleasure, etc. However, as we discussed in the previous section, one can simply redraw the agent-environment interface such that these parts of the body are considered to be outside of the agent (and thus part of the agent's environment). For example, if the goal concerns a robot's "internal" energy reservoirs, then these are considered part of the environment; if the goal concerns the positions of the robot's limbs, then these too are considered part of the environment—the boundary is drawn at the interface between the limbs and their control systems.

Roughly speaking, structures and processes are considered part of the agent if they are completely, directly, and with certainty, controllable; otherwise they are considered part of the environment. The ultimate goal is always something over which the reinforcement-learning agent has imperfect control: It cannot, for example, simply decree that the goal has been achieved (in the same way that it can arbitrarily set an internal parameter of its decision-making process). Therefore, we place the reward source outside of the agent. Note that this does not preclude the agent from defining for itself an internal goal, or a sequence of internal goals. Indeed, the commonly used method for temporal-credit assignment, based on Samuel's approach described above, does just that: It effectively defines internal goals.

Until this point, we have been imprecise when we spoke of the goal of learning as maximizing reward over the long run. How might this be formally defined? If the sequence of rewards received after time step t is denoted r_{t+1}, r_{t+2}, r_{t+3}, ..., then what aspect of this sequence do we wish to maximize? There are several useful answers to this question. The simplest is to maximize the *total reward*:

$$r_{t+1} + r_{t+2} + r_{t+3} + \dots + r_T \quad (1)$$

where T is a final time step. This approach makes sense in applications in which there is a natural notion of final time step in a trial, that is, when the agent-environment interaction breaks naturally into subsequences, such as plays of a game, trips through a maze, or any sort of repeated attempt where each repetition ends with a reset to a standard state. In these cases Equation 1 defines the *return* for time step t, i.e., the return that accumulates *after* time step t.

On the other hand, suppose that the agent-environment interaction does not naturally break into identifiable subsequences but simply goes on without limit. This would be the natural way to characterize a continuous process-control application, or an application to a robot with a long expected lifespan. The total-reward formulation then becomes problematic because the final time step becomes T approaches ∞, and the return as given by Equation 1 becomes a sum of an infinite number of terms. Thus the return, which is what the agent is trying to maximize, could itself be infinite (e.g, if the agent receives a reward of +1 at each time step).

The additional concept we need is that of *discounted return*. According to this approach, the agent's objective is to learn how to select actions so that, at every time step, the discounted sum of the rewards received over the future is maximized. That is, the objective is to learn to maximize the following definition of return for each time step t:

$$r_{t+1} + \gamma r_{t+2} + \gamma^2 r_{t+3} + \dots = \sum_{k=1}^{\infty} \gamma^{k-1} r_{t+k} \quad (2)$$

where γ is a positive number called the *discount factor*.

The discount factor determines the present value of future rewards: A reward received k time steps in the future is worth γ^{k-1} times what it would be worth if it were received immediately. If $0 \leq \gamma < 1$, this infinite discounted sum is finite as long as each individual reward is finite. If $\gamma = 0$, the agent is "myopic," i.e., only concerned with maximizing immediate rewards. Its objective in this case would be to learn how to act at each time step t so as to maximize only r_{t+1}. If each of the agent's actions happened only to influence the immediate reward, not future rewards as well, then a myopic agent could maximize Equation 2 by separately maximizing each immediate reward. But, in general, acting to maximize immediate reward can reduce access to future rewards so that the total reward may actually be reduced. As γ approaches one, the objective takes future rewards into account more strongly: The agent becomes more farsighted. Other definitions of return for infinite-duration problems are possible (Mahadevan, 1996), but the discounted return is the simplest mathematically.

Example: A problem that served as an early illustration of reinforcement learning is the problem of *pole-balancing* (Michie & Chambers, 1968). The objective here was to apply forces to a cart moving along a track so as to keep a pole hinged to the cart from falling over (**Figure 4**). We define a *balancing failure* as the fall of the pole past a given angle from vertical or the cart's exceeding the limits of the track. The pole is reset to vertical after each balancing failure. This problem could be treated as a total-reward problem, where the natural subsequences are the repeated attempts to balance the pole. The reward

FIGURE 4. The Pole-Balancing Problem. The objective is to apply forces to a cart moving along a track so as to keep a pole hinged to the cart from falling over.

in this case would be +1 for every time step on which failure did not occur, so that the return at each time would be the number of steps before failure. Alternatively, a punisher of -1 could be given for each failure and zero reward at all other times. The return at each time would then be related to $-\gamma^k$, where k is the number of time steps before failure. In either case, the return is maximized by keeping the pole balanced for as long as possible.

Situations and states

The *state* of a system with respect to an external observer is a summary of the observer's past experience with the system. The summary need not be a complete history of every observed input and output, but it must contain all the information that makes a difference as far as the system's future behavior is concerned. In particular, the observer must be able (in principle) to predict the system's future behavior just as well from knowledge of its current state as from knowledge of its complete history. For example, the state of a cannonball in flight is its current position vector and velocity vector. It doesn't matter how its current position and velocity came about.

In reinforcement learning, the external observer is the agent, and the state of interest is the state of the environment. In fact, the notion of "situation" is meant to be an approximation of the environment's state. What exactly is the state of the environment? The agent's past experience with the environment consists of all of the previous situations, actions, and rewards. Assuming that interaction began at $t = 0$, the complete *history* at time t is

$$H_t = \{s_t, r_t, a_{t-1}, s_{t-1}, r_{t-1}, a_{t-2}, s_{t-2}, \ldots, r_1, a_0, s_0\}. \quad (3)$$

Any signal $x_t \in X$, $t \geq 0$, gives the state of the environment at time step t if and only if the joint probability of the state, situation, and reward at time step $t+1$, given x_t and a_t, the action at time step t, is the same as their joint probability given H_t and a_t. When the number of possible states, situations, rewards, and actions are finite, this can be written simply as follows:

$$P\{x_{t+1}=x, s_{t+1}=s, r_{t+1}=r \mid x_t, a_t\} = P\{x_{t+1}=x, s_{t+1}=s, r_{t+1}=r \mid H_t, a_t\}, \quad (4)$$

for all $t \geq 0$, $x \in X$, $s \in S$, $r \in \Re$, $a \in A(s_t)$, and all possible H_t, where X, S, $\Re$, and $A(s_t)$ are finite sets of possible states, situations, rewards, and actions, respectively. In reinforcement learning, situations are intended to approximate the environment's states. The situations are in fact true states if and only if

$$P\{s_{t+1}=s, r_{t+1}=r \mid s_t, a_t\} = P\{s_{t+1}=s, r_{t+1}=r \mid H_t, a_t\} \quad (5)$$

for all $t \geq 0$, $s \in S$, $r \in \Re$, $a \in A(s_t)$, and all possible H_t. (If any of these sets are not finite, e.g., if a reward can be any real number, then the same conditions can be written in terms of probability density functions.) In this important special case, the environment and its interface define a Markov Decision Process, or MDP. If an MDP has a finite number of states, and a finite number of actions are available for each state, then it is a *finite MDP*. Because it is particularly easy to conceptualize and to prove theorems about finite MDPs, they play a central role in the theoretical analysis of reinforcement learning.

The conditional probability distributions given by Equation 5 constitute a complete description of the *dynamics* of the MDP. As far as the agent is concerned, the dynamics specify how the environment changes over time in response to its actions. If a reinforcement-learning agent has complete knowledge of its environment's dynamics, then it faces a reinforcement-learning problem under conditions of *complete information*. Most problem-solving methods in AI have addressed problems of complete information, whereas reinforcement learning focuses primarily on problems of incomplete information. The reader should be careful not to confuse complete and incomplete *information* with complete and incomplete *observation* of the environmental state. We refer to the case of complete observation by saying that the environment has the Markov property.

One can show that by iterating Equation 4 or Equation 5 an agent can predict *any* future state and reward from knowledge only of the current state and its proposed course of action (together with knowledge of the dynamics) as well as would be possible given the complete history. It also follows that the situations in MDPs provide the best possible basis for choosing actions. That is, the best policy for choosing actions as a function of situations is just as good as the best policy for choosing actions as a function of complete histories.

Even when the situations are not technically states in the sense of exactly satisfying Equation 5, it may still be appropriate to think of the situation in reinforcement learning as an approximation to the environment's state. In particular, we always want the situation to be a good basis for predicting future rewards and for selecting actions. For some purposes, it is also desirable to use present situations to accurately predict following situations. States provide an unsurpassed basis for doing all of these things. To the extent that situations approximate states in these ways, one can obtain better performance from reinforcement-learning systems. For all of these reasons, it is useful to think of the situation at each time step as an approximation to an MDP's state, although one should remember that a situation is often not precisely a state. Although most reinforcement-learning algorithms can be applied when the situations are not states, sometimes with good results, almost all of the formal theory rests on the assumption that situations are actual states.

Example: In the pole-balancing problem introduced in the previous section, a situation would be a state if it exactly specified, or made it possible to exactly reconstruct, the position and velocity of the cart along the track, the angle between the cart and the pole, and the rate at which this angle is changing (the angular velocity). In an idealized cart-pole system, this information would be sufficient to exactly predict the future behavior of the cart and pole, given the actions taken by the controller. In practice, however, it is never possible to know this information exactly because any real sensor would introduce some distortion and delay in its measurements. Furthermore, in any real cart-pole system there are always other components of the state, such as the bending of the pole, the temperatures of the wheel and pole bearings, and various forms of backlash, which slightly affect the behavior of the system. These factors would cause violations of Equation 5 if the role of state were played by only the positions and velocities of the cart and the pole.

However, often the situations of the positions and velocities serve quite well as approximate states. In several of the early studies of learning the pole-balancing problem, in fact, learning was successful despite the fact that each situation provided only a very coarse representation of the true state. For example, in our work (Barto, Sutton, & Anderson, 1983), the possible cart positions were divided into three regions: right, left, and middle. The situations indicated only in which of these three large regions the cart was located (and there were similarly rough quantizations of the other three intrinsic state variables). These rough approximations to the state were sufficient to easily solve the problem using reinforcement learning. In fact, this coarse representation of the state probably facilitated learning because it forced the learning agent to ignore fine distinctions that would not have been particularly useful in solving the problem.

Example: In draw poker, each player is dealt a hand of five cards. There is a round of betting in which each player exchanges some of his cards for new ones, and then there is a final round of betting. At each round of betting, a player must match the highest bets of the other players or else drop out (fold). After the second round of betting, the player with the best hand and who has not folded is the winner and collects all the bets.

The relevant state in draw poker is different for each player. Each player knows the cards in his own hand, but can only guess at those in the other players' hands. A common mistake is to think that the state must include the contents of all the players' hands and the cards remaining in the deck. However, this would provide more information than the state. In a fair game, one assumes that the players are in principle unable to determine these things from their past observations. If a player did have such information, some future events (such as the cards one could exchange for) could be *better* predicted than by remembering all past observations.

In addition to knowledge of one's own cards, the state in draw poker includes knowledge of the bets and the numbers of cards drawn by the other players. For example, if a player draws three new cards, you may suspect he retained a pair and adjust your estimate of the strength of his hand accordingly. The players' bets also influence your assessment of their hands. In fact, all of your past history with these particular players is part of the state. Does Ellen like to bluff, or does she play conservatively? Does her face or demeanor provide clues to the strength of her hand? How does Joe's play change when it is late at night, or when he has already won a lot of money?

Although everything ever observed about the other players may have an effect on the probabilities that they are holding various kinds of hands, in practice this is far too much to remember and analyze, and most of it will have no clear effect on one's predictions and decisions. Very good poker players are adept at remembering just the key clues and at sizing up new players quickly, but no one remembers everything that may be relevant. As a result, the situations people use to make their poker decisions are imperfect state models, and the decisions themselves are presumably imperfect. Nevertheless, people can still make very good decisions in such problems. The inability to have access to a *perfect* representation of the environment's state is probably not a severe problem for an AI agent.

Value Functions

Almost all reinforcement-learning algorithms are based on estimating *value functions*—functions of situations, or of situation-action pairs, that estimate *how good* it is for the agent to be in that situation. The notion of "how good" is defined in terms of expected future rewards or, to be precise, as the expected return given, by Equation 1 or Equation 2, for example. Of course, the rewards an agent can expect to receive in the future depend on what actions it takes. Accordingly, value functions are defined with respect to a particular policy. Recall that a policy, let us call it π, is a mapping from situations $s \in S$ to probability distributions over possible actions $a \in A(s)$. Informally, the value of a situation under a policy π, denoted $V^\pi(s)$, is the expected return when starting in s and following π. For MDPs, we can define $V^\pi(s)$ formally as:

$$V^\pi(s) = E_\pi \{\sum_{k=t}^{T} r_k \mid s_t = s\}, \tag{6}$$

for the total-reward case, where the return is defined by Equation 1, and where $E_\pi\{\}$ denotes the expected value given that the agent follows policy π. For the discounted case, in which the return is given by Equation 2, $V^\pi(s)$ is defined as:

$$V^\pi(s) = E_\pi \{\sum_{k=0}^{\infty} \gamma^k r_{t+k+1} \mid s_t = s\}. \tag{7}$$

Similarly, following Watkins (1989), we define the *action-value*, or *quality*, of taking action a in situation s under a policy π, denoted $Q^\pi(s,a)$, as the expected return starting from s, taking the action a, and thereafter following policy π:

$$Q^{\pi}(s,a) = E_{\pi}\{\sum_{k=t}^{T} r_k \mid s_t=s,\ a_t=a\}, \tag{8}$$

for the total-reward case, and

$$Q^{\pi}(s,a) = E_{\pi}\{\sum_{k=0}^{\infty} \gamma^k r_{t+k+1} \mid s_t=s,\ a_t=a\}, \tag{9}$$

for the discounted-reward case. In either case, the (situation) value function is related to the action value function by

$$V^{\pi}(s) = E\{Q^{\pi}(s,a_{\pi(s)})\}, \tag{10}$$

where $a_{\pi(s)}$ is the action selected according to the probability distribution over actions given by $\pi(s)$.

For MDPs, the value functions $V^{\pi}(s)$ and $Q^{\pi}(s,a)$ can be estimated from experience. For example, if an agent follows policy π and maintains an average, for each situation encountered, of the actual returns that have followed that situation, then the averages will converge to the situation's value, $V^{\pi}(s)$, as the number of times that situation is encountered approaches infinity. If separate averages are kept for each action taken in a situation, then these averages will similarly converge to the action values, $Q^{\pi}(s,a)$. Estimation methods of this kind are often called *Monte Carlo methods* because they involve averaging over many random samples of actual returns. Of course, if there are very many situations, or very many actions possible in each situation, then it may not be practical to keep separate averages for each situation individually. Instead, the agent may maintain V^{π} and Q^{π} as parameterized functions and adjust the parameters to better match the observed returns. This can also produce accurate estimates, although much depends on the nature of the parameterized function approximation.

A fundamental property of value functions used throughout reinforcement learning and dynamic programming is that they satisfy particular recursive relationships if the situations are genuine states. For any policy π and any state s the following consistency condition holds between the values of any "neighboring" states:

$$\begin{aligned} V^{\pi}(s) &= E_{\pi}\{\sum_{k=0}^{\infty} \gamma^k r_{t+k+1} \mid s_t=s\} \\ &= E_{\pi}\{r_{t+1} + \gamma \sum_{k=0}^{\infty} \gamma^k r_{t+k+2} \mid s_t=s\} \\ &= E_{\pi}\{r_{t+1} + \gamma V^{\pi}(s_{t+1}) \mid s_t=s\}, \end{aligned} \tag{11}$$

for the discounted-reward case, and

$$V^{\pi}(s) = E_{\pi}\{r_{t+1} + V^{\pi}(s_{t+1}) \mid s_t = s\} \tag{12}$$

for the total-reward case for $t \neq T$. Similar consistency conditions hold for Q^{π}. These conditions can be used in several different ways to compute or approximate V_{π} and Q^{π}.

Example: **Figure 5a** uses a rectangular grid to illustrate a simple finite MDP. The cells of the grid correspond to the states (situations) of the problem. At each cell, four actions are possible: NORTH, SOUTH, EAST, and WEST, which deterministically cause the agent to move one cell in the respective direction in the grid. Actions that would take the agent off the grid leave its location unchanged, but also result in a reward of -1. Other actions result in a reward of 0, except those that move the agent out of the special states A and B. From state A, all four actions yield a reward of +10 and take the system to A'. From state B, all actions yield a reward of +5 and take the system to B'.

Suppose the agent selects the four actions with equal probabilities in all states. **Figure 5b** shows the value function, V^{π}, for this policy, for the discounted-reward case with $\gamma = 0.9$. This value function was computed by solving the system of equations given by Equation 11. Notice the negative values near the lower edge; these are the result of the high probability of hitting the edge of the grid there under the random policy. Notice that A is the best state to be in under this policy, but that its expected return is less than 10, its immediate reward. The expected return is reduced because from A the agent is taken to A', from which it is likely to run into the edge of the grid. State B, on the other hand, is valued more than 5, its immediate reward, because from B the agent is taken to B', which has a positive value. From B'

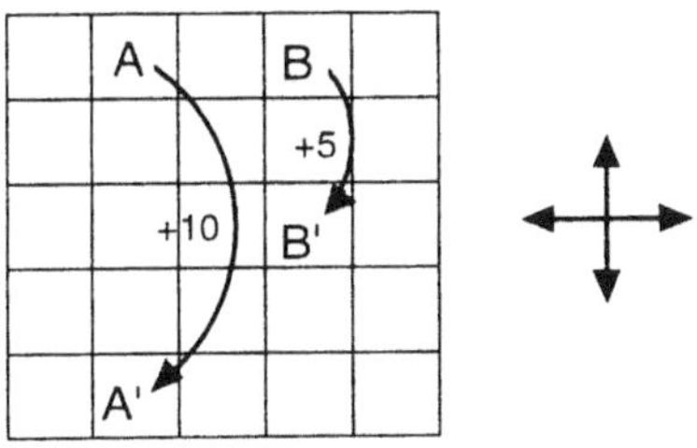

(a)

3.3	8.8	4.4	5.3	1.5
1.5	3.0	2.3	1.9	0.5
0.1	0.7	0.7	0.4	-0.4
-1.0	-0.4	-0.4	-0.6	-1.2
-1.9	-1.3	-1.2	-1.4	-2.0

(b)

FIGURE 5. Rectangular-Grid Illustration of a Value Function. **a.** Each cell is a state, and the agent can move one cell in any of the cardinal directions. Cells A, A', B, and B' are special states. Rewards are generated as described in the text. **b.** The value function for the policy of selecting the four actions with equal probabilities in all states.

the expected penalty (negative reward) for possibly running into an edge is more than compensated for by the expected gain for possibly stumbling onto A or B.

The Optimality Equation

Solving a reinforcement-learning problem means finding a policy that maximizes the expected return for each situation. For finite MDPs, we can precisely define the optimal solution in the following simple way. Value functions define a partial ordering over policies. A policy π is better than or equal to a policy π' if its expected return is greater than or equal to that of π' for all states. In other words, $\pi \geq \pi'$ if and only if $V^{\pi}(s) \geq V^{\pi'}(s)$ for all $s \in S$. There is always at least one policy that is better than or equal to all other policies. This is an *optimal policy*. Although there may be more than one, we denote all the optimal policies by π^*. They share the same value function, denoted V^*, defined as

$$V^*(s) = V^{\pi^*}(s) = \max_{\pi} V^{\pi}(s) \quad \text{for all } s \in S \tag{13}$$

and the same action-value function, denoted Q^*, defined as

$$Q^*(s,a) = Q^{\pi^*}(s,a) = \max_{\pi} Q^{\pi}(s,a) \quad \text{for all } s \in S, \text{ for all } a \in A(s). \tag{14}$$

Because V^* is the value function for a policy, it must satisfy a consistency condition such as Equation 11 or Equation 12. Because it is the optimal value function, however, V^*'s consistency condition can be written in a special form, often called the *optimality equation*, which is independent of the policy. Intuitively, the optimality equation is based on the fact that the value of a state under an optimal policy must equal the expected return for the best action from that state:

$$V^*(s) = \max_{a} E_{\pi^*} \{\sum_{k=0}^{\infty} \gamma^k r_{t+k+1} \mid s_t=s, a_t=a\}, \tag{15}$$

for the discounted-reward case. From this follows the optimality equation for the discounted-reward case:

$$V^*(s) = \max_{a} E\, \{r_{t+1} + \gamma V^*(s_{t+1}) \mid s_t=s, a_t=a\}, \tag{16}$$

for all $s \in S$. For finite MDPs, the optimality equation has a unique solution independent of the policy, unlike the ordinary consistency condition (11), whose solution depends on the policy. The optimality equation is actually a system of equations, one for each state, so that there are $|S|$ equations in $|S|$ unknowns. If the dynamics of the environment are known, then in principle one can solve this system of equations for V^* using one of a variety of methods for solving systems of nonlinear equations.

Once one has V^*, it is relatively easy to determine an optimal policy. For each state s, there will be one or more actions at which the maximum is obtained in the optimality equation. These are all equally good actions. Any policy that selects only from among these is an optimal policy. Another way of saying this is that any policy that is *greedy* with respect to the optimal evaluation function V^* is an optimal policy. The term greedy is used in computer science to describe any search or decision procedure that selects alternatives based only on local or immediate considerations, without considering the possibility that such a selection may prevent future access to even better alternatives (Pearl, 1984). Consequently, it is descriptive of policies that select actions based only on their short-term consequences. The beauty of V^* is that if one uses it to evaluate the short-term consequences of actions, specifically the one-step consequences, then a greedy policy is actually optimal in the long-term sense in which we are interested because V^* already takes into account the reward consequences of all possible future behavior. By means of V^*, the optimal expected long-term return is turned into a quantity that is locally and immediately available for each state.

The optimality equation therefore provides one route for finding an optimal policy, and thus for solving a reinforcement-learning problem. Unfortunately, the solution outlined above is almost never directly useful. This solution relies on three assumptions that are rarely true in practice: (1) Situations are actual states, i.e., the agent-environment interaction can be modeled as an MDP; (2) We accurately know the complete dynamics of the environment, required to even obtain the optimality equation; and (3) We need enough computational resources to complete the computation of the solution. For the kinds of problems in which we are interested, one is generally not able to implement this solution exactly because various combinations of these assumptions are violated. For example, although the first two assumptions present no problems for the game of backgammon, the third is a major impediment. Since the game has about 10^{20} states, it would take thousands of years on today's fastest computers to solve the optimality equation for V^*. Unless there is some special additional mathematical structure that can be exploited, one has to settle for approximate solutions.

Many different decision-making methods can be viewed as ways of approximately solving the optimality equation. For example, heuristic search methods of AI (Pearl, 1984) can be viewed as expanding the right-hand side of Equation 16 several times, up to some depth, forming a "tree" of possibilities, and then using a heuristic evaluation function to approximate V^* at the "leaf" nodes. (Heuristic search methods such as A^* are almost always based on the total-reward case, e.g., when the rewards are negative costs.) The methods of dynamic programming can be related even more closely to the optimality equation (Bertsekas, 1987). Many primitive reinforcement-learning methods are well understood as approximately solving the optimality equation, using actual experienced transitions in place of knowledge of the expected transitions.

Example: Suppose we solve the optimality equation for the simple grid problem introduced in the previous example and shown again in **Figure 6a**. Recall that state A is followed by a reward of +10 and transition to state A', while state B is followed by a reward of +5 and transition to state B'. **Figure 6b** shows the optimal value function, and **Figure 6c** shows the corresponding optimal policies. Where there are multiple arrows in a cell, either action is optimal.

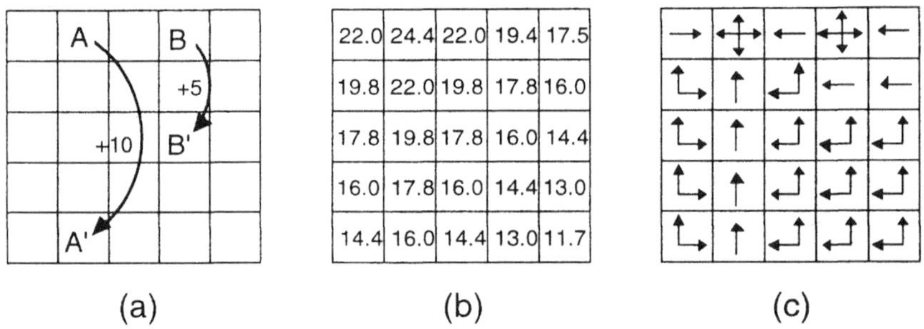

22.0	24.4	22.0	19.4	17.5
19.8	22.0	19.8	17.8	16.0
17.8	19.8	17.8	16.0	14.4
16.0	17.8	16.0	14.4	13.0
14.4	16.0	14.4	13.0	11.7

FIGURE 6. Rectangular-Grid Solution. **a.** The rectangular-grid illustration. **b.** The optimal value function, V^*. **c.** Optimal policies, π^*. Where there are multiple arrows in a cell, either action is optimal.

Learning

We have said that an agent's objective in a reinforcement-learning problem is, roughly speaking, to find a policy that maximizes the amount of reward it receives over the long run. We can now be more precise about what this means. The agent may be thought of as interacting with its environment over an infinite number of time steps. If the agent is trying to maximize total reward given by Equation 1 over subsequences of finite length, then we can imagine that it experiences an infinite number of such subsequences (e.g., can play an infinite number of games of backgammon). A reinforcement-learning agent has successfully completed learning when all the actions that it selects are optimal, i.e., when all of its actions are given by some optimal policy. This means that *after* successful learning, the agent always acts to maximize the expected return from each situation it encounters. Obviously, this ideal situation is usually only achievable in the limit, if at all, as the time step of interaction goes to infinity. Most of the algorithms we consider have been designed to achieve, under idealized circumstances, this kind of learning-in-the-limit, or *asymptotic learning*. This is what we mean by saying that an agent should *eventually* learn to act optimally.

Optimal learning may be contrasted with asymptotic learning. One might imagine that the agent should not only achieve optimal behavior in the limit, but should also improve its behavior as quickly as possible. That is, it should *optimally learn how to behave optimally* by making the best use of all of the experience it accumulates during its lifetime, as well as any prior knowledge it might bring to the task. An agent that learns optimally would maximize the total amount of reward it receives over its entire lifetime, not just over some time period in the infinite future. Although an ideal reinforcement-learning agent would be capable of optimal learning, we do not regard optimal learning as a realistic goal in designing reinforcement-learning agents. For the kinds of problems in which we are interested, optimal learning strategies can be generated only with extreme computational cost. We are, however, interested in algorithms by which an agent can improve its performance efficiently over time but without undertaking the complex process of designing an optimal learning strategy.

Even the ability to asymptotically learn to act optimally is usually beyond what is possible for a reinforcement-learning agent. We do not realistically expect that a reinforcement-learning agent would ever really achieve optimality even if it could learn over an infinite time period. A well-defined notion of optimality organizes the approach to learning we describe in this chapter and provides a way to understand the theoretical properties of various learning algorithms. However, it is an ideal that reinforcement-learning agents can only approximate to varying degrees. We noted previously that even if the reinforcement-learning agent has a complete and accurate model of its environment's dynamics, it is usually impossible for the agent to simply compute an optimal policy by solving the optimality equation. For example, board games such as chess are a tiny fraction of human experience, yet large, custom-designed computers still cannot compute the optimal moves. A critical aspect of the problem facing a reinforcement-learning agent will always be the computational power available to it, in particular, the amount of computation it can perform in a single time step. Although it is unclear how to quantify computational demands, they must be recognized.

The memory available to the agent is also an important constraint. The agent may require memory to build up approximations of value functions, policies, and models. In problems with small, finite situation sets, it is often possible to form these approximations using arrays or *tables*. In most cases of practical interest, however, there are far more situations than could possibly be entries in a table. In these cases the functions must be approximated using some sort of *compact representation*. Most of the theory of reinforcement learning applies to the tabular case, but many practical applications have used more compact representations.

So, our model of the reinforcement-learning problem forces us to settle for approximations. However, it also presents some unique opportunities for achieving useful approximations. For instance, we said above that a reinforcement-learning agent has successfully completed learning when all the actions it selects are given by some optimal policy. But, this does not mean that the agent's policy has to be optimal. An optimal policy specifies an optimal action for every possible situation, but for an agent to behave optimally its policy has to be optimal only for the situations it actually encounters. How the agent would act in situations it never encounters has no impact on the total amount of reward it will receive. Similarly, in approximating optimal behavior, there may be many situations that the agent will encounter with such a low probability that selecting suboptimal actions for them will have little impact on the amount of reward it receives. Tesauro's backgammon player, for example, plays with exceptional skill even though it might make very bad decisions on board configurations that occur rarely in games against experts. In fact, such rare configurations may make up a very large fraction of the game's state set. The interactive nature of reinforcement learning makes it possible to approximate optimal policies in ways that put more effort into learning to make good decisions for frequently encountered situations, at the expense of less effort for infrequently encountered situations. This is a key property that distinguishes reinforcement learning from other approaches to approximately solving MDPs.

Summary

Let us summarize the elements of the model of a reinforcement-learning problem that we have presented. Reinforcement learning is about learning how to behave in order to achieve a goal. The reinforcement-learning *agent* and its *environment* interact over a sequence of discrete time steps. The specification of their interface defines a particular problem: The *actions* are the choices made by the agent; the *situations* provide the agent's basis for making the choices; and the *rewards* are the basis for evaluating these choices. Everything inside the agent is completely known and controllable by the agent; everything outside is incompletely controllable but may or may not be completely known. A *policy* is a stochastic rule by which the agent selects actions as a function of situations. Roughly, the agent's objective is to learn a policy that maximizes the amount of reward it receives over the long run.

The *state* of the environment is a summary of the history of its situations, inputs (agent actions), and rewards that is sufficient to determine how it will behave in the future. The situation is meant to approximate the state. The *dynamics* of the environment are the stochastic relationships between the state and action at one time step and the situation and reward at the next. If an environment has the *Markov property*, then knowledge of the situation is sufficient to predict the environment's future behavior given a proposed course of action; the situation is a sufficient proxy for the state. This is rarely exactly true, but often nearly so, and situations should be chosen or constructed so that the Markov property approximately holds. If the Markov property does hold, then the environment is called a *Markov Decision Process*, or MDP. A *finite MDP* is an MDP with finite situation and action sets. Most of the current theory of reinforcement learning is restricted to finite MDPs.

The *return* is the function of future rewards that the agent seeks to maximize. It has several different definitions depending upon whether one is interested in *total reward* or *discounted reward*. A policy's *value function* assigns to each situation the expected return from that situation given that the agent uses the policy. The *optimal value function* assigns to each state the largest expected return from that state achievable by any policy. A policy whose evaluation function is the optimal evaluation function is an *optimal policy*. Whereas there is only one optimal evaluation function for a given MDP, there may be many optimal policies. Any policy that is greedy with respect to the optimal evaluation function is an optimal policy. The *optimality equation* is a special consistency condition that the optimal value function must satisfy and that can, in principle, be solved for the optimal value function, from which an optimal policy can be determined with relative ease.

Most of the algorithms that have been developed for reinforcement learning were designed for *asymptotic learning*, which means that, under ideal circumstances, as learning continues indefinitely, all the agent's actions approach optimal actions. We pointed out that this does not mean that the agent's policy must become an optimal policy; it only has to be optimal for the situations the agent actually encounters. We contrasted this with *optimal learning*, in which the agent should improve its behavior as quickly as possible by making the best possible use of all of the experience it accumulates during its lifetime, as well as any prior knowledge it might bring to the task. Although the rate of learning is a central issue in reinforcement learning, we do not regard optimal learning as a realistic goal in designing reinforcement-learning algorithms due to the extreme computational cost of obtaining optimal learning strategies for the kinds of problems that interest us.

The ability to asymptotically learn to act optimally is also usually impossible for a realistic reinforcement-learning agent due to limitations in computational resources. Even if the agent has a complete and accurate model of its environment, the agent may not be able to perform enough computation per time step to fully use it. The memory available to the agent is also an important constraint. The agent may require memory to build up approximations of value functions, policies, and models. In most cases of practical interest, there are far more situations than could possibly be entries in a look-up table, and the functions must be approximated using some sort of *compact representation*.

Although most of the theory of reinforcement learning is restricted to the tabular case, many practical applications have used more compact representations.

Reinforcement-learning problems differ according to the level of knowledge initially available to the agent. In problems of *complete information*, the agent has a complete and accurate model of its environment's dynamics. In problems of *incomplete information*, this level of knowledge is not available. It is important not to confuse complete and incomplete *information* with complete and incomplete *observation* of the environmental state. We refer to the case of complete observation by saying that the environment has the Markov property. For problems of incomplete information, *model-based* reinforcement-learning methods attempt to make up for the lack of a model by learning a model on line based on experience with the environment. *Primitive* methods attempt to optimize the policy without constructing a model of the environment's dynamics. Intermediate cases are possible as well.

A well-defined notion of optimality organizes the model of reinforcement learning we have described in this chapter. Optimality provides a way to understand the theoretical properties of various learning algorithms, but it is an ideal that reinforcement-learning agents can only approximate to varying degrees. In reinforcement learning, we are concerned with cases in which optimal solutions cannot be found but can be approximated in some way.

Neural-Networks Models of Cognition
J.W. Donahoe and V. Packard Dorsel (Editors)

CHAPTER 20

THE TD MODEL OF CLASSICAL CONDITIONING: RESPONSE TOPOGRAPHY AND BRAIN IMPLEMENTATION

J. W. Moore and J-S Choi
Department of Psychology
Neuroscience and Behavior Program
University of Massachusetts, Amherst

ABSTRACT

Classical-conditioning procedures instill knowledge about the temporal relationships between conditioned stimuli, which are regarded as predictive signals and triggers for action, and the unconditioned stimulus, the event to be timed. This knowledge is expressed in the temporal features of the conditioned response, which typically develops such that its peak amplitude occurs at times when the unconditioned stimulus is expected. A simple connectionist network based on Sutton and Barto's Time Derivative Model of Pavlovian Reinforcement provides a mechanism that can account for and simulate virtually all known aspects of conditioned-response timing in a variety of protocols, including delay and trace conditioning and conditioning under temporal uncertainty. The network is expressed in terms of equations that operate in real time according to Hebbian competitive-learning rules. The unfolding of time from the onsets and offsets of events such as conditioned stimuli is represented by the propagation of activity along delay lines. Inputs to the processing unit from conditioned stimuli arise from collateral taps off of each sequential element of these delay lines. The model can be aligned with anatomical circuits of the cerebellum and brainstem that are essential for learning and performance of eyeblink conditioned responses.

Overview

Classical conditioning is a form of reinforcement learning in which the behavioral output, the conditioned response (CR), carries information about the imminence and timing of the reinforcing event, the unconditioned stimulus (US). Conditioned responses typically increase progressively in amplitude over the interval spanned by the onset of the conditioned stimulus (CS) and the US, peaking at the time of US and decreasing afterwards. This feature of CR topography and timing has guided the development of learning rules that have found application in domains of prediction and control of artificial systems. One widely adopted learning rule is that employed by adaptive critics (Barto, 1995). Adaptive critics provide an actor with immediate evaluative feedback derived from predictions of future reinforcement. For example, Barto, Sutton, and Anderson (1983) applied an adaptive critic to the cart-pole balancing problem. The critic forewarns the agent responsible for generating control signals with information about the likelihood that the system's current state will result in a penalty without ameliorating action.

Sutton and Barto (1990) showed that the critic of the pole balancer can simulate a wide variety of classical-conditioning phenomena. In the context of classical conditioning, the adaptive critic is known as the Time Derivative Model of Pavlovian Reinforcement, or simply the TD model. The critic in the cart-pole system is mathematically identical to the TD learning rule (see Barto, Sutton, & Anderson, 1983, Equation 7, p. 841).

The TD model is a member of a broader class of models that Sutton and Barto (1990) refer to as $\dot{Y}$ theories of reinforcement learning. Such theories take the following form:

$$\Delta V_i = \alpha_i \overline{X}_i \times \beta \dot{Y} \quad (1)$$

As with Hebbian learning rules generally, changes in associative value, ΔV_i for CS_i, are computed as the product of two factors. The coefficients α_i and ß are rate parameters ($0 < \alpha_i, \beta \leq 1$). The factor $\overline{X}_i$ represents the salience and associability of CS_i, what Sutton and Barto (1990) call *eligibility*. Eligibility is a weighted average of previous and current strengths of CS_i. The other factor, $\dot{Y}$, represents *reinforcement*. Reinforcement in time-derivative models is a function of the difference (time derivative) between the response or output at time t, $Y(t)$, and the response or output on the previous time step, $Y(t - \Delta t)$ (Equation 2). Any system or device that would implement a time-derivative learning rule must be capable of monitoring the actor's output on both current and immediately preceding time steps.

$$\dot{Y} = Y(t) - Y(t - \Delta t) \quad (2)$$

Models that conform to the basic structure of Equation 1 have been applied with uneven success to data from classical eyeblink conditioning. As reviewed by Sutton and Barto (1990), the TD model is superior to most $\dot{Y}$ theories of reinforcement. Furthermore, it can describe the appropriate timing and topography of eyeblink CRs, if one assumes that the CS-US interval is segmented into a sequence of time-tagged units, each of which develops its own associative value over training. This representation of the CS is referred to as a complete serial compound (CSC). It resembles the approach to conditioned-response timing and topography employed by Desmond and Moore's VET model (Desmond, 1990; Desmond & Moore, 1988, 1991b, 1992; Moore, 1991, 1992; Moore, Desmond, & Berthier, 1989). VET is an acronym derived from associative *values* based on *expectations* of reinforcement *timing*. VET's advantage over other models lies in the wide range of phenomena it can encompass, but perhaps its most striking vindication is data illustrating that CR topography can be controlled by both the onset and offset of CS events.

The success of the VET model comes at some cost in parsimony. Unlike the TD model, it assumes the existence of two neuron-like processors. One processor, the E unit, "instructs" the other processor, the V unit, which generates the CR. In this role, the E unit resembles the critic in the pole-balancing problem. Where the critic provides a system's controller with predictions of penalties, the E unit provides the V unit with information about the expected timing of the US.

Goals

The purpose of this chapter is to show that the TD model can be extended to classical-conditioning protocols in which the offset as well as the onset of a CS controls the topography of the CR. One such case is trace conditioning. Another case involves training with varying CS-US intervals. In order to simulate appropriate CR topographies for these protocols, Desmond and Moore (1988) suggested in their VET model that CS offset, like CS onset, initiates a cascade of activation over a set of time-tagged serial components. We shall show that the TD model with a similar representation of both CS onset and offset is capable of simulating CRs from trace conditioning and from protocols with a varying CS-US interval. Furthermore, unlike VET, it can do this with one processing unit instead of two, although not without costs of its own.

Another purpose of this chapter is to suggest a scheme whereby the TD model might be implemented in the brain. There are a number of brain systems that express TD learning, such as the basal ganglia (Houk, Adams, & Barto, 1995), but we shall confine our attention to the cerebellum, as this part of the brain mediates the learning and performance of CRs such as the eyeblink.

The TD Model

The following equation expresses the TD learning rule for classical conditioning.

$$\Delta V_i(t) = \beta[\lambda(t) + \gamma Y(t) - Y(t-1)] \times \alpha \overline{X}_i(t) \tag{3}$$

where

$$Y(t) = \Sigma_j V_j(t) X_j(t). \tag{4}$$

The subscript j includes all serial CS components, and $X_j(t)$ indicates the on-off status of the jth component at time t. $Y(t)$ corresponds to CR amplitude at time t. It cannot take on negative value. $\lambda(t)$ represents the strength of the US at time t. α and β are rate parameters. Notice that we have dropped the subscript from α, so that $\alpha_i = \alpha$ for all i. $\overline{X}_i(t)$ is the eligibility of the ith CS component for modification at time t, given by the following expression.

$$\overline{X}_i(t + 1) = \overline{X}_i(t) + \delta[X_i(t) - \overline{X}_i(t)] \quad (5)$$

where $0 < \delta \leq 1$.

The parameter $\gamma(0 < \gamma \leq 1)$ is the "discount" factor (see Barto, 1995), a key feature of the TD model which primarily determines the rate of increase of CR amplitude, $Y(t)$, as the US becomes increasingly imminent over the CS-US interval. With the CSC representation of CSs, the TD model generates realistic portraits of CRs as they unfold in time. Realistic CRs resemble the classic goal gradients of traditional S-R reinforcement theory: The CR ramps upward to the predicted onset of the US. This aspect of CR waveforms reflects imminence-weighted (discounted) predictions of the US. Imminence weighting is a crucial feature of adaptive critics in reinforcement learning.

Figure 1 shows a family of asymptotic CR waveforms with different values of γ and δ. (Details regarding implementation of the TD learning rule for simulations can be found in Sutton & Barto, 1990.) The figure shows that CR topography depends primarily on γ: The smaller the value of γ, the lower the peak value of CR amplitude, $Y(t)$. Lower values of γ also increase the positive acceleration of CR amplitude, $\Delta\ \dot{Y}(t)$, without compromising the accuracy of $Y(t)$'s prediction of the timing of the US.

In practice, CR topography depends on the physical characteristics of CSs and their serial components. These characteristics, such as acoustic frequency and intensity, can be captured by the variables $X_j(t)$ in Equation 4, as suggested by Kehoe, Schreurs, Macrae, and Gormezano (1995), and by physical constraints of the motor system. Physical constraints include such things as the limitations on the positions that the effectors can assume. In the case of classically conditioned eyelid movements, the eyelids are normally open. In this position, CR amplitude has a value of 0. A fully developed CR is one in which the eyelid's position moves from open to completely closed. Yet, no matter how strong the prediction that the US will occur, the eyelids can only close so far. This constraint implies that the progressive closure of the lids in the course of CR production can saturate before the US's anticipated time of occurrence. In addition, these constraints on eyelid position render it impossible for negative predictions of the US to be expressed directly in eyelid movement—predictions that the US will not occur at some time when it would otherwise be anticipated. No matter how strong the prediction that the US will not occur, the eyelids can only open so far and no farther.

TD Model Differs from VET

Although they both aim to describe intra-trial as well as intertrial real-time aspects of classical conditioning, the TD model differs from the VET model in a number of ways. The TD model is simpler than the VET model because it consists of one learning rule that can be mapped directly onto behavior. The VET model has two learning rules: one for computing the expected time of the US, the other for computing the associative weights that map onto behavior.

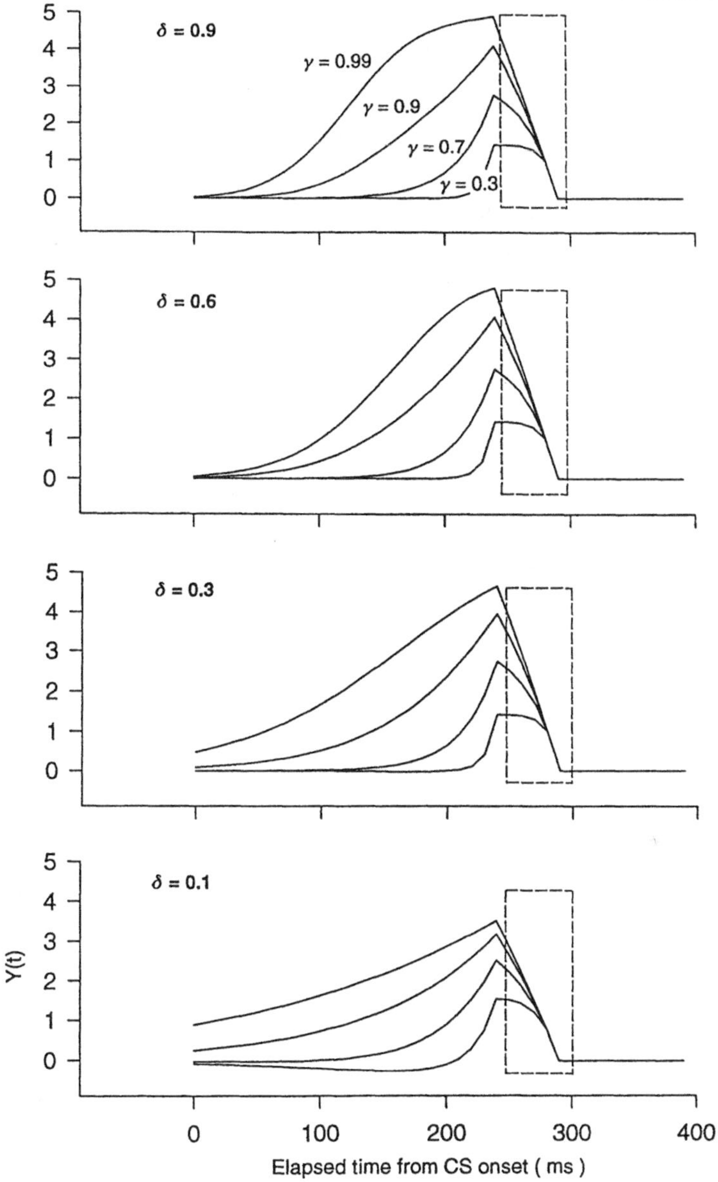

FIGURE 1. Simulated CRs, $Y(t)$, after 200 trials as a function of γ and δ. Time steps in this and other simulations are 10 ms, $\alpha = 0.05$, $\beta = 1.0$, and $\lambda = 1.0$. The rectangle in each panel indicates the duration of the US, which is 50 ms. Note that CR timing and amplitude are determined primarily by the discount factor, γ.

As noted above, the VET model employs a representation of the CS which assumes that CS onsets and offsets initiate cascades of time-tagged serial components. Although not crucial, published simulations with VET assumed that next-in-line serial components are activated with a latency of 10 ms and that they remain active for 100 ms. Although serial components nearer in time to the US are more eligible for modification than those that are temporally remote, and therefore acquire greater synaptic weight, the overlapping pattern of activation fosters the development of robust CRs that increase sharply in amplitude as the US becomes imminent. Unlike VET, the TD model does not assume overlapping activation of serial components. The amplitude of the CR increases over the CS-US interval simply because components nearer the US gain greater synaptic weight than those that are more remote.

The Sutton-Barto-Desmond Model

Our interest in the TD model arose from a long-standing goal of developing computational models capable of describing CR waveforms in a variety of classical-conditioning paradigms. We began this quest with the TD model's predecessor, the Sutton-Barto (SB) model, which is another $\dot{Y}$ learning theory developed by Sutton and Barto (1981). The Sutton-Barto (SB) model is given by the following equation.

$$\Delta V_i(t) = \beta[Y(t) - Y(t-1)] \times \alpha \overline{X}_i(t) \tag{6}$$

The resemblance of the SB model to the TD model is obvious, yet they differ fundamentally. In the SB model, $Y(t)$ can include the primary reinforcement term, λ. That is, $Y(t) = \Sigma_j V_j(t)X_j(t) + \lambda$. In the TD model, $Y(t)$ excludes k, as indicated by Equation 4. Instead, λ is explicitly represented in the learning rule, Equation 3.

Although the SB model does not generate realistic CRs, an elaboration of the basic model known as the Sutton-Barto-Desmond (SBD) model can generate rudimentary features of CR topography for a limited number of protocols. The SBD model has been described elsewhere (Blazis & Moore, 1991; Moore, 1991; Moore, Berthier, & Blazis, 1990; Moore & Blazis, 1989a,b,c; Moore, Desmond, Berthier, Blazis, Sutton, & Barto, 1986). Limitations of the SBD model (Desmond, 1990; Sutton & Barto, 1990) prompted development of the VET model.

The SBD model assumes that CR topography is guided by a template. Some agency exists that has a priori knowledge of what a fully developed CR is supposed to look like. CR topography depends on the assumed form of the template. In the VET model, CR topography is not guided by a template. Instead, CR topography arises from low-level mechanisms derived from the spread of activation from one serial component to the next and the timing of the US. In short, the form of the CR is selected by the timing relationship between the CS and the US, subject to physical constraints of the motor system. The TD model resembles the VET model in this regard.

The template approach to CR topography presents several difficulties which have been reviewed by Desmond (1990). The main difficulty is one of inflexibility. For example, the template assumed by the SBD model does not predict appropriate CR waveforms in a number of standard training protocols, including trace conditioning, long CS-US intervals, and mixed CS-US intervals. The VET and TD models overcome these limitations of flexibility because CR timing and topography are not predetermined. They are selected from ingredients contributed by the salience and timing of CS components with respect to the US.

Second-Order Conditioning

Although the VET model has proven to be superior to the SBD model in providing flexibility in CR topography and timing, it has limitations of its own. The main limitation is that it cannot generate second-order conditioning. The VET model is incapable of generating second-order conditioning because, like the Rescorla-Wagner model, its learning rule assumes that learning occurs only on time steps or trials where a discrepancy exists between the magnitude of the US predicted by CS elements and the magnitude of the US as represented by the scalar value of the US, λ. In second-order conditioning, the US does not occur, so there is no mechanism that permits the formation and modification of connections between the second-order CS and the CR.

In contrast, $\dot{Y}$ learning models such as SB are capable of generating second-order conditioning, so long as some portion of the second-order CS precedes the first-order CS (see Barto & Sutton, 1982, Figure 5, p. 230). If the would-be second-order CS occupies precisely the same time steps as the first-order CS, the SB model predicts blocking of conditioning to the would-be second-order CS. If the would-be second-order CS follows the first-order CS, its connection weights to the response become negative in value. The would-be second-order CS becomes a conditioned inhibitor. The TD model possesses these same attributes.

S-S versus S-R Learning

Experimental psychologists and animal-learning theorists distinguish between two broad classes of associative theory. They differ on the question: What is learned? For some, the answer is an S-R relationship. For others, the answer is an S-S relationship. S-R theories assume that connections are learned between stimuli and responses. S-S theories assume that connections are learned between stimuli or their representations in the brain. The theories of Thorndike, Guthrie, and Hull are examples of S-R theories. The theories of Tolman, Lewin, Mowrer, and Pavlov are examples of S-S theories. The distinction between the two types of associative theory becomes important for understanding why models behave as they do.

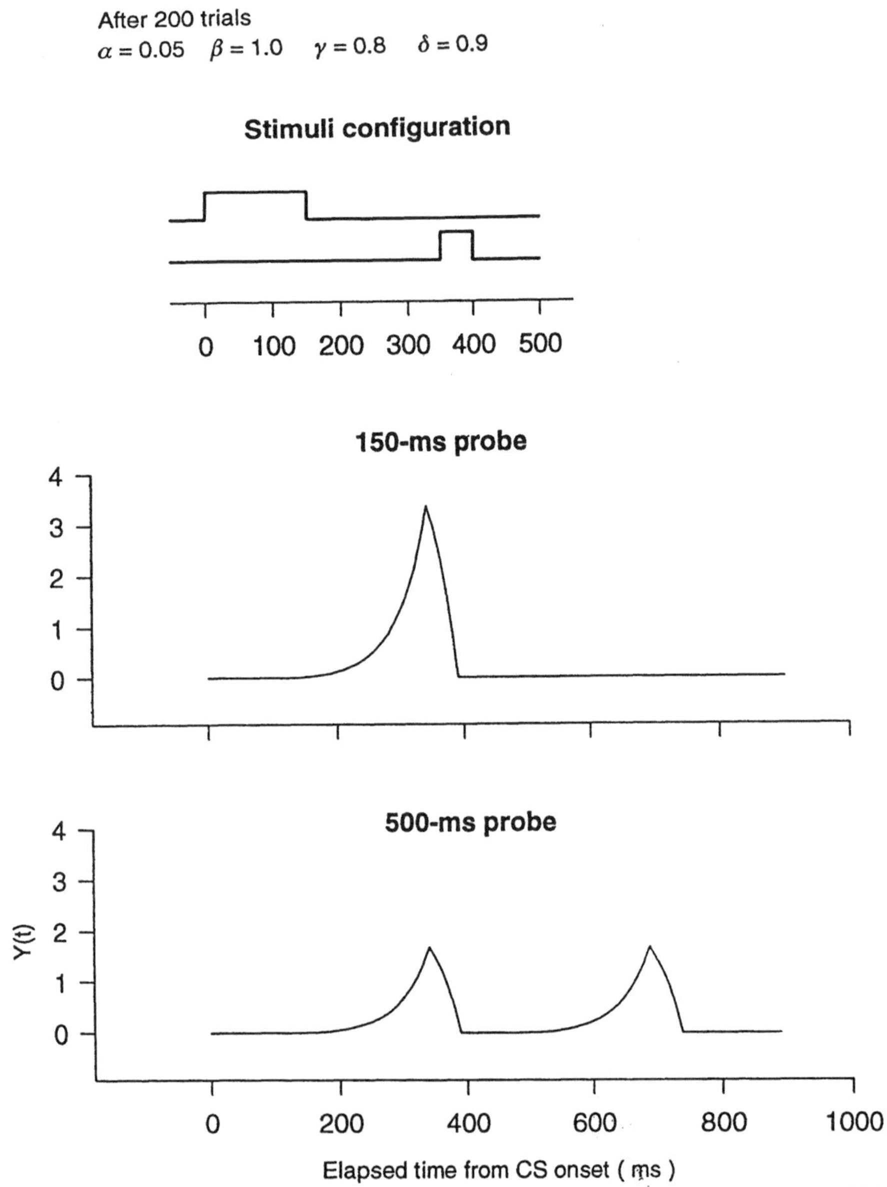

FIGURE 2. Simulated CRs to CS-alone probes of 150 and 500 ms, following 1,000 trials of trace conditioning with a 150-ns CS and a 200-ms trace interval, after Desmond and Moore (1991b).

The VET model is S-S because changes in connection weights are computed with respect to the discrepancy between (1) the magnitude of the representation of the US provided by CS elements active at the time and eligible for modification, and (2) the magnitude of the US provided by the US:

$$\Delta V_i(t) = \beta\,[\lambda(t) - Y(t)] \times \alpha \overline{X}_i \quad (7)$$

The SB model is S-R because changes in connection weights are computed with respect to the discrepancy between the current output of the system and preceding outputs, as indicated by Equation 6.

The TD model is a hybrid theory because it employs both types of discrepancy to compute changes in connection weights. The US dominates on time steps when it is present; the discrepancy between current and preceding output dominates on time steps when the US is not present.

TD Model Simulations with Onset and Offset CSC

As noted previously, the VET model requires two processing units. The TD model with the CSC assumption allows it to simulate the key features of CR topography with the same flexibility and precision as the VET model but with one processing unit instead of two. Hence, only one learning rule, Equation 3, is necessary to generate appropriate output, as illustrated by simulations of two benchmark protocols.

CS onset-offset synchrony in trace conditioning

Desmond and Moore (1991b) showed that CS onset and offset both contribute to the topography and timing of eyeblink CRs. To show this, rabbits were trained in trace conditioning with a CS-US interval of 350 ms. The CS duration was 150 ms, and the trace interval, the time between CS offset and the US, was 200 ms. After training, the rabbits were probed with CSs of varying duration. Probes of 150 ms elicited unimodal CRs with peaks at the temporal locus of the US, i.e., about 350 ms after CS onset. Longer probes resulted in bimodal CRs. For example, when probed with a CS duration of 400 ms, two peaks appeared. The first occurred 350 ms after CS onset, which was attributed to the onset cascade of serial components. The second peak occurred 200 ms later, which was attributed to asynchronous displacement of the offset cascade.

Figure 2 shows a simulation of a similar experiment using the TD model with the CSC assumption applied to onset and offset cascades. Like the VET model, the TD model predicts the two CRs that arise when the CS duration used in trace conditioning is lengthened on probe trials. In agreement with Desmond and Moore's (1991b) trace-conditioning study and predictions of the VET model, the amplitude of these peaks is not as great of that of the single peak in 150-ms probes. This is because the single peak represents the synchronized summation of associative values from onset and offset cascades.

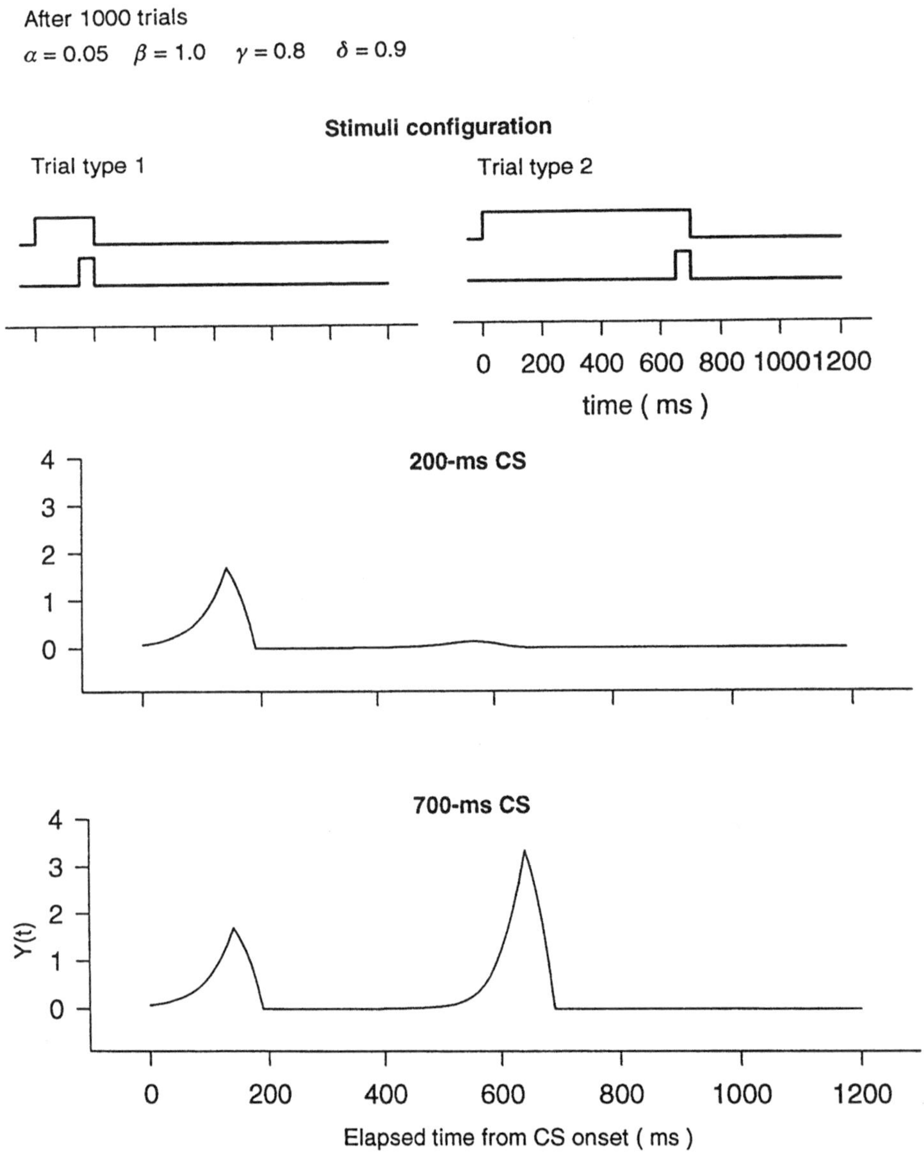

FIGURE 3. Simulated CRs to CS-alone probes of 200 and 700 ms, following 1,000 delay-conditioning trials with a 200-ms CS-US interval alternated with a 700-ms CS-US interval, after Millinsen et al (1977).

With longer probes, the associative values of these two cascades are no longer synchronized. Hence, the first peak is due to the onset cascade; the second peak is due to the offset cascade. The two peaks are equally high. This does not agree with data from Desmond and Moore (1991b) showing that the second peak is not as high as the first peak. The VET model captures this difference by including a global parameter that takes account of the inverted-U nature of optimal CS-US-interval functions. The function specifies that training at the optimal CS-US interval, 250 ms for the rabbit eyeblink, is more efficacious than training at other CS-US intervals. The TD model does not include this parameter.

Training with mixed CS-US intervals

Millenson, Kehoe, and Gormezano (1977) trained rabbits in delay eyeblink conditioning with a mixture of two CS-US intervals, 200 and 700 ms. After training, the animals were probed with CS durations of 200 and 700 ms. On 200-ms probes, CRs were unimodal with peaks at 200 ms. On 700-ms probes, CRs were bimodal with peaks at 200 and 700 ms. These results are predicted by the VET model (Desmond and Moore, 1988). They are also predicted by the TD model. As in the VET model, the TD model predicts only one CR peak on 200-ms probes because components of the offset cascade acquire negative associative values. This happens because offset serial components are systematically not reinforced by the US in the presence of positively valued onset components that occupy the interval between 200 and 700 ms. Training with a mixture of two CS-US intervals is tantamount to a conditioned-inhibition procedure for the offset components. The negatively valued offset components cancel the effect of the positively valued onset components in this interval, thereby reducing or eliminating the second amplitude peak on 200-ms probes. On 700-ms probes, the negatively valued offset components are activated too late to affect the CR.

Figure 3 is a simulation of the CS-US mixture experiment just described. It shows a single peak on 200-ms probes and two appropriately timed peaks on 700-ms probes. The second peak is higher than the first because extinction of onset components in the 200-700-ms interval is blocked on 200-ms training trials by the presence of negatively valued offset components. In contrast, onset components in the 0-200-ms interval undergo unblocked extinction during 700-ms training trials. This extinction is not blocked by the presence of negatively valued offset components.

The difference in the amplitude of the first and second peaks is not supported by data. Millenson et al (1977) found that the two peaks had about the same amplitude, and this is the prediction of the VET model. The VET model does not predict a smaller first peak because extinction, like acquisition, can only occur at times when the US is predicted by the E unit. The associative values of serial components active at other times do not change from one trial to the next.

Although not supported by data, the greater amplitude of the second peak on 700-ms probes makes sense from the CR-as-prediction viewpoint espoused by Sutton and Barto (1981). The amplitude of the second peak can be regarded as a measure of the likelihood of the US *given* that it did not occur at the point in time predicted by the first peak. Hence, by incorporating CS-offset cascades, the TD model computes *conditional* likelihoods of US timing.

Implementation of TD Learning in the Cerebellum

TD learning can be implemented in the cerebellum by aligning known anatomical arrangements with elements of the learning rule. In TD learning, we assume that each computational time step after the onset or offset of a CS is represented by an anatomically distinct input to the cerebellum. The easiest way to think about this is to imagine that the onset of a CS initiates a spreading pattern of activation among neurons tied to whatever sense modality is involved. This spreading of activation might occur within the brainstem or among cerebellar mossy fibers and their associated granule cells. A reasonable physical model of the process would be a tapped delay line, as in the VET model, but other plausible schemes have been suggested (e.g., Buonomano & Mauk, 1994).

Consistent with numerous neurobiological studies, the implementation scheme relies on evidence from rabbit eyeblink conditioning that CR topography is formed in cerebellar cortex through converging contiguous action of parallel-fiber and climbing-fiber input to Purkinje cells. This action produces synaptic changes known as long-term depression (LTD). Experimental evidence from a variety of techniques supports this view, and it has guided implementation schemes for other computational models such as VET (Moore & Desmond, 1992; Moore et al, 1989).

Figure 4, adapted from Rosenfield and Moore (1995), summarizes the neural circuits that are likely involved in rabbit eyeblink conditioning. The figure shows that CS information ascends to granule cells in the cerebellar cortex (Larsell's lobule H-VI) via mossy fibers originating in the pontine nuclei (PN). Information about the US ascends to cerebellar cortex by two routes, mossy-fiber (MF) projections from the sensory trigeminal complex, spinal oralis (SpO) in the figure, and climbing-fiber (CF) projections from the inferior olive (IO) nucleus. A CR is generated within deep cerebellar nucleus interpositus (IP), where the CR is formed by modulation from Purkinje cells (PCs). A full-blown CR is expressed as an increased rate of firing among IP neurons (e.g., Berthier & Moore, 1990; Berthier, Barto, & Moore, 1991). This activity is projected to the contralateral red nucleus (RN). From RN, activity is projected to motoneurons (MN) that innervate the peripheral musculature controlling the position and movements of the eyelids and eyeball (Desmond & Moore, 1991a). The RN also projects to SpO, giving rise to CR-related activity among these neurons (Richards, Ricciardi, & Moore, 1991).

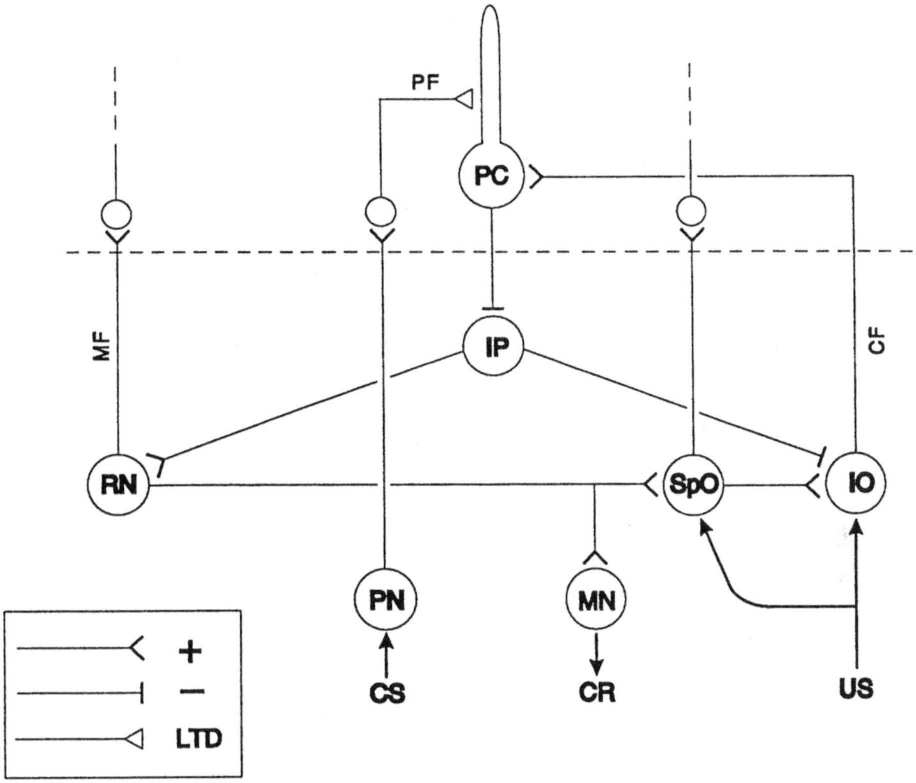

FIGURE 4. Cerebellar and brainstem circuits underlying eyeblink conditioning, after Rosenfield and Moore (1995).

Figure 4 depicts an inhibitory projection from IP to IO. The consequence of this arrangement is that olivary signals to PCs are suppressed when the CR representation within IP is robust. This anatomical feature suggests that climbing fibers are only excited when the US occurs *and* the CR is weak or absent, implementing a simple *delta* learning rule such as Equation 7. The TD learning rule is not a simple delta rule because of the $\gamma\ Y(t)$ term in Equation 3.

The TD learning rule is implemented by a combination of two reinforcement components. The first is donated by the US, λ in the model's learning rule. λ can be interpreted as the S-S or primary reinforcement component. The implementation scheme assumes that λ can be aligned with climbing-fiber activation of PCs, which functions to produce LTD among coactive parallel-fiber (PF) synapses, as depicted in the figure. The second reinforcement operator is donated by the $\dot{Y}(t)$ terms in the learning rule, $\gamma\ Y(t) - Y(t-1)$. $\dot{Y}(t)$ can be interpreted either as the S-R component or as the second-order conditioning component.

Figure 5 shows circuit elements, not shown in **Figure 4**, for implementing the $Y(t)$ component of the learning rule. These components include the projections to cerebellar cortex from the RN and SpO indicated in **Figure 4**. We hypothesize that the RN projection carries information about $Y(t)$ to cerebellar cortex as an efference copy. Parallel fibers project this information to PCs that have collaterals to a set of Golgi cells (Go). Because these projections are inhibitory (Ito, 1984), these PCs invert the efference signal from the RN. In addition, the interpositioning of the PCs between the RN and Golgi cells attenuates the signal and implements the TD model's discount factor, γ.

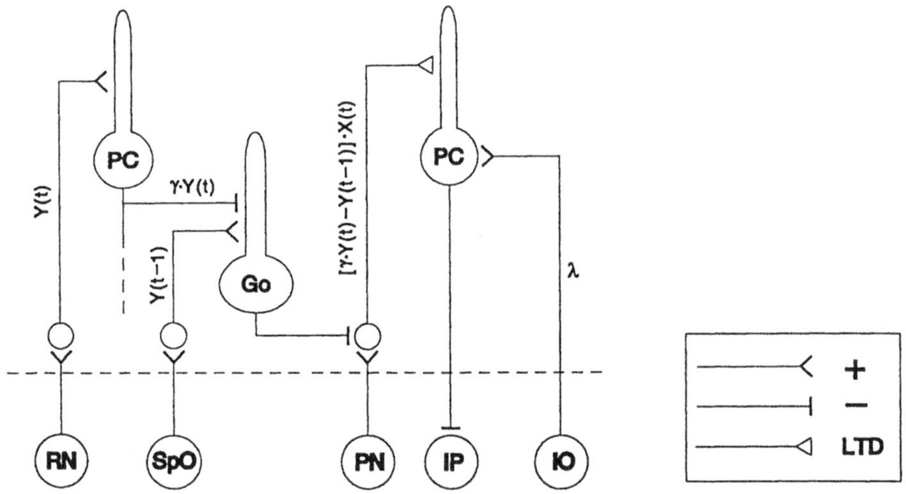

FIGURE 5. Neural circuits implementing $\gamma\ Y(t)$ and other variables of the TD learning rule.

Because Golgi cells are inhibitory on granule cells, the consequence of their inhibition by PCs receiving efference from the RN would be to disinhibit activity of granule cells. In other words, since granule cells relay CS information from the PN to PCs involved in LTD and CR generation, disinhibition of granule cells by Golgi cells enhances the information flow from active CS components. Mathematically, the implementation scheme assumes that the variables X_j in Equation 4 engage granule cells. The PFs arising from these granule cells engage LTD PCs in proportion to $Y(t) \times X_j$.

In this scheme, PCs driven by projections from the RN would increase their firing rate so as to mimic the representation of the CR as it passes through the RN en route to MN and SpO. Berthier and Moore (1986) recorded from several H-VI PCs with CR-mimicking increases in firing. Since increases in firing during a CS are incompatible with CR formation through LTD, it is likely that these PCs were inhibiting motor programs incompatible with CR generation, e.g., eyelid and eyeball musculature that would lead to eyelid opening instead of eyelid closure. Here, we are suggesting an additional function of these PCs, that of projecting inverted and discounted CR efference from the RN to Golgi cells.

The implementation scheme assumes that the Golgi cells that receive the inverted efference from the RN also receive a direct, non-inverted, excitatory projection from SpO. This projection carries information about the CR at time $t - \Delta t$. Therefore, the Golgi cell in **Figure 5** fires at a rate determined by the differential between two inputs: $\gamma\, Y(t)$ donated by the RN and $Y(t - \Delta t)$ donated by SpO. Hence, Golgi cells act as $\dot{Y}(t)$ detectors. In terms of Equation 3, $Y(t)$ is transmitted to cerebellar granule cells by the RN, and $Y(t - 1)$ is transmitted to granule cells from SpO. The RN input engages PCs that inhibit Golgi cells responsible for gating inputs from CSs to PCs. Efference from SpO engages the same Golgi cells directly. Because Golgi cells are inhibitory on granule cells, the bigger the RN input relative to SpO input, the bigger the signal from CS serial components active at that time, be they from onset or offset cascades. Enhanced throughput from active CS elements in the granular layer would lead to local recruitment of other active PFs that synapse on PCs involved in LTD and CR generation.

In this way, the Golgi cells that implement $\dot{Y}(t)$ reinforce and maintain the down-regulated state of active PF/PC synapses subject to LTD. Mechanisms that allow this to occur have been spelled out in recent articles (Eilers, Augustine, & Konnerth, 1995; Ghosh & Greenberg, 1995; Kano, Rexhausen, Dreesen, & Konnerth, 1992; Konnerth, Dreesen, & Augustine, 1992). Parallel-fiber/PC synapses that are activated by a CS element are down-regulated by the contiguous activation of climbing-fiber input from the inferior olive nucleus, triggered by the US. As CS elements earlier in the sequence of elements become capable of evoking an output that anticipates the US, inhibition is relayed to the olive and the US loses its capacity to trigger a climbing-fiber volley, as shown is **Figure 4**. However, the down-regulation of these synapses is maintained, and still earlier CS elements are recruited, by PFs carrying $\dot{Y} \times X_j$ to LTD PCs, as indicated in **Figure 5**.

In a single-unit recording study, Desmond and Moore (1991a) observed an average lead time of 36 ms from the onset of activity in RN cells having highly CR-related firing patterns and the peripherally observed CR. The average lead time of SpO cells with CR-related activity was 20 ms. Therefore, the time difference in CR-related efference arising from the two structures is probably on the order of 15-20 ms. This difference spans one 10-ms time step used in simulations employing the TD model. This temporal difference is consistent with a conduction velocity of 2 m/s for the 10-mm trajectory of unmyelinated axons from the RN to rostral portions of SpO. The 10-ms grain also ensures high-fidelity resolution of fast transients. The fastest transients in eyeblink conditioning occur during unconditioned responses (URs). At their fastest, the eyelids require 80 ms to move from completely open to completely closed, with a peak velocity of approximately 4-5 mm/20 ms.

Efference from SpO neurons recorded among H-VI PCs would tend to lag behind the peripherally observed CR, if it arises from more caudal portions of the structure. Berthier and Moore (1986) observed a continuum of lead and lag times among PCs that increased their firing to the CS. Purkinje cells that receive projections from SpO (not shown in the figure) would be expected to increase their firing, but with a lag relative to those receiving projections from the RN. It makes sense that the proportion of CR-leading PCs observed by Berthier and Moore (1986) matched the number of CR-lagging PCs, since these two populations would merely be reflecting CR efference from two temporal vantage points.

Figure 6 is an expanded version of **Figure 5** showing three sets of granule cells associated with three serial CS components. In the expanded model, these serial CS components arise from CS onsets and offsets, just as in Desmond and Moore's VET model. The degree to which information from any of these serial CS components reaches the PCs to which they project is determined by Golgi cells firing in proportion to $\dot{Y}(t)$, as just described. **Figure 1** shows that, depending on γ, TD-simulated CRs are positively accelerating in time up to the occurrence of the US, so $\dot{Y}(t)$ increases progressively over the CS-US interval. Therefore, those PF/PC synapses activated near the time of the climbing-fiber signal from the US would have the greatest impact in establishing and maintaining LTD of PF/PC synapses that ensure the appropriate form and timing of CRs. The spatial arrangement of PF/PC synapses has no significance for CR timing, nor would their arrangement have an effect on the responsiveness of PCs to PF input patterns (De Schutter & Bower, 1994).

Implications of the Implementation

The implementation scheme has several testable implications. One that has already been mentioned is that the firing pattern of most H-VI PCs with CR-related firing resembles the CR in form, in that their moment-to-moment rate of firing mirrors topographical features of the response. We maintain that this pattern of firing reflects CR efference. Since this efference cannot arise from proprioceptors, which are absent in muscles controlling the eyeblink, and since the axons of motoneurons innervating these muscles do not possess recurrent collaterals, this efference must arise from premotor centers. The RN and SpO are the prime candidates. One can only speculate as to the functions of this efference. We suggest that one function is to activate Golgi cells that modulate information flow through the granule cells. Another function would be to excite those PCs that project to deep nuclear cells which engage motor systems incompatible with CRs, such as those involved in eye opening or saccadic eye movements.

The implementation scheme also requires that Golgi cells—which modulate information flow from the time-tagged components of CS onsets and offsets—fire in relation to *changes* in eyelid position, i.e., they fire in relation to $\dot{Y}$. This property of Golgi-cell firing patterns has been reported by Van Kan, Gibson, and Houk (1993), in a study of monkey-limb movements, and Edgley and Lidierth (1987), in a study of cat locomotion.

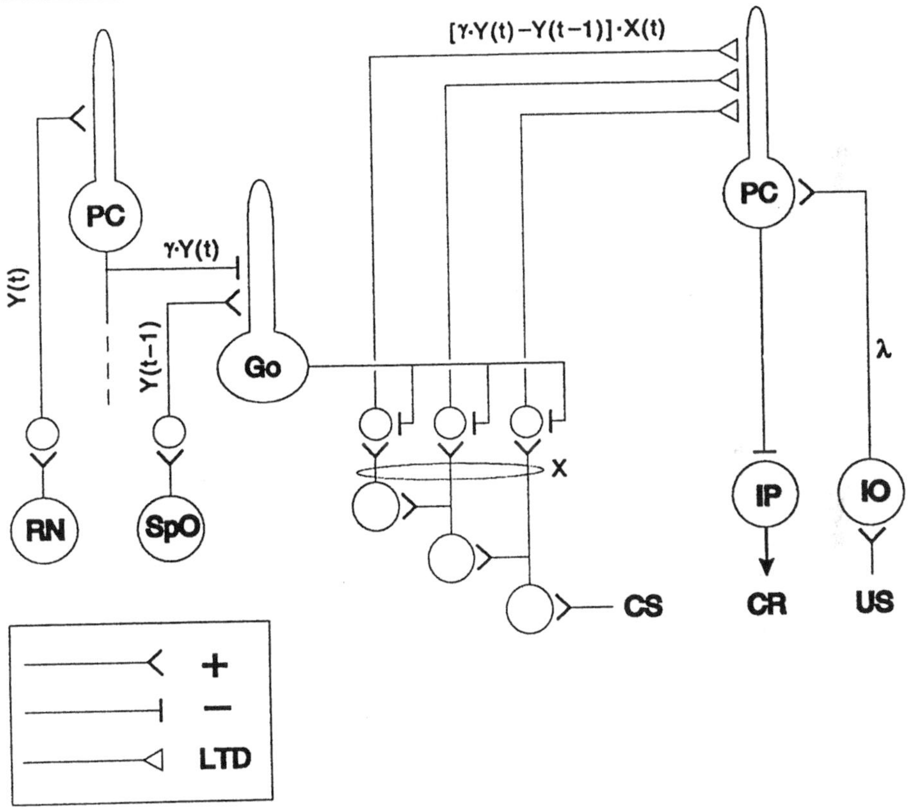

FIGURE 6. The complete TD implementation scheme showing three sequentially activated CS components, representing both onset and offset cascades in the manner of Desmond and Moore's (1988) VET model.

A third implication of the implementation scheme concerns the effects of inactivation of the RN through the use of pharmacological agents or cryostatic probes. Although activation of the RN would cause a temporary interruption of the information flow that results in a conditioned response, it would not prevent learning the primary association between components of the CS and the US. This association would proceed with little disruption because the PN and the IO would still be able to convey CS and US information to cerebellar cortex. Evidence for this proposition comes from a study of rabbit-eyeblink conditioning by Clark and Lavond (1993). They demonstrated that inactivation of the RN by cooling did not prevent learning, as CR magnitude recovered immediately upon reactivation of the RN.

Inactivation of the RN would, however, interrupt efference about the position of the eyelid at times t and $t - \Delta t$ from the RN and SpO. Thus, $\dot{Y}$ would not be available to cerebellar cortex. According to the TD model, $\dot{Y}$ allows for increments of predictive associations in the absence of the US, as would occur in second-order conditioning (Kehoe, Feyer, & Moses, 1981). This being the case, inactivation of the RN would interfere with second-order conditioning. Animals trained simultaneously in first- and second-order conditioning with the RN inactivated would be expected to show first-order learning, as in the Clark and Lavond (1993) study, but little or no second-order learning.

Interpretations of $\dot{Y}$: Efference or Afference

The implementation of the TD model interprets $\dot{Y}$ as efference. From a mathematical perspective, it is possible to interpret $\dot{Y}$ in terms of afference from CS components. This becomes apparent from the form of the TD learning rule actually employed in simulations (Sutton & Barto, 1990).

$$\Delta\ V_i(t) = \beta\ [\lambda\ (t) + \gamma \Sigma_j X_j(t)\ V_j(t-1) - \Sigma_j X_j(t-1)\ V(t-1)] \times \alpha \overline{X}_i(t) \qquad (8)$$

A recent study by Ramnani, Hardiman, and Yeo (1995) suggests that the efference interpretation of $\dot{Y}$ is correct. This experiment shows that temporary inactivation of IP by muscimol application prevents extinction of the CR. That is, CS-alone trials that would normally lead to a gradual elimination of the CR had no effect whatsoever. When tested later, after the muscimol blockade had been removed, the previously established CR was at full strength. It did extinguish with continuing presentation of CS-alone trials. This finding is consistent with the efference interpretation of the model because inactivation of IP eliminates the CR and therefore prevents efference from the RN and SpO from affecting learning. Under an afference interpretation, inactivation of IP would not prevent CS information from ascending to cerebellar cortex, where extinction would proceed normally, as this information arises from the PN.

Summary and Conclusions

This chapter considered how the TD theory of reinforcement learning, which lies at the heart of promising applications in adaptive control in both real and artificial systems, might be adapted to training protocols in which behavior is controlled by both the onset and offset of CSs. The TD model with the CSC assumption does a reasonably good job of generating appropriate CR waveforms in such protocols.

The chapter also reviewed the similarities and differences between the TD model (CSC) and earlier attempts to model CR topography and timing by Moore et al (1986) and Desmond and Moore (1988) in terms of concepts about associative learning developed and refined by experimental psychologists over the course of decades, emphasizing the TD model's relationship to S-R and S-S theories of learning.

The chapter also suggests an implementation scheme for TD learning within the cerebellum. The implementation draws on neurobiological evidence regarding how LTD is established, reinforced, and maintained among Purkinje cells that form the CR. The implementation incorporates recent anatomical findings, reviewed by Rosenfield and Moore (1995), that allow these Purkinje cells to receive both components of the TD model's reinforcement operator—the primary component donated by the US and the secondary component donated by $\dot{Y}(t) = Y(t) - Y(t - \Delta t)$. The implementation scheme lays the foundation for network simulations at the cellular level.

The entire exercise reinforces the synergy that has enlightened and invigorated behavioral and neurobiological studies of reinforcement learning. In particular, the TD model appears to be the most comprehensive rendering of classical conditioning that has been proposed to date. As far as applications to adaptive control in artificial systems are concerned, the TD model's role in reinforcement learning is becoming ever more ubiquitous (Sutton, 1992).

Neural-Networks Models of Cognition
J.W. Donahoe and V. Packard Dorsel (Editors)

CHAPTER 21

BIOLOGICAL SUBSTRATES OF PREDICTIVE MECHANISMS IN LEARNING AND ACTION CHOICE

P. Read Montague
Division of Neuroscience
Baylor College of Medicine
Houston, TX

ABSTRACT

A number of experimental studies have shown that diffuse monoaminergic systems influence neural plasticity and ongoing neural activity. In the vertebrate, these systems also play a role in activity-dependent development. In this chapter, I address the kinds of information that such systems could construct and deliver to their widespread target structures. I suggest that these diffuse systems may be positioned to construct prediction errors about the amount of reinforcement expected in the future. The framework is motivated by an established body of computational theory built around the method of temporal differences (TD). I demonstrate that this way of viewing the output of diffuse systems can account for the firing-rate changes in mesencephalic dopamine neurons during simple learning tasks, and provide a good account for the decision-making behavior exhibited by bees and humans on simple choice tasks.

Introduction

The ability of an animal to anticipate future salient stimuli requires *prediction.* Sensory events and their representations in the nervous system must reliably predict the likelihood, time, and magnitude of future important events such as food, danger, destructive stimuli, and mates. Over the years, psychological experiments have established that both vertebrates and invertebrates are capable of such prediction and are capable of selecting appropriate actions based on these predictions (for reviews see Dickinson, 1980; Mackintosh, 1983; Gallistel, 1990; Gluck & Thompson, 1987). Although many classical and instrumental conditioning phenomena are well understood from a psychological point of view, the neural mechanisms that generate, evaluate, and utilize predictions are in general unknown.

There are, however, systems of neurons in both vertebrates and invertebrates whose activity clearly relates to reward and salience processing. In the vertebrate brain, neuromodulatory systems have long been thought of as systems that report on the salience and emotional valence of events in the world. Although a number of subcortical structures are involved in reporting on reward and salience (e.g., see Ledoux, 1992), this information is thought to be carried *in part* by activity in widespread systems of axons which deliver neurotransmitters such as norepinephrine, acetylcholine, dopamine, and serotonin to their targets. These axonal systems originate in small nuclei in the midbrain and basal forebrain and are sometimes collectively called *diffuse ascending systems*. Invertebrates have similar sets of neurons that respond to rewarding stimuli and deliver neuromodulators to widespread target regions (Hammer, 1993). Diffuse systems differ in their neurotransmitters, afferent input, and exact pattern of output connections. However, they all have the common property of collecting information from a wide array of structures and sending information out to widespread targets.

We have previously proposed that diffuse neuromodulatory systems meet a number of general requirements for neural systems that could construct, distribute, and use information about predictions (Quartz, Dayan, Montague, & Sejnowski, 1992; Montague, Dayan, Nowlan, Pouget, & Sejnowski, 1993; Montague, Dayan, & Sejnowski, 1994; Egelman, Person, & Montague, 1995). We have further suggested a computational explanation for how diffuse systems play different roles in development, learning, and decision making.

In this chapter, I briefly summarize a set of algorithms—called temporal difference or TD algorithms—that learn the sort of temporal dependencies discussed above. These algorithms have been applied successfully as models of classical conditioning. I show how various forms of these algorithms provide a computational framework for understanding the kinds of signals that diffuse neuromodulatory systems may construct and use. I further show how the same framework captures the decision-making behavior displayed by bees and humans on simple choice tasks.

Temporal Difference Models of Conditioning

Any system that learns to predict must have the capacity to generate predictions, assay the error in the predictions, and make learning contingent on the errors. These requirements are equivalent to many features of most adaptation rules in psychology and engineering (Kalman, 1960; Rescorla & Wagner, 1972; Widrow & Stearns, 1985). The rule on which this chapter focuses is the temporal difference or TD rule. The TD rule is an adaptation of the Rescorla-Wagner rule for conditioning (Rescorla & Wagner, 1972) which takes proper account of the detailed time course of learning during training episodes (Sutton, 1988; Sutton & Barto, 1981, 1987, 1989, 1990). TD is also closely related to a learning rule suggested early by Konorski (1948) and has been used to model the behavior of the cerebellum at the cellular level (Moore, Berthier, & Blazis, 1990).

The computational problem

We briefly summarize the assumptions that underlie temporal difference (TD) models of classical conditioning (Sutton & Barto, 1987, 1989, 1990). At time t, let $x(t)$ be a vector with components $x_i(t) = 1$ or 0, denoting the presence or absence of stimulus i at time t. As a reminder that we will make a connection with real biological data, we will consider the components of $x(t)$ as populations of neurons in the cerebral cortex that represent the state of the environment. At time t, the animal also receives a scalar reward $r(t)$. Under TD, the computational goal of learning is to use the current stimuli $x(t)$ (as represented in the cortex) to predict a measure of the long term *discounted* reward $V(t)$ that will be available from time t onward:

$$V(t) = \sum_{s>t} \gamma^{s-t-1} r(s) = \gamma^0 r(t+1) + \gamma^1 r(t+2) + \gamma^2 r(t+3) + \ldots \quad (1)$$

where $0 \leq \gamma < 1$ is a discount factor that models the fact that future rewards may be worth less than current ones. In non-deterministic problems, the computational goal is to predict the expected value of this quantity. For this framework to work, it is important that $V(t)$ be a function of time solely through the inputs $x(t)$. This assumption requires that future rewards do not depend on past rewards except through the current stimulus state $x(t)$. Under this assumption, $V(x)$ satisfies the simple recursive relationship:

$$V(x(t-1)) = r(t) + \gamma V(x(t)) \quad (2)$$

The job for the cortex is to construct an estimate $\hat{V}(x(t))$ of the actual $V(x)$. In the simplest case, actual estimates of the predictions $\hat{V}(x(t))$ are constructed as:

$$\hat{V}(x(t)) = x(t) \cdot w(t) \quad (3)$$

where $w(t)$ is the weight of stimuli in the estimate $\hat{V}(x)$. Using the estimate $\hat{V}(x(t))$, the difference between the two sides of Equation 2, $\delta(t)$, is defined to be the error in the current estimates:

$$\delta(t) = r(t) + \gamma \hat{V}(x(t)) - \hat{V}(x(t-1)) \quad (4)$$

$\delta(t)$ is called the *TD error* since it involves the difference between the predictions at two successive time steps.

Weight changes are specified as:

$$\Delta w_i(t-1) = \eta x_i(t-1)\delta(t) \quad (5)$$

As in the Rescorla-Wagner rule or the engineering delta rule (Widrow & Stearns, 1985), the associated weights are adjusted to reduce the error. In Equation 5, η is a fixed learning rate.

As described above, learning in the TD rule is driven by any inconsistencies between the predictions made at one time and those made at later times, and specifies how adjustable weights should be changed to minimize these inconsistencies. Weight changes depend on presynaptic activity and a measure of ongoing prediction error (Equation 5). Any biological model based on this framework would require the prediction error to be communicated to widespread targets so that it can affect plasticity at all appropriate synaptic sites (see below).

Representing a stimulus through time

This description does not specify one critical detail: How is a sensory stimulus represented through time? It is necessary to specify such a representation so that when the prediction error $\delta(t)$ is large, there is some kind of

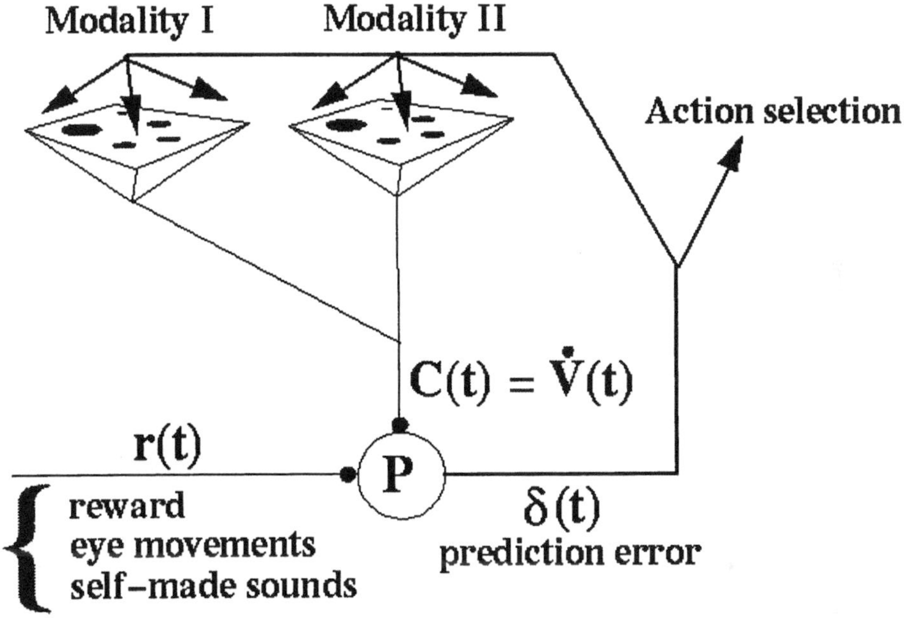

FIGURE 1. Constructing scalar predictions through convergence and divergence. Modality I and Modality II represent cortical regions or patches of cortex within a single region. Neuron P is a linear unit that collects highly convergent input $C(t)$ from these cortical representations in the form $\dot{V}(t) = \Sigma_i \dot{V}(i,t)$ where $\dot{V}(i,t)$ is a temporal derivative of the net excitatory input to domain i in the cortex. I use $V(t)$-$V(t$-$1)$ for $\dot{V}(t)$. In the simplest interpretation, the convergence onto P forces the summation $\Sigma_i \dot{V}(i,t)$, so that the input to P is a scalar. The output of P diverges to widespread targets, ensuring that the output is also a scalar. In this arrangement, any information encoded topographically in the cortical layers would be discarded. P also receives input from representations of salient events in the world and within the organism ($r(t)$). This arrangement permits the linear output of P, $\delta(t) = r(t) + V(t)$-$V(t$-$1)$, to act as a *prediction error* of future reward and expectations of reward. Fluctuations in neuromodulator delivery about some baseline level represent these prediction errors. The general features of this arrangement are supported by anatomical and physiological data (Schultz, 1992; Schultz et al, 1993). As indicated, the same error signal can be used to bias action selection (see **Figures 3, 5**).

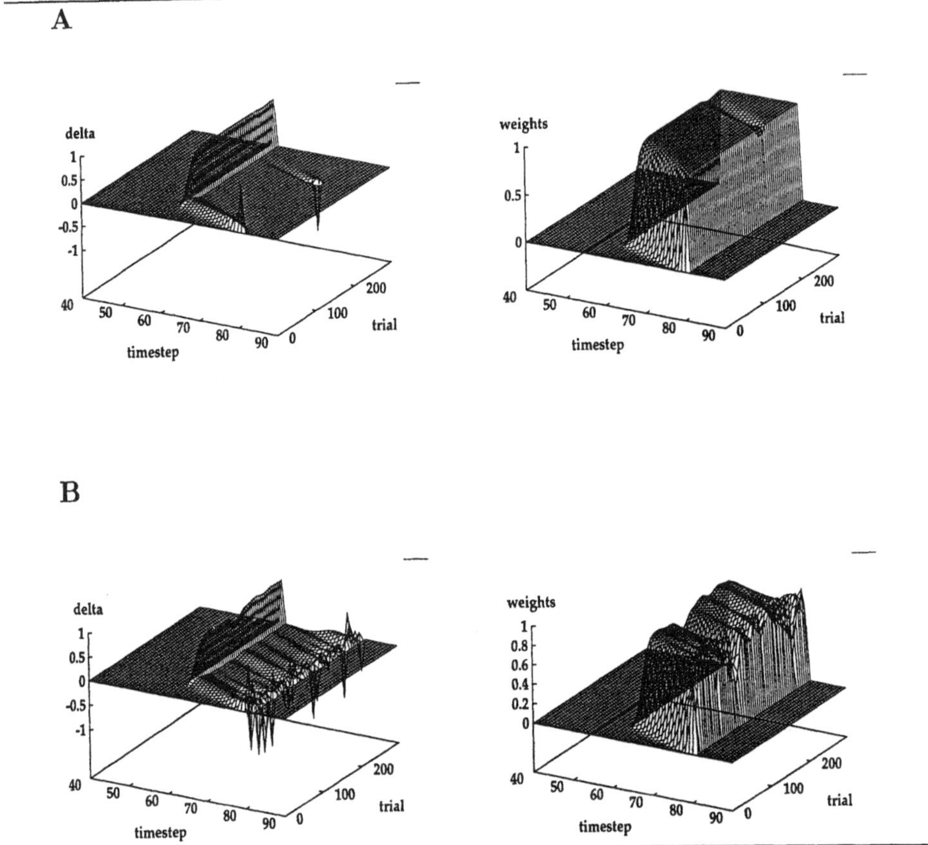

FIGURE 2. Conditioning using TD error signal. **A.** $\delta(t)$ plotted versus timestep and trial. At timestep 60, a sensory cue was presented and was followed 20 timesteps later by a reward ($r(t)=1$ for $t=80$; $r(t)=0$ for $t\neq 80$). The sensory cue has adjustable weights associated with the 40 timesteps following its initial onset at timestep 20. These weights are adjusted according to Equation 5. The evolution of $\delta(t)$ (left panel) and the weights (right panel) is shown. On trial 200, the reward is withheld and $\delta(t)$ is negative on $t=80$. If the plot of $\delta(t)$ were a physiological recording from a mesencephalic dopamine neuron (Schultz et al, 1993), one would see a transient change in firing at the onset of the predictive sensory cue, followed by a transient reduction in firing rate at the time that the reward had previously been delivered ($t=80$). **B.** These panels display the same information as in A, however, the consistency of reward delivery has changed. Reward is still delivered 20 timesteps after the onset of the sensory cue, but only on 80% of the trials. This inconsistency has obvious effects on the prediction error $\delta(t)$ at specific times. In a physiological experiment after training, such a profile for $\delta(t)$ could be seen as a transient increase in firing rate at the sensory cue onset, followed by changes in firing rate at the time the reward is delivered. These latter changes might be increases or decreases in firing rate (learning rate=0.3 for both A and B).

"temporal pointer" to the stimulus and time that consistently precede the fluctuations in $\delta(t)$.

We assume that the presentation of a sensory cue, say a light, elicits multiple representations of the light for a number of succeeding timesteps. In essence, we assume that different synapses are devoted to different times, although all of them represent the light from the point of view of the rest of the brain. I use the simplest form of such a representation: dividing the epoch following the stimulus into a finite number of equal time steps, each associated with a separate weight. For example, in **Figure 2**, the sensory cue is presented at timestep 60 and has weights associated with the succeeding 40 timesteps. In general, one would want to have a set of representations over continuous time rather than discrete time. This form of temporal representation is what Sutton and Barto (1990) call a complete serial-compound stimulus and is a form of Grossberg and Schmajuk's (1989) spectral timing model in which a learning rule selects from a spectrum of timed processes. How the brain generates analogous temporal representations is an important but unanswered question.

Neurons that compute and distribute prediction errors

We have recently proposed that the diffuse ascending systems are reporting information about *predictions of future stimuli* simultaneously to widespread targets (Quartz et al, 1992; Montague et al, 1993; Pouget et al, 1993; Montague et al, 1994). Specifically, we have suggested that descending connections from cortical representations to various subcortical nuclei in the midbrain and basal forebrain can make predictions about future stimuli through sets of adaptable weights (**Figure 1**). This work has suggested that errors in predictions analogous to those expressed in Equation 4 could be distributed to widespread target regions through diffusely projecting axonal systems, to influence activity-dependent map development in the neonate as well as learning and behavioral decisions in the adult (Quartz et al, 1992; Montague et al, 1993; Pouget et al, 1993; Montague & Sejnowski, 1994; Montague et al, 1994; Egelman et al, 1995).

Schultz and colleagues have recently discovered neurons in the primate midbrain whose activity is consistent with a computation of prediction error (Ljungberg, Apicella, & Schultz, 1992; Schultz, 1992; Schultz, Apicella, & Ljungberg, 1993). This area, called the ventral tegmental area (VTA), contains dopaminergic neurons which send axons to widespread target regions, including various limbic structures involved in reward-dependent learning (Oades & Halliday, 1987). Schultz and colleagues have recorded from these neurons in alert primates learning simple reaction-time and choice tasks. In one of these tasks, a sensory stimulus (light) consistently precedes the delivery of reward (mechanically delivered juice). Early in training, most of the neurons increase their firing rate to the delivery of the juice and do not respond to the onset of the light. Later in training, the neurons transiently increase their firing rate at the onset of the light and fire at baseline levels at the delivery of the reward. Using the temporal representation described above, we have shown that this change in firing is well described by a TD model in which the fluctuating output of a diffuse system is viewed as the prediction error (**Figure 1**; Quartz et al, 1992; Montague et al, 1993; Montague et al, 1994).

It has been suggested that diffuse neuromodulatory systems act as "gates" for synaptic plasticity, defining the epochs during which synaptic modification can and cannot occur (Rauschecker, 1991). Our proposal for diffuse ascending systems is consistent with this interpretation; however, the information carried in the "gating" signal is not a simple "print now" signal that defines those epochs when synaptic strengths can be changed. Instead, the activity levels in diffuse axons carry information about *prediction errors* in future rewards through fluctuating changes in neuromodulator release. The output $\delta(t)$ of diffusely projecting neurons carries a sign that permits it to select the direction of synaptic change. I view this sign as representing modulation of the output of P above ($\delta(t)>0$) and below ($\delta(t)<0$) some basal firing rate, thus incurring increased and decreased release of neuromodulator about some basal level.

Classical conditioning: predicting the time and magnitude of future states

In **Figure 2** I show how the TD model accounts for classical conditioning to a sensory cue that predicts the delivery of reward at a consistent time in the future. In these examples, a sensory cue is presented at timestep 60 and has weights in the succeeding 40 timesteps. At timestep 80, a reward ($r(t)=1.0$) is presented for one timestep. After training, if the reward is not delivered, there is a large negative fluctuation in $\delta(t)$ that would be viewed as a large reduction in firing rate in the diffusely projecting neuron (trial 200, **Figure 2A**). Such an effect has been observed by Schultz in recordings from dopamine neurons (Schultz et al, 1993).

As illustrated in panel **B** of **Figure 2**, the algorithm is sensitive to the consistency of the reward delivery. In this example, the reward was again delivered at timestep 80, but only on 80 of the trials. In a real experiment, peristimulus time averages of neuronal firing rates are often compiled for analysis. In this case (**Figure 2B**), such an average would show initially brisk responses only to reward delivery in early trials. In later trials, the sensory cue would cause a large response and the average response to reward would be variable—perhaps a fraction of the number of spikes resulting from presentation of the sensory cue. Effects similar to this latter suggestion are reported and discussed in Schultz (1993).

Translating Prediction Errors into Behavioral Decisions

One interesting aspect of TD models of conditioning is that the same signal $\delta(t)$ used to improve predictions can also be used to choose actions in an appropriate manner. As originally demonstrated by Barto, Sutton, and Ander-

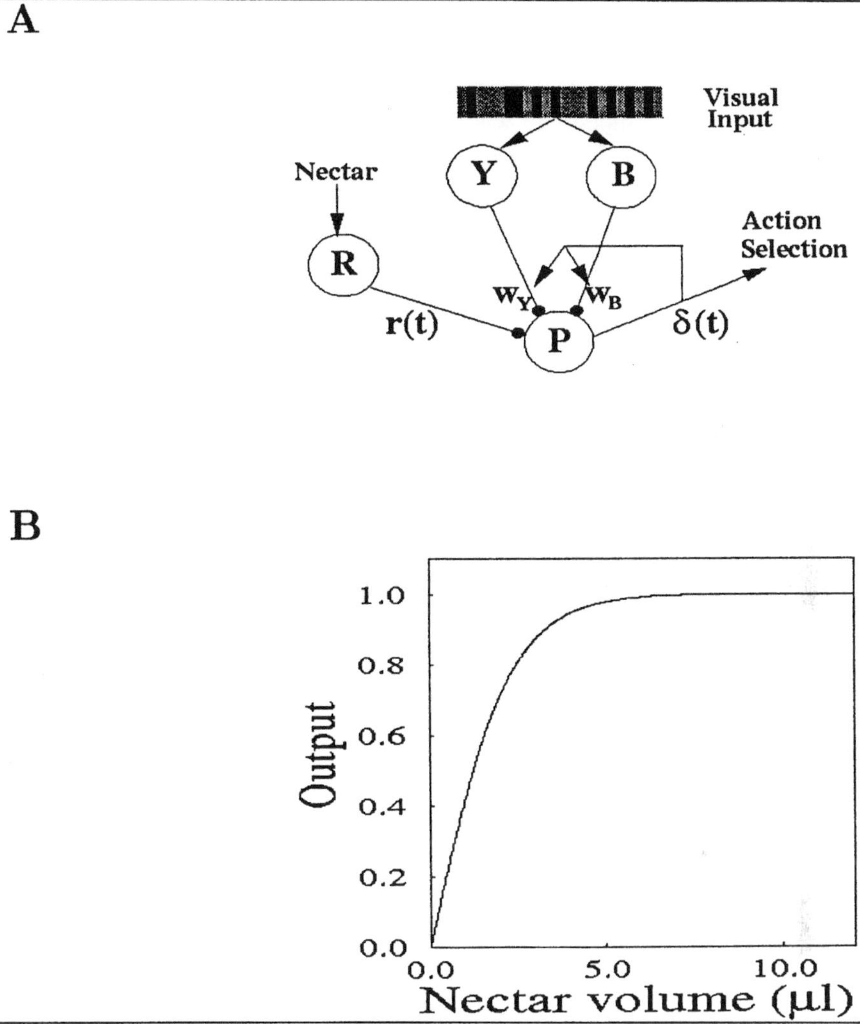

FIGURE 3. Neural architecture of model bee. **A.** Two units (B,Y) that receive input from a one-dimensional retina converge on linear unit *P*. The activity of these units represents the fraction of blue and yellow (black and gray here) in the visual field of the model bee. The model bee moves about in a circular arena and views a wall covered in stripes colored blue or yellow. Changes in the field of view change the input to B and Y, and this is communicated to *P* as temporal derivatives (one-timestep temporal differences). During movement through the simulated arena, $\delta(t)$ influences the decision to randomly reorient (Equation 7). On encountering a flower, $\delta(t)$ is used to update the weight associated with the selected flower according to Equation 9. **B.** Empirically derived utility curve for nectar volumes (Real, 1991). This curve was used as the saturating response of neuron *R* and acted as $r(t)$ at each flower landing.

son (1983), this use of the TD error signal can allow a system to learn difficult control problems (also see Barto, Sutton, & Watkins, 1989). In this section, I demonstrate how $\delta(t)$ can be used to select actions in a manner consistent with decision-making behavior demonstrated by bees and humans in simple tasks.

We have explored models in which the fluctuating output $\delta(t)$ of the diffuse system (output of P in **Figure 2**) could control neural activity that relates to behavioral decisions. These efforts require that the widely broadcast output of P directly influence the activity of target neurons involved in changing some kind of ongoing behavior. The models I present below are not necessarily meant to map directly onto specific neuroanatomy; however, they do suggest how the diffuse systems could sensibly influence behavioral choices in a manner consistent with real data. First, I present a simple model showing how this framework could control foraging behavior in a bee. Bees also possess analogous diffuse axonal systems that are important for reporting information about rewards to the rest of the bee brain (Hammer, 1993). Second, I apply a similar model to decision making in a simple card-choice task, and compare the results of the model to those obtained in humans performing the same task.

Risk aversion in foraging bees

Real and colleagues (Real, 1991) performed a series of experiments on bumblebees foraging on artificial flowers whose colors—blue and yellow—predicted the delivery of nectar. They examined how bees respond to the mean and variability of this reward delivery. The experiment was equivalent to a stochastic two-armed-bandit problem (Bush & Mosteller, 1955) except that the bees had to fly from choice to choice so that constraints other than reward delivery also played a role in their decision making. In one experiment, blue and yellow flowers yielded the same mean reward except that the variance of reward was 0 for blue flowers and >0 for yellow. In practice, 83% of the bees' visits were to the constant-yield blue flowers. Reversing the statistics of nectar delivery for blue and yellow caused the bees to switch their sampling accordingly. Again, they sampled approximately 83% from the constant-yield flower. In a second series of experiments, Real and colleagues demonstrated that the bees could be induced to visit the variable and constant flowers with equal frequency if the mean reward from the variable flower type was made sufficiently high.

These results suggest that the uncertainty of the reward is an important variable in the bees' decision making. This foraging strategy can be viewed as risk aversion, since the bees choose to avoid the more uncertain predictor (yellow or blue) despite the equivalent mean return from each flower type.

We used the model architecture shown in **Figure 3** to address the experiments described above (Montague et al, 1994). The existence of a diffusely projecting neuron P that would deliver prediction errors to targets has been suggested by the recent physiological and anatomical work of Hammer (1993).

In the model architecture, sensory units report on the percentage of blue and yellow in the visual field. Similar to the model in **Figure 1**, the weighted temporal derivatives of sensory-unit activity provide input to the linear unit P, along with information about reward ($r(t)$). Activity along $r(t)$ follows the empirically derived (utility) curve in **Figure 3B**. The output of P, $\delta(t)$, is used to update weights (w_B, w_Y) during flower encounters. In addition, the output was used to select actions. I have examined two models employing $\delta(t)$ in action selection; both give qualitatively similar results as shown in **Figure 4**.

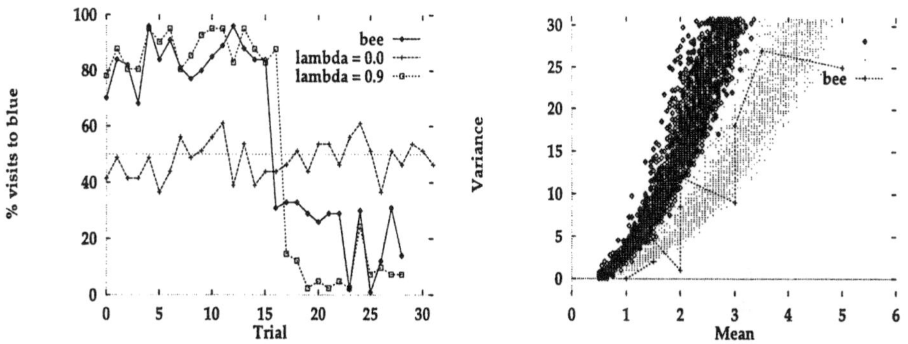

FIGURE 4. Risk-averse foraging by real bees and model bees. The left panel shows the behavior of real and model bees on a foraging task. The real bees were presented with an artificial field of flowers containing 100 yellow and 100 blue flowers. All the blue flowers contained 2 μl of nectar. One-third of the yellow flowers contained 6 μl, with the remainder containing no nectar. In practice, the real bee preferred the constant-yield blue flowers, sampling them on approximately 83% of landings. At a learning rate of 0.9 (λ in Equation 9), the model bee matched the real bee behavior (see Equations 7 and 9). Each trial represents 40 flower visits. The right panel demonstrates that real bees can be induced to forage equally from the constant-yield flowers and the variable-yield flowers if the mean return from the variable flowers is made sufficiently high. Each point represents an indifference point: the mean-variance pair at which the real or model bee foraged equally from each flower type. The points connected by a line represent pooled data from real bees. The spray of small dots indicates results for the model bee at learning rate $\lambda=0.9$. The larger diamonds are results for the model bee at learning rate $\lambda=0.1$.

In one case, we modeled the biasing of actions such as steering and landing with a probabilistic algorithm:

$$q(Y) = \frac{exp(\mu(w_Y x_Y))}{exp(\mu(w_B x_B)) + exp(\mu(w_Y x_Y))} \tag{6}$$

where $q(Y)$ is the probability of choosing a yellow flower. Values of $\mu > 0$ amplify the difference between the two predictions, so that larger values of μ

make it more likely that the larger prediction will result in choice toward the associated flower color. μ was varied from 2.8 to 6.0 and comparable results obtained. In this case, the predictions are used directly to choose an action.

In the second case, we gave the bee a cyclopean eye and allowed it to move about in a circular arena with flowers distributed along the wall of the arena. As the model bee changed direction, the output of P, $\delta(t)$, was used to bias actions by choosing the probability of randomly reorienting (tumbling) in the next step. $\delta(t)$ would fluctuate depending on the current value of the weights and the changes in sensory input. Using this error signal, the bee randomly reorients according to:

$$P_r = \frac{1}{1 + \exp(m\delta(t) + b)} \tag{7}$$

The latter model permits the bee to perform a biased random walk in a manner similar to klinokinesis displayed by bacteria moving up nutrient gradients (Spudich & Koshland, 1975).

In both models of action selection, the weights are updated according to the Rescorla-Wagner rule (Rescorla & Wagner, 1972) and only upon flower encounters. As shown in **Figure 4**, the model bee displays risk-averse foraging and the mean-variance tradeoff exhibited by real bees. The tradeoff between mean reward and its variance (uncertainty) results from the nonlinear utility of reward for increasing nectar volumes, i.e., the saturating response of the unit R reporting nectar volumes to P (**Figure 3**).

Although both models are extreme simplifications of the constraints that impinge on a foraging bee, they do demonstrate how bottom-up neural constraints can produce decision-making strategies exhibited by real biological systems. Since such strategies are usually analyzed in cognitive or economic terms, the possibility of relating these descriptions to testable neural constraints opens up a number of interesting questions.

Human choice behavior: a substrate for matching behavior

The preceding examples of bee foraging show how a simple use of TD error can capture bee foraging in an environment where the predictors of reward are initially uncertain but the reward distributions are stationary. One job of the foraging bee is to decide when and how to build up a model of its world through exploration and when and how to exploit the model to obtain reward. In short, there is a necessary tradeoff between exploration and exploitation.

I address the exploration-exploitation conflict below, in a model that uses the TD error ($\delta(t)$) in a fashion analogous to that employed with the model bee. In the following example, the reward distributions are not stationary but vary with the history of choices. In this case, I show how a use of $\delta(t)$ analogous to that above causes a system to sample so as to match the average rate of return from the two alternative choices. The results obtained with the network are then compared to some preliminary data from human subjects performing the same task.

Card-choice experiment

Figure 5 illustrates a card-choice task given to a model similar to the bee model above. The task was also given to human subjects (n=13). The task is to select a card from one of two decks, after which a reward is delivered along $r(t)$. As cards are selected, the amount of reward from each deck changes as a function of the fraction of choices made from Deck A as computed over a 30-card window. The reward functions are shown in **Figure 5**. This model amounts to a game where the environment chooses a fixed strategy and reacts to the opponent's choices. The rate at which rewards from either deck are changed depends on how fast the fraction of choices from Deck A changes. Hence, the speed with which the environment reacts is defined by the window over which the fraction of Deck A choices is computed. The model or human plays against this strategy. These reward curves were adapted from an experiment performed by Herrnstein to examine issues relating to rational choice theory (Herrnstein, 1990, 1991; Herrnstein, Loewenstein, Prelec, & Vaughan, 1993).

Notice that the reward functions cross at one point. After this crossing, the reward function for Deck B continues to grow and the reward function for Deck A stays approximately the same. At each card selection, the subject (or model) is thus given a choice of whether to switch decks or stay on the last deck chosen.

The model and its behavior

In this model, two sensory units (analogous to B and Y in **Figure 3A**) represented the two decks of cards and provided weighted input to unit P. The output of P, $\delta(t)$, was used to decide whether the current deck was selected. The model made choices by making random transitions between Deck A and Deck B, thus inducing fluctuations in $\delta(t)$ as before. The probability that the current deck was selected was:

$$P_s = \frac{1}{1 + \exp(m\delta(t) + b)} \tag{8}$$

where $r(t) = 0$ before a card is actually chosen. The model randomly chose one deck as a starting point and "looked back and forth" between decks; the fluctuations in $\delta(t)$ assigned a value to the transitions between choices; and P_s determined the probability that a given deck was selected after a transition (see **Figure 5A**). Analogous to the bee example, the sensory weights (w_A and w_B) determined the sign and magnitude of fluctuations in $\delta(t)$, and thus influenced

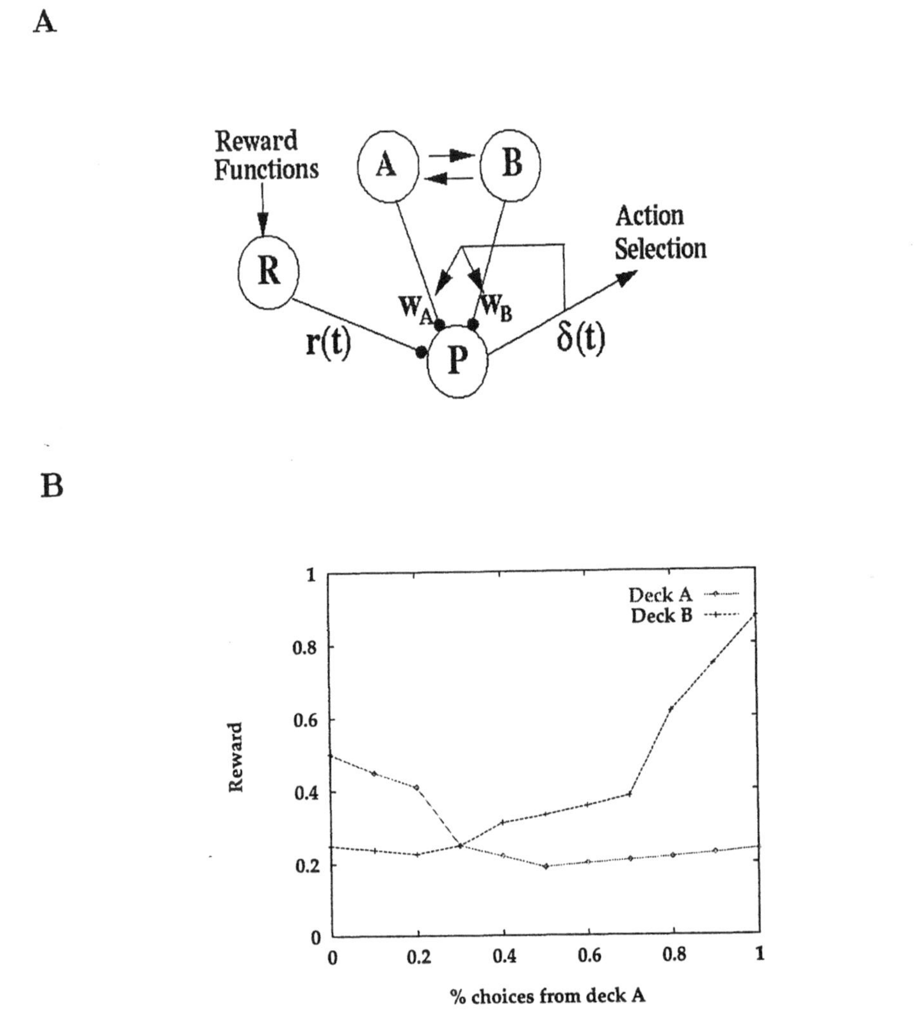

FIGURE 5. Card-choice task. **A.** A model analogous to the bee architecture above was given a simple card-choice task. The model made transitions between two decks of cards, A and B, which resulted in fluctuations in $\delta(t)$. $\delta(t)$ was used to determine whether the current deck was chosen using a decision function that depended on $\delta(t)$ (sigmoid in Equation 8). Once a deck was selected, a reward was received according to the reward functions shown in panel B. On reward presentation, the weights associated with each deck were updated according to a Rescorla-Wagner rule (Equation 9). **B.** Unlike the bee model, the reward distributions were not stationary but changed as a function of the fraction of choices made from Deck A. Human subjects were given the same task with identical reward distributions (**Figure 6**).

the choices between the decks. Once a deck was selected, the weight for the selected deck was updated according to a Rescorla-Wagner rule.

$$\Delta w_s(f_n) = \lambda(r_s(f_{n-1}) - w_s(f_{n-1})) \tag{9}$$

f_n is the fraction of choices from Deck A at iteration n, $w_s(f_{n-1})$ is the weight associated with the selected deck at iteration n-1, $r_s(f_{n-1})$ is the reward associated with the deck selected, and λ is a learning rate (or equivalently a forgetting rate). For the networks, the fraction of choices from Deck A converged to the zone 0.31-0.41 (**Figure 5B**). This range includes the crossing point of the reward functions. The slope of the linear portion of the decision function (*m*) was varied from (-0.1,-5.0) and *b* was varied from (0.0,15.0). In preliminary experiments, shown in **Figure 6A**, human subjects tended to fluctuate near the crossing point of the reward functions or remain near the optimal fraction of choice from Deck A (Egelman et al, 1995). These experimental results are compared to results with the model using various initial conditions, learning rates, and parameters for the decision function.

The humans and the networks tend to fluctuate around the crossing point of the reward functions, so that the average rate of return from the two choices is approximately equal. In experiments where an animal is given multiple behavioral alternatives, each of which yields rewards of various sizes or strengths, the animal tends to adjust its sampling of alternatives so as to match the relative rewards obtained from each alternative. In view of the importance of certain diffuse systems (e.g., dopamine) for reward-dependent behavior, our use of the diffuse-system output to constrain action choices provides one bottom-up explanation of how diffuse systems may establish constraints that favor matching.

Conclusions

We have seen that temporal difference algorithms provide a mathematical framework for describing one role that diffuse neuromodulatory systems could play during ongoing learning and action choice. In this framework the same error signal is used to improve predictions and choose actions. Diffuse neuromodulatory systems are known to influence both synaptic plasticity and ongoing neural activity, making these TD models a reasonable starting point for theories of their role during learning and decision making. There is one remaining domain in which diffuse ascending systems are known to play a role: activity-dependent stages of cortical development.

Preliminary work has suggested how a TD error signal could influence the initial self-organization of sensory representations (Quartz et al, 1992; Montague et al, 1993). This is an important possibility, since it would allow the system to adjust its sensory representations so as to better account for the constraints communicated by reward and punishment signals.

A

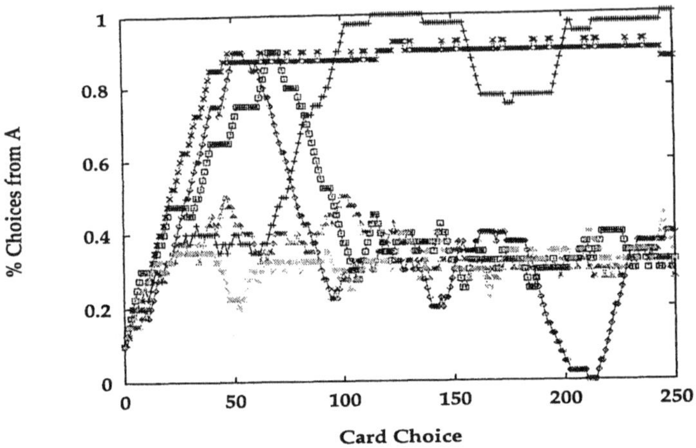

B

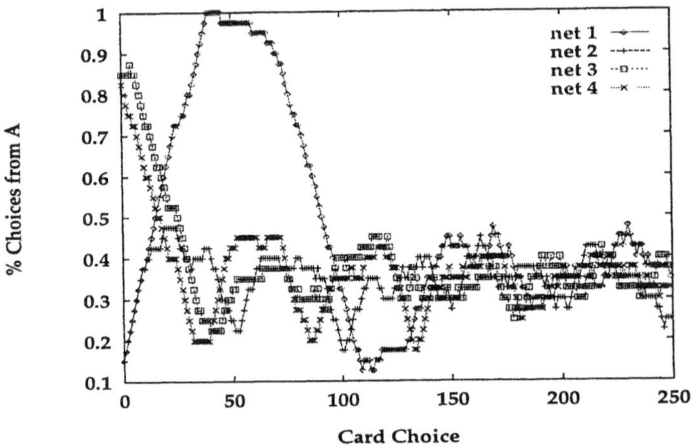

FIGURE 6. Human and model performance on card task. **A.** Raw data from 6 human subjects performing the same card task as the model. These data represent the trend observed in preliminary experiments: Subjects tend to fluctuate near the crossing point of the two reward functions, with a minority of subjects discovering the strategy for achieving better long-term returns (two subjects fluctuating around 0.8). **B.** Performance of 4 incarnations of the model (shown as net1-net4). In each case, some parameter is different: starting point, learning rate, slope of the linear portion of the decision function, inflection point of the decision function. The crossing point of the reward functions is a stable point for the network; as such, the networks sample so that the average rate of return from the two alternatives is approximately matched.

There are a number of interesting issues raised by our attempts to connect computational theories to biological substrates: How is time represented by neurons? How does the brain avoid picking up spurious correlations in the absence of sufficient data to average away noise? How are coherent sensory representations constructed? These questions are of course not answered here. However, connecting the physiological data with both behavioral and computational theories builds testable links between the biological hardware and its outputs.

ACKNOWLEDGMENTS

I thank David Egelman and Tony Reiser for helpful comments and criticisms. I thank Christophe Person and Paul Yates for help with computer simulations and robotic experiments. This work is supported in part by NIMH Grant R01MH52797-01 and the Center for Theoretical Neuroscience at Baylor College of Medicine.

Neural-Networks Models of Cognition
J.W. Donahoe and V. Packard Dorsel (Editors)

CHAPTER 22

THE ROLE OF TRAINING IN REINFORCEMENT LEARNING

Jeffery A. Clouse
Department of Computer Science
University of Massachusetts, Amherst

ABSTRACT

In reinforcement learning, an automated learning agent acquires the ability to perform tasks based on evaluative feedback from a critic. Unfortunately, in multi-step tasks this feedback is sparse and only weakly informative. In preliminary research reported here, feedback from a critic was augmented with additional information from a human training agent. Such information provides the learner with richer feedback and allows the learning agent to acquire its task much more quickly. In addition to reporting preliminary results with a human trainer, this chapter reviews other work of this type and identifies general issues in the design of automated training agents to complement reinforcement learning.

Introduction

In reinforcement learning, a learning agent attempts to improve its performance on a task in which it must choose a series of actions. The agent gains knowledge about the task through trial-and-error experience of the consequences of its actions. The source of that information is an automated critic and the information comes in the form of occasional feedback that indicates the desirability of the current state in which the learner finds itself.

At each successive time step of a multi-step task, the learning agent selects an action based on its current policy, which is a mapping between the current state of the task and an action. It then executes the chosen action, which changes the task state. At this point, the critic may provide the learning agent with a scalar reinforcement signal that indicates the value of the new state. The signal may arise from a mechanism internal to the critic or externally from the environmental consequences of the current action. The goal of the learning agent is to change its action-selection policy so that it will visit undesirable states less often and desirable states more often as learning progresses. That is, the agent adapts its policy to increase the frequency of positive signals and decrease that of negative signals.

Reinforcement-learning methods, such as actor/critic methods (Sutton, 1984) and Q-Learning (Watkins, 1989), were designed to allow a learning agent to develop a policy based solely on the critic's scalar feedback. Many researchers have demonstrated success with reinforcement-learning methods (Gullapalli, 1992; Lin, 1992; Whitehead, 1992; Barto, Bradtke, & Singh, 1993). However, the training information provided by the critic may be weak and sparse, often occurring only upon completion of the task. For example, after executing many actions, the learner may be informed by the critic that the state it has reached is not desirable. The critic does not give the learner any other information about the task—not which action to perform, or which of its many actions led to the receipt of the feedback, or which sequence of actions might have been better. The critic provides only simple, scalar feedback that is often delayed.

This chapter describes the effects of providing the learner with another source of information—a training agent. **Figure 1** presents the components of such a learning scenario and the interfaces between the components. The unshaded region of the figure depicts the basic components required for reinforcement learning: the *task*, the *learning agent*, and the *critic*. The task receives actions and produces state information; the learning agent receives state information and reinforcement, and produces actions; and the critic receives state information and produces reinforcement.

The shaded region of the figure represents the addition of a *training agent* to the scenario. As the learner engages in trial-and-error behavior, the training agent observes the state of the task and provides the learner with additional information. Note that, in reinforcement learning, the critic is also a training agent, but one whose interactions with the learner are constrained to provide scalar feedback. The training agent that is added to the reinforcement-learning scenario does not necessarily have such constraints placed upon it. A major goal of the present chapter is to identify some of the procedures whereby a training agent may best benefit the performance of the learning agent.

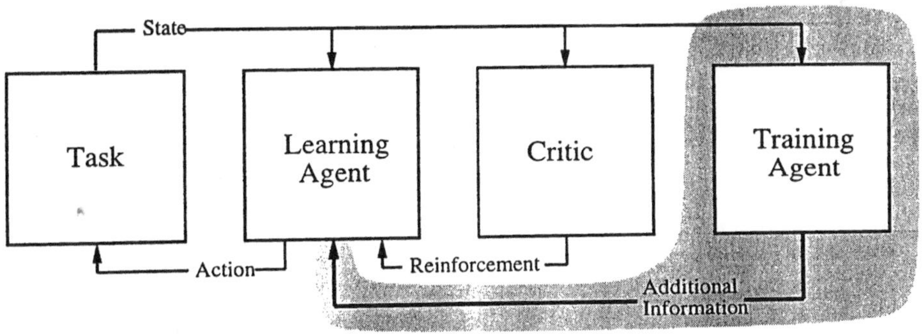

FIGURE 1. Additional Training Information.

Overview

Clouse and Utgoff (1992) implemented an instantiation of the scenario in **Figure 1**, wherein the critic's training information was augmented with training information from a *human* trainer. While observing the learning agent carry out the task, the human trainer provided the learning agent with actions to perform. Details of this system are presented in the next section of the chapter.

The remaining sections briefly review other learning systems in which the critic's feedback is augmented by information from a trainer, and then identify issues that arise when a training agent is added to a reinforcement-learning scenario. These issues include the form of the information supplied, the methods by which the learner incorporates the additional information, and the time at which the trainer provides the information. The chapter concludes by describing two types of multi-step tasks in which reinforcement learning does not, by itself, produce an effective policy but in which the addition of a training agent may alleviate the problem.

Learning Using a Human Training Agent

In a system first studied by Clouse and Utgoff (1992), a human served as the training agent that provided the automated learner with additional information about the task. While the automated agent learned the task by means of reinforcement learning, the human monitored the learner's performance in real time. When the trainer identified situations in which he believed that the learner needed assistance, he indicated an action that the learner should perform. Although the task was learnable without a trainer, the trainer was introduced to determine whether this could hasten learning. (See Clouse & Utgoff, 1992 for specific details of the implementation.)

The learning agent

To learn to choose the correct actions, the learning agent employed an actor/critic version of reinforcement learning, a method that is closely related to the dynamic programming method of policy iteration (Bellman, 1957; Barto, Bradtke, & Singh, 1993). The reinforcement-learning method is called "actor/critic" because of the two components of the adaptive mechanism. The learner maintains both a policy—which is an actor because it produces actions—and an evaluation function—which is a critic because it provides internal feedback for changing the policy.

The development of the evaluation function is based solely on the feedback that the environmental critic provides. Given the current state and the critic's current feedback, the evaluation function is trained to predict the future values of the feedback received in that state. The policy, on the other hand, is changed—or adapted— based on the output from the evaluation function, regardless of whether the critic provided any feedback. Because the evaluation function learns to predict future feedback, the current action taken by the learner receives credit or blame for that predicted feedback. The alternative is for the learner to perform actions until the critic provides feedback, and then attempt to allocate that feedback across the previous actions. The problem of deciding how to allocate current feedback across previous actions is known as the *temporal credit-assignment problem*. In the experiments described below, the evaluation function and policy are each implemented as a linear network (Nilsson, 1965).

In addition to the actor/critic method, another mechanism for alleviating the credit-assignment problem involves eligibility traces of previous actions. The trace mechanism slightly adjusts the evaluation of previous states and the choice of action in those states, thereby assigning partial credit or blame for the critic's current feedback to the learner's previous actions. For example, if the critic provides a negative scalar signal for the current state, then the evaluation of previous states is adjusted to make the choice of previous actions also less likely. The policy and evaluation function each have associated eligibility traces, which are vectors that record a decaying average of previous states and actions. The changes, or updates, to the policy and evaluation function are based on their associated eligibility traces. Thus, previous actions receive partial credit or blame for the critic's current feedback, because changes in the values of previous states are based on the evaluation of the current state and on the remaining contribution of previous states to the trace.

Incorporating the trainer's actions

When a learning agent adapts without a trainer, the learner operates within the standard reinforcement-learning cycle. In the first step, the agent chooses an action based on its current policy and then performs the action. Before receiving feedback from the critic, the learner updates its eligibility traces, recording the action it performed and the state of the task. In the final step after receiving the critic's feedback, the learner updates its evaluation function and policy based on the eligibility traces and the feedback received from the critic. The agent continues this cycle until training is stopped. These steps are presented in the first column of **Table 1**, labeled Reinforcement Learning.

So that the learner may adapt based on information received by the training agent, each of the steps mentioned above is changed. Before performing any of the learning steps, though, the learner determines whether the trainer has supplied an action. If not, the learner performs the standard reinforcement-learning steps, as if there were no trainer.

Each of the steps that the learner performs when the trainer provides an action differs from that of reinforcement learning alone because the actions provided by the trainer are not simple scalar values. These changes are summarized in the second column of **Table 1**, labeled Learning from Trainer's Action, and are highlighted with italics.

Reinforcement Learning (without trainer feedback)	**Learning from Trainer's Action** (with trainer feedback)
choose an action and perform it	perform the *trainer's* action
update eligibility traces	*zero* eligibility traces and *add* trainer's action
receive critic's feedback	*ignore* critic's feedback
adapt based on critic's feedback	adapt based on *positive* feedback

TABLE 1. Learning Steps.

The first change is that the learner performs the action provided by the trainer, not the action chosen by its own policy. Secondly, the trainer may provide the learner with an action when the learner's most recent actions were judged inappropriate by the trainer. This requires each eligibility trace to be treated in a special way: The traces are reset when the trainer provides an action. If the traces were not reset, then the learner's most recent actions would receive credit for having reached the state at which the trainer intervened. The effect of not resetting the traces would, therefore, be for the learner to increasingly perform those inappropriate actions. To prevent this, the learner's actions are expunged from the eligibility traces and do not receive credit. Note, the assumption that a "good" trainer always provides correct actions need not inevitably hold: An effective trainer may give the learner an action to emphasize that action *and then* reward the actions that lead up to it. Nevertheless, because the trainer can volunteer an action when the learner's recent actions were inappropriate, the traces are reset whenever the trainer provides the action.

After the eligibility traces have been reset, they are updated with the trainer's action and the current state. Then, any feedback from the critic is ignored and the learner associates positive feedback with the new state resulting from the trainer's action. In the last step, the evaluation function and policy are updated based on the new eligibility traces and the reinforcement signal. Thus, the trainer's action—and only the trainer's action—is rewarded. However, since the trainer's action may not necessarily be the best action, it too becomes part of the eligibility traces where it can be blamed (or credited) for future feedback received by the learner.

The multiple-step task

The multiple-step task employed in the experiments is the inherently unstable, dynamic cart-pole task (see Anderson & Miller, 1990). The task consists of a simulated 4.8-meter one-dimensional track, a wheeled cart that can move freely on the track, and a pole that is hinged to the top of the cart and can swing freely to the left or right (see **Figure 2**). The objective of the task is to keep the pole within twelve degrees of vertical and the cart within the boundaries of the track by pushing the cart left or right at each discrete time step. To achieve this end, the learner must develop a policy that specifies which of these two actions, push left or push right, to perform at each time step.

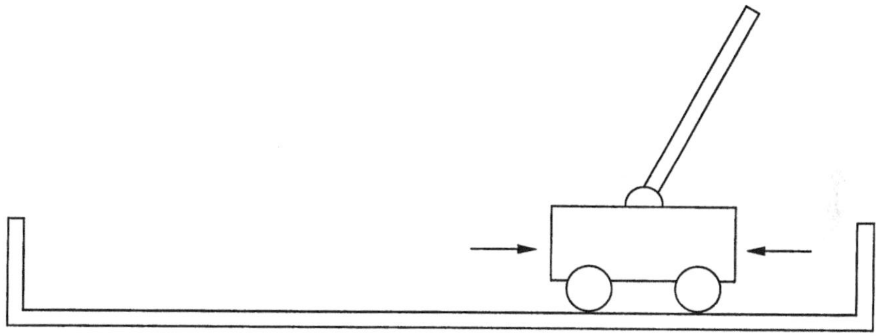

FIGURE 2. The Cart-Pole Task.

Experiments

To evaluate the learner's performance with and without access to a human trainer, we performed two sets of experiments. In the first set, the learner received training information from only the critic, which provided a failure signal (-1.0) when the pole was no longer within twelve degrees of vertical or the cart had hit the edge of the track. In the second set of experiments, a human trainer provided the learner with additional information through the use of a real-time, interactive graphics interface that allowed the trainer to observe the state of the cart-pole task as the learner manipulated it. The trainer provided the learner with training information by pressing either of two keys on a keyboard, one associated with a push to the left and the other with a push to the right. Each keystroke represented an action for the learner to perform at the time the key was pressed.

Both sets of experiments consisted of twenty runs, each of which began with the learning agent's policy and evaluation function initialized, and ended either when the learner failed to balance the pole or had kept it balanced successfully for 10,000 time steps. (The actions of the trainer, although performed by the learner, did not count towards this total). Each run consisted of several trials, each beginning with the cart-pole system in a random, solvable state.

	Without Training $\overline{X} \pm s$	With Training $\overline{X} \pm s$
Trials	992.0 ± 723.5	8.3 ± 6.1
Total Actions	181,529.2 ± 113,420.7	10,767.1 ± 866.5
Trainer's Actions	0.0 ± 0.0	7.3 ± 4.3

TABLE 2. Results of Experiments.

Results

The results of the first and second sets of experiments are reported in **Table 2** under the column headings Without Training and With Training, respectively. The table presents the means and standard deviations of the number of trials per run (top row), the total number of actions per run (middle row), and the number of trainer-supplied actions per run (bottom row).

As seen in the top row, the information provided by the human trainer reduced the total number of trials required to learn to keep the pole balanced. With only the critic's feedback, the learner needed an average of 992 trials to complete a run successfully. That is, the learner failed at the task an average of 991 times before developing a successful policy. In contrast, with the help of a human trainer, the learner achieved the same performance in an average of 8 trials—a decrease of *two orders of magnitude*. Furthermore, as shown in the bottom row of the table, this improvement required an average of only about seven actions by the human trainer. Thus, very little additional information was needed from the trainer to achieve a drastic reduction in the number of failures before the task was solved.

The addition of a trainer also reduced the total number of actions, or weight updates, for the two linear networks. As shown in the middle row of the table, the learning agent that developed its policy without benefit of a trainer needed to perform an average of 171,529 actions before beginning to satisfy the criterion of 10,000 successful actions in a row. Achieving this performance required approximately one hour of real-time simulation. With the trainer's help, the same criterion began to be satisfied in an average of 767 actions, which consumed about 3.5 minutes of real-time simulation. This also represents an improvement of two orders of magnitude—from 171,529 to 767 actions. Finally, notice that the variability of performance was much greater without a trainer. The large standard deviations without a trainer were partially due to the trial-and-error nature of reinforcement learning: The exploration choices made by the learning agent sometimes find an effective policy quickly, but at other times cause a delay in finding a policy. The standard deviations with a trainer were relatively small. Even though the learner was still exploring the task-space, the input from the training agent constrained the learner to explore those portions of the space in which a solution was more likely.

Discussion

In the foregoing experiment, a great reduction in the amount of training was achieved by allowing a human trainer to interact with the automated learning agent through interventions in real time. Similar results were achieved in a related set of experiments (Clouse & Utgoff, 1992) using a different method of reinforcement learning (Q-learning) and a different multiple-step task (a vehicle navigation problem). A human trainer helped the automated learner achieve a reduction in the number of training trials of one order of magnitude. Thus, this experiment also demonstrated that a learner can acquire a task in many fewer training trials when it receives actions from a training agent than when it is restricted to the critic's feedback alone.

Other Simulations of Training

Other researchers have taken related approaches to providing an automated learning agent with additional information. Utgoff and Clouse (1991) employed the simple reinforcement-learning method used by Samuel in his ground-breaking checkers program (Samuel, 1963) to learn an evaluation function over the state space of the problem. During training, the estimated value of the current state was changed to reflect the estimated value of the successor state chosen by the learning agent. Also, when the learner determined that its evaluation function was incorrect, it queried an automated trainer. An incorrect evaluation function was defined as one in which the difference between the estimated value of a state and the estimated value of its successor exceeded a criterion. The training agent provided the learner with the correct value for the current state. With the trainer's help, the learner was able to perform the task in *only one training trial*. Without a trainer, the agent required almost 500 trials.

Lin (1992) employed an approach he referred to as *teaching*. This approach required the human trainer to develop prior to the experiment complete sequences of actions for teaching the learning agent. In teaching, the human trainer led the learning agent from start states to goal states, with the learner recording the trainer's actions into a *lesson*. The learning agent then repeatedly replayed the lessons, updating its policy as if it were performing the actions in the lesson, while also performing trial-and-error experiments on its own. The results of the experiments indicated that the learning agents were able to learn *much* more quickly with training. In one case, the learning agent was able to learn a task with training that it did not learn with reinforcement learning alone.

More recently, researchers have studied providing the learning agent with information in the form of *advice*, represented as IF-THEN rules (Maclin & Shavlik, 1994). Because the learner's policy was represented in a feed-forward neural network, the advice could be incorporated using techniques from knowledge-based neural networks (Towell, Shavlik, & Noordewier, 1990) that allow IF-THEN rules to be added into the network. Although the advice could not be developed and presented in real time, the presentation of a single piece of advice allowed the learner to improve significantly its performance on the task.

Issues To Be Addressed When Adding a Training Agent

The possibility of adding a trainer to the learning scenario raises three general questions:

What information does the training agent provide?

How does the learner incorporate that information into its policy?

When does the trainer provide the training information?

These questions also apply to the feedback provided by the critic in reinforcement learning, and are answered as follows. First, the information provided by the critic takes the form of scalar signals. Second, these signals are incorporated into the learner's policy via a reinforcement-learning algorithm. Finally, the times when the information is given are largely determined by the nature of the problem. For example, the critic in the cart-pole task gives the learner negative scalar feedback when the pole is no longer balanced, thus indicating that the learner has not met the objective of the task. In other problems, the critic gives the learner positive feedback when the learner has achieved a particular goal or subgoal of the problem.

Unlike the critic, the trainer that is added to reinforcement learning is not constrained to provide only a single scalar feedback signal. The trainer can provide various forms of information, such as actions, scalar feedback, and advice. However, because reinforcement-learning methods are not designed to deal with non-scalar training information, the underlying learning mechanisms of the learner must be modified. Another complicating factor is that, while there are guiding principles for deciding when a critic should provide feedback, it is unclear when a training agent should supply the learner with training information.

The following sections examine these issues, point out how the systems presented above might deal with them, and present alternatives to be investigated in future simulation research concerning the effects of training on learning.

The form of additional information

In two of the training systems (Clouse & Utgoff, 1992; Lin, 1992), the trainer provided actions to the learner. These actions yielded rich knowledge in that a particular state of the problem was linked to the action that should be performed in that state. Moreover, the training agent's action might also have been useful because it caused the learner to explore a different part of the state space, thereby obtaining important knowledge about how to perform in the task that might not otherwise have been gained.

In Utgoff and Clouse (1991), the additional information given to the learner took the form of the direct value of particular states. Since this was exactly what the learner was attempting to establish, these values represented clearly useful training information. In other work (Maclin & Shavlik, 1994), the information supplied by the trainer took the form of IF-THEN rules, and this advice allowed the agent to perform better.

In addition to the forms of information discussed above (actions, state values, and advice), the information supplied by the critic can also be varied. The critic supplies criticism in the form of scalar values that allow the training agent to employ the operant conditioning technique of *shaping*. In shaping, the training agent slowly changes the criterion conditions for the presentation of reward so as to expose the learning agent to progressively more difficult problems. Using shaping, the trainer can reward the learner when it performs an approximation to the requisite task. After the learning agent is able to perform this simpler task, the training agent can then change its criterion for supplying scalar feedback to make the learner perform a slightly more complicated problem, but one that is still simpler than the target task. After several such approximations, the learning agent is able to perform the requisite task. Such a technique has been successfully implemented with a robot in a reaching task (Gullapalli, 1992; see also Gullapalli, this volume).

Another form of information, *sets* of actions, represents a new approach to giving task knowledge to a learner (Clouse, 1995). The set of actions can either be one from which the learning agent should choose an action, or one from which an action should not be chosen. When the set contains actions from which the learning agent should choose, the trainer can present a set with a limited number of actions early in the training. As the learner's abilities progress, the trainer can relax control over the learning agent, allowing more actions to be members of the set. Conversely, if the set represents actions the learning agent should *not* take, then the set may be large initially and slowly decrease as the learning agent improves its skill. By constraining the subset of possible actions, these two types of additional information allow the training agent to prevent the learner from making costly mistakes as it learns.

Because the training agent is not as constrained as the critic, there are many forms of information that he can supply to the learning agent. Findings from the study of a few systems indicate that advice, actions, scalar feedback, and exact values can all serve as beneficial forms of feedback. However, it is unknown which forms of information provide the most valuable feedback. It has also not been determined whether specific aspects of a particular task might influence which form of the trainer's information would most benefit the learning agent.

Incorporating additional information from the trainer

In addition to investigating what type of information is given the learning agent, we must also determine how the agent is to utilize the information, i.e., how the agent adapts its policy. Reinforcement-learning methods have been proven to produce policies that approach optimal policies (Watkins & Dayan, 1992), and they also appear to most closely approximate the methods whereby experience changes the behavior of living organisms. For these reasons, we assume below that the learner's underlying adaptive mechanism is reinforcement learning. Nevertheless, one must still determine how best to provide scalar feedback to the reinforcement-learning algorithm based on information that the trainer supplies, and how the learning agent chooses its actions. Depending on which form the information takes, the means of incorporating the information into the reinforcement-learning cycle must change in corresponding ways.

Basic reinforcement-learning methods produce adaptation based on scalar feedback from the critic. To accommodate learning with an automated trainer that gives scalar feedback, one must decide how to deal with simultaneous feedback from both the critic and the trainer. For example, should the trainer's concurrent feedback take precedence over the critic's, or be ignored? A more complicated solution would be to adapt based on a weighted average of the two scalar signals. Although incorporating another source of scalar feedback into reinforcement learning may seem a straightforward matter, important and perhaps unintended consequences flow from how the feedback signals are combined.

As we reported previously (see **Table 1**), Clouse and Utgoff (1992) introduced a new set of steps into the learning cycle in order for the learner to incorporate the trainer's action: The learning agent was required to perform the trainer's action. However, many plausible procedures exist for incorporating training information. For example, the learner might perform whatever action its policy dictates, but associate positive feedback with the action only if it is the same action given by the trainer. Also, in earlier work, the critic's feedback was ignored by the learning agent when the trainer provided the action. Alternatively, a weighted sum of the two scalar values might be used in the adaptation process. Finally, the eligibility traces were also treated in a special manner in previous work, being reset when the trainer provided an action. As an alternative, the traces might be left intact but training with them would only occur when the learner has produced the action.

In the case where the trainer provides sets of actions, the learner might be constrained to perform only those actions in the trainer's action set. That is, when the learner must choose an action to perform, its actions are confined to the choices in the trainer's specified action set. Another option would be for the learner to perform whatever action it chose, and then to associate a positive scalar signal with the action if it was one from the trainer's set. Many options exist for incorporating feedback from the trainer, and they remain to be explored in future simulation research.

When to provide training information

In addition to questions concerning what information to provide and how to provide it, questions arise about when the trainer should supply the information to the learning agent. Again, many possibilities exist. At one extreme the trainer might never volunteer information (pure reinforcement learning), while at the other the trainer might always provide information. Below, we identify some of the intermediate possibilities.

In most of the training systems described earlier, the additional information was provided at the whim of a human trainer. In Lin's (1992) work, the trainer provided the learning agent with entire sequences of actions, but the criteria for providing the sequences were unclear. In Clouse and Utgoff's (1992) work, the criteria employed by the human trainers were similarly unclear. Training was given whenever the human deemed it helpful. Similarly, Maclin and Shavlik (1994) allowed human trainers to provide advice at their discretion. None of these systems addressed the issue of *when* the training information should be provided, but simply allowed a human trainer to guide the system in an unprincipled manner.

Automating the trainer. In order to evaluate different criteria for deciding when to provide additional information, automated trainers must be designed that follow prescribed policies. The long and unknowable pre-experimental histories of human trainers and the variability inherent in living organisms make the pursuit of these questions difficult unless the trainer is automated (i.e., simulated).

A policy for deciding when to provide the learner with additional information can vary as a function of time, the output changing with the number of task actions that the learning agent performs. For example, the training agent might give help frequently at the beginning of training and less frequently as training progressed. More complicated functions of time can also be employed. For example, a policy might have the training agent help the learner for a period of time, stop helping for some period, and then revert to helping after yet another set period of time.

The trainer's policy need not be based on time alone; it can vary with the quality of the learner's performance, which can be measured by the actual scalar feedback. Thus, the trainer might help the learner only when the learner received negative feedback. However, because feedback can be scarce, a policy based on environmental feedback alone may be insufficient. For that reason, a policy based on both time and performance could be more effective, or—in the case of certain tasks—necessary. As an example of this type of policy, a decaying average of the environmental feedback could control whether the trainer helps the learner. As long as the decaying average remains below a specified threshold, the training agent continues to help the learning agent. As soon as the average of the environmental feedback surpasses this threshold, the training agent stops helping.

The training agent's policy for deciding when to provide information to the learner can also be a function of the current state of the learner's task. The current state of the task provides important information if the automated trainer knows which task states are difficult or dangerous or particularly significant for other reasons. This form of automated policy can be based on a set of heuristic rules for determining when to help.

Another possible technique for determining when the trainer should intervene involves active-exploration policies (cf. Thrun & Möller, 1992). Active exploration is designed to help reinforcement-learning agents develop effective trial-and-error experiences. Instead of employing these techniques to determine when to explore, they can be used to decide when the learner should be given help. For example, the trainer might develop a map of the task space that indicates the learning agent's competence at performing the task in each state. If this *competence map* indicates that the learner's acquired knowledge of the current state is inadequate, then the training agent can intervene.

Deciding when a trainer should help a learning agent is clearly a question whose answer lies in the future. The most well-studied cases are those in which the trainer always provides feedback (supervised learning) and those in which the trainer never provides feedback (reinforcement learning) about the specific characteristics of the learner's performance. We have also considered here some of the alternatives between these two extremes. For other related work, see research in Intelligent Tutoring Systems (Woolf, 1988), which concerns providing information by an automated trainer to human learners, and research in Distributed Artificial Intelligence (Durfee, Lesser, & Corkill, 1989), which concerns interactions between two automated agents.

Conclusions

Reinforcement learning is an effective means for adapting neural networks to the demands of many tasks. However, reinforcement-learning algorithms become much more powerful when they can take advantage of the contributions of a trainer. To take advantage of a training agent's knowledge of a task, a number of issues must be resolved about how the trainer interacts with the learning agent and how the learning agent incorporates the trainer's information. This chapter has presented several options for dealing with these issues.

Preliminary work clearly demonstrates the benefits of providing the learner with access to a training agent. In one system (Clouse & Utgoff, 1992), the trainer's presence allowed the learning agent to perform its task with two orders of magnitude fewer training trials. Another system (Utgoff & Clouse, 1991) demonstrated that a learner could perform a task after only one training trial with input from a trainer. A third system (Lin, 1992) determined that a task that could not be acquired with a reinforcement-learning algorithm alone could be mastered with the help of a trainer.

Multi-step tasks benefit particularly when reinforcement learning is supplemented with a training agent. Two major potential benefits are apparent. For some multi-step tasks, the learning agent employing only reinforcement learning takes a prohibitively long time to develop a policy. With the help of a training agent, the learner can arrive at a policy much more quickly. For other multiple-step tasks, failure may be highly undesirable or even catastrophic. A trainer can circumvent such consequences by disallowing a learning agent from performing whatever action is dictated by its own transient policy. The addition of an effective automated trainer provides a fail-safe means by which the learning agent can acquire the ability to perform the task. As examples, when the form of knowledge employed by the training agent is either individual task actions or sets of task actions, the training agent can supply state-action pairs that prevent the learner from failing and, by so doing, enable the learner to acquire those actions. By using scalar criticism (shaping), the trainer can begin by implementing a task at which the learner cannot fail catastrophically, and then progress slowly to the final target task. Through such means, the addition of an automated training agent to the learning scenario gives promise of developing automated learning agents that are competent to solve problems currently deemed beyond the reach of reinforcement learning.

ACKNOWLEDGMENTS

Many thanks go to the members of the Machine Learning Lunch group who commented on the contents and organization of a draft of this chapter. The work of Clouse & Utgoff (1992) was supported originally by the National Aeronautics and Space Administration under Grant NCC 2-658.

PART SIX: COMPLEX BEHAVIOR—LANGUAGE

A problematic trend resurfaced in the preceding section on Reinforcement Learning: As the complexity of the simulated phenomena increased, the degree to which the simulations were informed by the relevant biobehavioral science decreased. For example, reinforcement learning was sometimes motivated primarily by experimental findings about nonspecifically projecting, neuromodulatory systems (e.g., Donahoe and Montague) and at other times by mathematical arguments, as with dynamic programming (e.g., Barto & Sutton). Indeed, both experimental and mathematical considerations informed the same simulation, as with motor control using the biobehaviorally grounded concepts of reinforcement and shaping together with the mathematically inspired technique of backpropagation (Gullapalli). Moreover, different simulations by the same researcher drew upon different literatures at different times—experimental science on some occasions and formal (logical/mathematical) considerations on others (cf. Barto & Sutton in this volume and Houk, Adams, & Barto, 1995; see also Gluck & Myers, this volume).

The treatment of language in this section shows the same intermixing of formal considerations with biobehavioral findings and, further, the failure of relevant biobehavioral findings to fully inform the simulations. Using modern imaging techniques, **Raichle** identifies multiple neural systems for processing verbal stimuli, which **Gullapalli and Gelfand** then use to design their network architecture. However, each feature of a word in the simulations corresponds to a single unit, whereas ensembles of neurons are activated by single features in the real nervous system (e.g., Tanaka, this volume). Similarly, experimental research in phonetics provides the foundation for **Jordan**'s simulations of the parallel transmission of phonemes. But, different articulatory gestures correspond to single units, whereas experimental research indicates that responses reflect the concerted action of many motor neurons (Georgopoulos, this volume). Further, connection weights between units are modified using back propagation instead of reinforcement learning or other biologically plausible algorithms (e.g., Rumelhart & Zipser, 1985). Continuing, **Barnes and Hampson**'s simulations are motivated by behavioral research on the development of equivalence relations among verbal stimuli, but connection weights are again modified by backpropagation, and single units now correspond to whole words, not features. Finally, **Van Orden, Bosman, Goldinger and Farrar**'s simulations rely upon neuropsychological research on dyslexia and neuroanatomical work on recurrent connections, but the units in the neural network now represent letters, phonemes, and meanings instead of features or words. What are we to make of these differences (even inconsistencies) among the simulations with regard to their reliance upon formal rather than experimental constraints? And, what are we to make of differences in the extent to which relevant biobehavioral science informs the simulations?

The short answer is: We are at the very beginning of efforts to achieve biobehavioral interpretations of complex behavior and, hence, most current simulations are best regarded as preliminary forays into the domain of complex behavior whose primary purpose is to establish the *competence* of neural networks to produce complex behavior as their emergent product (cf. Elman, 1995). Viewed in this light, **Gullapalli and Gelfand**'s simulations successfully demonstrate that training can automatically shift verbal processing from one subnetwork to another (see short-circuiting, Chapter 18). **Jordan**'s simulations successfully demonstrate that parallel transmission is a general and emergent characteristic of neural networks, and not a language-specific phenomenon. **Barnes and Hampson**'s simulations successfully demonstrate that equivalence classes can be mediated by neural networks *if* the architecture is appropriate. And, the simulations of **Van Orden** and his colleagues successfully demonstrate that neural networks with recursive connections can mediate a variety of verbal relations that might otherwise be considered beyond their competence.

The longer answer is too long to be given here. However, it would include the following: The dominant approach to complex human behavior—cognitive psychology—has traditionally adopted a "top-down" approach to explanation. That is, explanation takes the form of special-purpose, high-level processes and structures—e.g., deep vs. surface processing, short-term vs. long-term memory, etc.—that arise as *inferences* from behavioral observations. This strategy contrasts with a biobehavioral approach in which high-level processes and structures are seen as emergent products of general-purpose, low-level processes and structures that are known through direct experimental observation (see Palmer & Donahoe, 1992; Donahoe & Palmer, 1994). This state of affairs is rapidly changing as cognitive science increasingly appreciates the unexpectedly powerful experimental and conceptual armamentarium of modern biobehavioral science. In the near future, it may be hoped that cognitive psychology will concur with its erstwhile bugaboo, the behaviorist B.F. Skinner: "The physiologist of the future will tell us all that can be known about what is happening inside the behaving organism. His account will be an important advance over a behavioral analysis, because the latter is necessarily 'historical'—that is to say, it is confined to functional relations showing temporal gaps. ... [Neuroscience] will make the picture of human action more nearly complete." (Skinner, 1974, pp. 236-237).

Neural-Networks Models of Cognition
J.W. Donahoe and V. Packard Dorsel (Editors)

CHAPTER 23

FUNCTIONAL BRAIN IMAGING AND VERBAL BEHAVIOR

Marcus E. Raichle
Department of Radiology and Neurology
Washington University School of Medicine
St Louis, MO

ABSTRACT

Recent advances in brain imaging with positron emission tomography (PET) and functional magnetic resonance imaging (fMRI) coupled with studies of event-related potentials (ERPs) now permit us to examine safely the normal human brain while it functions. These studies provide a much clearer understanding of the architecture of language systems in the brain and exciting new insights into how the normal human brain produces observable behavior. Neural-network models should yield important conceptual insights into the interpretation of the emerging data on functional brain imaging.

Introduction

Early discussions of the "mind-brain problem" largely treated the brain as a black box. Then, in 1861, French surgeon and anthropologist Pierre Paula Broca described a clear relationship between a patient's difficulty in speaking and an injury to a specific part of the patient's brain due to a stroke. Since this seminal observation, a vast body of scientific literature has accumulated implicating various parts of the human brain in specific aspects of human behavior including language. The remarkable level of sophistication to which this work has risen is detailed in several recent books (e.g., Damasio & Damasio, 1989; Shallice, 1988).

The view of brain organization arising from the study of patients with brain injury does raise some questions of interpretation. The size and location of brain injury varies greatly from patient to patient, making a precise correlation between damage to a particular area of the brain and the function normally served by that area sometimes difficult to determine. Furthermore, each patient may be assumed to have some features of brain organization that are unique to the patient. And, finally, it remains uncertain whether one can simply attribute a lost or disrupted function to a particular area of injury. Because of the interconnected nature of brain areas, injury in one area is likely to have effects on other areas that cannot necessarily be predicted from the location and size of the injury itself. Thus, although the study of patients with brain injury has provided much valuable information concerning brain organization, exactly how this information relates to the normal functioning of the human brain remains an open question.

Only recently have scientists interested in brain function had the opportunity to explore it analytically—to peer inside the black box during its normal activity. This ability stems from developments in imaging technology over the past 20 years, most notably positron emission tomography, usually referred to as PET, and magnetic resonance imaging, usually referred to as MRI. These techniques can now capture precisely localized physiological changes in the normal human brain associated with behaviorally induced changes in neuronal activity (Posner & Raichle, 1994).

It is important to point out that the underlying assumptions of current brain-mapping studies using PET and functional MRI (fMRI) are not modern versions of phrenology. The phrenologists of the past century posited that single areas of the brain, often identified by bumps on the skull, uniquely represented specific thought processes and emotions. In contrast, modern thinking posits that each area of the brain contributes quite simple mental operations that form the elementary components of observable behaviors. Observable behavior and thought processes emerge through the cooperative interactions of many such areas. Just as diverse instruments of a large orchestra play in a coordinated fashion to produce a symphony, so a group of diverse brain areas, each performing quite elementary and unique mental operations, work together in a coordinated fashion to produce human behavior. The prerequisite for such analyses is the conviction that complex behaviors can be broken down into a set of constituent mental operations.

The History of Functional Brain Imaging

The modern era of brain imaging began in the early 1970s with the introduction of a remarkable technique called X-ray computed tomography, now known as X-ray CT or just CT. South African physicist Allan M. Cormack and British engineer Sir Godfrey Hounsfield independently developed its principles, with Hounsfield constructing the first CT instrument in England. Both investigators received the Nobel Prize in 1979 for their contributions.

X-ray CT takes advantage of the fact that different tissues absorb different amounts of X-ray energy. The denser the tissue, the more energy it absorbs. A highly focused beam of X-rays traversing the body will exit at a reduced energy level depending on the tissues and organs through which it passes. A beam of X-rays passed through the body at many different angles through a plane collects sufficient information to reconstruct a picture of that body section. It was crucial to the application of X-ray CT that clever computing and mathematical techniques were developed to process the vast amount of information needed to create the images. Without the availability of sophisticated computers, the task would have been impossible.

The emergence of X-ray CT had two immediate consequences. First, it changed the practice of medicine, because X-ray CT was much superior to standard X-rays. For the first time, physicians could safely and effectively view living human tissue, such as the brain, with no discomfort or risk to the patient. Standard X-rays had revealed only bone and some surrounding soft tissue. Second, X-ray CT stimulated scientists and engineers to consider alternative ways of creating images of the body's interior using other mathematical and computer strategies for image construction. These efforts went beyond the picture of human anatomy provided by X-ray CT in that they began to focus on function.

Among the first groups to be intrigued by the possibilities opened by X-ray CT were experts in tissue autoradiography. This method had been used for many years in animal studies to investigate organ metabolism, biochemistry and blood flow. In tissue autoradiography, a radioactively labeled compound is injected into a vein. After the compound accumulates in the organ under investigation, the animal is sacrificed and the organ (e.g., the brain) removed for study. The organ is then carefully sectioned, and the individual slices are laid on a piece of film sensitive to radioactivity. Much as the film in a camera records a scene, the X-ray film records the distribution of radioactively labeled compound in each slice of tissue.

When the X-ray film is developed, scientists have a picture of the distribution of radioactivity within the organ and, hence, can deduce the organ's specific functions. The type of information provided by the picture is determined by the radioactive compound injected. For example, a radioactively labeled form of glucose measures brain metabolism, because glucose is the primary source of energy for cells of the brain. Central to functional brain imaging with PET is the measurement of brain blood flow, which is achieved through the injection of radioactively labeled water.

Investigators adept at tissue autoradiography were fascinated when X-ray CT was introduced. They realized that if the anatomy of an organ could be reconstructed by passing an X-ray beam through it, then they could also reconstruct the distribution of a previously administered radioisotope. One had simply to measure the emission of radioactivity from the body section. This insight was the birth of autoradiography with living human subjects.

One crucial element in the evolution of human autoradiography was the choice of radioisotope. Workers in the field selected a class of radioisotopes that emit positrons, which otherwise resemble electrons but carry a positive charge. Positrons produced within tissue almost immediately combine with nearby electrons. A positron and electron annihilate one another in this interaction, emitting two high-energy gamma rays in the process. Since the gamma rays travel in nearly opposite directions, radiation-detection devices arrayed in a circle around the organ of interest can detect the pairs of gamma rays and, with the aid of computers, locate their origin with remarkable precision. The crucial role of positrons in human autoradiography gave rise to the name positron emission tomography or PET.

More recently, another imaging technique has been developed to take its place alongside PET in revealing the function of the human brain. This technique is known as magnetic resonance imaging, or MRI. MRI is derived from the potent laboratory technique of nuclear magnetic resonance (NMR), which was designed to explore detailed chemical features of molecules. The technique garnered a Nobel Prize in 1972 for its developers, Felix Bloch of Stanford University and Edward Purcell of Harvard University. The method exploits the fact that many atoms behave as tiny compass needles when placed in a magnetic field. By skillfully manipulating the magnetic field, scientists can align the atoms. Applying radio-wave pulses to the sample perturbs the atoms in a precise manner and, as a result, they emit detectable radio signals unique to the number and state of the atoms in the sample. Careful adjustments of the magnetic field and the radio-wave pulses yield specific information about the sample under study.

NMR moved from the laboratory to the clinic when Paul C. Lauterbur of the University of Illinois found that NMR can form images when detecting protons. Protons are abundant in the human body, being found primarily in water and fat. Using mathematical techniques borrowed from X-ray CT, but later modified extensively, images of the anatomy of organs of the living human body were produced that far surpassed those produced by X-ray CT in their detail. Because the term "nuclear" made the procedure sound dangerous to some, NMR soon became known as magnetic resonance imaging, or MRI. The current excitement over both PET and MRI for imaging of normal brain function stems from their ability to detect signals associated with changes in neuronal activity through changes in local brain blood flow. I turn briefly to the nature of these changes in blood flow, and to their relation with changes in neuronal activity.

Measuring Brain Function with Imaging

Measurements of blood flow to local areas of the brain are at the heart of assessing brain function with both PET and fMRI (Posner & Raichle, 1994). The idea that blood flow is intimately related to brain function is a surprisingly old one. The English physiologists Charles S. Roy and Charles S. Sherrington formally proposed the idea in 1890. (For a detailed review of this history see Posner & Raichle, 1994 and Raichle, 1987.) They suggested that an automatic mechanism regulated the blood supply to the brain, with the amount of blood depending on local variations in activity. Although subsequent experiments have amply confirmed the existence of such an automatic mechanism, its exact nature remains somewhat unclear. Obviously, this remains a challenging area for research.

PET measures blood flow in the human brain through the use of an autoradiographic technique developed in the late 1940s by Seymour S. Kety and his colleagues for use with laboratory animals. (For a detailed review, see Raichle, 1987.) PET relies on radioactively labeled water, specifically hydrogen combined with oxygen-15, a radioactive isotope of oxygen. The labeled water emits copious numbers of positrons as it decays. The labeled water is administered into a vein in the arm and, in less than a minute, the radioactivity accumulates in the brain, providing the basis for an image of blood flow.

fMRI measures a complex function of blood flow: When blood flow increases during normal brain function, the amount of oxygen consumed by the brain does not (Fox, Raichle, Mintun, & Dence, 1988). Under these circumstances, more oxygen is present locally in the tissue because the blood flow has been increased but the demand for oxygen has not. Since the amount of oxygen in the tissue affects its magnetic properties, a fact first noted in 1935 by Linus Pauling (Pauling & Coryell, 1936), fMRI can detect the change. This change is often referred to as the BOLD—or **B**lood **O**xygen **L**evel **D**ependent—effect (Ogawa, Lee, & Tank, 1990; Ogawa, Tank, Menon, Ellermann, Kim, Merkile, & Ugurbil, 1992; Kwong, Belliveau, Chesler, Goldberg, Weiskoff, Poncelet, Kennedy, Hoppel, Cohen, Turner, Chen, Brady, & Rosen, 1992). A measure of blood flow equivalent to that measured by PET with oxygen-15-labeled water has proven difficult with fMRI, primarily because of the short half life (i.e., the T1 relaxation time) of the water protons in the fMRI experiment. (The T1 relaxation time of the water proton in brain tissue is approximately one second, whereas the half life of oxygen-15-labeled water is 123 seconds—more suitable for the measurement of blood flow in the human brain.)

The Imaging Strategy

A distinct strategy for the functional mapping of neuronal activity has emerged during the past 15 years. Initially developed for PET, it has been extended, with modifications, to fMRI.

This approach extends an idea first introduced into psychology in 1868 by Dutch physiologist Franciscus C. Donders. Donders proposed a general method to measure thought processes based on a simple logic. He subtracted the time needed to respond to the onset of any light (say, by pressing a key) from the time needed to respond only to the onset of a light of a particular color. He found that discriminating color required about 50 milliseconds more than responding to any light onset. In this way, Donders sought to isolate and measure a mental process by subtracting a *control state* (i.e., responding to any light) from a *task state* (i.e., responding to a particular color of light).

The current functional-imaging strategy is designed to accomplish a comparable subtraction regarding information about the areas of the brain that distinguish the task state from the control state. In particular, images of blood flow, or blood flow-related changes in the case of fMRI (i.e., the BOLD signal), obtained in a control state are subtracted from those obtained when the brain is engaged in the task. The control state and the task state are carefully chosen so as to isolate as well as possible a limited number of mental operations when the images of the two states are subtracted. Subtracting blood flow-

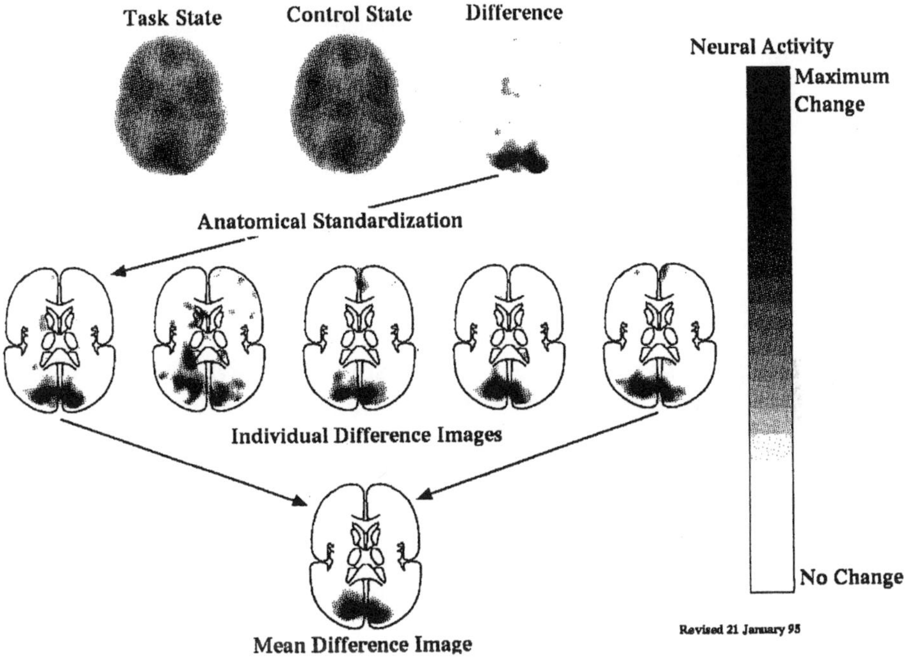

FIGURE 1. Illustration of image subtraction and image averaging in the development of positron emission tomographic (PET) images of brain function. In the top row are two PET blood-flow images in a normal human subject, one labeled Task State and the other labeled Control State. These images represent horizontal "slices" through the center of the brain. The front of the brain is toward the top and the left side is to the reader's left. Darker areas have higher blood flow than lighter areas. During the task state, the subject passively viewed a flashing annular checkerboard. During the control state the subject simply maintained fixation. (See text for details.) The difference in blood flow between the two states is shown in the difference image on the right in the top row. Once such a difference image is obtained, computer techniques transform it to a standard brain so that comparisons can be made with other individuals (middle row of difference images). From these individual difference images an averaged or mean difference image (bottom image) is made. Because the changes in blood flow are small, individual difference images tend to be somewhat variable due to the presence of statistical noise in the images and to variation among the subjects. Significant average changes reflect changes common to the sample of individuals. All of the images appearing subsequently in this chapter are mean difference images formed in the above manner.

dependent measurements made in the control state from the task state is intended to isolate those parts of the brain uniquely responsible for performing the task.

To obtain reliable data, the average of many responses is taken across individual subjects (usually the case with PET) or experimental trials from the same person (usually the case with fMRI). Averaging enables researchers to detect changes in blood flow associated with mental activity that would otherwise easily be confused with spurious shifts resulting from statistical noise in the resulting images. Averaging results across individuals has another important potential advantage in that it may indicate what common features of brain organization are shared by different subjects. Knowledge of such common features is essential for understanding the species-unique and universal capacity of humans for language, for example. An illustration of the image-subtraction and averaging strategy used for PET is shown in **Figure 1**.

For the remainder of the chapter, I focus on studies of language, largely from our own laboratory. This is done for several reasons: First, the work nicely illustrates the implementation of the strategies described above. Second, the work reflects the concerted effort of cognitive scientists and neuroscientists. Third, the results are pertinent with regard to cognitive theories of brain function. And fourth, these studies reveal important lessons about how to conduct functional-imaging research and what conclusions to draw from such studies. The following discussion is not intended as a complete review of functional-imaging studies of language but, rather, as an overview of functional brain-imaging strategies, using our studies of language as illustrations. These findings have important implications for neural-network modeling. (For a more in-depth review of human cognition and functional brain imaging, see Posner & Raichle, 1994.)

The Study of Language: An Example

The manner in which language skills are acquired and organized in the human brain has been the subject of intense investigation for more than a century. Work began in earnest in 1861 when Pierre Paula Broca described a patient for whom a damaged left frontal lobe destroyed the ability to speak. Broca's studies of language localization were complemented by those of Carl Wernicke, a German neurologist. In 1874, Wernicke told of people who had difficulty comprehending language following damage to the left temporal lobe. From these beginnings has emerged a concept of language organization in the human brain that, in broad outline, posits the following: Information flows from visual and auditory reception areas to areas at the junction between the left temporal and parietal lobes for comprehension, and then on to frontal areas for verbal response selection and speech production. Almost all of this information was gleaned from patients with brain damage. Could this organization represent the actual functioning of the normal brain as revealed through modern neural-imaging techniques?

In the mid-1980s my colleagues Steven E. Petersen, Michael I. Posner, Peter T. Fox, and Mark A. Mintun and I began a series of experiments (Petersen, Fox, Posner, Mintun, & Raichle, 1988, 1989; Petersen, Fox, Snyder, & Raichle, 1990; Raichle, Fiez, Videen, MacLeod, Pardo, Fox, & Petersen, 1994) to begin to answer this question. We elected to begin our work with an analysis of the manner in which the normal human brain processes single words from perception to speaking. The initial experiments were designed in a hierarchical manner in which levels of information processing of increasing complexity were employed. This design is in keeping with the subtractive model presented previously. In our experiments, words were presented to the subjects either on a television monitor or through earphones. In the presentation to follow, I will focus on those aspects of the study involving the *visual presentation* of words.

Opening the eyes

Regardless of the task to be performed by our subjects, they were always asked to fix their gaze on a small fixation point in the middle of a television monitor. This was done to prevent the unwanted activation of brain areas involved in saccadic eye movements from complicating the analysis. When one compares the simple act of opening one's eyes and fixing on a small dot in the middle of a television monitor with lying quietly when one's eyes are closed (**Figure 2**, top row) a significant increase in brain activity occurs in the back of the brain in those areas known to respond to visual stimuli. Subsequent changes in the strength of the image add to this already-present activity produced by simply opening the eyes and fixing the gaze.

Words as passive visual stimuli

Common English nouns appeared on the television monitor at the rate of one word a second while the subjects continued to fix their gaze on the fixation point. This added stimulation produced a marked increase in the extent and complexity of brain activity in the visual areas of the brain as shown in **Figure 2** (second row). The subjects were not instructed to respond to the words in any way, but simply to fix their gaze on the fixation point. These results suggested that words had special properties as visual stimuli that had powerful effects on the visual system of the human brain. What properties of the stimuli might produce these effects? Further analysis of words as visual stimuli was needed (Petersen et al, 1990).

Subsequent research indicates that a very important factor is the orthographic regularity of words and word-like symbols (Petersen et al, 1990). Two levels of analysis appear to be taking place in the visual system as we passively view words. At one level, the brain analyzes the visual features of the stimuli regardless of their relationships to letters and words. These visual features appear to be processed in multiple areas of the visual system on both sides of the brain. Responses to false fonts containing only meaningless features are particularly strong on the right side of the brain.

At a second level, the brain analyzes the visual word form. Visual stimuli consistent with the pronunciation rules of the English language uniquely activate a group of areas in the visual system of fluent readers of English. This coordinated response among a group of areas clustered in one part of the visual system is acquired as we learn to read, and is probably critical to the facility with which skilled readers handle words.

Reading words aloud

Following the hierarchical design of the original experiment, the subjects were next asked to read aloud words (i.e., common English nouns) when each

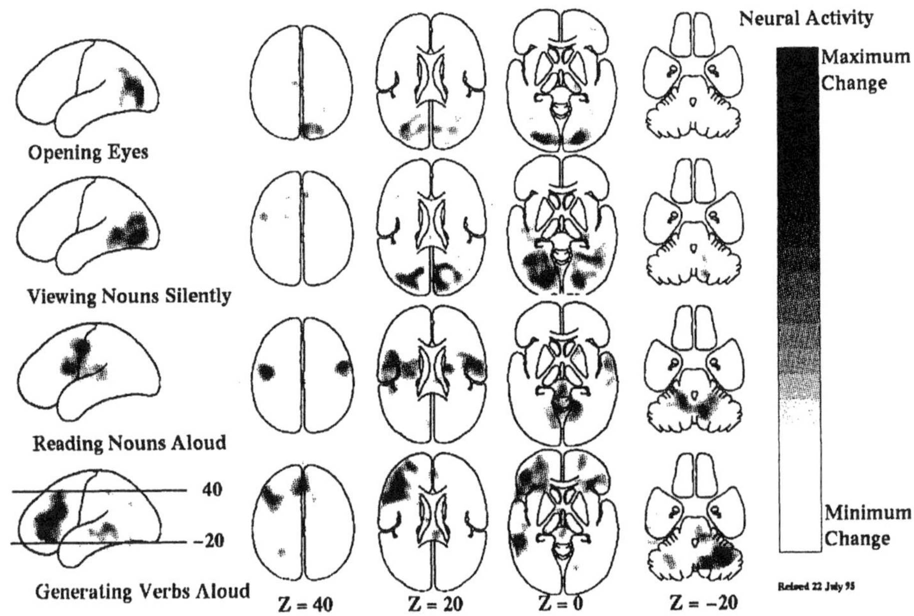

FIGURE 2. Mean difference PET images obtained from four different task states. A group of normal subjects performed a hierarchically designed series of language tasks. Each row represents the mean difference between images of the designated task state and a control state. The images on the left represent projections of the changes as seen on the lateral surface of the brain with the front of the brain to the reader's left. The horizontal lines through the bottom left image denote the orientation of the horizontal slices seen to the right of the images on the left. These horizontal images are oriented with the front of the brain toward the top and the left side to the reader's left. The Z values indicate millimeters above (+) or below (-) a horizontal plane through the brain at Z = 0. Four task states are indicated under the leftmost images. (See text for description of the conditions during the various task states.)

appeared on the television monitor. For most individuals fluent in their native language, this is an easy task to perform. One might easily envision such a task being performed effortlessly at the same time as another, unrelated task—the point being that such a task requires little of our conscious attention.

Not surprisingly, the motor areas of the brain were activated when individuals spoke the words as they were read from the television monitor. As shown in **Figure 2** (third row), these areas included the primary motor ccrtices in both cerebral hemispheres (**Figure 2**, third row, Z = 40). In addition, other motor areas buried more deeply within the cerebral hemispheres (**Figure 2**, third row, Z = 20) were also activated—including the cerebellum (**Figure 2**, third row, Z = -20).

At this point it should be noted that the act of speaking words, occurring in its simplest form during the reading aloud of single words, did not produce activity in the classic areas of the left cerebral hemisphere known as Broca's and Wernicke's areas. This was probably one of the first surprises to emerge from the studies of language in normal people using modern imaging techniques. The classical theories of language organization based on more than a century of research on patients with brain injury would have predicted clear-cut activity in these areas.

Although not mentioned above, *listening* to words does produce activity in posterior left temporal cortex at the temporo-parietal junction in the region classically thought of as Wernicke's area. Furthermore, this area remains active when normal individuals repeat aloud the words they hear. However, it is important for the standard theory of language organization that this area was not activated when the same individuals read aloud words that were presented visually. Clearly, Wernicke's area, as classically defined from studies of brain-injured patients, is not active in an obligatory fashion when we speak. Critics of this new view were quick to point out that the new imaging techniques, such as PET and fMRI, might simply not have been sensitive enough to detect a change in Wernicke's area when people read aloud. Were it not for the findings in the next stage of the experiment (generating verbs aloud for visually presented nouns), this criticism would have been difficult to evaluate.

Generating verbs aloud

The next stage of the experiment required the subjects to generate a verb aloud when a noun was presented visually (e.g., see *car*, say *drive*). This task may seem an unnecessarily complex next step in the hierarchical design of the experiment, and linguists have been particularly critical! After all, the task involves a number of complex mental operations such as determining the meaning of the presented word (semantics) and the relation between the meaning of the word and the choice of an appropriate verb (syntax). Additionally, it is a very powerful episodic-memory encoding task as well as a semantic retrieval task. Concerns about such issues obscure a most important difference between this task and reading or repeating words aloud. This difference lies in the requirement that subjects must devote considerable conscious attention to the task and, among other things, suppress the tendency to speak the word they see and, instead, utter an appropriate verb. Furthermore, the verbal response must occur rapidly, because the nouns were being presented at the rate of 40 to 60 per minute. It was clear to us that subjects found this task difficult when they first attempted it; they often fell behind and occasionally skipped responding to nouns.

The changes in neural activity we observed in the brain (**Figure 2**, bottom row) confirmed that the task of generating verbs for visually presented nouns placed a significant additional burden on processing resources. In addition to areas previously activated, new areas within the left frontal (**Figure 2**, bottom row, Z = 40, 20, and 0) and temporal (**Figure 2**, bottom row, Z = 0) lobes were activated together with an area along the anterior midline (**Figure 2**, bottom row, Z = 40). Areas qualifying as Broca's and Wernicke's areas were clearly activated, and there was a surprising additional involvement of the right cerebellum (**Figure 2**, bottom row, Z = -20). Recall that portions of the cerebellum were active during reading aloud (**Figure 2**, third row, Z = -20), but this additional activation in the right cerebellar hemisphere during verb generation took place in a distinctly separate area. Whatever else might have been predicted about the neural substrate of this task, the involvement of the right cerebellar hemisphere was completely unanticipated prior to obtaining these results.

The fact that some areas of the motor system active during word reading (**Figure 2**, third row, Z = 20) were actually inactive during verb generation was an additional surprise. Because of the nature of the intensity scale used in **Figure 2**, where only positive differences between the task state and the control state are shown, this inactivation is not shown. Suffice it to say that these areas, which were very active during reading aloud, were mysteriously inactive during verb generation. This pattern of results hints at the possibility that the task of verb generation actually requires *different* brain circuits rather than simply additional brain circuits for speech production. Our thinking in this regard was dramatically affected by the following entirely serendipitous event.

While we were studying an additional group of subjects on the verb-generation task, one subject was given practice on the task to ensure that he could do it with less difficulty and greater accuracy. Recall that subjects had never practiced the verb-generation task prior to performing it in the PET scanner. The additional practice had totally unexpected effects on our results. Practice not only improved performance but also led to a failure to activate any of the areas seen in our previous study of verb generation with naive subjects (**Figure 2**, bottom row). Practice on the verb-generation task appeared to allow the brain to perform the task with the same circuits used for simply reading a word aloud (**Figure 2**, row 3). If reliable, this was indeed a surprising finding. Therefore we set about studying the effect of practice on the brain circuits used for speech production.

Practice effects

The first study examined in greater detail the effect of practice on the verb-generation task. We learned (Raichle et al, 1994) that when normal subjects practiced generating verbs for the *same* list of 40 common English nouns their reaction times became significantly shorter over a period of about 10 minutes. During this time they went through the same word list 10 times, being encouraged by the examiner to proceed as quickly as possible. A critical feature of the learning process was the fact that as they practiced, their responses became quite stereotyped. Although each of the words on the list could be associated with several verbs, practice led to the repeated selection of just one verb for each noun. In a sense, an automated, stimulus-response pattern of behavior was established. If, after learning had occurred, a new word list was substituted, their behavior returned to the unpracticed state; i.e., responses were significantly slower and unstereotyped. It should also be noted that, regardless of the amount of practice or whether the subjects were speaking verbs or nouns, the actual time needed to say the word did not change. What did change was the time necessary to begin the response and the nature of the response (stereotyped versus unstereotyped). Armed with this more complete information concerning the behavioral effects of practice, we were now in a position to evaluate the effect of practice on the brain circuits in a new imaging experiment.

In the new imaging experiment (Raichle et al, 1994), normal subjects performed the verb-generation task with visually presented nouns naively and after 10 minutes of practice. The control task was simply reading aloud the same words as they were presented on the television monitor. Consistent with our earlier experiments, the naive generation of verbs in response to visually presented nouns again showed involvement of the same brain areas. This finding was reassuring and supported our confidence in the imaging method. With practice on the task, the changes were dramatic. The areas in the frontal and temporal cortex in the left hemisphere and the right side of the cerebellum, which were activated during the verb-generation task in the naive state (top row, **Figure 3**), were completely replaced by active areas deep within the brain hemispheres after only a few minutes' practice (bottom row, **Figure 3**). It should be noted that these latter areas were also used for the far simpler task of reading nouns aloud. In addition, areas in the left occipital cortex significantly increased their activity after practice on the verb-generation task (**Figure 4**). These same areas were active during the passive visual presentation of words (**Figure 2**, second row, Z = 0) and unchanged during reading nouns aloud (**Figure 2**, third row, Z = 0) and naively generating verbs for visually presented nouns (**Figure 2**, fourth row, Z = 0).

These results strongly support the hypothesis that different brain circuits are activated when performing verb generation for the first time than when practice has perfected or automated the task. Why should such an arrangement of brain circuitry be necessary? Why two circuits? Why not just do a better job of utilizing existing brain circuits as we learn? The answer may not simply be that

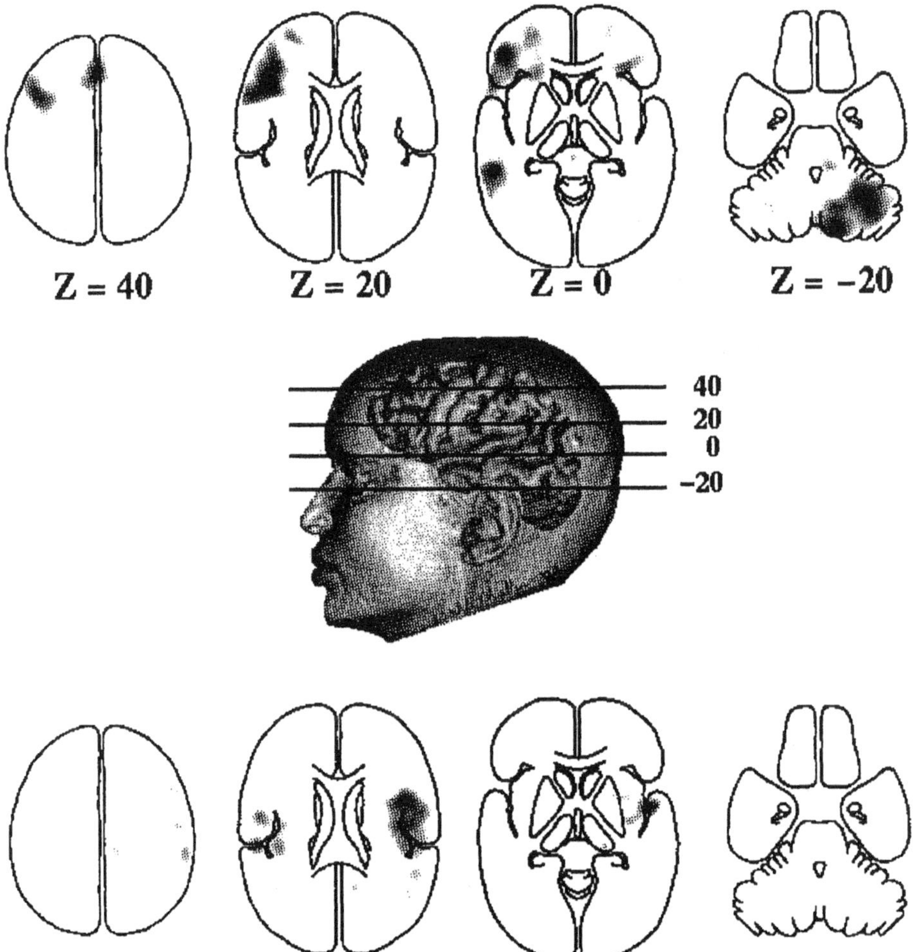

FIGURE 3. Neural activity shown by PET difference images during the verb-generation task in the early (upper row) and late (lower row) stages of practice. The control state for these subtraction images was simply reading aloud the same visually presented nouns.

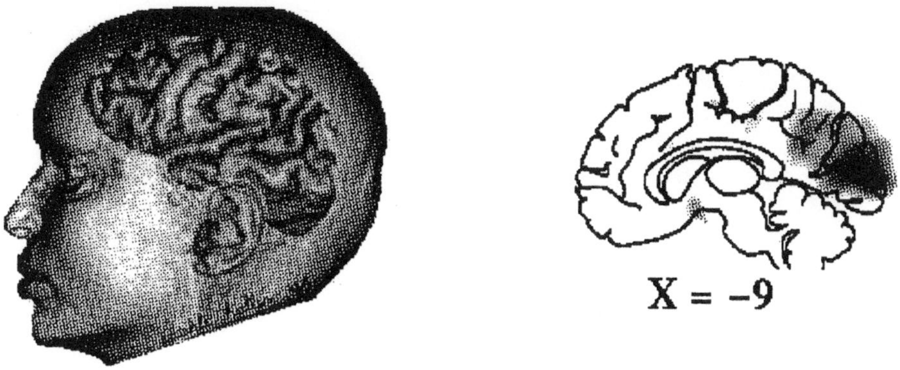

FIGURE 4. Sagittal view of neural activity shown by a PET difference image during a verb-generation task. The control state for this subtraction image was simply reading aloud the same visually presented nouns. The difference image to the right represents a sagittal slice of the brain 9 mm to the left of the midline. Note the marked increase in blood flow in the medial occipital cortices occurring after practice on the verb-generation task.

the brain needs two circuits, one for the non-conscious performance of highly automated tasks and the other for the performance of novel, non-automated tasks. The answer may instead be related to our need to strike a balance between the efficiency conferred by automation of much of our behavior and the occasional need to modify our programmed behavior in accordance with unexpected contingencies in our environment. Only further research can clarify such issues. What is clear is that functional brain imaging adds a remarkable new dimension to our thinking about how language and other cognitive activities are implemented in the human brain.

The Temporal Dimension

While functional imaging studies with PET provide new insights into the functional brain *anatomy* of the neuronal circuits underlying various cognitive activities, they do not provide any information on the *temporal sequence* of information processing within these circuits. Metaphorically, one might think of a network of brain areas (see **Figure 2**) as a group of individuals in the midst of a conference call. The temporal information sought would be equivalent to knowing who was speaking to whom and when. Such information is critical to understanding how specific brain areas are coordinated as a network to produce observable behavior.

The temporally varying changes in brain activity revealed by fMRI have raised the possibility that this technique, with its speed of data acquisition

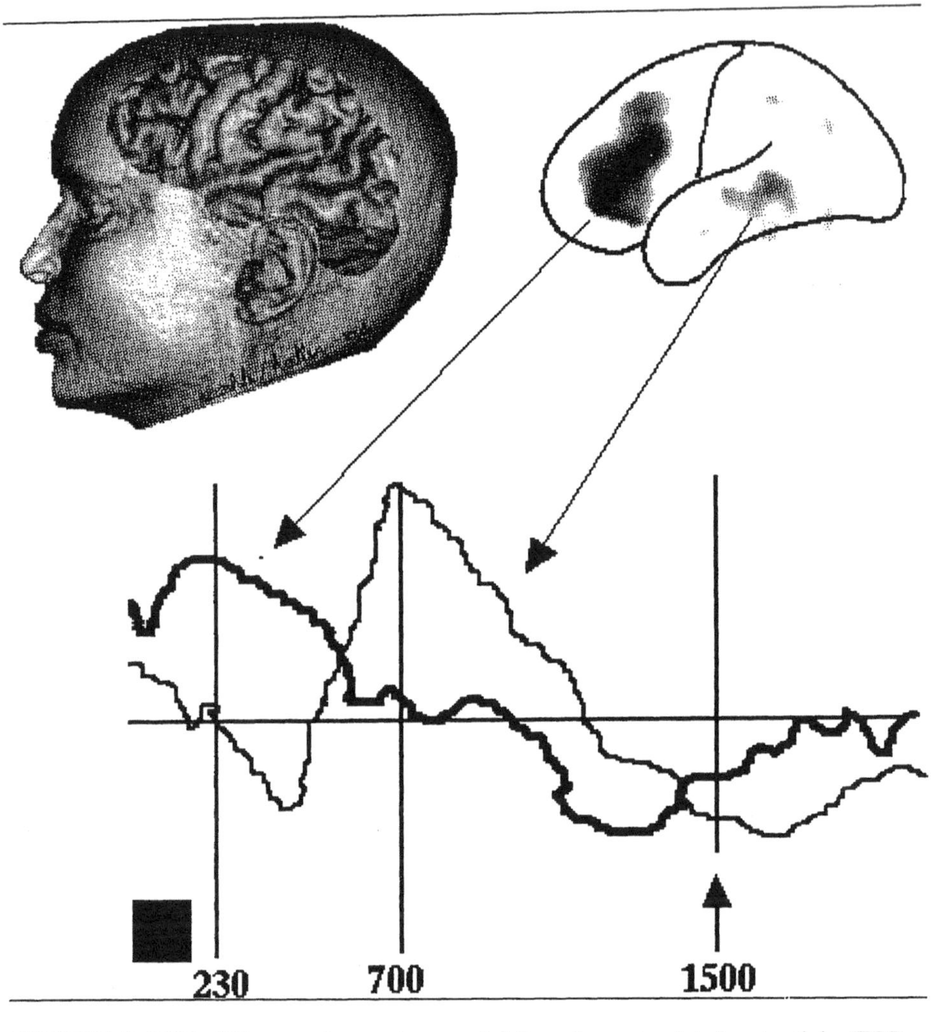

FIGURE 5. PET difference image (upper right) and event-related potentials (ERPs, lower row) during the verb-generation task. Note the increase in blood flow in the PET image occurring in left frontal and temporal cortices during naive performance of the verb-generation task. The ERP difference records were obtained by subtracting the waveforms recorded when reading nouns aloud from those obtained during the verb-generation task. The arrows connect the PET blood-flow responses with difference ERP waveforms recorded at the nearest overlying electrode (heavy line = frontal 7 electrode; light line = temporal 5 electrode). The horizontal bar below the ERP waveforms indicates the visual presentation of the noun and the vertical arrow indicates the cue to respond with the verb. Note that activity in frontal cortex precedes that in temporal cortex by over 400 msec. The numbers below the vertical lines indicate the times in msec after presentation of the word.

approaching a few tens of milliseconds, might provide both the anatomy and the sequence of information processing within functional brain circuits. The stumbling block, however, is the difference between the speed of neuronal activity and the rate of change of the fMRI signal. Signals from one part of the brain can travel to another in as little as a few milliseconds. Unfortunately, changes in blood flow and blood oxygenation (which is dependent on changes in blood flow) often require several seconds to occur after the onset of a change in neuronal activity. In all likelihood, the only methods that respond quickly enough are the electrical recording techniques such as electroencephalography (EEG) and magnetoencephalography (MEG).

One might reasonably ask why these latter techniques have not been used to address the type of experimental problems now studied with PET and fMRI. Put briefly, EEG and MEG have limitations in their spatial resolution and sensitivity. Even though great strides have been made, particularly with MEG, the accurate localization of the source of brain activity remains difficult when electrical recording is used in isolation. Furthermore, the resolution becomes poorer the deeper the source of brain activity.

Neither PET nor fMRI suffers from difficulty in localizing the anatomical source of a signal. Both techniques can sample all parts of the brain with equal spatial resolution and sensitivity. Recently, several successful attempts have been made to combine the spatial information of functional-imaging techniques with the temporal information from electrical techniques (Snyder, Abdullaev, Posner, & Raichle, 1995). One such attempt brought together investigators from our laboratory and the University of Oregon to study the temporal dynamics of the naive verb-generation task (Snyder et al, 1995). Event-related potentials (ERPs) were recorded during both verb generation and reading nouns aloud. Differences in the ERPs were then computed. Based on the findings of this study, it became clear that information processing unique to naive verb generation began in midline frontal cortices between 180 and 200 milliseconds after the noun was presented. This was followed by a spread of activity laterally to the left prefrontal cortex between 220 and 240 milliseconds after stimulus onset. Only later did activity arise in the left posterior temporal cortex, peaking between 620 and 640 milliseconds after the noun was presented (**Figure 5**).

These data, admittedly preliminary, provide important information about the role of specific areas of the brain in information processing during the verb-generation task. For example, the activation in posterior temporal cortex (**Figure 2**, fourth row, Z = 0 and **Figure 5**) in the vicinity of the classical Wernicke's area is involved rather late in many rapid semantic tasks that produce reaction times much faster than 600 milliseconds. The later activation of Wernicke's area may be more closely related to the integration of word meanings for the overall meaning of phrases, sentences or other more complex verbal units. Regardless of their final interpretation, data *combining* information from electrical and imaging studies are likely to benefit our models of information processing and to represent an important future direction in functional brain research.

Summing Up

Modern functional brain imaging with PET and fMRI, complemented by ERPs, will play an important role in our understanding of the functional organization of the normal human brain. These new tools, guided by the principles of cognitive science, have the potential to dissect the basic mental operations underlying behavior and to correlate them with specific neural circuitry. Referring to **Figure 2**, even relatively simple cognitive tasks recruit extensive areas of the brain to mediate performance. Although we are beginning to make statements about the roles of such groups of areas or circuits, further studies must determine the basic mental operations assigned to the individual brain areas within a given circuit. The studies of the neural processing of words that I have presented in this chapter give some indication of the manner in which such an analysis might proceed. This information, coupled with findings from studies of patients with brain injury and basic studies in laboratory animals using a variety of sophisticated techniques, bodes well for our future understanding of human brain function. Armed with such information, we will be in a much better position to appreciate the neural basis of human behavior.

Neural-Networks Models of Cognition
J.W. Donahoe and V. Packard Dorsel (Editors)

CHAPTER 24

NEURAL MODELING OF LEARNING IN VERBAL RESPONSE-SELECTION TASKS

Vijaykumar Gullapalli and Jack J. Gelfand
Princeton University
Princeton, NJ

ABSTRACT

We present a dynamic neuronal model of the diverse brain areas involved in learning verbal response selection that replicates observed human cognitive behavior. This model is based on recent positron emission tomography (PET), functional magnetic resonance imaging (fMRI), and event-related potential (ERP) data as well as the neuroanatomy and physiology of the brain areas involved. The model also captures the postulated dynamics of cortico-thalamo-basal ganglionic loops in the prefrontal cortex.

Introduction

Cognitive models of learning with practice have been studied for some time (Anderson, 1987; Crossman, 1959; Logan, 1988; MacKay, 1982; Newell & Rosenbloom, 1981; Schneider, 1985), and several different mechanisms for learning with practice have been proposed. Because of the wide latitude in devising cognitive models, these mechanisms—e.g., chunking (Newell & Rosenbloom, 1981), priority and association learning (Schneider, 1985), composition and strengthening (Anderson, 1987), instance recall (Logan, 1988), and strengthening of nodal connections (MacKay, 1982)—are based on widely different functional architectures. However, recent studies of the functional anatomy of the brain obtained through the positron emission tomography (PET), functional magnetic resonance imaging (fMRI) and event-related potential (ERP) techniques have shed some light on the actual brain mechanisms involved in learning with practice. Information obtained from such studies is crucial, because it yields strong modeling constraints over and above those obtained from cognitive studies alone.

The focus of this chapter is on modeling the brain mechanisms involved in iterated verbal response-selection tasks while satisfying neuroanatomical and functional constraints, as well as data from brain-imaging and ERP studies (Raichle, this volume). Human cognitive behavior in iterated verbal response-selection tasks is a simple yet interesting example of learning with practice. When subjects are asked to respond, for example, with appropriate verbs for a visually presented list of nouns, repeated presentation of the list initially elicits varying responses to each noun; with practice, however, stereotypic responses develop. Practiced responses also are produced faster than responses by naive subjects. In addition to these basic characteristics, verbal response selection shows dependence on several cognitive variables that, when manipulated, can give rise to priming, masking, interference, and other cognitive phenomena.

We present a heterogeneous dynamic neuronal model that integrates at a systems level the diverse areas of the brain involved in these tasks. Our model replicates both the high-level cognitive behavior and the neuronal-level characteristics of the brain circuits involved. This has necessitated the modeling of a significant portion of the language understanding and generation system on the neuronal level including the representation of words, direction of attention, response-selection processes and output processes. Much is unknown about some of these components. In some cases we have included the known neuronal circuits and hypothesized dynamics, and in others we have captured only the gross characteristics of their hypothesized functionality. One might ask "Why attempt such an undertaking at this stage?" There are several reasons. First, we have had to take into account data from a number of different fields to ensure that the model is consistent with what is known. The model requires the integration of such knowledge in an organized framework. In addition, assumptions are stated explicitly enough in the model to be verified or refuted by future experimental work. Our choice of the components of the model and the roles ascribed to them can also result in specific predictions about how the disruption of these components can affect overall behavior. These predictions can also be verified through future experimental work.

Many of the cognitive models proposed for learning with practice posit a single stream of processing and involve improving the efficiency or the focus of attention of the process with multiple exposures to the task. In contrast, other investigators have suggested that cognitive processing utilizes multiple processing streams and that the observed properties of learning are a manifestation of changes in one of those streams. (e.g., Kounios, Osman, & Meyer, 1987; Kounios, 1993; Logan, 1988, 1992; Schweikert & Wang, 1993). Kounios (Kounios et al, 1987; Kounios, 1993) has used a speed-accuracy decomposition method to infer that multiple simultaneous processes may be involved in semantic memory operations. He has postulated that a continuous computational process operates in parallel with discrete all-or-none search in a sentence-verification paradigm. Logan (1988; 1992) has postulated that learning of automatic responses involves the formation of a knowledge base of instances of each exposure to the task. A response in a particular task results from a race between two separate processing streams, one algorithmic and the other a pure recall of instances. With this approach, he has been able to reproduce the power law of speed-up in response time with practice and other observed changes with practice—such as changes in the distribution of response times, the shape of the learning curve, and the distribution of stereotypic responses with practice.

Multiple processing streams were also found in a recent study by Raichle, Fiez, Videen, MacLeod, Pardo, Fox, and Petersen (1994), who used the PET technique to examine the differences in the functional anatomy of the brain during naive and practiced performance of verbal response selection. PET scans over repeated presentations showed a shift in brain activity from the anterior cingulate, the left prefrontal and left posterior temporal cortices, and the right cerebellar hemisphere in the naive condition to the sylvian-insular cortex bilaterally and the left medial extrastriate cortex after practice. Introduction of a novel stimulus after practice reactivated the regions active in the naive condition. Raichle et al (1994) concluded that two distinct brain circuits were employed in verbal response generation, one for controlled selection of responses and the other for the production of learned or automatic responses.

In addition to the PET study of Raichle et al, there is a great deal of information to guide us in our modeling effort. Numerous studies have indicated that the left prefrontal cortex plays a role in language processing (Demb, Desmond, Wagner, Vaidya, Glover, & Gabrieli, 1995; Kapur, Rose, Liddle, Zipursky, Brown, Stuss, Houle, & Tulving, 1994; Petersen, Fox, Snyder, & Raichle, 1990; Raichle et al, 1994; Snyder, Abdullaev, Posner, & Raichle, 1995). There is evidence to suggest that temporal-association areas are responsible for accessing cognitive or semantic representations of words. This can include Wernicke's area as well as other left posterior temporal areas (Dronkers, Redfern, & Lucy, 1995; Wise, Chollet, Hadar, Friston, Hoffner, & Frackowiak, 1991). Furthermore, it is believed that anterior cingulate cortex specifically represents information pertaining to the task that is currently being performed (Pardo, Pardo, Janer, & Raichle, 1990; Janer & Pardo, 1991; Vogt, Finch, & Olson, 1992). The role of the sylvian-insular cortex is not well understood. Raichle et al (1994) have postulated that it serves as an associative store for learned responses to stimuli in a fashion similar to premotor cortex (Mitz, Godschalk, & Wise, 1991). In addition to the roles played by the cortical areas discussed above, it is believed that subcortical structures such as the thalamus and basal ganglia participate in language function (Crosson, 1992; Wallesch & Papagno, 1988).

Neural-Modeling Methods

Although verbal response selection is a high-level cognitive behavior, our model is implemented at the neuronal level to enable us to incorporate our knowledge of the neurophysiology of the brain areas participating in the task. We have used grouped models of the neurons in our implementation, primarily because little is known about the impact of sub-neuronal dynamics on the high-level cognitive behavior. However, one of our goals is to determine the appropriate level of detail needed to model all the relevant aspects of verbal response selection in humans.

The most important feature of our model is that it incorporates the dynamic characteristics of the brain circuits involved. This allows us to replicate the temporal characteristics of the observed cognitive behavior, which is essential for any meaningful study of the phenomenon of learning. It also enables us to compare the dynamic behavior of the model in simulations with the temporal characteristics of neuronal activations revealed by ERP data.

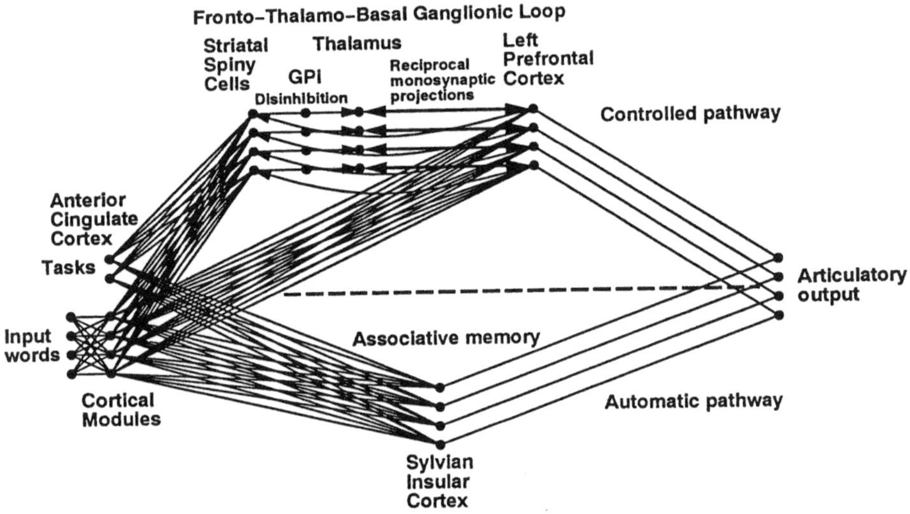

FIGURE 1. The architecture of the verbal response-selection model.

System Architecture

The architecture of the model is depicted in **Figure 1**. A primary feature of this architecture is the presence of two processing streams. The controlled stream consists of the frontal cortex, the basal ganglia, and the thalamus. There is also an automatic stream, which consists of the sylvian-insular cortex. Both receive inputs from the anterior cingulate cortex representing the task and cortical modules in left medial temporal cortex representing the input words. In the PET studies, these were the areas whose excitation differed significantly in the naive and practiced performance of the response-selection task (e.g., Raichle et al, 1994). Both these streams receive common inputs from the sensory areas and send outputs to the motor areas.

The cortical modules representing the anterior cingulate and the sensory and language areas are organized as columns of neurons that correspond to cortical columns, each functioning as a relatively coherent information-processing unit as discussed by Mountcastle (1978). Each cortical module forms a distributed representation of some internal state of the organism or external state of the environment. Cortico-cortical interconnections formed through Hebbian learning (Hebb, 1949) make it possible for the cortical modules to develop robust representations.

The representations used clearly have a major impact on the functioning of a model. In the current model, we used a distributed representation over the sensory and language cortical modules to encode stimulus words, with each module denoting a "feature" or "category" of words (e.g., "colors" or "verbs"). The rationale for this representation is presented in the next section. These cortical representations are input to the basal ganglia, the sylvian-insular cortex, and the frontal cortex. Additionally, the cingulate module stores a representation of the task (e.g., "generate a color response"), and provides it to the basal ganglia and the sylvian-insular cortex.

A more detailed diagram of the input-representation configuration and the projections involving the dorsolateral left prefrontal cortex and subcortical structures is shown in **Figure 2**. In addition to receiving projections from the

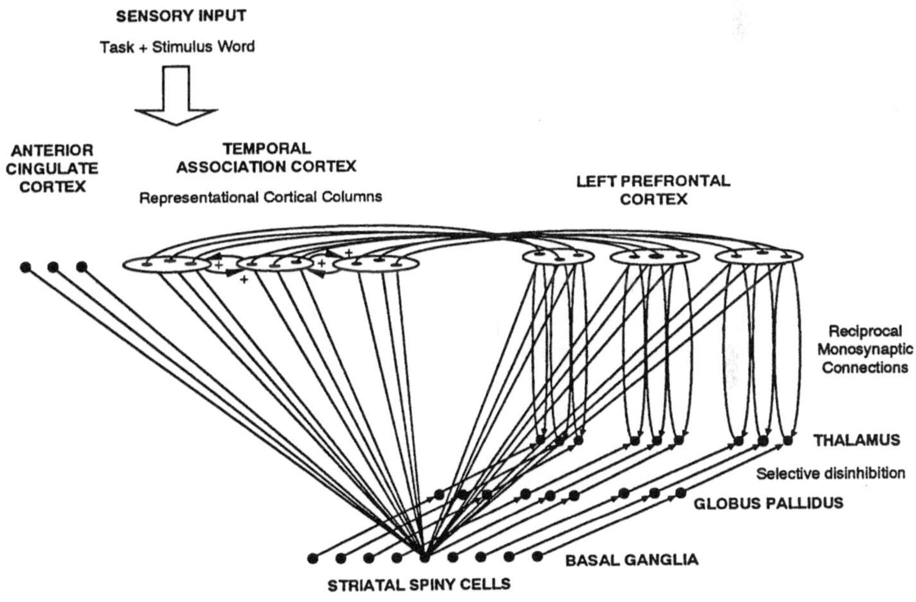

FIGURE 2. Architecture of the cortico-thalamo-basal ganglionic loop that makes up the controlled response-selection pathway. For clarity, only projections to one striatal spiny cell are shown. Similar convergent projections go to each striatal neuron. The lateral inhibitory connections between neurons within each association-cortex column are also not shown.

sensory/language cortical modules, the frontal cortex also has highly specific reciprocal projections with the thalamus, resulting in local cortico-thalamic loops that, when active, may sustain activity in frontal-cortex neurons (Houk & Wise, 1993; Alexander, Crutcher, & DeLong, 1990; Groenewegen & Berendse, 1994; Houk, 1995; Selemon & Goldman-Rakic, 1985). These loops could be activated through selective disinhibition by the basal ganglia (Chevalier & Deniau, 1990). In this model, striatal spiny cells in the basal ganglia function as a pattern recognizers providing a contextual set for the prefrontal cortex. Based on inputs from the cortical modules and the cingulate, the basal ganglia selectively disinhibit the frontal cortex-thalamus loops corresponding to the word features appropriate for the task, thereby latching these features in the frontal cortex. Thus the output of the controlled circuit is an appropriate word represented by the selected features in the frontal cortex.

In parallel with the controlled circuit, the sylvian-insular cortex module, which forms the automatic circuit, produces a response associated with the cortical inputs. The structure and function of the insular lobe has been reviewed by Augustine (1985). Because there is a lack of concrete anatomical evidence at this point for the mechanisms that mediate this associative response, we simply modeled the insular learning as a linear associative network. Although this is not biologically faithful, it does have the property of incrementally learning the correct output response based on examples given by the performance of the controlled pathway, and thus allows us to observe the overall dynamics of the model.

The output of the overall system is determined by the earliest information to arrive from the two parallel pathways. This is similar to a "race model" in a multi-process cognitive system as suggested by Kounios (1993) and Logan (1988, 1992). At each time step, we determined if any neuron in the pathway was selected based on whether the ratio of its activity to that of the other neurons in the pathway exceeded a fixed threshold. We then compared the activity levels of the maximally active neurons in each pathway and designated the pathway with the higher level of activation as the winner. Because the peak activity of neurons in both pathways was of comparable magnitude, the pathway that reached its peak activity sooner won the race to determine the output.

Representation of Words

Our use of a distributed representation over the sensory and language cortical modular array (Wise & Houk, 1994) to encode stimuli is inspired by the functional anatomy of the cortex (e.g., Penfield & Rasmussen, 1950; Mountcastle, 1978; Asanuma, 1975). The general organization of cortical circuits appears to be in the form of a distributed set of functionally specific regions or columns interactively involved in the execution of a given task. Each functionally specific region extracts from its inputs higher-level information regarding a particular aspect of the task.

Sharing of processed information through reciprocal cortico-cortical projections between regions enables information extracted in one region to influence the processing of information in other regions concerned with the execution of the task. Cortical organization in columns with reciprocal projections between columns has been observed, for example, in the primary and secondary visual areas (Mountcastle, 1978), as well as in the motor cortex (Asanuma, 1975).

Modules in the cortical modular array (Wise & Houk, 1994) in our implementation (**Figure 3**) correspond to local information-processing regions of the cortex, with each module concerned with the representation of a "feature" or "category" of words. For example, a module might represent a color or colors associated with the stimulus word, or the fact that it is a verb. As a result, each word is represented as a distributed activation of the features associated with that word.

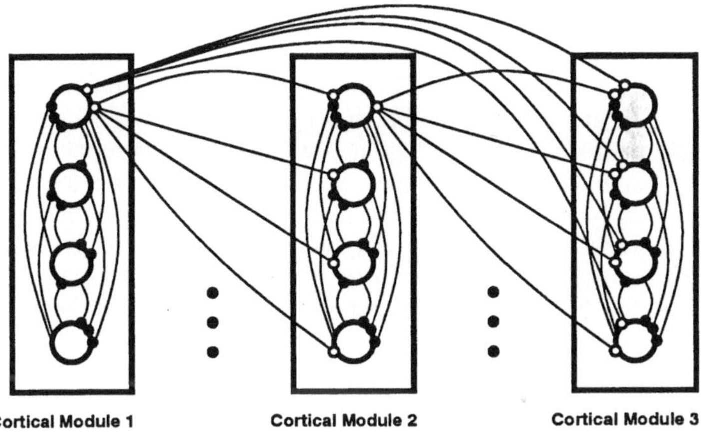

FIGURE 3. Block diagram of the cortical modular array used in our implementation to represent words. Lines ending in open circles denote reciprocal excitatory projections between neurons in different modules, while those ending in filled circles denote inhibitory projections between neurons within a module. For clarity, not all the excitatory projections are shown.

In our preliminary model, the nominal level of this distributed activation is predetermined for each word. The levels of activation are given in **Table 1**. Thus, presentation of a stimulus word is brought about by adding the activations indicated in the table to the corresponding neurons in the cortical modules. If no value is specified in the table for a neuron, its activity is not changed.

For example, when the stimulus APPLE is presented, (among others) the activations of the neurons representing EAT are increased by 0.5 and those of the neurons representing SWEET by 0.7, while the activations of the neurons representing PET, for instance, are not changed at all. These changes in activation values were selected to reflect the degree of association between the stimulus and the corresponding word. Thus, for instance, while APPLE is highly associated with both RED (1.0) and FRUIT (1.0), it is associated to a lesser extent with EAT (0.5), and not at all with BLACK (0.0).

Two factors influence the temporal dynamics of these distributed activations. First, *shunting-type* lateral inhibition (e.g., Pinter, 1983; Grossberg, 1973; Furman, 1965) among the neurons in a cortical module, depicted by filled circles in **Figure 3**, results in a winner-take-all type selection of the feature represented by that module. The equation governing this lateral inhibition within a module (see, for example, Pinter, 1983) is

$$dy_n/dt = I(y_n/T) - (y_n \sum_{m<>n} a_m f_m(y_m)), \quad (1)$$

where T is the cell time constant, I is the current weighted sum of inputs to the cell, a_m is the weight of inhibition, and $f_m(y_m)$is the conductance between neighboring neurons in a module. In our current implementation, we used $T = 1.0$, $a_m = 10.0$, and the identity function $f_m(y_m) = y_m$.

Modeling neurons as leaky integrators with shunting-type inhibition leads to dynamic interactions between the new activity due to presentation of a stimulus and the previously existing activity of neurons in a module. These interactions play a significant role in determining what is represented in each module: Previous strong activity of other features might inhibit weak new activity of a feature, resulting in suppression of a feature in the predetermined representation of the stimulus word. Alternatively, previous activation of the same feature might lead to priming of the new feature, increasing its prominence in the representation.

In contrast, the cortical modules attempt to maintain a coherent set of features in a representation by filling in missing features that were often active in the past in conjunction with those that are currently active. This is accomplished by mutually excitatory projections between neurons in different modules of the cortical modular array, as shown in **Figure 3**. The strengths of these projections are adapted via Hebbian learning (Hebb, 1949) according to the following equation:

$$w_{mn}(t+1) = \delta w_{mn}(t) + \alpha y_m(t) y_n(t), \quad (2)$$

where $w_{mn}(t)$ is the weight or synaptic strength between neurons n and m, δ is the weight-decay factor, α is the learning rate, and y_m and y_n are activations of neurons in different cortical modules.

Due to the dynamic nature of the representation of stimulus words, presenting the same word in different historical contexts can elicit different responses due to priming effects, much as occurs with human subjects.

Simulation Results

In this section, we present simulation results that demonstrate the features of the model. For these simulations, we selected a list of 20 words to represent. Of these, six were used as stimuli, while the responses could be selected from all 20. The words were classified into four groups: stimuli, color names, verbs, and miscellaneous. A separate cortical module was used to represent words in each group. **Table 1** contains the (predetermined) activations of the words in each sensory/language cortical module when each of the stimulus words is presented.

Stimulus:	**APPLE**	**BANANA**	**GRAPE**	**CAT**	**DOG**	**MOUSE**
Potential Responses						
APPLE	1.0					
BANANA		1.0				
GRAPE			1.0			
CAT				1.0		0.2
DOG				0.5	1.0	
MOUSE				0.6		1.0
BLACK				0.8	0.7	0.9
BROWN			0.5	0.5	0.9	0.7
RED	1.0		0.3			
YELLOW		1.0				
BUY	0.9	0.9	0.9			
EAT	0.5	0.2				
FALL	0.8	0.8	0.7	0.2	0.3	0.1
RUN				0.8	0.9	0.9
HOUSE				0.3	0.3	0.2
FRUIT	1.0	0.8	0.9			
PET				1.0	1.0	1.0
SWEET	0.7	0.4	0.5			
SOUR	0.5	0.1	0.9			
TREE	0.4	0.1				

TABLE 1. Representation of the stimulus words in the preliminary model as activations of cortical modular array neurons representing the features that constitute potential responses.

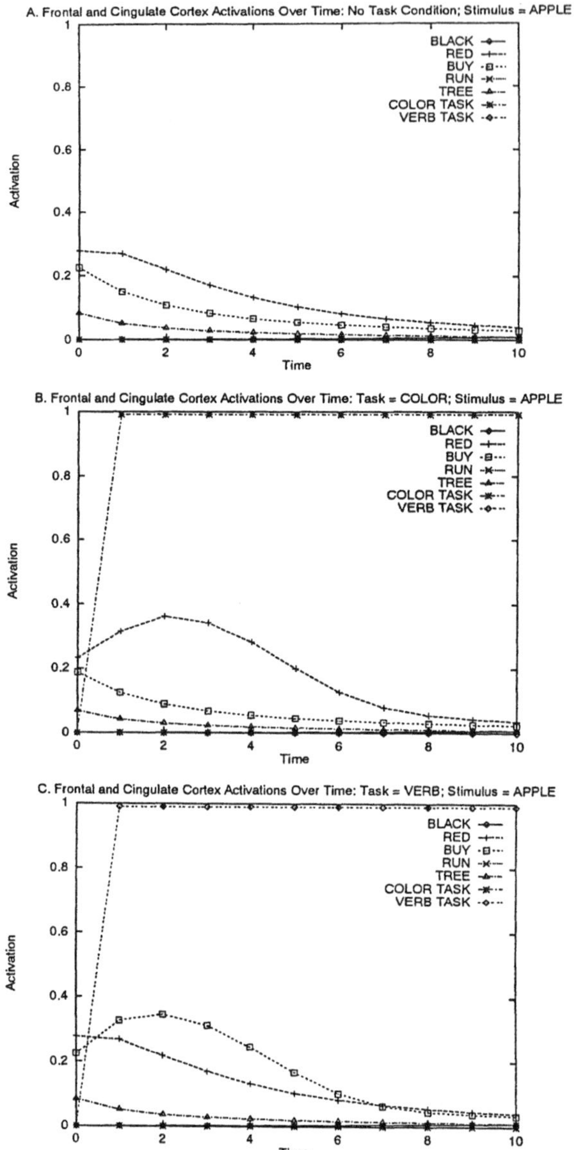

FIGURE 4. Example of selection of an appropriate response for different tasks by the controlled circuit. The COLOR and VERB task plots are activations of neurons in the cingulate cortex representing those tasks. The response activations (e.g., RED, BUY, TREE) are of neurons in the prefrontal cortex.

The model captures the dynamics of cortico-thalamo-basal ganglionic loops suggested by Houk and Wise (1993) based on neurophysiological evidence gathered by several researchers (Alexander et al, 1990; Chevalier & Deniau, 1990; Wang, Rinzel, & Rogawski, 1991; Fuster & Alexander, 1973; Goldman-Rakic & Friedman, 1991). An example of this is presented in **Figure 4**. The first plot shows the activity of frontal-cortex neurons when no task is specified. As can be seen, the initial activity of the neurons due to stimulus-word presentation decays with time. When a COLOR task is specified, the activity of neurons representing a color associated with the stimulus word is sustained through selective activation of the corresponding loops, while the activity of the other neurons decays away. Similarly, when a VERB task is specified, activity of neurons representing actions associated with the stimulus word is selectively sustained.

As reported by Raichle et al (1994), the median response times of human subjects decreases significantly over successive blocks of presentation of the same set of stimuli. If a novel stimulus set is presented immediately following these repeated blocks, the response time returns to about the same level as in the naive condition for the original stimulus set. We ran a similar experiment with our system. We presented three stimuli, APPLE, DOG, and MOUSE, for 10 successive blocks, followed by 10 more blocks with the stimuli BANANA, GRAPE, and CAT. The task (COLOR) remained the same. As illustrated in

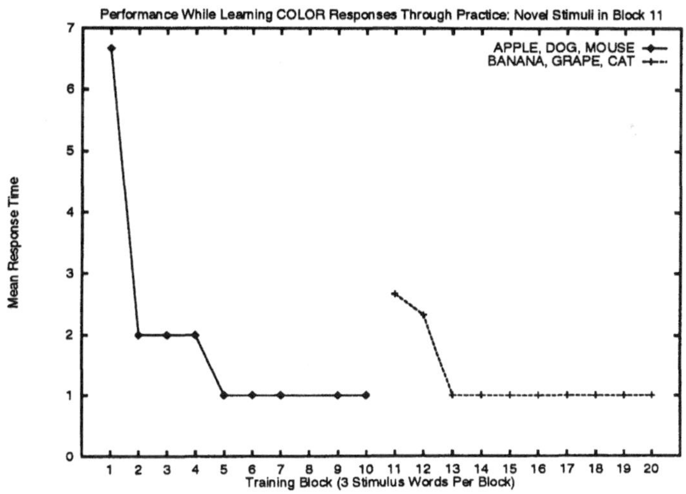

FIGURE 5. Performance while learning COLOR responses through practice. The stimuli were switched from (APPLE, DOG, MOUSE) to (BANANA, GRAPE, CAT) in block 11.

Figure 5, the response time of the system also decreased with repeated presentations of the same stimulus set for 10 blocks. Moreover, as with human subjects, presentation of a novel stimulus set resulted in a significant increase in the response time, which decreased with further repetitions.

A similar experiment, in which we switched the task instead of the stimuli in the 11th presentation, also produced interesting results. In this experiment, all six stimuli were presented for 10 blocks of the COLOR task. As shown in **Figure 6**, the response time of the model decreased with repeated presentations, as was expected. When the task was switched to VERB in the 11th block, the response time returned to the naive performance level and then began to decrease with further presentations.

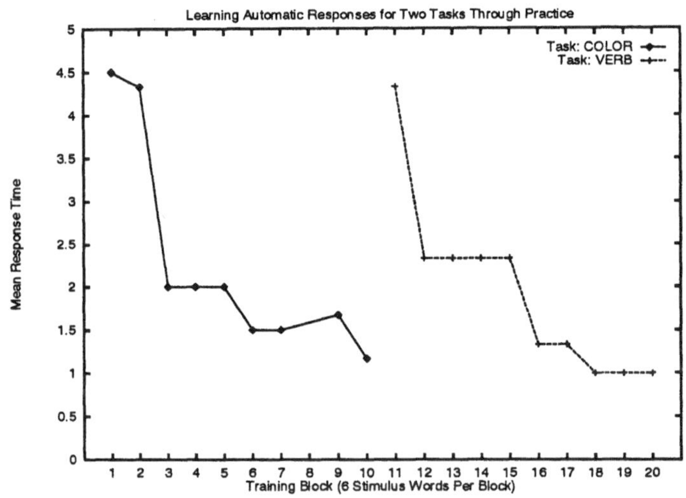

FIGURE 6. Performance while learning verbal responses in two different tasks. Task switched from COLOR to VERB in block 11.

Discussion

The results given in the last section indicate that a model based on the anatomical and functional features of the brain areas participating in learning word-association tasks does indeed simulate the cognitive behavior observed in humans. For this initial test of the overall model, we necessarily made many simplifications and assumptions concerning individual system components. Some of the important questions and issues raised by this modeling effort, along with possible directions of future work, are discussed below.

Comparison of the modeling that we have done here with regional cerebral blood flow assumed a direct relationship between neural activity and local blood flow. This simple relationship may not be generally true (Horwitz & Sporns, 1994). For example, Ackerman, Finch, Babb, and Engel (1984) have found that long-duration suppression of hippocampal pyramidal-cell firing was accompanied by increased 2-deoxyglucose uptake, indicating an increase in local regional cerebral blood flow. The mechanisms modeled in this paper do not include any suppression of activity as part of the operation of the circuit. It is possible, however, to envision a multi-stream architecture in which one stream might be actively suppressed while the other stream was in the process of generating a response. In this case, an increase in activity measured by PET or fMRI would indicate a decrease in excitation rather than an increase.

A primary issue for investigation is the existence of dual—controlled and automatic—pathways for verbal response selection. Demb et al (1995) have suggested that the decrease in fMRI and PET activation of the left prefrontal cortex with repeated semantic processing may not be an indication of a shift in processing between streams in a two-stream model, but rather a decreased response to the processing of repeated stimuli, as has been reported at the single-cell level in monkeys performing an object working-memory task (Miller & Desimone, 1994; Miller, Li, & Desimone, 1993). To address this issue, more information is needed to test the hypothesis that the sylvian-insular cortex serves as an associative store of the learned automatic response (Raichle et al, 1994). Although the PET data from the study of Raichle et al support such a model, there are other nearby areas in the Sylvian fissure that participate in language processing and generation. Higher-resolution data from fMRI is needed to pinpoint the precise area involved in the learning process. Additional functional evidence—especially temporal data from ERP—is also needed to evaluate the conclusions of the PET study. The model predicts that the shortening of response time is due to faster output from the insula. This could easily be tested with ERP data obtained from a learning protocol similar to that in Raichle's PET study. The ERP data of Snyder et al (1995) show that excitation of the insula occur after the cingulate, frontal, and temporal cortices for initial exposures to nouns in a use-generation task. If this two-stream model is correct, the excitation of the insula should occur more quickly as the excitation in the other areas diminishes with practice.

Raichle et al (1994) found that the insula were inhibited bilaterally with respect to simple word repetition during naive word-association tasks, but became more excited bilaterally after practice. Bilateral participation makes it difficult to use lesion studies of patients to uncover the function of the insula because of the small chance of finding a patient with an appropriate lesion on both the left and right sides. One might expect that patients with damage to any one side will recover all or nearly all of their language function. However, patients with focal damage to the language areas of the left insula are mildly aphasic and exhibit an apraxia of speech characterized by substituting one sound for another or omitting, adding, transposing, and distorting sounds (Dronkers, 1995; Miller, 1989). This occurs variably, with subjects sometime demonstrating normal production of the proper sound sequence for a given word. Thus, programming of word generation is disrupted. It is not, however, a Broca's aphasia where motor commands for speech are not generated. It is a disturbance of the motor plan. Based on these observations and the fact that the insula project to premotor areas (Augustine, 1985), we believe it worthwhile to investigate a recurrent-feedback network model of insular participation whose dynamics result in the ordered presentation of output as a series of phonemes or other linguistic primitives to the premotor (probably Broca's) area. This role for the insula is not necessarily contrary to a two-stream model of controlled and automatic response generation. It is possible that the associative-memory role of the insula is to associate an input stimulus with an output of the appropriate string of phonemes to be projected to the motor system.

At present, we have only modeled those areas that were observed to change their level of excitation during the learning process. In the future, we plan to include other known language-related areas in greater detail along with their projections to the motor system. Also, the architecture we describe here uses distributed, modular arrays for the representation and processing of information (Goldman-Rakic, 1988; Wise & Houk, 1994; Mesulam, 1990). Little is known about the specific information encoded by the cortical modules in word-association areas. In our current implementation, we postulate that cortical columns encode information regarding features or categories of words. Several alternative word representations organized around lexical, orthographic, or phonetic attributes of words are possible, and are being investigated at the present time. Because the representation of words must be a component of a more complex system that processes more complex linguistic structures such as sentences, a hierarchical system architecture (cf. MacKay, 1987) would probably be more suitable than the simple representation scheme used here.

The behavior of the neurons in the left prefrontal cortex in the model is similar to the sustained activity of prefrontal-cortex neurons observed by Fuster & Alexander (1973) and Goldman-Rakic (1994) in delayed-response tasks. Thus, the prefrontal area could serve as a working memory where task-specific representations are maintained for use by other cognitive and motor areas involved in the execution of a verbal response task. We would predict that lesions that cause a lack of inhibition of the cortico-thalamic loops would result in the generation of inappropriate responses. Lesions that cause a partial or global inhibition of the thalamo-cortical loops would result in deficits in initiating responses. Deficits in shifting set would result from lesions that damage the ability to alter the modulation of thalamo-cortical loops. Similar deficits have been observed in prefrontal- and subcortical-lesion patients. However, more specific correlations are needed between the lesion sites and the resulting effects on the dynamics of these neuronal circuits to confirm the features of this model.

Another important aspect of the system that requires further study is the functional role of the cerebellum. Raichle et al (1994) found that the right cerebellar hemisphere is activated in the naive condition of the word-association task. Leiner, Leiner, and Dow (1991, 1993), and others (Akshoomoff, Courchesne, Pres, & Iragui, 1992; Berntson & Torello, 1982; Ito, 1993; Schmahmann, 1991) have pointed out that the cerebellum has a significant role in cognitive and language functions, a fact that has long been overshadowed by its prominent motor role. It is possible that the cerebellum has a role in error correction for potential response words, especially when action-related responses are to be produced. This hypothesis is based on data from Fiez, Petersen, Cheney, and Raichle (1992), who found that damage to the right cerebellar hemisphere of a human subject due to stroke resulted in the patient's inability to generate appropriate word-association responses, especially verb responses. In this regard, the cerebellum may act as a language area for action-oriented aspects of language. This role may be a naturally arising complement of the well-established motor role of the cerebellum. The principal feature of Fiez's patient RC-1 was the absence of feedback concerning his performance in the word-association task. He made errors but was unaware of these errors. The mechanism of cerebellar participation may be similar to that for using feedback to correct errors in motor control. Berthier, Singh, Barto, and Houk (1993), for example, postulate such a mechanism that uses Purkinje cells in cortico-cerebellar modules to recognize complex states of cell firing in the cortex to select appropriate responses.

Finally, the cerebellum may play an important role in facilitating learning in the automatic pathway. In the above-mentioned study by Fiez et al (1992), the cerebellar patient was also impaired in learning new word associations with practice. Here again, the cerebellum could be acting as a pattern-recognizing system that detects when the type of response produced by the controlled pathway matches that required by the task. In case of a match, the cerebellum enables learning of the appropriate response. Simulations to test the authenticity of this mechanism and to determine possible alternative mechanisms for facilitating learning are currently being conducted.

ACKNOWLEDGMENTS

We would like to thank Nina Dronkers, James Houk, Philip Johnson-Laird, John Kounios, George Miller, and Marc Raichle for their guidance and many valuable suggestions. This work was partially supported by a grant from the James S. McDonnell Foundation to the Human Information Processing Group at Princeton University.

Earlier versions of this chapter appeared in the Proceedings of the 17th Cognitive Science Conference and in the Proceedings of the New York Academy of Sciences workshop on Structure and Functions of the Human Prefrontal Cortex.

Neural-Networks Models of Cognition
J.W. Donahoe and V. Packard Dorsel (Editors)

CHAPTER 25

SERIAL ORDER: A PARALLEL DISTRIBUTED PROCESSING APPROACH

Michael I. Jordan
Department of Brain and Cognitive Sciences
Massachusetts Institute of Technology

ABSTRACT

A theory of learned sequential behavior is presented, with a focus on coarticulatory phenomena in speech. The theory is implemented as a recurrent parallel distributed processing network that is trained via a generalized error-correcting algorithm. The basic idea underlying the theory is that both serial order and coarticulatory overlap can be represented in terms of relative levels of activation in a network if a clear distinction is made between the *state* of the network and the *output* of the network.

Introduction

Even the most cursory examination of human behavior reveals a variety of serially ordered action sequences. Our limb movements, our speech, and even our internal train of thought involve sequences of events that follow one another in time. We are capable of performing an enormous number of sequences, and we can perform the same actions in a variety of different contexts and orderings. Furthermore, most of the sequences that we perform were learned through experience.

A theory of serial order in behavior should clearly be able to account for these basic data. However, no such general theory has emerged, and an important reason for this is the failure of current formalisms to deal adequately with the parallel aspects of serially ordered behavior. We can tentatively distinguish two forms of parallelism. The first is parallelism that arises when actions in a sequence overlap in their execution. In speech research, such parallelism is referred to as *coarticulation* (Kent & Minifie, 1977; Moll & Daniloff, 1971; Öhman, 1966), and it greatly complicates the traditional description of sequential speech processes. The second form of parallelism occurs when two actions are required to be performed in parallel by the demands of the task or by implicit constraints. Such is the case, for example, in the *dual-task* paradigm, in which actions that have been learned separately must be performed together. This differs from the case of coarticulation, in which actions that are nominally separate in time are allowed to be performed in parallel. It is important to characterize both how such parallelism can arise within a sequential process and how it can be constrained so that unwanted parallel interactions are avoided.

In this paper, I present a theory of serial order that describes how sequences of actions might be learned and performed. The theory is embodied in the form of a parallel distributed processing network (Rumelhart & McClelland, 1986). Such networks are composed of a large number of simple processing units that are connected through weighted links. In various forms, such networks have been used as models of phenomena such as stereopsis (Marr & Poggio, 1976), word recognition (McClelland & Rumelhart, 1981), and reaching (Hinton, 1984). The success of these models has been due in large part to their high degree of parallelism, their ability to bring multiple interacting constraints to bear in solving complex problems, and their use of distributed representations. However, none of these properties seems particularly well suited to the problem of serial order. Indeed, a criticism of this class of models has been their inability to show interesting sequential behavior, whereas the more traditional symbolic approaches—typically by assuming a sequential processor as a primitive—deal with serial order in a much more straightforward manner. This criticism is challenged in this chapter, in the context of a theory of serial order that takes advantage of the underlying primitives provided by parallel distributed processing.

Serial Order

Many of the problems encountered in developing a parallel distributed processing approach to the serial-order problem were anticipated by Lashley (1951). Lashley pointed out the insufficiency of the *associative-chaining* solution to the serial-order problem. The associative-chaining solution assumes that serial ordering is encoded by directed links between control elements representing the actions to be ordered, and that the performance of a sequence involves following a path through the network of control elements. Lashley argued that this solution fails to allow different orderings of the same actions because there is no mechanism for specifying which link should be followed from an element having more than one outgoing link. He also argued that serial behavior shows anticipatory effects of future actions upon the current action, and that such context effects are not accounted for within the associationist framework.

Buffer approaches to serial order

Lashley's arguments have had an impact on those seeking to understand the role of feedback in a theory of motor behavior, but have been less influential on those interested in the structure of *motor programs*. This is in all likelihood due to the impact on theorists of the development of the digital computer, which made it possible to see how arbitrary sequential programs can be executed. Theories based explicitly on the computer metaphor have invoked the notion of a *buffer* which is loaded with the actions to be performed, and a program counter which steps through the buffer (Shaffer, 1976; Sternberg, Monsell, Knoll, & Wright, 1978). Despite the generality of such a theory, simple buffer theories are known to have several problems, including accounting for error patterns (Kent & Minifie, 1977; MacKay, 1981). It is also true that coarticulation is not well handled by buffer theories. One approach is to assume that buffer positions can interact with each other (Henke, 1966). However, this interaction, which must occur when successive actions are simultaneously present in the buffer, takes time, as does the process of reloading the buffer once a set of related actions has been executed. This approach implies the presence of delays at certain times in the production of long sequences, but such delays are not observed in fluent sequential behavior (cf. Shaffer & Hardwick, 1970). Another problem is that interactions between actions should depend on their relative positions in the buffer, not their absolute positions. For example, the interactions between the phonemes /i/ and /n/ should presumably be the same when saying "print" and "sprint." This would seem to imply the need for a complex mechanism whereby learned interactions can automatically generalize to all buffer positions. Such issues, which arise due to the explicit spatial representation of order in buffer theories, seem to be better handled within an associationist framework.

Associationist approach to serial order

Wickelgren (1969) revived the associationist approach by assuming that serial order was indeed encoded by directed links between control elements, but that the control elements were different for different orderings of the same actions. The control element for the action B in the sequence ABC can be represented by the form ${}_{A}B_{C}$ whereas the control element for B in the sequence CBA is represented as ${}_{C}B_{A}$. These control elements are distinct elements in the network, thus there is no problem with representing both the sequences ABC and CBA in the same network. In this account, actions are different in different contexts, not because they are executed in parallel, but because they are produced by different control elements.

Wickelgren's theory provides a solution to the problems posed by Lashley but it has several shortcomings. First, it requires a large number of elements, yet has difficulty with the pronunciation of words, such as "barnyard," that have repeated subsequences of length two or more (Wickelgren, 1969). Second, effects of context in speech have been shown to extend up to four or five phonemes forward in an utterance (Benguerel & Cowan, 1974). Extension of the theory to account for such effects would require an impossibly large number of control elements. Finally, note that there are only representations for *tokens* in the theory, and no representations for *types*. There is nothing in the theory to tie together the contextual variations of a given action. This means that there is no way to account for the linguistic and phonetic regularities that are observed when similar actions occur in similar contexts (Halwes & Jenkins, 1971).

Parallel-processing approaches to serial order

A different approach is to assume that actions are to some extent produced in parallel (Fowler, 1980; Rumelhart & Norman, 1982). The parallelism allows several control elements to influence behavior at a particular point in time, and therefore provides an account of coarticulatory effects, even though actions are represented in terms of context-free types. Rumelhart and Norman (1982) have shown that a model of typing incorporating parallelism can produce overlapping keystrokes much like those observed in transcription typing.

Allowing parallel activation of control elements accounts for context sensitivity; however, the problem of temporal ordering remains. Rumelhart and Norman achieved temporal ordering by assuming that elements suppress other elements through lateral inhibitory connections if they precede those elements in the sequence. This particular scheme is susceptible to Lashley's critique because all possible inhibitory connections must be present to allow the performance of the same elements in different orders, and a mechanism is needed for selecting the particular inhibitory connections used in the performance of a particular sequence. However, there are other ways of achieving the same effect that are not open to Lashley's critique (Grossberg, 1978; Grudin, 1981). Essentially, all of these schemes produce temporal order by inducing a graded activation pattern across the elements in the sequence, such that elements more distant in the future are activated less than earlier elements. Elements are assumed to influence behavior in proportion to their level of activation. Because the next action in the sequence is the most highly activated, it has the most influence on behavior. Once the activation of an element reaches a threshold, it is inhibited, allowing the performance of other items in the sequence.

A problem with these parallel-activation theories is that they have difficulty with sequences in which there are repeated occurrences of actions. In a pure type representation, there is simply no way to represent the repeated action. Rumelhart and Norman used a modified type representation in which they introduced special operators for doublings (e.g., AA) and alternations (e.g., ABA). However, they provided no general mechanism. For example, sequences such as ABCA invoked a parser to break the sequence into pieces, thus allowing no parallel influences across the break. This is not a satisfactory solution, in general, because data in speech show that coarticulatory influences can extend across sequences like ABCA (Benguerel & Cowan, 1974). Another possibility is to assume that repeated occurrences of actions are represented by separate control elements (representation by tokens). However, the combined effects of partially activated control elements will cause the first occurrence of a repeated action to move forward in time, whether or not this is actually desirable. Indeed, in a sequence such as ABBB, the B may overwhelm the A and be executed first. These problems are enhanced in featural representations of the kind that are often posited for actions (Grudin, 1983; Perkell, 1980; Rosenbaum, 1980) because the total activation from elements representing the repeated features will be greater than the activation levels for features that only occur once in the sequence, irrespective of the order of the features. Such problems arise because the single quantity of activation is being used to represent two distinct things: the parallel influences of actions and the temporal order of actions.

It is my view that many of these problems disappear when a clear distinction is made between the *state* of the system and the *output* of the system. Explicitly distinguishing between the state and the output means that the system has two activation vectors, which allows both temporal order and parallel influences to be represented in terms of activation. In the theory developed in this paper, the state and the output are assumed to be represented as patterns of activation on separate sets of processing units. These sets of units are linked by connections defining an *output function* for the system. Serial order is encoded both in the output function and in recurrent connections impinging on units representing the state; there is no attempt to encode order information in direct connections between the output units.

Coarticulation

In this section, I briefly introduce some of the parallel aspects of sequential behavior that have been considered important in the development of the current theory.

Several studies involving the recording of articulator trajectories have shown that speech gestures associated with distinct phonemes can occur in parallel. Moll and Daniloff (1971) showed that in an utterance such as "freon," the velar opening for the nasal */n/* can begin as early as the first vowel, thereby nasalizing the vowels. Benguerel and Cowan (1974) studied phrases such as "une sinistre structure," in which there is a string of the six consonants */strstr/* followed by the rounded vowel */y/*. They showed that lip-rounding for the */y/* can begin as early as the first */s/*. This is presumably allowable because the articulation of the consonants does not involve the lips.

These examples suggest that the speech system is able to take advantage of "free" articulators and use them in anticipating future actions. This results in parallel performance and allows speech to proceed faster and more smoothly than would otherwise be possible. Such parallelism clearly must be constrained by the abilities of the articulators. However, there are other constraints involved as well. In the case of "freon," for example, the velum is allowed to open during the production of the vowels because the language being spoken is English. In a language such as French, in which nasal vowels are different phonemically from non-nasal vowels, the velum would not be allowed to coarticulate with the vowels. Thus the articulatory control system cannot blindly anticipate articulations, but must be sensitive to phonemic distinctions in the language being spoken by only allowing certain coarticulations.

The situation is more complicated still if we note that constraints on parallelism may be specific to particular features. For example, in the case of */strstry/*, only the *rounding* of the */y/* can be anticipated. The *voicing* of the */y/* cannot be anticipated because that would change the phonemic identities of the consonants (for example, the */s/* would become a */z/*). Again, such knowledge cannot come from consideration of strategies of articulation, but must reflect higher-level phonemic constraints.

Thus, speech presents a difficult distributed-control problem in which constraints of various kinds are imposed on the particular patternings of parallelism and sequentiality that can be obtained in an utterance. What I wish to show in the remainder of this paper is how this problem can be approached with a theory based on parallel distributed processing networks.

A Theory of Serial Order

Let there be some sequence of *actions* $x_1, x_2, \ldots, x_r$, which is to be produced in the presence of a *plan* **p**. Each action is a vector in a parameter or feature space, and the plan can be treated as an action produced by a higher level of the system. The plan is assumed to remain constant during the production of the sequence, and serves primarily to designate the particular sequence that is to be performed.

In general, we would like the system to be able to produce many different sequences. Thus, different vectors **p** are assumed to be associated with different sequences of actions. A particular sequence is produced when a particular vector **p** is presented as input to the system. Note that, in principle, there need be no relationship between the form of plan vectors and the sequences that they evoke. Rather, a plan vector evokes a particular sequence because it was present as input to the system when the sequence was learned. Thus, plans may simply be arbitrary patterns of activation that serve to key particular sequences; they need not be scripts for the system to follow.

Actions are produced in a temporal context composed of actions nearby in time. This context entirely determines the desired action, in the sense that knowing the context makes it possible to specify what the current action should be. It is proposed that the system explicitly represents the temporal context of actions in the form of a state vector and chooses the current action by evaluating a function from states to actions. At each moment in time, an action is chosen based on the current state, and the state is then updated to allow the next action to be chosen. Serial order does not arise from direct connections between units representing the actions; rather, it arises from two functions that are evaluated at each time step: a function f which determines the output action x_n at time n,

$$\mathbf{x}_n = f(\mathbf{s}_n, \mathbf{p}) \qquad (1)$$

and a function g which determines the state $\mathbf{s}_{n+1}$,

$$\mathbf{s}_{n+1} = g(\mathbf{S}_N, \mathbf{p}), \qquad (2)$$

where both functions depend on the constant-plan vector as well as the current-state vector. Following the terminology of automata theory (Booth, 1967), f will be referred to as the *output* function, and g will be referred to as the *next-state* function. (From the definition, it can be seen that the plan **p** plays the role of the input symbol in a sequential machine. The use of the term "plan" is to emphasize the assumption that **p** remains constant during the production of the sequence. That is, we are not allowed to assume temporal order in the input to the system.)

Assumptions are made in the theory about the form of these functions. The output function f is assumed to arise through learned associations from state

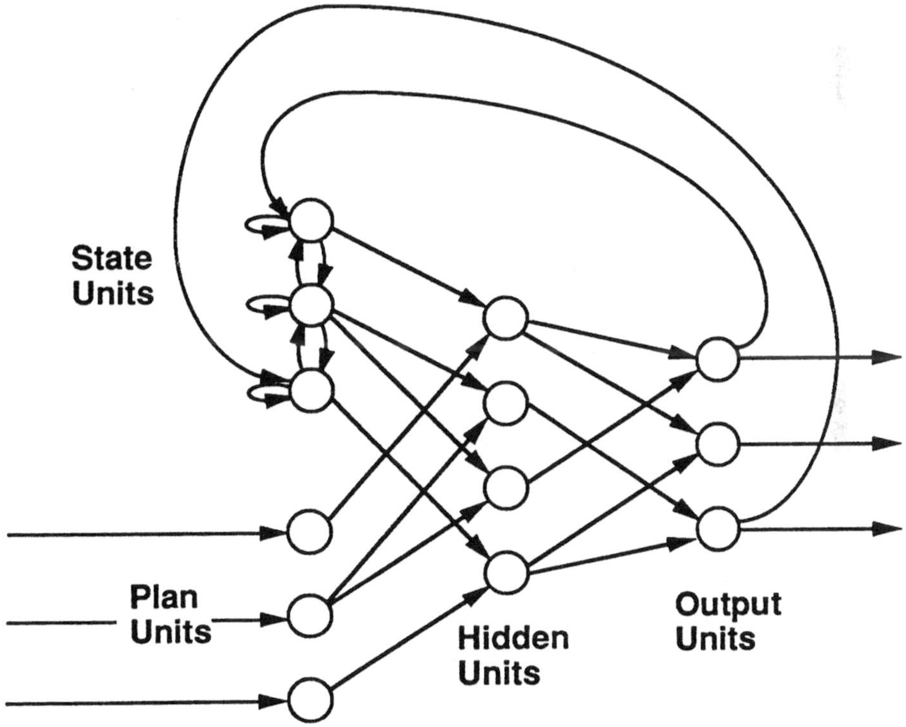

FIGURE 1. The processing units and basic interconnection scheme (not all connections are shown). The plan and state units together constitute the input units for the network.

and plan vectors to output vectors. These learned associations are assumed to generalize so that similar states and plans tend to lead to similar outputs. The major requirement for the next-state function g is that it have a continuity property: State vectors at nearby points in time are assumed to be similar. This requirement makes sense if the state is thought of as representing the temporal context of actions; intuitively, it seems appropriate that the temporal context should evolve continuously in time. Note that if the continuity property holds, then the generalizations made by the output function are such as to spread actions in time and, as learning proceeds, there is a tendency towards the increasing parallel execution of actions nearby in time. This process is discussed below in detail, where it is also shown how the generalizations leading to parallelism can be constrained.

A basic network architecture that embodies the theory is shown in **Figure 1**. The entities of the theory—plans, states, and outputs—are all assumed to be represented as distributed patterns of activation on three separate pools of processing units. The plan units and the state units together serve as the input units for a network that implements the output function f through weighted connections from the plan and state units to the output units. There are hidden units in the path from the plan and state units to the output units to allow for nonlinear output functions. Finally, the next-state function is implemented with recurrent connections from the state units to themselves and from the output units to the state units. This allows the current state to depend on the previous state and on the previous output (which is itself a function of the previous state and the plan).

In the proposed network, there is no explicit representation of temporal order and no explicit representation of action sequences. This is because there is only one set of output units for the network so that, at any point in time, only one output vector is present. Output vectors arise as a dynamic process, rather than being prepared in advance in a static buffer and then serially executed. Representing actions as distributed patterns on a common set of processing units has the virtue that partial activations blend together in a simple way to produce the output of the system.

Although it is possible that the next-state function as well as the output function arises through learning, this is not necessary for the system as a whole to be able to learn to produce sequences. Furthermore, given that the next-state function is set up in such a way that the continuity property holds, little is lost in the current framework if the recurrent connections necessary for the next-state function are taken as fixed and only the output function is learned. This latter approach is taken in the remainder of the chapter.

One choice of values for the fixed recurrent connections is based on the conception of the state as a temporal context. Consider the case of a sequence with a repeated subsequence or a pair of sequences with a common subsequence. It seems appropriate, given the positive transfer that can occur in such situations as well as the phenomena of capture errors (Norman, 1981), that the state should be similar during the performance of similar subsequences. This suggests defining the state in terms of the actions being produced. However, the representation must provide a sufficiently extensive temporal context that no ambiguities arise in cases involving repeated subsequences. If the state were to be defined as a function of the last n outputs, for example, then the system would be unable to perform sequences with repeated subsequences of length n, or to distinguish between pairs of sequences with a common subsequence of length n. To avoid such problems, the state can be defined as an exponentially weighted average of past outputs, so that the arbitrarily distant past has some representation in the state, albeit with ever-diminishing strength. This representation of the state is achieved if each output unit feeds back to a state unit with a weight of one, if each state unit feeds back to itself with a weight μ, and if the state units are linear. In this case, the state at time n is given by

$$s_N = \mu s_{n-1} + x_{n-1} \quad (3)$$

$$= \sum_{\tau=1}^{n-1} \mu^{\tau-1} x_{n-\tau} \quad (4)$$

The similarity between states depends on the particular actions that are added at each time step and on the value of μ. In general, with sufficiently large values of μ, the similarity extends forward and backward in time, growing weaker with increasing distance.

Learning and parallelism

In the network, learning is realized as an error-correcting process in which the weights of the network are incrementally adjusted based on the difference between the actual output of the network and a desired output. Essentially, the next-state function provides a time-varying state vector, and the error information drives changes in the mapping from this state vector and the plan vector to the output. The form that desired output vectors are assumed to take is a generalization of the approach used in traditional error-correction schemes (Rumelhart, Hinton, & Williams, 1986). Rather than assuming that a value is specified for each output unit, it is assumed that, in general, there are *constraints* specified on the values of the output units. Constraints may specify a range of values that an output unit may have, a particular value, or no value at all. This latter case is referred to as a "don't-care condition." It is also possible to consider constraints that are defined among output units; for example, the sum of the activations of a set of units might be required to take on a particular value. Constraints enter into the learning process in the following way: If the activation of an output unit fits the constraints on that unit, then no error corrections are instigated from that unit. If, however, a constraint is not met, then the error is defined as a proportion of the degree to which that constraint is not met, and this error is used in changing system parameters towards a configuration in which the constraint is met.

In many realistic sequence-learning problems, it would seem that desired outputs cannot be assumed to be directly available at the output units of the network. For example, in the case of speech production, the information provided to the learner is auditory or perceptual, whereas desired output information for the production module must be specified in terms of articulator motion. A related problem is that target information may be delayed in time relative to performance. Such problems of a "distal teacher" have been addressed in recent work that shows how the constraints may themselves be learned (Jordan & Rumelhart, 1992). The constraints are implemented in an auxiliary network that models the mapping from the network outputs to the distal results. Once the model is learned, backpropagation through the model converts distal error vectors into error vectors for the output units. For example, if the auxiliary network models the mapping from articulatory events to auditory events, then backpropagation can be used to convert auditory errors backward into articulatory errors. The error vectors that are computed by this process can be thought of as providing target outputs for the underlying sequential network. Thus, for current purposes, we can make the simplifying assumption that desired outputs are provided directly by an external agent. There is a caveat, however: When the auxiliary network models a many-to-one function, then the error vectors computed by backpropagation implicitly specify a region in output space, rather than a point. Of course, it is precisely this underspecification that is of interest, because it allows actions in a sequence to have an effect on one another. Here, I use don't-care conditions in the specification of desired output vectors to allow consideration of a particularly simple case: regions that are rectangular and parallel to the axes of the output space. For further discussion of the general case, see Jordan (1990).

Consider first the case in which desired output vectors specify values for only a single output unit. Suppose that a network with three output units is learning the sequence

$$\begin{bmatrix} .9 \\ * \\ * \end{bmatrix}, \begin{bmatrix} * \\ .9 \\ * \end{bmatrix}, \begin{bmatrix} * \\ * \\ .9 \end{bmatrix}.$$

The network is essentially being instructed to activate its output units in a particular order, and this case can be thought of as involving local representations for actions. At each time step, errors are propagated from only a single output unit, so that activation of that unit becomes associated to the current state s_i. Associations are learned from s_1 to activation of the first output unit, from s_2 to activation of the second output unit, and from s_3 to activation of the third output unit. These associations also generalize so that, for example, s_1 tends to produce partial activations of the second and third output units. This occurs because s_1 is similar to s_2 and s_3, and—by the assumption of continuity of the next-state function—similar inputs produce similar outputs in these networks. After learning, the network will likely produce a sequence such as

$$\begin{bmatrix} .9 \\ .7 \\ .5 \end{bmatrix}, \begin{bmatrix} .7 \\ .9 \\ .7 \end{bmatrix}, \begin{bmatrix} .5 \\ .7 \\ .9 \end{bmatrix},$$

where at each time step, there are parallel activations of all output units. If the network is driving a set of articulators that must travel a certain distance, or have a certain inertia, then it will be possible to go faster with these parallel control signals than with signals where only one output unit can be active at a time.

The foregoing example is simply the least constrained case and further constraints can be added. Suppose, for example, that the second output unit is not allowed to be active during the first action. This can be encoded in the target vector for the first action so that the network is instructed to learn the sequence

$$\begin{bmatrix} .9 \\ .1 \\ * \end{bmatrix}, \begin{bmatrix} * \\ .9 \\ * \end{bmatrix}, \begin{bmatrix} * \\ * \\ .9 \end{bmatrix}.$$

After learning, the output sequence will likely be as follows:

$$\begin{bmatrix} .9 \\ .1 \\ .5 \end{bmatrix}, \begin{bmatrix} .7 \\ .9 \\ .7 \end{bmatrix}, \begin{bmatrix} .5 \\ .6 \\ .9 \end{bmatrix},$$

where the added constraint is now met. In this example, the network must block the generalization that is made from s_2 to s_1.

As further constraints are added, and fewer generalizations across nearby states are allowed, performance becomes less parallel. Minimal parallelism will arise when neighboring actions specify conflicting values on all output units, in which case the performance will be strictly sequential. Maximal parallelism should be expected when neighboring actions specify values on nonoverlapping sets of output units. Note that there is no need to invoke a special process to introduce parallelism into the system. Parallelism arises from the ability of the system to generalize, and is a manifestation of the normal functioning of the system. Indeed, in most cases, it will be more difficult for the system to learn in the strictly sequential case when there are more constraints imposed on the system.

Serial order

Before turning to a more detailed discussion of coarticulation, it is worth considering how the current theory fares with respect to some of the general requirements of a theory of serial order. It should be clear that the theory can deal with the problem of converting a static input into a time-varying output, given that the state changes over time, and given that an appropriate output function can be constructed. Different orderings of the same actions can be achieved, both because the state trajectories may differ between the sequences and because the output function depends on the plan, and the plan can distinguish the different orderings. The theory has no problem with repeated actions; the existence of repeated actions simply indicates that the output function is not one-to-one, but that two or more state, plan pairs can map to the same output vector. Finally, sequences such as ABAC, which cause problems for an associative-chaining theory because of the transitions to distinct actions after a repeated action, are possible because the state after the first A is not the same as the state after the second A.

The theory is able, in principle, to account for a variety of regularities that occur within and between sequences. This is because outputs and states are represented as types; that is, there is only one set of output units and one set of state units. The same weights underlie the activation of actions, in whatever position in the sequence, and in whatever sequence. Thus, particular weights underlie the regularities observed for similar actions in similar contexts. For example, the fact in English that voiceless stops are aspirated following */s/* (e.g., */spIn/* is pronounced [*sbIn*]), would be encoded by inhibitory connections from state units encoding the recent occurrence of a voiceless alveolar fricative to output units controlling glottal and labial movements. In the sequential-network architecture, this encoding allows the allophonic regularity to generalize immediately in contexts other than the initial portion of the word. Such a sensitivity to relative position, rather than absolute position, is difficult to obtain in architectures using spatial buffers (Sejnowski & Rosenberg, 1986), and problematic to obtain (in full generality) in schemes using context-sensitive allophones (Rumelhart & McClelland, 1986; Wickelgren, 1969).

One of the more important tests of a theory of serial order is that it account for interactions both forward and backward in time. In the current theory, time is represented implicitly by the configuration of the state vector. Interactions in time are due to the similarity of the state vector at nearby points in time. There is no time arrow associated with this similarity, thus, forward and backward interactions are equally possible.

Limitations on the structure of the functions f and g will lead to some sequences being more difficult to learn and perform than others. For example, the temporal context cannot extend indefinitely far in time; thus, the repetition of lengthy subsequences that make transitions to different actions can be difficult to learn and perform. Also, similarity between action transitions in different plans can cause interference, as can similarity between plan representations. The interference can lead to errors and to the learning of one sequence causing negative transfer on another sequence. Such interactions can also have a positive side, of course, in the form of positive transfer.

Dynamic properties of the networks

When a network learns to perform a sequence, it essentially learns to follow a trajectory through a state space. The state space consists of the ensemble of possible vectors of activation of the output units. An important fact about the learned trajectories is that they tend to influence points nearby in the state space. Indeed, the learned trajectories tend to be *attractors*.

Consider, for example, a network taught to perform the cyclic sequence

$$\begin{bmatrix} .25 \\ .25 \end{bmatrix}, \begin{bmatrix} .75 \\ .25 \end{bmatrix}, \begin{bmatrix} .75 \\ .75 \end{bmatrix}, \begin{bmatrix} .25 \\ .75 \end{bmatrix}, \begin{bmatrix} .25 \\ .25 \end{bmatrix}.$$

The trajectory of the network is on the four corners of a square in the first quadrant of the plane. The trajectory will repeatedly move around this square if the initial vector of activations of the output units is one of the corners of the square. It is also possible to set the initial activations of the output units to other values, thereby starting the network at points in the space other than the four corners of the square. **Figure 2** (left panel) shows the results of a simulation experiment in which the network was started at the point (.4,.4). As can be seen, the trajectory spirals outward and begins to approximate the square more and more closely. When the network is started at a point outside of the square, the trajectory is found to spiral inward towards the square. A sample trajectory starting from the point (.05,.05) is shown in the right panel of **Figure 2**. When the network was initialized at each of 100 points in the state space, it was found that all trajectories reached the square in the limit, demonstrating that the square is a *periodic attractor*. Note that trajectories starting inside the square approach the limit cycle less rapidly than do trajectories starting outside the square. At a point inside the square, the trajectory is subject to influences associated with all four corners, and these influences are in conflicting directions and therefore tend to cancel one another. At a point outside the square, however, only a pair of adjacent corners tend to influence the trajectory, and adjacent influences do not conflict in this example.

The dynamics exhibited by the networks described above has several useful properties. The system tends to be noise-resistant, because perturbed trajectories return to the attractor trajectory. The system is also relatively insensitive to initial conditions. Finally, the learning of a particular trajectory automatically generalizes to nearby trajectories, which is what is desired in many situations. The relevance of these properties to motor control has been recognized by several authors (Kelso, Saltzman, & Tuller, 1987; Saltzman & Kelso, 1987). I wish to suggest that such dynamics may also characterize the higher-level dynamic system that is responsible for serial ordering.

Application of the theory to coarticulation

The theory presented in this paper involves a dynamic system that is constrained through a learning process to follow particular trajectories. The learning process relies on constraints on the output of the system. These constraints implicitly define regions in output space through which trajectories must pass, and thereby delimit the possible range of effects of temporal context.

In the case of speech, the form of the constraints on articulation depends on inter-articulator organization, both kinematic and dynamic, and on the function

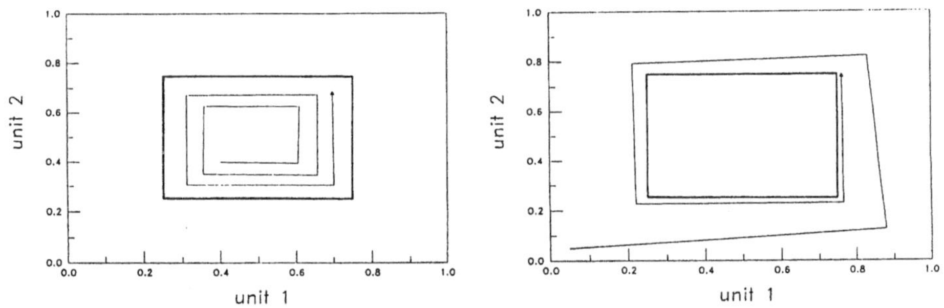

FIGURE 2. Two examples of the activations of the two output units plotted with time as a parameter. In each case, the square is the trajectory that the network learned. Left panel: The spiral trajectory is the path that the network followed when started at the point (.4,.4). Right panel: The spiral trajectory is the path that the network followed when started at the point (.05,.05).

that relates articulatory events to perceptual events. This latter function includes at least two kinds of mappings—one that relates articulator motion to pre-categorical auditory representations, and one that relates pre-categorical representations to post-categorical representations. After preliminary learning, both of these mappings can be assumed to be represented internally and thereby available to compute articulatory constraints from perceptual data as discussed in Jordan (1990). The salient characteristic of both of these mappings is that they are many-to-one (cf. Atal, Chang, Mathews, & Tukey, 1978). Thus, during imitative learning, the error vectors that are computed from the back-propagation of perceptual information implicitly specify *regions* of articulatory space rather than points. As described previously, the underlying dynamic system will form trajectories that pass smoothly through these regions. This yields contextually dependent variants of a given articulatory equivalence class. In summary, coarticulation is hypothesized to be a form of smoothness in articulatory space that is subject to perceptual constraints.

The perceptual information that provides target vectors for imitative learning may be either pre-categorical or post-categorical. Clearly, children's ability to acquire accent and other non-distinctive aspects of speech suggests that learning must be at least partially based on pre-categorical target information. It is tempting to hypothesize that the locus of target information evolves as post-categorical representations are formed over the course of development: Using a post-categorical target specifies a larger region of articulatory space, and therefore allows more flexibility in the choice of an articulatory trajectory. Of course, this flexibility is obtained with a corresponding loss in the ability to acquire articulation that reflects pre-categorical details.

In this section, I present some simple simulations of a system learning phonetic sequences. It should be emphasized that I am not proposing a realistic model of speech production in this section. A major simplification is that I have defined desired output vectors directly in articulatory terms using target values and don't-care conditions. This representation ignores the problem of converting perceptual information into articulatory information as well as the effects of articulatory dynamics. (Both of these issues can be addressed, however, within the framework of the forward-modeling approach; see Jordan & Rumelhart, 1992.) Nonetheless, the simulations are useful in elucidating the network algorithms hypothesized to underlie coarticulation. Also, they allow some qualitative predictions to be made.

The problem of serial ordering in speech is typically treated in discrete terms, and the relation between discrete higher-level processes and continuous lower-level articulatory processes has provoked much debate in the literature on speech production (Fowler, 1980; Hammarberg, 1982; Perkell, 1980). In the current theory, however, such issues are not particularly problematic because the entire system can be thought of as operating in continuous time. It is consistent with the current theory to assume that the defining state equations are simply a discrete version of a continuous-time dynamic system. In the continuous case, learning involves imposing constraints intermittently on the system at various points in time. In geometric terms, constraints appear as regions through which continuous-network trajectories must pass, with trajectories between regions unconstrained. To approximate the continuous system in the simulation, I have inserted several time steps between steps at which constraints are imposed. During these intermediate time steps, the network is free running (these intermediate steps can be thought of as having don't-care conditions on all of the output units). By conducting the simulation in this manner, it is possible to demonstrate the differences between the current approach and an assimilatory model in which different allophones are produced at each time step and interactions must begin and end at allophonic boundaries (cf. Fowler, 1980).

Feature	*i*	*s*	*t*	*r*	*s*	*t*	*r*	*y*
voice	8	1	1	*	1	1	*	8
place	7	9	9	2	9	9	2	7
sonorant	8	2	1	5	2	1	5	8
sibilant	1	9	2	4	9	2	4	1
nasal	*	*	1	*	*	1	*	*
height	9	9	9	9	9	9	9	9
back	1	*	*	2	*	*	2	1
round	1	*	*	*	*	*	*	9

TABLE 1. Target vectors for the string */istrstry/*.

Simulation experiments

For the purposes of describing the simulations section, I use the term "phoneme" to refer to a vector of target values and don't-care conditions. Representations for the phonemes were adapted from a list of real-valued features proposed by Ladefoged (1982). Eight features were selected that provided adequate discriminations between the particular phonemes used in the simulations. The feature values were all between 0.1 and 0.9. Choices for don't-care conditions were based on known allophonic variations (for example, the rounding for the French */s/* was taken to be a don't-care condition, because it is possible to have a rounded or an unrounded */s/*).

The network used in the simulations had 8 output units, 10 hidden units, 6 plan units, and 8 state units. The state units had recurrent connections onto themselves with weights of $\mu = 0.5$.

The procedure used in the simulation was essentially that of the preceding section, with the following modification. During learning trials, target vectors (i.e., phonemes) were presented to the network every fourth time step. Learning occurred only on these time steps. During the intermediate three time steps, the units were updated normally with no learning occurring.

In the first experiment, the network was taught to perform the sequence "sinistre structure." The phonemes that were used are shown in **Table 1** for the embedded sequence */istrstry/* only. The learning process involved repeated trials in which the phonemes in the sequence were used as target vectors for the network. The plan was a particular constant vector whose composition is irrelevant here because the network learned only this one sequence. The results for the embedded sequence */istrstry/* are shown in **Figure 3**, which displays the output trajectories actually produced by the network once the sequence was learned to criterion. The network learned to produce the specified values, as can be seen by comparing the values produced at every fourth time step with the values in the table. The network also produced values for the don't-care conditions and for unconstrained parts of the trajectories. In particular, the value of .9 for the rounding feature of the rounded vowel */y/* was anticipated as early as the third time step. In a control experiment, the sequence "sinistre stricture," in which the same consonant sequence is followed by the unrounded vowel */i/*, was taught to the network. As shown in **Figure 4**, there was now no rounding during the entire utterance. These results parallel the data obtained by Benguerel and Cowan (1974).

In a third experiment, the network learned the sequence "freon," where the feature of interest was the nasal feature associated with the terminal */n/*. In the phoneme vectors, the */f/* was specified as 0.1 for the nasal feature, the */n/* was specified as 0.9, and the intervening three phonemes had don't-care values for the nasal feature. Thus, this experiment is analogous to the previous experiment, with the interest in the anticipation of the nasal feature rather than the rounding feature. The results are shown in **Figure 5** in terms of the activation of the nasal feature at every fourth time step. As in the data of Moll and Daniloff (1971), there was substantial anticipation of the nasal value of */n/* before and during the two vowels. Note that a steeper dropoff in the amount of anticipation occurred in this sequence than in the sequence */istrstry/*. An investigation of the weights learned during these sequences revealed that the extensive coarticulation in the latter sequence arose from the repetition of phonemes. The rounding of */y/* was produced in a temporal context in which

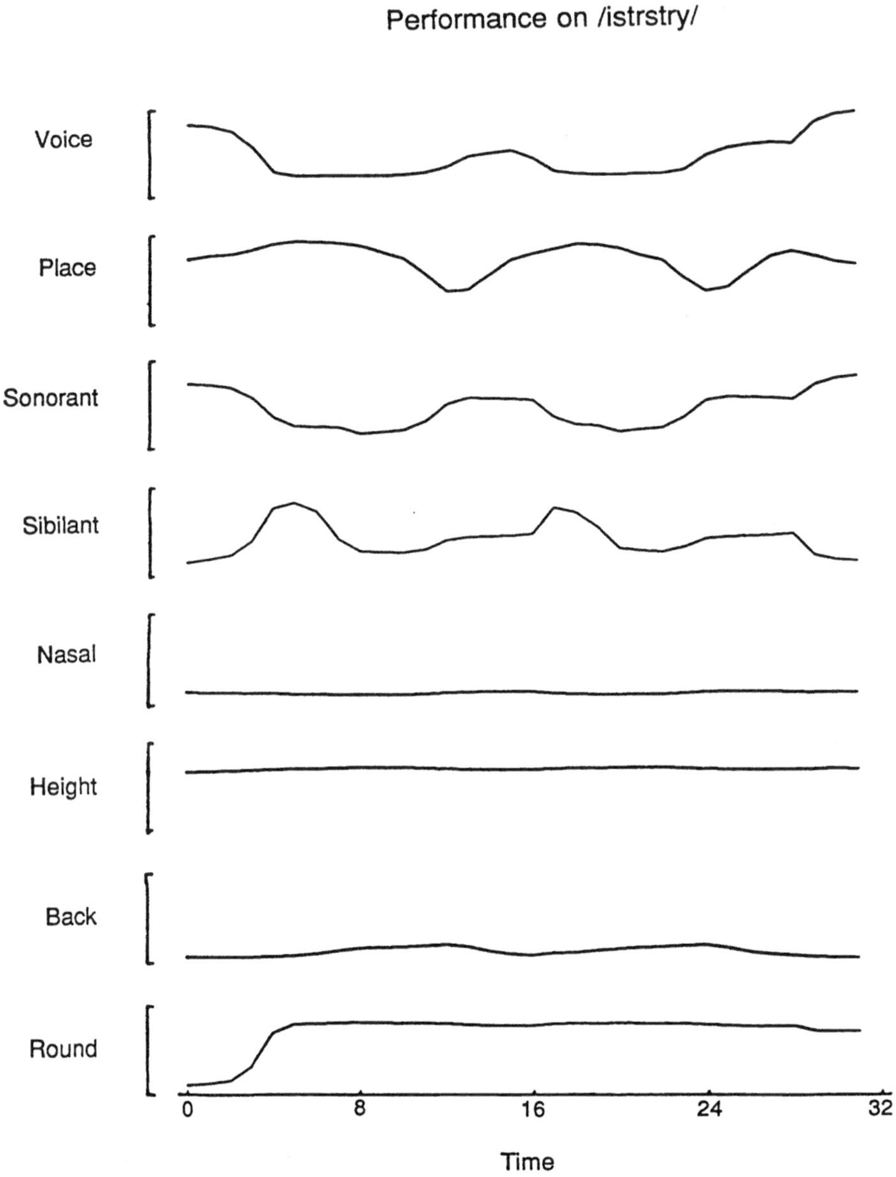

FIGURE 3. Output trajectories for the sequence /*istrstry*/.

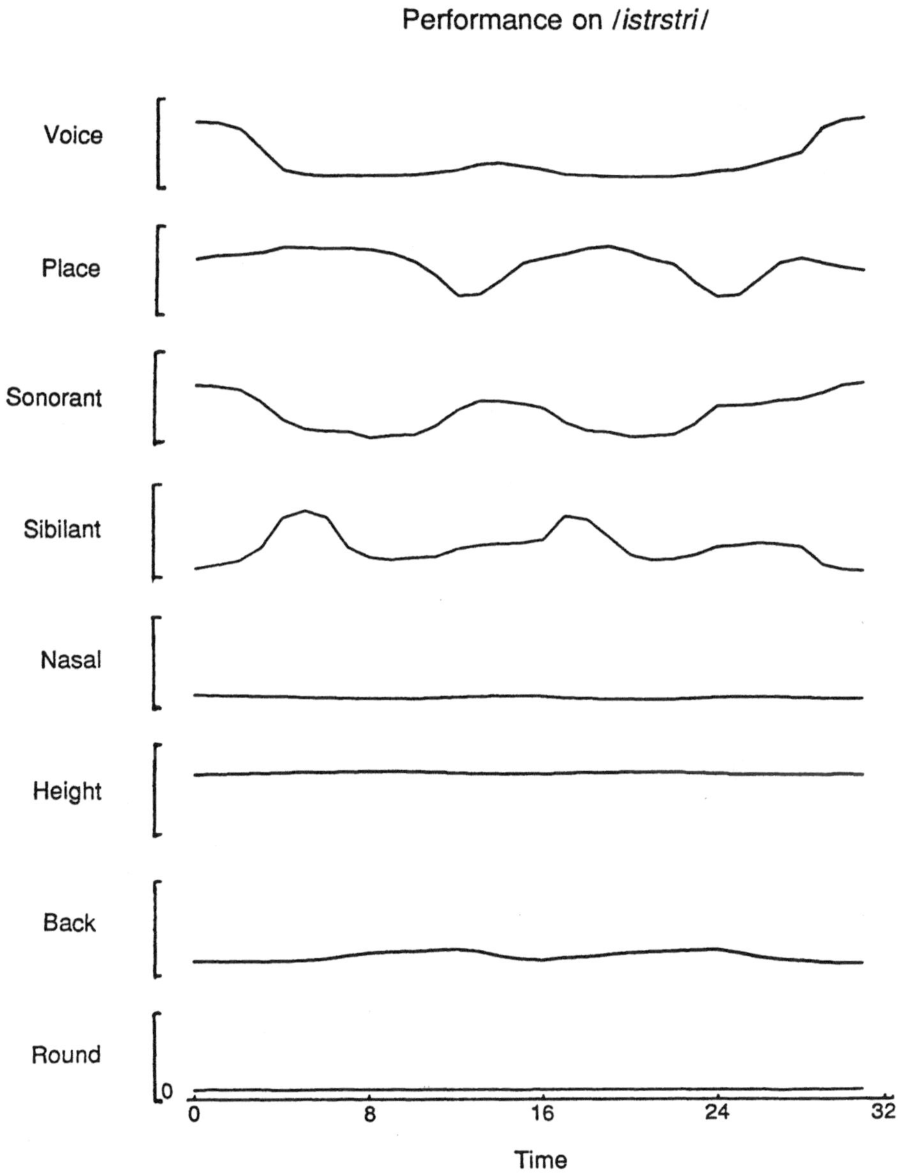

FIGURE 4. Output trajectories for the sequence */istrstri/*.

/str/ was the preceding subsequence. A very similar context occurred after the first */r/*, thus, there was necessarily coarticulation into the first repetition of */str/*. These considerations suggest that, in general, more forward coarticulation should occur over strings that have homogeneous phonemic structure than over strings with heterogeneous phonemes.

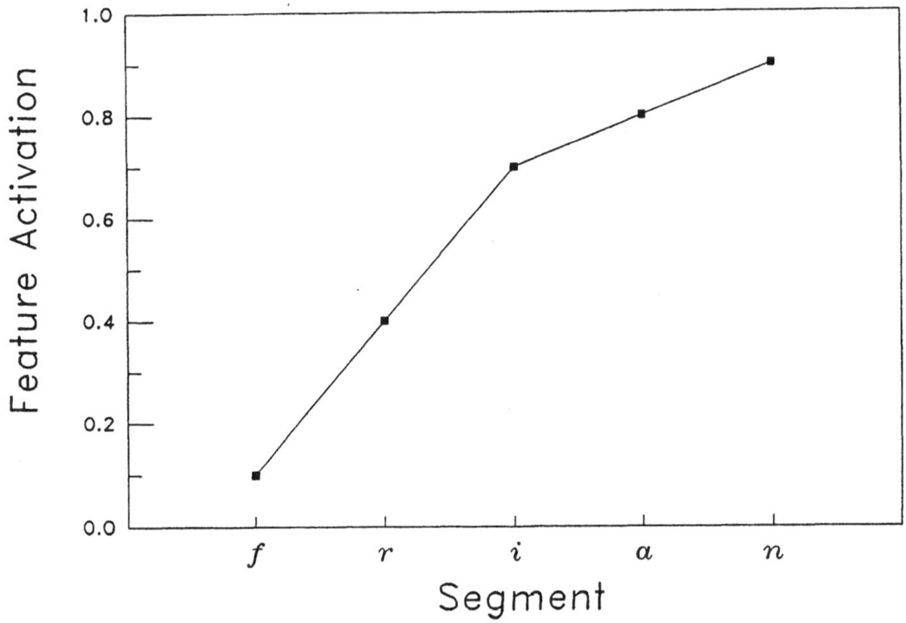

FIGURE 5. Activation of the nasal feature at every fourth time step during performance of the word "freon."

Another interesting aspect of the simulated coarticulation can be seen by considering the voicing feature in **Figure 5**. This feature is unspecified for the phoneme */r/* (the */r/* in French can be voiced or unvoiced depending on the context; compare "rouge" and "lettre"), but is specified as a 0.1 for the directly adjacent features */t/* and */s/*. Nevertheless, the first */r/* receives a small amount of voicing, which comes from the positive value of voicing for the nearby, but not adjacent, phonemes */i/* and */y/*. This result emphasizes the underlying mechanism of activation of the output units: Units are activated to the extent that the current state is similar to the state in which they were learned. This means that units with don't-care conditions take on values that are, in general, a compromise involving the values of several nearby phonemes, and not merely the nearest specified value. Typically, however, the nearest phoneme has the most influence.

These considerations suggest that the amount of forward coarticulation should depend not only on the preceding phonemes, but also on the following phoneme. If the phoneme following /y/ is unrounded, for example, then rounding of the /y/ should be anticipated less than when the following phoneme is rounded or unspecified on the rounding feature (as in the example of "structure"). This prediction was borne out in simulation. The French pseudowords "virtuo," "virtui," and "virtud," in which the rounded phoneme /y/ is followed by the rounded phoneme /o/, the unrounded phoneme /i/, or the "don't-care" phoneme /d/, were taught to the network. The results are shown in **Figure 6** in terms of the activation of the rounding feature at successive points in time. The figure shows that forward coarticulation in the network clearly depends on the following context.

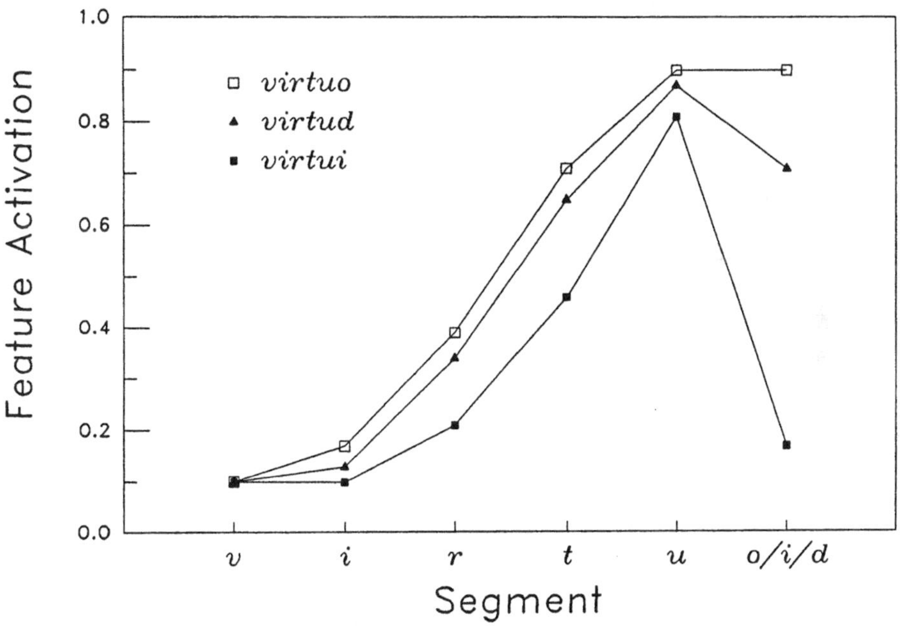

FIGURE 6. Activation of the rounding feature at every fourth time step during performance of three French pseudowords.

Discussion

In their review on coarticulation, Kent and Minifie (1977) distinguish between submovements in an articulatory sequence that have "immediate successional impact," that is, those that "must follow one another in a prescribed sequence," and submovements without immediate successional impact, that are "accommodated within the sequential pattern defined by the locally critical articulatory transitions." The model presented in this section obeys this distinction, where constraints specify the locally critical articulatory transitions. The model also provides a mechanism for the process of "accommodation," by which features without immediate successional impact can be integrated into the articulatory program.

It is worthwhile to compare the current simulation to a feature-spreading model such as that proposed by Henke (1966). Henke's model is essentially a buffer model, in which positions in the buffer are loaded with the phonemes to be produced. Phonemes are lists of trinary features, each of which can have the value +, -, or 0. When a buffer position is to be executed, features having value 0 are filled in by an operator that serially inspects "future" buffer positions until a plus or a minus is found. Once all features are filled in, the allophonic variation thus created can be executed. Although this model is similar to the current simulations in the sense that both rely on context-independent representations of phonemes that specify dimensions along which the phonemes can be altered, there are important differences.

From a conceptual point of view, the underlying mechanisms that determine output values are quite different and have different empirical consequences. In the current approach, parallel performance arises automatically, without the need for a special process to program in the parallelism. This occurs because the current state is similar to the state in which nearby phonemes were learned, and similar states tend to produce similar activations of the output units. There is therefore no implication that features can spread indefinitely in time, which is true of a strict interpretation of Henke's model (Gelfer, Harris, & Hilt, 1981). Rather, the spread of a feature in time diminishes due to the dropoff in similarity of the state. For related reasons, there is no implication that feature vectors change discretely in time. As the state evolves continuously in time, the components of the output vector also evolve continuously in time, with no necessary coherence between anticipated or perseverated features and adjacent segments (cf. Fowler, 1980). Indeed, there is really no notion of a segment in the output of the network. Also, whereas Henke's model is an assimilatory model of coarticulation, the current model is best thought of as a model of parallelism in speech production. As shown in the simulations, the parallel model predicts nonadjacent interactions: For example, the amount of forward coarticulation of a feature in a phoneme depends on what follows the phoneme. Although an assimilatory model could be constructed to mimic such behavior, it would seem better accounted for within the parallel approach. However, I know of no empirical evidence relevant to deciding this issue. (The following experiment would constitute a critical test. Consider forward vowel-to-vowel coarticulation, such as the raising of */a/* when it is followed by */i/* in the sequence */papi/* (Manuel & Krakow, 1984). When this sequence is followed by */e/* (e.g., in the sequence */papipe/*), the */e/* acts to lower the */i/*. The question is what happens to the */a/* in */papi/* vs. */papipe/*. Under an assimilation hypothesis, the */a/* should be lower in the latter case because the source for its raising (the */i/*) has been lowered. Under the hypothesis of parallelism, on the other hand, the */a/* should be at least as high in */papipe/* as in */papi/*, because both the */e/* and the */i/* act to raise */a/*.) Finally, it should be noted that in the current model, utterances are not explicitly represented (i.e., in a buffer) before being produced. Rather, the process is truly dynamic; utterances are implicit in the weights of the network, and become explicit only as the network evolves in time.

The simulations presented above relied only on the simplest constraints on the output units. However, much could be gained by considering more complex constraints such as inequality constraints, range constraints, or constraints between units. For example, certain low-level effects of context, such as the dentalization of the */d/* in "width," are often treated as phonological in origin, rather than resulting from coarticulation. This is presumably because the place of articulation jumps discretely to dental, rather than moving somewhere between alveolar and dental. In the current model, however, the */d/* could be represented as having a range constraint on the place of constriction feature (i.e., a constraint that the place be between a pair of values). The actual value chosen for the place feature will be dependent on the neighboring context, and a context such as the dental fricative could well drive this value against a boundary of the range constraint. Similarly, constraints between units can determine which articulatory configuration is chosen out of several possibilities. For example, if the sum of the activations of three output units must be a particular value, then it is possible to trade off the activations among the units if particular units are further constrained by the neighboring context. Finally, the general case of learned, nonlinear constraints allows modeling of the role of the nonlinear mapping from articulation to acoustics in determining the way in which articulatory components trade off (Jordan, 1990).

There are two possible versions of a parallel model of coarticulation. The first assumes that parallelism is feature-specific, that is, that particular features of a phoneme can be anticipated or perseverated. This approach is consistent with the distinction of Kent and Minifie (1977) discussed above, and is the approach that I have emphasized. However, it is also possible to assume that all of the components of a phoneme must be activated together. This is the approach favored by Fowler (1980), who claims that coarticulation results from the coproduction of "canonical forms." In the current framework, such phoneme-specific parallelism occurs when phonemes specify constraints on nonoverlapping sets of output units. In the limiting case, each phoneme can constrain a unique output unit, in which case the partial activations of output units lead to the partial production of entire phonemes rather than specific features. It is still possible to represent phonemes by features, but this must be done at a lower level in the system, below the level at which parallelism arises.

However, it would appear that feature-specific parallelism is necessary. For example, in the production of a sequence of vowels followed by an */n/*, it would seem important that only the velar movement associated with the nasal be anticipated, and not the alveolar tongue position. There is some evidence for this in the data of Kent, Carney, and Severeid (1974). In recordings of the articulatory movements during the utterance "contract," they found that the movement towards the alveolar tongue position for the */n/* began 120 milliseconds *after* the onset of velar lowering for the */n/*. This suggests that the features of the */n/* are not being controlled synchronously.

To summarize, the current proposal is that coarticulation results from the similarity structure of the state at nearby points in time. The dropoff in similarity of the state defines the zone in which the features of a phoneme can possibly be present in the output. Within this zone, the pattern of coarticulation that is obtained depends on the constraints that are imposed by the features corresponding to nearby phonemes.

Conclusions

The current theory provides an alternative to the traditional motor-program approach to the serial-order problem. The traditional approach, based on the von Neumann conception of a stored program, assumes that motor actions are instructions that are assembled into a structure that is then scanned by a sequential processor. The parallelism and interactiveness of real behavior prove burdensome to such an approach, and typically, extra mechanisms must be invoked. In the current approach, on the other hand, parallelism is a primitive, arising directly from the continuity of the mappings defining the system. Strictly sequential performance is simply the limiting, most highly constrained case.

This chapter has concentrated on only certain aspects of the serial-order problem, namely those involving learning and coarticulation. Jordan (1986) discusses other issues, including rate, errors, hierarchies, and dual-task parallelism.

The concept of state is central to the current theory. Time is represented implicitly by the configuration of the state vector, and it is the assumption of a continuously varying state that relates nearby moments in time and provides a natural way for behavior to be parallel and interactive locally while still broadly sequential. The similarity structure of the underlying state provides a theoretical point of convergence for many kinds of behavioral data. The pattern of coarticulation depends on this similarity structure, errors are more likely when discriminations must be made between similar states, dual-task interference is a function of similarity, and learning is faster when similar actions are associated to similar states. Thus, if the theory is to prove useful, elucidation of the similarity structure of the states underlying sequential behavior becomes an overriding theoretical and empirical concern.

ACKNOWLEDGMENTS

This chapter is an abridged version of Jordan (1986). My sincere thanks to Donald Norman and David Rumelhart for their support of many years. I also wish to acknowledge the help of Eileen Conway, Kathy Farrelly, Jonathan Grudin, Bernhard Keller, Michael Mozer, David Navon, and Stanley Parkinson. This research was supported by Contract N00014-85-K-0450, NR 667-548 with the Personnel and Training Research Programs of the Office of Naval Research.

Neural-Networks Models of Cognition
J.W. Donahoe and V. Packard Dorsel (Editors)

CHAPTER 26

CONNECTIONIST MODELS OF ARBITRARILY APPLICABLE RELATIONAL RESPONDING: A POSSIBLE ROLE FOR THE HIPPOCAMPAL SYSTEM

Dermot Barnes and Peter J. Hampson
Department of Applied Psychology
University College Cork
Cork, Ireland

ABSTRACT

Connectionist models have been developed that can simulate some of the extended histories of reinforcement that are thought to be largely responsible for complex human functioning. In particular, a number of models have successfully simulated equivalence-class formation and other forms of arbitrarily applicable relational responding. A recent example of one such model is described. We then consider the major weaknesses of the model, and focus on the idea that future models should be neurophysiologically constrained. Finally, we examine the hypothesis that hippocampal functioning may be critical to the emergence of arbitrarily applicable relational responding.

Introduction

> *The physiologist of the future will tell us all that can be known about what is happening inside the behaving organism. His account will be an important advance over a behavioral analysis, because the latter is necessarily "historical"—that is to say, it is confined to functional relations showing temporal gaps. Something is done today which affects the behavior of an organism tomorrow. No matter how clearly that fact can be established, a step is missing, and we must wait for the physiologist to supply it. He will be able to show how an organism is changed when exposed to contingencies of reinforcement and why the changed organism then behaves in a different way, possibly at a much later date. What he discovers cannot invalidate the laws of a science of behavior, but it will make the picture of human action more nearly complete. (Skinner, 1974, p. 215).*

Skinner clearly argued that the study of neurophysiology will play an important role in developing a complete science of behavior. We presume few behavior analysts would disagree with Skinner on this point. Nevertheless, we question the idea that physiologists will one day provide behavior analysis with the necessary data to fill those temporal gaps between stimuli and responses. In other words, it would be a mistake to assume that an imaginary neurophysiologist will one day knock on the behavior-analytic door and say: "I hear you've been waiting some time for the necessary data to fill that temporal gap between stimulus and response—well here it is, and feel free to call me if you have any further questions!" Unfortunately for behavior analysis, the scientific community simply doesn't work like that. The questions and issues that interest neurophysiological researchers are unlikely to coincide or overlap at a specific point in time, at least in any meaningful way, with the research interests and questions of the behavior-analytic community. Thus, if behavior analysis wants to fill those temporal gaps with neurophysiological data, then its researchers will have to become more aware, and possibly involved, in the necessary research activity—a daunting prospect for many if not most behavioral researchers, who often do not have the appropriate scientific training in the biological sciences. Nevertheless, all is not lost. Fortunately, the development of connectionist science has provided the behavior-analytic community with an opportunity to forge those all-important links with those involved in the study of neurophysiology. The current chapter represents a small step in this direction.

Connectionism and Neurophysiological Plausibility

Much of the connectionist research conducted during the 1980s was of the demonstration variety. In effect, connectionist scientists were content to develop models that successfully simulated a certain type of behavior (cf. Bechtel & Abrahamsen, 1991). More recently, however, connectionist science has become more interested in constraining its models with neurophysiological data. In effect, a connectionist model, it is argued, should not only simulate a particular performance, but should also be designed and operate in accordance with what is known about neurophysiological structures and processes (see Donahoe, Burgos, & Palmer, 1993).

Connectionist scientists with a background in neurophysiology are clearly best placed to design neurophysiologically plausible connectionist networks. Presumably, however, their knowledge of the processes of nonhuman and human learning is somewhat limited. It is feasible, therefore, that a connectionist researcher with a background in neurophysiology could develop a neurophysiologically accurate network, but fail to simulate the appropriate performance, not because the network is faulty, but because the researcher has failed to provide the necessary conditions for learning. This is where the behavior analyst can be of use, and in the process he or she can learn something about the neurophysiological processes that fill that temporal gap between stimuli and responses. Clearly, if this is to happen, however, it is important for behavior analysis to make a serious contribution to the development of neurophysiologically plausible connectionist simulations of nonhuman and human behavior.

Simulating Complex Human Behavior

A first step in the process of contributing towards the development of neurophysiologically plausible models of behavior may be to put aside, initially, the requirements imposed by neurophysiological constraints and simply develop networks that successfully simulate the data from behavior-analytic studies. Once a number of working but non-neurophysiologically constrained models have been developed, the behavior analyst can then refine and modify the models in accordance with the more neurophysiologically constrained models of the modern connectionist researcher.

In fact, a number of behaviorally oriented researchers have developed connectionist models of nonhuman behavior that have not been explicitly constrained by neurophysiology (see Commons, Grossberg, & Staddon, 1991, for some relevant examples), whereas others have developed very tightly neurophysiologically constrained models of basic nonhuman conditioning effects (e.g., Donahoe et al, 1993). As yet, however, there are very few connectionist simulations of data obtained from behavior-analytic studies of complex human behavior. Although some might argue that we should deal with the more basic nonhuman behaviors before even attempting to simulate human performances, there appears to be no good reason why some researchers cannot work from the "bottom up" (i.e., starting with basic nonhuman behavior) while others work from the "top down" (i.e., starting with complex human behavior). In fact, it could be argued that this "pincer movement" might be a more effective research strategy. For example, top-down research may encourage and help bottom-up researchers to extend their basic models to complex human behavior, and bottom-up research may encourage and help top-down researchers to develop models that produce complex human behavior by building upon the simpler units of nonhuman behavior (see Thompson & Lubinski, 1986). Of course, as suggested previously, behavioral researchers (both bottom-up and top-down) should be aiming towards neurophysiologically plausible models, be they of simple nonhuman behavior or complex human performances. It is with this overall research strategy in mind that we turn our attention to complex human behavior.

Stimulus Equivalence and Relational Responding

Behavior analysis has always been interested in human behavior, but until relatively recently most human research in behavior analysis was of the applied variety. In recent years, however, basic research into human behavior has flourished. This is due, in part, to the behavioral effect known as stimulus equivalence. This phenomenon is normally produced in the following way. Subjects are first trained in a series of related conditional discriminations, using a matching-to-sample format and arbitrary stimuli such as nonsense syllables or abstract shapes (in the interests of clarity, authors reporting on equivalence studies normally label the syllables or shapes with alphanumerics, such as A1, B1, C1, and so forth). For example, a subject might be exposed to a training procedure in which reinforcers (e.g., points exchangeable for money) are delivered when the subject chooses the comparison stimuli B1 and C1 in the presence of the sample stimulus A1, and chooses the comparisons B2 and C2 in the presence of the sample A2 (choosing the "incorrect" comparison, such as B2 in the presence of A1, is not reinforced). If the subject then selects, without further training, C1 in the presence of B1 (and vice versa), and C2 in the presence of B2 (and vice versa), the stimuli are said to participate in two equivalence relations (i.e., A1-B1-C1 and A2-B2-C2). The concept of conditional discrimination, as traditionally defined, does not predict the emergence of this untaught performance. For example, neither the B1 nor the C1 stimulus has a history of differential reinforcement as a conditional discriminative stimulus with regard to the other, and therefore neither stimulus should reliably control selection of the other. Remember also that the stimuli used in equivalence studies are arbitrary (i.e., they are not related to each other along any consistent physical dimension, such as color, shape, or size) and thus primary stimulus generalization is also unable to account for equivalence responding.

One explanation for the equivalence effect is based on the idea that equivalence is a form of generalized, or overarching class, of operant behavior (see Barnes, 1994; Barnes & Holmes, 1991; Hayes, 1991; Hayes, 1992; Hayes & Hayes, 1989; Hayes & Hayes, 1992). According to this account, known as relational frame theory, equivalence and other forms of derived relational responding emerge because the human verbal community *explicitly* reinforces equivalence responding in young children. For example, during early language development a child might be exposed to the following three types of parent-child interaction; (i) reinforcers in the form of praise and/or cuddling may be delivered for uttering two different names (e.g., "seat" and "chair") for an object (i.e., A1-B1 and A1-C1), (ii) reinforcers may be delivered for correctly identifying an appropriate object (an actual chair/seat) when someone else provides either of the two names (B1-A1 and C1-A1), and (iii) reinforcers may be given for uttering "chair" when asked for another word that means "seat," and vice versa (i.e., B1-C1 and C1-B1). In effect, this sequence of verbal interactions, and others like it, provides a history of explicit reinforcement for responding in accordance with the relation of equivalence. Consequently, if a child with this history of reinforcement is told on one occasion, while looking at a picture of a motor vehicle, that it is a picture of a "car," and on another occasion that it is a picture of an "automobile," the child may, *without any further teaching*, derive, in an appropriate context, the equivalence relation between "car" and "automobile" (e.g., the child will utter "automobile" when asked for another word that means "car," or vice versa). According to relational frame theory, a functionally similar behavioral effect is obtained during the typical equivalence experiment. In other words, the context of the equivalence experiment is discriminative for responding that is functionally similar to the responding that was reinforced when the subject was taught that some words "mean the same thing." (See Barnes, 1994, for a detailed discussion of this issue.) In fact, the matching-to-sample format may be particularly likely to produce equivalence because this format is often used in preschool education exercises in which children have to look at a picture and point to the appropriate word or words that "mean" the picture.

It is important to understand that relational frame theory also addresses other types of relational responding that cannot readily be categorized as equivalence (Dymond & Barnes, in press, a; Dymond & Barnes, in press, b; Steele & Hayes, 1991.) Consider, for example, a subject who is taught that A is the opposite of B, and B is the opposite of C, and is then asked what is the relation between A and C. According to the relational frame account, for a subject to provide the correct answer (i.e., A is the same as C) he or she would require an appropriate history of explicit reinforcement for responding in accordance with the relational frame of opposition (e.g., explicit reinforcement would have been provided for responding to cold as the *opposite* to hot, to hot as the *opposite* to chilly, and to cold and chilly as the *same*).

It should be clear by now that one of the most important features of the relational frame account of stimulus equivalence, and other forms of relational responding more generally, is the emphasis it places on extended and sometimes complex histories of explicit reinforcement as one of the main sources of untaught or derived behavior. One problem with this approach, however, is that these histories are often so prolonged and complex that it would be very difficult, and in some cases highly unethical, to synthesize them in the behavioral laboratory, and thereby test directly the relational frame account. Consider, for example, the basic idea (outlined previously) that the verbal community provides a young child with an appropriate reinforcement history for equivalence responding when the community reinforces uttering the names for stimuli, and also reinforces the following of simple instructions containing those names. Clearly, it would be highly unethical to test this form of history effect by confining all verbal interactions with a newborn child to those types of interaction that are supposed to be necessary for equivalence responding, and preventing or suppressing all other forms of interaction that are supposedly *not* causally related to equivalence.

Of course, researchers interested in examining the role of extended reinforcement histories on equivalence responding do not have to confine themselves to newborn children as subjects. In one recent study, for example, a California sea lion was provided with a history of explicit reinforcement for responding in accordance with twelve different sets of equivalence relations, and then demonstrated untaught or derived equivalence responding across a further eighteen sets of stimuli (Schusterman & Kastak, 1993). This finding clearly supports relational frame theory, because equivalence responding only emerged after the subject had been provided with a history of explicit reinforcement for equivalence responding (i.e., a history that is functionally similar to the "naming history" that is presumed to be largely responsible for equivalence in humans).

Although it may be possible to demonstrate relational frame history effects on equivalence responding using nonhuman subjects, it is important to remember that relational frame theory does not confine itself to explaining only equivalence; it also represents an attempt to explain far more complex forms of human behavior (e.g., complex instructional control, analogical reasoning, clinical disorders). The important point here is that these complex human behaviors are presumed to be produced by verbal histories that are far more extended and complex than the "basic" naming history that is thought to be largely responsible for equivalence. The problem, therefore, for the behavioral researcher is that any attempt to simulate these histories using nonhuman subjects will become increasingly difficult as the histories grow in length and complexity. For example, a child's ability to respond appropriately to a simple instruction that he or she has not been explicitly taught to follow (e.g., "When the bell rings, get the cake from the oven.") normally requires, from a relational frame perspective, a behavioral history that spans a number of years (see Hayes & Hayes, 1989, pp. 179-180). The difficulty involved in synthesizing such a history should not, of course, discourage researchers from attempting to simulate these histories with nonhuman subjects. Nevertheless, the relatively recent developments in connectionist science provide the behavior analyst with another alternative to nonhuman research when he or she wishes to examine complex history effects that would be difficult or unethical to simulate accurately using human subjects. In fact, it will be shown here that the highly controllable and extremely precise methodology provided by the connectionist approach may allow the behavior-analytic researcher to examine the types of prolonged and often complex histories of reinforcement that, from a relational frame point of view, characterize many aspects of human behavior. The next section focuses on this very issue.

A Connectionist Simulation of Complex Human Behavior

The following model was designed to simulate complex human behaviors, as viewed from the relational frame perspective. Having described this connectionist simulation, we will then examine some of the important issues arising from this work, including our key concern that future models of complex human behavior should strive for neurophysiological plausibility.

Responding in accordance with sameness, difference, and opposition

As outlined previously, relational frame theory represents an attempt to explain a great deal more than stimulus equivalence. For example, the theory explicitly incorporates forms of relational responding that cannot be categorized as equivalence. In fact, the first empirical investigation of relational frame theory (Steele & Hayes, 1991) aimed to demonstrate that teenage human subjects could respond in accordance with the derived relations of coordination (or sameness), opposition, and difference.

In Steele and Hayes' study, subjects were first provided with nonarbitrary relational pretraining that was used to establish three contextual stimuli (abstract shapes) as functionally equivalent to the words "same," "different," and "opposite" (**Figure 1**, top panel). For instance, subjects were trained to match a long line with a long line when presented with the SAME contextual stimulus, a long line with a short line when presented with the OPPOSITE contextual stimulus, and a long line with a square when presented with the DIFFERENT contextual stimulus. After the subjects had successfully completed the pretraining, they were trained in six matching-to-sample tasks (the stimuli were abstract shapes) in the presence of the three contextual stimuli (**Figure 1**, second and third panels from top). When the SAME contextual stimulus was presented, matching B1 and C1 to A1 was reinforced. When the OPPOSITE stimulus was presented, matching B3 and C3 to A1 was reinforced. When the DIFFERENT stimulus was presented, matching B2 and C2 to A1 was reinforced.

After subjects had successfully completed the foregoing training, they were repeatedly exposed to 15 test tasks in the absence of any feedback. The first four of these tasks tested for the mutually entailed relations of sameness and opposition (**Figure 1**, lower four panels). For example, subjects were trained on Task 2 to match B3 to A1 in the presence of the OPPOSITE contextual stimulus; if, on Task 8, they then consistently matched A1 to B3 in the presence of the OPPOSITE stimulus, this demonstrated responding in accordance with the mutually entailed relation of opposition (i.e., train OPPOSITE/A1-B3 and obtain without further training OPPOSITE/B3-A1). The remaining 11 tasks, 11 to 21 (**Figure 1**, lower three panels) examined combinatorially entailed relational responding. Tasks 17 to 21 involved the more complex of these derived relations and will, therefore, be examined individually. On Task 17, subjects were presented with the SAME contextual stimulus, C1 as the sample, two familiar comparisons, B1 and B2, and a novel comparison, N3 (i.e., S [C1] B1-B2-N3). As predicted, subjects consistently chose B1; this demonstrated responding in accordance with the combinatorially entailed relation of sameness (i.e., train SAME/A1-B1 and SAME/A1-C1, and obtain without further training SAME/B1-C1). The subjects' choice of B2 on Task 18 (i.e., D [C1] B1-*B2*-N3) demonstrated responding in accordance with the relations of sameness and difference (i.e., train SAME/A1-C1 and DIFFERENT/A1-B2 and obtain without further training DIFFERENT/C1-B2). Choosing N3 on Task 19 (i.e., O [C1] B1-B2-*N3*) was based on the relations of sameness, difference, and opposition. In effect, neither B1 nor B2 could be the correct choice on Task 19 because subjects had been trained that B1 and C1 are the same as A1, and B2 is different from A1, but not the opposite to A1, and thus not the opposite to C1. By choosing N3, the only remaining option, subjects thereby demonstrated three different types of relational control, none of which were directly trained. The subjects' performance on Task 19 indicated that the forced choice of N3 should cause it to enter into the network of relations. In effect, the forced choice of N3 in the presence of OPPOSITE and C1 might be sufficient to establish N3 as the opposite to C1 and therefore the same as C3. The responses on Task 20 (O [N3] *C1*-C2-C3) and on Task 21 (S [N3] C1-C2-*C3*) to C1 and C3 respectively, supported this prediction.

The connectionist simulation

Our first connectionist model, called RELNET, was designed to simulate the pattern of arbitrary relational responding obtained by Steele and Hayes, but with one important difference. Steele and Hayes used normally developing adolescent subjects, who had already been exposed to the verbal community for many years and had therefore (it was assumed) been provided with an extended history of explicit reinforcement for responding in accordance with the relations of sameness, difference, and opposition. A key prediction of relational frame theory is that derived responding in accordance with arbitrary stimulus relations should only emerge in subjects who have received explicit training to respond in accordance with those relations (i.e., derived responding should not emerge if subjects have received insufficient explicit training in the relevant stimulus relations). A further prediction is that the accuracy on tests of derived responding should increase as a function of prior explicit training in the relevant stimulus relations (see Lipkens, Hayes, & Hayes, 1993). Accordingly, a complete connectionist simulation of the relational responding reported by Steele and Hayes must demonstrate both the training and testing performances produced in the laboratory, and the inferred effects of prior explicit training by the verbal community on the laboratory-produced performances.

To simulate prior explicit training, RELNET was designed so that seven different sets of stimuli, each composed of eight stimuli, could be used (i.e., Set 1: A1, B1, C1, B2, C2, B3, C3, N3/1; Set 2: D1, E1, F1, E2, F2, E3, F3, N3/2, and so on). The seven stimulus sets were used to produce eight levels of explicit training (i.e., no explicit training, exposure to one explicit training set, and so on up to seven explicit training sets). Exposure to Training Set 1 involved training all 21 tasks illustrated in **Figure 1**. Exposure to Training Sets 1 and 2 involved training all 21 tasks shown in **Figure 1**, and also training the same 21 tasks *but using Stimulus Set 2*. Thus, a total of 147 tasks were explicitly trained when subjects were exposed to all seven sets of stimuli (i.e., 21 multiplied by 7).

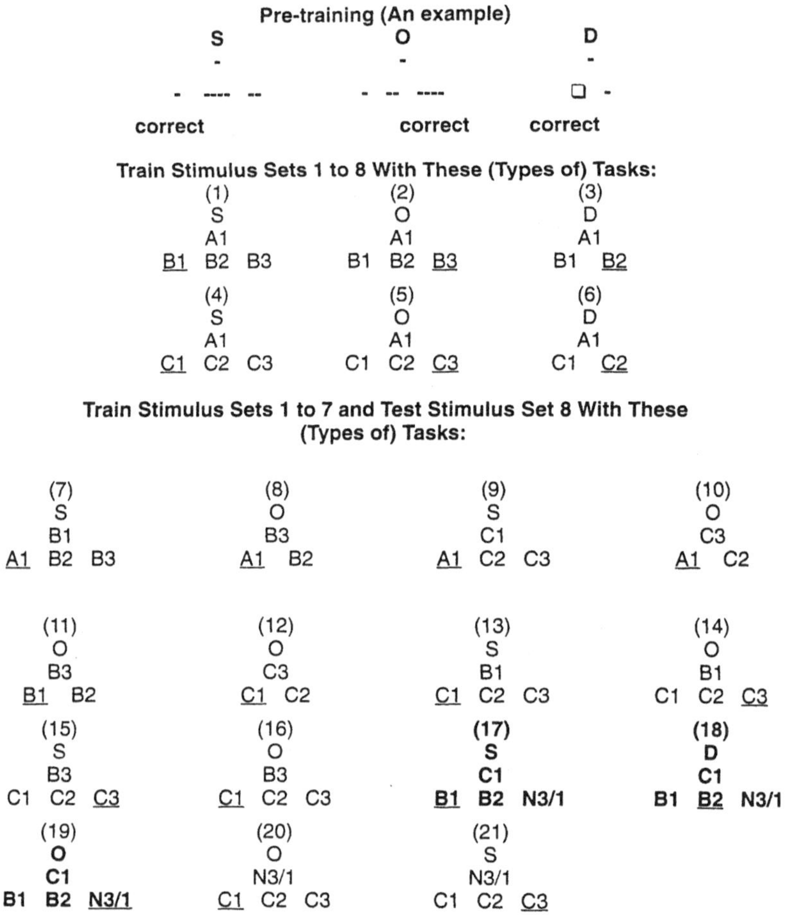

FIGURE 1. Relational frame tasks from Steele and Hayes (1991). Each task presents a pretrained contextual cue, a sample, and two or three comparisons. In the Steele and Hayes study, subjects were trained on the first six tasks and were then tested on the remaining fifteen tasks, the assumption being that adolescent subjects would have sufficient exposure to the contingencies of reinforcement provided by the verbal community to allow them to derive the predicted relationships without explicit training on similar tasks with different stimuli. In contrast, the connectionist network was trained on all twenty-one types of task, using up to seven different stimulus sets (i.e., twenty-one tasks per set), each set containing completely new stimuli (thus simulating the prior exposure to the verbal contingencies experienced by the Steele and Hayes subjects). The network was also trained on the first six tasks only from an eighth stimulus set, and then tested using the remaining fifteen tasks from this eighth set.

In conjunction with this explicit training across 0 to 7 stimulus sets, an eighth and completely novel set of stimuli (i.e., W1, X1, Y1... N3/8) was used to train RELNET on Tasks 1 to 6 only (i.e., limited training), before it was tested on Tasks 7 to 21 (see **Figure 1**, and substitute the A, B ,C, N3/1 stimuli with the W, X, Y, N3/8 stimuli). In effect, training with Stimulus Sets 1 to 7 simulates the preexperimental history of explicit reinforcement for responding in accordance with the relations of sameness, difference, and opposition, whereas the training and testing on Stimulus Set 8 directly simulates the Steele and Hayes experimental procedures with adolescent subjects.

The complete RELNET model consists of an encoder, a central relational responding machine, and a decoder. The three modules were implemented separately in the simulation. The encoder and decoder simply preprocess stimuli for and decode outputs from the relational responding machine, and thus we will focus upon the latter.

The relational responding machine is a three-layered network with 83 inputs, 8 hidden units, and 19 outputs. The input layer and the hidden layer are fully interconnected and the hidden and output layers are connected as shown in **Figure 2**. The stimulus identity element (input units 1-64, labeled A1 to N3/8) represents the stimuli that function as samples and comparisons across the eight stimulus sets. If, for example, we wished to present a task to the network in which the stimuli A1, B1, and B2 were used, input units 1, 2, and 3 would be activated. The sample-marking duplicator (input units 65-80, labeled Z1 to Z8/s) simply copies the activation from each task (as represented in the stimulus identity element), and marks one of the stimuli as a sample from that task. For example, when A1 is activated as a sample with B1 and B2 as comparisons, the Z1, Z1/s, Z2, and Z3 input units are activated. If B1 was the sample, then Z2/s would be activated and Z1/s would be turned off. The sample-marking duplicator mirrors activation in exactly the same way for each individual task across each of the eight stimulus sets (e.g., when W1 is activated as a sample with X1 and X2 as comparisons, then Z1, Z1/s, Z2, and Z3 are activated in the sample-marking duplicator). The element labeled SAME/DIFF/OPP contains the inputs for contextual stimuli. In effect, the three input units 81, 82, and 83 represent SAME, DIFFERENT, and OPPOSITE, respectively (e.g., when the SAME cue is presented on a relational task, unit 81 is activated, and 82 and 83 are turned off).

The outputs are representations of the stimulus set identity (output units 1-8), the output stimulus identity (output units 9-16), and the sample-comparison relations of sameness, difference, and opposition (output units 17-19). The output from the stimulus set identity classifies the chosen stimulus as belonging to one of the 8 stimulus sets (i.e., output unit 1 represents Set 1, output unit 2 represents Set 2, and so on). The output stimulus identity classifies the chosen stimulus within each set. For example, depending on the output from the stimulus set identity, output 9 represents the first stimulus within each stimulus set (i.e., A1, D1, and so forth to W1), output unit 10 represents the second stimulus in each set (i.e., B1, E1, and so forth to X1), and so on (i.e., output unit 16 represents the eighth stimulus in each set; N3/1, N3/2, and so forth to N3/8). Thus, if output units 1 and 9 are activated this identifies stimulus A1 as a chosen comparison. If, however, output units 2 and 10 are activated this identifies stimulus E1 as the chosen comparison. Finally, the three units 17, 18, and 19 represent the three relations of sameness, different, and opposite, respectively (e.g., if unit 17 is activated, but 18 and 19 are not, sameness is the controlling relation).

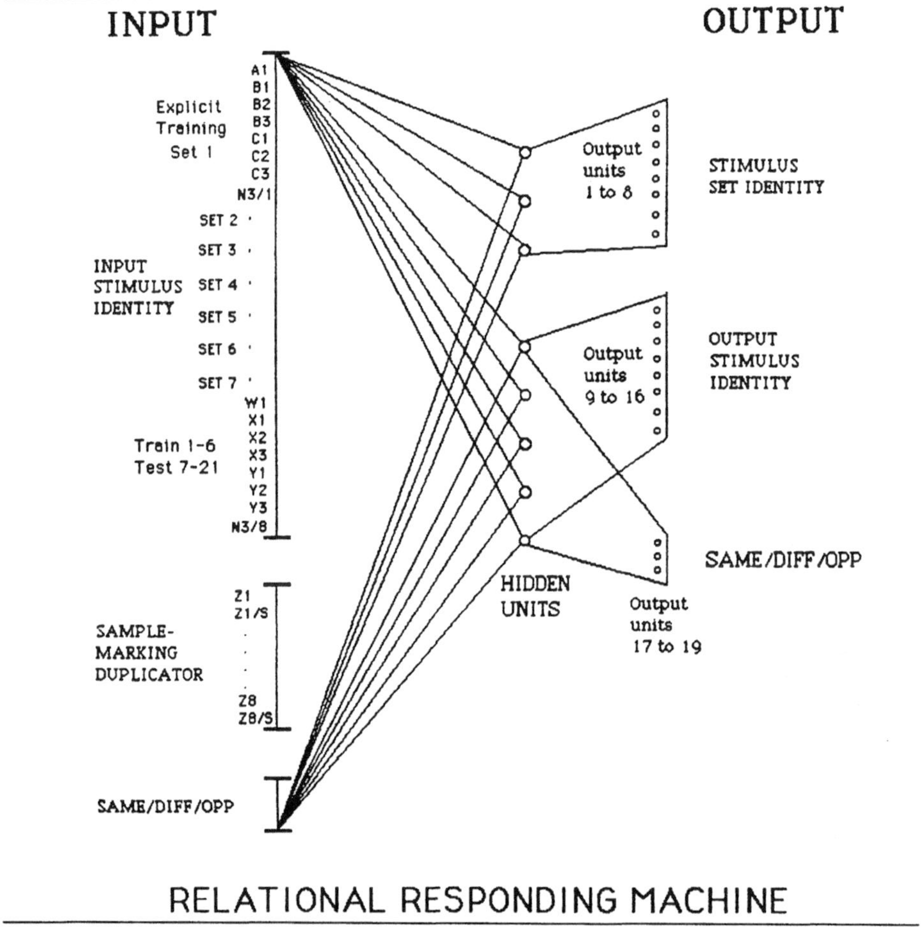

FIGURE 2. A schematic representation of the relational responding machine that shows how the training and testing was implemented.

RELNET was trained using the standard backward-error propagation algorithm (Rumelhart, Hinton, & Williams, 1986), and the output format described above, to output the correct comparison, and the controlling relation (sameness, difference, or opposition) on each task. There were 8 levels of explicit training, involving either no exposure to any of the 7 stimulus sets, or exposure to from 1 to 7 sets of stimuli. Limited training was also given at each level of explicit training on Tasks 1 to 6 using Stimulus Set 8. The same randomly generated starting weights were employed across all 8 levels of training to simulate increasing amounts of exposure to the verbal community. This entire procedure was repeated 10 times using different randomly generated starting weights, yielding a total of 80 runs. All training was conducted to the same error criterion (ecrit < .05 difference between the total sum of squares of the actual output and the target output; see McClelland & Rumelhart, 1988, pp. 140-141). Testing RELNET involved presenting the 15 untrained tasks from Set 8 to the model and recording any differences between the obtained and predicted outputs.

The effects of explicit training across Stimulus Sets 0 to 7, together with limited training on Tasks 1 to 6 of Stimulus Set 8, on the overall performances of the relational responding machine on the untrained Tasks 7-21 from Stimulus Set 8, are shown in **Figure 3**. In **Figure 3** (upper graph), performance is represented as the total error sums of squares (TSS) calculated across the 15 test tasks, for Set 8, and averaged across 10 runs. **Figure 3** (lower graph) shows performance in discrete error scores (DES). A DES was defined as (i) an output unit that failed to acquire at least 50% of its predicted activation (i.e., it failed to identify the correct comparison, relation, or both), or (ii) an unpredicted response by an output unit of greater than 50% (i.e., it identified either a sample [always an incorrect response], an incorrect comparison, an incorrect relation, or some combination thereof). **Figure 3** (lower graph) displays DESs calculated across the 15 test tasks averaged across the 10 runs. Analysis of the TSS and DES scores produced similar results. Training and testing on Stimulus Set 8, after explicit training on Set 1, vastly improved the performance of the relational responding machine compared with no explicit training on Set 1. We also predicted that performance should depend on the amount (in addition to the mere presence) of previous explicit training. To examine this, repeated-measures analyses of variance were conducted for explicit training with 1 to 7 stimulus sets for TSS and DES data. The effects of amount of training were highly significant; error scores were an inverse function of the amount of explicit training, $F(6, 54) = 14.50$, $p < .0001$ for TSS and $F(6, 54) = 5.66$, $p < .0001$ for DES. Posttests revealed that for both measures, increasing the amount of training from 1 set to 3 sets reliably decreased error scores, whereas training with additional sets did not yield significant increases in performance, though the trend was in the predicted direction,

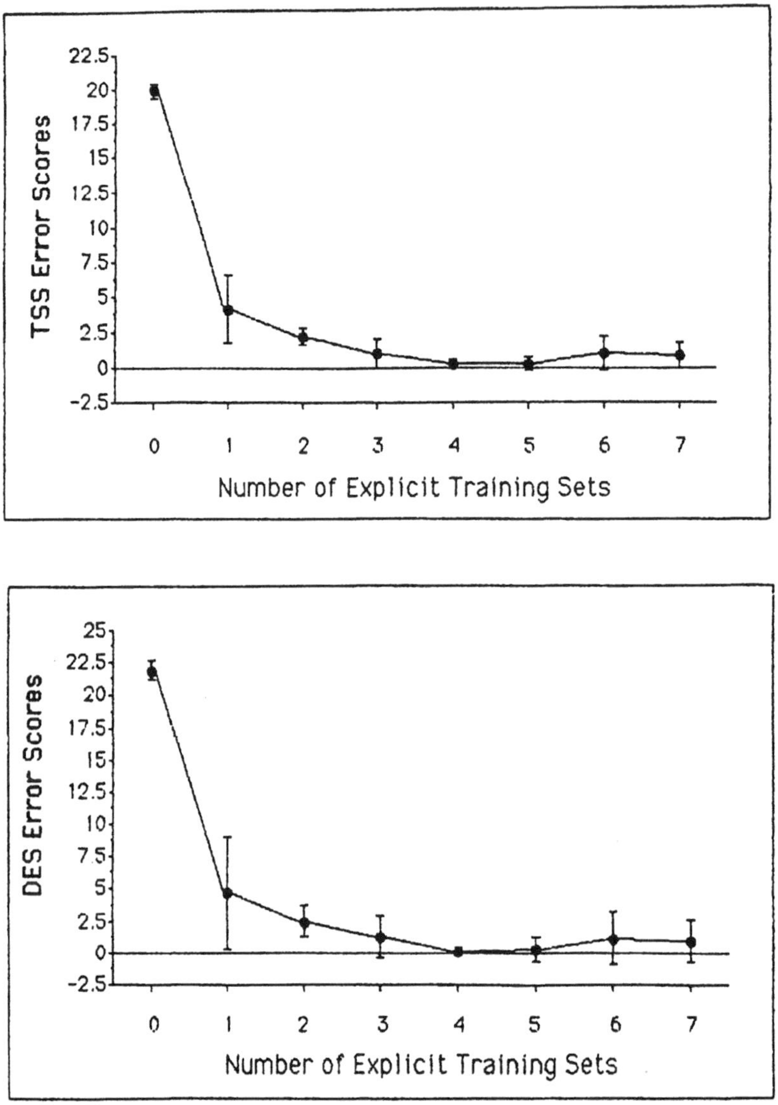

FIGURE 3. Upper graph: Mean total sum of squares error scores (TSS) produced by the trained relational responding machine, summed across 15 test tasks from Stimulus Set 8, as a function of the number of explicit training sets. Lower graph: Mean discrete error scores (DES) produced by the trained relational responding machine, summed across fifteen test tasks from Stimulus Set 8, as a function of the number of explicit training sets.

with some suggestion, though nonsignificant, that accuracy reached its maximum at four training sets.

In summary, these data show that it is possible to construct a network that will perform well on tasks that require arbitrarily applicable relational responding. In effect, the network responded in accordance with the relations of sameness, difference, and opposition on 15 untrained tasks following training on only 6 tasks (i.e., Stimulus Set 8). Furthermore, it was shown that this performance required that explicit training be provided on at least three of seven stimulus sets, each of which employed the same 21 types of task as Stimulus Set 8. Overall, the accuracy of the network's performance on the eighth stimulus set depended on both the presence and amount of explicit training on Sets 1 to 7. Explicit training, therefore, is a necessary precursor if the network is to respond in accordance with derived sameness, difference, and opposition relations. Finally, the fact that some explicit training is essential is shown by the finding that limited training on Set 8 alone reduced noise in the network, but did not lead to an accurate performance.

The foregoing demonstration model clearly shows that it is possible to simulate adult human performances obtained in the behavioral laboratory using a relatively simple connectionist network. We have also simulated other examples of derived relational responding with modified versions of RELNET. These simulations have been reported elsewhere (see Barnes & Hampson, 1992; Barnes & Hampson, 1993; Cullinan, Barnes, Hampson, & Lyddy, 1994), and will not be addressed here because they did not involve any substantial changes to the basic design concept of RELNET, or to the procedures used to train the network. These simulations of human performance observed in the behavioral laboratory represent an important step for behavior analysis, and perhaps, in the long run, for connectionist science. Nevertheless, it is only one step in the right direction. There is a major problem (with many sub-problems) that must be resolved before connectionism can hope to provide a thoroughly adequate domain for modeling human behavior. In the remaining half of this chapter we will examine this problem (including many of the inherent sub-problems), and in so doing help to highlight the more important directions for future research in this area.

Problems with the Current Models

The major problem with our current family of models is that they were designed as "in-principle" demonstrations to show that connectionist techniques are applicable to issues of complex learning. They were not intended to represent neurophysiologically plausible simulations of the phenomena in question, nor were they intended as complete accounts. When we began our project, issues of neural plausibility were less pressing for connectionist researchers than they are today. It was important, then, to show that connectionism was not ruled out from the start as a way of modeling relational frame effects, given the doubts expressed by Fodor and Pylyshyn (1988) about its general relevance for complex combinatorial phenomena. Today, however, simply demonstrating that connectionism can simulate relational framing behaviors is no longer an appropriate research strategy. In the intervening period, approaches have emerged in both cognitive and behavioral psychology that emphasize the neurological variables involved in behavior. Cognitive neuroscience is now attempting to explain cognitive phenomena in terms of data and theories drawn from the brain sciences as well as data derived solely from psychological experimentation (e.g., Kosslyn, 1994; Kosslyn & Koening, 1992; Posner & Raichle, 1994). Its models, too, are no longer as underconstrained as they were previously. Nowadays, ideally, cognitive models should do justice to what is known about the structure and function of relevant brain systems as well as produce a pattern of responses to inputs analogous to that of a human subject in a psychological experiment. In the behavioral camp, biobehaviorism now seeks to supplement standard descriptions of behavioral interactions, couched in the language of the experimental analysis of behavior, with accounts of the internal neurophysiological processes thought to support such interactions. Connectionist modeling is now uniquely poised to pull together these two research strands, particularly where phenomena are modeled which are of interest to both cognitive and behavioral workers, such as the interaction between old and new learning (e.g., exactly how, at the neurological level, is a generalized, overarching class of operant behavior [or relational frame] established, and exactly how is new learning incorporated into this operant class?).

The basic limitation of our earlier approach—its lack of neural plausibility— produces several specific problems that we shall now outline. Several of these problems are related and their solutions intertwined, but in the interests of clarity we shall first consider them in isolation before looking at ways to resolve them.

Problem 1: The simulations were too global, and hence the issue of task decomposition into simpler subtasks was not sufficiently addressed, nor was any attempt made to model, in more detail, possible subcomponents of the tasks

Our model was designed to simulate all of the important behavior obtained in the Steele and Hayes relational frame study, and the explicit history of reinforcement (from the verbal community) that supposedly allowed the teenage subjects to demonstrate relational framing. In effect, we aimed to mimic (i) the acquisition of the relevant overarching operant classes (i.e., relational frames) across a number of different stimulus sets or stimulus domains (referred to as explicit training or old learning), (ii) training on the to-be-tested set of stimuli (referred to as limited training or new learning), and (iii) the demonstration of derived relational responding (referred to as successfully integrating the new learning with the old learning). Upon closer inspection, however, the relational frame procedures involve at least the following subtasks:

(i) The learning of several input-output pairs in succession without interference. In a typical stimulus-equivalence or relational frame experiment, participants must learn the appropriate responses to a series of related conditional discriminations or stimulus configurations. We refer to this as *learning multiple input-output pairs*.

(ii) The learning of the entailed relations across or between trials. An important, if obvious, point about a relational frame or equivalence experiment is that stimuli recur across a series of trials in different configurations. For example, suppose a participant learns that VEK is the same as JUF on Trial 1, but different from YUG on Trial 2, then the relationships between Trials 1 and 2 have implications for the relationship between JUF and YUG which might be tested at some later time. For correct responding to occur, this relationship must either be learned at the time that Trials 1 and 2 are mastered, or derived at the time of testing. Either way, what we call *inter-trial learning* must take place.

(iii) The retention of learned relationships. Learned relationships from a given domain must be retained (have a durable effect) since derived responding at some future time is possible. The effects of learning must endure (though for how long is, of course, an empirical question). We refer to this as *learning retention*.

(iv) The use of previous learning. Previous learning, from other domains, is used to interpret current learning. For example, to exploit relations such as "same," "different," or "opposite," participants must be able to use previous experience with these types of relations. However, for old learning to modulate new learning in this way, its effects will need to be recovered or elicited at the appropriate time, presumably being activated by certain contextual cues inherent in the new learning. This is what we refer to as the *recovery of old by current learning*.

(v) Derivation of new responses. New responses are derived by combining old and new learning. This we refer to as the *bringing to bear of old on new learning*.

While it is not inconceivable that all five of these subtasks could be carried out by one undifferentiated network, it is clear that they cannot all be carried out satisfactorily by ours. Our networks were designed to show how relationships derived from extensive training on other domains (or stimulus sets) could be brought to bear and used to produce new (untaught) responses following limited training on a new set of pattern pairs (i.e., the eighth stimulus set). Thus we focused on the issues of inter-trial learning and the bringing to bear of old on new learning. The three other subtasks—learning multiple input-output pairs, learning retention, and the recovery of old by current learning— were heavily finessed, but as we shall see shortly, these issues must be faced if neurophysiologically plausible models are to be constructed.

Problem 2: The learning of several input-output pairs in sequence without interference was finessed

Human and nonhuman animals are constantly dealing with new learning situations. In many cases, new learning can be acquired without interfering with or being disrupted by old learning, though of course there are conditions under which interference effects do occur in human learning and memory (see Baddeley, 1976, for a discussion). Connectionist models, of the type we used, are not so powerful. They suffer from catastrophic interference in which previous learning is disrupted by current input (McCloskey & Cohen, 1989; Ratcliff, 1990). We avoided this problem by presenting all pattern sets together in a given training epoch (equivalent to old learning always being refreshed at the same time as new). In reality, of course, the world is not so obliging; new environments do not routinely include copies of old environments to assist the learner! Humans learn from new input, without necessarily being disrupted by it. In effect, they cope with the "sensitivity-stability" problem (Hebb, 1949).

Problem 3: Pattern sets used to represent stimuli were overspecified

One important aspect of relational framing is the (presumably learned) ability to respond to a set of stimuli as participating in a particular relational frame, and for this ability to be produced, at least in part, by previous learning in other domains (i.e., with different stimulus sets). In other words, the system must exhibit inter-trial learning as well as bringing to bear relevant old learning from other domains. To a large extent we finessed this aspect of the problem too. RELNET overspecified the relational frames, insofar as the sample-marking duplicator provided a set of inputs to the network *that remained constant* across each of the different stimulus sets. Learning, in the case of our model, entailed the gradual strengthening of the relational frames by the sample-marking duplicator, across a set of domains (or stimulus sets). The relational frames were not, therefore, fully derived.

Problem 4: The networks were designed with unprincipled architectures; we made no attempt to map networks onto known aspects of neuroanatomical structure or function

All of our simulations have made use of simple three-layer networks whose architecture was motivated as much by considerations of parsimony as by any known neuroanatomy. The scheme of connections was used because it permitted a clear separation between the domain-variant aspects of the input-output mappings (i.e., the explicit training sets) and the domain-invariant aspects (i.e., the sample-marking duplicator). No attempt was made to make contact with relevant neurophysiological knowledge.

Problem 5: A biologically dubious learning rule was used

It is highly debatable whether backward error propagation is a biologically plausible learning rule (cf. Crick & Asunama, 1986). The algorithm assumes that an error signal is sent back through the network and used to assess the amount by which connection weights need be altered to produce the desired response. Despite the fact that many brain areas are known to be reciprocally connected, with as many backward connections as there are forward (Kosslyn & Koenig, 1992), the evidence for such a learning scheme is quite tenuous. A further obvious problem is that backprop is a global learning mechanism which operates over the entire network, whereas there are reasons for thinking that synaptic changes are generally local.

Toward More Plausible Simulations

In this section we outline our revised strategy to work toward more neurophysiologically plausible simulations, by offering some potential solutions to the problems we have just considered.

Problems 1, 2, 3, and 4: one or two learning systems?

On grounds of parsimony we initially approached the relational frame task as a whole, finessed the problem of sequential learning, chose an architecture consisting of one network, and used highly specified pattern sets, but as we have just seen, there are grounds for thinking that learning tasks often decompose into simpler subtasks. As we shall now explain, there are also grounds for arguing that the neural processes of learning are correspondingly complex and involve at least two learning subsystems.

In this spirit, we are now exploring the idea that complex human learning, of the sort that we are investigating, involves the interaction between at least two neural subsystems. The first, the cortical-response (S-R) subsystem, is sensitive to moment-to-moment changes in the environment and the appropriate responses to them, and is ultimately capable of long-term retention of the stable products of previous learning. While this subsystem is capable of dealing rapidly and flexibly with changing environmental demands, if unassisted it is limited in its ability to support inter-trial learning and is potentially vulnerable to catastrophic interference between old and new learning. These difficulties can, however, be overcome with the help of the hippocampal system, the second learning subsystem. This second subsystem relies on the hippocampus and associated structures and has a variety of functions which support those aspects of learning which go beyond the learning of single trials. These functions include (i) extracting regularities across sets of trials, (ii) retaining the results of several environment-behavior interactions as captured by the cortical-response system over intermediate time periods, (iii) interleaving new and old learning in the cortex, and (iv) the reinstatement and bringing to bear of old on new learning. We use the term *hippocampal system* in the same way as Cohen and Eichenbaum (1993) to refer to the *hippocampus* or *hippocampal formation* itself together with the *para-hippocampal system*. According to this usage, the para-hippocampal system is taken to include the various cortical areas surrounding the hippocampus, whereas the term hippocampal formation applies only to the dentate gyrus, areas CA3 and CA1 of Ammon's horn, and the subiculum.

Our appeal to two learning systems is in line with recent work on cognitive neuroscience (e.g., Cohen & Eichenbaum, 1993; Eichenbaum, Otto, & Cohen, 1994; Squire, 1992), connectionist modeling (e.g., French, 1995; Gluck, Chapter 21, this volume; McClelland, McNaughton & O'Reilly, 1994; O'Reilly & McClelland, 1994), and biobehavioral approaches (e.g., Donahoe, Burgos & Palmer, 1993; Donahoe & Palmer, 1994; Donahoe, Chapter 19, this volume).

We now summarize what we see as three major roles of the hippocampal learning system, and show how these, when properly understood, will allow us to resolve some of the problems with our earlier simulations. We do not pretend to offer a complete review of what is by now an extensive literature on the hippocampal system; instead we survey its likely computational functions and indicate their relevance to relational framing behaviors.

Hippocampus as a special type of learning system. A reasonable body of evidence can be adduced to support the proposition that the hippocampal system is involved in a qualitatively distinct form of learning. The cognitive researchers Cohen and Eichenbaum (1993) have recently described this learning as relational or declarative. According to these researchers, such learning can take many forms. It might, for instance, involve the links between an animal's movements through a maze, the relative positions of cues in the environment, and the availability of reinforcement at a given location, or it might involve learning a list of paired associates. Despite these variations, all declarative learning, according to Cohen and Eichenbaum, shares the common characteristic that it captures the outcome of various processing modules, which are bound or chunked together by the hippocampal system. Its representations are therefore relational and compositional. Its stored memories can be accessed by a wide variety of other processors and expressed flexibly in novel contexts (cf. Cohen & Eichenbaum, 1993, p. 73).

We assume that relational framing involves what Cohen and Eichenbaum call declarative information processing, since whenever responses in the matching-to-sample task go beyond simple configural learning, a wide variety of relata are involved. In the case of derived responding, in particular, the current input-output mappings (new learning) must be related with those previous mappings that combined to yield neural patterns that support responding in accordance with relational frames (old learning). If the new and old mappings are successfully related, then the new mappings are incorporated into a set of already established mappings, and the neural patterns for responding to a set of stimuli as participants in a specific relational frame are established.

To investigate the tasks which interest us we are now using the working assumption that the hippocampus, acting as a relational processor, *and in cooperation with the cortex*, produces the more important neural patterns necessary for relational frame behaviors to emerge from multiple exemplars of explicitly taught stimulus relations. In other words, the hippocampus is essential for the formation and activation of relational frames.

In drawing attention to the role of the hippocampus as a special type of processing system, we go beyond the views expressed on its role by Donahoe and Palmer (1994), who concentrate on its role as an enhancer or integrator of existing cortically co-occurrent or "polysensory" events. They chiefly focus on the interface between the hippocampus and the cortex and effectively treat the hippocampus as a black box. We, on the other hand, suggest that a full understanding of complex learning will be possible only following an examination of the special contribution made by hippocampal activity. We appreciate and value their parsimonious account, but defend our own breach of parsimony on the grounds that it seems to be required both by our analysis of the relational frame task and from relevant work on the neurobiology of learning.

Hippocampal involvement in the coordination of old and new learning and temporal aspects of learning. Together with other structures, the hippocampus appears to coordinate the neural effects of learning. According to a range of cognitive researchers, there is good evidence that it can act as an intermediate store of recently acquired and not fully established memories (Eichenbaum et al, 1994), that it exhibits a phenomenon known as long-term potentiation involving changes in synaptic plasticity (Lynch, 1986; Lynch & Baudry, 1988), that it is involved in the consolidation of memories (Kim & Fanselow, 1992; Squire, 1992; Zola, Morgan, & Squire, 1990), and that it is not the final repository of information storage or locus in which the effects of learning are neurally represented (Cohen & Eichenbaum, 1993; Squire, Shimamura, & Amaral, 1989).

Many of these facets of hippocampal activity are brought together by the idea that the hippocampus acts as a buffer device for retaining information prior to its eventual transfer to more durable cortical locations (McClelland, McNaughton, & O'Reilly, 1994). The use of such a dual-learning system, according to McClelland et al, permits new learning to be gradually meshed or interleaved with old, thus avoiding catastrophic interference. Such interleaving can, in humans at least, take place over time intervals as long as fifteen years. McClelland et al have constructed connectionist models of these key hippocampal-cortical interactions. In their simulations, the hippocampus is designed to acquire rapidly several input-output mappings without interfering with previously acquired structure, and to aid or "teach" the cortex to extract new structure from inputs.

With somewhat different emphasis, in a two-stage theory of hippocampal processing, Eichenbaum et al (1994) have argued that the para-hippocampal region is capable of retaining individual learning experiences over intermediate time periods in a passive form, thus protecting individual traces from the moment-to-moment interference and overwriting experienced by short-term, active cortical representations. The hippocampus proper is then used for relational comparisons between new learning and old learning, and is finally assumed to exert its long-term cortical effects by modifying, or making connections between, relevant cortical areas.

Whatever the final outcome of this debate, there is sufficient agreement about hippocampal functions to conclude that for relational framing to occur, the hippocampal system (i) is needed for the learning of multiple input-output pairs without interference, (ii) permits inter-trial learning, and (iii) acts cooperatively with the cortex in learning retention.

Hippocampus for use of/interpretation of learning. Although a large proportion of the current literature on hippocampal-system activity is devoted to its role during learning or encoding, there is, however, also a role for the hippocampus at retrieval. There are two major ways in which the system is thought to be involved at the stage that memories (prior learning) are used. First, as a temporary, intermediate locus of storage, it will allow the appropriate responses to be acquired and re-emitted across sets of input-output pairs. Thus, as we have just pointed out, Eichenbaum et al (1994) have shown that the para-hippocampal system may be specialized for this. Second, and more relevant for our current concerns, the hippocampal system seems to have a role in the reinstatement or reconstruction of cortically retained learning. Again, accounts of this differ, but O'Reilly and McClelland (1994) have used formal modeling methods to show that the hippocampus is well designed to accomplish pattern completion at retrieval.

The important point here is that the act of relational framing can be thought of as a type of cueing by contextual stimuli in which new (current) inputs are discriminative for old learning and, *during a hippocampal-dependent pattern completion process*, untaught relational responses emerge.

An example of hippocampal support. We can illustrate some of these hypothesized functions with a simple example and toy network. Suppose a child has already learned the names and ages of two individuals—Fred, aged 40, and George, aged 10. We assume that cortical neuronal groups already signify these names and ages and their co-occurrence and will refer to these groups as name, age, and person nodes respectively, such that name and age nodes are mutually facilitatory and interconnected through person nodes. Suppose, now, the child learns that Fred is the father of George and, *on a later occasion*, that George is the son of Fred; that is, in the presence of George and Fred, she

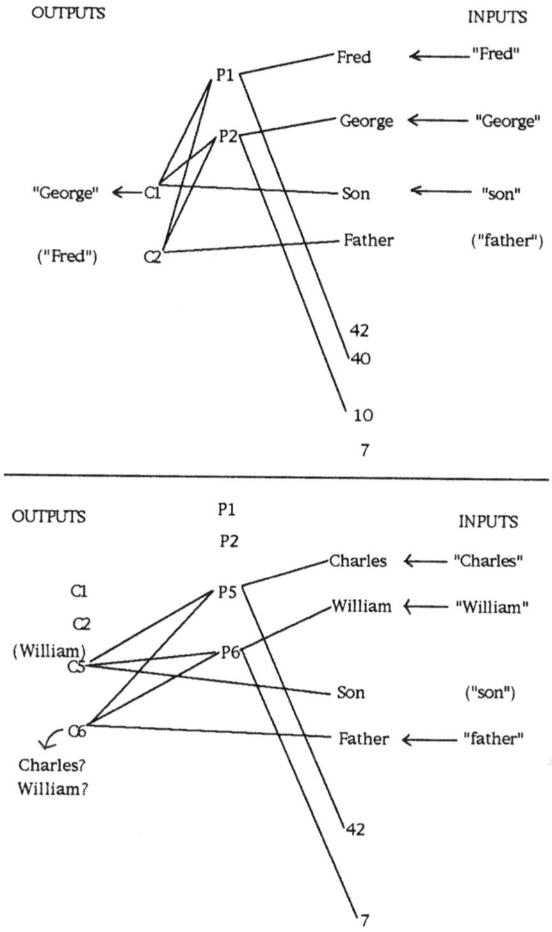

FIGURE 4. Simple networks for learning to select the appropriate response given the names of two individuals as input and the specification of the relationship, "son" or "father." Upper section: the network response "George" is guided by the configuration node C1, which represents the co-occurrence of Person 1, (P1), Person 2 (P2) and "son;" (the response "Fred" is guided by the corresponding co-occurrence at C2 of P1, P2, and "father"). Lower section: previous selective training on "Charles," "William," and "son" as inputs and "William" as output (and other similar mappings) is insufficient to permit derived responding of "Charles" as response, given "Charles," "William," and "father" as inputs.

learns to orient or respond to George when the word "son" is used, and to Fred when the word "father" is used. One simple way in which this could occur is for a response to be shaped up to the co-occurrence of person, age, and the new status information. In this way, it is theoretically possible for the child to learn directly about a number of father-son relationships, say that Peter is the father of Joe, Joe is Peter's son, and so on. An illustration of this for the implicative relationships between Fred and George is illustrated in **Figure 4** (upper section). Notice that so far all we are assuming is that the child is responding to the co-occurrence of information, and that all learning is cortical. We make no strong assumptions about learning rules at this stage, though we do assume that connected nodes are mutually excitatory.

However, this simple scheme is likely to fail as learning progresses for two reasons. First, as other father-son pairs are learned, catastrophic interference will occur as new learning disrupts old. For this reason alone some complementary learning and gradual interleaving of new learning is likely to be necessary. Second, and crucially for our understanding of relational frame effects, the system is incapable of derived responding since no inter-trial learning has taken place. The network may well "know," having been previously instructed, that, say, Charles is aged 42 and has a son, aged 7, named William, and be able to respond with "William" to the input configuration "Charles," "William," "son." Thus constituted, however, the network will be unable to respond "Charles" when presented with "Charles," "William," "father" (**Figure 4**, lower section). As it stands, the network has not acquired the necessary neural patterns to demonstrate responding in accordance with a relational frame or frames in which sons and fathers may participate. A simplified neural pattern that would, however, support an appropriate relational frame is represented in **Figure 5**. Here the links between the "father" node and the set of ages around the "42" node (within the age range of the prototypical father) permit additional activation to flow from the "42" node to node P5, which is strongly associated to the name "Charles."

We showed in our previous models that bidirectional training alone, without suitable architecture to support it, cannot give rise to relational framing behaviors. We now postulate that relational framing can be accomplished with hippocampal support. Suppose that the hippocampus, acting as a relational processor, coordinates information from the stimulus side with information regarding successful responses. Neuroanatomically it is known that responding to both stimulus and response information is possible since the hippocampal system is bidirectionally connected to a range of areas including the motor cortex (Cohen & Eichenbaum, 1993; Amaral, 1987). Also, work on nonhuman learning suggests that so-called hippocampal place maps are response valenced in the sense that they indicate the behavioral significance of places visited, and do not merely offer an epistemologically neutral survey of the world (see

Cohen & Eichenbaum, 1993, for a discussion). Given an accumulated set of mappings involving father-son relationships (i.e., links between named individuals specified as "father" or "son"), and the appropriate responses to them, the hippocampus now has the "raw material" to relay back to the cortex the common characteristics of all father-son mappings encountered to date. One way in which this could take place is for it to categorize mappings as either "father" or "son," and in the process extract their prototypical features. Thus, in our simple example, father mappings are characterized by the selection of individuals aged around 40 years, while son mappings typically involve individuals of 10 years or less. In other words, the hippocampus establishes a neural pattern that supports a relational frame in which all fathers and sons, and their prototypical features, may participate and transmits this to the cortex (see French, 1995, for details of a simple network architecture that can accomplish this). Alternatively, the neural patterns that support individual relational frames may gradually take place in the cortex itself. In either case, new linkages will eventually arise in the cortex that allow for responding in accordance with relational frames.

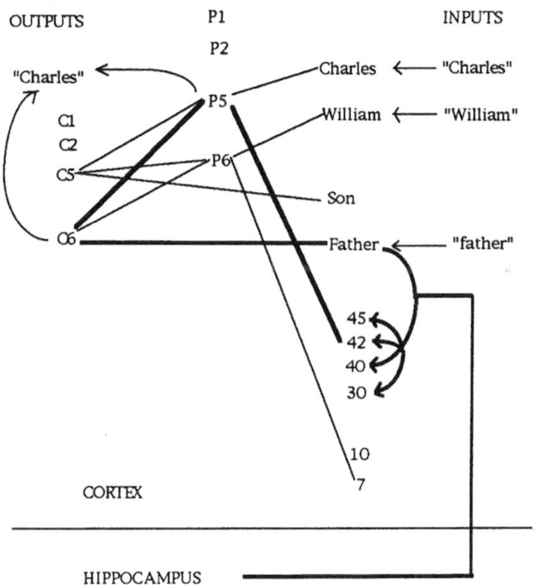

FIGURE 5. Prototypical information about a series of father and son mappings, supplied with the assistance of the hippocampus, permits new linkages between "father" and the set of typical ages of fathers. The latter, in turn, increases activation in co-occurrent nodes, in this case P5, which then guides the response "Charles." (See text for full explanation).

These new linkages can then be used to support derived relational responding. To return to our example, recall that the two mappings between father and son are characterized by the selection of older and younger individuals respectively. Persons with ages closer to the relevant prototypical age will thus be more likely to be selected as a father or a son, as appropriate. Thus the inputs "Charles," "William," "father" will now be able to activate the response "Charles," since the crucial additional linkage, between nodes "father" and "42," has been interleaved into the cortex from the results of previous mappings distilled in the hippocampus.

To summarize: the hippocampus provides the cortex with the information needed to relate the specific aspects of new learning with the general, prototypical aspects of earlier learning. Using our simple example again, the terms "father" and "son" eventually trigger general characteristics of father and son mappings rather than being tied to any single individual. In our discussion, we have assumed, probably naively, that this transfer is all-or-none and relatively direct. In practice, the process of information transfer is likely to be much more interactive than this with the prototypical information slowly growing in the cortex, and the hippocampus, in turn, having progressively more and more prototypical information available from the cortex to join with the interpretation of specific mappings.

The details of this process are far from fully specified, but we can hypothesize that it involves the following elements: (i) the hippocampus combines mappings from a series of learning trials, within and between domains (or stimulus sets), (ii) it helps in the separation of these mappings in the cortex during initial learning (it avoids catastrophic interference), (iii) as learning progresses, it helps the cortex to develop the neurological patterns required for responding in accordance with relational frames, (iv) it helps bring to bear the effects of old learning on new learning.

This analysis goes some way towards dealing with Problems 1 to 4, since the subtasks of relational framing are addressed. In effect, the problem of multiple input-output pairs is faced; the neurological patterns required for relational framing are extracted, across pattern sets, by the hippocampus in conjunction with the cortex; it is not supplied *ab initio*; and the architecture needed to carry out these activities is neurologically principled.

Problem 5: alternative learning rules

A major problem with our earlier simulations was the use of backward error propagation as the learning rule for training our networks, the biological implausibility of which has already been noted.

There are by now in the connectionist literature a variety of other learning rules which could serve as alternatives, but space does not permit us to review all of these here. We do, however, wish to note that one recent learning rule, in which neuronal connections are selectively strengthened by broadcast, reinforcement-triggered signals, offers one currently very attractive alternative (cf. Donahoe & Palmer, 1994). As currently constituted, Donahoe and Palmer present a powerful, parsimonious, and plausible account of single-trial input-output learning that relies on straightforward cortical-motor links. It remains to be seen, however, whether their learning rule can be further expanded to cope with multiple-trial learning without interference and inter-trial learning. Donahoe and Palmer argue that output from CA1 neurons can similarly modulate and strengthen stimulus co-occurrences at the site(s) of original learning, and argue that this may be sufficient to explain equivalence responding.

We, on the other hand, maintain that the hippocampus needs to combine the co-relations of input-output neural *mappings* within and across domains (i.e., within and across stimulus sets) in order to produce relational frame behaviors. Whether or not this process in its entirety can be accomplished using the learning rule developed by Donahoe and Palmer is not clear. It is possible, for example, that hippocampal-cortical interactions are governed by the type of learning rule suggested by Donahoe and Palmer, but it might also be the case that the unique and hard-wired architecture of the hippocampus has been naturally selected to fulfill its "special" role as a relational processor. We also assume that hippocampal "teaching" of the cortex eventually results in the formation/connection of new neuronal groups in association areas not necessarily directly involved in the early stages of learning to respond in accordance with relational frames (e.g., when learning to name), and these groups then contribute towards the neural patterns that support increasingly complex relational framing behaviors (e.g., following instructions and analogical reasoning).

Conclusions

In this chapter we have outlined some of the phenomena associated with relational framing behaviors, discussed some of our initial attempts to model them using connectionism, criticized our models on various grounds, and offered ways forward to a more neurologically plausible family of models. In doing so, we were prompted to examine relational framing as composed of several subtasks, and argued for the existence of two complementary learning systems at the neural level.

There are benefits to be gained from this type of local and detailed analysis of complex learning behaviors, and from submitting ourselves, as psychological scientists, to the constraints of neurophysiologically plausible models. When we do so, the issues revealed have a global relevance for both behavioral and cognitive neuroscience.

Neural-Networks Models of Cognition
J.W. Donahoe and V. Packard Dorsel (Editors)

CHAPTER 27

A RECURRENT-NETWORK ACCOUNT OF READING, SPELLING, AND DYSLEXIA

Guy C. Van Orden, Anna M.T. Bosman,
Stephen D. Goldinger, and William T. Farrar, IV
Arizona State University

ABSTRACT

We present a recurrent-network account of printed word perception, spelling, and dyslexia. Our account explains why phonology is fundamental to reading and spelling, and why spelling is more difficult than reading. It also provides a basis for simulating the behavior displayed by children with reading problems (developmental dyslexia) and by adults with acquired reading problems (acquired dyslexia).

Introduction

Recurrent networks are connectionist models in which activation flows from input to output and back again, creating feedback loops. *Behavior* is modeled in self-organizing patterns of activation, but activation in any part of the network is always reflected throughout the network. Bidirectional flow of activation binds activation at each part to activation at every other part. This holism is the basis of our claim to a *mind as embodied action* metaphor: Perception and action are emergent in the coupling (resonance) of an organism and its environment (Varela, Thompson, & Rosch, 1991). One entailment of this metaphor is that an organism and its environment are themselves interdependent (cf. Turvey & Carello, 1981). In this chapter, we illustrate these claims in an explanation of reading and spelling that assumes interdependence between readers and the printed word. The empirical basis of our account is performance in laboratory reading and spelling tasks.

The description presented in this chapter derives from a general theoretical framework proposed by us and our colleagues (Grossberg & Stone, 1986; Stone & Van Orden, 1994; Van Orden & Goldinger, 1994; Van Orden, Pennington, & Stone, 1990, 1996), and is rooted in mathematical dynamic-systems theory (cf. Thelen & Smith, 1994). Clarifying the metaphor of *mind as embodied action* will be easier once we have presented our account in more detail. We begin by discussing a classic cognitive phenomenon: phonologic mediation in reading and spelling. Then, we describe a resonance theory of word perception and spelling. At its heart, we offer an explanation of why phonology is fundamental to reading and spelling, and why spelling is more difficult than reading. This account predicts non-intuitive "feedback" phenomena that have been corroborated in laboratory studies. We describe these studies, and then suggest how a recurrent network might accommodate the anomalous reading behavior of developmental and acquired dyslexics. In the final section, we return to the metaphor that underlies our approach.

Phonologic Mediation in Reading and Spelling

The emphasis in much reading research is on the perception of single words, as this is the most important aspect of reading skill. Poor word perception severely limits the development of skilled reading and reading comprehension (Perfetti, 1985). A perennial question in such research is whether a word's phonology (loosely, the "sound" of a word) influences visual word perception. A recent accumulation of empirical findings forcefully suggests that phonology's role in word perception is fundamental. Numerous experiments have shown that the phonology of a letter string affects its perception in simple reading tasks. (For overviews, see Berent & Perfetti, 1995; Van Orden et al, 1990.) For example, subjects tend to mistake ROZE for an exemplar of FLOWER in a categorization task, or they misclassify the letter string SUTE as a word in a lexical decision task, or they overlook misspellings such as MUNKEY in a proofreading task (Van Orden, Stone, Garlington, Markson, Pinnt, Simonfy, & Brichetto, 1992). And in writing, systematic misspellings, such as substituting ROZE or ROWS for ROSE are common. (For an overview, see Bosman & Van Orden, in press.) These errors indicate that phonology is central to reading and spelling. With respect to phonology, a ROZE is a ROWS is a ROSE. Note, however, that phonology is not explicit in these printed forms; it is only implicit with respect to the knowledge that readers bring to reading and spelling.

The effects of phonology are apparent with readers and writers spanning the full range of reading skill (i.e., beginning, skilled, and disabled readers). They are found across languages (in both alphabetic and non-alphabetic writing systems) and across laboratory tasks. These tasks include naming (quickly reading words aloud), lexical decision (quickly classifying words versus nonwords), semantic categorization (quickly determining whether words belong to designated categories), and proofreading (carefully checking a document for spelling errors). Why is phonology so involved in reading or spelling? It clearly is not always helpful, often leading to errors in these experimental tasks. In the next section, we describe a recurrent-network account of word perception and spelling. Our account pertains to a very simple recurrent network that has been implemented (Farrar & Van Orden, 1994), but the principled basis of our account is not tied to the specifics of our simulation. No claim is made with respect to a "correct" architecture (see Stone & Van Orden, 1994; Van Orden & Goldinger, 1994; Van Orden et al, 1990, 1996). We return to this issue in the final section of the chapter.

Reading and Spelling are Fundamentally Related

Imagine a fictitious nervous system that perceives printed words. This system consists of three families of neurons: letter neurons, phoneme neurons, and semantic neurons. Every neuron in each family is (potentially) bidirectionally connected to every neuron of the other two families. Bidirectionally connected means that if a feedforward connection exists from neuron "x" to neuron "y," there is also a feedback connection from neuron "y" to "x." Now, imagine a specific pattern of activation across the letter neurons, due to the presentation of a printed word. This letter pattern feeds activation forward through a matrix of "synaptic" connections, creating patterns of activation across phoneme and semantic neurons. The phoneme and semantic neurons, in turn, feed activation back through a top-down matrix of connections, transforming their patterns back into letter patterns. Whenever the feedback patterns match the original letter pattern, top-down activation *conserves* bottom-up activation. Consequently, the "matched" letter neurons conserve their capacity to reactivate matching phoneme and semantic neurons that, in turn, reactivate the letter neurons, and so on. This feedback cycle is temporarily stable, resulting in a coherent dynamic whole: a *resonance*.

This simple neural network is only for exposition. It is helpful to consider word perception in terms of artificial neural activity, but analogies between cognitive systems and actual nervous systems, albeit compelling, are limited. We conceive of word perception in cognitive terms. **Figure 1** illustrates cogni-

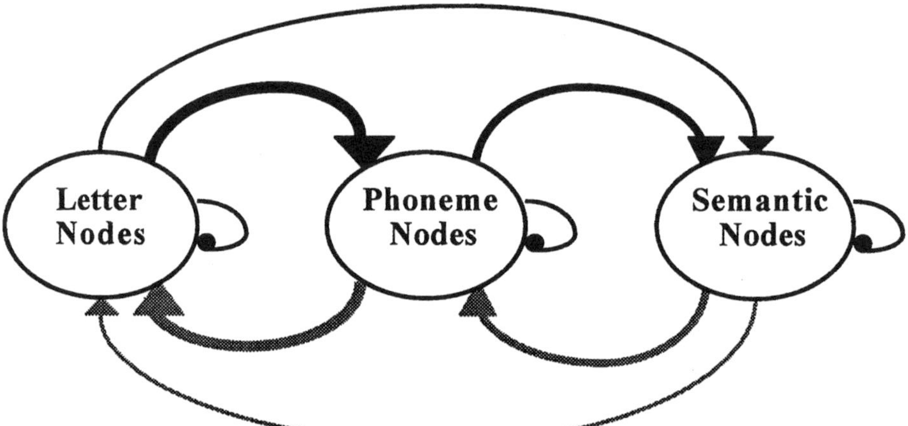

FIGURE 1. Macrodynamics of reading and spelling performance emergent in a recurrent network. The boldness of the arrows indicates the overall strength of the relations between letter, phoneme, and semantic node families (see text).

tive macrodynamics of word perception (Van Orden & Goldinger, 1994) and spelling (Bosman & Van Orden, in press), and **Figure 2** illustrates microdynamics.

Figure 1 portrays a recurrent network with three families of fully interdependent nodes (letter nodes, phoneme nodes, and semantic nodes). On average, the connections between node families differ in strength; the rank order of overall strength is illustrated by the relative boldness of arrows in the figure. In alphabetic languages, letters and phonemes correlate quite strongly. For example, the letter B is almost always pronounced as /b/, and the phoneme /b/ is always written with a B. Correlations between phonemes and semantic features, or letters and semantic features, are far weaker than correlations between letters and phonemes. Knowing that a word begins with the letter B indicates almost nothing about its meaning, but much about its initial pronunciation.

Notice also that phoneme-semantic connections are depicted as stronger than letter-semantic connections, primarily because we speak before and more often than we read. Moreover, once in place, this asymmetry is self-perpetuating. Reading strengthens phoneme-semantic connections, because phonology functions in every instance of printed-word perception. Thus, even the exceptional condition of people who read more than they speak would support phoneme-semantic connections that are at least as strong as letter-semantic connections. Also, if a coherent positive-feedback loop forms from semantic to phoneme nodes before the feedback loop from semantic to letter nodes, then printed or spoken discourse may proceed without the contribution of the feedback loop from semantic to letter nodes. The absence of resonance in the latter feedback loop may preclude strengthening the connections between letter and semantic nodes (see discussion below, and Grossberg & Stone, 1986). Thus, at this macro-level of description, families of nodes differ in the overall strength of relations with other families. These differences in overall correlational structure are illustrated in the relative boldness of the arrows in **Figure 1**.

The strong bidirectional connections between letter and phoneme nodes, as compared to those with semantic nodes, causes the letter-phoneme dynamic to cohere (resonate) before all others. These strong connections between letters and phonemes explain why phonology is so fundamental to reading and spelling. Stated differently, it explains why sound-alike words (ROSE and ROWS) may be confused in reading (Van Orden, 1987), and explains why the majority of spelling errors (ROZE instead of ROSE) are phonologically acceptable. (Van Orden & Goldinger, 1994, 1996 describe various other phenomena that derive from the powerful correlations of spelling and phonology.)

In a model analogous to **Figure 1**, presentation of a printed word activates letter nodes that, in turn, activate phoneme and semantic nodes. Following initial activation, recurrent feedback begins among all these families of nodes. Similarly, presentation of a spoken word activates phoneme nodes that, in turn, activate semantic and letter nodes. (And, word production would begin with activation of semantic nodes that, in turn, activate phoneme and letter nodes.) In all these cases, initial activation leads to recurrent feedback among all families of nodes. However, the strongest recurrent dynamic is between letter and phoneme nodes, which creates the common basis of reading and spelling.

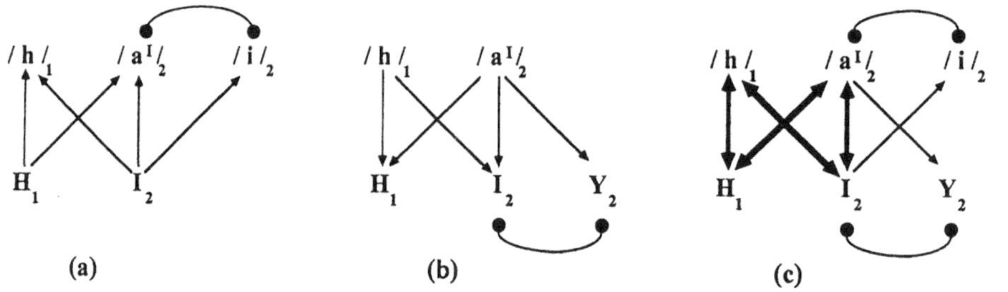

FIGURE 2. A simplified illustration of microdynamics that describe reading and spelling performance for the word HI. **a.** Presented with HI, activation feeds forward from letter nodes to phoneme nodes. **b.** In turn, phoneme nodes feed activation back to letter nodes. **c.** A resonance that emerges between letter and phoneme nodes corresponding to HI. (To reduce the number of lines in the figure, bidirectional connections are depicted with double-headed arrows.)

Figure 2 illustrates microdynamics. We zoom in on the connectivity between letter and phoneme nodes (and ignore, for now, phoneme-semantic and letter-semantic connectivity). In **Figure 2A**, reading the printed word HI activates the letter nodes H_1 and I_2, which activate the phoneme nodes $/h/_1$ and $/a^I/_2$ but also competing nodes such as $/I/_2$ (as in /hIt/) which must be inhibited. (The subscripts refer to the positions of the letters or phonemes within words.) **Figure 2B** shows how, in turn, phoneme nodes feed activation back to letter nodes (illustrated for the phoneme nodes $/h/_1$ and $/a^I/_2$). The phoneme node $/a^I/_2$ activates the correct letter nodes H_1 and I_2 and also competing letter nodes such as the letter node Y_2 in MY or BY. Thus, early patterns of activation are loosely structured, entailing activation of correct, but also many incorrect, candidates for resonance (Van Orden et al, 1990). Interactions between nodes then select combinations of nodes through cooperative-competitive dynamics (Grossberg & Stone, 1986; Stone & Van Orden, 1994).

Reliable performance emerges if the overall bidirectional configuration of connections favors mutual activation between the letter nodes H_1 and I_2 and the phoneme nodes $/h/_1$ and $/a^I/_2$. This advantage grows over time as the "strong grow stronger" and the "weak grow weaker" (McClelland & Rumelhart,

1981). This is illustrated in **Figure 2C**, which combines the flow of activation from letter nodes to phoneme nodes and from phoneme nodes back to letter nodes, as assumed in a recurrent network. Presentation of the spoken word /ha^I/ to the network (as in a spelling task) leads to a similar dynamic between phoneme and letter nodes. Thus, activation initiated in phoneme nodes may generate a coherent pattern of activity across letter nodes.

Why Spelling is More Difficult than Reading

In the course of writing, everybody experiences occasional doubts about how to spell a word, but we almost never forget how to read a word. This asymmetry between reading and spelling is evident at all levels of skill (Bosman & Van Orden, in press). The account we offer explains why people find spelling more difficult than reading. It may be simply described with respect to the previous illustrations of microdynamics (letter-phoneme dynamics), and macrodynamics (dynamics among families of nodes).

Returning to **Figure 2**, reading the word HI not only activates the phoneme nodes $/h/_1$ and $/a^I/_2$, and the letter nodes H_1 and I_2, but also all possible pronunciations of H_1 and I_2 and all possible spellings of $/h/_1$ and $/a^I/_2$. Thus, correctly reading a word requires inhibition of incorrect phoneme nodes, and correctly spelling a word requires inhibition of incorrect letter nodes. In the case of reading, the letters are presented to the model (or reader) such that phoneme→letter ambiguity is quite unlikely to produce full activation of incorrect letter nodes, because the persistent and stable input at letter nodes accelerates the formation of feedback loops with phoneme and semantic nodes (as illustrated by bold arrows in **Figure 2C**). However, in the case of spelling, this resonant pattern must be generated from phonologic and semantic activation alone. There is no environmental support for correct letter nodes.

The crux of spelling is that English orthography, generally, has more possible spellings for any given word than possible readings, and this is true for most (but not all) alphabetic writing systems (e.g., Stone, Vanhoy, & Van Orden, in press). Consider, for example, the multiple possible "spelling bodies" that may correspond to the "rime" /_ûrch/: _IRCH as in BIRCH, _ERCH as in PERCH, _URCH as in LURCH, and _EARCH as in SEARCH. Stone et al (in press) estimated that 69% of low-frequency English one-syllable words are spelling→phonology consistent (at the grain size of spelling bodies and rimes), but only 28% are phonology→spelling consistent (at the same grain size). In a larger sample, including both low- and high-frequency one-syllable words, 72% of all spelling→phonology consistent words were phonology→spelling inconsistent. These linguistic analyses clearly indicate that phonology→spelling inconsistency is the rule for English.

Although both reading and spelling are powerfully constrained by the strong correlational structure of the letter-phoneme dynamic, the occasional inconsistencies in these relations are resolved by different sources of constraint. Now, we refer again to the illustration of macrodynamics in **Figure 1**. When a model "reads" a low-frequency, spelling→phonology *inconsistent word* such as PINT, the more consistent letter-phoneme relation would rhyme with MINT (and HINT, LINT, TINT, etc.; Kawamoto & Zemblidge, 1992). Similarly, the letter-phoneme dynamic would yield two correct pronunciations for words like WIND (although it would typically favor the more regular pronunciation, Kawamoto & Zemblidge, 1992). In both these cases, relatively strong semantic-phoneme relations may supply sufficient secondary constraints to encourage the appropriate letter-phoneme dynamic. In the case of WIND, semantic constraints may also be due to context, and contextual sources of semantic activation contribute via the relatively strong connections between semantic and phoneme nodes.

In the case of spelling, however, a model must resolve the inverted patterns of ambiguity in the phoneme-letter dynamic. To spell a low-frequency phonology→spelling inconsistent word such as HEAP, the rime /_ip/'s correct spelling would compete with a more strongly correlated incorrect spelling body _EEP (as in DEEP, BEEP, KEEP, PEEP, SEEP, and WEEP). Additionally, the phoneme-letter dynamic yields two correct spellings for homophones (e.g., ROSE/ROWS). In either case, correct spelling must rely on relatively weak semantic-letter dynamics to sufficiently activate the appropriate letter nodes (as illustrated in **Figure 1**); even contextual support is filtered through the weak letter-semantic connections. This weaker support for spelling, compared to the strong support for reading (i.e., phoneme-semantic dynamics) is the "macro-basis" for the asymmetry between reading and spelling. Thus, spelling is more difficult than reading for two reasons: (1) Microdynamic phoneme→letter relations are more inconsistent than letter→phoneme relations (Stone et al, in press), and (2) macrodynamic support for spelling (i.e., letter-semantic connections) is weaker than macrodynamic support for reading (i.e., phoneme-semantic connections).

Productive Use of This Simple Model

The recurrent dynamic system described above predicts a rather non-intuitive micro-effect. This effect concerns the consistency of relations between letters and phonemes. Until recently, all discussion of consistency has concerned a classic "feedforward," spelling→phonology effect. Inconsistent words such as PINT are named more slowly than consistent words such as DUCK. (_INT in PINT may be pronounced as in MINT; _UCK is only pronounced as in DUCK.) The feedforward consistency effect answers the question: Does it matter in word perception that a spelling may have more than one pronunciation? From most theoretical perspectives, this is the only sensible question. In a naming task, the letter string is unambiguous to subjects (it is right in front of their eyes); the only potential ambiguity arises with respect to derived phonology. However, our "feedback hypothesis" generalizes perceptual ambiguity in the phonology→spelling direction as well. We were led to ask the feedback question: Does it matter in *visual* word perception that a *pronunciation* may have more than one *spelling*?

Recently, Stone et al (in press) tested for the effects of both feedforward and feedback consistency on performance in a lexical-decision task. They used a factorial design that included four types of words. In bidirectionally consistent words such as DUCK, the spelling body (_UCK) can only be pronounced one way, and the pronunciation rime (/_uk/) is only spelled one way. In spelling→phonology inconsistent words such as MOTH, the spelling body can be pronounced in multiple ways (BOTH), but the pronunciation rime (/_ôth/) is only spelled one way. In phonology→spelling inconsistent words such as HURL, the spelling body is pronounced in only one way, but the pronunciation rime can be spelled in more than one way (GIRL). In bidirectionally inconsistent words such as WORM, the spelling body can be pronounced in multiple ways (DORM), and the pronunciation rime can be spelled in multiple ways (FIRM). Stone et al found strong evidence for perception as a "two-way street;" correct response times were equally (and strongly) slowed by both feedforward and feedback inconsistency. Additionally, they found a reliable interaction; all inconsistent words produced approximately equal response times, even those that were inconsistent in both directions. Only words that were bidirectionally consistent produced better performance. Recently, Patrice Gibbs (personal communication, May 1995) found a similar effect of phonology→spelling consistency in a naming task.

Again, note the non-intuitive nature of this phenomenon. The letter string is clearly visible to the subject, and it remains visible until a response is recorded. However, if feedback from phonology suggests that some *other* letter string could have been presented, performance is slower. Ziegler and Jacobs (1995) reported a similar counter-intuitive finding in a letter-search task. Subjects in their experiment were briefly presented with a letter string such as BRANE (a "pseudohomophone" of the word "Brain"), followed by a pattern mask (#####). The subjects were instructed to indicate whether a predesignated letter, for example the letter "i," was present in the masked letter string. In the case of BRANE, they *mis*reported having seen the letter "i" more often than in a control stimulus. Similarly, they *mis*reported not having seen the letter "i" in the letter string TAIP (a pseudohomophone of the word "Tape"). Presumably, the phonology of the pseudohomophones BRANE or TAIP suggested that "Brain" or "Tape" was presented, causing subjects to misreport the presence or absence of the letter "i."

Developmental Dyslexia

Dyslexic children read poorly relative to non-dyslexic children of the same age, background, intelligence, and instructional level. No skill is more essential than reading for normal functioning within literate cultures; developmental dyslexia can broadly undercut a child's potential for success and happiness (Bryant & Bradley, 1985). Dyslexics typically show qualitative differences from non-dyslexics in simple reading and language tasks. Moreover, their performance is impaired even relative to younger non-dyslexic children who successfully read at the same level (i.e., *reading-age* control subjects; Bosman, van Leerdam, & de Gelder, 1995; Pennington, Van Orden, Smith, Green, & Haith, 1990; see Rack, Snowling, & Olson, 1992, for review). These developmental dyslexics show specific deficits on tasks that require constructive use of phonology (e.g., phonological awareness and pseudoword-naming tasks, described shortly). We have offered an account of developmental dyslexia that derives from our account of phonologic mediation in skilled reading (Van Orden & Goldinger, in press).

A pseudoword-naming task requires fine-grain "phoneme-size" knowledge of how letter strings translate into phonology. In this task, a subject is shown a letter string, such as the pseudoword BINT, that shares spelling structure with actual words. (Consider MINT, BIN, etc. Skilled readers pronounce pseudowords analogously to words; Seidenberg, Plaut, Petersen, McClelland, McRae, 1994). Dyslexic readers name pseudowords much more slowly and produce more "unacceptable" pronunciations than reading-age controls (Rack et al, 1992). They may also perform poorly when judging whether *someone else* has given an "acceptable" pronunciation of a pseudoword (Snowling, 1980). Poor performance in pseudoword naming is a primary symptom of dyslexia.

Correct performance in phonological-awareness tasks also depends upon fine-grain, phoneme-size knowledge of the phonology of a word. These tasks typically require subjects to manipulate or judge the phonology of words. In a "pig Latin" task, for example, the first phoneme of a word must be moved to the end and pronounced with /AY/ (e.g., /dog/ becomes /og·day/). Dyslexics perform very poorly on this task relative to reading-age control subjects, even when they need only recognize whether someone else has produced correct pig Latin (Pennington et al, 1990). Deficits in phonological awareness are typically correlated with deficits in pseudoword naming, and both deficits appear to be influenced by heredity (Olson, Wise, Conners, Rack, & Fulker, 1989; Pennington et al, 1990).

These findings all motivate the hypothesis that dyslexia is a deficit in fine-grain knowledge of phonology and its relation to print in alphabetic languages. The importance of phonology in dyslexia agrees with our account of skilled reading in which phonology also plays a crucial role. The crux of reading is perception of individual printed words (Perfetti, 1985), and the crux of word perception is coherent phonology. Accordingly, developmental dyslexia might be explained by an absence of phonology in reading. It turns out, however, that *absent* phonology is far too simple a hypothesis (Bruck, 1988). Dyslexic subjects, who show a pronounced deficit in pig-Latin performance (Pennington et al, 1990), nevertheless produce a very high proportion of categorization errors to homophonic foils (e.g., ROWS or ROZE categorized as FLOWERS, Van Orden et al, 1990; Van Orden & Goldinger, 1996).

The paradox for the absent-phonology hypothesis is that the same dyslexic subjects show *both* negative and positive phonology effects. Demonstrations of negative phonology (such as pseudoword naming and pig-Latin deficits) are consistent with the absent-phonology hypothesis (Bruck, 1988). However, categorization errors to homophonic foils are not. A key difference between pseudoword (BINT) naming versus categorization may be the added constraints in categorization produced by category names. This source of constraint may exaggerate the dyslexics' susceptibility to phonology in the categorization task, especially with pseudoword homophones such as ROZE. It does so, however, by enhancing ("cleaning up") letter-phoneme dynamics, which explains their very high error rates to homophonic foils. Thus, our ability to observe phonology effects in dyslexics is partly a function of the task examined. We propose that different tasks emphasize different grain sizes of phonology, and these contribute to the respective positive and negative effects. To better understand our proposal, it is first necessary to understand how *covariant learning* serves as a basis for non-dyslexic reading.

Crosstalk is the basis of covariant learning. Crosstalk extracts positive correlations between families of nodes (cf. Reeke & Edelman, 1984). Reading performance is enhanced by consistent crosstalk whenever a letter-phoneme or letter-phoneme-semantic correspondence is shared across a neighborhood of words. In word naming, consistent crosstalk is the source of many common effects, such as rule-strength and word-frequency effects. Rule strength is estimated by a count of all words that share a particular letter-phoneme correspondence. Strong-rule words are composed of letter-phoneme relations that appear in many words (K-/k/). Weak-rule words have at least one letter-phoneme relation that is relatively rare (ZZ-/z/). Strong-rule words (DESK) and pseudowords (DASK) are named faster and more accurately than weak-rule words (FIZZ) and pseudowords (NOZZ; Rosson, 1985). Also, high-frequency words are named faster and more accurately than low-frequency words (Forster & Chambers, 1973). **Figure 3** illustrates how these effects would emerge via covariant learning in a very simple model. BE and BY share a relatively strong rule (B-/b/), and BE is the more frequent word (in the figure).

In **Figure 3A**, a BE learning trial brings four pairs of nodes into collective resonance: $B_1 \longleftrightarrow /b/_1$, $B_1 \longleftrightarrow /i/_2$, $E_2 \longleftrightarrow /b/_1$, and $E_2 \longleftrightarrow /i/_2$. Such resonance increases the connection weights between all the nodes involved. At this point in the model's development, the resonance $B_1E_2 \longleftrightarrow /b_1i_2/$ is an "encapsulated" whole. Although we can anticipate potential subresonances in the *a*

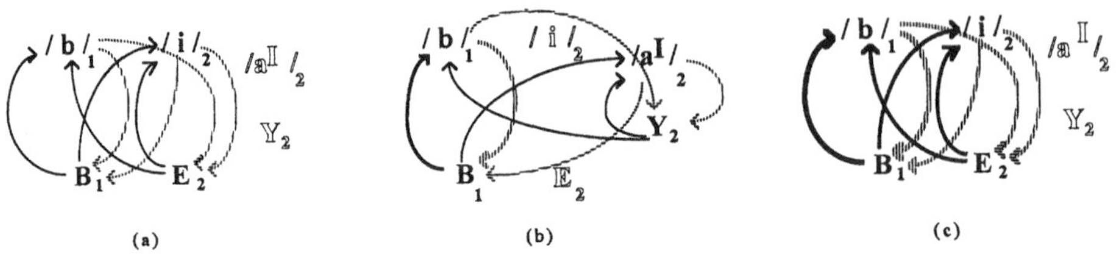

FIGURE 3. Consistent crosstalk in a recurrent network capable of covariant learning. a. The consequences for the connections between letter nodes and phoneme nodes of a learning trial for the word BE. b. Likewise for the word BY. c. A second learning trial for the word BE. The presence of a line between two nodes indicates an increase in the connection weights between them. The width of the lines ranks the strength of the relations that accumulates across learning trials. Notice across the figures that the width of the lines between B_1 and $/b/_1$ increases faster than those of other relations. This is due to consistent crosstalk.

priori relations between letter and phoneme nodes, these are not reflected in the model's behavior. If the network were presented again with the word BE, the activation values of the same four nodes would grow symmetrically toward resonance. Thus, the model's behavior would reflect only coarse-grain, word-size knowledge. This simple figure illustrates how dimensionally nonspecific (holistic) relations can emerge behaviorally prior to relational rule-like knowledge (Thelen & Smith, 1994). The later emergence of rule-like knowledge is illustrated in panels **3B** and **3C**.

A subsequent BY trial, shown in **3B**, brings its four pairs of nodes into collective resonance ($B_1 \longleftrightarrow /b/_1$, $B_1 \longleftrightarrow /a^I/_2$, $Y_2 \longleftrightarrow /b/_1$, and $Y_2 \longleftrightarrow /a^I/_2$), thus adjusting all the connection weights involved. Notice at this point that the connections between B_1 and $/b/_1$ have been adjusted more often than any other connections. If, in turn, another BE learning trial occurs (**3C**), then the four pairs of connection weights: $B_1 \longleftrightarrow /b/_1$, $B_1 \longleftrightarrow /i/_2$, $E_2 \longleftrightarrow /b/_1$, and $E_2 \longleftrightarrow /i/_2$ are adjusted again by its four component resonances. Because B correlates with the same pronunciation in BY and BE, the configuration of weights in the resonance $B_1 \longleftrightarrow /b/_1$ is tuned toward this strong subword resonance more often than configurations promoting other component resonances. The bidirectional connections between B_1 and $/b/_1$ emerge as a strong rule via consistent crosstalk.

Strong-rule resonances, such as B $\longleftrightarrow$ /b/, are examples of local (fine-grain) dynamics exhibiting relatively high *self-consistency*. After learning, the component resonances of a strong-rule word show themselves behaviorally because they coalesce quickly and thereby facilitate naming. Consequently, even relatively unfamiliar words are named quickly if they are composed of strong rules (Rosson, 1985). Additionally, the naming of pseudowords (e.g., BINT) is primarily constrained by these same fine-grain resonances, as will be seen in our account of developmental dyslexia.

Earlier in this chapter, we described a fictitious nervous system to introduce the construct *resonance*. We also oversimplify the nervous system in this section to introduce our account of dyslexia. Postmortem studies have found anatomical anomalies in the brains of dyslexics (e.g., see Galaburda, Rosen, & Sherman, 1989) that may be due to subtle anomalies in neuronal migration. Small deviations in neural positioning may cause large changes in patterns of connectivity between neurons in different brain regions. A rough analogy with connectivity in network models inspired our "haphazard-connections" hypothesis concerning the performance deficits of dyslexics. (Please do not interpret this rough analogy as a claim to anatomical plausibility. We merely wish to acknowledge the inspiration for our behavioral account.)

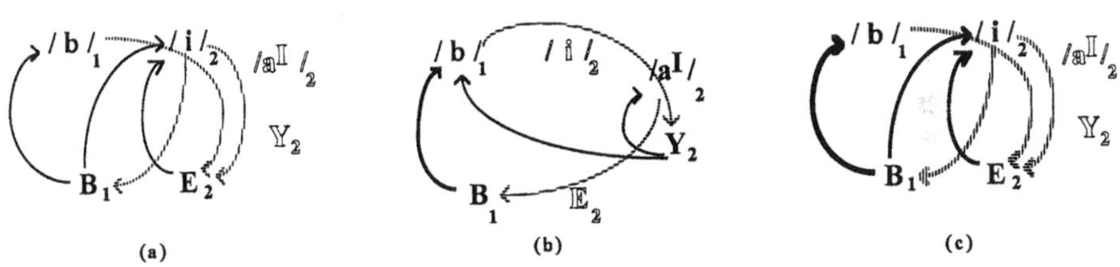

FIGURE 4. An illustration of the failure to develop fine-grain dimensions of word perception. **a.** The consequences for the connections between the letter nodes and phoneme nodes of a learning trial for the word BE when letter and phoneme nodes are haphazardly connected. **b.** Likewise for the word BY. **c.** A second learning trial for the word BE. The presence of a line between two nodes indicates an increase in the connection weights between them. Notice that the bidirectional fine-grain relation between B_1 and $/b/_1$ does not emerge in this haphazardly connected illustration.

Figure 4 illustrates a haphazard pattern of connectivity between letter and phoneme nodes. (Compare this to **Figure 3**, in which all connections are symmetrical.) We can easily track the outcome of covariant learning, given this haphazard connectivity. Once again, consistent crosstalk extracts positive correlations between letter and phoneme nodes. However, although BE and BY share a relatively strong rule (B-/b/), it does not emerge as a self-consistent subresonance in the behavior of the model. This is the key to our account.

In **Figure 4A**, a BE learning trial brings four nodes into collective resonance ($B_1 \rightarrow /b/_1$, $B_1 \longleftrightarrow /i/_2$, $E_2 \leftarrow /b/_1$, and $E_2 \longleftrightarrow /i/_2$), and adjusts six connection weights. As in **Figure 3A**, at this point in the model's development the resonance $B_1E_2 \longleftrightarrow /b_1i_2/$ is an encapsulated whole. The model would correctly produce the whole-word phonology of BE, but its behavior would be opaque to substructures such as $B_1 \longleftrightarrow /b/_1$. A subsequent BY trial, shown in **4B**, brings four pairs of nodes into collective resonance ($B_1 \longleftrightarrow /b/_1$, $B_1 \leftarrow /a^I/_2$, $Y_2 \longleftrightarrow /b/_1$, and $Y_2 \rightarrow /a^I/_2$) and adjusts five connection weights. Notice the difference between panel **B** in **Figure 3** and panel **B** in **Figure 4**. The feedforward connection weight $b_1 \rightarrow /b/_1$ grows faster than other connections in the resonance $B_1E_2 \longleftrightarrow /b_1i_2/$, due to covariant learning. However, the sub-word relation $B_1 \longleftrightarrow /b/_1$ does not grow in self-consistency due to the absence of feedback connections. The haphazard connectivity between B_1 and $/b/_1$ in **Figure 4B** does not allow the emergence of the component resonance $B_1 \longleftrightarrow /b/_1$, and this state of affairs is unchanged by additional BE trials (**4C**).

Consider the subword dynamics for the word BE following the learning trials depicted in **Figure 4**. An advantage due to covariant learning between B_1 and $/b/_1$ may still affect performance. The relatively strong connection $B_1 \rightarrow /b/_1$ would promote a faster overall time to resonance for the word BE. The node $/b/_1$ is strongly activated by B_1, and $/b/_1$ conserves this strong activation for the whole-word resonance when it feeds activation back to E_2. In turn, E_2 feeds this activation forward to $/i/_2$, $/i/_2$ feeds it back to B_1 and E_2, and the activation has been conserved in a *word-size feedback loop*. At no time, however, does a *subresonance* $B_1 \longleftrightarrow /b/_1$ emerge in the behavior of the model.

We noted previously that pseudoword (BINT) naming is primarily dependent upon fine-grain resonances such as $B_1 \longleftrightarrow /b/_1$. These are necessary to insure integrity of pseudoword pronunciation, but they do not reliably emerge in a model with haphazard connectivity. This translates behaviorally into a deficit in pseudoword naming (as well as reading proper names, unfamiliar words, etc.). Skilled pseudoword naming is a fairly predictable function of the statistical relation between words' spellings and their pronunciations (Seidenberg et al, 1994). With haphazard connectivity, a model is robbed of its strongest source of information about how pseudoword spellings relate to phonology. A model analogous to our simple illustration would fail to derive a full complement of the fine-grain letter-phoneme relations necessary to mimic skilled pseudoword naming. Thus, dyslexia may be a failure of perceptual development to derive these fine-grain dimensions of word perception.

Remember, however, that strong positive phonology effects are found in categorization performance with both word (ROWS) and pseudoword (ROZE) homophones. The word effect is explained by the coarse-grain (word-size) phonology that emerges in the haphazardly connected model. However, the effect of pseudoword (ROZE) phonology requires the added contextual constraints on phoneme-semantic dynamics due to availability of the category name (FLOWER). This context is strong enough to cause skilled readers to misinterpret highly familiar homophone words in a categorization task (Jared

& Seidenberg, 1991). We suggest that context acts to compensate for the noisy letter-phoneme dynamics of dyslexic readers via semantic-phoneme dynamics, as illustrated in **Figure 1**.

The essential point for implementing our scheme is that haphazard connectivity blocks the emergence of letter-phoneme resonances (i.e., attractors; see Jordan, this volume) that would normally act relatively independently of the coarse-grain resonances from which they derive. These fine-grain dynamic structures are necessary to mimic the full range of intact performance to printed language. We could even zoom in on the phoneme nodes, revealing their finer-grain resonances between acoustic and articulatory nodes. Then, we could propose haphazard connectivity between acoustic and articulatory nodes, thus precluding the development of proper "phoneme resonances." In this way, the haphazard-connectivity hypothesis might be extended to explain poor performance on phonological-awareness tasks.

Acquired Dyslexia

Patients with *acquired dyslexia* have reading difficulties as a consequence of brain trauma. In a seminal article, Marshall and Newcombe (1973) described two apparently distinct syndromes of acquired dyslexia: *surface* and *deep* dyslexia. Both are defined by characteristic profiles of errors in the naming task (Shallice, 1988). The utility and reliability of syndrome categories is highly controversial (Caramazza, 1986), but we need not endorse such distinctions for our purpose. We merely focus on theoretically important patterns of naming errors associated with each syndrome. Our goal is to produce similar errors in "lesioned" models that previously produced skilled patterns of naming performance. Here, we describe briefly our simulations of two error types identified by Marshall and Newcombe: the regularization error and the semantic error.

Regularization errors are characteristic of surface-dyslexic patients. These errors occur when words such as PINT, with irregular pronunciations, are incorrectly read aloud to rhyme with similar regular words (e.g., MINT, HINT, and LINT). Although skilled readers also occasionally make regularization errors (Kawamoto & Zemblidge, 1992), surface-dyslexic patients make many more. *Semantic errors* are characteristic of deep-dyslexic patients, occurring when words are incorrectly read aloud as semantically related words. For example, the word BUSH might be read aloud incorrectly as TREE. The separate occurrence of semantic and regularization errors is sometimes interpreted as evidence against recurrent-network models (Shallice, 1988; see Van Orden et al, 1996 for a review and counter-argument). Farrar and Van Orden (1994) recently simulated these two error types.

We began with a recurrent-network architecture very similar in structure to the simple illustrations presented earlier in this chapter. Three families of nodes (see **Figure 1**) were "taught" a sample of English words using a Hebbian-type learning algorithm, until the model produced patterns of naming performance similar to those of skilled readers. (Specifically, we implemented 10 learning trials for each "high-frequency" word and one for each "low-frequency" word. The "naming response" was taken from the pattern of the most active phoneme units, and "naming time" was defined as the number of cycles required to generate a coherent pattern of activation of the phoneme units.) Of particular present relevance, the model produced an interaction of frequency with consistency. Low-frequency inconsistent words such as PINT were named more slowly than low-frequency consistent words such as DUCK, whereas all high-frequency words were named quickly (see Waters & Seidenberg, 1985).

To simulate the regularization error, we added noise to our intact network. Noise was implemented as a uniform distribution of small positive or negative changes in activation that were added in each cycle to the activation values of randomly chosen nodes. The effect was to erode the network's capacity for correct naming of words having highly inconsistent pronunciations, such as PINT. Instead of the correct phonemes for PINT, the network activated phoneme nodes that regularized PINT to rhyme with MINT. In effect, the noise eroded the (already weak) phoneme-semantic constraints, such that the model expressed only the powerful constraints of letter-phoneme dynamics. Because letter-phoneme dynamics primarily reflect the strongest correlations between letters and phonemes, these dynamics lead to regularization errors (Kawamoto & Zemblidge, 1992).

It is important to note that we could have implemented noise in many ways with the same consequences. For example, we could have introduced small changes in randomly chosen connection strengths. Similarly, the *locus* of noise is not crucial. Bidirectional flow of activation makes the system highly interdependent. Consequently, noise introduced anywhere in the system spreads throughout the system, in the next time step.

To simulate semantic errors, we further "lesioned" the noisy network that was producing regularization errors. We set all of the letter-phoneme connection weights to zero, effectively "cutting" those connections. (We could have cut fewer connections with the same effect; the minimum proportion of disconnections that produces semantic errors is interdependent with other modeling choices.) Subsequently, the network produced semantic errors; when presented with BUSH the network generated a pattern of activity across phoneme nodes corresponding to TREE. Setting the letter-phoneme connections to zero creates a highly unstable network, causing it to rely on semantic-phoneme dynamics, the most reliable remaining source of constraints. However, in the absence of letter-phoneme constraints, semantic-phoneme dynamics are sometimes misled into a semantic error from which the normally weak letter-semantic dynamics cannot rescue the network. Semantic errors are especially likely when semantic nodes of one word (BUSH) are strongly correlated with phoneme nodes of a different word (TREE).

Mind as Embodied Action

The attentive reader has probably noticed that we refrained from calling our cognitive account a neural network. We chose the term recurrent network instead. This choice was not made simply for aesthetic reasons. The term *neural* in "neural network" has a connotation we wish to avoid. It suggests too strong an analogy with the nervous system, or (worse yet) that cognition should be explained in terms of the nervous system. For all we know, there may be more differences than similarities between cognitive behavior and the behavior of nervous systems (cf. Freeman, 1995). Thus, "nodes" in our recurrent network do not refer to neurons, nor do their interconnections refer to synapses.

The previous caveat resurrects the issue of the metaphor underlying our account, specifically, *mind as embodied action* (see Introduction). This view of cognition, as described by Varela et al (1991), means, first, that

> ...cognition depends upon the kinds of experience that come from having a body with various sensorimotor capacities, and second, that these individual sensorimotor capacities are themselves embedded in a more encompassing biological, psychological, and cultural context. (pp. 172-173)

The term *action* emphasizes that perception and action are fundamentally interrelated (see also Turvey & Carello, 1981). This proposed interdependence is obvious in the behavior of a recurrent network. When our network is presented with a printed word, the initial activations of the letter nodes feed activation forward to phonologic and semantic nodes that, in turn, feed activation back to the letter nodes. In a strongly nonlinear system, feedforward and feedback activation undergo successive nonlinear transformations, eventually producing a resonant whole. This illustrates, in model systems, the irreducible interdependence of input and output (see also van Leeuwen, Steyvers, & Nooter, 1995).

It is easy to confuse nodes in a network (or *subsymbols*, Van Orden et al 1990; Van Orden & Goldinger, 1994) with traditional symbolic representations. This is another confusion that we wish to avoid. Nodes are not mental representations. They are pragmatic notations for purposes of modeling or illustration, and serve a narrative function only. Thus, nodes are not psychologically real (atomic) units of cognition, and they should not be attributed causal or explanatory properties independent of the dynamics in which they participate (cf. Putnam, 1981; Turvey & Carello, 1981). The network models that we propose refer only to observed patterns of intercorrelation between laboratory manipulations and performance. This pragmatic approach to modeling implies that nodes chosen at one grain size are no more "real" than nodes that might have been chosen at other grain sizes. To rephrase, modeling a particular behavior or performance requires a smart (or pragmatic) choice of nodes (see also Putnam, 1981). We choose nodes at the finest grain size of reliable covariation between laboratory manipulations and performance (Van Orden et al, 1990; Van Orden & Goldinger, 1994). A fairly good description of results obtained from reading and spelling research can be achieved using letter, phoneme, and semantic (feature) nodes. Note that a degree of arbitrariness is unavoidable; other research problems may lead to other choices. The crux is the ability to mimic, in as parsimonious a model as possible, the observed complexity in laboratory performance.

The pragmatic constraints on the choice of nodes are further revealed by the following example. Choosing to model letter perception using a grain size of letter nodes would be too "coarse" a choice, because it ignores reliable effects of font, handwriting, and other episodic variables (e.g., Sanocki, 1987). A finer grain size (e.g., letter-feature nodes) would be necessary. Conversely, choosing to model sentence comprehension using a grain size of letter nodes is a too "fine" a choice of grain. Using letter nodes to model the phenomena of sentence comprehension would build in unnecessary detail. These performance phenomena are typically word- or morpheme-size, and nodes chosen at these grain sizes would be more appropriate. The strength of our model comes not from the discovery of true nodes, but from the generality and simplicity of its behavioral account. Dynamic interactions among small families of nodes can account for a vast literature of performance in laboratory reading tasks, and the entailed principles extend to cognition at large (Stone & Van Orden, 1994; Van Orden et al, 1990; Van Orden & Goldinger, 1994).

In summary, we have shown that a simple recurrent network has great utility for mimicking phenomena in reading and spelling. Our approach motivates principled explanations for why phonology is fundamental to reading and spelling, why spelling is more difficult than reading, why words with multiple rime spellings are more slowly read, and why developmental and acquired dyslexics have difficulties in reading.

ACKNOWLEDGMENTS

Preparation of this chapter was funded by research grants 5 R29 NS26247-05 from the National Institute of Neurological Disorders and Stroke, National Institutes of Health, awarded to Guy Van Orden; 1 R29 DC02629-01 from the National Institute on Deafness and Other Communication Disorders, National Institutes of Health, awarded to Steve Goldinger; 1 F32 DC 00187-01 from the National Institute on Deafness and Other Communication Disorders, National Institutes of Health, awarded to Bill Farrar, and a NATO-Fellowship N58-92 from the Netherlands Organization for Scientific Research (NWO) awarded to Anny Bosman. Correspondence should be addressed to the senior author.

REFERENCES

Note: The chapter numbers in which each reference is cited are indicated in parentheses following the reference.

Abbott, L.F., Varela, J.A., Sen, K., & Nelson, S.B. (1997). Synaptic depression and cortical gain control. *Science*, **275**, 220-224. (Pt 5)

Abeles, M. (1991). *Corticonics: Neural circuits of the cerebral cortex.* Cambridge, UK: Cambridge University Press. (8)

Abeles, M., Bergman, H., Margalit, E., & Vaadia, E. (1993). Spatiotemporal firing patterns in the frontal cortex of behaving monkeys. *J. of Neurophysiology*, **70**, 1629-1638. (8)

Abraham, W.C., Bliss, T.V.P., & Goddard, G.V. (1985). Heterosynaptic changes accompany long-term but not short-term potentiation of the perforant path in the anaesthetized rat. *J. of Physiology*, **363**, 335-349. (6)

Abraham, W.C., Dragunow, M., & Tate, W.P. (1991). The role of immediate early genes in the stabilization of long-term potentiation. *Molecular Neurobiology*, **5**, 297-314. (5)

Abraham, W.C., Mason, S.E., Demmer, J., Williams, J.M., Richardson, C.L., Tate, W.P., Lawlor, P.A., & Dragunow, M. (1993). Correlations between immediate early gene induction and the persistence of long-term potentiation. *Neuroscience*, **56**, 717- 727. (6)

Ackerman, R., Finch, D., Babb, T., & Engel, J. (1984). Increased glucose metabolism during long-duration recurrent inhibition of hippocampal pyramidal cells. *J. of Neuroscience*, **4**, 251-264. (24)

Ackley, D., & Littman, M. (1992). Interactions between learning and evolution. In C.G. Langton, C. Taylor, J.D. Farmer, & S. Rasmussen (Eds.), *Artificial Life II: Proceedings of the Workshop on Artificial Life* (pp. 487-509). Redwood City: Addison-Wesley. (4)

Agnati, L.F., Zoli, M., Pich, E.M., Benfenati, F., & Fuxe, K. (1990). Aspects of neural plasticity in the central nervous system VII: Theoretical aspects of brain communication and computation. *Neurochemistry International*, **16**, 479-500. (5)

Agranoff, B.W., & Klinger, P.D. (1964). Puromycin effect on memory formation in the goldfish. *Science*, **146**, 952-953. (6)

Ahissar, E., Vaadia, E., Ahissar, M., Bergman, H., Arieli, A., & Abeles, M. (1992). Dependence of cortical plasticity on correlated activity of single neurons and on behavioral context. *Science*, **257**, 1412-1415. (8)

Aiple, F., & Krüger, J. (1988). Neuronal synchrony in monkey striate cortex: Interocular signal flow and dependency on spike rates. *Experimental Brain Research*, **72**, 141-149. (8)

Akshoomoff, N.A., Courchesne, E., Pres, G.A., & Iragui, V. (1992). Contribution of the cerebellum to neuropsychological functioning: Evidence from a case of cerebellar degenerative disorder. *Neuropsychologia*, **30**, 315-328. (24)

Albright, T.D., & Gross, C.G. (1990). Do inferior temporal cortex neurons encode shape by acting as Fourier descriptor filters? *Proceedings of International Conference on Fuzzy Logic and Neural Networks*, Izuka, Japan, 375-378. (9)

Albus, J. (1981). *Brains, behavior, and robotics.* Peterborough, NH: Byte Books. (15)

Alexander, G.E., Crutcher, M.D., & DeLong, M.R. (1990). Basal ganglia-thalamocortical circuits: Parallel substrates for motor, oculomotor, "prefrontal" and "limbic" functions. In H.B.M. Uylings, C.G. VanEden, J.P.C. De Bruin, M.A. Corner, & M.G.P. Feenstra (Eds.), *The prefrontal cortex: Its structure, function and pathology* (pp. 119-146). Amsterdam: Elsevier.

Alexander, G.E., DeLong, M.R., & Strick, P.L. (1986) Parallel organization of functionally segregated circuits linking basal ganglia and cortex. *Annual Review of Neuroscience*, **9**, 357-381. (17)

Alkon, D.L., Amaral, D.G., Bear, M.F., Black, J., Carew, T.J., Cohen, N.J., Disterhoft, J.F., Eichenbaum, H., Golski, S., Gorman, L.K., Lynch, G., McNaughton, B.L., Mishkin, M., Moyer, J.R.J., Olds, J.L., Olton, D.S., Otto, T., Squire, L.R., Staubli, U., Thompson, L.T., & Wible, C. (1991). Learning and memory. *Brain Research Review*, **16**, 193-220. (6)

Allen, R.B. (1989). Adaptive training of connectionist state machines. Paper presented at ACM Computer Science Conference, Louisville, KY. (16)

Allman, J. (1990). Evolution of the cortex. In A. Peters & E.G. Jones (Eds.), *Cerebral cortex, Vol. 8a: Comparative structure and evolution of cerebral cortex, Part I* (pp. 269-283). New York: Plenum Press. (4)

Altman, J. (1969). Autoradiographic and histological studies of postnatal neurogenesis IV: Cell proliferation and migration in the anterior forebrain, with special reference to persisting neurogenesis in the olfactory bulb. *J. of Comparative Neurology*, **137**, 433-458. (2)

Altman, J., & Bayer, S.A. (1990). Mosaic organization of the hippocampal neuroepithelium and the multiple germinal sources of dentate granule cells. *J. of Comparative Neurology*, **301**, 325-342. (3)

Altman, J., & Bayer, S.A. (1995). *Atlas of prenatal rat brain development.* Boca Raton, FL: CRC Press. (3)

Alvarado, M., & Rudy, J. (1995). A comparison of kainic acid plus colchicine and ibotenic acid-induced hippocampal formation damage on four configural tasks in rats. *Behavioral Neuroscience*, **109**, 1052-1062. (12)

Amaral, D.G. (1987). Memory: Anatomical organization of candidate brain regions. N. Plum (Ed.,), *Handbook of physiology: Section I. Neurophysiology, Vol. 5: Higher functions of the brain* (pp. 211-294). Bethesda, MD: American Physiological Society. (18) (26)

Amaral, D.G., & Witter, M.P. (1989). The three-dimensional organization of the hippocampal

formation: A review of anatomical data. *Neuroscience*, **31**, 571-591. (12)
Ambros-Ingerson, J., Granger, R., & Lynch, G. (1990). Simulation of paleocortex performs hierarchical clustering. *Science*, **247**, 1344-1348. (12)
Amit, D.J. (1989). *Modeling brain function: The world of attractor neural networks*. Cambridge, UK: Cambridge University Press. (4)
Anderson, C.W., & Miller, T.M., III. (1990). Challenging control problems. In T. Miller, R.S. Sutton, & P.J. Werbos (Eds.), *Neural networks for control*. Cambridge, MA: MIT Press. (22)
Anderson, J. (1987). Skill acquisition: Compilation of weak-method problem solutions. *Psychological Review*, **94**, 192-210. (24)
Anderson, J.A., & Rosenfeld, E. (1988). *Neurocomputing*. Cambridge, MA: MIT Press. (7)
Anderson, J.R. (1983). *The architecture of cognition*. Cambridge, MA: Harvard University Press. (1)
Angenstein, F., Matthies, H., Jr., Staeck, S., Reymann, K.G., & Staak, S. (1992). The maintenance of hippocampal long-term potentiation is paralleled by a dopamine-dependent increase in glycoprotein fucosylation. *Neurochemistry International*, **21**, 403-408. (6)
Antón, P.S., Lynch, G., & Granger, R. (1991). Computation of frequency-to-spatial transform by olfactory bulb glomeruli. *Biological Cybernetics*, **65**, 407-414. (7)
Antonini, A., & Stryker, M.P. (1993). Development of individual geniculocortical arbors in cat striate cortex and effects of binocular impulse blockade. *J. of Neuroscience*, **13**, 3549-3573. (8)
Apicella, P., Ljungberg, T., Scarnati, E., & Schultz, W. (1991) Responses to reward in monkey dorsal and ventral striatum. *Experimental Brain Research*, **85**, 491-500. (17)
Arbas, E.A., Meinertzhagen, I.A., & Shaw, S.R. (1991). Evolution in nervous systems. *Annual Review of Neuroscience*, **14**, 9-38. (4)
Arbib, M.A., & Dominey, P.F. (1995) Modeling the roles of the basal ganglia in timing and sequencing saccadic eye movements. In J.C. Houk, J.L. Davis, & D.G. Beiser (Eds.), *Models of information processing in the basal ganglia* (pp. 149-162). Cambridge, MA: MIT Press. (17)
Armstrong, R.C., & Montminy, M.R. (1993). Transsynaptic control of gene expression. *Annual Review of Neuroscience*, **16**, 17-29. (5)
Artola, A., Bröcher, S., & Singer, W. (1990). Different voltage-dependent thresholds for inducing long-term depression and long-term potentiation in slices of rat visual cortex. *Nature*, **347**, 69-72. (8)
Artola, A., & Singer, W. (1987). Long-term potentiation and NMDA receptors in rat visual cortex. *Nature*, **330**, 649-652. (8)
Artola, A., & Singer, W. (1993). Long-term depression of excitatory synaptic transmission and its relationship to long-term potentiation. *Trends in Neuroscience*, **16**, 480-487. (8)
Asanuma, H. (1975). Recent developments in the study of the columnar arrangements of neurons in the motor cortex. *Physiological Review*, **55**, 143-156. (24)
Asanuma, H., & Brooks, V.B. (1965). Recurrent cortical effects following stimulation of internal capsule. *Archives Italiennes de Biologie*, **103**, 220-246. (13)
Ashe, J., Taira, M., Smyrnis, N., Pellizzer, G., Georgakopoulos, T., Lurito, J.T., & Georgopoulos, A.P. (1993). Motor cortical activity preceding a memorized movement trajectory with an orthogonal bend. *Experimental Brain Research*, **95**, 118-130. (13)
Ashmead, D.H., McCarty, M.E., Lucas, L.S., & Belvedere, M.C. (1993). Visual guidance in infants' reaching toward suddenly displaced targets. *Child Development*, **64**, 1111-1127. (15)
Aston-Jones, G. (1985). Behavioral functions of locus coeruleus derived from cellular attributes. *Physiological Psychology*, **13**, 118-126. (5)
Atal, B.S., Chang, J.J., Mathews, M.V., & Tukey, J.W. (1978). Inversion of articulatory-to-acoustic transformation in the vocal tract by a computer sorting technique. *J. of the Acoustic Society of America*, **63**, 1535-1555. (25)
Augustine, J. (1985). The insular lobe in primates including humans. *Neurological Research*, **7**, 2-8. (24)
Austin, C., & Cepko, C. (1989). Lineage analysis of the mouse cerebral cortex using retrovirus vectors. *Society of Neuroscience Abstracts*, **15**, 599. (4)
Austin, C.P., & Cepko, C.L. (1990). Cellular migration patterns in the developing mouse cerebral cortex. *Development*, **110**, 713-732. (2)
Azrin, N.H., Hutchinson, R.R., & McLaughlin, R. (1965). The opportunity for aggression is an operant reinforcer during aversive stimulation. *J. of the Experimental Analysis of Behavior*, **8**, 171-180. (14)
Baddeley, A.D. (1976). *The psychology of memory*. New York: Basic Books. (26)
Baldwin, J.M. (1896). A new factor in evolution. *American Naturalist*, **30**, 441-451. (4)
Baldwin, J.M. (1895). *Mental development in the child and race*. New York: Macmillan. (14)
Barlow, H.B. (1972). Single units and sensation: A neuron doctrine for perceptual psychology? *Perception*, **1**, 371-394. (10)
Barlow, H.B. (1985). The twelfth Bartlett memorial lecture: The role of single neurons in the psychology of perception. *Quarterly J. of Experimental Psychology*, **37A**, 121-145. (10)
Barnes, C.A. (1979). Memory deficits associated with senescence: A neurophysiological and behavioural study in the rat. *J. of Comparative and Physiological Psychology*, **93**, 74-104. (6)
Barnes, D. (1994). Stimulus equivalence and relational frame theory. *Psychological Record*, **43**, 91-124. (26)
Barnes, D., & Hampson, P.J. (1992). Learning to learn: The contribution of behaviour analysis to connectionist models of inferential skill in humans. In G. Orchard (Ed.), *Neural computing research and applications: Part one* (pp. 129-138). Bristol, UK: IOP Publishing. (26)
Barnes, D., & Hampson, P.J. (1993). Stimulus equivalence and connectionism: Implications for behavior analysis and cognitive science. *Psychological Record*, **43**, 617-638. (26)
Barnes, D., & Holmes, Y. (1991). Radical behaviorism, stimulus equivalence, and human cognition. *Psychological Record*, **41**, 19-31. (26)
Barondes, S.M., & Cohen, H.D. (1967). Comparative effects of cycloheximide and puromycin on cerebral protein synthesis and consolidation of memory in mice. *Brain Research*, **4**, 44-51. (6)

Barsalou, L.W. (1993). Flexibility, structure and linguistic vagary in concepts: Manifestations of a compositional system of perceptual symbols. In A.F. Collins, S.E. Gathercole, M.A. Conway, & P.E. Morris (Eds.), *Theories of memory*. Hillsdale, NJ: Lawrence Erlbaum. (11)

Barto, A.G. (1990). Connectionist learning for control: An overview. In T. Miller, R.S. Sutton, & P.J. Werbos (Eds.), *Neural networks for control*. Cambridge, MA: MIT Press. (15)

Barto, A.G. (1992). Reinforcement learning and adaptive critic methods. In D.A. White & D.A. Sofge (Eds.), *Handbook of intelligent control: Neural, fuzzy, and adaptive approaches* (pp. 469-491). New York: Van Nostrand Reinhold. (19)

Barto, A.G. (1994). Reinforcement learning control. *Current Opinion in Neurobiology*, **4**, 888-893. (19)

Barto, A.G. (1995). Adaptive critics and the basal ganglia. In J.C. Houk, J.L. Davis, & D.C. Beiser (Eds.), *Models of information processing in the basal ganglia* (pp. 215-232). Cambridge, MA: MIT Press. (17) (20)

Barto, A.G., Bradtke, S.J., & Singh, S.P. (1995). Learning to act using real-time dynamic programming. *Artificial Intelligence*, **72**, 81-138. (19) (22)

Barto, A.G., & Sutton, R.S. (1982). Simulation of anticipatory responses in classical conditioning by a neuron-like adaptive element. *Behavioral Neuroscience*, **4**, 221-235. (20)

Barto, A.G., Sutton, R.S., & Anderson, C.W. (1983). Neuronlike elements that can solve difficult learning control problems. *IEEE Transactions on Systems, Man, and Cybernetics*, **13**, 835-846. Reprinted in J.A. Anderson & E. Rosenfeld (Eds.), *Neurocomputing: Foundations of research*. Cambridge, MA: MIT Press, 1988. (19) (20) (21)

Barto, A.G., Sutton, R.S., & Watkins, C.J.C.H. (1989). *Learning and sequential decision making*. Technical Report 89-95, Computer and Information Science, University of Massachusetts, Amherst, MA. (21)

Bashir, Z.I., Bortolotto, Z.A., Davies, C.H., Berretta, N., Irving, A.J., Seal, A.J., Henley, J.M., Jane, D.E., Watkins, J.C., & Collingridge, G.L. (1993). Induction of LTP in the hippocampus needs synaptic activation of glutamate metabotropic receptors. *Nature*, **363**, 347-350. (6)

Baskys, A., & Malenka, R.C. (1991). Agonists at metabotropic glutamate receptors presynaptically inhibit EPSCs in neonatal rat hippocampus. *J. of Physiology (London)*, **444**, 687-701. (6)

Baudry, M. (1986). Long-term potentiation and kindling: Similar biochemical mechanisms? *Advances in Neurophysiology*, **44**, 401-410. (6)

Baulac, M., Verney, C., & Berger, B. (1986). Dopaminergic innervation of parahippocampal and hippocampal regions in the rat. *Review of Neurology*, **142**, 895-905. (6)

Bayer, S.A. (1983). 3H-thymidine-radiographic studies of neurogenesis in the rat olfactory bulb. *Experimental Brain Research*, **50**, 329-340. (2)

Baylis, G.C., Rolls, E.T., & Leonard, C.M. (1985). Selectivity between faces in the responses of a population of neurons in the cortex in the superior temporal sulcus of the monkey. *Brain Research*, **342**, 91-102. (9) (10)

Baylis, G.C., Rolls, E.T., & Leonard, C.M. (1987). Functional subdivisions of the temporal lobe neocortex. *J. of Neuroscience*, **7**, 330-342. (9)

Bear, M.F., Cooper, L.N., & Ebner, F.F. (1987). A physiological basis for a theory of synapse modification. *Science* , **237**, 42-48. Reprinted in K.L. Kelner & D.E. Koshland, Jr. (Eds.), *Molecules to models: Advances in neuroscience* (pp. 168-181). Washington, DC: American Association for the Advancement of Science, 1989. (8)

Bear, M.F., Kleinschmidt, A., Gu, Q., & Singer, W. (1990). Disruption of experience-dependent synaptic modifications in striate cortex by infusion of an NMDA receptor antagonist. *J. of Neuroscience* , **10**, 909-925. (8) (18)

Bear, M.F., & Singer, W. (1986). Modulation of visual cortical plasticity by acetylcholine and noradrenaline. *Nature* , **320**, 172-176. (8) (18)

Bechtel, W., & Abrahamsen, A. (1991). *Connectionism and the mind*. Oxford: Basil Blackwell. (26)

Beecher, M.D. (1988). Some comments on the adaptationist approach to learning. In R.C. Bolles & M. D. Beecher (Eds.), *Evolution and learning*. Hillsdale, NJ: Lawrence Erlbaum. (4)

Bekkers, J.M., & Stevens, C.F. (1990). Presynaptic mechanism for long-term potentiation in the hippocampus. *Nature*, **346**, 724- 729. (5)

Belew, R.K., McInerney, J., & Schraudolph, N.N. (1992). Evolving networks: Using the genetic algorithm with connectionist learning. In C.G. Langton, C. Taylor, J.D. Farmer, & S. Rasmussen (Eds.), *Artificial Life II: Proceedings of the Workshop on Artificial Life* (pp. 511-547). Redwood City, CA: Addison-Wesley. (4)

Bellman, R.E. (1957). *Dynamic programming*. Princeton, NJ: Princeton University Press. (22)

Benguerel, A.P., & Cowan, H.A. (1974). Coarticulation of upper lip protrusion in French. *Phonetica*, **30**, 41-55. (25)

Beninger, R.J. (1983). The role of dopamine in locomotor activity and learning. *Brain Research Review*, **6**, 173-196. (6)

Beninger, R.J., & Hahn, B.L. (1983) Pimozide blocks establishment but not expression of amphetamine-produced environment-specific conditioning. *Science*, **220**, 1304-1306. (17)

Berardi, N., Cellerino, A., Comenici, L., Fagiolini, M., Pizzorusso, T., Cattaneo, A., & Maffei, L. (1994). Monoclonal antibodies to nerve growth factor affect the postnatal development of the visual system. *Proceedings of the National Academy of Sciences USA*, **91**, 684-688. (8)

Berent, I., & Perfetti, C.A. (1995). A rose is a REEZ: The two-cycles model of phonology assembly in reading English. *Psychological Review*, **102**, 146-184. (27)

Berger, A.J., Gaspar, P.L., & Verney, R.N. (1991). Dopaminergic innervation of the cerebral cortex: Unexpected differences between rodents and primates. *Trends in Neurosciences*, **14**, 21-27. (18)

Bergevin, R., & Levine, M.D. (1993). Generic object recognition: Building and matching course descriptions from line drawings. *IEEE Transactions on Pattern Analysis and Machine Intelligence*, **15**, 19-36. (11)

Berntson, G.G., & Torello, M.W. (1982). The paleocerebellum and the integration of behavioral function. *Physiological Psychology*, **10**, 2-12. (24)

Berridge, M. (1986). Second messenger dualism in neuromodulation and memory. *Nature*, **323**, 294-295. (6)

Berry, S., & Thompson, R. (1979). Medial septal

lesions retard classical conditioning of the nictitating membrane response in rabbits. *Science*, **205**, 209-211. (12)

Bersini, H., & Detours, V. (1994). Asynchrony induces stability in cellular automata based models. In R.A. Brooks & P. Maes (Eds.), *Artificial life IV: Proceedings of the Fourth International Workshop on the Synthesis and Simulation of Living Systems* (pp. 382-387). Cambridge, MA: MIT Press. (4)

Berthier, N.E. (1996). Learning to reach: A mathematical model. *Developmental Psychology*, **32**, 811-823. (15)

Berthier, N.E., Barto, A.G., & Moore, J.W. (1991). Linear systems analysis of the relationship between firing of deep cerebellar neurons and the classically conditioned nictitating membrane response in rabbits. *Biological Cybernetics*, **65**, 99-105. (20)

Berthier, N.E., & McCarty, M.E. (1996). Speed of infant reaching during the first year: Confirmation of a prediction. *Infant Behavior and Development*, **19**, 331. (15)

Berthier, N.E., & Moore, J.W. (1986). Cerebellar Purkinje cell activity related to the classically conditioned nictitating membrane response. *Experimental Brain Research*, **63**, 341-350. (20)

Berthier, N.E., & Moore, J.W. (1990). Activity of deep cerebellar nuclear cells during classical conditioning of nictitating membrane extension in rabbits. *Experimental Brain Research*, **83**, 44-54. (20)

Berthier, N.E., Singh, S.P., Barto, A.G., & Houk, J.C. (1993). Distributed representation of limb motor programs in arrays of adjustable pattern generators. *J. of Cognitive Neuroscience*, **5**, 56-78. (15) (24)

Bertsekas, D.P. (1987). *Dynamic programming: Deterministic and stochastic models*. Englewood Cliffs, NJ: Prentice-Hall. (19)

Betarbet, R., Zigova, T., Bakay, R.A.E., & Luskin, M.B. (in press). Dopaminergic and GABAergic interneurons of the olfactory bulb are derived from the neonatal subventricular zone. *International J. of Developmental Neuroscience*. (2)

Betarbet R., Zigova T., Bakay, R.A.E., & Luskin, M.B. (1995). Transmitter phenotype of olfactory bulb interneurons derived from neuronal progenitor cells of the neonatal subventricular zone. *Abstracts of the XVIth Meeting of the Association for Chemoreception Sciences*, 210-213. (2)

Biederman, I. (1987). Recognition-by-components: A theory of human image understanding. *Psychological Review*, **94**, 115-147. (11)

Biederman, I., & Cooper, E.E. (1991a). Evidence for complete translational and reflectional invariance in visual object priming. *Perception*, **20**, 595-593. (11)

Biederman, I., & Cooper, E.E. (1991b). Priming contour deleted images: Evidence for intermediate representations in visual object recognition. *Cognitive Psychology*, **23**, 393-419. (11)

Biederman, I., & Cooper, E.E. (1992). Size invariance in visual object priming. *J. of Experimental Psychology: Human Perception and Performance*, **18**, 121-133. (11)

Biederman, I., & Gerhardstein, P.C. (1993). Recognizing depth-rotated objects: Evidence and conditions for 3-dimensional viewpoint invariance. *J. of Experimental Psychology: Human Perception and Performance*, **19**, 1162-1182. (11)

Biederman, I., & Gerhardstein, P.C. (1995). Viewpoint-dependent mechanisms in visual object recognition: A critical analysis. *J. of Experimental Psychology: Human Perception and Performance*. (in press). (11)

Bindra, D. (1968) Neuropsychological interpretation of the effects of drive and incentive-motivation on general activity and instrumental behavior. *Psychological Review*, **75**, 1-22. (17)

Bischoff, S. (1986). Mesohippocampal dopamine system: characterization, functional and clinical implications. In R.L. Isaacson & K.H. Pribram (Eds.), *The hippocampus* (pp. 1-32). New York: Plenum Press. (6)

Bitterman, M.E. (1965). Phyletic differences in learning. *American Psychologist*, **20**, 396-410. (4)

Blakemore, C., & Cooper, G.F. (1970). Development of the brain depends on the visual environment. *Nature*, **228**, 477-478. (8)

Blakemore, C., & Tobin, E.A. (1972). Lateral inhibition between orientation detectors in the cat's visual cortex. *Experimental Brain Research* , **15**, 439-440. (8)

Blakemore, C., van Sluyters, R.C., & Movshon, J.A. (1976). Synaptic competition in the kitten's visual cortex. *Cold Spring Harbor Symposia on Quantitative Biology XL*, **40**, 601-609. (8)

Blazis, D.E.J., & Moore, J.W. (1991). Conditioned stimulus duration in classical trace conditioning: Test of a real-time computational model. *Behavioral Brain Research*, **43**, 73-78. (20)

Bliss, T.V.P., & Collingridge, G.L. (1993). A synaptic model of memory: Long-term potentiation in the hippocampus. *Nature*, **361**, 31-39. (5) (6)

Bliss, T.V., & Gardner-Medwin, A.R. (1973). Long-lasting potentiation of synaptic transmission in the dentate area of the unanaesthetized rabbit following stimulation of the perforant path. *J. of Physiology (London)*, **232**, 357-374. (6)

Bliss, T.V.P., Goddard, G.V., & Riives, M. (1983). Reduction of long-term potentiation in the dentate gyrus of the rat following selective depletion of monoamines. *J. of Physiology*, **334**, 475-491. (6)

Bliss, T.V., & Lomo, T. (1973). Long-lasting potentiation of synaptic transmission in the dentate area of the anaesthetized rabbit following stimulation of the perforant path. *J. of Physiology (London)*, **232**, 331-356. (6)

Blitzer, R.D., Gil, O., & Landau, E.M. (1990). Cholinergic stimulation enhances long-term potentiation in the CA1 region of rat hippocampus. *Neuroscience Letters*, **119**, 207-210. (6)

Blough, D.S. (1975). Steady state data and a quantitative model of operant generalization and discrimination. *J. of Experimental Psychology: Animal Behavior Processes*, **1**, 3-21. (18)

Bode-Greuel, K., & Singer, W. (1989). The development of N-methyl-D-aspartate receptors in cat visual cortex. *Developmental Brain Research* , **46**, 197-204. (8)

Bolles, R.C. (1970). Species-specific defense reactions and avoidance learning. *Psychological Review*, **77**, 32-48. (4)

Bortolotto, Z.A., & Collingridge, G.L. (1993). Characterisation of LTP induced by the activation of glutamate metabotropic receptors in area CAl of the hippocampus. *Neuropharmacology*, **32**, 1-9. (6)

Bosman, A.M.T., & Van Orden, G.C. (in press). Why spelling is more difficult than reading. In C.A. Perfetti, M. Fayol, & L. Rieben (Eds.),

Learning to spell. Hillsdale, NJ: Lawrence Erlbaum. (27)

Bosman, A.M.T., van Leerdam, M., & de Gelder, B. (1995). The /A/ in AMEN is different from the /A/ in ARK: Phonologic effects in normal-reading and reading-disabled Dutch children. Unpublished manuscript. (27)

Boussaoud, D., Desimone, R., & Ungerleider, L.G. (1991). Visual topography of area TEO in the macaque. *J. of Comparative Neurology*, **306**, 554-575. (9)

Bouton, M., & Peck, C. (1989). Context effects on conditioning, extinction and the reinstatement in an appetitive conditioning paradigm. *Animal Learning & Behavior*, **17**, 188-198. (12)

Braitenberg, V. (1967). Is the cerebellar cortex a biological clock in the millisecond range? *Progress in Brain Research*, **25**, 334-336. (7)

Braitenberg, V. (1978). In R. Heim & G. Palm, (Eds.), *Lecture notes in Biomathematics 21: Theoretical approaches in complex systems* (pp. 171-188). Berlin: Springer. (8)

Bramham, C.R. (1992). Opioid receptor dependent long-term potentiation: Peptidergic regulation of synaptic plasticity in the hippocampus. *Neurochemistry International*, **20**, 441-455. (6)

Bramham, C.R., Errington, M.L., & Bliss, T.V. (1988). Naloxone blocks the induction of long-term potentiation in the lateral but not in the medial perforant pathway in the anesthetized rat. *Brain Research*, **449**, 352-356. (6)

Bramham, C.R., Milgram, N.W., & Srebro, B. (1991). Delta opioid receptor activation is required to induce LTP of synaptic transmission in the lateral perforant path in vivo. *Brain Research*, **567**, 42-50. (6)

Bressler, S.L., Coppola, R., & Nakamura, R. (1993). Episodic multiregional cortical coherence at multiple frequencies during visual task performance. *Nature* , **366**, 153-156. (8)

Briggs, C.A., & McAfee, D.A. (1988). Long-term potentiation at nicotinic synapses in the rat superior cervical ganglion. *J. of Physiology (London)*, **404**, 129-144. (6)

Bröcher, S., Artola, A., & Singer, W. (1992). Agonists of cholinergic and noradrenergic receptors facilitate synergistically the induction of long-term potentiation in slices of rat visual cortex. *Brain Research* , **573**, 27-36. (8)

Brown, M.C., Hopkins, W.G., & Keynes, R.J. (1991). *Essentials of neural development*. Cambridge, UK: Cambridge University Press. (4)

Brown, R., & Hanlon, C. (1970). Derivational complexity and order of acquisition in child speech. In J.R. Hayes (Ed.), *Cognition and the development of language*. New York: Wiley. (18)

Brown, T., Kariss, E., & Keenan, C. (1990). Hebbian synapses: Biophysical mechanisms and algorithms. *Annual Review of Neuroscience*, **13**, 475-511. (12)

Bruce, C., Desimone, R., & Gross, C.G. (1981). Visual properties of neurons in a polysensory area in superior temporal sulcus of the macaque. *J. of Neurophysiology*, **46**, 369-384. (9)

Bruce, V., Cowey, A., Ellis, A.W., & Perrett, D.I. (Eds.) (1992). Processing the facial image. *Philosophical Transactions of the Royal Society of London B*, **335**, 1-128. (10)

Bruck, M. (1988). The word recognition and spelling of dyslexic children. *Reading Research Quarterly*, **23**, 51-69. (27)

Bryant, P., & Bradley, L. (1985). *Children's reading problems*. Oxford, UK: Basil Blackwell Ltd. (27)

Buhl, E., Halasy, K., & Somogyi, P. (1994). Diverse sources of hippocampal unitary inhibitory postsynaptic potentials and the number of synaptic release sites. *Nature* **368**, 823-828. (7)

Bullier, J., Munk, M.H.J., & Nowak, L.G. (1992). Synchronization of neuronal firing in areas V1 and V2 of the monkey. *Society Neuroscience Abstracts* , **18**, 11.7. (8)

Bullock, S., Lössner, B., Krug, M., Frey, S., Rose, S.P., & Matthies, H. (1990). Posttetanic long-term potentiation in rat dentate area increases postsynaptic 411B immunoreactivity. *J. of Neurochemistry*, **55**, 708-713. (6)

Buonomano, D.V., & Byrne, J.H. (1990). Long-term synaptic changes produced by a cellular analog of classical conditioning in Aplysia. *Science*, **249**, 420-423. (4)

Buonomano, D.V., & Mauk, M.D. (1994). Neural network model of the cerebellum: Temporal discrimination and the timing of motor responses. *Neural Computation*, **6**, 38-55. (7) (20)

Buonomano, D.V., & Merzenich, M.M. (1995). Temporal information transformed into a spatial code by a neural network with realistic properties. *Science*, **267**, 1026-1028. (7)

Burgard, E.C., & Sarvey, J.M. (1990). Muscarinic receptor activation facilitates the induction of long-term potentiation (LTP) in the rat dentate gyrus. *Neuroscience Letters*, **116**, 34-39. (6)

Burke, R.E., Marks, W.B., & Ulfhake, B. (1992). A parsimonious description of motoneuron dendritic morphology using computer simulation. *J. of Neuroscience*, **12**, 2403-2406. (3)

Bush, R.R., & Mosteller, F. (1955). *Stochastic models for learning*. New York: Wiley. (21)

Buzsaki, G., & Gage, F.H. (1989). Absence of long-term potentiation in the subcortically deafferented dentate gyrus. *Brain Research*, **484**, 94-101. (6)

Byrne, J.H., Gingrich, K.J., & Baxter, D.A. (1989). Computational capabilities of single neurons: Relationship to simple forms of associative and nonassociative learning in Aplysia. In R.D. Hawkins & G.H. Bower (Eds.), *Computational models of learning in simple neural models* (pp. 31-63). (4)

Cabelli, R.J., Hohn, A., & Shatz, C.J. (1995). Inhibition of ocular dominance column formation by infusion of NT-4/5 or BDNF. *Science*, **267**, 1662-1666. (8)

Cajal, S.R. (1911). *Hystologie du systeme nerveux de l'homme et des vertebres*. Paris: Maloine. (6)

Callaway, E.M., & Katz, L.C. (1990). Emergence and refinement of clustered horizontal connections in cat striate cortex. *J. of Neuroscience* , **10**, 1134-1153. (8)

Callaway, E.M., & Katz, L.C. (1991). Effects of binocular deprivation on the development of clustered horizontal connection in cat striate cortex. *Proceedings of the National Academy of Sciences USA*, **88**, 745-749. (8)

Calverley, R.K.S., & Jones, D.G. (1990). Contributions of dendritic spines and perforated synapses to synaptic plasticity. *Brain Research Reviews*, **15**, 215-249. (5)

Caminiti, R., Johnson, P.B., Galli, C., Ferraina, S., Burnod, Y., & Urbano, A. (1991). Making arm movements within different parts of space: The premotor and motor cortical representation of a coordinate system for reaching at visual targets. *J. of Neuroscience*, **11**, 1182-1197. (13)

Caminiti, R., Johnson, P.B., & Urbano, A. (1990). Making arm movements within different parts of space: Dynamic aspects in the primate motor cortex. *J. of Neuroscience*, **10**, 2039-2058. (13)

Campbell, D.T. (1975). On the conflicts between biological and social evolution and between psychology and moral tradition. *American Psychologist*, **30**, 1103-1126. (14)

Canevari, L., Richter-Levin, G., & Bliss, T.V.P. (1994). LTP in the dentate gyrus is associated with a persistent NMDA receptor-dependent enhancement of synaptosomal glutamate release. *Brain Research*, **667**, 115-117. (6)

Caramazza, A. (1986). On drawing inferences about the structure of normal cognitive systems from the analysis of patterns of impaired performance: The case for single-patient studies. *Brain and Cognition*, **5**, 41-66. (27)

Carlin, R.K., & Siekevitz, P. (1983). Plasticity in the central nervous system: Do synapses divide? *Proceedings of the National Academy of Sciences USA*, **80**, 3517-3521. (5)

Carlson-Radvanski, L., & Irwin, D.E. (in press). Memory for structural descriptions across eye movements. *Journal of Experimental Psychology: Learning, Memory, and Cognition*. (11)

Carmeliet, P., Schoonjans, L., Kieckens, L., Ream, B., Degen, J., Bronson, R., De Vos, R., Van den Oord, J.J., Collen, D., & Mulligan, R.C. (1994). Physiological consequences of loss of plasminogen activator gene function in mice. *Nature*, **368**, 419-424. (6)

Carmignoto, G., Canella, R., Candeo, P., Comelli, M.C., & Maffei, L. (1993). Effects of nerve growth factor on neuronal plasticity of the kitten visual cortex. *J. of Physiology* , **464**, 343-360. (8)

Carriquiry, A.L., Ireland, W.P., Kliemann, W., & Uemura, E. (1991). Statistical evaluation of dendritic growth models. *Bulletin of Mathematical Biology*, **53**, 579-589. (3)

Caserta, F., Eldred, W.D., Fernandez, E., Hausman, R.E., Stanford, L.R., Bulderev, S.V., Schwarzer, S., & Stanley, H.E. (1995). Determination of fractal dimension of physiologically characterized neurons in two and three dimensions. *J. of Neuroscience Methods*, **56**, 133-144. (3)

Castellucci, V.F., Frost, W.N., Goelet, P., Montarolo, P.G., Schacher, S., Morgan, J.A., Blumenfeld, H., & Kandel, E.R. (1986). Cell and molecular analysis of long-term sensitization in Aplysia. *J. of Physiology (Paris)*, **81**, 349-357. (6)

Castro, C.A., Silbert, L.H., McNaughton, B.L., & Barnes, C.A. (1989). Recovery of spatial learning deficits after decay of electrically induced synaptic enhancement in the hippocampus. *Nature*, **342**, 545-548. (6)

Catania, A.C. (1973). The concept of the operant in the analysis of behavior. *Behaviorism*, **1**, 103-116. (14)

Cavicchio, D.J. (1970). Adaptive search using simulated evolution. Unpublished doctoral dissertation, University of Michigan. (4)

Caviness, V.S., Jr., & Sidman, R.L. (1973). Time of origin of corresponding cell classes in the cerebral cortex of normal and reeler mutant mice: An autoradiographic analysis. *J. of Comparative Neurology*, **148**, 141-151. (4)

Chapman, B., Zahs, K.R., & Stryker, M.P. (1991). Relation of cortical cell orientation selectivity to alignment of receptive fields of the geniculo-cortical column in ferret visual cortex afferents that arborize within a single orientation. *J. of Neuroscience*, **11**, 1347-1358. (2)

Chatterjee, A., Morris, M., Bowers, D., Williamson, D., Doty, L., & Heilman, K. (1993). Cholinergic treatment of an amnestic man with a basal forebrain lesion: Theoretical implications. *J. of Neurology, Neurosurgery and Psychiatry*, **56**, 1282-1289. (12)

Chelazzi, L., Miller, E.K., Duncan, J., & Desimone, R. (1993). A neural basis for visual search in inferior temporal cortex. *Nature*, **363**, 345-347. (9)

Chenn, A., & McConnell, S.K. (1995). Cleavage orientation and the asymmetric inheritance of Notch1 immunoreactivity in mammalian neurogenesis. *Cell*, **82**, 631-641. (2)

Chetkovich, D.M., Gray, R., Johnston, D., & Sweatt, J.D. (1991). N-methyl-D-aspartate receptor activation increases cAMP levels and voltage-gated Ca2+ channel activity in area CA1 of hippocampus. *Proceedings of the National Academy of Sciences USA*, **88**, 6467-6471. (6)

Chetkovich, D.M., & Sweatt, J.D. (1993). NMDA receptor activation increases cyclic AMP in area CA1 of the hippocampus via calcium/calmodulin stimulation of adenylyl cyclase. *J. of Neurochemistry*, **61**, 1933-1942. (6)

Chevalier, G., & Deniau, J.M. (1990). Disinhibition as a basic process in the expression of striatal functions. *Trends in Neuroscience*, **13**, 277-280. (24)

Chino, Y.M., Smith, E.L., Kaas, J.H., Sasaki, Y., & Cheng, H. (1995). Receptive-field properties of differentiated visual cortical neurons after topographic map reorganization in adult cats. *J. of Neuroscience*, **15**, 2417-2433. (8)

Chioda, L.A., & Berger, T.W. (1986). Interactions between dopamine and amino-acid induced excitation and inhibition in the striatum. *Brain Research*, **375**, 198-203. (18)

Church, R.M. (1984). Properties of the internal clock. In J. Gibbon & L.G. Allan (Eds.), *Timing and time perception* (pp. 566- 582). New York: New York Academy of Sciences. (7)

Clark, K.A., Randall, A.D., & Collingridge, G.L. (1994). A comparison of paired-pulse facilitation of AMPA and NMDA receptor-mediated excitatory postsynaptic currents in the hippocampus. *Experimental Brain Research*, **101**, 272-278. (7)

Clark, R.E., & Lavond, D.G. (1993). Reversible lesions of the red nucleus during acquisition and retention of a classically conditioned behavior in rabbits. *Behavioral Neuroscience*, **107**, 264-270. (20)

Cliff, D. (1991). Computational neuroethology: A provisional manifesto. In J.A. Meyer & S.W. Wilson (Eds.), *From animals to animats: Proceedings of the First International Conference on Simulation of Adaptive Behavior* (pp. 263-272). Cambridge, MA: MIT Press. (4)

Clouse, J.A. (1995). *On training automated agents*. Unpublished doctoral dissertation. Department of Computer and Information Science, University of Massachusetts, Amherst, MA. (22)

Clouse, J.A., & Utgoff, P.E. (1992). A teaching method for reinforcement learning. *Machine Learning: Proceedings of the Ninth International Conference* (pp. 92-101). San Mateo, CA: Morgan Kaufmann. (19) (22)

Clowes, M.B. (1967). Perception, picture processing and computers. In N.L. Collins & D. Michie (Eds.), *Machine intelligence* (Vol. 1, pp. 181-197). Edinburgh, Scotland: Oliver & Boyd. (11)

Coan, E.J., & Collingridge, G.L. (1987). Characterization of an N-methyl-D-aspartate receptor component of synaptic transmission in rat hippocampal slices. *Neuroscience*, **22**, 1-8. (6)

Coates, P.W., Fermini, B., Strahlendorf, J.C., & Strahlendorf, H.K. (1992). Utilization of three-dimensional culture for early morphometric and electrophysiological analyses of solitary cerebellar neurons. *Developmental Neuroscience*, **14**, 35-43. (3)

Cohen, N. (1984). Preserved learning capacity in amnesia: Evidence for multiple learning systems. In L. Squire & N. Butters (Eds.), *Neuropsychology of memory* (pp. 83-103). New York: Guilford Press. (12)

Cohen, N.J., & Eichenbaum, H. (1993). *Memory, amnesia and the hippocampal system*. Cambridge, MA: MIT Press. (26)

Cole, A.J., Saffen, D.W., Baraban, J.M., & Worley, P.F. (1989). Rapid increase of an immediate early gene messenger RNA in hippocampal neurons by synaptic NMDA receptor activation. *Nature*, **340**, 474-476. (6)

Colley, P.A., & Routtenberg, A. (1993). Long-term potentiation as synaptic dialogue. *Brain Research Review*, **18**, 115-122. (6)

Collingridge, G.L. (1992). The mechanism of induction of NMDA receptor-dependent long-term potentiation in the hippocampus. *Experimental Physiology*, **77**, 771-797. (6)

Collingridge, G.L., & Bliss, T.V.P. (1987). NMDA receptors—their role in long-term potentiation. *Trends in Neuroscience*, **10**, 288-293. (6)

Commons, M.L., Grossberg, S., & Staddon, J.E.R. (Eds.) (1991). *Neural network models of conditioning and action*. Hillsdale, NJ: Lawrence Erlbaum. (26)

Conel, J. (1939). *The postnatal development of the human cerebral cortex: Vol. I, Cortex of the newborn*. Cambridge, MA: Harvard University Press. (15)

Cooper, M.L., & Rakic, P. (1983). Gradients of cellular maturation and synaptogenesis in the superior colliculus of the fetal rhesus monkey. *J. of Comparative Neurology*, **215**, 165- 186. (4)

Corner, M.A., & Ramakers, G.J.A. (1992). Spontaneous firing as an epigenetic factor in brain development—physiological consequences of chronic tetrodotoxin and picrotoxin exposure on cultured rat neocortex neurons. *Developmental Brain Research*, **65**, 57-64. (4)

Crair, M.C., & Malenka, R.C. (1995). A critical period for long-term potentiation at thalamocortical synapses. *Nature*, **375**, 325-328. (8)

Creager, R., Dunwiddie, T., & Lynch, G. (1980). Paired-pulse facilitation in the CA1 region of the in vitro rat hippocampus. *J. of Physiology*, **299**, 409-424. (7)

Crick, F. (1984). Function of the thalamic reticular complex: the searchlight hypothesis. *Proceedings of the National Academy of Sciences USA*, **81**, 4586-4590. (9)

Crick, F.H.C., & Asunama, C. (1986). Certain aspects of the anatomy and physiology of the cerebral cortex. In J.L. McClelland, D.E. Rumelhart, & The PDP Research Group (Eds.), *Parallel distributed processing* (Vol. 2, pp. 333-371). Cambridge, MA: MIT Press. (18) (26)

Crites, R.H., & Barto, A.G. (1996). Improving elevator performance using reinforcement learning. In D. S. Touretzky, M.C. Mozer & M.E. Hasselmo (Eds.), *Advances in neural information processing systems 8* (pp. 1017-1023). Cambridge, MA: MIT Press. (19)

Crossman, E. (1959). A theory of the acquisition of speed skill. *Ergonomics*, **2**, 153-166. (24)

Crosson, B. (1992). Theories of subcortical structures in language. In B. Crosson (Ed.), *Subcortical functions in language and memory* (pp. 111-144). New York: Guildford. (24)

Cullinan, V., Barnes, D., Hampson, P.J., & Lyddy, F. (1994). A transfer of explicitly and nonexplicitly trained sequence responses through equivalence relations: An experimental demonstration and connectionist model. *Psychological Record*, **44**, 559-585. (26)

Damasio, H., & Damasio, A.R. (1989). *Lesion analysis in neuropsychology*. New York: Oxford University Press. (23)

Darian-Smith, C., & Gilbert, C.D. (1995). Topographic reorganization in the striate cortex of the adult cat and monkey is cortically mediated. *J. of Neuroscience*, **15**, 1631-1647. (8)

Das, A., & Gilbert, C.D. (1995). Long-range horizontal connections and their role in cortical reorganization revealed by optical recording of cat primary visual cortex. *Nature* , **375**, 780-784. (8)

Davies, C.H., Davies, S.N., & Collingridge, G.L. (1990). Paired-pulse depression of monosynaptic GABA-mediated inhibitory postsynaptic responses in rat hippocampus. *J. of Physiology*, **424**, 513-531. (7)

Davies, S.N., Lester, R.A., Reymann, K.G., & Collingridge, G.L. (1989). Temporally distinct pre- and post-synaptic mechanisms maintain long-term potentiation. *Nature*, **338**, 500-503. (6)

Davis, K., & Mohs, R. (1982). Enhancement of memory processes in Alzheimer's disease with multiple-dose intravenous physostigmine. *American J. of Psychiatry*, **139**, 1421-1424. (12)

Davis, M. (1989). Neural systems involved in fear-potentiated startle. *Annals of the New York Academy of Sciences*, **563**, 165- 183. (4)

Dawkins, R. (1986). *The blind watchmaker*. New York: Norton. (4) (14)

Dayan, P., & Hinton, G.E. (1993). Feudal reinforcement learning. In S.J. Hanson, J.D. Cohen, & C.L. Giles (Eds.), *Advances in neural information processing systems 5* (pp. 271-278). San Mateo, CA: Morgan Kaufmann. (19)

Dean, P. (1976). Effects of inferotemporal lesions on the behavior of monkeys. *Psychological Bulletin*, **83**, 41-71. (9)

de Curtis, M., & Llinas, R. (1993). Entorhinal cortex long-term potentiation evoked by theta-patterned stimulation of associative fibers in the isolated in vitro guinea pig brain. *Brain Research*, **600**, 327-330. (12)

Deisz, R.A., & Prince, D.A. (1989). Frequency-dependent depression of inhibition in guinea-pig neocortex in vitro by $GABA_B$ receptor feed-back on GABA release. *J. of Physiology*, **412**, 513-541. (7)

De Jong, K.A. (1975). An analysis of the behavior of a class of genetic adaptive systems. Unpublished doctoral dissertation, University of Michigan. (4)

de Jonge, M., & Racine, R.J. (1985). The effects of repeated induction of long-term potentiation in the dentate gyrus. *Brain Research*, **328**, 181-185. (6)

Demb, J., Desmond, J., Wagner, A., Vaidya, C., Glover, G., & Gabrieli, J. (1995). Semantic encoding and retrieval in the left inferior prefrontal cortex: A functional MRI study of task difficulty and process specificity. *J. of Neuroscience*, in press. (24)

Demmer, J., Dragunow, M., Lawlor, P.A., Mason, S.E., Leah, J.D., Abraham, W.C., & Tate, W.P. (1993). Differential expression of immediate early genes after hippocampal long-term potentiation in awake rats. *Molecular Brain Research*, **17**, 279-286. (6)

Dennett, D.C. (1981). *Brainstorms*. Cambridge, MA: MIT Press. (Pt 5)

Dennett, D.C. (1995). *Darwin's dangerous idea: Evolution and the meanings of life*. New York: Simon & Schuster. (4)

Descarries, L., Séguéla, P., & Watkins, K.C. (1991). Nonjunctional relationships of monoamine axon terminals in the cerebral cortex of adult rat. In K. Fuxe & L.F. Agnati (Eds.), *Volume transmission in the brain: Novel mechanisms for neural transmission* (pp. 53-62). New York: Raven Press. (5)

DeShutter, E. (1992). A consumer guide to neuronal modeling software. *Trends in Neuroscience*, **15**, 462-465. (3)

Desimone, R., Albright, T.D., Gross, C.G., & Bruce, C. (1984). Stimulus-selective properties of inferior temporal neurons in the macaque. *J. of Neuroscience*, **4**, 2051-2062. (9) (10)

Desimone, R., Miller, E.K., & Chelazzi, L. (1994). The interaction of neural systems for attention and memory. In C. Koch (Ed.), *Large scale theories of the brain* (pp. 75-91). Cambridge/London: MIT Press. (9)

Desmond, J.E. (1990). Temporally adaptive responses in neural models: The stimulus trace. In M. Gabriel & J. Moore (Eds.), *Learning and computational neuroscience: Foundations of adaptive networks* (pp. 421-461). Cambridge, MA: MIT Press. (20)

Desmond, J.E., & Moore, J.W. (1988). Adaptive timing in neural networks: The conditioned response. *Biological Cybernetics*, **58**, 405-415. (20)

Desmond, J.E., & Moore, J.W. (1991a). Activity of red nucleus neurons during the classically conditioned rabbit nictitating membrane response. *Neuroscience Research*, **10**, 260-279. (20)

Desmond, J.E., & Moore, J.W. (1991b). Altering the synchrony of stimulus trace processes: Tests of a neural-network model. *Biological Cybernetics*, **65**, 161-169. (20)

De Valois, R.L., & De Valois, K.K. (1988). *Spatial vision*. New York: Oxford University Press. (18)

Devane, W.A., Hanuš, L., Breuer, A., Pertwee, R.G., Stevenson, L.A., Griffin, G., Gibson, D., Mandelbaum, A., Etinger, A., & Mechoulam, R. (1992). Isolation and structure of a brain constituent that binds to the cannabinoid receptor. *Science*, **258**, 1946-1949. (5)

Dickinson, A. (1980) *Contemporary animal learning theory*. Cambridge, UK: Cambridge University Press. (17) (21)

Dickinson, A., Hall, G., & Mackintosh, N.J. (1976) Surprise and the attenuation of blocking. *J. of Experimental Psychology: Animal Behavior Processes*, **2**, 313-322. (17)

Dickinson, S.J., Pentland, A.P., & Rosenfeld, A. (1992). 3-D shape recovery using distributed aspect matching. *IEEE Transactions on Pattern Analysis and Machine Intelligence*, **14**, 174-198. (11)

Dinsmoor, J.A. (1995). The role of observing and attention in establishing stimulus control. *J. of the Experimental Analysis of Behavior*, **43**, 365-381. (14)

Domenici, L., Cellerino, A., Berardi, N., Cattaneo, A., & Maffei, L. (1994). Antibodies to nerve growth factor (NGF) prolong the sensitive period for monocular deprivation in the rat. *Neuroreport*, **5**, 2041-2044. (8)

Domjan, M. (1983). Biological constraints on instrumental and classical conditioning: Implications for general process theory. *Psychology of learning and motivation* (Vol. 17, pp. 215-277). New York: Academic Press. (1)

Domjan, M., & Hollis, K.L. (1988). Reproductive behavior: A potential model for adaptive specializations in learning. In R.C. Bolles & M.D. Beecher (Eds.), *Learning and evolution*. Hillsdale, NJ: Lawrence Erlbaum. (4)

Donahoe, J.W., Burgos, J.E., & Palmer, D.C. (1993). A selectionist approach to reinforcement. *J. of the Experimental Analysis of Behavior*, **60**, 17-40. (4) (6) (18) (26)

Donahoe, J.W., Crowley, M.A., Millard, W.J., & Stickney, K.A. (1982). A unified principle of reinforcement. *Quantitative models of behavior* (Vol. 2, pp. 493-521). Cambridge, MA: Ballinger. (1) (14) (18)

Donahoe, J.W., & Palmer, D.C. (1989). The interpretation of complex behavior: Some reactions to *Parallel Distributed Processing*. *J. of the Experimental Analysis of Behavior*, **51**, 399-416. (18)

Donahoe, J.W., & Palmer, D.C. (1994). *Learning and complex behavior*, Boston: Allyn & Bacon. (1) (4) (16) (14) (18) (26)

Donahoe, J.W., Palmer, D.C., & Burgos, J.E. (1996, in press). The S-R Issue: Its status in behavior analysis and in Donahoe and Palmer's *Learning and Complex Behavior*. *J. of the Experimental Analysis of Behavior*. (14) (18)

Donegan, N.H., Gluck, M.A., & Thompson, R.F. (1989). Integrating behavioral and biological models of conditioning. In R.D. Hawkins & G.H. Bower (Eds.), *Psychology of learning and motivation* (Vol. 22, pp. 109-156). Academic Press. (12)

Doucet, G., Descarries, L., & Garcia, S. (1986) Quantification of the dopamine innervation in adult rat neostriatum. *Neuroscience*, **19**, 427-445. (17)

Douglas, R.J., & Martin, K.A.C. (1991). A functional microcircuit for cat visual cortex. *J. of Physiology*, **440**, 735-769. (7)

Douglas, R.J., Martin, K.A.C., & Whitteridge, D. (1989). A canonical microcircuit for neocortex. *Neural Computation*, **1**, 480-488. (8)

Doyère, V., Burette, F., Rédini-Del Negro, C., & Laroche, S. (1993a). Long-term potentiation of hippocampal afferents and efferents to prefrontal cortex: Implications for associative learning. *Neuropsychologia*, **31**, 1031-1053. (6)

Doyère, V., Rédini-Del Negro, C., & Laroche, S. (1993b). Changes in synaptic function of an LTP form in associative learning and memory. *Neuroscience Research Communications*, **13** Suppl. **1**, S23-S26. (6)

Dragunow, M., Abraham, W.C., Goulding, M., Mason, S.E., Robertson, H.A., & Faull, R.L. (1989a). Long-term potentiation and the induction of c-fos mRNA and proteins in the dentate gyrus of unanesthetized rats. *Neuroscience Letters*, **101**, 274-280. (6)

Dragunow, M., Currie, R.W., Faull, R.L.M., Robertson, H.A., & Jansen, K. (1989b). Immediate-early genes, kindling and long-term potentiation. *Neuroscience Biobehavioral Reviews*, **13**, 301-313. (6)

Dronkers, N. (1995). Unpublished observations. (24)

Dronkers, N., Redfern, B., & Lucy, C. (1995). Lesion localization in chronic Wernicke's aphasia. *Proceedings of the 1995 conference of the society for cognitive neuroscience*, San Francisco, CA. (24)

Duffy, C., Teyler, T.J., & Shashoua, V.E. (1981). Long-term potentiation in the hippocampal slice: Evidence for stimulated secretion of newly synthesized proteins. *Science*, **212**, 1148-1151. (6)

Dumery, V., Derer, P., & Blozovski, D. (1988). Enhancement of passive avoidance learning through small doses of intra-amygdaloid physostigmine in the young rat: Its relation to the development of acetylcholinesterase. *Developmental Psychobiology*, **21**, 553-565. (12)

Dunwiddie, T.V., Roberson, N.L., & Worth, T. (1982). Modulation of long-term potentiation: Effects of adrenergic and neuroleptic drugs. *Pharmacology Biochemistry and Behavior*, **17**, 1257-1264. (6)

Durfee, E.H., Lesser, V.R., & Corkill, D.D. (1989). Cooperative distributed problem solving. In A. Barr, L. Cohen, & E.A. Feigenbaum (Eds.), *The handbook of artificial intelligence: Volume IV*. Reading, MA: Addison-Wesley. (22)

Dymond, S.O., & Barnes, D. (1995). A transformation of self-discrimination response functions in accordance with the arbitrarily applicable relations of sameness, more-than, and less-than. *J. of the Experimental Analysis of Behavior*, **64**, 163-184. (26)

Dymond, S.O., & Barnes, D. (1996). A transformation of self-discrimination response functions in accordance with the arbitrarily applicable relations of sameness and opposition. *Psychological Record*, in press. (26)

Dyson, S.E., & Jones, D.G. (1984). Synaptic remodelling during development and maturation: Junction differentiation and splitting as a mechanism for modifying connectivity. *Developmental Brain Research*, **13**, 125-137. (5)

Eacott, M.J., Gaffan, D., & Murray, E.A. (1994). Preserved recognition memory for small sets, and impaired stimulus identification for large sets, following rhinal cortex ablations in monkeys. *European J. of Neuroscience*, **6**, 1466-1478. (9)

Easter, S.S., Jr., Ross, L.S., & Frankfurter, A. (1993). Initial tract formation in the mouse brain. *J. of Neuroscience*, **13**, 285-299. (2)

Eccles, J.C. (1964). *The physiology of synapses*. Berlin: Springer. (6)

Eccles, J.C. (1989). *Evolution of the brain: Creation of the self*. London: Routledge. (4)

Eckenhoff, M.F., & Rakic, P. (1991). A quantitative analysis of synaptogenesis in the molecular layer of the dentate gyrus in the rhesus monkey. *Developmental Brain Research*, **64**, 129-135. (4)

Eckhorn, R., Bauer, R., Jordan, W., Brosch, M., Kruse, W., Munk, M., & Reitboek, H.J. (1988). Coherent oscillations: a mechanism of feature linking in the visual cortex? Multiple electrode and correlation analyses in the cat. *Biological Cybernetics*, **60**, 121-130. (8)

Eckhorn, R., Frien, A., Bauer, R., Woelbern, T., & Kehr, H. (1993). High frequency (60-90 Hz) oscillations in primary visual cortex of awake monkey. *NeuroReport*, **4**, 243-246. (8)

Eckhorn, R., Schanze, T., Brosch, M., Salem, W., & Bauer, R. (1992). In E. Basar & T.H. Bullock (Eds.), *Induced Rhythms in the Brain* (pp.47-82). Boston: Birkhauser. (8)

Edelman, G.M. (1987). *Neural Darwinism: The theory of neuronal group selection*. New York: Basic Books. (8) (4)

Edelman, G.M. (1989). *The remembered present: A biological theory of consciousness*. New York: Basic Books. (8)

Edelman G.M. (1993). Neural Darwinism: Selection and reentrant signaling in higher brain function. *Neuron*, **10**, 115-125. (3)

Edelman, S. (1995). Representation, similarity and the chorus of prototypes. *Minds and Machines*, **5**, 45-68. (9)

Edgley, S.A., & Lidierth, M. (1987). Discharges of cerebellar Golgi cells during locomotion in cats. *J. of Physiology (London)*, **392**, 315-332. (20)

Egelman, D.M., Person, C., & Montague, P.R. (1995). A predictive model for diffuse systems matches human choices in a simple decision-making task. *Society of Neuroscience Abstracts*, in press. (21)

Eichenbaum, H., Cohen, N.J., Otto, T., & Wible, C. (1991). Learning and memory: A snapshot without the album. *Brain Research Reviews*, **16**, 209-215. (12)

Eichenbaum, H., Cohen, N.J., Otto, T., & Wible, C. (1992b). Memory representation in the hippocampus: Functional domain and functional organization. In L.R. Squire, G. Lynch, N.M. Weinberger, & J.L. McGaugh (Eds.), *Memory organization and locus of change* (pp. 163-204). Oxford, UK: Oxford University Press. (12)

Eichenbaum, H., Fagan, A., & Cohen, N.J. (1986). Normal olfactory discrimination learning set and facilitation of reversal learning after medial-temporal damage in rats: Implication for an account of preserved learning abilities in amnesia. *J. of Neuroscience*, **6**, 1876-1884. (12)

Eichenbaum, H., Fagan, A., Mathews, P., & Cohen, N.J. (1988). Hippocampal system dysfunction and odor discrimination learning in rats: Impairment or facilitation depending on representational demands. *Behavioral Neuroscience*, **102**, 331-339. (12)

Eichenbaum, H., Otto, T., & Cohen, N.J. (1992a). The hippocampus—What does it do? *Behavioral and Neural Biology*, **57**, 2-36. (12)

Eichenbaum, H., Otto, T., & Cohen, N.J. (1994). Two functional components of the hippocampal

memory system. *Behavioral and Brain Sciences*, **17**, 449-465. (26)
Eichenbaum, H., Otto, T., Wible, C., & Piper, J. (1991). Building a model of the hippocampus in olfaction and memory. In J. Davis & H. Eichenbaum (Eds.), *Olfaction as a model for computational neuroscience* (pp. 167-210). Cambridge, MA: MIT Press. (12)
Eilers, J., Augustine, G.J., & Konnerth, A. (1995). Subthreshold synaptic Ca^{2+} signaling in fine dendrites and spines of cerebellar Purkinje neurons. *Nature*, **373**, 155-158. (20)
Eliot, L.S., Dudai, Y., Kandel, E.R., & Abrams, T.W. (1989). Ca2+/calmodulin sensitivity may be common to all forms of neural adenylate cyclase. *Proceedings of the National Academy of Sciences USA*, **86**, 9564-9568. (6)
Ellis, R., & Allport, D.A. (1986). Multiple levels of representation for visual objects: A behavioural study (pp. 245-257). In A.G. Cohen & J.R. Thomas (Eds.), *Artificial Intelligence and its Applications*. New York: Wiley. (11)
Ellis, R., Allport, D.A., Humphreys, G.W., & Collis, J. (1989). Varieties of object constancy. *Quarterly J. of Experimental Psychology*, **4**, 775-796. (11)
Elman, J.L. (1995). Language as a dynamical system. In R. Port & T. Gelder (Eds.), *Mind as motion*. Cambridge: MIT Press. (Pt 6)
Emmeche, C. (1991). *The garden in the machine: The emerging science of Artificial Life*. Translated by S. Sampson (1994). Princeton, NJ: Princeton University Press. (4)
Engel, A.K., König, P., Gray, C.M., & Singer, W. (1990). *European J. of Neuroscience*, **2**, 588-592. (8)
Engel, A.K., König, P., Kreiter, A.K., Schillen, T.B., & Singer, W. (1992). Temporal coding in the visual cortex: New vistas on integration in the nervous system. *Trends in Neuroscience*, **15**, 218-226. (9)
Engel, A.K., König, P., Kreiter, A.K., & Singer, W. (1991). Interhemispheric synchronization of oscillatory neuronal responses in cat visual cortex. *Science*, **252**, 1177-1179. (8) (11)
Engel, A.K., König, P., & Singer, W. (1991). Direct physiological evidence for scene segmentation by temporal coding. *Proceedings of the National Academy of Sciences USA*, **88**, 9136-9140. (8)
Engel, A.K., Kreiter, A.K., König, P., & Singer, W. (1991). Synchronization of oscillatory neuronal responses between striate and extrastriate visual cortical areas of the cat. *Proceedings of the National Academy of Sciences USA*, **88**, 6048-6052. (8) (18)
Ennaceur, A., & Meliani, K. (1992). Effects of physostigmine and scopolamine on rats' performances in object-recognition and radial-maze tests. *Psychopharmacology*, **109**, 321-330. (12)
Ennouri, K., Dubon, C., Notides, C., & Bloch, H. (1994). Visual control upon hand reaching movements in neonates. *Infant Behavior and Development* (Special ICIS Issue), **17**, 626. (15)
Enns, J.T. (1992). Sensitivity of early human vision to 3-D orientation in line-drawings. *Canadian J. of Psychology*, **46**, 143-169. (11)
Erickson, R.P. (1968). Stimulus coding in topographic and nontopographic afferent modalities: On the significance of the activity of individual neurons. *Psychological Review*, **75**, 447-465. (9)
Errington, M.L., Lynch, M.A., & Bliss, T.V.P. (1987). Long-term potentiation in the dentate gyrus: Induction and increased glutamate release are blocked by D⁻ aminophosphonovalerate. *Neuroscience*, **20**, 279-284. (6)
Etscorn, F., & Stephens, R. (1973). Establishment of conditioned taste aversions with a 24-hour CS-US interval. *Physiological Psychology*, **1**, 251-253. (4)
Fallon, J.H., & Laughlin, S.E. (1987). Monoamine innervation of cerebral cortex and a theory of the role of monoamines in cerebral cortex and basal ganglia. In E.G. Jones & A. Peters (Eds.), *Cerebral cortex. Vol. 6: Further aspects of cortical function, including hippocampus* (pp. 41-127). New York: Plenum. (18)
Farah, M. (1992). Is an object an object an object: Cognitive and neuropsychological investigations of domain-specificity in visual object recognition. *Current Directions in Psychological Science*, **1**, 164-169. (11)
Farrar, W.T., & Van Orden, G.C. (1994). Simulation of surface and deep dyslexia in a unified network. Poster presented at the Annual Meeting of the Psychonomic Society, November, St. Louis, MO. (27)
Fazeli, M.S., Breen, K., Errington, M.L., & Bliss, T.V.P. (1994). Increase in extracellular NCAM and amyloid precursor protein following induction of long-term potentiation in the dentate gyrus of anaesthetized rats. *Neuroscience Letters*, **169**, 77-80. (6)
Fazeli, M.S., Corbet, J., Dunn, M.J., Dolphin, A.C., & Bliss, T.V.P. (1993). Changes in protein synthesis accompanying long-term potentiation in the dentate gyrus in vivo. *J. of Neuroscience*, **13**, 1346-1353. (6)
Fazeli, M.S., Errington, M.L., Dolphin, A.C., & Bliss, T.V. (1988). Long-term potentiation in the dentate gyrus of the anaesthetized rat is accompanied by an increase in protein efflux into push-pull cannula perfusates. *Brain Research*, **473**, 51-59. (6)
Fazeli, M.S., Errington, M.L., Dolphin, A.C., & Bliss, T.V. (1990). Increased efflux of a haemoglobin-like protein and an 80 kDa protease into push-pull perfusates following the induction of long-term potentiation in the dentate gyrus. *Brain Research*, **521**, 247-253. (6)
Feasey, K.J., Lynch, M.A., & Bliss, T.V. (1986). Long-term potentiation is associated with an increase in calcium-dependent, potassium-stimulated release of [14C]glutamate from hippocampal slices: An ex vivo study in the rat. *Brain Research*, **364**, 39-44. (6)
Feig, S., & Lipton, P. (1993). Pairing the cholinergic agonist carbachol with patterned Schaffer collateral stimulation initiates protein synthesis in hippocampal CA1 pyramidal cell dendrites via a muscarinic, NMDA-dependent mechanism. *J. of Neuroscience*, **13**, 1010-1021. (6)
Fetters, L., & Todd, J. (1987). Quantitative assessment of infant reaching movements. *J. of Motor Behavior*, **19**, 147-166. (15)
Fibiger, H.C., & Phillips, A.G. (1986) Reward, motivation, cognition: Psychobiology of mesotelencephalic dopamine systems. In F.E. Bloom (Ed.) *Handbook of physiology: The nervous system* (Vol.

IV, pp. 647-675). Baltimore, MD: Williams and Wilkins. (17)
Fiez, J.A., Petersen, S.E., Cheney, M.K., & Raichle, M.E. (1992). Impaired non-motor learning and error detection associated with cerebellar damage. *Brain*, **115**, 155-178. (24)
Fishell, G., Mason, C.A., & Hatten, M.E. (1993). Dispersion of neural progenitors within the germinal zones of the forebrain. *Nature*, **362**, 636-638. (2)
Fitts, P.M. (1954). The information capacity of the human motor system in controlling the amplitude of movement. *J. of Experimental Psychology*, **47**, 381-391. (15)
Flaherty, A.W., & Graybiel, A. (1993) Two input systems for body representations in the primate striatal matrix: Experimental evidence in the squirrel monkey. *J. of Neuroscience*, **13**, 1120-1137. (17)
Flaherty, A.W., & Graybiel, A. (1994) Input-output organization of the sensorimotor striatum in the squirrel monkey. *J. of Neuroscience*, **14**, 599-610. (17)
Flash, T., & Henis, E. (1991). Arm trajectory modifications during reaching towards visual targets. *J. of Cognitive Neuroscience*, **3**, 220-230. (15)
Flash, T., & Hogan, N. (1985). The coordination of arm movements: An experimentally confirmed mathematical model. *J. of Neuroscience*, **7**, 1688-1703. (15)
Flexner, J.B., Flexner, L.B., & Stellar, E. (1963). Memory in mice is affected by intracerebral puromycin. *Science*, **141**, 51-59. (6)
Flood, J.F., Smith, G.E., & Jarvik, M.E. (1980). A comparison of the effects of localized brain administration of catecholamine and protein synthesis inhibitors on memory processing. *Brain Research*, **197**, 153-165. (6)
Fodor, J.A., & Pylyshyn, Z.W. (1981). Connectionism and cognitive architecture: A critical analysis. *Cognition*, **28**, 3-71. (26)
Fodor, J.A., & Pylyshyn, Z.W. (1988). Connectionism and cognitive architecture: A critical analysis. In Pinker, S., & Mehler, J. (Eds.) *Connections and symbols*. MIT Press: Cambridge, MA. (11)
Forster, K.I., & Chambers, S.M. (1973). Lexical access and naming time. *J. of Verbal Learning and Verbal Behavior*, **12**, 627-635. (27)
Fortier, P.A., Kalaska, J.F., & Smith, A.M. (1989). Cerebellar neuronal activity related to whole-arm reaching movements in the monkey. *J. of Neurophysiology*, **62**, 198-211. (13)
Fowler, C.A. (1980). Coarticulation and theories of extrinsic timing. *J. of Phonetics*, **8**, 113-133. (25)
Fox, K., & Daw, N.W. (1993). Do NMDA receptors have a critical function in visual cortical plasticity? *Trends in Neurosciences*, **16**, 116-122. (8)
Fox, K., Daw, N., Sato, H., & Czepita, D. (1992). The effect of visual experience on development of NMDA receptor synaptic transmission in kitten visual cortex. *J. of Neuroscience*, **12**, 2672-2684. (8)
Fox, K., Sato, H., & Daw, N. (1989). The location and function of NMDA receptors in cat and kitten visual cortex. *J. of Neuroscience*, **9**, 2443-2454. (8)
Fox, P.T., Raichle, M.E., Mintun, M.A., & Dence, C. (1988). Nonoxidative glucose consumption during focal physiologic neural activity. *Science*, **241**, 462-464. (23)
Frazier-Cierpial, L., & Brunjes, P. C. (1989). Early postnatal cellular proliferation and survival in the olfactory bulb and rostral migratory stream of normal and unilaterally odor-deprived rats. *J. of Comparative Neurology*, **289**, 481-492. (2)
Freeman, W.J. (1995). *Societies of brains: A study in the neuroscience of love and hate*. Hillsdale, NJ: Erlbaum. (27)
Frégnac, Y., & Imbert, M. (1984). Development of neuronal selectivity in primary visual cortex of cat. *Physiological Review*, **64**, 325-434. (8)
Frégnac, Y., Shulz, D., Thorpe, S., & Bienenstock, E. (1988). A cellular analogue of visual cortical plasticity. *Nature*, **333**, 367-370. (8)
French, R.M., & Messinger, A. (1994). Genes, phenes and the Baldwin Effect: Learning and evolution in a simulated population. In R.A. Brooks & P. Maes (Eds.), *Artificial life IV: Proceedings of the Fourth International Workshop on the Synthesis and Simulation of Living Systems* (pp. 277-282). Cambridge, MA: MIT Press. (4)
French, R.M. (1995). Interactive tandem networks and the sequential learning problem. Research MS: Center for Research on Concepts and Cognition. Indiana University. (ftp: //cogsci. indiana.edu. /pub/french. tandem-stm-ltm.ps.Z) (26)
Freund, T.T., Powell, J.F., & Smith, A.D. (1984) Tyrosine hydroxylase-immunoreactive boutons in synaptic contact with identified striatonigral neurons, with particular reference to dendritic spines. *Neuroscience*, **13**, 1189-1215. (17)
Frey, P.W., & Ross, L.E. (1968). Classical conditioning of the rabbit eyelid response as a function of interstimulus interval. *J. of Comparative and Physiological Psychology*, **65**, 246-250. (4)
Frey, S., Schweigert, C., Krug, M., & Lossner, B. (1991). Long-term potentiation induced changes in protein synthesis of hippocampal subfields of freely moving rats: Time-course. *Biomedical Biochemical Acta*, **50**, 1231-1240. (6)
Frey, U., Frey, S., Schollmeier, F., & Krug, M. (1995a). Influence of actinomycin D, an RNA-synthesis inhibitor, on long-term potentiation in rat hippocampal neurons in vivo and in vitro. *J. of Physiology*, **490**, 703-711. (6)
Frey, U., Hartmann, S., & Matthies, H. (1989a). Domperidone, an inhibitor of the D_2-receptor, blocks a late phase of an electrically induced long-term potentiation in the CA1-region in rats. *Biomedical Biochemical Acta*, **48**, 473-476. (6)
Frey, U., Huang, Y-Y., & Kandel, E.R. (1993a). Effects of cAMP simulate a late stage of LTP in hippocampal CA1 neurons. *Science*, **260**, 1661-1664. (6)
Frey, U., Krug, M., Brödemann, R., Reymann, K., & Matthies, H. (1989b). Long-term potentiation induced in dendrites separated from rat's CA1 pyramidal somata does not establish a late phase. *Neuroscience Letters*, **97**, 135-139. (6)
Frey, U., Krug, M., Reymann, K.G., & Matthies, H. (1988). Anisomycin, an inhibitor of protein synthesis, blocks late phases of LTP phenomena in the hippocampal CA1 region in vitro. *Brain Research*, **452**, 57-65. (6)
Frey, U., Matthies, H., & Reymann, K.G. (1991). The effect of dopaminergic D_1 receptor blockade during tetanization on the expression of long-term

potentiation in the rat CA1 region in vitro. *Neuroscience Letters*, **129**, 111-114. (6)
Frey, U., & Morris, R.G.M. (in press). Synaptic tagging: Synapse specificity during protein synthesis-dependent long-term potentiation. *Nature*. (Pt 2)
Frey, U., Müller, M., & Kuhl, D. (1996). A novel form of long-lasting potentiation revealed in tissue plasminogen activator mutant mice. *Fourth IBRO World Congress of Neuroscience, Kyoto* (abstract). (6)
Frey, U., Schollmeier, K., Reymann, K.G., & Seidenbecher, T. (1995b). Asymptotic hippocampal long-term potentiation in rats does not preclude additional potentiation at later phases. *Neuroscience*, **67**, 799-807. (6)
Frey, U., Schroeder, H., & Matthies, H. (1990). Dopaminergic antagonists prevent long-term maintenance of posttetanic LTP in the CA1 region of rat hippocampal slices. *Brain Research*, **522**, 69-75. (6)
Frey, U., Seidenbecher, T., & Krug, M. (1993b). The action of the RNA-synthesis inhibitor actinomycin D on a late stage of LTP in hippocampal neurons in vivo and in vitro. *Society of Neuroscience Abstracts*, **19**, 912. (6)
Freyd, J.J. (1987). Dynamic mental representation. *Psychological Review*, **94**, 427-438. (13)
Friedländer, M.J., Martin, K.A.C., & Wassenhove-McCarthy, D. (1991). Effects of monocular visual deprivation on geniculocortical innervation of area 18 in cat. *J. of Neuroscience*, **11**, 3268-3288. (8)
Friston, K.J., Tononi, G., Reeke, G.N., Jr., Sporns, O., & Edelman, G.M. (1994) Value-dependent selection in the brain: Simulation in a synthetic neural model. *Neuroscience*, **59**, 229-243. (17)
Frostig, R.D., Lieke, D.E., Ts'o, D.Y., & Grinvald, A. (1990). Cortical functional architecture and local coupling between neuronal activity and the microstimulation revealed by in vivo high-resolution optical imaging of intrinsic signals. *Proceedings of the National Academy of Sciences USA*, **87**, 6082-6086. (9)
Fu, Q-G., Flament, D., Coltz, J.D., & Ebner, T.J. (1995). Temporal encoding of movement kinematics in the discharge of primate primary motor and premotor neurons. *J. of Neurophysiology*, **73**, 836-854. (13)
Fujita, I., Tanaka, K., Ito, M., & Cheng, K. (1992). Columns for visual features of objects in monkey inferotemporal cortex. *Nature*, **360**, 343-346. (9)
Fujita, S. (1990). Morphogenesis of the brain as studied by 3-D computer graphics. *J. of Microscopy*, **157**, 259-269. (3)
Fukuda, A., Mody, I., & Prince, D.A. (1993). Differential ontogenesis of presynaptic and postsynaptic GABA inhibition in rat somatosensory cortex. *J. of Neurophysiology*, **70**, 448-452. (7)
Funauchi, M., Haruta, H., & Tsumoto, T. (1994). Effects of an inhibitor for calcium/calmodulin-dependent protein phosphatase, calcineurin, on induction of long-term potentiation in rat visual cortex. *Neuroscience Research*, **19**, 269-278. (6)
Funka-Lea, G.D., & Schwaber, J.S. (1994). A digital brain atlas and its application to the visceral neuraxis. *J. of Neuroscience Methods*, **54**, 253-260. (3)
Furman, G.G. (1965). Comparison of models for subtractive and shunting lateral inhibition in receptor-neuron fields. *Kybernetic*, **2**, 257-274. (24)
Fuster, J.M. (1989). *The prefrontal cortex*. New York: Raven. (18)
Fuster, J.M., & Alexander, G.E. (1973). Firing changes in cells of the nucleus medialis dorsalis associated with the delayed response behavior. *Brain Research*, **61**, 79-81. (24)
Fuster, J.M., Bauer, R.H., & Jervey, J.P. (1982). Cellular discharge in the dorsolateral prefrontal cortex of the monkey in cognitive tasks. *Experimental Neurology*, **77**, 679-694. (9)
Fuxe, K., & Agnati, L.F. (1985). Receptor-receptor interactions in the central nervous system: A new integrative mechanism in synapses. *Medicinal Research Reviews*, **5**, 441-482. (5)
Gabrieli, J., McGlinchey-Berroth, R., Carrillo, M., Gluck, M., Cermack, L., & Disterhoft, J. (1995). Intact delay-eyeblink classical conditioning in amnesia. *Behavioral Neuroscience*, **109**, 819-827. (12)
Gaffan, D., & Murray, E.A. (1992). Monkeys (*Macaca fascicularis*) with rhinal cortex ablations succeed in object discrimination learning despite 24-hr intertrial intervals and fail at matching to sample despite double sample presentations. *Behavioral Neuroscience*, **1**, 30-38. (9)
Gaffan, D. (1994). Dissociated effects of perirhinal cortex ablation, fornix transection and amygdalectomy: Evidence for multiple memory systems in the primate temporal lobe. *Experimental Brain Research*, **99**, 411-432. (9)
Galaburda, A.M., Rosen, G.D., & Sherman, G.F. (1989). The neural origin of developmental dyslexia: Implications for medicine, neurology, and cognition. In A.M. Galaburda (Ed.), *From reading to neurons* (pp. 377-388). Cambridge, MA: MIT Press. (27)
Gall, C.M., Hendry, S.H.C., Seroogy, K.B., Jones, E.G., & Haycock, J.W. (1987). Evidence for coexistence of GABA and dopamine in neurons of the rat olfactory bulb. *J. of Comparative Neurology*, **266**, 307-318. (2)
Gallant, J.L., Braun, J., & Van Essen, D.C. (1993). Selectivity for polar, hyperbolic, and Cartesian gratings in macaque visual cortex. *Science*, **259**, 100-103. (10)
Galli, L., & Maffei, U. (1988). Spontaneous impulse activity of rat retinal ganglion cells in prenatal life. *Science*, **242**, 90-91. (8)
Gallistel, C.R. (1990). *The organization of learning*. Cambridge, MA: MIT Press. (21)
Gally, J.A., Montague, P.R., Reeke, G.N., Jr., & Edelman, G.M. (1990). The NO hypothesis: Possible effects on a short-lived, rapidly diffusible signal in the development and function of the nervous system. *Proceedings of the National Academy of Sciences USA*, **87**, 3547-3551. (3)
Galuske, R.A.W., Kim, D.S., Castren, E., Thoenen, H., & Singer, W. (1996). Brain derived neurotrophic factor reverses experience-dependent synaptic modifications in kitten visual cortex. *European J. of Neuroscience*, **8**, 1554-1559. (8)
Galuske, R.A.W., & Singer, W. (1996). The origin and topography of long range intrinsic projections in cat visual cortex: A developmental study. *Cerebral Cortex*, **6**, 417-430. (8)
Gamzu, E.R., & Williams, D.R. (1973). Associative factors underlying the pigeon's key pecking in autoshaping procedures. *J. of the Experimental Analysis of Behavior*, **19**, 225-232. (4)

Garcia, J., & Koelling, R. (1966). Relation of cue to consequence in avoidance learning. *Psychonomic Science*, **4**, 123- 124. (4)

Garris, P.A., Ciolkowski, E.L., Pastore, P., & Wightman, R.M. (1994). Efflux of dopamine from the synaptic cleft in the nucleus accumbens of the rat brain. *J. of Neuroscience*, **14**, 6084-6093. (5)

Garthwaite, J. (1991). Glutamate, nitric oxide and cell-cell signalling in the nervous system. *Trends in Neurosciences*, **14**, 60-67. (5)

Gawne, T.J., & Richmond, B.J. (1993). How independent are the messages carried by adjacent inferior temporal cortical neurons? *J. of Neurosciences*, **137**, 2758-2771. (8)

Geinisman, Y., deToledo-Morrell, L., Morrell, F., Heller, R.E., Rossi, M., & Parshall, R.F. (1993). Structural synaptic correlate of long-term potentiation: Formation of axospinous synapses with multiple, completely partitioned transmission zones. *Hippocampus*, **3**, 435-446. (5)

Gelfer, C.E., Harris, K.S., & Hilt, G. (1981). Temporal constraints on anticipatory coarticulation. *Status Report on Speech Research SR-77/78*. New Haven, CT: Haskins Laboratories. (25)

Geman, S., Bienenstock, E., & Doursat, R. (1992). Neural networks and the bias/variance dilemma. *Neural Computation*, **4**, 1-58. (15)

Gentner, D. (1983). Structure-mapping: A theoretical framework for analogy. *Cognitive Science*, **7**, 155-170. (11)

Gentner, D.R., Grudin, J., & Conway, E. (1981). Finger movements in transcription typing. *Center for Human Information Processing, Technical Report 8001*. San Diego: University of California. (25)

Georgopoulos, A.P. (1986). On reaching. *Annual Review of Neuroscience*, **9**, 147-170. (15)

Georgopoulos, A.P. (1988). Neural integration of movement: Role of motor cortex in reaching. *FASEB J.*, **2**, 2849-2857. (13)

Georgopoulos, A.P. (1990). Neurophysiology of reaching. In M. Jeannerod (Ed.), *Attention and performance XIII* (pp. 227-263). Hillsdale, NJ: Lawrence Erlbaum. (13)

Georgopoulos, A.P. (1995). On the translation of directional motor cortical commands to activation of muscles via spinal interneuronal systems. *Cognitive Brain Research*, in press. (13)

Georgopoulos, A.P., Ashe, J., Smyrnis, N., & Taira, M. (1992). Motor cortex and the coding of force. *Science*, **256**, 1692-1695. (13)

Georgopoulos, A.P., Caminiti, R., Kalaska, J.F., & Massey, J.T. (1983). Spatial coding of movement: A hypothesis concerning the coding of movement direction by motor cortical populations. *Experimental Brain Research Supplement*, **7**, 327-336. (13)

Georgopoulos, A.P., Crutcher, M.D., & Schwartz, A.B. (1989b). Cognitive spatial-motor processes 3: Motor cortical prediction of movement direction during an instructed delay period. *Experimental Brain Research*, **75**, 183-194. (13)

Georgopoulos, A.P., Kalaska, J.F., & Caminiti, R. (1985). Relations between two-dimensional arm movements and single cell discharge in motor cortex and area 5: Movement direction versus movement endpoint. *Experimental Brain Research Supplement*, **10**, 176-183. (13)

Georgopoulos, A.P., Kalaska, J.F., Caminiti, R., & Massey, J.T. (1982). On the relations between the direction of two-dimensional arm movements and cell discharge in primate motor cortex. *J. of Neuroscience*, **2**, 1527-1537. (13)

Georgopoulos, A.P., Kalaska, J.F., Crutcher, M.D., Caminiti, R., & Massey, J.T. (1984). The representation of movement direction in the motor cortex: Single cell and population studies. In G.M. Edelman, W.E. Gall, & W.M. Cowan (Eds.), *Dynamic aspects of neocortical function* (pp. 501-524). New York: Wiley. (13)

Georgopoulos, A.P., Kettner, R.E., & Schwartz, A.B. (1988). Primate motor cortex and free arm movements to visual targets in three-dimensional space II: Coding of the direction of movement by a neuronal population. *J. of Neuroscience*, **8**, 2928-2937. (13)

Georgopoulos, A.P., & Lurito, J.T. (1991). Cognitive spatial-motor processes 6: Visuomotor memory scanning. *Experimental Brain Research*, **83**, 453-458. (13)

Georgopoulos, A.P., Lurito, J.T., Petrides, M., Schwartz, A.B., & Massey, J.T. (1989a). Mental rotation of the neuronal population vector. *Science*, **243**, 234-236. (13)

Georgopoulos, A.P., & Massey, J.T. (1987). Cognitive spatial-motor processes 1: The making of movements at various angles from a stimulus direction. *Experimental Brain Research*, **65**, 361-370. (13)

Georgopoulos, A.P., & Massey, J.T. (1988). Cognitive spatial motor processes 2: Information transmitted by the direction of two-dimensional arm movements and by neuronal populations in primate motor cortex and area 5. *Experimental Brain Research*, **69**, 315-326. (13)

Georgopoulos, A.P., Schwartz, A.B., & Kettner, R.E. (1986). Neuronal population coding of movement direction. *Science*, **233**, 1416-1419. (13)

Georgopoulos, A.P., Taira, M., & Lukashin, A. (1993). Cognitive neurophysiology of the motor cortex. *Science*, **260**, 47-52. (13)

Gerstein, G.L., & Gochin, P.M. (1992). In A. Aensen & V. Braitenberg (Eds.), *Information processing in the cortex: Experiments and theory* (pp. 139-173). Berlin/Heidelberg/New York: Springer. (8)

Ghosh, A., & Greenberg, M.E. (1995). Calcium signaling in neurons: Molecular mechanisms and cellular consequences. *Science*, **268**, 239-247. (20)

Gibbon, J., & Allan, L. (1984). *Timing and time perception*. New York: New York Academy of Sciences. (7)

Gibbon, J., Baldock, M.D., Locurto, C., Gold, L., & Terrace, H.S. (1977). Trial and intertrial durations in autoshaping. *J. of Experimental Psychology: Animal Behavior Processes*, **3**, 264-284. (4)

Gibson, E.J. (1988). Exploratory behavior in the development of perceiving, acting, and the acquiring of knowledge. *Annual Review of Psychology*, **39**, 1-41. (15)

Gibson, E.J., & Spelke, E.S. (1983). The development of perception. In J.H. Flavell & E.M. Markman (Eds.), *Handbook of child psychology* (Vol. III). New York: Wiley. (15)

Gick, M.L., & Holyoak, K.J. (1983). Schema induction and analogical transfer. *Cognitive Psychology*, **15**, 1-38. (11)

Gilbert, C.D., & Wiesel, T.N. (1989). Columnar specificity of intrinsic horizontal and corticocortical

connection in cat visual cortex. *J. of Neuroscience*, **97**, 2432-2442. (8)
Gilbert, C.D., & Wiesel, T.N. (1992). Receptive field dynamics in adult primary visual cortex. *Nature*, **356**, 150-152. (8)
Gluck, M.A., & Granger, R. (1993). Computational models of the neural bases of learning and memory. *Annual Review of Neuroscience*, **16**, 667-706. (12)
Gluck, M., & Myers, C. (1993). Hippocampal mediation of stimulus representation: A computational theory. *Hippocampus*, **3**, 491-516. (12)
Gluck, M., Myers, C., & Goebel, J. (1994). A computational perspective on dissociating hippocampal and entorhinal function (Response to Eichenbaum et al). *Behavioral and Brain Sciences*, **17**, 478-479. (12)
Gluck, M.A., & Thompson, R.F. (1987). Modeling the neural substrates of associative learning and memory: A computational approach. *Psychological Review*, **94**, 176-191. (21)
Gochin, P.M., Miller, E.K., Gross, C.G., & Gerstein, G.L. (1991). Functional interactions among neurons in inferior temporal cortex of the awake macaque. *Experimental Brain Research*, **84**, 505-516. (8)
Goelet, P., Castellucci, V.F., Schacher, S., & Kandel, E.R. (1986). The long and the short of long-term memory—a molecular framework. *Nature*, **322**, 419-422. (6)
Goldberg, D.E. (1989). *Genetic algorithms in search, optimization, and machine learning*. Reading, MA: Addison-Wesley. (19)
Goldberg, D.E., & Deb, K. (1992). A comparative analysis of selection schemes used in genetic algorithms. In G.J.E. Rawlins (Ed.), *Foundations of genetic algorithms* (pp. 69-93). San Mateo, CA: Morgan Kaufmann. (4)
Goldfield, E.C., Kay, B.A., & Warren, W.H. (1993). Infant bouncing: The assembly and tuning of action systems. *Child Development*, **64**, 1128-1142. (15)
Goldman-Rakic, P.S. (1987). Circuitry of primate prefrontal cortex and regulation of behavior by representational memory. In V.B. Mountcastle, F. Plum, & S.R. Geiger (Eds.), *Handbook of physiology*, Vol. I, section 1 (pp. 373-417). New York: Oxford University Press. (18)
Goldman-Rakic, P. (1988). Topography of cognition: parallel distributed networks in primate association cortex. *Annual Review of Neuroscience*, **11**, 137-156. (24)
Goldman-Rakic, P. (1994). The issue of memory in the study of the prefrontal cortex. In A. Thierry, J. Glowinski, P. Goldman-Rakic, & Y. Christen (Eds.), *Motor and cognitive functions of the prefrontal cortex*. Berlin: Springer-Verlag. (24)
Goldman-Rakic, P.E., Chafee, M., & Friedman, H. (1993) Allocation of function in distributed circuits. In T. Ono, L.R. Squire, M.E. Raichle, D.I. Perrett, & M. Fukada (Eds.), *Brain mechanisms of perception and memory: From neuron to behavior* (pp.445-456). New York: Oxford University Press. (18)
Goldman-Rakic, P.S., & Friedman, H.R. (1991). The circuitry of working memory revealed by anatomy and metabolic imaging. In H.S. Levin, H.M. Eisenberg, & A.L. Benton (Eds.), *Frontal lobe function and dysfunction* (pp. 72-91). New York: Oxford University Press. (24)
Goldman-Rakic, P.S., Leranth, C., Williams, M.S., Mons, N., & Geffard, M. (1989) Dopamine synaptic complex with pyramidal neurons in primate cerebral cortex. *Proceedings of the National Academy of Sciences USA*, **86**, 9015-9019. (17)
Goldmeier, E. (1972). Similarity in perceived visual forms. *Psychological Issues*, **8**, 1-135. (11)
Gonon, F.G. (1988) Nonlinear relationship between impulse flow and dopamine released by rat midbrain dopaminergic neurons as studied by in vivo electrochemistry. *Neuroscience*, **24**, 19-28. (17)
Gonon, F.G., Suaud-Chagny, M.F., Mermet, C.C., & Buda, M. (1991). Relation between impulse flow and extracellular catecholamine levels as studied by *in vivo* electrochemistry in CNS. In K. Fuxe & L.F. Agnati (Eds.), *Volume transmission in the brain: Novel mechanisms for neural transmission* (pp. 337-350). New York: Raven Press. (5)
Good, M., & Honey, R.C. (1991). Conditioning and contextual retrieval in hippocampal rats. *Behavioral Neuroscience*, **105**, 499-509. (12)
Goodale, M.A., & Milner, A.D. (1992). Separate visual pathways for perception and action. *Trends in Neuroscience*, **15**, 20-25. (9)
Goodale, M.A., Milner, A.D., Jakobson, L.S., & Carey, D.P. (1991). A neurological dissociation between perceiving objects and grasping them. *Nature*, **349**, 154-156. (9)
Goodman, C.S., & Shatz, C.J. (1993). Developmental mechanisms that generate precise patterns of neuronal connectivity. *Neuron (Suppl.)*,**10**, 77-98. (8)
Gormezano, I., & Kehoe, E.J. (1981). Classical conditioning and the law of contiguity. In P. Harzem & M.D. Zeiler (Eds.), *Predictability, correlation, and contiguity* (pp. 1-45). New York: Wiley. (Pt 5)
Grace, A.A. (1991). Phasic versus tonic dopamine release & the modulation of dopamine system responsivity: A hypothesis for the etiology of schizophrenia. *Neuroscience*, **41**, 1-24. (6)
Grant, S.G.N., O'Dell, T.J., Karl, K.A., Stein, P.L., Soriano, P., & Kandel, E.R. (1992). Impaired long-term potentiation, spatial learning, and hippocampal development in *fyn* mutant mice. *Science*, **258**, 1903-1910. (6)
Gray, C.M., König, P., Engel, A.K., & Singer, W. (1989). Oscillatory responses in cat visual cortex exhibit inter-columnar synchronization which reflects global stimulus properties. *Nature*, **338**, 334-347. (8) (11)
Gray, C.M., & Singer, W. (1989). Stimulus specific neuronal oscillations in orientation columns of cat visual cortex. *Proceedings of the National Academy of Sciences USA*, **86**, 1698-1702. (11)
Gray, C.M., & Viana Di Prisco, G. (1993). Properties of stimulus-dependent rhythmic activity of visual cortical neurons in the alert cat. *Society of Neuroscience Abstracts*, **19**, 359.8. (8)
Gray, G.E., & Sanes, J.R. (1991). Migratory paths and phenotypic choices of clonally related cells in the avian optic tectum. *Neuron*, **6**, 211-225. (2)
Grecksch, G., & Matthies, H. (1980). Two sensitive periods for the amnesic effect of anisomycin. *Pharmacology Biochemistry and Behavior*, **12**, 663-665. (6)
Grecksch, G., Ott, T., & Matthies, H. (1980). The effect of intrahippocampally applied anisomycin on

the retention of brightness discrimination in rats. *Behavioral Neural Biology*, **29**, 281-288. (6)

Greengard, P., Jen, J., Nairn, A.C., & Stevens, C.F. (1991). Enhancement of the glutamate response by cAMP-dependent protein kinase in hippocampal neurons. *Science*, **253**, 1135-1138. (6)

Greenough, W.T., Armstrong, K.E., Comery, T.A., Hawrylak, N., Humphreys, A.G., Kleim, J., Swain, R.A., & Wang, X. (1994). Plasticity-related changes in synapse morphology. In A.I. Selverston & P. Ascher (Eds.), *Cellular and molecular mechanisms underlying higher neural functions* (pp. 211-220). New York: Wiley. (6)

Greuel, J.M., Luhmann, H.J., & Singer, W. (1988). Pharmacological induction of use-dependent receptive field modifications in the visual cortex. *Science* , **242**, 74-77. (8)

Gribkoff, V., & Ashe, J. (1984). Modulation by dopamine of population responses and cell membrane properties of hippocampal CA1 neurons in vitro. *Brain Research*, **292**, 327-338. (6)

Grice, G.R. (1948). The relation of secondary reinforcement to delayed reward in visual discrimination learning. *J. of Experimental Psychology*, **38**, 1-16. (14)

Griffith, J.S. (1966). A theory of the nature of memory. *Nature*, **211**, 1160-1163. (6)

Groenewegen, H., & Berendse, H. (1994). Anatomical relationships between the prefrontal cortex and basal ganglia in the rat. In A. Thierry, J. Glowinski, P. Goldman-Rakic, & Y. Christen (Eds.), *Motor and cognitive functions of the prefrontal cortex*. Berlin: Springer-Verlag. (24)

Gross, C.G. (1973). Visual functions of inferotemporal cortex. In R. Jung (Ed.), *Handbook of sensory physiology* (Vol. 7, part 3B, pp. 451-482). Berlin: Springer-Verlag. (9)

Gross, C.G. (1992). Representation of visual stimuli in inferior temporal cortex. *Philosophical Transactions of the Royal Society of London B*, **335**, 3-10. (10)

Gross, C.G. (1994). How inferior temporal cortex became a visual area. *Cerebral Cortex*, **5**, 455-469. (9)

Gross, C.G., Bender, D.B., & Rocha-Miranda, C.E. (1969). Visual receptive fields of neurons in inferotemporal cortex of the monkey. *Science*, **166**, 1303-1306. (9)

Gross, C.G., Rocha-Miranda, C.E., & Bender, D.B. (1972). Visual properties of neurons in inferotemporal cortex of the macaque. *J. of Neurophysiology*, **35**, 96-111. (9) (10)

Gross, C.G., & Sergent, J. (1992). Face recognition. *Current Opinion in Neurobiology*, **2**, 156-161. (10)

Grossberg, S. (1973). Contour enhancement, STM, and constancies in reverberating neural nets. *Studies in Applied Mathematics*, **52**, 217-257. (24)

Grossberg, S. (1976). Adaptive pattern classification and recoding: Part I. *Biological Cybernetics*, **23**, 121-134. (12)

Grossberg, S. (1978). A theory of human memory: Self-organization and performance of sensory-motor codes, maps, and plans. *Progress in Theoretical Biology*, **5**, 233-302. (25)

Grossberg, S. (1980). How does a brain build a cognitive code? *Psychological Review*, **87**, 1-51. (8)

Grossberg, S. (1991). A neural network architecture for Pavlovian conditioning: Reinforcement, attention, forgetting, timing. In M.L. Commons, S. Grossberg, & J.E.R. Staddon (Eds.), *Neural network models of conditioning and action* (pp. 69-122). Hillsdale, NJ: Lawrence Erlbaum. (4)

Grossberg, S., & Schmajuk, N.A. (1989). Neural dynamics of adaptive timing and temporal discrimination during associative learning. *Neural Networks*, **2**, 79-102. (7) (21)

Grossberg, S., & Stone, G.O. (1986). Neural dynamics of word recognition and recall: Priming, learning, and resonance. *Psychological Review*, **93**, 46-74. (27)

Grove, E.A., Williams, B.P., Li, D-Q., Hajhosseini, M., Friedrich, A., & Price, J. (1993). Multiple restricted lineages in the embryonic rat cerebral cortex. *Development*, **117**, 553-561. (2)

Groves, P.M., Garcia-Munoz, M., Linder, J.C., Manley, M.S., Martone, M.E., & Young, S.J. (1995). Elements of the intrinsic organization and information processing in the neostriatum. In J.C. Houk, J.L. Davis, & D.G. Beiser (Eds.), *Models of information processing in the basal ganglia* (pp. 51-96). Cambridge: MIT Press. (17) (18)

Groves, P.M., Linder, J.C., & Young, S.J. (1994). 5-Hydroxydopamine-labeled axons: Three dimensional reconstructions of axons, synapses, and postsynaptic targets in rat neostriatum. *Neuroscience*, **58**, 593-604. (18)

Gruau, F., & Whitley, D. (1993). Adding learning to the cellular development of neural networks: Evolution and the Baldwin Effect. *Evolutionary Computation*, **1**, 213-233. (4)

Grudin, J.T. (1981). *The organization of serial order in typing*. Unpublished doctoral dissertation, University of California, San Diego. (25)

Grudin, J.T. (1983). Error patterns in skilled and novice transcription typing. In W. E. Cooper (Ed.), *Cognitive aspects of skilled typewriting*. New York: Springer-Verlag. (25)

Gu, Q., Bear, M.F., & Singer, W. (1989). Blockade of NMDA-receptors prevents ocularity changes in kitten visual cortex after reversed monocular deprivation. *Developmental Brain Research*, **47**, 281-288. (8)

Gu, Q., & Singer, W. (1993). Effects of intracortical infusion of anticholinergic drugs on neuronal plasticity in kitten striate cortex. *European J. of Neuroscience*, **5**, 475-485. (8)

Gu, Q., & Singer, W. (1995). Involvement of serotonin in developmental plasticity of kitten visual cortex. *European J. of Neuroscience*, **7**, 1146-1153. (8)

Gullapalli, V. (1990). A stochastic reinforcement learning algorithm for learning real-valued functions. *Neural Networks*, **3**, 671-692. (16)

Gullapalli, V. (1992). Learning control under extreme uncertainty. In S.J. Hanson, J.D. Cowan, & C.L. Giles (Eds.) *Advances in neural information processing systems 5* (pp. 327-334). San Mateo, CA: Morgan Kaufmann. (15)

Gullapalli, V. (1992). *Reinforcement learning and its application to control*. Unpublished doctoral dissertation, Department of Computer and Information Science, University of Massachusetts, Amherst, MA. (22)

Haas, H.L., Jefferys, J.G., Slater, N.T., & Carpenter, D.O. (1984). Modulation of low calcium induced field bursts in the hippocampus by monoamines and cholinomimetics. *Pflugers Archives*,

400, 28-33. (6)
Haas, H.L., Sergueeva, O.A., Vorobjev, V.S., & Sharonova, I.N. (1995). Subcortical modulation of synaptic plasticity in the hippocampus. *Behavioral Brain Research*, **66**, 41-44. (6)
Hablitz, J.J., & Thalmann, R.H. (1987). Conductance changes underlying a late synaptic hyperpolarization in hippocampal CA3 neurons. *J. of Neurophysiology*, **58**, 160-179. (7)
Haith, M.M. (1980). *Rules babies look by: The organization of newborn visual activity*. Hillsdale, NJ: Erlbaum. (15)
Halasz, N., Ljungdahl, A., & Hökfelt, T. (1979). Transmitter histochemistry of the rat olfactory bulb III: Autoradiographic localization of 3H-GABA. *Brain Research*, **167**, 221-240. (2)
Halasz, N., & Shepherd, G.M. (1983). Neurochemistry of the vertebrate olfactory bulb. *Neuroscience*, **10**, 579-619. (2)
Hall, G., & Honey, R. (1990). Context-specific conditioning in the conditioned-emotional-response procedure. *J. of Experimental Psychology: Animal Behavior Processes*, **16**, 271-278. (12)
Halverson, H.M. (1931). Study of prehension in infants. *Genetic Psychology Monographs*, **10**, 107-285. (15)
Halverson, H.M. (1933). The acquisition of skill in infancy. *J. of Genetic Psychology*, **43**, 3-48. (15)
Halwes, T., & Jenkins, J.J. (1971). Problem of serial order in behavior is not resolved by context-sensitive associative memory models. *Psychological Review*, **78**, 122-129. (25)
Hammarberg, R. (1982). On redefining coarticulation. *J. of Phonetics*, **10**, 123-137. (25)
Hammer, M. (1993). An identified neuron mediates the unconditioned stimulus in associative learning in honeybees. *Nature*, **366**, 59-63. (21)
Hansel, C., Artola, A., & Singer, W. (1994). Calcium influx accompanies the induction of long-term depression in pyramidal cells of the rat visual cortex. *European J. of Neuroscience Supplement*, **7**, 46.09.
Harder, L.D., & Real, L.A. (1987). Why are bumble bees risk averse. *Ecology*, **68**, 1104-1108. (21)
Harnad, S., Hanson, S., & Lubin, J. (1994). Learned categorical perception in neural nets: Implications for symbol grounding. In V. Honovar and L. Uhr (Eds.), *Artificial intelligence and neural networks: Steps toward principled integration* (pp. 191-206). New York: Academic Press. (12)
Harp, S.A., & Samad, T. (1991). Genetic synthesis of neural network architecture. In L. Davies (Ed.), *Handbook of genetic algorithms* (pp. 202-221). New York: Van Nostrand Reinhold. (4)
Harris, K.S. (1984). Coarticulation as a component in articulatory description. In R.G. Daniloff (Ed.), *Articulatory assessment and treatment issues*. San Diego: College-Hill Press. (25)
Hartline, H.K., & Ratliffe, F. (1957). Inhibitory interaction of receptor units in the eye of Limulus. *J. of General Physiology*, **40**, 357-376. (18)
Harvey, I. (1991). The artificial evolution of behaviour. In J.A. Meyer & S.W. Wilson (Eds.), *From animals to animats: Proceedings of the First International Conference on Simulation of Adaptive Behavior* (pp. 263-272). Cambridge, MA: MIT Press. (4)
Harvey, J., & Collingridge, G.L. (1992). Thapsigargin blocks the induction of long-term potentiation in rat hippocampal slices. *Neuroscience Letters*, **139**, 197-200. (6)
Hasselmo, M., & Schnell, E. (1994). Feedback regulation of cholinergic modulation and hippocampal memory function. *Proceedings of World Congress on Neural Networks*, **2**, 729-734. San Diego: INNS Press. (12)
Hasselmo, M., Schnell, E., & Barkai, E. (1995). Dynamics of learning and recall at excitatory synapses and cholinergic modulation in hippocampal region CA3. *J. of Neuroscience* (Part 2), **15**, 5249-5262. (12)
Hata, Y., Tsumoto, T., Sato, H., Hagihara, K., & Tamura, H. (1988). Inhibition contributes to orientation selectivity in visual cortex of cat. *Nature*, **335**, 815-817. (8)
Hata, Y., Tsumoto, T., Sato, H., & Tamura, H. (1991). Horizontal interactions between visual cortical neurons studied by cross-correlational analysis in the cat. *J. of Physiology*, **441**, 593-614. (8)
Hawkins, R.D. (1989). A biologically based computational model for several simple forms of learning. In R.D. Hawkins & G.H. Bower (Eds.), *Computational models of learning in simple neural systems* (pp. 65-108). San Diego: Academic Press. (4)
Haydon, P.G., & Drapeau, P. (1995). From contact to connection: Early events during synaptogenesis. *Trends in Neuroscience*, **18**, 196-201. (3)
Hayes, C. (1951). *The ape in our house*. New York: Harper & Row. (Pt 1)
Hayes, S.C. (1991). A relational control theory of stimulus equivalence. In L.J. Hayes & P.N. Chase (Eds.), *Dialogues of verbal behavior: The first international institute on verbal relations* (pp. 19-40). Reno, NV: Context Press. (26)
Hayes, S.C. (1992). Verbal relations, time, and suicide. In S.C. Hayes & L.J. Hayes (Eds.), *Understanding verbal relations: The second and third international institute on verbal relations* (pp. 109-118). Reno, NV: Context Press. (26)
Hayes, S.C., & Hayes, L.J. (1989). The verbal action of the listener as a basis for rule-governance. In S.C. Hayes (Ed.), *Rule-governed behavior: Cognition, contingencies, and instructional control* (pp. 153-190). New York: Plenum. (26)
Hayes, S.C., & Hayes, L.J. (1992). Verbal relations and the evolution of behavior analysis. *American Psychologist*, **47**, 1383-1395. (26)
Hebb, D.O. (1949). *The organization of behavior: A neuropsychological theory*. New York: Wiley. (8) (5) (6) (24) (26)
Henke, W.L. (1966). *Dynamic articulatory model of speech production using computer simulation*. Unpublished doctoral dissertation, Massachusetts Institute of Technology. (25)
Herkenham, M., Lynn, A.B., Little, M.D., Johnson, M.R., Melvin, L.S., de Costa, B.R., & Rice, K.R. (1990). Cannabinoid receptor localization in brain. *Proceedings of the National Academy of Sciences USA*, **87**, 1932-1936. (5)
Herrnstein, R.J. (1990). Rational choice theory. *American Psychologist*, **45**, 356-367. (21)
Herrnstein, R.J. (1991). Experiments on stable suboptimality in individual behavior. *American Economic Review Papers Proceedings*, **83**, 360-364. (21)

Herrnstein, R.J., Loewenstein, G.F., Prelec, D., & Vaughan, W. (1993). Utility maximization and melioration: Internalities in individual choice. *J. of Behavioral Decision Making*, **6**, 149-185. (21)

Hestrin, S. (1992). Developmental regulation of NMDA receptor-mediated synaptic currents at a central synapse. *Nature*, **357**, 686-689. (8)

Hille, B. (1992). G protein-coupled mechanisms and nervous signaling. *Neuron*, **9**, 187-195. (5)

Hinde, R.A. (1973). Constraints on learning: An introduction to the problem. In R.A. Hinde & J. Stevenson-Hinde (Eds.), *Constraints on learning* (pp. 1-19). London: Academic Press. (4)

Hinds, J.W. (1968). Autoradiographic study of histogenesis in the mouse olfactory bulb II: Cell proliferation and migration. *J. of Comparative Neurology*, **134**, 305-322. (2)

Hinton, G.E. (1984). Parallel computations for controlling an arm. *J. of Motor Behavior*, **16**, 171-194. (25)

Hinton, G.E. (1987). Connectionist learning procedures. *Technical Report CMU-CS-87-115*. Department of Computer Science, Carnegie-Mellon University, Pittsburgh, PA. (16)

Hinton, G.E. (1989). Connectionist learning procedures. *Artificial Intelligence*, **40**, 185-234. (12)

Hinton, G.E., & Nowlan, S.J. (1987). How learning can guide evolution. *Complex Systems*, **1**, 495-502. (4)

Hirsch, H.V.B., & Spinelli, D.N. (1970). Visual experience modifies distribution of horizontally and vertically oriented receptive fields in cats. *Science*, **168**, 869-871. (8)

Hirsch, J., & Gilbert, C. (1993). Long-term changes in synaptic strength along specific intrinsic pathways in the cat visual cortex. *J. of Physiology* , **461**, 247-262. (8)

Hirsh, R. (1974). The hippocampus and contextual retrieval of information from memory: A theory. *Behavioral Biology*, **12**, 421-444. (12)

Hocherman, S., & Wise, S.P. (1991). Effects of hand movement path on motor cortical activity in awake, behaving rhesus monkeys. *Experimental Brain Research*, **83**, 285-302. (13)

Hoebel, B.G. (1988). Neuroscience and motivation: Pathways and peptides that define motivational systems. In R.A. Atkinson (Ed.), *Stevens' handbook of experimental psychology* (pp. 547-625). New York: Wiley. (18)

Hökfelt, T. (1991). Neuropeptides in perspective: The last ten years. *Neuron*, **7**, 867-879. (5)

Holland, J.H. (1975). *Adaptation in natural and artificial systems*. Ann Arbor: University of Michigan Press. (4) (19)

Hollis, K.L. (1984). The biological function of Pavlovian conditioning: The best defense is a good offense. *J. of Experimental Psychology: Animal Behavior Processes*, **10**, 413-425. (4)

Hollis, K.L. (1990). The role of Pavlovian conditioning in territorial aggression and reproduction. In D.A. Dewsbury (Ed.), *Contemporary issues in comparative psychology*. Sunderland, MA: Sinauer. (4)

Hollis, K.L., Cadieux, E.L., & Colbert, M.M. (1989). The biological function of Pavlovian conditioning: A mechanism for mating success in the blue gourami (*Trichogaster trichopterus*). *J. of Comparative Psychology*, **103**, 115-121. (4)

Hollis, K.L., Martin, K.A., Cadieux, E.L., & Colbert, M.M. (1984). The biological function of Pavlovian conditioning: Learned inhibition of aggressive behavior in territorial fish. *Learning and Motivation*, **15**, 459-478. (4)

Hollis, K.L., & Overmier, J.B. (1982). Effect of telencephalon ablation on the reinforcing and eliciting properties of species specific events in *Betta splendens*. *J. of Comparative and Physiological Psychology*, **96**, 574- 590. (4)

Hollmann, M., Hartley, M., & Heinemann, S. (1991). Ca^{2+} permeability of KA-AMPA-gated glutamate receptor channels depends on subunit composition. *Science*, **252**, 851-853. (5)

Holyoak, K.J., & Thagard, P. (1989). Analogical mapping by constraint satisfaction. *Cognitive Science*, **13**, 295-355. (11)

Honey, R., & Good, M. (1993). Selective hippocampal lesions abolish the contextual specificity of latent inhibition and conditioning. *Behavioral Neuroscience*, **107**, 23-33. (12)

Honig, W.K., & Staddon, J.E.R. (1977). *Handbook of operant behavior*. Englewood Cliffs, NJ: Prentice-Hall. (16)

Hopkins, W.F., & Johnston, D. (1984). Frequency-dependent noradrenergic modulation of long-term potentiation in the hippocampus. *Science*, **226**, 350-352. (6)

Horowitz, F.D. (1992). John B. Watson's legacy: Learning and environment. *Developmental Psychology*, **28**, 360-367. (15)

Horwitz, B., & Sporns, O. (1994). Neural modeling and functional neuroimaging. *Human Brain Mapping*, **1**, 269-283. (24)

Houk, J. (1995). Information processing in modular circuits linking basal ganglia. In J.C. Houk, J.L. Davis, & D. Beiser (Eds.), *Models of information processing in the basal ganglia* (pp. 3-9). Cambridge, MA: MIT Press. (24)

Houk, J.C., Adams, J.L., & Barto, A.G. (1995). A model of how the basal ganglia generate and use neural signals that predict reinforcement. In J.C. Houk, J.L. Davis, & D.C. Beiser (Eds.), *Models of information processing in the basal ganglia* (pp. 249-270). Cambridge, MA: MIT Press. (17) (18) (20)

Houk, J.C., & Wise, S.P. (1993). Outline for a theory of motor behavior: Involving cooperative actions of the cerebellum, basal ganglia, and cerebral cortex. In P. Rudomin, M.A. Arbib, F. Cervantes-Perez, & R. Romo (Eds.), *Neuroscience: From neural networks to artificial intelligence—Research notes in neural computation* (Vol. 4, pp. 471-493). Berlin: Springer-Verlag. (24)

Huang, Y-Y., & Kandel, E.R. (1994). Recruitment of long-lasting and protein kinase A-dependent long-term potentiation in the CA1 region of hippocampus requires repeated tetanization. *Learning & Memory*, **1**, 74-82. (6)

Huang, Y-Y., & Kandel, E.R. (1995). D_1/D_5 receptor agonists induce a protein synthesis-dependent late potentiation in the CA1 region of the hippocampus. *Proceedings of the National Academy of Sciences USA*, **92**, 2446-2450. (6)

Hubel, D.H. (1975). Physiological responses and functional architecture of the monkey visual cortex. In R. Galun, P. Hillman, I. Parnas, & R. Werman (Eds.), *Sensory physiology and behavior* (pp. 21-24). New York: Plenum Press. (8)

Hubel, D.H., & Wiesel, T.N. (1962). Receptive fields, binocular interaction and functional architecture in the cat's visual cortex. *J. of Physiology*, **160**, 106-154. (8)
Hubel, D.H., & Wiesel, T.N. (1963). Receptive fields of cells in striate cortex of very young, visually inexperienced kittens. *J. of Neurophysiology*, **26**, 994-1002. (8)
Hubel, D.H., & Wiesel, T.N. (1968). Receptive fields and functional architecture of monkey striate cortex. *J. of Physiology (London)*, **195**, 215-243. (18)
Huerta, M.F., Koslow, S.H., & Leshner, A.I. (1993). The Human Brain Project: an international resource. *Trends in Neuroscience*, **16**, 436-438. (3)
Hull, C.L. (1945). The place of innate individual and species differences in a natural-science theory of behavior. *Psychological Review*, **52**, 55-60. (4)
Hummel, J.E., & Biederman, I. (1990). Dynamic binding: A basis for the representation of shape by neural networks. *Proceedings of the 12th Annual Conference of the Cognitive Science Society*, 614-621. (11)
Hummel, J.E., & Biederman, I. (1992). Dynamic binding in a neural network for shape recognition. *Psychological Review*, **99**, 480-517. (11)
Hummel, J.E., & Holyoak, K.J. (1992). Indirect analogical mapping. *Proceedings of the 14th Annual Conference of the Cognitive Science Society*, 516-521. (11)
Hummel, J.E., & Holyoak, K.J. (1993). Distributing structure over time. Review of L. Shastri and V. Ajjenegadde, *From simple associations to systematic reasoning: A connectionist representation of rules, variables and dynamic bindings. Behavioral and Brain Sciences*, 464. (11)
Hummel, J.E., & Melz, E.R., Thompson, J., & Holyoak, K.J. (1994). Mapping hierarchical structures with synchrony for binding: Preliminary investigations. In A. Ram & K. Eiselt (Eds.), *Proceedings of the Sixteenth Annual Conference of the Cognitive Science Society* (pp. 433-438). Hillsdale, NJ: Erlbaum. (11)
Hummel, J.E., & Stankiewicz, B.J. (1995a). An architecture for rapid, hierarchical structural description. In T. Inui & J. McClelland (Eds.), *Attention and performance XVI*. (in press). (11)
Hummel, J.E., & Stankiewicz, B.J. (1995b). Coordinates and spatial relations in object memory. *Technical Report 95-01, UCLA Shape Perception Laboratory*. University of California, Los Angeles. (11)
Huntley, G.W., & Jones, E.G. (1991). Relationship of intrinsic connections to forelimb movement representations in monkey motor cortex: A correlative anatomic and physiological study. *J. of Neurophysiology*, **66**, 390-413. (13)
Innocenti, G.M., & Frost, D.O. (1979). Effects of visual experience on the maturation of the efferent system to the corpus callosum. *Nature*, **280**, 231-234. (8)
Irwin, D.E. (1992). Perceiving an integrated visual world. In D. E. Meyer & S. Kornblum (Eds.), *Attention and performance XIV: Synergies in experimental psychology, artificial intelligence, and cognitive neuroscience—A silver jubilee* (pp. 121-142). Cambridge, MA: MIT Press. (11)
Irwin, D.E., & Andrews, R.V. (1995). Integration and accumulation of information across saccadic eye movements. In T. Inui & J. McClelland (Eds.), *Attention and performance XVI: Information integration in perception and communication* (pp. 125-155). Cambridge, MA: MIT Press. (11)
Isaac, J.T.R., Nicoll, R.A., & Malenka, R.C. (1995). Evidence for silent synapses: Implications for the expression of LTP. *Neuron*, **15**, 427-434. (6)
Ito, I., Hidaka, H., & Sugiyama, H. (1991). Effects of KN-62, a specific inhibitor of calcium/calmodulin-dependent protein kinase II, on long-term potentiation in the rat hippocampus. *Neuroscience Letters*, **121**, 119-121. (6)
Ito, M. (1984). *The cerebellum and neural control*. New York: Raven Press. (20)
Ito, M. (1993). How does the cerebellum facilitate thought? In T. Ono, L.R. Squire, M.E. Raichle, D.I. Perrett, & M. Fukada (Eds.), *Brain mechanisms of perception and memory: From neuron to behavior* (pp. 651-658). New York: Oxford University Press. (24)
Ito, M., Fujita, I., Tamura, H., & Tanaka, K. (1994). Processing of contrast polarity of visual images in inferotemporal cortex of the macaque monkey. *Cerebral Cortex*, **5**, 499-508. (9)
Ito, M., Tamura, H., Fujita, I., & Tanaka, K. (1995). Size and position invariance of neuronal responses in monkey inferotemporal cortex. *J. of Neurophysiology*, **73**, 218-226. (9)
Izumi, Y., Clifford, D.B., & Zorumski, C.F. (1991). 2-Amino-3-phosphonopropionate blocks the induction and maintenance of long-term potentiation in rat hippocampal slices. *Neuroscience Letters*, **122**, 187-190. (6)
Jaakkola, T., Singh, S.P., & Jordan, M.I. (1996). Reinforcement learning algorithm for partially observable Markov decision problems. In G. Tesauro, D.S. Touretzky and T.K. Leen (Eds.), *Advances in neural information processing systems 7* (pp. 345-352). Cambridge, MA: MIT Press. (19)
Jackendoff, R., & Landau, B. (1991). Spatial language and spatial cognition. In D.J. Napoli & J. Kegl (Eds.), *Bridges between psychology and linguistics: A Swarthmore Festschrift for Lila Gleitman* (pp. 145-169). Hillsdale, NJ: Erlbaum. (11)
Jackson, D.M., & Westlind-Danielsson, A. (1994). Dopamine receptors: Molecular biology, biochemistry and behavioural aspects. *Pharmacological Therapeutics*, **64**, 291-369. (6)
Jacobs, B.L., & Fornal, C.A. (1993). 5-HT and motor control. *Trends in Neurosciences*, **16**, 346-352. (5)
Jacobs, R.A. (1988). Increased rates of convergence through learning rate adaptation. *Neural Networks*, **1**, 295-307. (12)
Jacobs, R.E., & Fraser, S.E. (1994). Magnetic resonance microscopy of embryonic cell lineages and movements. *Science*, **263**, 681-684. (3)
Jagadeesh, B., Wheat, H.S., & Ferster, D. (1993). Linearity of summation of synaptic potentials underlying direction selectivity in simple cells of the cat visual cortex. *Science*, **262**, 1901-1904. (8)
James, W. (1896). *Introduction to Psychology*. New York: Holt. (12)
Jan, Y.N., & Jan, L.Y. (1983). A LHRH-like peptidergic neurotransmitter capable of 'action at a distance' in autonomic ganglia. *Trends in Neurosciences*, **6**, 320-325. (5)
Janer, K.W., & Pardo, J.V. (1991). Deficits in selec-

tive attention following bilateral anterior cingulotomy. *J. of Cognitive Neuroscience*, **3**, 231-241. (24)
Jared, D., & Seidenberg, M.S. (1991). Does word identification proceed from spelling to sound to meaning? *J. of Experimental Psychology: General*, **120**, 358-394. (27)
Jarrard, L. (1993). On the role of the hippocampus in learning and memory in the rat. *Behavioral and Neural Biology*, **60**, 9-26. (12)
Jay, T.M., Burette, F., & Laroche, S. (1995). NMDA receptor-dependent long-term potentiation in the hippocampal afferent fiber system to the prefrontal cortex in the rat. *European J. of Neuroscience*, **7**, 247-250. (18)
Jeffery, K.J., Abraham, W.C., Dragunow, M., & Mason, S.E. (1990). Induction of Fos-like immunoreactivity and the maintenance of long-term potentiation in the dentate gyrus of unanesthetized rats. *Molecular Brain Research*, **8**, 267-274. (6)
Jerne, N.K. (1955). The natural selection theory of antibody formation. *Proceedings of the National Academy of Sciences USA*, **41**, 849-856. (14)
John, E.R. (1967). *Mechanisms of memory*. London: Academic Press. (6)
Jolicoeur, P. (1985). The time to name disoriented natural objects. *Memory & Cognition*, **13**, 289-303. (11)
Jolicoeur, P. (1990). Identification of disoriented objects: A dual systems theory. *Mind and Language*, **5**, 387-410. (11)
Jones, B., & Mishkin, M. (1972). Limbic lesions and the problem of stimulus-reinforcement association. *Experimental Neurology*, **36**, 362-377. (12)
Jones, E.G., & Wise, S.P. (1977). Size, laminar and columnar distribution of efferent cells in the sensory-motor cortex of monkeys. *J. of Comparative Neurology*, **175**, 391-438. (13)
Jordan, M.I. (1986). Serial order: A parallel distributed processing approach. *Institute for Cognitive Science Tech. Rep. 8604*. University of California, San Diego. (25)
Jordan, M.I. (1990). Motor learning and the degrees of freedom problem. In M. Jeannerod (Ed.), *Attention and performance, XIII*. Hillsdale, NJ: Erlbaum. (25)
Jordan, M.I., & Rumelhart, D.E. (1992). Supervised learning with a distal teacher. *Cognitive Science*, **16**, 307-354. (25)
Jork, R., Grecksch, G., & Matthies, H. (1982). Apomorphine and glycoprotein synthesis during consolidation. *Pharmacology Biochemistry and Behavior*, **17**, 11-13. (6)
Jork, R., Grecksch, G., & Matthies, H. (1986). Impairment of glycoprotein fucosylation in rat hippocampus and the consequences on memory formation. *Pharmacology Biochemistry and Behavior*, **25**, 1137-1144. (6)
Jork, R., Lössner, B., & Matthies, H. (1979). Dopamine and macromolecule synthesis in rat hippocampus. *Pharmacology Biochemistry and Behavior*, **11**, 247-249. (6)
Juraska, J.M., & Fifkova, E. (1979). A Golgi study of the early postnatal development of the visual cortex of the hooded rat. *J. of Comparative Neurology*, **183**, 247-256. (3)
Kaas, J.H. (1995). The reorganization of sensory and motor maps in adult mammals. In M.S. Gazzaniga (Ed.), *The cognitive neurosciences* (pp. 51-71). Cambridge, MA: MIT Press. (8)
Kaelbling, L.P. (1993). *Learning in embedded systems*. Cambridge MA: MIT Press. (19)
Kalaska, J.F., Caminiti, R., & Georgopoulos, A.P. (1983). Cortical mechanisms related to the direction of two-dimensional arm movements: Relations in parietal area 5 and comparison with motor cortex. *Experimental Brain Research*, **51**, 247-260. (13)
Kalaska, J.F., Cohen, D.A.D., Hyde, M.L., & Prud'homme, M. (1989). A comparison of movement direction-related versus load direction-related activity in primate motor cortex, using a two-dimensional reaching task. *J. of Neuroscience*, **9**, 2080-2102. (13)
Kalman, R.E. (1960). A new approach to linear filtering and prediction problems. *J. of Basic Engineering, Transactions of the ASME, Series D*, **82**, 35-45. (21)
Kalsner, S., & Westfall, T.C. (1990). Presynaptic receptors and the question of autoregulation of neurotransmitter release. *Annals of the New York Academy of Sciences*, **604**, 652-655. (5) (18)
Kamil, A.C. (1995). A synthetic approach to the study of animal intelligence. In L.A. Real (Ed.), *Behavioral mechanisms in evolutionary ecology* (pp. 11-45). Chicago, IL: University of Chicago Press. (4)
Kamin, L.J. (1968). Attention-like processes in classical conditioning. In M.R. Jones (Ed.), *Miami symposium on the prediction of behavior* (pp. 9-31). Miami, FL: University of Miami Press. (1) (14)
Kamin, L.J. (1969). Predictability, surprise, attention and conditioning. In B.A. Campbell & R.M. Church (Eds.), *Punishment and aversive behavior* (pp. 279-296). New York: Appleton-Century-Crofts. (14) (18)
Kamiya, H., Sawada, S., & Yamamoto, C. (1993). Long-lasting potentiation of synaptic transmission in the Schaffer collateral-commissural pathway of the guinea pig hippocampus by activation of postsynaptic N-methyl-D-aspartate receptor. *Synapse*, **13**, 186- 194. (6)
Kandel, E.R. (1985). Cellular mechanisms of learning and the biological basis of individuality. In E.R. Kandel & J.H. Schwartz (Eds.), *Principles of neural science* (pp. 816-833). New York: Raven Press. (4)
Kano, M., Rexhausen, U., Dreesen, J., & Konnerth, A. (1992). Synaptic excitation produces a long-lasting rebound potentiation of inhibitory synaptic signals in cerebellar Purkinje cells. *Nature*, **356**, 601-604. (20)
Kapur, S., Rose, R., Liddle, P.F., Zipursky, R.B., Brown, G.M., Stuss, D., Houle, S., & Tulving, E. (1994). The role of the left prefrontal cortex in verbal processing: Semantic processing or willed action? *NeuroReport*, **5**, 2193-2196. (24)
Kasai, H., & Petersen, O.H. (1994). Spatial dynamics of second messengers: IP_3 and cAMP as long-range and associative messengers. *Trends in Neurosciences*, **17**, 95-101. (5)
Kasamatsu, T., & Pettigrew, J.D. (1979). Preservation of binocularity after monocular deprivation in the striate cortex of kittens treated with 6-hydroxydopamine. *J. of Comparative Neurology*, **185**, 139-161. (8)
Kater, S.B., & Mills, L.R. (1991). Regulation of

growth cone behavior by calcium. *J. of Neuroscience*, **11**, 891-899. (3)

Kato, N., Artola, A., & Singer, W. (1991). Developmental changes in the susceptibility to long-term potentiation of neurones in rat visual cortex slices. *Developmental Brain Research*, **60**, 43-50. (8)

Kato, N., Brain, M.S., Artola, A., & Singer, W. (1991). NMDA receptor-mediated EPSP in developing rat neocortex. *Experimental Brain Research*, **20**, 81-84. (8)

Katsuki, H., Saito, H., & Satoh, M. (1992). The involvement of muscarinic, ß-adrenergic and metabotropic glutamate receptors in long-term potentiation in the fimbria-CA3 pathway of the hippocampus. *Neuroscience Letters*, **142**, 249-252. (6)

Kauer, J.A., Malenka, R.C., & Nicoll, R.A. (1988). NMDA application potentiates synaptic transmission in the hippocampus. *Nature*, **334**, 250-252. (6)

Kawagoe, K.T., Garris, P.A., Wiedemann, D.J., & Wightman, R.M. (1992) Regulation of transient dopamine concentration gradients in the microenvironment surrounding nerve terminals in the rat striatum. *Neuroscience*, **51**, 55-64. (17)

Kawamoto, A., & Zemblidge, J. (1992). Pronunciation of homographs. *J. of Memory and Language*, **31**, 349-374. (27)

Kawato, M., Hayakawa, H., & Inui, T. (1993). A forward-inverse optics model of reciprocal connections between visual cortical areas. *Network*, **4**, 415-422. (9)

Kawato, M., Inui, T., Hongo, S., & Hayakawa, H. (1991). Computational theory and neural network models of interaction between visual cortical areas. *ATR Technical Report, TR-A-0105*. Kyoto: ATR. (9)

Kaye, Y., & Pearce, J. (1987). Hippocampal lesions attenuate latent inhibition and the decline of the orienting response in rats. *Quarterly J. of Experimental Psychology*, 39B, 107-125. (12)

Keesing, R., & Stork, D.G. (1991). Evolution and learning in neural networks: The number and distribution of learning trials affect the rate of evolution. In R.P. Lippman, J.E. Moody, & D.S. Touretzky (Eds.), *Advances in neural information processing systems, vol. 3* (pp. 804-810). San Mateo, CA: Morgan Kaufmann. (4)

Kehoe, E.J., Feyer, A.M., & Moses, J.L. (1981). Second-order conditioning of the rabbit's nictitating membrane response as a function of the CS2-CS1 and CS1-US intervals. *Animal Learning & Behavior*, **9**, 304-315. (20)

Kehoe, E.J., Schreurs, B.G., Macrae, M., & Gormezano, I. (1995). Effects of modulating tone frequency, intensity, and duration on the classically conditioned rabbit nictitating membrane response. *Psychobiology*, **23**, 103-115. (20)

Keller, A., & White, E.L. (1987). Synaptic organization of GABAergic neurons in the mouse SmI cortex. *J. of Comparative Neurology*, **262**, 1-12. (3)

Keller, F.S., & Schoenfeld. W.N. (1950). *Principles of psychology*. New York: Appleton-Century-Crofts. (14)

Kelso, J.A.S., Saltzman, E.L., & Tuller, B. (1987). The dynamical perspective on speech production: Data and theory. *J. of Phonetics*, **14**, 29-60. (25)

Kent, R.D., Carney, P.J., & Severeid, L.R. (1974). Velar movement and timing: Evaluation of a model for binary control. *J. of Speech and Hearing Research*, **17**, 470-488. (25)

Kent, R.D., & Minifie, F.D. (1977). Coarticulation in recent speech production models. *J. of Phonetics*, **5**, 115-117. (25)

Kettner, R.E., Schwartz, A.B., & Georgopoulos, A.P. (1988). Primate motor cortex and free arm movements to visual targets in three-dimensional space III: Positional gradients and population coding of movement direction from various movement origins. *J. of Neuroscience*, **8**, 2938-2947. (13)

Kiani, M.F., & Hudetz, A.G. (1991). Computer simulation of growth of anastomosing microvascular networks. *J. of Theoretical Biology*, **150**, 547-560. (3)

Kim, J.J., & Fanselow, M.S. (1992). Modality-specific retrograde amnesia of fear. *Science*, **256**, 675-677. (26)

Kirkwood, A., & Bear, M.F. (1994). Homosynaptic long-term depression in the visual cortex. *J. of Neuroscience*, 3404-3412. (8)

Kirkwood, A., & Bear, M.F. (1994). Hebbian synapses in visual cortex. *J. of Neuroscience*, **14**, 1634-1645. (8)

Kirkwood, A., Dudek, S.M., Gold, J.T., Aizenman, C.D., & Bear, M.F. (1993). Common forms of synaptic plasticity in the hippocampus and neocortex in vitro. *Science*, **260**, 1518-1521. (8)

Kirkwood, A., Lee, H.K., & Bear, M.F. (1995). Co-regulation of long-term potentiation and experience-dependent synaptic plasticity in visual cortex by age and experience. *Nature*, **375**, 328-331. (8)

Kirkwood, B.L., Price, J., & Grove, E.A. (1992). Technical comment. *Science*, **258**, 317-320. (2)

Kishi, K. (1987). Golgi studies on the development of granule cells of the rat olfactory bulb with reference to migration in the subependymal layer. *J. of Comparative Neurology*, **258**, 112-124. (2)

Kishi, K., Peng, J.Y., Kakuta, S., Murakami, K., Kuruda, M., Yokota, S., Hayakawa, S., Kuge, T., & Asayama, T. (1990). Migration of bipolar cells, precursors of granule cells of the olfactory bulb, with reference to the arrangement of radial glial fibers. *Archives of Histology and Cytology*. **53**, 219-226. (2)

Kleinschmidt, A., Bear, M.F., & Singer, W. (1987). Blockade of "NMDA" receptors disrupts experience-dependent plasticity of kitten striate cortex. *Science*, **238**, 355-358. (8)

Klopf, A.H. (1988). A neuronal model of classical conditioning. *Psychobiology*, **16**, 85-125. (4)

Klopf, A.H., & Morgan, J.S. (1990). The role of time in natural intelligence: Implications of classical and instrumental conditioning for neuronal and neural-network modeling. In M. Gabriel & J. Moore (Eds.), *Learning and computational neuroscience: Foundations of adaptive networks* (pp. 463-495). Cambridge, MA: MIT Press. (4)

Knapp, M., Knopman, D., Solomon, P., Pendlebury, W., Davis, C., & Gracon, S. (1994). A 30-week randomized controlled trial of high-dose Tacrine in patients with Alzheimer's disease. *J. of the American Medical Association*, **271**, 985-991. (12)

König, P., Engel, A.K., Löwel, S., & Singer, W. (1993). Squint affects synchronization of oscillatory responses in cat visual cortex. *European J. of Neuroscience*, **5**, 501-508. (8)

Knowlton, B.J., Mangels, J.A., & Squire, L.R. (1996). A neostriatal habit learning system in humans. *Science*, **273**, 1399-1402. (18)
Knowlton, B., Squire, L., & Gluck, M. (1994). Probabilistic classification learning in amnesia. *Learning and Memory*, **1**, 106-120. (12)
Kobatake, E., & Tanaka, K. (1994). Neuronal selectivities to complex object features in the ventral visual pathway of the macaque cerebral cortex. *J. of Neurophysiology*, **71**, 856-867. (9)
Kobatake, E., Tanaka, K., & Tamori, Y. (1992). Long-term learning changes the stimulus selectivity of cells in the inferotemporal cortex of adult monkeys. *Neuroscience Research*, **S17**, S237. (9)
Kobatake, E., Tanaka, K., Wang, G., & Tamori, Y. (1993). Effects of adult learning on the stimulus selectivity of cells in the inferotemporal cortex. *Society of Neuroscience Abstract*, **19**, 975. (9)
Kohler, C. (1986). Intrinsic connections of the retrohippocampal region in the rat brain II: The medial entorhinal area. *J. of Comparative Neurology*, **246**, 149-169. (12)
Kohonen, T. (1984). *Self-organization and associative memory*. New York: Springer-Verlag. (12)
Konishi, M. (1991). Deciphering the brain's codes. *Neural Computation*, **3**, 1-18. (10)
Konnerth, A., Dressen, J., & Augustine, G.T. (1992). Brief dendritic signals initiate long-lasting synaptic depression in cerebellar Purkinje cells. *Proceedings of the National Academy of Sciences USA*, **89**, 7051-7055. (20)
Konorski, J. (1948). *Conditioned reflexes and neuron organization*. Cambridge, UK: Cambridge University Press. (6) (21)
Konorski, J. (1967). *Integrative activity of the brain*. Chicago: University of Chicago Press. (10)
Konorski, J., & Miller, S. (1937). On two types of conditioned reflex. *J. of Genetic Psychology*, **16**, 264-272. (14)
Korn, H., Sur, C., Charpier, S., Legendre, P., & Faber, D.S. (1994). The one-vesicle hypothesis and multivesicular release. In L. Stjärne, P. Greengard, S. Grillner, T. Hökfelt, & D. Ottoson (Eds.), *Molecular and cellular mechanisms of neurotransmitter release* (pp. 301-322). New York: Raven Press. (5)
Kornack, D.R., & Rakic, P. (1995). Radial and horizontal deployment of clonally related cells in the primate neocortex: Relationship to distinct mitotic lineages. *Neuron*, **15**, 311-321. (2)
Korol, D.L., Abel, T.W., Church, L.T., Barnes, C.A., & McNaughton, B.L. (1993). Hippocampal synaptic enhancement and spatial learning in the Morris swim task. *Hippocampus*, **3**, 127- 132. (6)
Kossel, A., Löwel, S., & Bolz, J. (1995). Relationships between dendritic fields and functional architecture in striate cortex of normal and visually deprived cats. *J. of Neuroscience*, **15**, 3913-3926. (8)
Kosslyn, S.M. (1994). *Image and brain*. Cambridge, MA: MIT Press. (26)
Kosslyn, S.M., & Koenig, O. (1992). *Wet mind: The new cognitive neuroscience*. New York: Free Press. (26)
Kosslyn, S.M., & Mumford, D. (Eds.). (1991). Special issue on face perception. *J. of Cognitive Neuroscience*, **3**, 1-88. (10)
Kounios, J. (1993). Process complexity in semantic memory. *J. of Experimental Psychology: Learning, Memory and Cognition*, **19**, 338-351. (24)
Kounios, J., Osman, A., & Meyer, D. (1987). Structure and process in semantic memory: New evidence based on speed-accuracy decomposition. *J. of Experimental Psychology: General*, **1**, 3-25. (24)
Krug, M., Chepkova, A.N., Geyer, C., & Ott, T. (1983). Aminergic blockade modulates long-term potentiation in the dentate gyrus of freely moving rats. *Brain Research Bulletin*, **11**, 1-6. (6)
Krug, M., Jork, R., Reymann, K., Wagner, M., & Matthies, H. (1991). The amnesic substance 2-deoxy-D-galactose suppresses the maintenance of hippocampal LTP. *Brain Research*, **540**, 237-242. (6)
Krug, M., Lössner, B., & Ott, T. (1984). Anisomycin blocks the late phase of long-term potentiation in the dentate gyrus of freely moving rats. *Brain Research Bulletin*, **13**, 39-42. (6)
Kruschke, J. (1992). Human category learning: Implications for backpropagation models. *Connection Science*, **5**, 3-36. (12)
Kugler, P.N., Kelso, J.A.S., & Turvey, M.T. (1980). On the concept of coordinative structures as dissipative structures: I, Theoretical lines of convergence. In G.E. Stelmach & J. Requin (Eds.), *Tutorials in motor behavior* (pp. 3-47). New York: North Holland. (15)
Kusunoki, M., Tanaka, Y., Ohtsuka, H., Ishiyama, K., & Sakata, H. (1993). Selectivity of the parietal visual neurons in the axis orientation of objects in space. *Society of Neuroscience Abstract*, **19**, 770. (9)
Kwong, K.K., Belliveau, J.W., Chesler, D.A., Goldberg, I.E., Weiskoff, R.M., Poncelet, B.P., Kennedy, D.N., Hoppel, B.E., Cohen, M.S., Turner, R., Cheng, H-M., Brady, T.J., & Rosen, B.R. (1992). Dynamic magnetic resonance imaging of human brain activity during primary sensory stimulation. *Proceedings of the National Academy of Sciences USA*, **89**, 5675-5679. (23)
Ladefoged, P. (1982). *A course in phonetics*. New York: Harcourt Brace Jovanovich. (25)
Lamme, V.A.F. (1995). The neurophysiology of figure-ground segregation in primary visual cortex. *J. of Neuroscience*, **15**, 1605-1615. (8)
Lamme, V.A.F., van Dijk, B.W., & Spekreijse, H. (1993). Contour from motion processing occurs in primary visual cortex. *Nature*, **363**, 541-543. (8)
Landry, P., Wilson, C.J., & Kitai, S.T. (1984). Morphological and electrophysiological characteristics of pyramidal tract neurons in the rat. *Experimental Brain Research*, **57**, 177-190. (7)
Langton, C.G. (1992). Preface. In C.G. Langton, C. Taylor, J.D. Farmer, & S. Rasmussen (Eds.), *Artificial life II* (xiii- xviii). Redwood City, CA: Addison-Wesley. (4)
Langton, C.G. (1994). Editor's introduction. *Artificial Life*, **1**, v-viii. (4)
Lashley, K.S. (1951). The problem of serial order in behavior. In L.A. Jeffress (Ed.), *Cerebral mechanisms in behavior*. New York: Wiley. (25)
Lattal, K.A., & Gleeson, S. (1990). Response acquisition with delayed reinforcement. *J. of Experimental Psychology: Animal Behavior Processes*, **16**, 27-39. (14)
Lattal, K.A., & Metzger, B. (1994). Response acquisition by Siamese fighting fish (*Betta splendens*) with delayed visual reinforcement. *J. of the Experimental Analysis of Behavior*, **61**, 35-44. (14)

Lawrence, D.H. (1952). The transfer of a discrimination along a continuum. *J. of Comparative and Physiological Psychology*, **45**, 511-516. (12)
LeDouarin, N.M. (1982). *The neural crest*. Cambridge, UK: Cambridge University Press. (2)
LeDoux, J.E. (1986). Sensory systems and emotion: A model of affective processing. *Integrative Psychiatry*, **4**, 237-243. (4)
Ledoux, J.E. (1992). Brain mechanisms of emotion and emotional learning. *Current Opinion in Neurobiology*, **2**, 191-197. (21)
LeDoux, J.E. (1993). Emotional memory systems in the brain. *Behavioural Brain Research*, **58**, 69-79. (4)
Lee, M.K., Rebhun, L.I., & Frankfurter, A. (1990b). Posttranslational modification of class III ß-tubulin. *Proceedings of the National Academy of Sciences USA*, **87**, 7195-7199. (2)
Lee, M.K., Tuttle, J.B., Rebhun, L.L., Cleveland, D.W., & Frankfurter, A. (1990a). The expression and posttranslational modification of a neuron-specific ß-tubulin isotype during chick embryogenesis. *Cell Motility and the Cytoskeleton*, **17**, 118-132. (2)
Lehky, S.R., Sejnowski, T.J., & Desimone, R. (1992). Predicting responses of nonlinear neurons in monkey striate cortex to complex patterns. *J. of Neuroscience*, **12**, 3568-3581. (9)
Leiner, H.C., Leiner, A.L., & Dow, R.S. (1991). The human cerebro-cerebellar system: Its computing, cognitive, and language skills. *Behavioral Brain Research*, **44**, 113-128. (24)
Leiner, H. C., Leiner, A. L., & Dow, R. S. (1993). Cognitive and language functions of the human cerebellum. *Trends in Neuroscience*, **16**, 444-454. (24)
Leonard, C.M., Rolls, E.T., Wilson, F.A.W., & Baylis, G.C. (1985). Neurons in the amygdala of the monkey with responses selective for faces. *Behavioral Brain Research*, **15**, 159-176. (9)
Levi, G., & Raiteri, M. (1993). Carrier-mediated release of neurotransmitters. *Trends in Neurosciences*, **16**, 415-419. (5)
LeVine, S.M., & Goldman, J.E. (1988). Embryonic divergence of oligodendrocyte and astrocyte lineages in developing rat cerebrum. *J. of Neuroscience*, **8**, 3992-4006. (2)
Levison, S.W., & Goldman, J.E. (1993). Both oligodendrocytes and astrocytes develop from progenitors in the subventricular zone of postnatal rat forebrain. *Neuron*, **10**, 201-212. (2)
Levy, W.B., Brassel, S.E., & Moore, S.D. (1983). Partial quantification of the associative synaptic learning rule of the dentate gyrus. *Neuroscience*, **8**, 799-808. (12)
Levy, W.B., & Steward, O. (1979). Synapses as associative memory elements in the hippocampal formation. *Brain Research*, **175**, 233- 245. (6)
Li, G-H., Qin, C-D., & Wang, L-W. (1995). Computer model of growth cone behavior and neuronal morphogenesis. *J. of Theoretical Biology*, **174**, 381-389. (3)
Li, L., Miller, E.K., & Desimone, R. (1993). The representation of stimulus familiarity in anterior inferior temporal cortex. *J. of Neurophysiology*, **69**, 1918-1929. (9)
Liao, D., Hessler, N.A., & Malinow, R. (1995). Activation of postsynaptically silent synapses during pairing-induced LTP in CA1 region of hippocampal slice. *Nature*, **375**, 400-404. (6)
Libby, A. (1951). Two variables in the acquisition of depressant properties by a stimulus. *J. of Experimental Psychology*, **42**, 100-107. (4)
Lübke, J., & Albus, K. (1992). Lack of exuberance in clustered intrinsic connections in the striate cortex of one-month-old kitten. *European J. of Neuroscience*, **14**, 189-192. (8)
Lin, L-J. (1992). Self-improving reactive agents based on reinforcement learning, planning and teaching. *Machine Learning*, **8**, 293-321. (19) (22)
Lindsay, R.D., & Scheibel, A.B. (1974). Quantitative analysis of the dendritic branching pattern of small pyramidal cells from adult rat somesthetic and visual cortex. *Experimental Neurology*, **45**, 424-434. (3)
Lindvall, O.A., Bjorklund, A., & Divac, I. (1978). Organization of catecholamine neurons projecting to frontal cortex in the rat. *Brain Research*, **142**, 1-24. (18)
Link, W., Konietzko, U., Kauselmann, G., Krug, M., Schwanke, B., Frey, U., & Kuhl, D. (1995). Somatodendritic expression of an immediate early gene is regulated by synaptic activity. *Proceedings of the National Academy of Sciences USA*, **92**, 5734- 5738. (6)
Lipkens, R., Hayes, S.C., & Hayes, L.J. (1993). Longitudinal study of the development of derived relational responding in an infant. *J. of Experimental Child Psychology*, **56**, 201-239. (26)
Lipp, H-P., Wolfer, D.P., Bozizevic, M., Carmeliet, P., Collen, D., & Mulligan, R.M. (1993). Swimming navigation learning in mice deficient for tissue plasminogen activator: Impaired acquisition but intact spatial memory. *Society of Neuroscience Abstracts*, **19**, 799. (6)
Lisman, J. (1989). A mechanism for the Hebb and the anti-Hebb processes underlying learning and memory. *Proceedings of the National Academy of Sciences USA*, **86**, 9574-9578. (8)
Livingstone, M.S. (1991). Visually evoked oscillations in monkey striate cortex. *Society of Neuroscience Abstracts*, **17**, 73.3. (8)
Ljungberg, T., Apicella, P., & Schultz, W. (1991) Responses of monkey midbrain dopamine neurons during delayed alternation performance. *Brain Research*, **586**, 337-341. (17)
Ljungberg, T., Apicella, P., & Schultz, W. (1992) Responses of monkey dopamine neurons during learning of behavioral reactions. *J. of Neurophysiology*, **67**, 145-163. (17) (21)
Llinas, R., & Muhlethaler, M. (1988). Electrophysiology of guinea-pig cerebellar nuclear cells in the in vitro brain stem-cerebellar preparation. *J. of Physiology*, **404**, 241-258. (7)
Loewi, O. (1921). Über humorale Übertragbarkeit der Herznervenwirkung. *Pflüger's Archiv für die gesamte Physiologie des Menschen und der Tiere*, **189**, 239-242. (5)
Logan, F.A. (1952). The role of delay of reinforcement in determining reaction potential. *J. of Experimental Psychology*, **43**, 393-399. (14)
Logan, G.D. (1988). Toward an instance theory of automatization. *Psychological Review*, **95**, 492-527. (24)
Logan, G.D. (1992). Shapes of reaction-time distributions and shapes of learning curves: A test of the instance theory of automaticity. *J. of Experimental Psychology*, **18**, 883-914. (24)

Logan, G.D. (1994). Spatial attention and the apprehension of spatial relations. *J. of Experimental Psychology: Human Perception and Performance*. (in press). (11)

Lössner, B., Jork, R., Krug, M., & Matthies, H. (1982). Protein synthesis in rat hippocampus during training and stimulation experiments. In C.A. Marsan & H. Matthies (Eds.), *Neural plasticity and memory formation* (pp. 183-191). New York: Ravens Press. (6)

Lössner, B., Schweigert, C., Pchalek, V., Krug, M., Frey, S., & Matthies, H. (1987). Training- and LTP-induced changes of protein synthesis in rat hippocampus. *Neuroscience Supplements*, **22**, 9512. (6)

Lovinger, D.M., Wong, K.L., Murakami, K., & Routtenberg, A. (1987). Protein kinase C inhibitors eliminate hippocampal long-term potentiation. *Brain Research*, **436**, 177-183. (6)

Lowe, D.G. (1987). The viewpoint consistency constraint. *International J. of Computer Vision*, **1**, 57-72. (11)

Löwel, S., & Singer, W. (1992). Selection of intrinsic horizontal connections in the visual cortex by correlated neuronal activity. *Science*, **255**, 209-212. (8)

Lubow, R. (1973). Latent inhibition. *Psychological Bulletin*, **79**, 398-407. (12)

Lubow, R.E., Rifkin, B., & Alek, M. (1976). The context effect: The relationship between stimulus pre-exposure and environmental pre-exposure determines subsequent learning. *J. of Experimental Psychology: Animal Behavior Processes*, **2**, 38-47. (12)

Luhmann, H.J., Singer, W., & Martinez-Millan, L. (1990). Horizontal interactions in cat striate cortex: I. Anatomical substrate and postnatal development. *European J. of Neuroscience*, **2**, 344-357. (8)

Lukashin, A.V. (1990). A learned neural network that simulates properties of the neuronal population vector. *Biological Cybernetics*, **63**, 377-382. (13)

Lukashin, A.V., Amirikian, B.R., & Georgopoulos, A.P. (1996) Neural computations underlying the exertion of force: A model. *Biological Cybernetics*, in press. (13)

Lukashin, A.V., & Georgopoulos, A.P. (1994a). A neural network for coding of trajectories by time series of neuronal population vectors. *Neural Computation*, **6**, 19-28. (13)

Lukashin, A.V., & Georgopoulos, A.P. (1994b). Directional operations in the motor cortex modeled by a neural network of spiking neurons. *Biological Cybernetics*, **71**, 79-85. (13)

Lukashin, A.V., Wilcox, G.L., & Georgopoulos, A.P. (1994). Overlapping neural networks for multiple motor engrams. *Proceedings of the National Academy of Sciences USA*, **91**, 8651-8654. (13)

Lundberg, J. M., & Hökfelt, T. (1986). Multiple coexistence of peptides and classical transmitters in peripheral autonomic and sensory neurons functional and pharmacological implications. In T. Hökfelt, K. Fuxe, & B. Pernow (Eds.), *Coexistence of neuronal messengers: A new principle in chemical transmission. Progress in Brain Research*, (Vol. 68, pp. 241-262). Amsterdam: Elsevier. (5)

Lurito, J.L., Georgakopoulos, T., & Georgopoulos, A.P. (1991). Cognitive spatial-motor processes 7: The making of movements at an angle from a stimulus direction: Studies of motor cortical activity at the single cell and population levels. *Experimental Brain Research*, **87**, 562-580. (13)

Luskin, M.B. (1992). Generation of cell diversity in the mammalian visual cortex: Lineage relationships. In R. Lent (Ed.), *The visual system from genesis to maturity* (pp. 111-130). Cambridge, MA: Birkhauser Press. (2)

Luskin, M.B. (1993a). Restricted proliferation and migration of postnatally generated neurons derived from the forebrain subventricular zone. *Neuron*, **11**, 173-189. (2)

Luskin, M.B. (1993b). Mammalian forebrain development: How cells find a permanent home and establish their identity. *J. of NIH Research*, **5**, 60-64. (2)

Luskin, M.B., & McDermott, K.W. (1994). Divergent lineages for oligodendrocytes and astrocytes originating in the forebrain subventricular zone. *Glia*, **11**, 211- 226. (2)

Luskin, M.B., Parnavelas, J.G., & Barfield, J.A. (1993). Neurons astrocytes and oligodendrocytes of the rat cerebral cortex originate from separate progenitor cells: An ultrastructural analysis of clonally related cells. *J. of Neuroscience*, **13**, 1730-1750. (2)

Luskin, M.B., Pearlman, A.L., & Sanes, J.R. (1988). Cell lineage in the cerebral cortex of the mouse studied in vivo and in vitro with a recombinant retrovirus. *Neuron*, **1**, 635-647. (2)

Luskin, M.B., & Shatz, C.J. (1985a). Neurogenesis of the cat's primary visual cortex. *J. of Comparative Neurology*, **216**, 611-631. (2)

Luskin, M.B., & Shatz, C.J. (1985b). Studies of the earliest generated cells of the cat's visual cortex: Cogeneration of subplate and marginal zone cells. *J. of Neuroscience*, **5**, 1062-1075. (2)

Lynch, G. (1986). *Synapses, circuits and the beginning of memory*. Cambridge, MA: MIT Press. (26)

Lynch, G., & Baudry, M. (1988). Structure-function relationships in the organization of memory. In M.S. Gazzaniga (Ed.), *Perspectives in memory research*. Cambridge, MA: MIT Press. (26)

Lynch, M.A., Errington, M.L., & Bliss, T.V. (1985). Long-term potentiation of synaptic transmission in the dentate gyrus: Increased release of [14C]glutamate without increase in receptor binding. *Neuroscience Letters*, **62**, 123-129. (6)

Lynch, M.A., Errington, M.L., Clements, M.P., Bliss, T.V., Rédini-Del Negro, C., & Laroche, S. (1990). Increases in glutamate release and phosphoinositide metabolism associated with long-term potentiation and classical conditioning. *Progress in Brain Research*, **83**, 251-256. (6)

Macar, F. (1985). Time psychophysics and related models. In J.A. Michon & J.L. Jackson (Eds.), *Time, mind and behavior* (pp. 112- 130). Berlin: Springer-Verlag. (7)

Macdonald, R.L., & Olsen, R.W. (1994). $GABA_A$ receptor channels. *Annual Review of Neuroscience*, **17**, 569-602. (5)

MacKay, D. (1982). The problem of flexibility, fluency and speed-accuracy trade-off in skilled behavior. *Psychological Review*, **89**, 483-506. (24)

MacKay, D. (1987). *The organization of perception and action: A theory for language and other cognitive skills*. Heidelberg: Springer-Verlag. (24)

MacKay, D.G. (1981). A general theory of serial ordering in behavior. Paper presented at the

Twenty-Second Annual Meeting of The Psychonomic Society, San Diego, CA. (25)
Mackintosh, N.J. (1975) A theory of attention: Variations in the associability of stimulus with reinforcement. *Psychological Review*, **82**, 276-298. (17)
Mackintosh, N.J. (1983). *Conditioning and associative learning*. Oxford, UK: Oxford University Press. (21)
Maclin, R., & Shavlik, J.W. (1994). Incorporating advice into agents that learn from reinforcements. *Proceedings of the Twelfth National Conference on Artificial Intelligence (AAAI-94)*, 694-699. (19) (22)
Madison, D.V., & Nicoll, R.A. (1986). Actions of noradrenaline recorded intracellularly in rat hippocampal CA1 pyramidal neurones in vitro. *J. of Physiology*, **372**, 245-259. (7)
Maeda, T., Kaneko, S., & Satoh, M. (1994). Roles of endogenous cholinergic neurons in the induction of long-term potentiation at hippocampal mossy fiber synapses. *Neuroscience Research*, **20**, 71-78. (6)
Maes, P., & Brooks, R. (1990). Learning to coordinate behavior. *Proceedings of the Eighth National Conference on Artificial Intelligence* (pp. 796-802). San Mateo, CA: Morgan Kaufmann. (19)
Maffei, L., Berardi, N., Domenici, L., Parisi, V., & Pizzorusso, T. (1992). Nerve growth factor (NGF) prevents the shift in ocular dominance distribution of visual cortical neurons in monocularly deprived rats. *J. of Neuroscience*, **12**, 4651-4662. (2)
Magee, J.C., & Johnston, D. (1997). A synaptically controlled, associative signal for Hebbian plasticity in hippocampal neurons. *Science*, **275**, 209-213. (Pt 5)
Mahadevan, S. (1966). Average reward reinforcement learning: Foundations, algorithms, and empirical results. *Machine Learning*, **22**, 159-196. (19)
Mahadevan, S., & Connell, J. (1991). Scaling reinforcement learning to robotics by exploiting the subsumption architecture. In L.A. Birnbaum & G.C. Collins (Eds.), *Machine learning: Proceedings of the eighth international workshop* (pp. 328-332). San Mateo, CA: Morgan Kaufmann. (19)
Mahowald, M.A., & Mead, C. (1991). The silicon retina. *Scientific American*, **264**, 76-82. (3)
Maines, M.D. (1993). Carbon monoxide: An emerging regulator of cGMP in the brain. *Molecular and Cellular Neuroscience*, **4**, 389- 397. (5)
Malach, R. (1994). Cortical columns as devices for maximizing neuronal diversity. *Trends in Neuroscience*, **17**, 101-104. (9)
Malach, R., Amir, Y., Harel, M., & Grinvald, A. (1993). Relationship between intrinsic connections and functional architecture revealed by optical imaging and in vivo targeted biocytin injections in primate striate cortex. *Proceedings of the National Academy of Sciences USA*, **90**, 10469-10473. (8)
Malenka, R.C. (1995). LTP and LTD: Dynamic and interactive processes of synaptic plasticity. *The Neuroscientist*, **1**, 35-42. (5)
Malenka, R.C., Kauer, J.A., Perkel, D.J., Mauk, M.D., Kelly, P.T., Nicoll, R.A., & Waxham, M.N. (1989a). An essential role for postsynaptic calmodulin and protein kinase activity in long term potentiation. *Nature*, **340**, 554-556. (6)
Malenka, R.C., Kauer, J.A., Perkel, D.J., & Nicoll, R.A. (1989b). The impact of postsynaptic calcium on synaptic transmission—its role in long-term potentiation. *Trends in Neuroscience*, **12**, 444-450. (6)
Malinow, R., Madison, D.V., & Tsien, R.W. (1988). Persistent protein kinase activity underlying long-term potentiation. *Nature*, **335**, 820-824. (6)
Malinow, R., Schulman, H., & Tsien, R.W. (1989). Inhibition of postsynaptic PKC or CaMKII blocks induction but not expression of LTP. *Science*, **245**, 862-866. (6)
Manahan-Vaughan, D., & Reymann, K.G. (1995). 1S, 3R-ACPD dose-dependently induces a slow-onset potentiation in the rat hippocampal CA1 region in vivo. *Neuropharmacology*, **34**, 1103-1105. (6)
Manning, F. (1994). Tacrine therapy for the dementia of Alzheimer's disease. *American Family Physician*, **50**, 819-826. (12)
Manuel, S.Y., & Krakow R.A. (1984). Universal and language particular aspects of vowel-to-vowel coarticulation. *Status Report on Speech Research SR-77/78*. New Haven, CT: Haskins Laboratories. (25)
Margoliash, D. (1983). Acoustic parameters underlying the responses of song-specific neurons in the white-crowned sparrow. *J. of Neuroscience*, **3**, 133-143. (7)
Margoliash, D., & Fortune, E.S. (1992). Temporal and harmonic combination-sensitive neurons in the zebra finch's HVc. *J. of Neuroscience*, **12**, 4309-4326. (7)
Markowska, A., Olton, D., & Givens, B. (1995). Cholinergic manipulations in the medial septal area: Age-related effects on working memory and hippocampal electrophysiology. *J. of Neuroscience* (3 Pt. 1), **15**, 2063-2073. (12)
Markram, H., Lübke, J., Frotscher, M., & Sakmann, B. (1997). Regulation of synaptic efficacy by coincidence of postsynaptic APs and EPSPs. *Science*, **275**, 213-215. (Pt 5)
Marr, D. (1969). A theory of cerebellar cortex. *J. of Physiology (London)*, **202**, 437-470. (6)
Marr, D., & Nishihara, H. K. (1978). Representation and recognition of three dimensional shapes. *Proceedings of the Royal Society of London B*, **200**, 269-294. (11)
Marr, D., & Poggio, T. (1976). Cooperative computation of stereo disparity. *Science*, **194**, 283-287. (25)
Marshall, J. C., & Newcombe, F. (1973). Patterns of paralexia: A psycholinguistic approach. *J. of Psycholinguistic Research*, **2**, 175-199. (27)
Marshall, J., Molloy, R., Moss, G.W.J., Howe, J.R., & Hughes, T.E. (1995). The jellyfish green fluorescent protein: A new tool for studying ion channel expression and function. *Neuron*, **14**, 211-215. (3)
Martin, K.A.C. (1988). From single cells to simple circuits in the cerebral cortex. *Quarterly J. of Experimental Physiology*, **73**, 637-702. (13)
Mathew, A., & Cook, M. (1990). The control of reaching movements by young infants. *Child Development*, **61**, 1238-1258. (15)
Matthies, H. (1974). The biochemical basis of learning and memory. *Life Science*, **15**, 2017-2031. (6)
Matthies, H. (1989a). Neurobiological aspects of learning and memory. *Annual Review of Psychology*, **40**, 381-404. (6)
Matthies, H. (1989b). In search of cellular mechanisms of memory. *Progress in Neurobiology*, **32**,

277-349. (6)
Matthies, H., Frey, U., Reymann, K., Krug, M., Jork, R., & Schroeder, H. (1990). Different mechanisms and multiple stages of LTP. *Advances in Experimental Medicine and Biology*, **268**, 359-368. (6)
Matthies, H., & Reymann, K.G. (1993). Protein kinase A inhibitors prevent the maintenance of hippocampal long-term potentiation. *NeuroReport*, **4**, 712-714. (6)
Matthies, H., Reymann, K.G., Krug, M., Frey, U., Loessner, B., & Popov, N. (1989). Multiple stages of posttetanic LTP and memory. In Rahmann (Ed.), *Fundamentals of memory formation: Neuronal plasticity and brain function* (pp. 395-396). Stuttgart, New York: Gustav Fischer Verlag. (6)
Matthies, H., Ruethrich, H., Ott, T., Matthies, H.K., & Matthies, R. (1986). Low frequency perforant path stimulation as a conditioned stimulus demonstrates correlations between long-term synaptic potentiation and learning. *Physiology and Behavior*, **36**, 811-821. (6)
Maunsell, J.H.R. (1996). The brain's visual world: Representation of visual targets in cerebral cortex. *Science*, **270**, 764-768. (8)
Mazzoni, P., Andersen, R.A., & Jordan, M.I. (1991) A more biologically plausible learning rule than backpropagation applied to a network model of cortical area 7. *Cerebral Cortex*, **1**, 293-307. (17)
McClelland, J.L., & Rumelhart, D.E. (1981). An interactive-activation model of context effects in letter perception. Part 1: An account of basic findings. *Psychological Review*, **88**, 375-407. (25) (27)
McClelland, J.L., McNaughton, B.L., & O'Reilly, R.C. (1994). Why there are complementary learning systems in the hippocampus and neo-cortex: Insights from the success and failures of connectionist models of learning and memory. *Technical Report, PDP.CNS.94.1*. (26)
McClelland, J.L., & Rumelhart, D.E. (1988). *Explorations in parallel distributed processing: A handbook of models, programs, and exercises*. London: MIT Press. (26)
McCloskey, M., & Cohen, N.J. (1989). Catastrophic interference in connectionist networks: The sequential learning problem. In G.H. Bower (Ed.), *The psychology of learning and motivation*. New York: Academic Press. (26)
McConnell, S.K. (1988). Development and decision-making in the mammalian cerebral cortex. *Brain Research Review*, **13**, 1-23. (2) (4)
McConnell, S.K. (1989). The determination of neuronal fate in the cerebral cortex. *Trends in Neuroscience*, **12**, 342-349. (4)
McConnell, S.K. (1991). The generation of neuronal diversity in the central nervous system. *Annual Review of Neuroscience*, **14**, 269-300. (4)
McCormick, D.A. (1989). GABA as an inhibitory neurotransmitter in human cerebral cortex. *J. of Neurophysiology*, **62**, 1018-1027. (7)
McCormick, D.A., & Williamson, A. (1989). Convergence and divergence of neurotransmitter action in human cerebral cortex. *Proceedings of the National Academy of Sciences USA*, **86**, 8098-8102. (5)
McDermott, K.W., & Luskin, M.B. (1995). The use of retroviral vectors in the study of cell lineage and migration during the development of the mammalian central nervous system. In M.G. Kaplitt and A.P. Loewy (Eds.), *Viral vectors: Tools for study and genetic manipulation of the nervous system* (pp.411-433). New York: Academic Press. (2)
McGraw, M.B. (1935). *Growth: A study of Johnny and Jimmy*. New York: Appleton-Century-Crofts. (15)
McGraw, M.B. (1941). Development of neuromuscular mechanisms as reflected in the crawling and creeping behavior of the human infant. *J. of Genetic Psychology*, **58**, 83-111. (15)
McLean, J.H., & Shipley, M.T. (1988). Postmitotic, postmigrational expression of tyrosine hydroxylase in olfactory bulb dopaminergic neurons. *J. of Neuroscience*, **8**, 3658-3669. (2)
McNaughton, B., Chen, L., & Markus, E. (1991). "Dead reckoning," landmark learning, and the sense of direction: A neurophysiological and computational hypothesis. *J. of Cognitive Neuroscience*, **3**, 190-202. (12)
McNaughton, B.L., Douglas, R.M., & Goddard, G.V. (1978). Synaptic enhancement in fasciadentata: Cooperativity among coactive afferents. *Brain Research*, **157**, 277-294. (6)
McNaughton, B.L., & Morris, R.G.M. (1987). Hippocampal synaptic enhancement and information storage within a distributed memory system. *Trends in Neuroscience*, **10**, 408-415. (6)
Mechner, F. (1958). Sequential dependencies of the lengths of consecutive response runs. *J. of the Experimental Analysis of Behavior*, **1**, 229-233. (14)
Medin, D.L., Goldstone, R.L., & Gentner, D. (1994). Respects for similarity. *Psychological Review*, **100**, 254-278. (11)
Meister, M., Wong, R.O.L., Baylor, D.A., & Shatz, C.J. (1991). Synchronous bursts of action potentials in ganglion cells of the developing mammalian retina. *Science*, **252**, 939-943. (8)
Melchers, B.P., Pennartz, C.M., & Lopes Da Silva, F.H. (1987). Differential effects of elevated extracellular calcium concentrations on field potentials in dentate gyrus and CA1 of the rat hippocampal slice preparation. *Neuroscience Letters*, **77**, 37-42. (6)
Memberg, S.P., & Hall, A.K. (1995). Dividing neuron precursors express neuron-specific tubulin. *J. of Neurobiology*, **27**, 26-43. (2)
Menezes, J.R.L., & Luskin, M.B. (1994). Expression of neuron-specific tubulin defines a novel population of neurons in the proliferative layers of the developing telencephalon. *J. of Neuroscience*, **14**, 5399-5416. (2)
Menezes, J.R.L., Smith, C.M., Nelson, K., & Luskin, M.B. (1995). The division of neuronal progenitor cells during migration in the neonatal mammalian forebrain. *Molecular and Cellular Neuroscience*, **6**, 496-508. (2)
Merzenich, M.M., & de Charms, C. (1996). Neural representations, experience, and change. In P. Churchland & R. Llinas (Eds.) *The mind and brain continuum: Sensory processes* (pp. 61-82). Boston: MIT Press. (8)
Merzenich, M.M., Kaas, J.H., Wall, J.T., Nelson, T.J., Sur, M., & Felleman, D.T. (1983). Topographic reorganization of somatosensory cortical areas 3b and 1 in adult monkeys following restricted deafferentation. *Neuroscience*, **8**, 33-55. (3)
Merzenich, M.M., & Sameshima, K. (1993). Corti-

cal plasticity and memory. *Current Opinion in Neurobiology*, **3**, 187-196. (8)
Mesulam, M. (1990). Large-scale neurocognitive networks and distributed processing for attention, language, and memory. *Annals of Neurology*, **28**, 597-613. (24)
Metherate, R., & Ashe, J.H. (1994). Facilitation of an NMDA receptor-mediated EPSP by paired-pulse stimulation in rat neocortex via depression of GABAergic IPSPs. *J. of Physiology*, **481**, 331-348. (7)
Meunier, M., Bachevalier, J., Mishkin, M., & Murray, E.A. (1993). Effects on visual recognition of combined and separate ablations of the entorhinal and perirhinal cortex in rhesus monkeys. *J. of Neuroscience*, **13**, 5418-5432. (9)
Meyer, D.E., Abrams, R.A., Kornblum, A., Wright, C.E., & Smith, J.E.K. (1988). Optimality in human motor performance: Ideal control of rapid aimed movements. *Psychological Review*, **95**, 340-370. (15)
Meyer, J.S., Feldman, R.S., & Quenzer, L. (1996). *Principles of neuropsychopharmacology*. Sunderland, MA: Sinauer. (5)
Michael, J. (1982). Distinguishing between discriminative and motivational functions of stimuli. *J. of the Experimental Analysis of Behavior*, **37**, 149-155. (14)
Michalski, A., Gerstein, G.L., Czarkowska, J., & Tarnecki, R. (1983). Interactions between cat striate cortex neurons. *Experimental Brain Research*, **51**, 97-107. (8)
Michie, D., & Chambers, R.A. (1968). BOXES: An experiment in adaptive control. In E. Dale & D. Michie (Eds.), *Machine intelligence 2* (pp. 137-152). Oliver and Boyd. (19)
Millenson, J.R., Kehoe, E.J., & Gormezano, I. (1977). Classical conditioning of the rabbit's nictitating membrane response under fixed and mixed CS-US intervals. *Learning and Motivation*, **8**, 351-366. (20)
Miller, E., & Desimone, R. (1994). Parallel neuronal mechanisms for short-term memory. *Science*, **263**, 520-522. (24)
Miller, E.K., Li, L., & Desimone, R. (1991). A neural mechanism for working and recognition memory in inferior temporal cortex. *Science*, **254**, 1377-1379. (9)
Miller, E., Li, L., & Desimone, R. (1993). Activity of neurons in anterior inferior temporal cortex during a short-term memory task. *J. of Neuroscience*, **12**, 1460-1478. (24)
Miller, G.F., Todd, P.M., & Hegde, S.U. (1989). Designing neural networks using genetic algorithms. *Proceedings of the Third International Conference on Genetic Algorithms* (pp. 379- 383). San Mateo, CA: Morgan Kaufmann. (4)
Miller, J.D., Sanghera, M.K., & German, D.C. (1981) Mesencephalic dopaminergic unit activity in the behaviorally conditioned rat. *Life Sciences*, **29**, 1255-1263. (17)
Miller, K.D., Keller, J.B., & Stryker, M.P. (1989). Ocular dominance column development: Analysis and simulation. *Science*, **245**, 605-615. (8)
Miller, M.W. (1988). Development of projection and local circuit neurons in neocortex. In A. Peters & E.G. Jones (Eds.), *Cerebral cortex, vol. 7: Development and maturation of cerebral cortex* (pp. 133-175). New York: Plenum Press. (4)
Miller, N. (1989). Apraxia of speech. In C. Code (Ed.), *The characteristics of aphasia*. London: Taylor & Francis. (24)
Miller, N.E., & Dworkin, B.R. (1974). Visceral learning: Recent difficulties with curarized rats and significant problems for human research. In P.A. Obrist, A.H. Black, J. Brener, & L.V. DiCara (Eds.), *Cardiovascular psychophysiology*. Chicago: Aldine. (14)
Miller, W.T., Kraft, L.G., & Glanz, F.H. (1992). Real time comparison of neural network and traditional adaptive controllers. *Proceedings of the Seventh Yale Workshop on Adaptive Systems*, 99-104. (15)
Mills, S.K., & Beatty, J.H. (1979). The propensity interpretation of fitness. *Philosophy of Science*, **46**, 263-286. (4)
Milner, B. (1972). Disorders of learning and memory after temporal lobe lesions in man. *Clinical Neurosurgery*, **19**, 421-446. (6)
Minsky, M.L. (1961). Steps toward artificial intelligence. *Proceedings of the Institute of Radio Engineers*, **49**, 8-30. Reprinted in E.A. Feigenbaum & J. Feldman (Eds.), *Computers and thought* (pp. 406-450). New York: McGraw-Hill, 1963. (19)
Mione, M.C., Danevic, C., Boardman, P., Harris, B., & Parnavelas, J.G. (1994). Lineage analysis reveals neurotransmitter (GABA or glutamate) but not calcium-binding protein homogeneity in clonally related cortical neurons. *J. of Neuroscience*, **14**, 107-123. (2)
Mirenowicz, J., & Schultz, W. (1994). Importance of unpredictedness for reward responses in primate dopamine neurons. *J. of Neurophysiology*, **72**, 1024-1027. (17)
Mirenowicz, J., & Schultz, W. (1996). Preferential activation of midbrain dopamine neurons by appetitive rather than aversive stimuli. *Nature*, **379**, 449-451. (17)
Mishkin, M., Ungerleider, L.G., & Macko, K.A. (1983). Object vision and spatial vision: Two cortical pathways. *Trends in Neuroscience*, **6**, 414-417. (9)
Mitchell, K., Oke, A.F., & Adams, R.N. (1994). In vivo dynamics of norepinephrine release-reuptake in multiple terminal field regions of rat brain. *J. of Neurochemistry*, **63**, 917-926. (5)
Mitchell, T.M., & Thrun, S.B. (1993). Explanation-based neural network learning for robot control. In S.J. Hanson, J.D. Cohen, & C.L. Giles (Eds.), *Advances in neural information processing systems 5* (pp. 287-294), San Mateo, CA: Morgan Kaufmann. (19)
Mitz, A.R., Godschalk, M., & Wise, S.P. (1991). Learning-dependent neuronal activity in the premotor cortex: Activity during the acquisition of conditional motor associations. *J. of Neuroscience*, **11**, 1855-1872. (24)
Miyamoto, M., Narumi, S., Nagaoka, A., & Coyle, J. (1989). Effects of continuous infusion of cholinergic drugs on memory impairment in rats with basal forebrain lesions. *J. of Pharmacology and Experimental Therapy*, **248**, 825-835. (12)
Miyashita, Y. (1988). Neuronal correlate of visual associative long-term memory in the primate temporal cortex. *Nature*, **335**, 817-820. (8)
Miyashita, Y. (1993). Inferior temporal cortex: Where visual perception meets memory. *Annual*

Review of Neuroscience, **16**, 245-263. (9)

Miyashita, Y., Higuchi, S., Zhou, X., Okuno, H., & Hasegawa, I. (1994). Disruption of backward connections from the rhinal cortex impairs visual associative coding of neurons in inferotemporal cortex of monkeys. *Society of Neuroscience Abstract*, **20**, 428. (9)

Mizumori, S.J.Y., Rosenzweig, M.R., & Bennett, E.L. (1985). Long-term working memory in the rat: Effects of hippocampally applied anisomycin. *Behavioral Neuroscience*, **99**, 220-232. (6)

Moll, K.L., & Daniloff, R.G. (1971). Investigation of the timing of velar movements during speech. *J. of the Acoustical Society of America*, **50**, 678-684. (25)

Montague, P.R., Dayan, P., Nowlan, S.J., Pouget, A., & Sejnowski, T.J. (1993). Using aperiodic reinforcement for directed self-organization. In Giles, C.L., Hanson, S.J., & Cowan, J.D. (Eds.), *Advances in neural information processing systems 5*. San Mateo, CA: Morgan Kaufmann. (21)

Montague, P.R., Dayan, P., Person, C., & Sejnowski, T.J. (1995) Bee foraging in uncertain environments using predictive hebbian learning. *Nature*, **377**, 725-728. (17) (21)

Montague, P.R., Dayan, P., & Sejnowski, T.J. (1994). Foraging in an uncertain environment using predictive Hebbian learning. *Advances in neural information processing systems 6* (pp. 598-605). San Mateo, CA: Morgan Kaufmann. (21)

Montague, P.R., & Sejnowksi, T.J. (1994). The predictive brain: Temporal coincidence and temporal order in synaptic learning mechanisms. *Learning and Memory*, **1**, 1-33. (21)

Montminy, M.R., Gonzalez, G.A., & Yamamoto, K.K. (1990). Regulation of cAMP-inducible genes by CREB. *Trends in Neuroscience*, **13**, 184-188. (6)

Moody, S.A., Quigg, M.S., & Frankfurter, A. (1989). Development of the peripheral trigeminal system in the chick revealed by an isotype-specific anti-beta-tubulin monoclonal antibody. *J. of Comparative Neurology*, **279**, 567-580. (2)

Moore, A.W., & Atkeson, C.G. (1993). Prioritized sweeping: Reinforcement learning with less data and less real time. *Machine Learning*, **13**, 103-130. (19)

Moore, J.W. (1991). Implementing connectionist algorithms for classical conditioning in the brain. In M. L. Commons, S.Grossberg, & J.E.R. Staddon (Eds.), *Neural network models of conditioning and action* (pp. 181-191). Hillsdale, NJ: Lawrence Erlbaum. (4) (20)

Moore, J.W. (1992). A mechanism for timing conditioned responses. In F. Macar, V. Pouthas, & W. J. Friedman (Eds.), *Time, action and cognition* (pp. 229-238). Dordrecht, The Netherlands: Kluwer Academic Publishers. (20)

Moore, J.W., & Berthier, N.E. (1987). Purkinje cell activity and the conditioned nictitating membrane response. In M. Glickstein, C. Yeo, & J. Stein (Eds.), *Cerebellum and neuronal plasticity* (pp. 338-352). New York: Plenum. (20)

Moore, J.W., & Blazis, D.E.J. (1989). Cerebellar implementation of a computational model of classical conditioning. In P. Strata (Ed.), *The olivocerebellar system in motor control*. New York: Springer-Verlag. (4)

Moore, J.W., & Blazis, D.E.J. (1989). Simulation of a classically conditioned response: A cerebellar neural network implementation of the Sutton-Barto-Desmond model. In J.H. Byrne & W.O. Berry (Eds.), *Neural models of plasticity* (pp. 187-207). San Diego, CA: Academic Press. (20)

Moore, J.W., Berthier, N.E., & Blazis, D.E.J. (1990). Classical eye blink conditioning: Brain systems and implementation of a computational model. In M. Gabriel & J. Moore (Eds.) *Learning and computational neuroscience: Foundations of adaptive networks* (pp. 359-387). Cambridge, MA: Bradford Books/MIT Press. (4) (20) (21)

Moore, J.W., & Desmond, J.E. (1992). A cerebellar neural network implementation of a temporally adaptive conditioned response. In I. Gormezano & E.A. Wasserman (Eds.), *Learning and memory: The behavioral and biological substrates* (pp. 347-368). Hillsdale, NJ: Lawrence Erlbaum. (20)

Moore, J.W., Desmond, J.E., & Berthier, N.E. (1982). The metencephalic basis of the conditioned nictitating membrane response. In C.D. Woody (Ed.), *Conditioning: Representation of involved neural function*. New York: Plenum Press. (4)

Moore, J.W., Desmond, J.E., & Berthier, N.E. (1989). Adaptively timed conditioned responses and the cerebellum: A neural network approach. *Biological Cybernetics*, **62**, 17-28. (20)

Moore, J.W., Yeo, C.H., Oakley, D.A., & Russell, S. (1980). Conditioned inhibition of the membrane nictitating response in decorticate rabbits. *Behavioural Brain Research*, **1**, 397-409. (4)

Moran, J., & Desimone, R. (1985) Selective attention gates visual processing in the extrastriate cortex. *Science*, **229**, 782-784. (8) (9)

Morgan, J.I., & Curran, T. (1989). Stimulus-transcription coupling in neurons: Role of cellular immediate-early genes. *Trends in Neurosciences*, **12**, 459-462. (5) (6)

Morris, R. (1983). An attempt to dissociate "spatial-mapping" and "working-memory" theories of hippocampal function. In W. Seifert (Ed.), *Neurobiology of the hippocampus* (pp. 405-432). London: Academic Press. (12)

Morris, R.G. (1989). Synaptic plasticity and learning: Selective impairment of learning rats and blockade of long-term potentiation in vivo by the N-methyl-D-aspartate receptor antagonist AP5. *J. of Neuroscience*, **9**, 3040-3057. (6)

Morris, R.G., Anderson, E., Lynch, G.S., & Baudry, M. (1986). Selective impairment of learning and blockade of long-term potentiation by an N-methyl-D-aspartate receptor antagonist, AP5. *Nature*, **319**, 774-776. (6)

Morrone, C., Burr, D.C., & Maffei, L. (1982). Functional implications of cross-orientation inhibition of cortical visual cells. I. Neurophysiological evidence. *Proceedings of the Royal Society of London B: Biological Sciences*, **216**, 335-354. (8)

Moser, E., Moser, M-B., & Andersen, P. (1993). Synaptic potentiation in the rat dentate gyrus during exploratory learning. *NeuroReport*, **5**, 317-320. (6)

Mountcastle, V.B. (1978). An organizing principle for cerebral function: The unit module and the distributed system. In V.B. Mountcastle & G.M. Edelman (Eds.), *The mindful brain* (pp. 7-50). Cambridge, MA: MIT Press. (9) (24)

Muller, D., Molinari, I., Soldati, L., & Bianchi, G. (1995). A genetic deficiency in calpastatin and isovalerylcarnitine treatment is associated with

enhanced hippocampal long-term potentiation. *Synapse*, **19**, 37-45. (6)
Murray, E.A., & Mishkin, M. (1986). Visual recognition in monkeys following rhinal cortical ablations combined with either amygdalectomy or hippocampectomy. *J. of Neuroscience*, **6**, 1991-2003. (9)
Murthy, V.N., & Fetz, E.E. (1992). Coherent 25-35-Hz oscillations in the sensorimotor cortex of awake behaving monkeys. *Proceedings of the National Academy of Sciences USA*, **89**, 5670-5674. (2)
Myers, C.E., Ermita, B., Harris, K., Hasselmo, M., Solomon, P., & Gluck, M.A. (1996). A computational model of the effects of septohippocampal disruption on classical eyeblink conditioning. *Neurobiology of Learning and Memory*, **66**, 51-66. (12)
Myers, C., & Gluck, M. (1994). Context, conditioning and hippocampal re-representation. *Behavioral Neuroscience*, **108**, 835-847. (12)
Myers, C., Gluck, M., & Granger, R. (1995). Dissociation of hippocampal and entorhinal function in associative learning: A computational approach. *Psychobiology*, **23**, 116-138. (12)
Nakamura, H., Matsumoto, K., Mikami, A., & Kubota, K. (1994) Visual response properties of single neurons in the temporal pole of behaving monkeys. *J. of Neurophysiology*, **71**, 1206-1221. (9)
Nakamura, H., Mikami, A., & Kubota, K. (1992) Activity of single neurons in the monkey amygdala during performance of a visual discrimination task. *J. of Neurophysiology*, **67**, 1447-1463. (9)
Nakatsuji, N., Kadokawa, Y., & Suemori, H. (1991). Radial columnar patches in the chimeric cerebral cortex visualized by use of mouse embryonic stem cells expressing ß-galactosidase. *Development, Growth and Differentiation*, **33**, 571-578. (2)
Nathan, T., & Lambert, J.D.C. (1991). Depression of the fast IPSP underlies paired-pulse facilitation in area CA1 of the rat hippocampus. *J. of Neurophysiology*, **66**, 1704-1715. (7)
Nedivi, E., Hevroni, D., Naot, D., Israeli, D., & Citri, Y. (1993). Numerous candidate plasticity-related genes revealed by differential cDNA cloning. *Nature*, **363**, 718-722. (6)
Nelson, J.I. (1985). In D. Rose & V.G. Dobson (Eds.), *Models of the visual cortex* (pp. 108-122). Chichester: Wiley & Sons. (8)
Nelson, J.I., Nowak, L.G., Chouvet, G., Munk, M.H.J., & Bullier, J.I. (1992). Synchronization between cortical neurons depends on activity in remote areas. *Society of Neuroscience Abstracts*, **18**, 11.8. (8)
Newell, A., & Rosenbloom, P. (1981). Mechanisms of skill acquisition and the law of practice. In J. Anderson (Ed.), *Cognitive skills and their acquisition* (pp. 1-55). Hillsdale, NJ: Erlbaum. (24)
Nguyen, P.V., Abel, T., & Kandel, E.R. (1994). Requirement of a critical period of transcription for induction of a late phase of LTP. *Science*, **265**, 1104-1107. (6)
Nicoll, R.A. (1988). The coupling of neurotransmitter receptors to ion channels in the brain. *Science*, **241**, 545-551. (5)
Niki, H., & Watanabe, M. (1979). Prefrontal and cingulate unit activity during timing behavior in the monkey. *Brain Research*, **171**, 213-224. (17)
Nikolaev, E., Tischmeyer, W., Krug, M., Matthies, H., & Kaczmarek, L. (1991). c-fos protooncogene expression in rat hippocampus and entorhinal cortex following tetanic stimulation of the perforant path. *Brain Research*, **560**, 346-349. (6)
Nilsson, N.J. (1965). *Learning machines*. New York: McGraw-Hill. (22)
Nishijo, H., Ono, T., & Nishino, H. (1988). Single neuron responses in amygdala of alert monkey during complex sensory stimulation with affective significance. *J. of Neuroscience*, **8**, 3570-3583. (17)
Nosofsky, R. (1984). Choice, similarity, and the context theory of classification. *J. of Experimental Psychology: Learning, Memory and Cognition*, **10**, 104-114. (12)
Nosofsky, R.M. (1991). Tests of an exemplar model for relating perceptual classification and recognition memory. *J. of Experimental Psychology: Human Perception and Performance*, **17**, 3-27. (10)
Notterman, J.M. (1959). Force emission during bar pressing. *J. of Experimental Psychology*, **58**, 341-347. (14)
Nowak, L.G., Munk, M.H.J., Chounlamountri, N., & Bullier, J. (1994). In C. Pantev et al (Eds.), *Oscillatory Event-Related Brain Dynamics*, (pp.85-98). New York: Plenum Press. (8)
Nowakowski, R.S., Hayes, N.L., & Egger, M.D. (1992). Competitive interactions during dendritic growth: A simple stochastic growth algorithm. *Brain Research*, **576**, 152-6. (3)
Nowlan, S.J. (1988). Gain variation in recurrent error propagation networks. *Complex Systems*, **2**, 305-320. (16)
Oades, R.D., & Halliday, G.M. (1987). Ventral tegmental (A10) system: Neurobiology. 1. Anatomy and connectivity. *Brain Research*, **434**, 117-165. (21)
Ochoa, E.L.M., Chattopadhyay, A., & McNamee, M.G. (1989). Desensitization of the nicotinic acetylcholine receptor: Molecular mechanisms and effect of modulators. *Cellular and Molecular Neurobiology*, **9**, 141-178. (5)
O'Dell, T.J., Kandel, E.R., & Grant, S.G. (1991). Long-term potentiation in the hippocampus is blocked by tyrosine kinase inhibitors. *Nature*, **353**, 558-560. (6)
O'Donohue, T.L., Millington, W.R., Handelmann, G.E., Contreras, P.C., & Chronwall, B.M. (1985). On the 50th anniversary of Dale's law: Multiple neurotransmitter neurons. *Trends in Pharmacological Sciences*, **6**, 305-308. (5)
Ogawa, S., Lee, T.M., & Tank, D.W. (1990). Brain magnetic resonance imaging with contrast dependent on blood oxygenation. *Proceedings of the National Academy of Sciences USA*, **87**, 9868-9872. (23)
Ogawa, S., Tank, D.W., Menon, R., Ellermann, J.M., Kim, S-G., Merkle, H., & Ugurbil, K. (1992). Intrinsic signal changes accompanying sensory stimulation: Functional brain mapping with magnetic resonance imaging. *Proceedings of the National Academy of Sciences USA*, **89**, 5951-5955. (23)
Ogura, H., & Aigner, T. (1993). MK-801 impairs recognition memory in Rhesus monkeys: Comparison with cholinergic drugs. *J. of Pharmacology*

and Experimental Therapeutics, **266**, 60-64. (12)
Ohman, S.E.G. (1966). Coarticulation in VCV utterances: Spectrographic measurements. *J. of the Acoustical Society of America*, **39**, 151-168. (25)
O'Keefe, J. (1979). A review of the hippocampal place cells. *Progress in Neurobiology*, **13**, 419-439. (12)
O'Keefe, J., & Nadel, L. (1978). *The hippocampus as a cognitive map*. Oxford, UK: Clarendon University Press. (6) (12)
Olds, J., & Milner, P. (1954). Positive reinforcement produced by electrical stimulation of the septal and other regions of rat brain. *J. of Comparative and Physiological Psychology*, **47**, 419-427. (18)
Olshausen, B.A., Anderson, C.H., & Van Essen, D.C. (1993). A neurobiological model of visual attention and invariant pattern recognition based on dynamic routing of information. *J. of Neuroscience*, **13**, 4700-4719. (11)
Olson, R.K., Wise, B., Conners, F.A., Rack, J.P., & Fulker, D. (1989). Specific deficits in component reading and language skills: Genetic and environmental influences. *J. of Learning Disabilities*, **22**, 339-348. (27)
Oram, M.W., & Perrett, D.I. (1994). Responses of anterior superior temporal polysensory (STPa) neurons to "biological motion" stimuli. *J. of Cognitive Neuroscience*, **6**, 99-116. (9)
O'Reilly, R.C., & McClelland, J.L. (1994). Hippocampal conjunctive encoding, storage and recall: Avoiding a trade off. *Technical Report, PDP.CNS.94.4*. (26)
O'Rourke, N.A., Dailey, M.E., Smith, S.J., & McConnell, S.K. (1992). Diverse migratory pathways in the developing cerebral cortex. *Science*, **258**, 299-302. (2)
Otani, S., & Abraham, W.C. (1989). Inhibition of protein synthesis in the dentate gyrus, but not the entorhinal cortex, blocks maintenance of long-term potentiation in rats. *Neuroscience Letters*, **106**, 175-180. (6)
Otani, S., Marshall, C.J., Tate, W.P., Goddard, G.V., & Abraham, W.C. (1989). Maintenance of long-term potentiation in rat dentate gyrus requires protein synthesis but not messenger RNA synthesis immediately post-tetanization. *Neuroscience*, **28**, 519-526. (6)
Otto, T., Schottler, F., Staubli, U., Eichenbaum, H., & Lynch, G. (1991). Hippocampus and olfactory discrimination learning: Effects of entorhinal cortex lesions on olfactory learning and memory in a successive-cue, go-no-go task. *Behavioral Neuroscience*, **105(1)**, 111-119. (12)
Packard, M.G., & White, N.M. (1989). Memory facilitation produced by dopamine agonists: Role of receptor subtype and mnemonic requirements. *Pharmacology Biochemistry and Behavior*, **33**, 511-518. (6)
Palm, G. (1990). Cell assemblies as a guideline for brain research. *Concepts in Neuroscience*, **1**, 133-147. (8)
Palm, G. (1993). Cell assemblies, coherence, and corticohippocampal interplay. *Hippocampus*, **3**, Spec. No. 219-225. (3)
Palmer, D.C. (1986). Chomsky's nativism: A critical review. In P.N. Chase & L.J. Parrott (Eds.), *Psychological aspects of language* (pp. 49-60). Springfield, IL: Charles Thomas. (18)
Palmer, D.C., & Donahoe, J.W. (1992). Selectionism and essentialism in behaviorism and cognitive science. *American Psychologist*, **47**, 1344-1358. (14)
Palmer, S.E. (1977). Hierarchical structure in perceptual representation. *Cognitive Psychology*, **9**, 441-474. (11)
Palmer, S.E. (1978a). Fundamental aspects of cognitive representation. In E. Rosch & B.B. Lloyd (Eds.), *Cognition and categorization* (pp. 259-303). Hillsdale, NJ: Lawrence Erlbaum. (11)
Palmer, S.E. (1978b). Structural aspects of similarity. *Memory and Cognition*, **6**, 91-97. (11)
Pardo, J.V., Pardo, P.J., Janer, K.W., & Raichle, M.E. (1990). The anterior cingulate cortex mediates processing selection in the Stroop attentional conflict paradigm. *Proceedings of the National Academy of Sciences USA*, **87**, 256-259. (24)
Paredis, J. (1991). The evolution of behavior: Some experiments. In J.A. Meyer & S.W. Wilson (Eds.), *From animals to animats: Proceedings of the First International Conference on Simulation of Adaptive Behavior* (pp. 263-272). Cambridge, MA: MIT Press. (4)
Paredis, J. (1994). Steps towards co-evolutionary classification neural networks. In R.A. Brooks & P. Maes (Eds.), *Artificial life IV: Proceedings of the Fourth International Workshop on the Synthesis and Simulation of Living Systems* (pp. 102-108). Cambridge, MA: MIT Press. (4)
Parnavelas, J.G., Barfield, J.A., Franke, E., & Luskin, M.B. (1991). Separate progenitor cells give rise to pyramidal and nonpyramidal neurons in the rat telencephalon. *Cerebral Cortex*, **1**, 463-468. (2)
Parnavelas, J.G., Luder, R., Pollard, S.G., Sullivan, K., Lieberman, A.R. (1983). A qualitative and quantitative ultrastructural study of glial cells in the developing visual cortex of the rat. *Philosophical Transactions of the Royal Society of London B*, **301**, 55-84. (2)
Patrylo, P.R., Schweitzer, J.S., & Dudek, F.E. (1994). Potassium-dependent prolonged field bursts in the dentate gyrus: Effects of extracellular calcium and amino acid receptor antagonists. *Neuroscience*, **61**, 13-19. (6)
Pauling, L., & Coryell, C.D. (1936). The magnetic properties and structure of hemoglobin, oxyhemoglobin and carbonmonoxyhemoglobin. *Proceedings of the National Academy of Sciences USA*, **33**, 210-216. (23)
Pavlov, I.P. (1927). *Conditioned reflexes*. London: Oxford University Press. Reprint: New York: Dover, 1960. (4) (18)
Pearce, J.M., & Hall, G. (1980) A model for Pavlovian conditioning: Variations in the effectiveness of conditioned but not of unconditioned stimuli. *Psychological Review*, **87**, 532-552. (17)
Pearl, J. (1984). *Heuristics: Intelligent search strategies for computer problem solving*. Addison-Wesley. (19)
Pearson, J.C., Norris, J.R., & Phelps, C.H. (1985). Subclassification of neurons in the subthalamic nucleus of the lesser Bushbaby (Galago senegalensis): A quantitative Golgi study using principal components analysis. *J. of Comparative Neurology*, **238**, 323-339. (3)
Pellizzer, G., & Georgopoulos, A.P. (1993). Common processing constraints for visuomotor and

visual mental rotation. *Experimental Brain Research*, **93**, 165-172. (13)
Pellizzer, G., Sargent, P., & Georgopoulos, A.P. (1995). Motor cortical activity in a context-recall task. *Science*, **269**, 702-705. (13)
Penfield, W., & Milner, B. (1958). Memory deficit produced by bilateral lesions in the hippocampal zone. *Archives of Neurology and Psychiatry*, **79**, 475-479. (6)
Penfield, W., & Rasmussen, T. (1950). *The cerebral cortex of man: A clinical study of localization of function*. New York: Macmillan. (24)
Penick, S., & Solomon, R. (1991). Hippocampus, context and conditioning. *Behavioral Neuroscience*, **105**, 611-617. (12)
Pennington, B.F., Van Orden, G.C., Smith, S.D., Green, P.A., & Haith, M.M. (1990). Phonological processing skills and deficits in adult dyslexics. *Child Development*, **61**, 1753-1778. (27)
Perfetti, C.A. (1985). *Reading ability*. New York: Oxford University Press. (27)
Perrett, D.I., Hietanen, J.K., Oram, M.W., & Benson, P.J. (1992). Organization and functions of cells responsive to faces in the temporal cortex. *Philosophical Transactions of the Royal Society of London B*, **335**, 23-30. (9)
Perrett, D.I., Harries, M.H., Bevan, R., Thomas, S., Benson, P.J., Mistlin, A.J., Chitty, J.K., Hietanen, J.K., & Ortega, J.E. (1989). Frameworks of analysis for the neural representation of animate objects and actions. *J. of Experimental Biology*, **146**, 87-113. (9)
Perrett, D.I., Mistlin, A.J., & Chitty, A.J. (1987). Visual neurones responsive to faces. *Trends in Neuroscience*, **10**, 358-364. (9) (10)
Perrett, D.I., Rolls, E.T., & Caan, W. (1982). Visual neurones responsive to faces in the monkey temporal cortex. *Experimental Brain Research*, **47**, 329-342. (9)
Perrett, D.I., Smith, P.A.J., Mistlin, A.J, Chitty, A.J., Head, A.S., Potter, D.D., Broennimann, R., Milner, A.D., & Jeeves, M.A. (1985). Visual analysis of body movements by neurons in the temporal cortex of the macaque monkey: A preliminary report. *Behavioral Brain Research*, **16**, 153-170. (9)
Peters, A., Palay, S.L., & Webster, de F. (1991). *The fine structure of the nervous system: Neurons and Their supporting cells, 3rd ed.* New York: Oxford University Press. (2)
Peters, A., & Yilmaz, E. (1993). Neuronal organization in area 17 of cat visual cortex. *Cerebral Cortex*, **3**, 49-68. (9)
Petersen, S.E., Fox, P.T., Posner, M.I., Mintun, M.A., & Raichle, M.E. (1988). Positron emission tomographic studies of the cortical anatomy of single word processing. *Nature*, **331**, 585-589. (23)
Petersen, S.E., Fox, P.T., Posner, M.I., Mintun, M.A., & Raichle, M.E. (1989). Positron emission tomographic studies of the processing of single words. *J. of Cognitive Neuroscience*, **1**, 153-170. (23)
Petersen, S.E., Fox, P.T., Snyder, A.Z., & Raichle, M.E. (1990). Activation of extrastriate and frontal cortical areas by visual words and word-like stimuli. *Science*, **249**, 1041-1044. (23) (24)
Pettet, M.W., & Gilbert, C.D. (1992). Dynamic changes in receptive-field size in cat primary visual cortex. *Proceedings of the National Academy of Sciences USA*, **89**, 8366-8370. (8)
Phillips, R., & LeDoux, J. (1994). Lesions of the dorsal hippocampal formation interfere with background but not foreground contextual fear conditioning. *Learning & Memory*, **1**, 34-44. (12)
Pinching, A.J., & Powell, T.P.S. (1971). The neuron types of the glomerular layer of the olfactory bulb. *J. of Cell Science*, **9**, 305-345. (2)
Pinter, R.B. (1983). The electrophysiological bases for linear and for non-linear product term lateral inhibition and the consequences for wide field textured stimuli. *J. of Theoretical Biology*, **105**, 233-243. (24)
Pockett, S., & Figurov, A. (1993). Long-term potentiation and depression in the ventral horn of rat spinal cord in vitro. *Neuroreport*, **4**, 97-99. (6)
Pockett, S., Slack, J.R., & Peacock, S. (1993). Cyclic AMP and long-term potentiation in the CA1 region of rat hippocampus. *Neuroscience*, **52**, 229-236. (6)
Poggio, T., & Edelman, S. (1990). A neural network that learns to recognize three-dimensional objects. *Nature*, **317**, 314-319. (11)
Poggio, T., & Girosi, F. (1990). Networks for approximation and learning. *Proceedings of the IEEE*, **78**. (15)
Pohle, W., & Matthies, H. (1974). Incorporation of 3H-leucine into brain cells after learning. *Pharmacology Biochemistry and Behavior*, **2**, 573-577. (6)
Popov, N., Rüthrich, H.L., Pohle, W., Schulzeck, S., & Matthies, H. (1976). Increased fucose incorporation into rat hippocampus during learning: A biochemical and microautoradiographic study. *Brain Research*, **101**, 295-304. (6)
Popov, N., Schulzeck, S., Pohle, W., & Matthies, H. (1980). Changes in the incorporation of [3H]fucose into rat hippocampus after acquisition of a brightness discrimination reaction: An electrophoretic study. *Neuroscience*, **5**, 161-167. (6)
Popper, K. (1972). *Objective knowledge: An evolutionary approach*. Oxford: Clarendon Press. (14)
Port, R., Mikhail, A., & Patterson, M. (1985). Differential effects of hippocampectomy on classically conditioned rabbit nictitating membrane response related to interstimulus interval. *Behavioral Neuroscience*, **99**, 200-208. (12)
Port, R., & Patterson, M. (1984). Fimbrial lesions and sensory preconditioning. *Behavioral Neuroscience*, **98**, 584-589. (12)
Port, R., Romano, A., & Patterson, M. (1986). Stimulus duration discrimination in the rabbit: Effects of hippocampectomy on discrimination and reversal learning. *Physiological Psychology*, **4**, 124-129. (12)
Posner, M., & Raichle, M.E. (1994). *Images of mind*. New York: Freeman. (23) (26)
Pouget, A., Montague, P.R., Dayan, P., & Sejnowski, T.J. (1993). A developmental model of map registration in the superior colliculus using predictive Hebbian learning. *Society of Neuroscience Abstracts*, **19**, 858. (21)
Price, D.J., & Blakemore, C. (1985). Retrogressive events in the postnatal development of association projections in the visual cortex. *Nature*, **316**, 721-724. (8)
Price, J.P. (1970). The morphology of granule cells of the olfactory bulb. *J. of Cell Science*, **7**, 91-123. (2)
Price, J. (1973). An autoradiographic study of

complementary laminar patterns of termination of afferent fiber to the olfactory cortex. *J. of Comparative Neurology*, **150**, 87-108. (12)

Price, J., & Thurlow, J. (1988). Cell lineage in the rat cerebral cortex: A study using retroviral-mediated gene transfer. *Development*, **104**, 473-482. (2)

Privat, A. (1975). Postnatal gliogenesis in the mammalian brain. *International Review of Cytology*, **40**, 281-323. (2)

Prokasy, W. (1972). Developments with the two-stage model applied to human eyelid conditioning. In A. Black & W. Prokasy (Eds.), *Classical conditioning II: Current research and theory* (pp. 119-147). New York: Appleton-Century-Crofts. (12)

Purves, D., & Hadley, R.D. (1985). Changes in the dendritic branching of adult mammalian neurones revealed by repeated imaging *in situ*. *Nature*, **315**, 404-406. (5)

Purves, D., Hadley, R.D., & Voyvodic, J.T. (1986). Dynamic changes in dendritic geometry of individual neurons visualized over periods of up to three months in the superior cervical ganglion of living mice. *J. of Neuroscience*, **6**, 1051-1060. (3)

Purves, D., & Lichtman, J.W. (1985). *Principles of neural development*. Sunderland, MA: Sinauer. (2) (4)

Purves, D., Riddle, D.R., & LaMantia, A-S. (1992). Iterated patterns of brain circuitry (or how the cortex gets its spots). *Trends in Neuroscience*, **15**, 362-368. (9)

Purves, D., Voyvodic, J.T., Magrassi, L., & Yawo, H. (1987). Nerve terminal remodeling visualized in living mice by repeated examination of the same neuron. *Science*, **238**, 1122-1126. (5)

Putnam, H. (1981). Reductionism and the nature of psychology. In J. Haugeland (Ed.), *Mind design* (pp. 205-219). Cambridge, MA: MIT Press. (27)

Qian, Z., Gilbert, M.E., Colicos, M.A., Kandel, E.R., & Kuhl, D. (1993). Tissue-plasminogen activator is induced as an immediate-early gene during seizure, kindling and long-term potentiation. *Nature*, **361**, 453-457. (6)

Quartz, S., Dayan, P., Montague, P.R., & Sejnowski, T.J. (1992). Expectation learning in the brain using diffuse ascending connections. *Society of Neuroscience Abstracts*, **18**, 1210. (21)

Quinlan, P.T. (1991). Differing approaches to two-dimensional shape recognition. *Psychological Bulletin*, **109**, 224-241. (11)

Racine, R.J., Milgram, N.W., & Hafner, S. (1983). Long-term potentiation phenomena in the rat limbic forebrain. *Brain Research*, **260**, 217-231. (6)

Rack, J.P., Snowling, M.J., & Olson, R.K. (1992). The nonword reading deficit in developmental dyslexia: A review. *Reading Research Quarterly*, **27**, 29-53. (27)

Raichle, M.E. (1987). Circulatory and metabolic correlates of brain function in normal humans. In *Handbook of physiology: The nervous system V, Higher functions of the brain* (pp. 643-674). Bethesda: The American Physiological Society. (23)

Raichle, M.E., Fiez, J.A., Videen, T.O., MacLeod, A.M.K., Pardo, J.V., Fox, P.T., & Petersen, S.E. (1994). Practice-related changes in human brain functional anatomy during non-motor learning. *Cerebral Cortex*, **4**, 8-26. (23) (24)

Rakic, P. (1974). Neurons in the rhesus monkey visual cortex: Systematic relation between time of origin and eventual disposition. *Science*, **183**, 425-427. (2)

Rakic, P. (1988a). Intrinsic and extrinsic determinants of neocortical parcellation: A radial unit model. In P. Rakic & W. Singer (Eds.), *Neurobiology of the neocortex* (pp. 5-27). New York: Wiley. (4)

Rakic, P. (1988b). Specification of cerebral cortical areas. *Science*, **241**, 170-176. (2) (4)

Rakic, P. (1990). Principles of neural cell migration. *Experientia*, **46**, 882-891. (2)

Rakic, P., Bourgeois, J.P., Eckenhoff, M.F., Zecevic, N., & Goldman-Rakic, P.S. (1986). Concurrent overproduction of synapses in diverse regions of the primate cerebral cortex. *Science*, **232**, 232-235. (4)

Ramnani, N., Hardiman, M.J., & Yeo, C.H. (1995). Temporary inactivation of the cerebellum prevents the extinction of conditioned nictitating membrane responses. *Society for Neuroscience Abstracts*, **21**, 1222. (20)

Ramon Molinar, E. (1962). An attempt at classifying nerve cells on the basis of their dendritic patterns. *J. of Comparative Neurology*, **119**, 211-227. (3)

Ramon y Cajal, S. (1911). *Histologie du systeme nerveux de l'homme et des vertebres*, 2 vols. Paris: Maloine, 1909, 1911; Madrid: Instituto Ramon y Cajal, 1952, 1955. (3)

Rasio, F.A., & Ford, E.B. (1996). Dynamical instabilities and the formation of extrasolar planetary systems. *Science*, **274**, 954-966. (Pt 1)

Ratcliff, R. (1990). Connectionist models of learning and memory: Constraints imposed by learning and forgetting functions. *Psychological Review*, **97**, 285-308. (26)

Rauschecker, J.P. (1991). Mechanisms of visual plasticity: Hebb synapses, NMDA receptors, and beyond. *Physiological Reviews*, **71**, 587-615. (8) (21)

Rauschecker, J.P., & Singer, W. (1979). Changes in the circuitry of the kitten visual cortex are gated by postsynaptic activity. *Nature*, **280**, 58-60. (8)

Rauschecker, J.P., & Singer, W. (1981). The effects of early visual experience on the cat's visual cortex and their possible explanation by Hebb synapses. *J. of Physiology*, **310**, 215-239. (8)

Real, L.A. (1991). Animal choice behavior and the evolution of cognitive architecture. *Science*, **253**, 980-986. (21)

Recanzone, G.H., Schreiner, C.E., & Merzenich, M.M. (1993). Plasticity in the frequency representation of primary auditory cortex following discrimination training in adult owl monkeys. *J. of Neuroscience*, **13**, 87-103. (8)

Reed, S.K., & Friedman, M.P. (1973). Perceptual vs. conceptual categorization. *Memory & Cognition*, **1**, 157-163. (10)

Reeke, G.N., & Edelman, G.M. (1984). Selective networks and recognition automata. *Annals of the New York Academy of Science*, **426**, 181-201. (27)

Reid, C.B., Liang, I., & Walsh, C. (1995). Systematic widespread clonal organization in cerebral cortex. *Neuron*, **15**, 299-310. (2)

Reilly, S., Harley, C., & Revusky, S. (1993). Ibotenate lesions of the hippocampus enhance latent inhibition in conditioned taste aversion and increase resistance to extinction in conditioned taste preference. *Behavioral Neuroscience*, **107**, 996-1004. (12)

Reiter, H.O., & Stryker, M.P. (1988). Neural plasticity without postsynaptic action potentials: Less-active inputs become dominant when kitten visual cortical cells are pharmacologically inhibited. *Proceedings of the National Academy of Sciences USA*, **85**, 3623-3627. (8)

Rescorla, R.A. (1966). Predictability and number of pairings in Pavlovian fear conditioning. *Psychonomic Science*, **4**, 383-385. (4)

Rescorla, R.A. (1968). Probability of shock in the presence and absence of CS in fear conditioning. *J. of Comparative and Physiological Psychology*, **66**, 1-5. (4)

Rescorla, R.A. (1968). Conditioned inhibition of fear. In N.J. Mackintosh & W.K. Honig (Eds.), *Fundamental issues in associative learning* (pp. 65-89). Halifax, Nova Scotia: University of Dalhousie Press. (18)

Rescorla, R.A., & Wagner, A.R. (1972). A theory of Pavlovian conditioning: Variations in the effectiveness of reinforcement and nonreinforcement. In A.H. Black & W.F. Prokasy (Eds.), *Classical conditioning II: Current research and theory* (pp. 64-99). New York: Appleton-Century-Crofts. (1) (12) (14) (17) (18) (21)

Reymann, K.G. (1993). Mechanisms underlying synaptic long-term potentiation in the hippocampus: Focus on postsynaptic glutamate receptors and protein kinases. *Functional Neurology*, **8**, 7-32. (6)

Reymann, K.G., Brödemann, R., Kase, H., & Matthies, H. (1988a). Inhibitors of calmodulin and protein kinase C block different phases of hippocampal long-term potentiation. *Brain Research*, **461**, 388-392. (6)

Reymann, K., Frey, U., & Matthies, H. (1988b). A multi-phase model of synaptic long-term potentiation in hippocampal CA1 neurones: Protein kinase C activation and protein synthesis are required for the maintenance of the trace. In H.L. Haas & G. Buzsaki (Eds.), *Synaptic plasticity in the hippocampus* (pp. 126- 129). Berlin/Heidelberg: Springer. (6)

Reymann, K.G., Frey, U., Jork, R., & Matthies, H. (1988c). Polymyxin B, an inhibitor of protein kinase C, prevents the maintenance of synaptic long-term potentiation in hippocampal CA1 neurons. *Brain Research*, **440**, 305-314. (6)

Reymann, K.G., Malisch, R., Schulzeck, K., Brödemann, R., Ott, T., & Matthies, H. (1985). The duration of long-term potentiation in the CA1 region of the hippocampal slice preparation. *Brain Research Bulletin*, **15**, 249-255. (6)

Reymann, K.G., & Matthies, H. (1989). 2-Amino-4-phosphonobutyrate selectively eliminates late phases of long-term potentiation in rat hippocampus. *Neuroscience Letters*, **98**, 166- 171. (6)

Reymann, K.G., Matthies, H.K., Frey, U., Vorobyev, V.S., & Matthies, H. (1986). Calcium-induced long-term potentiation in the hippocampal slice: Characterization of the time course and conditions. *Brain Research Bulletin*, **17**, 291-296. (6)

Richards, W.G., Ricciardi, T.N., & Moore, J.W. (1991). Activity of spinal trigeminal pars oralis and adjacent reticular formation units during differential conditioning of the rabbit nictitating membrane response. *Behavioral Brain Research*, **44**, 195-204. (20)

Richmond, B.J., & Sato, T. (1987). Enhancement of inferior temporal neurons during visual discrimination. *J. of Neurophysiology*, **58**, 1292-1306. (9)

Richter-Levin, G., Canevari, L., & Bliss, T.V.P. (1995). Long-term potentiation and glutamate release in the dentate gyrus: Links to spatial learning. *Behavior and Brain Research*, **66**, 37-40. (6)

Riley, D. (1968). *Discrimination learning*. Boston: Allyn & Bacon. (12)

Robbins, T.W., & Everitt, B.J. (1992) Functions of dopamine in the dorsal and ventral striatum. *Seminars in Neuroscience*, **4**, 119-128. (17)

Robin, D., Berthier, N.E., & Clifton, R.K. (1996). Infants' predictive reaching for moving objects in the dark. *Developmental Psychology*, **32**, 824-835. (15)

Robinson, G.B., & Racine, R.J. (1985). Long-term potentiation in the dentate gyrus: Effects of noradrenaline depletion in the awake rat. *Brain Research*, **325**, 71-78. (6)

Robinson, T.E., & Berridge, K.C. (1993) The neural basis for drug craving: An incentive-sensitization theory of addiction. *Brain Research Reviews*, **18**, 247-291. (17)

Rockland, K.S., Saleem, K.S., & Tanaka, K.. (1994). Divergent feedback connections from areas V4 and TEO in the macaque. *Visual Neuroscience*, **11**, 579-600. (9)

Rockland, K.S., & Van Hoesen, G.W. (1994). Direct temporal-occipital feedback connections to striate cortex (V1) in the macaque monkey. *Cerebral Cortex*, **4**, 300-313. (9)

Rodman, H.R., O'Scalaidhe, S.P., & Gross, C.G. (1993). Response properties of neurons in temporal cortical visual areas of infant monkeys. *J. of Neurophysiology*, **70**, 1115-1136. (8)

Roe, A.W., & Ts'o, D.Y. (1992). Functional connectivity between V1 and V2 in the primate. *Society of Neuroscience Abstracts* , **18**, 11.4. (8)

Roelfsema, P.R., König, P., Engel, A.K., & Singer, W. (1993). Interactions between visual cortical areas of awake behaving cats. *European J. of Neuroscience Supplement*, **7**, 13.04. (8)

Roelfsema, P.R., König, P., Engel, A.K., Sireteanu, R., & Singer, W. (1994). Reduced synchronization in the visual cortex of cats with strabismic amblyopia. *European J. of Neuroscience*, **6**, 1645-1655. (8)

Roland, P. E., & Ziles, K. (1994). Brain atlases—a new research tool. *Trends in Neuroscience*, **17**, 458-467. (3)

Rolls, E.T. (1991). Neural organization of higher visual functions. *Current Opinion in Neurobiology*, **1**, 274-278. (9)

Rolls, E.T. (1992). Neurophysiological mechanisms underlying face processing within and beyond the temporal cortical visual areas. *Philosophical Transactions of the Royal Society of London B*, **335**, 11-21. (9)

Rolls, E.T., & Tovee, M.J. (1995). Sparseness of the neuronal representation of stimuli in the primate temporal visual cortex. *J. of Neurophysiology*, **73**, 713-726. (9)

Romo, R., & Schultz, W. (1990) Dopamine neurons of the monkey midbrain: Contingencies of responses to active touch during self-initiated arm movements. *J. of Neurophysiology*, **63**, 592-606. (17)

Rosenbaum, D.A. (1980). Human movement initiation: Specification of arm, direction, and extent. *J. of Experimental Psychology: General*, **109**, 444-

474. (25)
Rosenbaum, D.A. (1991). *Human motor control.* San Diego: Academic Press. (15)
Rosenbaum, D.A., Loukopoulos, L., Meulenbroek, R., Vaughan, J., & Engelbrecht, S. (1995). Planning reaches by evaluating stored postures. *Psychological Review*, **102**, 28-67. (15)
Rosenblatt, F. (1962). *Principles of neurodynamics.* New York: Spartan. (1)
Rosenfield, M.E., & Moore, J.W. (1985). Red nucleus lesions impair acquisition of the classically conditioned nictitating membrane response but not eye-to-eye savings or conditioned response amplitude. *Behavioural Brain Research*, **17**, 77-81. (4)
Rosenfield, M.E., & Moore, J.W. (1995). Connections to cerebellar cortex (Larsell's HVI) in the rabbit: A WGA-HRP study with implications for classical eyeblink conditioning. *Behavioral Neuroscience*, **109**, in press. (20)
Ross, L.E. (1961). Conditioned fear as a function of CS-UCS and probe stimulus intervals. *J. of Experimental Psychology*, **61**, 265-273. (4)
Ross, R., Orr, W., Holland, P., & Berger, T. (1984). Hippocampectomy disrupts acquisition and retention of learned conditional responding. *Behavioral Neuroscience*, **98**, 211-225. (12)
Ross, S. (1983). *Introduction to stochastic dynamic programming.* New York: Academic Press. (19)
Rosson, M.B. (1985). The interaction of pronunciation rules and lexical representations in reading aloud. *Memory & Cognition*, **13**, 90-99. (27)
Rudy, J., & Sutherland, R. (1989). The hippocampal formation is necessary for rats to learn and remember configural discriminations. *Behavioral Brain Research*, **34**, 97-109. (12)
Rumelhart, D.E., Hinton, G.E., & Williams, R.J. (1986). Learning internal representations by error propagation. In D.E. Rumelhart & J.L. McClelland (Eds.), *Parallel distributed processing: Explorations in the microstructure of cognition*, (Vol 1, pp. 318-362). Cambridge, MA: Bradford Books/MIT Press. (1) (12) (16) (18) (19) (25)
Rumelhart, D.E., Hinton, G.E., & Williams, R.J. (1986). Learning representations by back-propagating errors. *Nature*, **323**, 533-536. (26)
Rumelhart, D.E., McClelland, J.L., & The PDP Group (Eds.). (1986). *Parallel distributed processing.* (Vols. 1 and 2). Cambridge, MA: MIT Press. (18)
Rumelhart, D.E., & Norman, D.A. (1982). Simulating a skilled typist: A study of skilled cognitive-motor performance. *Cognitive Science*, **6**, 1-36. (25)
Rumelhart, D., & Zipser, D. (1985). Feature discovery by competitive learning. *Cognitive Science*, **9**, 75-112. (12)
Sagar, S.M., & Sharp, F.R. (1993). Early response genes as markers of neuronal activity and growth factor action. In F.J. Seil (Ed.), *Advances in neurology, Vol. 59* (pp. 273-284). New York: Raven Press. (5)
Saiki, J., & Hummel, J.E. (1995a). Attribute conjunctions and the part configuration advantage in object category learning. Manuscript submitted for publication. (11)
Saiki, J., & Hummel, J.E. (1995b). Connectedness and structural relations in object category learning. Manuscript submitted for publication. (11)
Saiki, J., & Hummel, J.E. (1995c). Connectedness and the integration of parts with relations in shape perception. Manuscript submitted for publication. (11)
Sakai, K., & Miyashita, Y. (1991). Neural organization for the long-term memory of paired associates. *Nature*, **354**, 152-155. (8) (9)
Sakata, H., & Kusunoki, M. (1992). Organization of space perception: Neural representation of three-dimensional space in the posterior parietal cortex. *Current Opinion in Neurobiology*, **2**, 170-174. (9)
Saleem, K.S., Tanaka, K., & Rockland, K.S. (1993). Specific and columnar projection from area TEO to TE in the macaque inferotemporal cortex. *Cerebral Cortex*, **3**, 454-464. (9)
Salinas, E., & Abbott, L.F. (1994). Vector reconstruction from firing rates. *J. of Computational Neuroscience*, **1**, 89-107. (13)
Saltzman, E.L., & Kelso, J.A.S. (1987). Skilled actions: A task-dynamic approach. *Psychological Review*, **94**, 84-106. (25)
Samuel, A.L. (1959). Some studies in machine learning using the game of checkers. *IBM J. on Research and Development*, 210-229. Reprinted in E.A. Feigenbaum & J. Feldman (Eds.), *Computers and thought.* New York: McGraw-Hill, 1963. (19) (22)
Sanes, J.R. (1989). Analyzing cell lineage with a recombinant retrovirus. *Trends in Neurosciences*, **12**, 21-28. (2)
Sanocki, T. (1987). Visual knowledge underlying letter perception: Font-specific, schematic tuning. *J. of Experimental Psychology: Human Perception and Performance*, **13**, 267-278. (27)
Santucci, A., Kanof, P., & Haroutunian, V. (1989). Effect of physostigmine on memory consolidation and retrieval processes in intact and nucleus basalis-lesioned rats. *Psychopharmacology*, **99**, 70-74. (12)
Sato, M., Lopez-Mascaraque, L., Heffner, C.D., & O'Leary, D.D.M. (1994). Action of a diffusible target-derived chemoattractant on cortical axon branch induction and directed growth. *Neuron*, **13**, 791-803. (3)
Schaffer, J.D., Caruana, R.A., & Eshelman, L.J. (1990). Using genetic search to exploit the emergent behavior of neural networks. In S. Forrest (Ed.), *Emergent computation.* Amsterdam: North Holland. (4)
Schaffer, J.D., Eshelman, L.J., & Offutt, D. (1991). Spurious correlations and premature convergence in genetic algorithms. In G.J.E. Rawlins (Ed.), *Foundations of Genetic Algorithms* (pp. 102-112). San Mateo, CA: Morgan Kaufmann. (4)
Schlinger, H.D., & Blakely, E. (1994). The effects of delayed reinforcement and a response-produced auditory stimulus on the acquisition of operant behavior in rats. *Psychological Record*, **44**, 391-409. (14)
Schmahmann, J.D. (1991). An emerging concept: The cerebellar contribution to higher function. *Archives of Neurology*, **48**, 1178-1187. (24)
Schmajuk, N., & DiCarlo, J. (1990). Backpropagation, classical conditioning and hippocampal function. *Technical Report*, Northwestern University. (12)
Schmajuk, N.A., & DiCarlo, J.J. (1991). Neural dynamics of hippocampal modulation of classical conditioning. In M.L. Commons, S. Grossberg, & J.E.R. Staddon (Eds.), *Neural network models of*

conditioning and action (pp. 149-180). Hillsdale, NJ: Lawrence Erlbaum. (4)
Schmaltz, L., & Theios, J. (1972). Acquisition and extinction of a classically conditioned response in hippocampectomized rabbits (Oryctolagus cuniculus). *J. of Comparative and Physiological Psychology*, **79**, 328-333. (12)
Schmid, L.M., Rosa, M.G.P., & Calford, M.B. (1995). Retinal detachment induces massive immediate reorganization in visual cortex. *NeuroReport*, **6**, 1349-1353. (8)
Schmidt, R.A. (1982). *Motor control and learning: A behavioral emphasis*. Champaign, IL: Human Kinetics Publishers. (15)
Schmidt, S., Löwel, S., Goebel, R., & Singer, W. (in preparation). (2)
Schneider, W. (1985). Toward a model of attention and the development of automatic processing. In M. Posner, & O. Martin (Eds.), *Attention and performance* (Vol. XI, pp. 475-492). Hillsdale, NJ: Erlbaum. (24)
Schneiderman, N. (1966). Interstimulus interval function of the nictitating membrane response of the rabbit under delay versus trace conditioning. *J. of Comparative and Physiological Psychology*, **62**, 397-402. (4)
Schofield, P.R., Shivers, B.D., & Seeburg, P.H. (1990). The role of receptor subtype diversity in the CNS. *Trends in Neurosciences*, **13**, 8-11. (5)
Schultz, W. (1986) Responses of midbrain dopamine neurons to behavioral trigger stimuli in the monkey. *J. of Neurophysiology*, **56**, 1439-1462. (17)
Schultz, W. (1992). Activity of dopamine neurons in the behaving primate. *Seminars in the Neurosciences*, **4**, 129-138. (21)
Schultz, W. (1995). The primate basal ganglia between the intention and outcome of action. In M. Kimura & A.M. Graybiel (Eds.), *Functional linkages between the cerebral cortex and basal ganglia in the functional control of movement* (pp. 31-48). Tokyo: Springer. (17)
Schultz, W., Apicella, P., & Ljungberg, T. (1993) Responses of monkey dopamine neurons to reward and conditioned stimuli during successive steps of learning a delayed response task. *J. of Neuroscience*, **13**, 900-913. (17) (21)
Schultz, W., Apicella, P., Scarnati, E., & Ljungberg, T. (1992) Neuronal activity in monkey ventral striatum related to the expectation of reward. *J. of Neuroscience*, **12**, 4595-4610. (17)
Schultz, W., & Romo, R. (1990) Dopamine neurons of the monkey midbrain: Contingencies of responses to stimuli eliciting immediate behavioral reactions. *J. of Neurophysiology*, **63**, 607-624. (17)
Schultz, W., Romo, R., Ljungberg, T., Mirenowicz, J., Hollerman, J.R., & Dickinson, A. (1995). Reward-related signals carried by dopamine neurons. In J.C. Houk, J.L. Davis, & D.G. Beiser (Eds.), *Models of information processing in the basal ganglia* (pp. 233-248). Cambridge, MA: MIT Press. (17) (18)
Schusterman, R.J., & Kastak, D. (1993). A California sea lion (Zalophus californianus) is capable of forming equivalence relations. *Psychological Record*, **43**, 823-839. (26)
Schwartz, A.B. (1992) Motor cortical activity during drawing movements: Single unit activity during sinusoid tracing. *J. of Neurophysiology*, **68**, 528-541. (13)
Schwartz, A.B. (1993). Motor cortical activity during drawing movements: Population representation during sinusoid tracing. *J. of Neurophysiology*, **70**, 28-36. (13)
Schwartz, A.B. (1994). Direct cortical representation of drawing. *Science*, **265**, 540-542. (13)
Schwartz, A.B., Kettner, R.E., & Georgopoulos, A.P. (1988). Primate motor cortex and free arm movements to visual targets in three-dimensional space I: Relations between single cell discharge and direction of movement. *J. of Neuroscience*, **8**, 2913-2927. (13)
Schwartz, B. (1980). Development of complex, stereotyped behavior in pigeons. *J. of the Experimental Analysis of Behavior*, **33**, 153-166. (14)
Schwartz, B. (1982). Reinforcement-induced behavioral stereotypy: How not to teach people to discover rules. *J. of Experimental Psychology: General*, **111**, 23-59. (14)
Schwartz, E.I., Desimone, R., Albright, T.D., & Gross, C.G. (1983). Shape recognition and inferior temporal neurons. *Proceedings of the National Academy of Sciences USA*, **80**, 5776-5778. (9)
Schwarz, C., & Bolz, J. (1991). Functional specificity of a long-range horizontal connection in cat visual cortex: A cross-correlation study. *J. of Neuroscience* , **11**, 2995-3007. (8)
Schweikert, R., & Wang, Z. (1993). Effects on response time factors selectively influencing processes in acyclic task networks with or gates. *British J. of Mathematical and Statistical Psychology*, **46**, 1-30. (24)
Sears, L.L., & Steinmetz, J.E. (1990). Acquisition of classically conditioned-related activity in the hippocampus is affected by lesions of the cerebellar interpositus nucleus. *Behavioral Neuroscience*, **104**, 681-692. (12)
Seidenberg, M.S., Plaut, D.C., Petersen, A.S., McClelland, J.L., & McRae, K. (1994). Nonword pronunciation and models of word recognition. *J. of Experimental Psychology: Human Perception and Performance*, **20**, 1177-1196. (27)
Selemon, L., & Goldman-Rakic, P. (1985). Longitudinal topography and interdigitation of corticostriatal projections in the rhesus monkey. *J. of Neuroscience*, **5**, 776-794. (24)
Selfridge, O., Sutton, R.S., & Barto, A.G. (1985). Training and tracking in robotics. *Proceedings of the Ninth International Joint Conference on Artificial Intelligence*, Los Angeles, CA. (16)
Seligman, M.E.P., & Hager, J.L. (1972). *Biological boundaries on learning*. New York: Appleton-Century-Crofts. (4)
Seltzer, B., & Pandya, D.N. (1978). Afferent cortical connections and architectonics of the superior temporal sulcus and surrounding cortex in the rhesus monkey. *Brain Research*, **149**, 1-24. (9)
Seltzer, B., & Pandya, D.N. (1984). Further observations on pirate-temporal connections in the rhesus monkey. *Experimental Brain Research*, **55**, 301-312. (9)
Seltzer, B., & Pandya, D.N. (1989). Intrinsic connections and architectonics of the superior temporal sulcus in the rhesus monkey. *J. of Comparative Neurology*, **290**, 451-471. (9)
Seltzer, B., & Pandya, D.N. (1994). Parietal, temporal, and occipital projections to cortex of the superior temporal sulcus in the rhesus monkey: A retrograde tracer study. *J. of Comparative Neu-*

rology, **343**, 445-463. (9)

Senft, S.L. (1995). Derivation of neuron geometry from confocal scans. *Neuroscience Abstracts*, **25**, 1078. (3)

Senft, S.L., & Woolsey, T.A. (1991). Computer-aided analyses of thalamocortical afferent ingrowth. *Cerebral Cortex*, **1**, 336-347. (3)

Shaffer, L.H. (1976). Intention and performance. *Psychological Review*, **83**, 375-393. (25)

Shaffer, L.H., & Hardwick, J. (1970). The basis of transcription skill. *J. of Experimental Psychology*, **84**, 424-440. (25)

Shallice, T. (1988). *From neuropsychology to mental structure*. New York: Cambridge University Press. (23) (27)

Shashoua, V.E. (1976). Identification of specific changes in the pattern of brain protein synthesis after training. *Science*, **193**, 1264-1266. (6)

Shashoua, V.E. (1985). The role of brain extracellular proteins in neuroplasticity and learning. *Cellular and Molecular Neurobiology*, **5**, 183-207. (6)

Shashoua, V.E. (1988). The role of ependymin in the development of long lasting synaptic changes. *J. of Physiology (Paris)*, **83**, 232-239. (6)

Shashoua, V.E. (1990). The role of ependymin in neuronal plasticity and LTP. *Advances in Experimental Medicine and Biology*, **268**, 333-345. (6)

Shashoua, V.E. (1991). Ependymin, a brain extracellular glycoprotein, and CNS plasticity. *Annals of the New York Academcy of Science*, **627**, 94-114. (6)

Shastri, L., & Ajjanagadde, V. (1993). From simple associations to systematic reasoning: A connectionist representation of rules, variables and dynamic bindings. *Behavioral and Brain Sciences*, **16**, 417-494. (11)

Shaw, C., Needler, M.C., Wilkinson, M., Aoki, C., & Cynader, M. (1984). Alterations in receptor number, affinity and laminar distribution in cat visual cortex during critical period. *Progress in Neurobiological Psychology*, **8**, 627-634. (8)

Shaw, C.A., Lanius, R.A., & Van den Doel, K. (1994). The origin of synaptic neuroplasticity: Crucial molecules or a dynamical cascade. *Brain Research Review*, **19**, 241-263. (6)

Sheng, M., & Greenberg, M.E. (1990). The regulation and function of c-fos and other immediate early genes in the nervous system. *Neuron*, **4**, 477-485. (6)

Shepard, R. (1957). Stimulus and response generalization: A stochastic model relating generalization to distance in psychological space. *Psychometrika*, **22**, 325-345. (12)

Shepard, R. (1958). Stimulus and response generalization: Deduction of the generalization gradient from a trace model. *Psychological Review*, **65**, 242-256. (12)

Shepard, R.N., & Cooper, L.A. (1982). *Mental images and their transformations*. Cambridge: MIT Press. (13)

Shettleworth, S.J. (1973). Food reinforcement and the organization of behaviour in golden hamsters. In R.A. Hinde & J. Stevenson-Hinde (Eds.), *Constraints on learning* (pp. 243-263). New York: Academic Press. (4)

Shimizu, T., & Wolfe, L. S. (1990). Arachidonic acid cascade and signal transduction. *J. of Neurochemistry*, **55**, 1-15. (5)

Shizgal, P., Bielajew, C., & Rompre, P-P. (1988). Quantitative characteristics of the directly stimulated neurons subserving self-stimulation of the medial forebrain bundle: Psychophysical inference and electrophysiological measurement. In M.L. Commons, R.M. Church, J.R. Stellar, & A.R. Wagner (Eds.), *Quantitative analyses of behavior: Vol. 7, Biological determinants of reinforcement* (pp. 59-85). Hillsdale, NJ: Lawrence Erlbaum. (18)

Sillito, A.M., Jones, H.G., Gerstein, G.L., & West, D.C. (1994). Feature-linked synchronization of thalamic relay cell firing induced by feedback from the visual cortex. *Nature*, **369**, 479-482. (8)

Silva, A.J., Stevens, C.F., Tonegawa, S., & Wang, Y. (1992). Deficient hippocampal long-term potentiation in alpha-calcium-calmodulin kinase II mutant mice. *Science*, **257**, 201-206. (6)

Silveira, J., & Kimble, D. (1968). Brightness discrimination and reversal in hippocampally-lesioned rats. *Physiology & Behavior*, **3**, 625-630. (12)

Simpson, G.G. (1953). The Baldwin effect. *Evolution*, **7**, 110-117. (4)

Singer, W. (1985). Activity-dependent self-organization of the mammalian visual cortex. In D. Rose & V.G. Dobson (Eds.), *Models of the visual cortex* (pp. 123-136). Chichester: Wiley & Sons. (8)

Singer, W. (1990). The formation of cooperative cell assemblies in the visual cortex. *J. of Experimental Biology*, **153**, 177-197. (8)

Singer, W. (1993). Synchronization of cortical activity and its putative role in information processing and learning. *Annual Review of Physiology*, **55**, 349-374. (9)

Singer, W. (1993). Synchronization of cortical activity and its putative role in information processing and learning. *Annual Review of Physiology*, **55**, 349-374. (8)

Singer, W. (1994). Putative functions of temporal correlations in neocortical processing. In C. Koch & J.L. Davis (Eds.), *Large-scale neuronal theories of the brain* (pp. 201-237). Cambridge, MA: MIT Press. (8)

Singer, W., & Artola, A. (1995) The role of NMDA receptors in use-dependent synaptic plasticity of the visual cortex. In A. Thomson (Ed.), *Excitatory amino acids and synaptic function*, 2nd Edition (pp. 249-261). London: Academic Press. (8)

Singer, W., & Gray, C.M. (1995). Visual feature integration and the temporal correlation hypothesis. *Annual Review of Neuroscience*, **18**, 555-586. (8)

Singer, W., & Tretter, F. (1976). Unusually large receptive fields in cats with restricted visual experience. *Experimental Brain Research*, **26**, 171-184. (8)

Singh, S.P. (1991). Transfer of learning across compositions of sequential tasks. In L.A. Birnbaum & G.C. Collins (Eds.), *Machine learning: Proceedings of the Eighth International Workshop* (pp. 348-352), San Mateo, CA: Morgan Kaufmann. (19)

Singh, S.P. (1992). The efficient learning of multiple task sequences. In J.E. Moody, S.J. Hanson, & R.P. Lippman (Eds.), *Advances in neural information processing systems 4*, (pp. 251-258), San Mateo, CA: Morgan Kaufmann. (19)

Singh, S.P., Barto, A.G., Grupen, R., & Connolly, C. (1994). Robust reinforcement learning in motion planning. In J.D. Cohen, G. Tesauro, & J. Alspector (Eds.), *Advances in neural information process-*

ing systems 6 (pp. 655-662), San Francisco, CA: Morgan Kaufmann. (19)

Singh, S.P., Jaakkola, T., & Jordan, M.I. (1994). Learning without state-estimation in partially observable Markovian decision problems. In W.W. Cohen & H. Hirsch (Eds.), *Machine learning: Proceedings of the Eleventh International Conference* (pp. 284-292), San Francisco: Morgan Kaufmann. (19)

Skelton, R.W., Miller, J.J., & Phillips, A.G. (1985). Long-term potentiation facilitates behavioral responding to single-pulse stimulation of the perforant path. *Behavioral Neuroscience*, **99**, 603-620. (6)

Skinner, B.F. (1935). The generic nature of the concepts of stimulus and response. *J. of General Psychology*, **12**, 40-65. (14)

Skinner, B.F. (1938). *The behavior of organisms*. New York: Appleton-Century-Crofts. (14) (16) (18)

Skinner, B.F. (1948). "Superstition" in the pigeon. *J. of Experimental Psychology*, **38**, 168-172. (14) (18)

Skinner, B.F. (1950). Are theories of learning necessary? *Psychological Review*, **57**, 193-216. (Pt 2)

Skinner, B.F. (1953). *Science and human behavior*. New York: Macmillan. (16)

Skinner, B.F. (1974). *About behaviorism*. London: Jonathan Cape. (26)

Skinner, B.F. (1974). *About behaviorism*. New York: Random House. (Pt 6)

Skinner, B.F. (1975). The shaping of phylogenic behavior. *J. of the Experimental Analysis of Behavior*, **24**, 117-120. (18)

Slack, J.R., & Pockett, S. (1991). Cyclic AMP induces long-term increase in synaptic efficacy in CA1 region of rat hippocampus. *Neuroscience Letters*, **130**, 69-72. (6)

Smiley, J.F., Levey, A.I., Ciliax, B.J., & Goldman-Rakic, P.S. (1994). D dopamine receptor immunoreactivity in human and monkey cerebral cortex: Predominant and extrasynaptic localization in dendritic spines. *Proceedings of the National Academy of Sciences USA*, **91**, 5720-5724. (5)

Smirnova, T., Laroche, S., Errington, M.L., Hicks, A.A., Bliss, T.V.P., & Mallet, J. (1993). Transsynaptic expression of a presynaptic glutamate receptor during hippocampal long-term potentiation. *Science*, **262**, 433-436. (6)

Smith, A.D., & Bolam, J.P. (1990) The neural network of the basal ganglia as revealed by the study of synaptic connections of identified neurones. *Trends in Neuroscience*, **13**, 259-265. (17)

Smith, M.C., Coleman, S.R., & Gormezano, I. (1969). Classical conditioning of the rabbit's nictitating membrane response at backward, simultaneous, and forward CS-US intervals. *J. of Comparative and Physiological Psychology*, **69**, 226-231. (4)

Smith, Y., Bennett, B.D., Bolam, J.P., Parent, A., & Sadikot, A.F. (1993) Synaptic interactions between the dopaminergic afferents and the cortical or thalamic input at the single cell level in the striatum of monkey. *Society of Neuroscience Abstracts*, **19**, 977. (17)

Smolensky, P. (1986). Neural and conceptual interpretations of PDP models. In J. L. McClelland, D. E. Rumelhart, & The PDP Research Group (Eds.), *Parallel distributed processing: Vol. 2* (pp. 390-431). Cambridge, MA: MIT Press. (1)

Smolensky, P. (1990). Tensor product variable binding and the representation of symbolic structures in connectionist systems. *Artificial Intelligence*, **46**, 159-216. (11)

Smyrnis, N., Taira, M., Ashe, J., & Georgopoulos, A.P. (1992). Motor cortical activity in a memorized delay task. *Experimental Brain Research*, **92**, 139-151. (13)

Snippe, H.P., & Koenderink, J.J. (1992). Discrimination thresholds for channel-coded systems. *Biological Cybernetics*, **66**, 543-551. (9)

Snowling, M.J. (1980). The development of grapheme-phoneme correspondence in normal and dyslexic readers. *J. of Experimental Child Psychology*, **29**, 294-305. (27)

Snyder, A.Z., Abdullaev, Y.G., Posner, M.I., & Raichle, M.E. (1995). Scalp electrical potentials reflect regional cerebral blood flow responses during processing of written words. *Proceedings of the National Academy of Sciences USA*, **92**, 1689-1693. (23) (24)

Sokolov, M.V., & Kleschevnikov, A.M. (1995). Atropine suppresses associative LTP in the CA1 region of rat hippocampal slices. *Brain Research*, **672**, 281-284. (6)

Solomon, P., Groccia-Ellison, M., Flynn, D., Mirak, J., Edwards, K., Dunehew, A., & Stanton, M. (1993). Disruption of human eyeblink conditioning after central cholinergic blockade with scopolamine. *Behavioral Neuroscience*, **107**, 271-279. (12)

Solomon, P., & Moore, J. (1975). Latent inhibition and stimulus generalization of the classically conditioned nictitating membrane response in rabbits (Oryctolagus cuniculus) following dorsal hippocampal ablation. *J. of Comparative and Physiological Psychology*, **89**, 1192-1203. (12)

Solomon, P., Solomon, S., Van der Schaaf, E., & Perry, H. (1983). Altered activity in the hippocampus is more detrimental to classical conditioning than removing the structure. *Science*, **220**, 329-331. (12)

Sonnenberg, J.L., Rauscher, F.J., Morgan, J.I., & Curran, T. (1989). Regulation of proenkephalin by Fos and Jun. *Science*, **246**, 1622-1625. (6)

Southam, E., & Garthwaite, J. (1993). The nitric oxide-cyclic GMP signalling pathway in rat brain. *Neuropharmacology*, **32**, 1267-1277. (5)

Spence, K.W. (1947). The role of secondary reinforcement in delayed reward learning. *Psychological Review*, **54**, 1-8. (18)

Spitzer, H., Desimone, R., & Moran, J. (1988). Increased attention enhances both behavioral and neuronal performance. *Science*, **240**, 338-340. (9)

Spudich, J.L., & Koshland, D.E. (1975). Quantification of the sensory response in bacterial chemotaxis. *Proceedings of the National Academy of Sciences USA*, **72**, 710-713. (21)

Squire, L. (1987). *Memory and brain*. New York: Oxford University Press. (12)

Squire, L.R. (1992). Memory and the hippocampus: A synthesis from findings with rats, monkeys, and humans. *Psychological Review*, **99**, 191-231. (26)

Squire, L.R., Shimura, A.P., & Amaral, D.G. (1989). Memory and the hippocampus. In J. Byrne & W.O. Berry (Eds.), *Neural models of plasticity* (pp. 208-239). New York: Academic Press. (26)

Staddon, J.E.R. (1983). *Adaptive behavior and learning*. New York: Cambridge University Press.

(16)
Staddon, J.E.R., & Ettinger, R.H. (1989). Learning: An introduction to the principles of adaptive behavior. New York: Harcourt Brace Jovanovich. (16)
Stankiewicz, B.J., Hummel J.E., & Cooper, E.E. (1995). The role of attention in priming for left-right reflections of object images: Evidence for a dual representation of object shape. Manuscript submitted for publication. (11)
Stanton, P.K., Mody, I., & Heinemann, U. (1989). Down-regulation of norepinephrine sensitivity after induction of long-term neuronal plasticity (kindling) in the rat dentate gyrus. *Brain Research*, **476**, 367-372. (6)
Stanton, P.K., & Sarvey, J.M. (1984). Blockade of long-term potentiation in rat hippocampal CA1 region by inhibitors of protein synthesis. *J. of Neuroscience*, **4**, 3080-3088. (6)
Stanton, P.K., & Sarvey, J.M. (1985a). The effect of high-frequency electrical stimulation and norepinephrine on cyclic AMP levels in normal versus norepinephrine-depleted rat hippocampal slices. *Brain Research*, **358**, 343-348. (6)
Stanton, P.K., & Sarvey, J.M. (1985b). Blockade of norepinephrine-induced long-lasting potentiation in the hippocampal dentate gyrus by an inhibitor of protein synthesis. *Brain Research*, **361**, 276-283. (6)
Stanton, P.K., & Sarvey, J.M. (1985c). Depletion of norepinephrine, but not serotonin, reduces long-term potentiation in the dentate gyrus of rat hippocampal slices. *J. of Neuroscience*, **5**, 2169-2176. (6)
Stanton, P.K., & Sarvey, J.M. (1987). Norepinephrine regulates long-term potentiation of both the population spike and dendritic EPSP in hippocampal dentate gyrus. *Brain Research Bulletin*, **18**, 115-119. (6)
Steele, D.M., & Hayes, S.C. (1991). Stimulus equivalence and arbitrarily applicable relational responding. *J. of the Experimental Analysis of Behavior*, **56**, 519-555. (26)
Stefanis, C., & Jasper, H. (1964). Recurrent collateral inhibition in pyramidal tract neurons. *J. of Neurophysiology*, **27**, 855-877. (13)
Stein, L. (1975). Norepinephrine reward pathways: Role of self- stimulation, memory consolidation, and schizophrenia. *Nebraska Symposium on Motivation*, **22**, 113-159. (6)
Stein, L., & Belluzzi, J. D. (1989). Cellular investigations on behavioral reinforcement. *Neuroscience and Biobehavior Reviews*, **13**, 69-80. (18)
Stein, L., Xue, B.G., & Belluzzi, J.D. (1993). A cellular analogue of operant conditioning. *J. of the Experimental Analysis of Behavior*, **60**, 41-53. (18)
Stein, L., Xue, B.G., & Belluzzi, J.D. (1994). In vitro reinforcement of hippocampal bursting: A search for Skinner's atoms of behavior. *J. of the Experimental Analysis of Behavior*, **61**, 155-168. (14)
Steinfels, G.F., Heym, J., Strecker, R.E. & Jacobs, B.L. (1983) Behavioral correlates of dopaminergic unit activity in freely moving cats. *Brain Research*, **258**, 217-228. (17)
Steinmetz, M.A., Motter, B.C., Duffy, C.J., & Mountcastle, V.B. (1987). Functional properties of parietal visual neurons: Radial organization of directionalities within the visual field. *J. of Neuroscience*, **7**, 177-191. (13)
Stellar, J.R., & Stellar, E. (1985). *The neurobiology of motivation and reward*. Heidelberg: Springer-Verlag. (18)
Sternberg, S. (1969). Memory scanning: Mental processes revealed by reaction time experiments. *American Scientist*, **54**, 421-457. (13)
Sternberg, S., Monsell, S., Knoll, R.L., & Wright, C.E. (1978). The latency and duration of rapid movement sequences: Comparisons of speech and typewriting. In G.E. Stelmach (Ed.), *Information processing in motor control and learning*. New York: Academic Press. (25)
Sternweis, P.C., & Pang, I-H. (1990). The G protein-channel connection. *Trends in Neurosciences*, **13**, 122-126. (5)
Stevens, C.F., & Wang, Y. (1994). Changes in reliability of synaptic function as a mechanism for plasticity. *Nature*, **371**, 704-707. (5)
Steward, O., & Wallace, C.S. (1995). mRNA distribution within dendrites: Relationship to afferent innervation. *J. of Neurobiology*, **26**, 447-458. (6)
Stone, G.O., & Van Orden, G.C. (1994). Building a resonance framework for word recognition using design and system principles. *J. of Experimental Psychology: Human Perception and Performance*, **20**, 1248-1268. (27)
Stone, G.O., Vanhoy, M., & Van Orden, G.C. (in press). Perception is a two-way street: Feedforward and feedback phonology in visual word recognition. *J. of Memory and Language*. (27)
Stryker, M.P. (1991). Activity-dependent reorganization of afferents in the developing mammalian visual system. In D.M-K. Lau & C.J. Shatz (Eds.), *Development of the visual system* (pp.267-287). Cambridge, MA: MIT Press. (8)
Stryker, M.P., & Harris, W.A. (1986). Binocular impulse blockade prevents the formation of ocular dominance columns in cat visual cortex. *J. of Neuroscience*, **6**, 2117-2133. (8)
Stuart, G.J., & Sakmann, B. (1994). Active propagation of somatic action potentials into neocortical pyramidal cell dendrites. *Nature*, **367**, 69-72. (8)
Sunahara, R.K., Guan, H., O'Dowd, B.F., Seeman, P., Laurier, L.G., Ng, G., George, S.R., Torchia, J., Van Tol, H.H.M., & Niznik, H.B. (1991). Cloning of the gene of the human dopamine D_5 receptor with higher affinity to dopamine than D_1. *Nature*, **350**, 614-619. (6)
Sutcliffe, J.G., Milner, R.J., Gottesfeld, J.M., & Reynolds, W. (1984). Control of neuronal gene expression. *Science*, **225**, 1308-1315. (3)
Sutherland, N.S. (1968). Outlines of a theory of visual pattern recognition in animals and man. *Proceedings of the Royal Society of London B*, **171**, 95-103. (11)
Sutton, R.S. (1984). *Temporal credit assignment in reinforcement learning*. Unpublished doctoral dissertation, Department of Computer and Information Science, University of Massachusetts, Amherst, MA. (22)
Sutton, R.S. (1991). Planning by incremental dynamic programming. *Proceedings of the Eighth International Conference on Machine Learning* (pp. 353-357), San Mateo, CA: Morgan Kaufmann. (19)
Sutton, R.S. (Ed.). (1992). Special issue on reinforcement learning. *Machine Learning*, **8**. Also published as *Reinforcement learning*. Boston:

Kluwer Academic Press, 1992. (19) (20)
Sutton, R.S. (1988). Learning to predict by the method of temporal differences. *Machine Learning*, **3**, 9-44. (19) (21)
Sutton, R.S. (1990). Integrated architectures for learning, planning, and reacting based on approximating dynamic programming. *Proceedings of the Seventh International Conference on Machine Learning* (pp. 216-224), San Mateo, CA: Morgan Kaufmann. (19)
Sutton, R.S., & Barto, A. G. (1981). Toward a modern theory of adaptive networks: Expectation and prediction. *Psychological Review*, **88**, 135-170. (1) (4) (17) (20) (21)
Sutton, R.S., & Barto, A.G. (1987). A temporal-difference model of classical conditioning. *Proceedings of the Ninth Annual Conference of the Cognitive Science Society*, Seattle, WA. (21)
Sutton, R.S., & Barto, A.G. (1989). Time-derivative models of Pavlovian reinforcement. In M. Gabriel & J. Moore (Eds.), *Learning and computational neuroscience*. Cambridge, MA: MIT Press. (21)
Sutton, R.S., & Barto, A. G. (1990). Time-derivative models of Pavlovian reinforcement. In M. Gabriel & J. Moore (Eds.), *Learning and computational neuroscience: Foundations of adaptive networks* (pp. 497-537). Cambridge, MA: MIT Press. (20) (21)
Sutton, R.S., & Whitehead, S.D. (1993). Online learning with random representation. In *Proceedings of the Tenth International Conference on Machine Learning* (pp. 314-321). San Mateo, CA: Morgan Kaufmann. (15)
Suzuki, W.A., Zola-Morgan, S., Squire, L.R., & Amaral, D.G. (1993). Lesions of perirhinal and parahippocampal cortices in the monkey produce long-lasting memory impairment in the visual and tactual modalities. *J. of Neuroscience*, **13**, 2430-2451 (9)
Suzuki, W.A., & Amaral, D.G. (1995). Perirhinal and parahippocampal cortices of the macaque monkey: Cortical afferents. *J. of Comparative Neurology*, **349**, 1-36. (9)
Sweeney, J., Bachman, E., & Coyle, J. (1990). Effects of different doses of galanthamine, a long-acting acetylcholinesterase inhibitor, on memory in mice. *Psychopharmacology*, **102**, 191-200. (12)
Szentágothai, J. (1970). Local neuron circuits of the neocortex. In F.D. Schmitt & F.G. Worden (Eds.), *The neurosciences 4th study program.* (pp. 399-415). Cambridge, MA: MIT Press. (8)
Taira, M., Mine, S., Georgopoulos, A.P., Murata, A., & Sakata, H. (1990). Parietal cortex neurons of the monkey related to the visual guidance of hand movement. *Experimental Brain Research*, **83**, 29-36 (9)
Tallal, P., & Galaburda, A.M., Llinas, R.R., & von Euler, C. (1993). *Temporal information processing in the nervous system.* New York: New York Academy of Sciences. (7)
Tan, S-S., & Breen, S. (1993). Radial mosaicism and tangential cell dispersion both contribute to mouse neocortical development. *Nature*, **362**, 638-640. (2)
Tanaka, K. (1993). Neuronal mechanisms of object recognition. *Science*, **262**, 685-688. (10)
Tanaka, K., Saito, H., Fukada, Y., & Moriya, M. (1991). Coding visual images of objects in the inferotemporal cortex of the macaque monkey. *J. of Neurophysiology*, **66**, 170-189. (9) (10)
Tanaka, Y., Kusunoki, M., Ohtsuka, H., Takiura, K., & Sakata, H. (1992). Analysis of three dimensional directional selectivity of the monkey parietal depth-movement-sensitive neurons using a stereoscopic computer display system. *Neuroscience Research*, **S17**, S238. (9)
Tanaka, Y., Murata, A., Taira, M., Shikata, E., & Sakata, H. (1994). Responses of the parietal visual neurons to stereoscopic stimuli on the computer graphic display in alert monkeys. *Neuroscience Research*, **S19**, S200. (9)
Tank, D.W., & Hopfield, J.J. (1987). Neural computation by concentrating information in time. *Proceedings of the National Academy of Sciences USA*, **84**, 1896-1900. (7)
Tarr, M.J. (1995). Rotating objects to recognize them: A case study on the role of viewpoint dependency in the recognition of three-dimensional objects. *Psychonomic Bulletin & Review*, **2(1)**, 55-82. (11)
Tarr, M.J., & Bölthoff, H. (1995). Conditions for viewpoint-dependence and viewpoint-invariance: What mechanisms are used to recognize an object? *J. of Experimental Psychology: Human Perception and Performance*. (in press). (11)
Tarr, M.J., & Pinker. S. (1989). Mental rotation and orientation dependence in shape recognition. *Cognitive Psychology*, **21**, 233-283. (11)
Tarr, M.J., & Pinker. S. (1990). When does human object recognition use a viewer-centered reference frame? *Psychological Science*, **1**, 253-256. (11)
Taube, J. (1991). Space, the final hippocampal frontier? *Hippocampus*, **1**, 247-249. (12)
Temple, S. (1989). Division and differentiation of isolated CNS blast cells in microculture. *Nature*, **340**, 471-473. (2)
Terrace, H. (1963). Discrimination learning with and without "errors." *J. of Experimental Analysis of Behavior*, **6**, 1-27. (12)
Tesauro, G.J. (1994). TD-Gammon, a self-teaching backgammon program, achieves master-level play. *Neural Computation*, **6**, 215-219. (19)
Tesauro, G.J. (1995). Temporal difference learning and TD-Gammon. *Communications of the ACM*, **38**, 58-68. (19)
Teyler, T.J., & DiScenna, P. (1986). The hippocampal memory indexing theory. *Behavioral Neuroscience*, **100**, 147-154. (6)
Thal, L., Fuld, P., Masur, D., & Sharpless, N. (1983). Oral physostigmine and lecithin improves memory in Alzheimer's disease. *Annals of Neurology*, **13**, 491-496. (12)
Thelen, E. (1995). Motor development: A new synthesis. *American Psychologist*, **50**, 79-95. (15)
Thelen, E., & Adolph, K.E. (1992). Arnold L. Gesell: The paradox of nature and nurture. *Developmental Psychology*, **28**, 368-380. (15)
Thelen, E., Corbetta, D., Kamm, K., Spencer, J.P., Schneider, K., & Zernicke, R.F. (1993). The transition to reaching: Mapping intention to intrinsic dynamics. *Child Development*, **64**, 1058-1098. (15)
Thelen, E., Fisher, D.M., & Ridley-Johnson, R. (1984). The relationship between physical growth and a newborn reflex. *Infant Behavior and Development*, **7**, 479-493. (15)
Thelen, E., Kelso, J.A.S., & Fogel, A. (1987). Self-organizing systems and infant motor development. *Developmental Review*, **7**, 39-65. (15)

Thelen, E., & Smith, L. (1994). *A dynamic systems approach to the development of cognition and action*. Cambridge, MA: Harvard. (27)
Thompson, R. (1972). Sensory preconditioning. In R. Thompson & J. Voss (Eds.), *Topics in Learning and Performance* (pp. 105-129). New York: Academic Press. (12)
Thompson, R.F. (1986). The neurobiology of learning and memory. *Science*, **233**, 941-947. (4) (6)
Thompson, R., Berger, R., Berry, S., Hoehler, F., Kettner, R., & Weisz, D. (1980). Hippocampal substrate of classical conditioning. *Physiological Psychology*, **8**, 262-279. (12)
Thompson, R.F., Berger, T.W., & Madden, J., IV. (1983). Cellular processes of learning and memory in the mammalian CNS. *Annual Review of Neuroscience*, **6**, 447-491. (4)
Thompson, T., & Lubinski, D. (1986). Units of analysis and kinetic structure of behavioral repertoires. *J. of the Experimental Analysis of Behavior*, **46**, 219-242. (26)
Thomson, A. M., & Deuchars, J. (1994). Temporal and spatial properties of local circuits in neocortex. *Trends in Neuroscience*, **17**, 119-126. (9)
Thorpe, S.J., Rolls, E.T., & Maddison, S. (1983) The orbitofrontal cortex: Neuronal activity in the behaving monkey. *Experimental Brain Research*, **49**, 93-115. (17)
Thrun, S.B., & Möller, K. (1992). Active exploration in dynamic environments. In Moody, Hanson, & Lippman (Eds.), *Advances in neural information processing systems*. San Mateo, CA: Morgan Kaufmann. (22)
Tierney, A.J. (1986). The evolution of learned and innate behavior: Contributions from genetics and neurobiology to a theory of behavioral evolution. *Animal Learning & Behavior*, **14**, 339-348. (4)
Tipper, S.P. (1985). The negative priming effect: Inhibitory effects of ignored primes. *Quarterly J. of Experimental Psychology*, **37A**, 571-590. (11)
Tipper, S.P., & Driver, J. (1988). Negative priming between pictures and words in a selective attention task: Evidence for semantic processing of ignored stimuli. *Memory and Cognition*, **16**, 64-70. (11)
Todd, P.M., & Miller, G.F. (1991). Exploring adaptive agency II: Simulating the evolution of associative learning. In J.A. Meyer & S.W. Wilson (Eds.), *From animals to animats: Proceedings of the First International Conference on Simulation of Adaptive Behavior* (pp. 263-272). Cambridge, MA: MIT Press. (4)
Tovee, M.J., & Rolls, E.T. (1992). Oscillatory activity is not evident in the primate visual cortex with static stimuli. *NeuroReport*, **3**, 369-372. (9) (11)
Towell, G., Shavlik, J., & Noordewier, M. (1990). Refinement of approximate domain theories by knowledge-based neural networks. *Proceedings of the Eighth National Conference on Artificial Intelligence* (pp. 861-866). Boston: Morgan Kaufmann. (22)
Townsend, J.T. (1972). Some results on the identifiability of serial and parallel processes. *British J. of Mathematical and Statistical Psychology*, **25**, 168-199. (1)
Toyama, K., Kimura, M., & Tanaka, J. (1981). Cross-correlation analysis of interneuronal connectivity in cat visual cortex. *J. of Neurophysiology*, **6**, 191-201. (8)
Toyama, K., Kimura, M., & Tanaka, J. (1981). Organization of cat visual cortex as investigated by cross-correlation technique. *J. of Neurophysiology*, **6**, 202-214. (8)
Trehub, A. (1977). Neuronal models for cognitive processes: Networks for learning, perception and imagination. *J. of Theoretical Biology*, **65**, 141-169. (10)
Trehub, A. (1991). *The cognitive brain*. Cambridge: MIT Press. (10)
Treisman, A. (1993). The perception of features and objects. In A. Baddeley & L. Weiskrantz (Eds.), *Attention: selection, awareness and control: A tribute to Donald Broadbent* (pp. 5-35). Oxford: Clarendon Press. (11)
Treisman, A., & DeSchepper, B. (1995). Object tokens, attention, and visual memory. In T. Inui & J. McClelland (Eds.), *Attention and performance XVI*. In press. (11)
Treisman, A., & Gelade, G. (1980). A feature integration theory of attention. *Cognitive Psychology*, **12**, 97-136. (11)
Treisman, A.M., & Schmidt, H. (1982). Illusory conjunctions in the perception of objects. *Cognitive Psychology*, **14**, 107-141. (11)
Tretter, F., Cynader, M., & Singer, W. (1975). Modification of direction selectivity of neurons in the visual cortex of kittens. *Brain Research*, **84**, 143-149. (8)
Ts'o, D., & Gilbert, C. (1988). The organization of chromatic and spatial interaction in the primate striate cortex. *J. of Neuroscience*, **8**, 1712-1727. (8)
Ts'o, D., Gilbert, C., & Wiesel, T.N. (1986). Relationships between horizontal interactions and functional architecture in cat striate cortex as revealed by cross-correlation analysis. *J. of Neuroscience*, **6**, 1160-1170. (8)
Tsumoto, T. (1992). Long-term potentiation and long-term depression in the neocortex. *Progress in Neurobiology*, **39**, 209-228. (8)
Tsumoto, T. (1993). Long-term depression in cerebral cortex: A possible substrate of "forgetting" that should not be forgotten. *Neuroscience Research*, **16**, 263-270. (8)
Tsumoto, T., Hagihara, K., Sato, H., & Hata, Y. (1987). NMDA receptors in the visual cortex of young kittens are more effective than those of adult cats. *Nature* , **327**, 513-514. (8)
Turvey, M.T., & Carello, C. (1981). Cognition: The view from ecological realism. *Cognition*, **10**, 313-321. (27)
Tversky, A. (1977). Features of similarity. *Psychological Review*, **84**, 327-352. (12)
Ullman, S. (1989). Aligning pictoral descriptions: An approach to object recognition. *Cognition*, **32**, 193-254. (11)
Ullman, S., & Basri, R. (1991). Recognition by liner combinations of models. *IEEE Transactions on Pattern Analysis and Machine Intelligence*, **13**, 992-1006. (11)
Utgoff, P.E., & Clouse, J.A. (1991). Two kinds of training information for evaluation function learning. *Proceedings of the Ninth National Conference on Artificial Intelligence* (pp. 596-600). Anaheim, CA: MIT Press. (22)
Vaadia, E., Haalman, I., Abeles, M., Bergman, H., Prut, Y., Slovin, H., & Aertsen, A. (1995). Dynamics of neuronal interactions in monkey

cortex in relation to behavioral events. *Nature*, **373**, 515-518. (8)

Vallejo, M. (1994). Transcriptional control of gene expression by cAMP-response element binding proteins. *J. of Neuroendocrinology*, **6**, 587-596. (5)

Van Hoesen, G.W. (1982) The parahippocampal gyrus: New observations regarding its cortical connections in the monkey. *Trends in Neuroscience*, **5**, 345-350. (9)

Van Hoesen, G., & Pandya, D. (1975). Some connections of the entorhinal (area 28) and perirhinal (area 35) cortices of the rhesus monkey I: Temporal lobe afferents. *Brain Research*, **95**, 1-24. (12)

Van Kan, P.L.E., Gibson, A.R., & Houk, J.C. (1993). Movement-related inputs to intermediate cerebellum of the monkey. *J. of Neurophysiology*, **69**, 74-94. (20)

Van Leeuwen, C., Steyvers, M., & Nooter, M. (1995). Stability and intermittency in high-dimensional dynamical models for perceptual grouping. (Submitted for publication.) (27)

Van Orden, G.C. (1987). A ROWS is a ROSE: Spelling, sound, and reading. *Memory & Cognition*, **15**, 181-198. (27)

Van Orden, G.C., & Goldinger, S.D. (1994). Interdependence of form and function in cognitive systems explains perception of printed words. *J. of Experimental Psychology: Human Perception and Performance*, **20**, 1269-1291. (27)

Van Orden, G.C., & Goldinger, S.D. (1996). Phonologic mediation in skilled and dyslexic reading. In C. Chase, G. Rosen, & G. Sherman (Eds.), *Developmental dyslexia: Neuro, cognitive, and genetic mechanisms*. Timonium, MD: York Press. (27)

Van Orden, G.C., Pennington, B.F., & Stone, G.O. (1990). Word identification in reading and the promise of subsymbolic psycholinguistics. *Psychological Review*, **97**, 488-522. (27)

Van Orden, G.C., Pennington, B.F., & Stone, G. O. (1995). What do double dissociations prove? Inductive methods and theory in psychology. Manuscript submitted for publication. (27)

Van Orden, G.C., Stone, G.O., Garlington, K.L., Markson, L.R., Pinnt, G.S., Simonfy, C.M., & Brichetto, T. (1992). "Assembled" phonology and reading: A case study in how theoretical perspective shapes empirical investigation. In R. Frost & L. Katz (Eds.), *Orthography, phonology, morphology, and meaning* (pp. 249-292). Amsterdam: Elsevier. (27)

Van Veen, M.P., & Van Pelt, J. (1994). Neuritic growth rate described by modeling microtubule dynamics. *Bulletin of Mathematical Biology*, **56**, 249-273. (3)

Varela, F.J., Thompson, E., & Rosch, E. (1991). *The embodied mind*. Cambridge, MA: MIT Press. (27)

Verhage, M., Ghijsen, W.E.J.M., & Da Silva, F.H.L. (1994). Presynaptic plasticity: The regulation of Ca^{2+}-dependent transmitter release. *Progress in Neurobiology*, **42**, 539-574. (5)

Verma, A., Hirsch, D.J., Glatt, C.E., Ronnett, G.V., & Snyder, S.H. (1993). Carbon monoxide: A putative neural messenger. *Science*, **259**, 381-384. (5)

Villanueva, L., & Le Bars, D. (1993). Computer-assisted reconstruction of axonal arborizations anterogradely labeled with the Phaseolus vulgaris leucoagglutinin technique. *J. of Neuroscience Methods*, **50**, 217-224. (3)

Vincent, S.R., & Hope, B.T. (1992). Neurons that say NO. *Trends in Neurosciences*, **15**, 108-113. (5)

Vogt, B.A., Finch, D.M., & Olson, C.R. (1992). Functional heterogeneity in cingulate cortex: The anterior executive and posterior evaluative regions. *Cerebral Cortex*, **2**, 435-443. (24)

vom Saal, W., & Jenkins, H.M. (1970). Blocking the development of stimulus control. *Learning and Motivation*, **1**, 52-64. (14)

von der Malsburg, C. (1985). Nervous structures with dynamical links. *Physical Chemistry*, **89**, 703-710. (8)

von Hofsten, C. (1979). Development of visually directed reaching: The approach phase. *J. of Human Movement Studies*, **43**, 3-48. (15)

von Hofsten, C. (1980). Predictive reaching for moving objects by human infants. *J. of Experimental Child Psychology*, **30**, 369-382. (15)

von Hofsten, C. (1982). Eye-hand coordination in newborns. *Developmental Psychology*, **18**, 450-461. (15)

von Hofsten, C. (1983). Catching skills in infancy. *J. of Experimental Psychology: Human Perception and Performance*, **9**, 75-85. (15)

von Hofsten, C. (1991). Structuring of early reaching movements: A longitudinal study. *J. of Motor Behavior*, **23**, 280-292. (15)

von Hofsten, C. (1993). Prospective control: A basic aspect of action development. *Human Development*, **36**, 253-270. (15)

von Hofsten, C., & Lindhagen, K. (1979). Observations on the development of reaching for moving objects. *J. of Experimental Child Psychology*, **28**, 158-173. (15)

von Hofsten, C., & Ronnqvist, L. (1988). Preparation for grasping an object: A developmental study. *J. of Experimental Psychology: Human Perception and Performance*, **14**, 610-621. (15)

von Hofsten, C., Spelke, E., & Feng, Q. (1993). Principles of predictive reaching in 6-month-old infants. *34th Annual Meeting of the Psychonomic Society Abstracts*, **9**. (15)

Wallesch, C-W., & Papagno, C. (1988). Subcortical aphasia. In F. Rose, R. Whurr, & M. Wyse (Eds.), *Aphasia* (pp. 256-287). London: Whurr Publishers. (24)

Walsh, C., & Cepko, C.L. (1988). Clonally related cortical cells show migration patterns. *Science*, **241**, 1342-1345. (2)

Walsh, C., & Cepko, C.L. (1992). Widespread dispersion of neuronal clones across functional regions of the cerebral cortex. *Science*, **255**, 434-440. (2)

Wang, J-H., & Stelzer, A. (1994). Inhibition of phosphatase 2B prevents expression of hippocampal long-term potentiation. *NeuroReport*, **5**, 2377-2380. (6)

Wang, G., Tanaka, K., & Tanifuji, M. (1994). Optical imaging of functional organization in macaque inferotemporal cortex. *Society of Neuroscience Abstracts*, **20**, 316. (9)

Wang, L.Y., Salter, M.W., & MacDonald, J.F. (1991). Regulation of kainate receptors by cAMP-dependent protein kinase and phosphatases. *Science*, **253**, 1132-1135. (6)

Wang, X-J., Rinzel, J., & Rogawski, M. A. (1991). A model of the T-type calcium current and the low-threshold spike in thalamic neurons. *J. of Neurophysiology*, **66**, 839-850. (24)

Wann, D.F., Woolsey, T.A., Dierker, M.L., & Cowan, W.M. (1973). An on-line digital computer system for the semiautomatic analysis of Golgi-impregnated neurons. *IEEE transactions on biomedical engineering*, **20**, 233-247. (3)

Waters, G.S., & Seidenberg, M.S. (1985). Spelling-sound effects in reading: Time course and decision criteria. *Memory and Cognition*, **13**, 557-572. (27)

Watkins, C.J.C.H. (1989). Learning from delayed rewards. Unpublished PhD thesis, Cambridge University, Cambridge, England. (19) (22)

Watkins, C.J.C.H., & Dayan, P. (1992). Q-learning. *Machine Learning*, **8**, 279-292. (15) (22)

Watson, J.D., Gilman, M., Witkowski, J., & Zoller, M. (1992). *Recombinant DNA, second edition*. New York: William Freeman. (5)

Weikart, C., & Berger, T.W. (1986). Hippocampal lesions disrupt classical conditioning of crossmodality reversal learning of the rabbit nictitating membrane response. *Behavioral Brain Research*, **22**, 85-90. (4)

Weiland, A.P., & Leighton, R.R. (1988). Shaping schedules as a method for accelerating learning. In International Neural Network Society Meeting. (16)

Weiler, I.J., Hawrylak, N., & Greenough, W.T. (1995). Morphogenesis in memory formation: Synaptic and cellular mechanisms. *Behavioural Brain Research*, **66**, 1-6. (5)

Weinberger, N.M. (1993). Learning-induced changes of auditory receptive fields. *Current Opinion in Neurobiology*, **3**, 570-577. (8)

Werbos, P.J. (1974). Beyond regression: New tools for prediction and analysis in the behavioral sciences. Unpublished doctoral dissertation, Applied Mathematics. Harvard University, Cambridge, MA. (1)

Wertheim, S.L., & Sidman, R.L. (1991). Databases for Neuroscience. *Nature*, **354**, 88-89. (3)

White, B.L., Castle, P., & Held, R. (1964). Observations on the development of visually-guided reaching. *Child Development*, **35**, 349-364. (15)

Whitehead, S.D. (1992). *Reinforcement learning for the adaptive control of perception and action*. Unpublished doctoral dissertation, Computer Science, University of Rochester, Rochester, NY. (22)

Whitehead, S.D., & Ballard, D.H. (1990). Active perception and reinforcement learning. *Neural Computation*, **2**, 409-419. (19)

Whitehouse, P., Price, D., Struble, R., Clark, A., Coyle, J., & DeLong, M. (1982). Alzheimer's disease and senile dementia: Loss of neurons in the basal forebrain. *Science*, **215**, 1237-1239. (12)

Whitley, D., Starkweather, T., & Bogart, C. (1990). Genetic algorithms and neural networks: Optimizing connection weights and connectivity. *Parallel Computing*, **14**, 347-361. (4)

Wickelgren, W.A. (1969). Context-sensitive coding, associative memory, and serial order in (speech) behavior. *Psychological Review*, **76**, 1-15. (25)

Wickens, J., & Kötter, R. (1995). Cellular models of reinforcement. In J. C. Houk, J. L. Davis, & D. G. Beiser (Eds.). *Models of information processing in the basal ganglia* (pp. 187-214). Cambridge, MA: MIT Press. (17) (18)

Widrow, B., & Hoff, M.E. (1960). Adaptive switching circuits. *Institute of Radio Engineers, Western Electronic Show and Convention Record*, **4**, 96-104. (12) (17)

Widrow, B., & Stearns, S.D. (1985). *Adaptive signal processing*. Englewood Cliffs, NJ: Prentice-Hall. (21)

Wieland, A.P. (1991). Evolving controls for unstable systems. In D.S. Touretsky, J.L. Elman, T.J. Sejnowski, & G.E. Hinton, (Eds.), Connectionist Models: Proceedings of the 1990 Summer School. San Mateo, CA: Morgan Kaufmann. (16)

Wiesel, T.N., & Hubel, D.H. (1974). Ordered arrangement of orientation columns in monkeys lacking visual experience. *J. of Comparative Neurology*, **158**, 307-318. (8)

Wilkenfield, J., Nickel, M., & Blakely, E. (1992). Acquisition of lever-press responding in rats with delayed reinforcement: A comparison of three procedures. *J. of the Experimental Analysis of Behavior*, **58**, 431-443. (14)

Williams, B.P., Read, J., & Price, J. (1991). The generation of neurons and oligodendrocytes from a common precursor cell. *Neuron*, **7**, 685-693. (2)

Williams, J., Dragunow, M., Lawlor, P., Mason, S., Abraham, W.C., Leah, J., Bravo, R., Demmer, J., & Tate, W. (1995). Krox20 may play a key role in the stabilization of long-term potentiation. *Molecular Brain Research*, **28**, 87-93. (6)

Williams, J.H., Li, Y-G., Nayak, A., Errington, M.L., Murphy, K.P.S.J., & Bliss, T.V.P. (1993). The suppression of long-term potentiation in rat hippocampus by inhibitors of nitric oxide synthase is temperature and age dependent. *Neuron*, **11**, 877-884. (6)

Williams, R.W., & Herrup, K. (1988). The control of neuron number. *Annual Review of Neuroscience*, **11**, 423-453. (3)

Willis, W.D., & Koppe, J.A. (1991). Brain Browser: Hypercard application for the Macintosh. *Science*, **251**, 1500-1502. (3)

Wilson, C.J. (1995) The contribution of cortical neurons to the firing pattern of striatal spiny neurons. In J.C. Houk, J.L. Davis & D.G. Beiser (Eds.) *Models of information processing in the basal ganglia* (pp. 29-50). Cambridge, MA: MIT Press. (17)

Wilson, E.O. (1975). *Sociobiology*. Cambridge, MA: Harvard University Press. (1)

Winston, P. (1975). Learning structural descriptions from examples. In P. Winston (Ed.), *Psychology of computer vision* (pp. 157-209). New York: McGraw-Hill. (11)

Wisden, W., Errington, M.L., Williams, S., Dunnett, S.B., Waters, C., Hitchcock, D., Evan, G., Bliss, T.V., & Hunt, S.P. (1990). Differential expression of immediate early genes in the hippocampus and spinal cord. *Neuron*, **4**, 603-614. (6)

Wise, R., Chollet, F., Hadar, U., Friston, K., Hoffner, E., & Frackowiak,R. (1991). Distribution of cortical neural networks involved in word comprehension and word retrieval. *Brain*, **114**, 1803-1817. (24)

Wise, R.A., & Hoffman, D.C. (1992) Localization of drug reward mechanisms by intracranial injections. *Synapse*, **10**, 247-263. (17)

Wise, R.A., Spindler, J., de Wit, H., & Gerber, G.J. (1978) Neuroleptic-induced 'anhedonia' in rats: Pimozide blocks reward quality of food. *Science*, **201**, 262-264. (17)

Wise, S., & Houk, J.C. (1994). Modular neuronal architecture for planning and controlling motor behavior. *Biological Communications of the Danish Royal Academy of Sciences*, **43**, 21-33. (24)

Wolff, J.R., Laskawi, R., Spatz, W.B., & Missler, M. (1995). Structural dynamics of synapses and synaptic components. *Behavioural Brain Research*, **66**, 13-20. (5)

Wood, C. (1976). Discriminability, response bias, and phoneme categories in discrimination of voice onset time. *J. of the Acoustical Society of America*, **60**, 1381-1388. (7)

Wood, P.L., Tadimeta, S.R., Iyengar, S., Lanthorn, T., Monahan, J., Cordi, A., Sun, E., Vazquez, M., Gray, N., & Contreras, P. (1990). A review of the in vitro and in vivo neurochemical characterization of the NMDA/PCP/glycine/ion channel receptor macrocomplex. *Neurochemical Research*, **15**, 217-230. (5)

Woodhams, P., Celio, M., Ulfig, N., & Witter, M. (1993). Morphological and functional correlates of borders in the entorhinal cortex and hippocampus. In R. Nitsch & T. Ohm (Eds.), *Hippocampus*, **3**, *Special issue on entorhinal-hippocampal interaction* (pp. 313-312). (12)

Woolf, B. (1988). Intelligent tutoring systems: A survey. In H. Shrobe & American Association for Artificial Intelligence (Eds.), *Exploring artificial intelligence*. San Mateo, CA: Morgan Kaufmann. (22)

Worley, P.F., Bhat, R.V., Baraban, J.M., Erickson, C.A., McNaughton, B.L., & Barnes, C.A. (1993). Thresholds for synaptic activation of transcription factors in hippocampus: Correlation with long-term enhancement. *J. of Neuroscience*, **13**, 4776-4786. (6)

Wörgötter, F., & Koch, C. (1991). A detailed model of the primary visual pathway in the cat: Comparison of afferent excitatory and intracortical inhibitory connection schemes for orientation selectivity. *J. of Neuroscience*, **11**, 959-1979. (7)

Wu, L.G., & Saggau, P. (1994). Presynaptic calcium is increased during normal synaptic transmission and paired-pulse facilitation, but not long-term potentiation in area CA1 of hippocampus. *J. of Neuroscience*, **14**, 645-654. (7)

Wurtz, R.H., Goldberg, M.E., & Robinson, D.L. (1980). Behavioral modulation of visual responses in the monkey: Stimulus selection for attention and movement. *Progress in Psychobiology and Physiological Psychology*, **9**, 43-83. (8)

Yaginuma, S., Osawa, Y., Yamaguchi, K., & Iwai, E. (1993). Differential functions of central and peripheral visual field representations in monkey prestriate cortex. In T. Ono, L. R. Squire, M. E. Raichle, D. L. Perrett, & M. Fukuda (Eds.), *Brain mechanisms of perception and memory: From neuron to behavior* (pp. 3-15). New York/Oxford: Oxford University Press. (9)

Yamagata, K., Andreasson, K.I., Kaufmann, W.E., Barnes, C.A., & Worley, P.F. (1993). Expression of a mitogen-inducible cyclooxygenase in brain neurons: Regulation by synaptic activity and glucocorticoids. *Neuron*, **11**, 371-386. (6)

Yamagata, K., Sanders, L.K., Kaufmann, W.E., Yee, W., Barnes, C.A., Nathans, D., & Worley, P.F. (1994). Rheb, a growth factor- and synaptic activity-regulated gene, encodes a novel Ras-related protein. *J. of Biological Chemistry*, **269**, 16333-16339. (6)

Yamane, S., Kaji, S., & Kawano, K. (1988). What facial features activate face neurons in the inferotemporal cortex of the monkey? *Experimental Brain Research*, **73**, 209-214. (9)

Yeo, A.G. (1974). The acquisition of conditioned suppression as a function of interstimulus interval duration. *Quarterly J. of Experimental Psychology*, **26**, 405-416. (4)

Yeo, C.H., Hardiman, M.J., & Glickstein, M. (1984). Discrete lesions of the cerebellar cortex abolish the classically conditioned nictitating response of the rabbit. *Behavioural Brain Research*, **13**, 261-266. (4)

Young, M.P., Tanaka, K., & Yamane, S. (1992). On oscillating neuronal responses in the visual cortex of the monkey. *J. of Neurophysiology*, **67**, 1464-1474. (9)

Young, M.P., & Yamane, S. (1992). Sparse population coding of faces in the inferotemporal cortex. *Science*, **29**, 1327-1331. (8) (9) (10) (13)

Yuste, R., & Denk, W. (1995). Dendritic spines as basic functional units of neuronal integration. *Nature*, **375**, 682-684. (8)

Zelazo, P.R., Zelazo, N.A., & Kolb, S. (1972). "Walking" in the newborn. *Science*, **176**, 314-315. (15)

Zhang, W., & Dietterich, T. G. (1995). A reinforcement learning approach to job-shop scheduling. *Proceedings of the Fourteenth International Joint Conference on Artificial Intelligence* (pp. 1114-1120). (19)

Ziegler, J.C., & Jacobs, A.M. (1995). Phonological information provides early sources of constraint in the processing of letter strings. *J. of Memory and Language*, **34**, 567-593. (27)

Zigova, T., Betarbet, R., Soteres, B.J., Brock, S., Bakay, R.A.E., & Luskin, M.B. (1996). A comparison of the pattern of migration and destination of homotopically transplanted neonatal subventricular zone cells and heterotopically transplanted telencephalic ventricular zone cells. *Developmental Biology*, **173**, 459-474. (2)

Zola-Morgan, S., & Squire, L. (1986). Memory impairment in monkeys following lesions limited to the hippocampus. *Behavioral Neuroscience*, **100**, 155-160. (12)

Zola-Morgan, S., & Squire, L.R. (1990). The primate hippocampal formation: Evidence for a time-limited role in memory storage. *Science*, **250**, 288-290. (26)

Zola-Morgan, S., & Squire, L. (1992). Components of the medial temporal lobe memory system. In L. Squire & N. Butters (Eds.), *Neuropsychology of memory* (2nd edition, pp. 325-335). New York: Guilford Press. (12)

Zola-Morgan, S., Squire, L.R., Amaral, D.G., & Suzuki, W.A. (1989). Lesions of perirhinal and parahippocampal cortex that spare the amygdala and hippocampal formation produce severe memory impairment. *J. of Neuroscience*, **9**, 4355-4370. (9)

Zola-Morgan, S., Squire, L.R., Clower, R.P., & Rempel, N.L. (1993). Damage to the perirhinal cortex exacerbates memory impairment following lesions to the hippocampal formation. *J. of Neuroscience*, **13**, 251-265. (9)

Zola-Morgan, S., Squire, L., Rempel, N., Clower, R., & Amaral, D. (1992). Enduring memory

impairment in monkeys after ischemic damage to the hippocampus. *J. of Neuroscience*, **12**, 2582-2596. (12)
Zucker, R.S. (1989). Short-term synaptic plasticity. *Annual Review of Neuroscience*, **12**, 13-31. (7)

INDEX

FSC
www.fsc.org
MIX
Papier | Fördert gute Waldnutzung
FSC® C083411

Kolibri 360 GmbH
Ferdinand-Jühlke-Straße 7
99095 Erfurt, Deutschland
produktsicherheit@kolibri360.de